ENGINEE CHANICS

Dynamics

ENGINEERING MECHANICS SI EDITION

Anthony Bedford
and
Wallace Fowler
University of Texas at Austin

Adapted by Eddie Morris

ADDISON-WESLEY
Reading, Massachusetts · Menlo Park, California · New York · Don Mills, Ontario
Harlow, England · Amsterdam · Bonn · Sydney · Singapore · Tokyo
Madrid · San Juan · Milan

Addison Wesley Longman Limited
Edinburgh Gate
Harlow
Essex
CM20 2JE
England

and Associated Companies throughout the world.

Photo Credits
Cover: Medford Taylor/Superstock
Chapter 2: 2.11, The Harold E. Edgerton 1992 Trust, courtesty Palm, Press, Inc.;
2.44, courtesy of Intelsat.
Chapter 4: P4., U.S. Geological Survey.
Chapter 5: 5.4, The Harold Edgerton 1992 Trust, courtesy Palm Press, Inc.
Chapter 6: 6.44 (a & b), NASA.
Chapter 10: 10.20, US Geological Survey.

Cover designed by Designers & Partners, Oxford
Typset by Techset Composition Limited, Salisbury
Printed and bound in the United States of America

First printed 1996

ISBN 0-201-40341-2
British Library Cataloguing-in-Publication Data
A catalogue record for this book is available from the British Library.

Library of Congress Cataloging-in-Publication Data is available

About the Authors

Anthony Bedford is Professor of Aerospace Engineering and Engineering Mechanics at the University of Texas at Austin. He received his BS degree at the University of Texas at Austin, his MS degree at the California Institute of Technology, and his PhD degree at Rice University in 1967. He has industrial experience at the Douglas Aircraft Company and TRW Systems, and has been on the faculty of the University of Texas at Austin since 1968.

Dr Bedford's main professional activity has been education and research in engineering mechanics. He is author or co-author of many technical papers on the mechanics of composite materials and mixtures and of two books, *Hamilton's Principle in Continuum Mechanics* and *Introduction to Elastic Wave Propagation*. He has developed undergraduate and graduate courses in engineering mechanics and is the recipient of the General Dynamics Teaching Excellence Award.

He is a licensed professional engineer and a member of the Acoustical Society of America, the American Society for Engineering Education, the American Academy of Mechanics and the Society of Natural Philosophy.

Wallace Fowler is Paul D. and Betty Robertson Meek Professor of Engineering in the Department of Aerospace Engineering and Engineering Mechanics at the University of Texas at Austin. Dr Fowler received his BS, MS and PhD degrees at the University of Texas at Austin, and has been on the faculty since 1966. During 1976 he was on the staff of the United States Air Force Test Pilot School, Edwards Air Force Base, California, and in 1981–1982 he was a visiting professor at the United States Air Force Academy. Since 1991 he has been Director of the Texas Space Grant Consortium.

Dr Fowler's areas of teaching and research are dynamics, orbital mechanics and spacecraft mission design. He is author or co-author of many technical papers on trajectory optimization and attitude dynamics, and has also published many papers on the theory and practice of engineering teaching. He has received numerous teaching awards, including the Chancellor's Council Outstanding Teaching Award, the General Dynamics Teaching Excellence Award, the Halliburton Education Foundation Award of Excellence and the AIAA-ASEE Distinguished Aerospace Educator Award.

He is a licensed professional engineer, a member of many technical societies, and a fellow of the American Institute of Aeronautics and Astronautics and the American Society for Engineering Education.

Preface

For 25 years we have taught the two-semester introductory course in engineering mechanics. During that time, students often told us that they could understand our classroom presentation, but had difficulty understanding the textbook. This comment led us to examine what the instructor does in class that differs from the traditional textbook presentation, and eventually resulted in this book. Our approach is to present material the way we do in the classroom, using more sequences of figures and stressing the importance of careful visual analysis and conceptual understanding. Throughout the book, we keep the student foremost in mind as our audience.

Figure 7.16

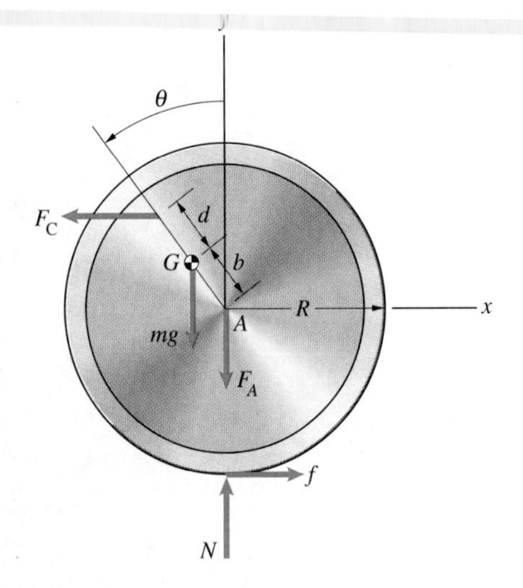

(a) Free-body diagram of the wheel.

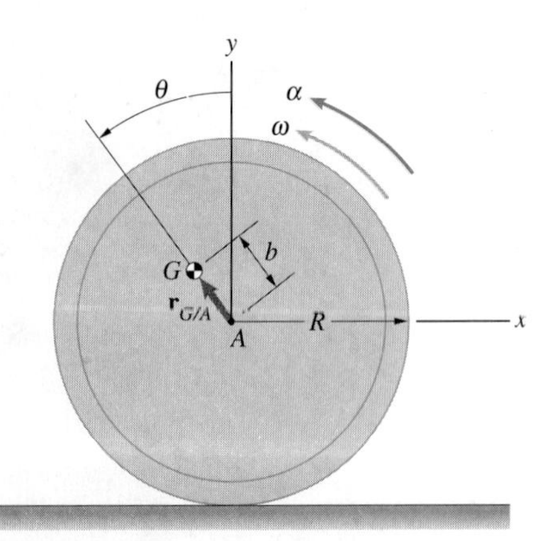

(b) Expressing the acceleration of the centre of mass G in terms of the acceleration of the centre A.

Goals and Themes

Problem Solving We emphasize the critical importance of good problem-solving skills. In our worked examples, we teach students to think about problems before they begin their solution. What principles apply? What must be determined, and in what order? Separate *Strategy* sections that precede most of the examples illustrate this preliminary analysis. Then we give a careful and complete description of the solution, often showing alternative methods. Finally, many examples conclude with *Discussion* sections that point out properties of the solution, or comment on and compare alternative solution methods, or point out ways to check the answers. (See, for instance, Example 3.2, pp. 106–7.) Our objective is to teach students how to approach problems and critically judge the results. In addition, for those students who tell us that they understand the material in class, but don't know how to begin their homework problems we also provide brief Strategy sections in selected homework problems.

Visualization One of the essential elements in successful problem solving is visualization, especially the use of free-body diagrams. In the classroom, the instructor can draw a diagram one step at a time, describing each step and developing the solution in parallel with the diagram. We have done the same thing in this book, showing the same sequence of diagrams we use in class and carefully indicating the relationships between them. In Example 8.2, pp. 378–9, instead of simply showing the free-body diagram, we repeat the initial figure with the isolated part highlighted and everything else shown as a pale ghosted image. In this way we show the student exactly how to isolate the part that will become the free-body diagram. In Example 9.8, p. 456, we use a ghosted image to indicate the motion of a rigid body about an axis. This helps students visualize the true motion of the object.

We use colour to help students distinguish and understand the various elements in figures. By using the same colours for particular elements consistently – such as blue for force vectors and green for accelerations – we have tried to make the book easier for students to read and understand. (See, for example, Figure 7.16 on the left.) In addition, the greater realism of colour illustrations helps motivate students. (See Figure 3.7, p. 117; Figure 5.13, p. 202; and problem illustrations throughout the book.)

Emphasizing Basic Principles Our primary goal for this book is to teach students fundamental concepts and methods. Instead of presenting dynamics as a sequence of independent methods, we emphasize its coherence by showing how energy and momentum techniques can be derived from Newton's second law. We apply the same approach to a system of particles to obtain the equations describing the dynamics of rigid bodies. In describing motions of rigid bodies, we consistently use the angular velocity vector and the vector equations describing the relative motions of points. Traditionally, dynamics texts have waited until they discuss rigid bodies to show that the sum of the external forces acting on an object is equal to the product of its mass and the acceleration of its centre of mass. We introduce this simple result as soon as we have discussed Newton's second law, in Chapter 3, because we find our students gain confidence in their solution. They don't need to be concerned about whether a given object can be modelled as a particle; they know they're determining the motion of its centre of mass. To help students identify important results, key equations are highlighted (see, for example, p. 18), and the concepts discussed in each chapter are reinforced in a chapter-end summary.

Thinking Like Engineers Engineering is an exciting discipline, requiring creativity and imagination as well as knowledge and systematic thinking. In this book we try to show the place of engineering mechanics within the larger context of engineering practice. Engineers in industry and the Accrediting Board for Engineering and Technology (ABET) are encouraging instructors to introduce design early in the engineering curriculum. We include simple design and safety ideas in many of our examples and problems without compromising emphasis on fundamental mechanics. Many problems are expressed in terms of design and safety considerations (for example, Problems 3.101 and 3.102, p. 136); in some cases, students are asked to choose a design parameter from a range of possible values based on stated criteria (for example, Problems 4.118, p. 180; and 4.125, p. 181). Our students have responded very positively to these motivational elements and have developed an awareness of how these essential ideas are applied in engineering.

Pedagogical Features

Based on our own teaching experiences and advice from many colleagues, we have included several features to help students learn and to broaden their perspective on engineering mechanics.

Problem-Solving Strategies Worked examples and homework problems are the heart of a course in engineering mechanics. Throughout the book, we provide descriptions of the approaches that we use in the examples and which students will find helpful in working problems. We do not provide recipes that students are intended to follow rigidly. Instead, we describe general lines of thought that apply to broad classes of problems and give useful advice and helpful warnings of common pitfalls, the kind of information we give to students during office hours. (See, for example, pp. 33, 242, 262 and 311.)

Applications Many of our examples and problems are derived from actual engineering practice, ranging from familiar household items to advanced engineering applications. In addition, examples labelled as 'Applications to Engineering' provide more detailed case studies from different engineering disciplines. These examples show how the principles learned in the text are directly applicable to current and future engineering problems. Our goal is to help students see the importance of engineering mechanics in these applications and so gain the motivation to learn it. (See, for example, pp. 79, 118 and 218.)

Computer Problems Surveys tell us that most instructors make some use of computers in engineering mechanics courses, but there is no consensus on how it should be done. We give the instructor the opportunity to introduce students to computer applications in dynamics, including the use of finite differences to integrate equations of motion, without imposing a particular approach. Optional sections called 'Computational Mechanics' contain examples and problems suitable for the use of a programmable calculator or computer. (See, for example, pp. 128 and 174.) The instructor can choose how students solve these problems, for example by using a programming language, a spreadsheet or a higher-level problem-solving environment. These sections are independent and self-contained.

Chapter Openings We begin each chapter with an illustration showing an application of the ideas in the chapter, often choosing objects that are familiar to students. By seeing how the concepts in this course relate to the design and function of familiar objects around them, students can begin to appreciate the importance and excitement of engineering as a career. (See pp. 98, 230 and 302.)

Commitment to Students and Instructors

We have taken precautions that ensure the accuracy of this book to the best of our ability. Reviewers examined each stage of the manuscript for errors. We have each solved the problems in an effort to be sure that their answers are correct and that the problems are of an appropriate level of difficulty. Eugene Davis, author of the Solutions Manual, further verified the answers while developing his solutions. As a further check, James Whitenton examined the entire text for errors that crept in during the typesetting process.

Any errors that remain are the responsibility of the authors. We welcome communication from students and instructors concerning errors or areas for improvement. Our mailing address is Department of Aerospace Engineering and Engineering Mechanics, University of Texas at Austin, Austin, Texas 78712, USA. Our electronic mail address is bedford@aw.com.

Printed Supplements

Instructor's Solutions Manual The manual for the instructor contains step-by-step solutions to all problems. Each solution includes the problem statement and the associated art.

Study Guide This guide reinforces the Strategy–Solution–Discussion process outlined in the text. Selected solutions are provided in great detail, accompanied by suggested strategies for approaching problems of that type.

Transparencies Approximately 100 figures from the text have been prepared in four colours on acetate for use on an overhead projector.

Software Supplements

Student Edition of Working Model® Working Model (Knowledge Revolution, Inc.) is a simulation and modelling program that allows the student to visualize engineering problems. The program calculates the effects of forces on an object (or objects), animates the results, and provides output data such as force, moment, velocity and acceleration in digital or graphical form. The Student Edition make this powerful program affordable for undergraduate students. It is available in both Windows and Macintosh versions.

Working Model® Simulations Approximately 100 problems and examples from the text have been re-created on disk as Working Model simulations. These simulations have been constructed to allow the student to change variables and see the results. The student can explore physical situations in a 'what if' manner and thereby develop deeper conceptual insights than possible through quantitative problem solving alone. Students can purchase these simulations combined with the text for a nominal additional charge.

Acknowledgements

We are grateful to our teachers, colleagues and students for what we have learned about mechanics and about teaching mechanics. Many colleagues reviewed the manuscript and generously shared their knowledge and experience to improve our book. They are

Nick Altiero
Michigan State University

James G. Andrews
University of Iowa

Gautam Batra
University of Nebraska, Lincoln

Rathi Bhatacharya
Bradley University

Clarence Calder
Oregon State University

Anthony DeLuzio
Merrimack College

Xioamin Deng
University of South Carolina

James Dent
Montana State University

Robert W. Fitzgerald
Worcester Polytechnic Institute

Mark Frisina
Wentworth Institute

Robert W. Fuessle
Bradley University

John Giger
Rose State College

Robert A. Howland
University of Notre Dame

David B. Johnson
Southern Methodist University

Charles M. Krousgrill
Purdue University

Richard Lewis
Louisiana Technological University

Brad C. Liebst
University of Minnesota

Bertram Long
Northeastern University

V. J. Lopardo
US Naval Academy

Frank K. Lu
University of Texas, Arlington

Donald L. Margolis
University of California, Davis

George Mase
Michigan State University

William W. Seto
San Jose University

Francis M. Thomas
University of Kansas

Mark R. Virkler
University of Missouri, Columbia

William H. Walston, Jr
University of Maryland

Julius Wong
University of Louisville

We particularly thank Eugene Davis, Serope Kalpakjian and Eric Sandgren for suggesting many problems based on their broad knowledge of applications of mechanics in engineering. We thank the people of Addison-Wesley for their friendship and generous help, especially Bette Aaronson, Jennifer Duggan, Don Fowley, Joyce Grandy, Stuart Johnson, Laurie McGuire and Jim Rigney. We are especially grateful to our gifted editor, David Chelton, and equally gifted artist, James Bryant, for transforming our work far beyond our modest conception. We thank our chairman Richard Miksad for unfailing support and relief from distractions that made the project possible. And of course we thank our families, for everything.

Anthony Bedford and Wallace Fowler
July 1994
Austin, Texas

Contents

The first space shuttle flight took place on 12 April 1981. The space shuttle *Columbia* went into orbit 271 km above the earth. To achieve orbit, it had to attain a velocity relative to the centre of the earth of approximately 8 km per second. After two days, with Commander John Young at the controls, it landed at Edwards Air Force Base, California.

Chapter 1

Introduction

THE Space Shuttle was conceived as an economical method to transport personnel and equipment to orbit. Throughout its development, engineers used the principles of dynamics to predict its motion during boost, in orbit and while landing. These predictions were essential for the design of its aerodynamic configuration and structure, rocket engines and control system. Dynamics is one of the sciences underlying the design of all vehicles and machines.

1.1 *Engineering and Mechanics*

How do engineers design complex systems and predict their characteristics before they are constructed? Engineers have always relied on their knowledge of previous designs, experiments, ingenuity and creativity to develop new designs. Modern engineers add a powerful technique: they develop mathematical equations based on the physical characteristics of the devices they design. With these **mathematical models**, engineers predict the behaviour of their designs, modify them, and test them prior to their actual construction. Aerospace engineers used mathematical models to predict the paths the space shuttle would follow in flight. Civil engineers used mathematical models to analyse the response to loads of the steel frame of the 443 m Sears Tower in Chicago.

Engineers are responsible for the design, construction and testing of the devices we use, from simple things such as chairs and pencil sharpeners to complicated ones such as dams, cars, airplanes and spacecraft. They must have a deep understanding of the physics underlying these devices and must be familiar with the use of mathematical models to predict system behaviour. Students of engineering begin to learn how to analyse and predict the behaviour of physical systems by studying mechanics.

At its most basic level, mechanics is the study of forces and their effects. Elementary mechanics is divided into **statics**, the study of objects in equilibrium, and **dynamics**, the study of objects in motion. The results obtained in elementary mechanics apply directly to many fields of engineering. Mechanical and civil engineers who design structures use the equilibrium equations derived in statics. Civil engineers who analyse the responses of buildings to earthquakes and aerospace engineers who determine the trajectories of satellites use the equations of motion derived in dynamics.

Mechanics was the first analytical science; consequently fundamental concepts, analytical methods and analogies from mechanics are found in virtually every field of engineering. For example, students of chemical and electrical engineering gain a deeper appreciation of basic concepts in their fields such as equilibrium, energy and stability by learning them in their original mechanical contexts. In fact, by studying mechanics they retrace the historical development of these ideas.

1.2 *Learning Mechanics*

Mechanics consists of broad principles that govern the behaviour of objects. In this book we describe these principles and provide you with examples that demonstrate some of their applications. Although it's essential that you practise working problems similar to these examples, and we include many problems of this kind, our objective is to help you understand the principles well enough to apply them to situations that are new to you. Each generation of engineers confronts new problems.

Problem Solving

In the study of mechanics you learn problem-solving procedures you will use in succeeding courses and throughout your career. Although different types of problems require different approaches, the following steps apply to many of them:

- Identify the information that is given and the information, or answer, you must determine. It's often helpful to restate the problem in your own words. When appropriate, make sure you understand the physical system or model involved.
- Develop a **strategy** for the problem. This means identifying the principles and equations that apply and deciding how you will use them to solve the problem. Whenever possible, draw diagrams to help visualize and solve the problem.
- Whenever you can, try to predict the answer. This will develop your intuition and will often help you recognize an incorrect answer.
- Solve the equations and, whenever possible, interpret your results and compare them with your prediction. The latter step is called a **reality check**. Is your answer reasonable?

Calculators and Computers

Most of the problems in this book are designed to lead to an algebraic expression with which to calculate the answer in terms of given quantities. A calculator with trigonometric and logarithmic functions is sufficient to determine the numerical value of such answers. The use of a programmable calculator or a computer with problem-solving software such as *Mathcad* or *TK! Solver* is convenient, but be careful not to become too reliant on tools you will not have during tests.

Sections called *Computational Mechanics* contain examples and problems that are suitable for solution with a programmable calculator or a computer.

Engineering Applications

Although the problems are designed primarily to help you learn mechanics, many of them illustrate uses of mechanics in engineering. Sections called *Application to Engineering* describe how mechanics is applied in various fields of engineering.

We also include problems that emphasize two essential aspects of engineering:

- *Design*. Some problems ask you to choose values of parameters to satisfy stated design criteria.
- *Safety*. Some problems ask you to evaluate the safety of devices and choose values of parameters to satisfy stated safety requirements.

1.3 *Fundamental Concepts*

Some topics in mechanics will be familiar to you from everyday experience or from previous exposure to them in physics courses. In this section we briefly review the foundations of elementary mechanics.

Space and Time

Space simply refers to the three-dimensional universe in which we live. Our daily experiences give us an intuitive notion of space and the locations, or positions, of points in space. The distance between two points in space is the length of the straight line joining them.

Measuring the distance between points in space requires a unit of length. We use both the International System of units, or SI units, and US Customary units. In SI units, the unit of length is the metre (abbreviated to m). In US Customary units, the unit of length is the foot (ft).

Time is, of course, familiar – our lives are measured by it. The daily cycles of light and darkness and the hours, minutes and seconds measured by our clocks and watches give us an intuitive notion of time. Time is measured by the intervals between repeatable events, such as the swings of a clock pendulum or the vibrations of a quartz crystal in a watch. In both SI units and US Customary units, the unit of time is the second (abbreviated to s). Minutes (min), hours (h) and days are also frequently used.

If the position of a point in space relative to some reference point changes with time, the rate of change of its position is called its **velocity**, and the rate of change of its velocity is called its **acceleration**. In SI units, the velocity is expressed in metres per second (m/s) and the acceleration is expressed in metres per second per second, or metres per second squared (m/s^2). In US Customary units, the velocity is expressed in feet per second (ft/s) and the acceleration is expressed in feet per second squared (ft/s^2).

Newton's Laws

Elementary mechanics was established on a firm basis with the publication, in 1687, of *Philosophiae naturalis principia mathematica*, by Isaac Newton. Although highly original, it built upon fundamental concepts developed by many others during a long and difficult struggle towards understanding. Newton stated three 'laws' of motion, which we express in modern terms:

(1) *When the sum of the forces acting on a particle is zero, its velocity is constant. In particular, if the particle is initially stationary, it will remain stationary.*
(2) *When the sum of the forces acting on a particle is not zero, the sum of the forces is equal to the rate of change of the linear momentum of the particle. If the mass is constant, the sum of the forces is equal to the product of the mass of the particle and its acceleration.*
(3) *The forces exerted by two particles on each other are equal in magnitude and opposite in direction.*

Notice that we did not define force and mass before stating Newton's laws. The modern view is that these terms are defined by the second law. To demonstrate, suppose that we choose an arbitrary object and define it to have unit mass. Then we define a unit of force to be the force that gives our unit mass an acceleration of unit magnitude. In principle, we can then determine the mass of any object: we apply a unit force to it, measure the resulting acceleration and use the second law to determine the mass. We can also determine the magnitude of any force: we apply it to our unit mass, measure the resulting acceleration and use the second law to determine the force.

Thus Newton's second law gives precise meanings to the terms **mass** and **force**. In SI units, the unit of mass is the kilogram (kg). The unit of force is the newton (N), which is the force required to give a mass of one kilogram an acceleration of one metre per second squared. In US Customary units, the unit of force is the pound (lb). The unit of mass is the slug, which is the amount of mass accelerated at one foot per second squared by a force of one pound.

Although the results we discuss in this book are applicable to many of the problems met in engineering practice, there are limits to the validity of Newton's laws. For example, they don't give accurate results if a problem involves velocities that are not small compared with the velocity of light (3×10^8 m/s). Einstein's special theory of relativity applies to such problems. Elementary mechanics also fails in problems involving dimensions that are not large compared with atomic dimensions. Quantum mechanics must be used to describe phenomena on the atomic scale.

Newtonian Gravitation

Another of Newton's fundamental contributions to mechanics is his postulate for the gravitational force between two particles in terms of their masses m_1 and m_2 and the distance r between them (Figure 1.1). His expression for the magnitude of the force is

$$F = \frac{Gm_1m_2}{r^2} \tag{1.1}$$

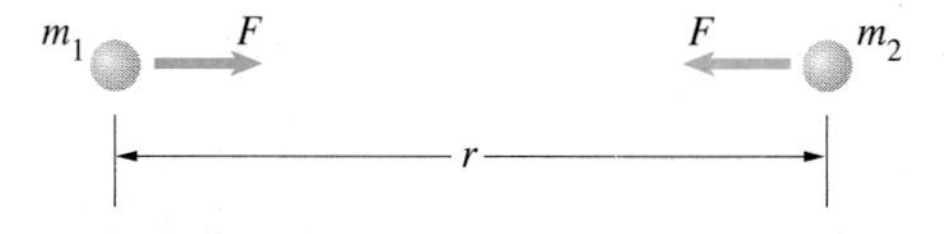

Figure 1.1

The gravitational forces between two particles are equal in magnitude and directed along the line between them.

where G is called the **universal gravitational constant**.

Newton calculated the gravitational force between a particle of mass m_1 and a homogeneous sphere of mass m_2 and found that it is also given by Equation (1.1), with r denoting the distance from the particle to the centre of the sphere. Although the earth is not a homogeneous sphere, we can use this result to approximate the weight of an object of mass m due to the gravitational attraction of the earth.

$$W = \frac{Gmm_E}{r^2} \tag{1.2}$$

where m_E is the mass of the earth and r is the distance from the centre of the earth to the object. Notice that the weight of an object depends on its location relative to the centre of the earth, whereas the mass of the object is a measure of the amount of matter it contains, and doesn't depend on its position.

When an object's weight is the only force acting on it, the resulting acceleration is called the **acceleration due to gravity**. In this case, Newton's second law states that $W = ma$, and from Equation (1.2) we see that the acceleration due to gravity is

$$a = \frac{gm_E}{r^2} \tag{1.3}$$

The **acceleration due to gravity at sea level** is denoted by g. Denoting the radius of the earth by R_E, we see from Equation (1.3) that $Gm_E = gR_E^2$. Substituting this result into Equation (1.3), we obtain an expression for the acceleration due to gravity at a distance r from the centre of the earth in terms of the acceleration due to gravity at sea level:

$$a = g\frac{R_E^2}{r^2} \tag{1.4}$$

Since the weight of the object $W = ma$, the weight of an object at a distance r from the centre of the earth is

$$W = mg\frac{R_E^2}{r^2} \tag{1.5}$$

At sea level, the weight of an object is given in terms of its mass by the simple relation

$$W = mg \tag{1.6}$$

The value of g varies from location to location on the surface of the earth. The values we use in examples and problems are $g = 9.81\ \text{m/s}^2$ in SI units and $g = 32.2\ \text{ft/s}^2$ in US Customary units.

Numbers

Engineering measurements, calculations and results are expressed in numbers. You need to know how we express numbers in the examples and problems and how to express the results of your own calculations.

Significant Digits This term refers to the number of meaningful (that is, accurate) digits in a number, counting to the right starting with the first nonzero digit. The two numbers 7.630 and 0.007 630 are each stated to four significant digits. If only the first four digits in the number 7 630 000 are known to be accurate, this can be indicated by writing the number in scientific notation as 7.630×10^6.

If a number is the result of a measurement, the significant digits it contains are limited by the accuracy of the measurement. If the result of a measurement is stated to be 2.43, this means that the actual value is believed to be closer to 2.43 than to 2.42 or 2.44.

Numbers may be **rounded off** to a certain number of significant digits. For example, we can express the value of π to three significant digits, 3.14, or we can express it to six significant digits, 3.14159. When you use a calculator or computer, the number of significant digits is limited by the number of digits the machine is designed to carry.

Use of Numbers in This Book You should treat numbers given in problems as exact values and not be concerned about how many significant digits they contain. If a problem states that a quantity equals 32.2, you can assume its value is 32.200 . . . . We express intermediate results and answers in the examples and the answers to the problems to at least three significant digits. If you use a calculator, your results should be that accurate. Be sure to avoid round-off errors that occur if you round off intermediate results when making a series of calculations. Instead, carry through your calculations with as much accuracy as you can by retaining values in your calculator.

1.4 Units

The SI system of units has become nearly standard throughout the world. In the USA, US Customary units are also used. In this section we summarize these two systems of units and explain how to convert units from one system to another.

International System of Units

In SI units, length is measured in metres (m) and mass in kilograms (kg). Time is measured in seconds (s), although other familiar measures such as minutes (min), hours (hr) and days are also used when convenient. Metres, kilograms and seconds are called the **base units** of the SI system. Force is measured in newtons (N). Recall that these units are related by Newton's second law: one newton is the force required to give an object of one kilogram mass an acceleration of one metre per second squared.

$$1\,\mathrm{N} = (1\,\mathrm{kg})(1\,\mathrm{m/s^2}) = 1\,\mathrm{kg.m/s^2}$$

Since the newton can be expressed in terms of the base units, it is called a **derived unit**.

To express quantities by numbers of convenient size, multiples of units are indicated by prefixes. The most common prefixes, their abbreviations and the multiples they represent are shown in Table 1.1. For example, 1 km is 1 kilometre, which is 1000 m, and 1 Mg is 1 megagram, which is 10^6 g or 1000 kg. We frequently use kilonewtons (kN).

Table 1.1 The common prefixes used in SI units and the multiples they represent.

Prefix	Abbreviation	Multiple
nano-	n	10^{-9}
micro	μ	10^{-6}
milli-	m	10^{-3}
kilo-	k	10^{3}
mega-	M	10^{6}
giga-	G	10^{9}

US Customary Units

In US Customary units, length is measured in feet (ft) and force is measured in pounds (lb). Time is measured in seconds (s). These are the base units of the US Customary system. In this system of units, mass is a derived unit. The unit of mass is the **slug**, which is the mass of material accelerated at one foot per second squared by a force of one pound. Newton's second law states that

$$1 \text{ lb} = (1 \text{ slug})(1 \text{ ft/s}^2)$$

From this expression we obtain

$$1 \text{ slug} = 1 \text{ lb.s}^2/\text{ft}$$

We use other US Customary units such as the mile (1 mi = 5280 ft) and the inch (1 ft = 12 in.). We also use the kilopound (kip), which is 1000 lb.

In some engineering applications, an alternative unit of mass called the pound mass (lbm) is used, which is the mass of material having a weight of one pound at sea level. The weight at sea level of an object that has a mass of one slug is

$$W = mg = (1 \text{ slug})(32.2 \text{ ft/s}^2) = 32.2 \text{ lb}$$

so 1 lbm = (1/32.2) slug. When the pound mass is used, a pound of force is usually denoted by the abbreviation lbf.

Angular Units

In both SI and US Customary units, angles are normally expressed in radians (rad). We show the value of an angle θ in radians in Figure 1.2. It is defined to be the ratio of the part of the circumference subtended by θ to the radius of the circle. Angles are also expressed in degrees. Since there are 360 degrees (360°) in a complete circle, and the complete circumference of the circle is $2\pi R$, 360° equals 2π rad.

Equations containing angles are nearly always derived under the assumption that angles are expressed in radians. Therefore when you want to substitute the value of an angle expressed in degrees into an equation, you should first convert is into radians. A notable exception to this rule is that many calculators are designed to accept angles expressed in either degrees or radians when you use them to evaluate functions such as $\sin\theta$.

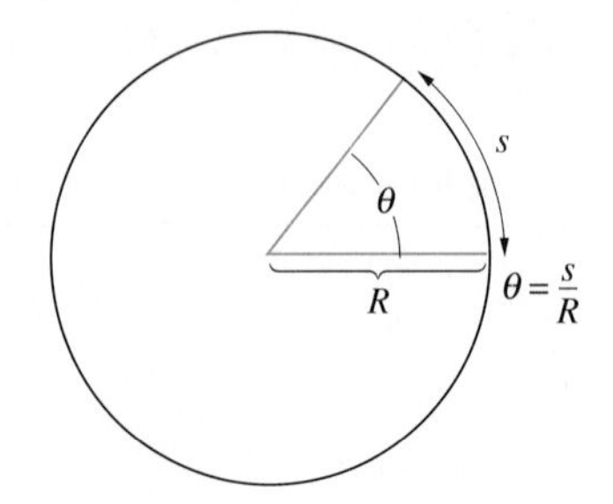

Figure 1.2
Definition of an angle in radians.

Conversion of Units

Many situations arise in engineering practice that require you to convert values expressed in units of one kind into values in other units. If some data in a problem are given in terms of SI units and some are given in terms of US Customary units, you must express all of the data in terms of one system of units. In problems expressed in terms of SI units, you will occasionally be given data in terms of units other than the base units of seconds, metres, kilograms and newtons. You should convert these data into the base units before working the problem. Similarly, in problems involving US Customary units you should convert terms into the base units of seconds, feet, slugs and pounds. After you gain some experience, you will recognize situations in which these rules can be relaxed, but for now they are the safest procedure.

Converting units is straightforward, although you must do it with care. Suppose that we want to express 1 mi/hr in terms of ft/s. Since one mile equals 5280 ft and 1 hour equals 3600 seconds, we can treat the expressions

$$\left(\frac{5280\ \text{ft}}{1\ \text{mi}}\right) \quad \text{and} \quad \left(\frac{1\ \text{hr}}{3600\ \text{s}}\right)$$

as ratios whose values are 1. In this way we obtain

$$1\ \text{mi/hr} = 1\ \text{mi/hr} \times \left(\frac{5280\ \text{ft}}{1\ \text{mi}}\right) \times \left(\frac{1\ \text{hr}}{3600\ \text{s}}\right) = 1.47\ \text{ft/s}$$

We give some useful conversions in Table 1.2.

Table 1.2 Unit conversions.

Time	1 minute	=	60 second
	1 hour	=	60 minutes
	1 day	=	24 hours
Length	1 foot	=	12 inches
	1 mile	=	5280 feet
	1 inch	=	25.4 millimetres
	1 foot	=	0.3048 metres
Angle	2π radians	=	360 degrees
Mass	1 slug	=	14.59 kilograms
Force	1 pound	=	4.448 newtons

Example 1.1

If an Olympic sprinter (Figure 1.3) runs 100 metres in 10 seconds, his average velocity is 10 m/s. What is his average velocity in mi/hr?

Figure 1.3

SOLUTION

$$10\,\text{m/s} = 10\,\text{m/s} \times \left(\frac{1\,\text{ft}}{0.3048\,\text{m}}\right) \times \left(\frac{1\,\text{mi}}{5280\,\text{ft}}\right) \times \left(\frac{3600\,\text{s}}{1\,\text{hr}}\right)$$
$$= 22.4\,\text{mi/hr}$$

Example 1.2

Suppose that in Einstein's equation

$$E = mc^2$$

the mass m is in kg and the velocity of light c is in m/s.
(a) What are the SI units of E?
(b) If the value of E in SI units is 20, what is its value in US Customary base units?

STRATEGY

(a) Since we know the units of the terms m and c, we can deduce the units of E from the given equation.
(b) We can use the unit conversions for mass and length from Table 1.2 to convert E from SI units to US Customary units.

SOLUTION

(a) From the equation for E,

$$E = (m\,\text{kg})(c\,\text{m/s})^2$$

the SI units of E are $\text{kg.m}^2/\text{s}^2$.

(b) From Table 1.2, 1 slug = 14.59 kg and 1 ft = 0.3048 m. Therefore

$$1\,\text{kg.m}^2/\text{s}^2 = 1\,\text{kg.m}^2/\text{s}^2 \times \left(\frac{1\,\text{slug}}{14.59\,\text{kg}}\right) \times \left(\frac{1\,\text{ft}}{0.3048\,\text{m}}\right)^2$$
$$= 0.738\,\text{slug.ft}^2/\text{s}^2$$

The value of E in US Customary units is

$$E = (20)(0.738) = 14.8\,\text{slug-ft}^2/\text{s}^2$$

Example 1.3

George Stephenson's *Rocket* (Figure 1.4), an early steam locomotive, weighed about 7 tons with its tender. (A ton is 2000 lb.) What was its approximate mass in kilograms?

Figure 1.4

STRATEGY

We can use Equation (1.6) to obtain the mass in slugs and then use the conversion given in Table 1.2 to determine the mass in kilograms.

SOLUTION

The mass in slugs is

$$m = \frac{W}{g} = \frac{15\,680\,\text{lb}}{32.2\,\text{ft/s}^2} = 487.0\,\text{slugs}$$

From Table 1.2, 1 slug equals 14.59 kg, so the mass in kilograms is (to three significant digits)

$$m = (487.0)(14.59) = 7105\,\text{kg}$$

Problems

1.1 The value of π is 3.141 592 654. . . . What is its value to four significant digits?

1.2 What is the value of e (the base of natural logarithms) to five significant digits?

1.3 Determine the value of the expression $1/(2 - \pi)$ to three significant digits.

1.4 If $x = 3$, what is the value of the expression $1 - e^{-x}$ to three significant digits?

1.5 Suppose that you have just purchased a Ferrari Dino 246GT coupe and you want to know whether you can use your set of wrenches to work on it. You have wrenches with widths $w = 1/4$ in., 1/2 in., 3/4 in. and 1 in., and the car has nuts with dimensions $n = 5$ mm, 10 mm, 15 mm, 20 mm and 25 mm. Defining a wrench to fit if w is no more that 2 per cent larger than n, which of your wrenches can you use?

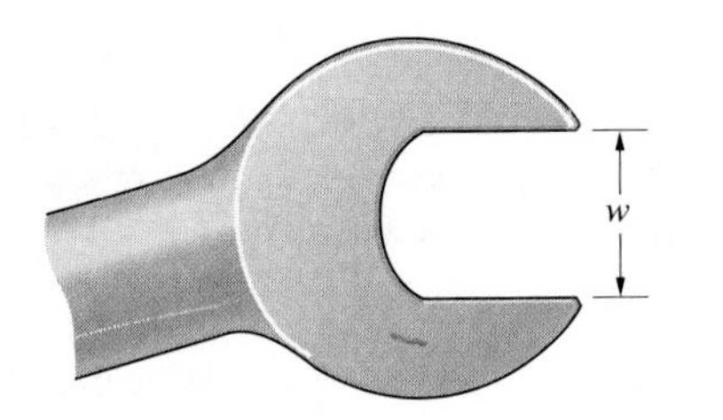

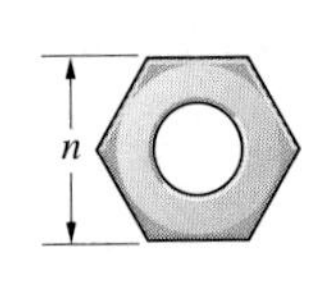

P1.5

1.6 The 1829 *Rocket*, shown in Example 1.3, could draw a carriage with 30 passengers at 25 mi/hr. Determine its velocity to three significant digits: (a) in ft/s; (b) in km/hr.

1.7 High-speed 'bullet trains' began running between Tokyo and Osaka, Japan, in 1964. If a bullet train travels at 240 km/hr, what is its velocity in mi/hr to three significant digits?

1.8 Engineers who study shock waves sometimes express velocity in millimetres per microsecond (mm/μs). Suppose the velocity of a wavefront is measured and determined to be 5 mm/μs. Determine its velocity: (a) in m/s; (b) in mi/s.

1.9 Geophysicists measure the motion of a glacier and discover it is moving at 80 mm/year. What is its velocity in m/s?

1.10 The acceleration due to gravity at sea level in SI units is $g = 9.81$ m/s^2. By converting units, use this value to determine the acceleration due to gravity at sea level in US Customary units.

1.11 A *furlong per fortnight* is a facetious unit of velocity, perhaps made up by a student as a satirical comment on the bewildering variety of units engineers must deal with. A furlong is 660 ft (1/8 mi). A fortnight is two weeks (14 nights). If you walk to class at 5 ft/s, what is your velocity in furlongs per fortnight to three significant digits?

1.12 The cross-sectional area of a beam is 480 in^2. What is its cross-sectional area in m^2?

1.13 A truck can carry 15 cubic yards of gravel. A yard equals 3 feet. How many cubic metres of gravel can the truck carry?

1.14 A pressure transducer measures a value of 300 lb/in^2. Determine the value of the pressure in pascals. A pascal (Pa) is 1 N/m^2.

1.15 A horsepower is 550 ft.lb/s. A watt is 1 N.m/s. Determine the number of watts generated by (a) the Wright brothers' 1903 aeroplane, which had a 12-horsepower engine; (b) a modern passenger jet with a power of 100 000 horsepower at cruising speed.

P1.15

1.16 In SI units, the universal gravitational constant $G = 6.67 \times 10^{-11}$ N.m^2/kg^2. Determine the value of G in US Customary base units.

1.17 If the earth is modelled as a homogeneous sphere, the velocity of a satellite in a circular orbit is

$$v = \sqrt{\frac{gR_E^2}{r}}$$

where R_E is the radius of the earth and r is the radius of the orbit.
(a) If g is in m/s^2 and R_E and r are in metres, what are the units of v?
(b) If $R_E = 6370$ km and $r = 6670$ km, what is the value of v to three significant digits?

(c) For the orbit described in part (b), what is the value of v in mi/s to three significant digits?

1.18 In the equation

$$T = \frac{1}{2} I \omega^2$$

the term I is in kg-m^2 and ω is in s^{-1}.
(a) What are the SI units of T?
(b) If the value of T is 100 when I is in kg-m^2 and ω is in s^{-1}, what is the value of T when it is expressed in terms of US Customary base units?

1.19 The 'crawler' developed to transport the *Saturn V* launch vehicle from the vehicle assembly building to the launch pad is the largest land vehicle ever built, weighing 4.9×10^6 lb at sea level.
(a) What is its mass in slugs?
(b) What is its mass in kilograms?
(c) A typical car has a mass of about 1000 kg. How many such cars does it take to have the same weight as the crawler at sea level?

1.20 The acceleration due to gravity is 13.2 ft/s^2 on the surface of Mars and 32.2 ft/s^2 on the surface of earth. If a woman weighs 125 lb on earth, what would she weigh on Mars?

1.21 The acceleration due to gravity is 13.2 ft/s^2 on the surface of Mars and 32.2 ft/s^2 on the surface of the earth. A woman weighs 125 lb on earth. To survive and work on the surface of Mars, she must wear life-support equipment and carry tools. What is the maximum allowable weight on earth of the woman's clothing, equipment and tools if the engineers don't want the total weight on Mars of the woman and her clothing, equipment and tools to exceed 125 lb?

1.22 A person has a mass of 50 kg.
(a) The acceleration due to gravity at sea level is $g = 9.81$ m/s^2. What is the person's weight at sea level?
(b) The acceleration due to gravity on the surface of the moon is 1.62 m/s^2. What would the person weigh on the moon?

1.23 The acceleration due to gravity at sea level is $g = 9.81$ m/s^2. The radius of the earth is 6370 km. The universal gravitational constant $G = 6.67 \times 10^{-11}$ N-m^2/kg^2. Use this information to determine the mass of the earth.

1.24 A person weighs 180 lb at sea level. The radius of the earth is 3960 mi. What force is exerted on the person by the gravitational attraction of the earth if he is in a space station in near-earth orbit 200 mi above the surface of the earth?

1.25 The acceleration due to gravity on the surface of the moon is 1.62 m/s^2. The radius of the moon is $R_M = 1738$ km. Determine the acceleration due to gravity of the moon at a point 1738 km above its surface.

Strategy: Write an equation equivalent to Equation (1.4) for the acceleration due to gravity of the moon.

1.26 If an object is near the surface of the earth, the variation of its weight with distance from the centre of the earth can often be neglected. The acceleration due to gravity at sea level is $g = 9.81$ m/s^2. The radius of the earth is 6370 km. The weight of an object at sea level is mg, where m is its mass. At what height above the surface of the earth does the height of the object decrease to $0.99\,mg$?

1.27 The centres of two oranges are 1 m apart. The mass of each orange is 0.2 kg. What gravitational force do they exert on each other? (The universal gravitational constant $G = 6.67 \times 10^{-11}$ N.m^2/kg^2.)

1.28 One inch equals 25.4 mm. The mass of one cubic metre of water is 1000 kg. The acceleration due to gravity at sea level is $g = 9.81$ m/s^2. The weight of one cubic foot of water at sea level is approximately 62.4 lb. By using this information, determine how many newtons equal one pound.

P1.19

The position and velocity of the *Voyager 2* space probe at the time of its release near Earth determined the trajectory (path) it followed to reach the planet Jupiter. The gravitational field of Jupiter altered the trajectory of *Voyager 2* so that it could pass near Saturn, which altered its trajectory again so that it could pass near Uranus, and so on to Neptune. In this chapter you will determine trajectories of objects and analyse their positions, velocities and accelerations using different types of coordinate systems.

Chapter 2

Motion of a Point

ENGINEERS designing a vehicle, whether a bicycle or a spacecraft, must be able to analyse and predict its motion. To design an engine, they must analyse the motions of each of its moving parts. Even when designing 'static' structures such as buildings, bridges and dams, they must often analyse motions resulting from wind loads and potential earthquakes.

In this chapter we begin the study of motion. We are not concerned here with the properties of objects or the causes of their motions – we merely want to describe and analyse the motion of a point in space. However, keep in mind that the point can represent some point (such as the centre of mass) of a moving object. After defining the position, velocity and acceleration of a point, we consider the simplest example: motion along a straight line. We then show how motion of a point along an arbitrary path, or **trajectory**, is expressed and analysed in various coordinate systems.

2.1 Position, Velocity and Acceleration

We can describe the position of a point P by choosing a **reference point** O and introducing the **position vector r** from O to P (Figure 2.1(a)). Suppose that P is in motion relative to O, so that $\mathbf{r}$ is a function of time t (Figure 2.1(b)). We express this by the notation

$$\mathbf{r} = \mathbf{r}(t)$$

The **velocity** of P relative to O at time t is defined by

$$\mathbf{v} = \frac{d\mathbf{r}}{dt} = \lim_{\Delta t \to 0} \frac{\mathbf{r}(t + \Delta t) - \mathbf{r}(t)}{\Delta t} \tag{2.1}$$

where the vector $\mathbf{r}(t + \Delta t) - \mathbf{r}(t)$ is the change in position, or **displacement** of P, during the interval of time Δt (Figure 2.1(c)). Thus the velocity is the rate of change of the position of P relative to O.

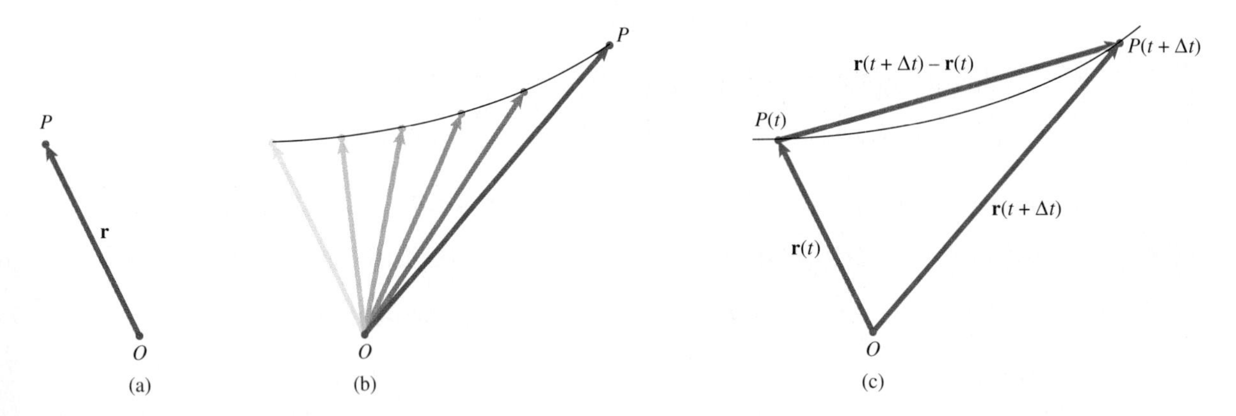

Figure 2.1
(a) The position vector $\mathbf{r}$ of P relative to O.
(b) Motion of P relative to O.
(c) Change in the position of P from t to $t + \Delta t$.

The dimensions of a derivative are determined just as if it were a ratio, so the dimensions of $\mathbf{v}$ are (distance)/(time). The reference point being used is often obvious, and we simply call $\mathbf{v}$ the velocity of P. However, you must remember that the position and velocity of a point can be specified only relative to some reference point.

Notice in Equation (2.1) that the derivative of a vector with respect to time is defined in exactly the same way as is the derivative of a scalar function. As a result, it shares some of the properties of the derivative of a scalar function. We will use two of these properties. The time derivative of the sum of two vector functions $\mathbf{u}$ and $\mathbf{w}$ is

$$\frac{d}{dt}(\mathbf{u} + \mathbf{w}) = \frac{d\mathbf{u}}{dt} + \frac{d\mathbf{w}}{dt}$$

and the time derivative of the product of a scalar function f and a vector function $\mathbf{u}$ is

$$\frac{d(f\mathbf{u})}{dt} = \frac{df}{dt}\mathbf{u} + f\frac{d\mathbf{u}}{dt}$$

The **acceleration** of P relative to O at time t is defined by

$$\mathbf{a} = \frac{d\mathbf{v}}{dt} = \lim_{\Delta t \to 0} \frac{\mathbf{v}(t+\Delta t) - \mathbf{v}(t)}{\Delta t} \tag{2.2}$$

where $\mathbf{v}(t+\Delta t) - \mathbf{v}(t)$ is the change in the velocity of P during the interval of time Δt (Figure 2.2). The acceleration is the rate of change of the velocity of P at time t (the second time derivative of the displacement), and its dimensions are (distance)/(time)^2.

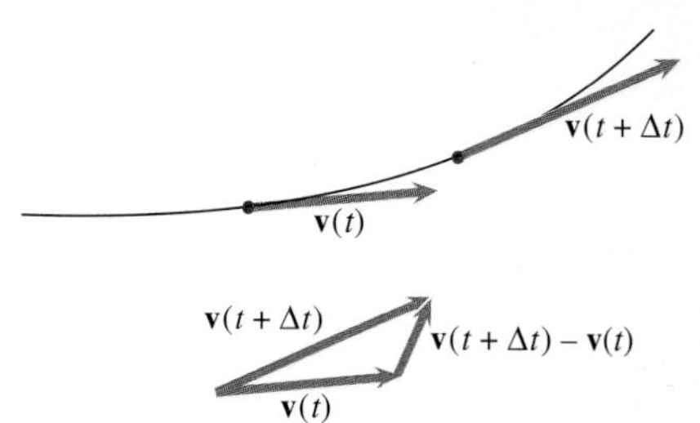

Figure 2.2
Change in the velocity of P from t to $t + \Delta t$.

2.2 Straight-Line Motion

We discuss this simple type of motion primarily so that you can gain experience and insight before proceeding to the general case of motion of a point. But engineers must analyse straight-line motions in many practical situations, such as the motion of a vehicle on a straight road or track or the motion of a piston in an internal combustion engine.

Description of the Motion

We can specify the position of a point P on a straight line relative to a reference point O by the coordinate s measured along the line from O to P (Figure 2.3(a)). In this case we define s to be positive to the right, so s is positive when P is to the right of O and negative when P is to the left of O. The **displacement** Δs relative to O during an interval time of time from t_0 to t is the change in the position, $\Delta s = s(t) - s(t_0)$.

By introducing a unit vector $\mathbf{e}$ that is parallel to the line and points in the positive s direction (Figure 2.3(b)), we can write the position vector of P relative to O as

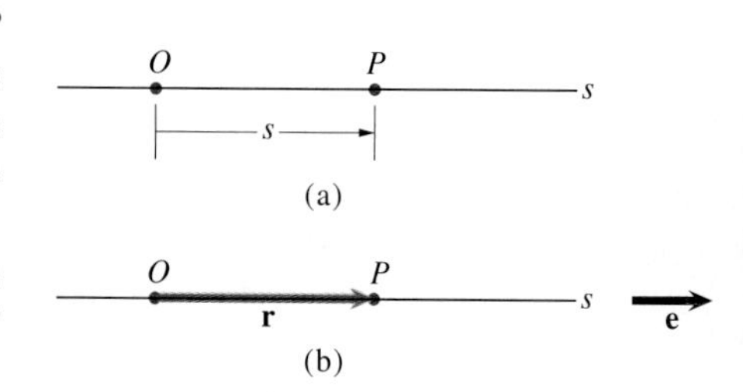

Figure 2.3
(a) The coordinate s from O to P.
(b) The unit vector $\mathbf{e}$ and position vector $\mathbf{r}$.

$$\mathbf{r} = s\,\mathbf{e}$$

If the line does not rotate, the unit vector $\mathbf{e}$ is constant and the velocity of P relative to O is

$$\mathbf{v} = \frac{d\mathbf{r}}{dt} = \frac{ds}{dt}\mathbf{e}$$

We can write the velocity vector as $\mathbf{v} = v\,\mathbf{e}$, obtaining the scalar equation

$$v = \frac{ds}{dt}$$

The velocity v of point P along the straight line is the rate of change of its position s. Notice that v is equal to the slope at time t of the line tangent to the graph of s as a function of time (Figure 2.4).

The acceleration of P relative to O is

$$\mathbf{a} = \frac{d\mathbf{v}}{dt} = \frac{d}{dt}(v\,\mathbf{e}) = \frac{dv}{dt}\mathbf{e}$$

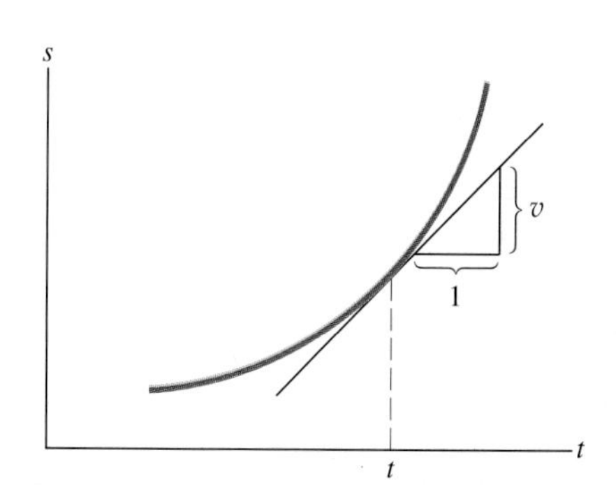

Figure 2.4
The slope of the straight line tangent to the graph of s versus t is the velocity at time t.

Writing the acceleration vector as $\mathbf{a} = a\,\mathbf{e}$, we obtain the scalar equation

$$a = \frac{dv}{dt} = \frac{d^2s}{dt^2}$$

The acceleration a is equal to the slope at time t of the line tangent to the graph of v as a function of time (Figure 2.5).

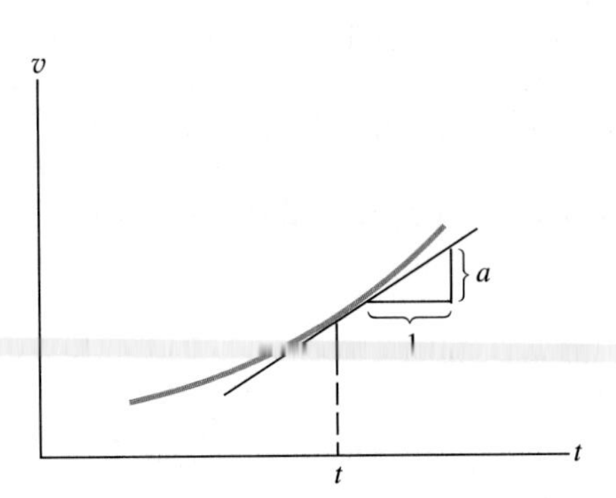

Figure 2.5
The slope of the straight line tangent to the graph of v versus t is the acceleration at time t.

By introducing the unit vector **e**, we have obtained scalar equations describing the motion of P. The position is specified by the coordinate s, and the velocity and acceleration are governed by the equations

$$v = \frac{ds}{dt} \tag{2.3}$$

$$a = \frac{dv}{dt} \tag{2.4}$$

Analysis of the Motion

In some situations, you will know the position s of some point of an object as a function of time. Engineers use methods such as radar and laser-doppler interferometry to measure positions as functions of time. In this case, you can obtain the velocity and acceleration as functions of time from Equations (2.3) and (2.4) by differentiation. For example, if the position of the truck in Figure 2.6 during the interval of time from $t = 2$ s to $t = 4$ s is given by the equation

$$s = \left(6 + \frac{1}{3}t^3\right)\text{m}$$

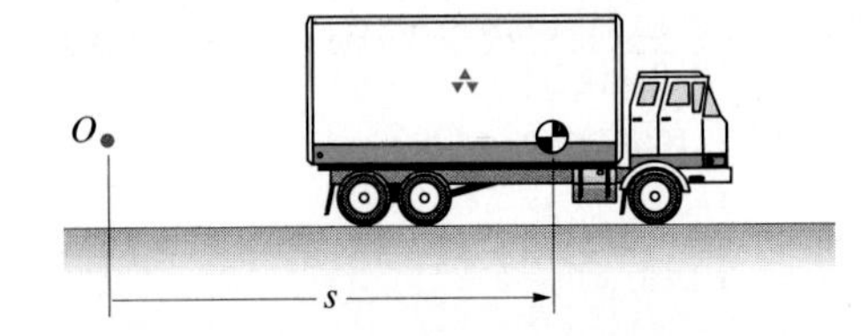

Figure 2.6
The coordinate s measures the position of the centre of mass of the track relative to a reference point.

its velocity and acceleration during that interval of time are

$$v = \frac{ds}{dt} = t^2 \text{ m/s}$$

$$a = \frac{dv}{dt} = 2t \text{ m/s}^2$$

However, it is more common to know an object's acceleration than to know its position, because the acceleration of an object can be determined by Newton's second law when the forces acting on it are known. When the acceleration is known, you can determine the velocity and position from Equations (2.3) and (2.4) by integration. We discuss three important cases in the following sections.

Acceleration Specified as a Function of Time If the acceleration is a known function of time $a(t)$, we can integrate the relation

$$\frac{dv}{dt} = a(t) \tag{2.5}$$

with respect to time to determine the velocity as a function of time

$$v = \int a(t)\, dt + A \tag{2.6}$$

where A is an integration constant. Then we can integrate the relation

$$\frac{ds}{dt} = v \tag{2.7}$$

to determine the position as a function of time

$$s = \int v\, dt + B \tag{2.8}$$

where B is another integration constant. We would need additional information about the motion, such as the values of v and s at a given time, to determine the constants A and B.

Instead of using indefinite integrals, we can write Equation (2.5) as

$$dv = a(t)\, dt$$

and integrate in terms of definite integrals:

$$\int_{v_0}^{v} dv = \int_{t_0}^{t} a(t)\, dt$$

The lower limit v_0 is the velocity at time t_0, and the upper limit v is the velocity at an arbitrary time t. Evaluating the left integral, we obtain an expression for the velocity as a function of time:

$$v = v_0 + \int_{t_0}^{t} a(t)\, dt \tag{2.9}$$

We can then write Equation (2.7) as

$$ds = v\,dt$$

and integrate in terms of definite integrals

$$\int_{s_0}^{t} ds = \int_{t-0}^{t} v\,dt$$

where the lower limit s_0 is the position at time t_0 and the upper limit s is the position at an arbitrary time t. Evaluating the left integral, we obtain the position as a function of time:

$$s = s_0 + \int_{t_0}^{t} v\,dt \tag{2.10}$$

Although we have shown how to determine the velocity and position when you know the acceleration as a function of time, you shouldn't try to remember results such as Equations (2.9) and (2.10). As we will demonstrate in the examples, we recommend that you solve straight-line motion problems by beginning with Equations (2.3) and (2.4).

We can make some useful observations from Equations (2.9) and (2.10):

- The area defined by the graph of the acceleration of P as a function of time from t_0 to t is equal to the change in the velocity from t_0 to t (Figure 2.7(a)).
- The area defined by the graph of the velocity of P as a function of time from t_0 to t is equal to the displacement, or change in position, from t_0 to t (Figure 2.7(b)).

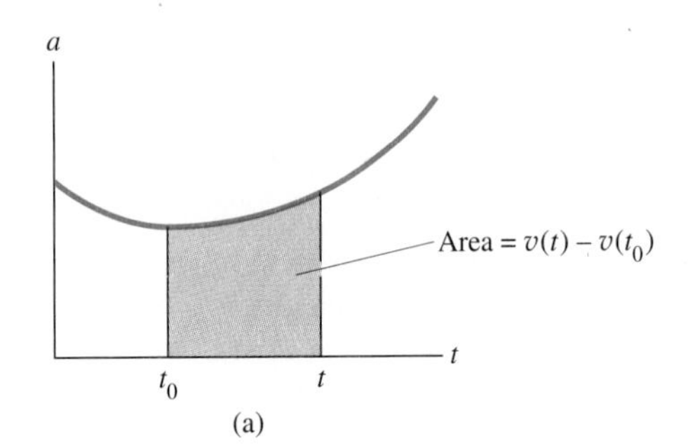

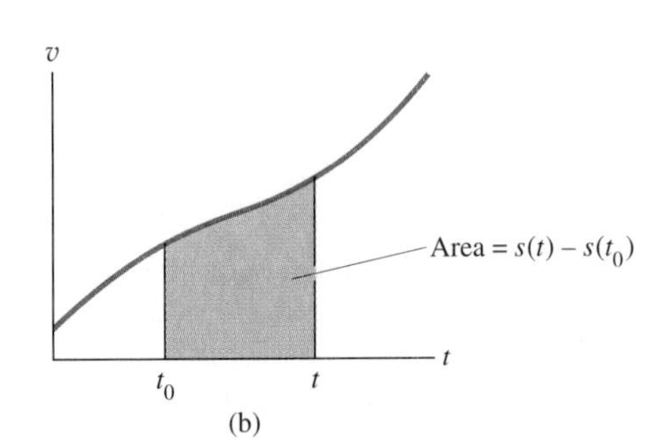

Figure 2.7
Relations between areas defined by the graphs of the acceleration and velocity of P and changes in its velocity and position.

You can often use these relationships to obtain a qualitative understanding of an object's motion, and in some cases you can even use them to determine its motion.

In some situations, the acceleration of an object is constant, or nearly constant. For example, if you drop a dense object such as a golf ball or rock and it doesn't fall too far, you can neglect aerodynamic drag and assume that its acceleration is equal to the acceleration of gravity at sea level.

Let the acceleration be a known constant a_0. From Equations (2.9) and (2.10), the velocity and position as functions of time are

$$v = v_0 + a_0(t - t_0) \tag{2.11}$$

$$s = s_0 + v_0(t - t_0) + \frac{1}{2}a_0(t - t_0)^2 \tag{2.12}$$

where s_0 and v_0 are the position and velocity, respectively, at time t_0. Notice that *if the acceleration is constant, the velocity is a linear function of time.*

We can use the **chain rule** to express the acceleration in terms of a derivative with respect to s:

$$a_0 = \frac{dv}{dt} = \frac{dv}{ds}\frac{ds}{dt} = \frac{dv}{ds}v$$

Writing this expression as $v dv = a_0 ds$ and integrating,

$$\int_{v_0}^{v} v dv = \int_{s_0}^{s} a_0 \, ds$$

we obtain an equation for the velocity as a function of position:

$$v^2 = v_0^2 + 2a_0(s - s_0) \tag{2.13}$$

You are probably familiar with Equations (2.11)–(2.13). Although these results can be useful *when you know that the acceleration is constant,* you must be careful not to use them otherwise.

The following examples illustrate how you can use Equations (2.3) and (2.4) to obtain information about straight-line motions of objects. You may need to choose the reference point and the positive direction for s. When you know the acceleration as a function of time, you can integrate Equation (2.4) to determine the velocity and then integrate Equation (2.3) to determine the position.

Example 2.1

Engineers testing a vehicle that will be dropped by parachute estimate that its vertical velocity when it reaches the ground will be 6.1 m/s. If they drop the vehicle from the test rig in Figure 2.8, from what height h should they drop it to simulate the parachute drop?

Figure 2.8

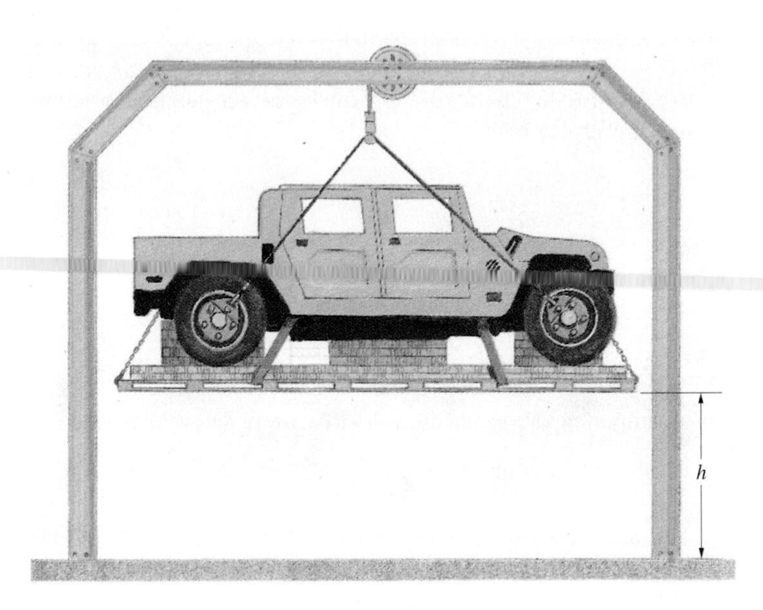

STRATEGY

We can assume that the vehicle's acceleration during its short fall is $g = 9.81\ \mathrm{m/s^2}$. We can determine the height h in two ways:

- *First method.* We can integrate Equations (2.3) and (2.4) to determine the vehicle's motion.
- *Second method.* We can use Equation (2.13), which relates the velocity and position when the acceleration is constant.

SOLUTION

We let s be the position of the bottom of the platform supporting the vehicle relative to its initial position (Figure (a)). The vehicle's acceleration is $a = 9.8\ \mathrm{m/s^2}$.

First Method From Equation (2.4),

$$\frac{dv}{dt} = a = 9.81\ m/s^2$$

Integrating, we obtain

$$v = 9.81t + A$$

where A is an integration constant. If we let $t = 0$ be the instant the vehicle is

(a) The coordinate s measures the position of the bottom of the platform relative to its position.

dropped, $v = 0$ when $t = 0$, so $A = 0$ and the velocity as a function of time is

$$v = 9.81 \text{ m/s}$$

Then by integrating Equation (2.3),

$$\frac{ds}{dt} = v = 9.81t$$

we obtain

$$s = 4.905t^2 + B$$

where B is a second integration constant. The position $x = 0$ when $t = 0$, so $B = 0$ and the position as a function of time is

$$s = 4.905t^2$$

From the equation for the velocity, the time of fall necessary for the vehicle to reach 6.1 m/s is $t = 6.1/9.81 = 0.622$ s. Substituting this time into the equation for the position, the height h needed to simulate the parachute drop is

$$h = 4.905(0.622)^2 = 1.90 \text{ m}$$

Second Method Because the acceleration is constant, we can use Equation (2.13) to determine the distance necessary for the velocity to increase to 6.1 m/s:

$$v^2 = v_0^2 + 2a_0(s - s_0)$$
$$(6.1)^2 = 0 + (9.81)(s - 0)$$

Solving for s, we obtain $h = 1.90$ m.

Example 2.2

The cheetah, *Acinonyx jubatus* (Figure 2.9), can run as fast as 75 mi/hr. If you assume that the animal's acceleration is constant and that it reaches top speed in 4 s, what distance can it cover in 10 s?

Figure 2.9

STRATEGY

The acceleration has a constant value for the first 4 s and is then zero. We can determine the distance travelled during each of these 'phases' of the motion and sum them to obtain the total distance covered. We do so both analytically and graphically.

SOLUTION

The top speed in terms of feet per second is

$$75\,\text{mi/hr} = 75\,\text{mi/hr} \times \left(\frac{5280\,\text{ft}}{1\,\text{mi}}\right) \times \left(\frac{1\,\text{hr}}{3600\,\text{s}}\right) = 110\,\text{ft/s}$$

First Method Let a_0 be the acceleration during the first 4 s. We integrate Equation (2.4),

$$\int_0^v dv = \int_0^t a_0\,dt$$

obtaining the velocity as a function of time during the first 4 s:

$$v = a_0 t\ \text{ft/s}$$

When $t = 4$ s, $v = 110$ ft/s, so $a_0 = 110/4 = 27.5$ ft/s^2. Now we integrate Equation (2.3),

$$\int_0^s ds = \int_0^t 27.5t\,dt$$

obtaining the position as a function of time during the first 4 s:

$$s = 13.75t^2 \text{ ft}$$

At $t = 4$ s, the position is $s = 13.75(4)^2 = 220$ ft.
From $t = 4$ to $t = 10$ s, the velocity is constant. The distance travelled is

$$(110 \text{ ft/s})(6 \text{ s}) = 660 \text{ ft}$$

The total distance the animal travels is $220 + 660 = 880$ ft, or 268.2 m in 10 s.

Second Method We draw a graph of the animal's velocity as a function of time in Figure (a). The acceleration is constant during the first 4 s of motion, so the velocity is a linear function of time from $v = 0$ at $t = 0$ to $v = 110$ ft/s at $t = 4$ s. The velocity is constant during the last 6 s. The total distance covered is the sum of the areas during the two phases of motion:

$$\frac{1}{2}(4 \text{ s})(110 \text{ ft/s}) + (6 \text{ s})(110 \text{ ft/s}) = 220 \text{ ft} = 880 \text{ ft}$$

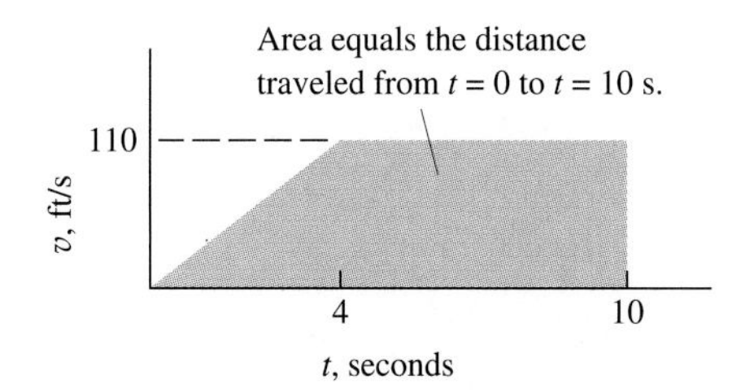

(a) The cheetah's velocity as a function of time.

DISCUSSION

Notice that in the first method we used definite, rather than indefinite, integrals to determine the cheetah's velocity and position as functions of time. You should rework the example using indefinite integrals and compare your results with ours. Whether to use indefinite or definite integrals is primarily a matter of taste, but you need to be familiar with both procedures.

Example 2.3

Figure 2.10

Suppose that the acceleration of the train in Figure 2.10 during the interval of time from $t = 2$ s to $t = 4$ s is $a = 2t\,\text{m/s}^2$, and at $t = 2$ s its velocity is $v = 180$ km/hr. What is the train's velocity at $t = 4$ s, and what is its displacement (change in position) from $t = 2$ s to $t = 4$ s?

STRATEGY

We can integrate Equations (2.3) and (2.4) to determine the train's velocity and position as functions of time.

SOLUTION

The velocity at $t = 2$ s in terms of m/s is

$$180\,\text{km/hr} = 180\,\text{km/hr} \times \left(\frac{1000\,\text{m}}{1\,\text{km}}\right) \times \left(\frac{1\,\text{hr}}{3600\,\text{s}}\right) = 50\,\text{m/s}$$

We write Equation (2.4) as

$$dv = a\,dt = 2t\,dt$$

and integrate, introducing the condition $v = 50$ m/s at $t = 2$ s:

$$\int_{50}^{v} dv = \int_{2}^{t} 2t\,dt$$

Evaluating the integrals, we obtain

$$v = t^2 + 46\,\text{m/s}$$

Now that we know the velocity as a function of time, we write Equation (2.3) as

$$ds = v\,dt = (t^2 + 46)\,dt$$

and integrate, defining the position of the train at $t = 2$ s to be $s = 0$:

$$\int_{0}^{2} ds = \int_{2}^{1} (t^2 + 46)\,dt$$

The position as a function of time is

$$s = \left(\frac{1}{3}t^3 + 46t - 94.7\right)\text{m}$$

Using our equations for the velocity and position, the velocity at $t = 4$ s is

$$v = (4)^2 + 46 = 62\,\text{m/s}$$

and the displacement from $t = 2$ s to $t = 4$ s is

$$\Delta s = \left[\frac{1}{3}(4)^3 + 46(4) - 94.7\right] - 0 = 110.7\,\text{m}$$

DISCUSSION

The acceleration in this example is not constant. You must not try to solve such problems by using equations that are valid only when the acceleration is constant. To convince yourself, try applying Equation (2.11) to this example: set $a_0 = 2t\,\mathrm{m/s^2}$, $t_0 = 2$ s, and $v_0 = 50$ m/s, and solve for the velocity at $t = 4$ s.

Problems

The following problems involve straight-line motion. The time t is in seconds unless otherwise stated.

2.1 The graph of the position s of a point as a function of time is a straight line. When $t = 4$ s, $s = 24$ m, and when $t = 20$ s, $s = 72$ m.

(a) Determine the velocity of the point by calculating the slope of the straight line.

(b) Obtain the equation for s as a function of time and use it to determine the velocity of the point.

2.2 The graph of the position s of a point of a milling machine as a function of time is a straight line. When $t = 0.2$ s, $s = 90$ mm. During the interval of time from $t = 0.6$ s to $t = 1.2$ s, the displacement of the point is $\Delta s = -180$ mm.

(a) Determine the equation for s as a function of time.

(b) What is the velocity of the point?

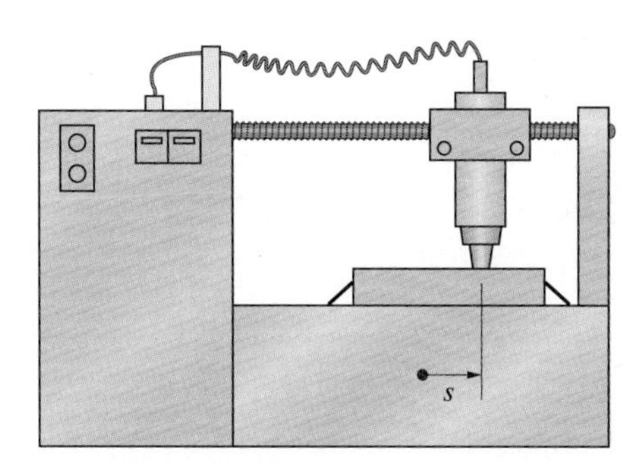

P2.2

2.3 The graph of the velocity v of a point as a function of time is a straight line. When $t = 2$ s, $v = 4$ m/s, and when $t = 4$ s, $v = -10$ m/s.

(a) Determine the acceleration of the point by calculating the slope of the straight line.

(b) Obtain the equation for v as a function of time and use it to determine the acceleration of the point.

2.4 The position of a point is $s = (2t^2 - 10)$ m.

(a) What is the displacement of the point from $t = 0$ to $t = 4$ s?

(b) What are the velocity and acceleration at $t = 0$?

(c) What are the velocity and acceleration at $t = 4$ s?

2.5 A rocket starts from rest and travels straight up. Its height above the ground is measured by radar from $t = 0$ to $t = 4$ s and is found to be approximated by the function $s = 10t^2$ m.

(a) What is the displacement during this interval of time?

(b) What is the velocity at $t = 4$ s?

(c) What is the acceleration during the first 4 s?

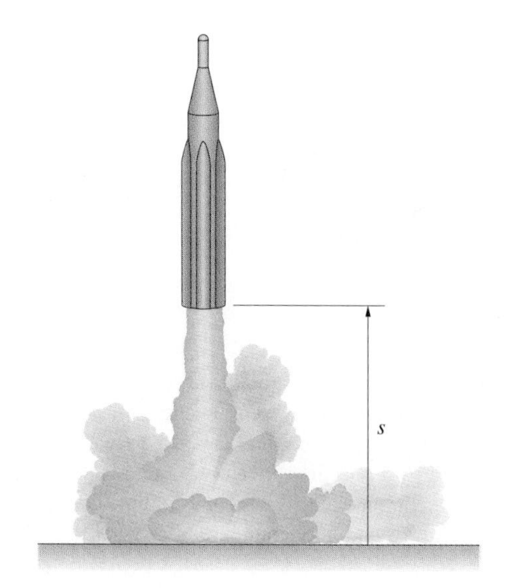

P2.5

2.6 The position of a point during the interval of time from $t = 0$ to $t = 6$ s is $s = (-\frac{1}{2}t^3 + 6t^2 + 4t)$ m.

(a) What is the displacement of the point during this interval of time?

(b) What is the maximum velocity during this interval of time, and at what time does it occur?

(c) What is the acceleration when the velocity is a maximum?

2.7 The position of a point during the interval of time from $t = 0$ to $t = 3$ s is $s = (12 + 5t^2 - t^3)$ m.
(a) What is the maximum velocity during this interval of time, and at what time does it occur?
(b) What is the acceleration when the velocity is a maximum?

2.8 A seismograph measures the horizontal motion of the ground during an earthquake. An engineer analysing the data determines that for a 10 s interval of time beginning at $t = 0$, the position is approximated by $s = 100\cos(2\pi t)$ mm. What are the (a) maximum velocity and (b) maximum acceleration of the ground during the 10 s interval?

2.9 During an assembly operation, a robot's arm moves along a straight line. During an interval of time from $t = 0$ to $t = 1$ s, its position is given by $s = (75t^2 - 50t^3)$ mm. Determine, during this 1 s interval: (a) the displacement of the arm; (b) the maximum and minimum values of the velocity; (c) the maximum and minimum values of the acceleration.

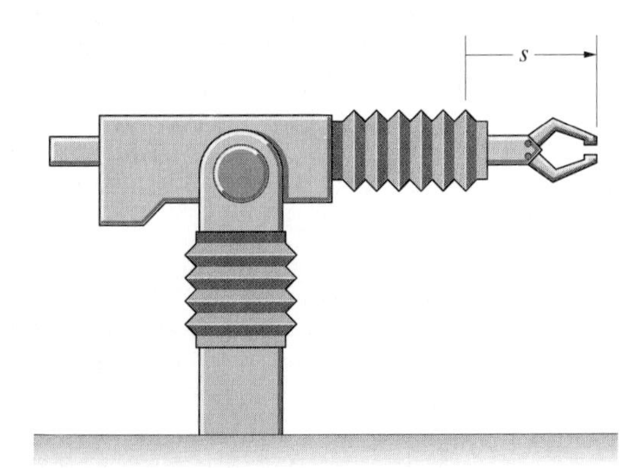

P2.9

2.10 In a test of a prototype car, the driver starts the car from rest at $t = 0$, accelerates, and then applies the brakes. Engineers measuring the position of the car find that from $t = 0$ to $t = 18$ s it is approximated by $s = (1.5t^2 + 0.1t^3 - 0.006t^4)$ m.
(a) What is the maximum velocity, and at what time does it occur?
(b) What is the maximum acceleration, and at what time does it occur?

P2.10

2.11 Suppose you want to approximate the position of a vehicle you are testing by the power series $s = A + Bt + Ct^2 + Dt^3$, where A, B, C and D are constants. The vehicle starts from rest at $t = 0$ and $s = 0$. At $t = 4$ s, $s = 54$ m and at $t = 8$ s, $s = 136$ m.
(a) Determine A, B, C and D.
(b) What are the approximate velocity and acceleration of the vehicle at $t = 8$ s?

2.12 The acceleration of a point is $a = 20t\ \text{m/s}^2$. When $t = 0$, $s = 40$ m and $v = -10$ m/s. What are the position and velocity at $t = 3$ s?

2.13 The acceleration of a point is $a = (60t - 36t^2)\ \text{m/s}^2$. When $t = 0$, $s = 0$ and $v = 20$ m/s. What are the position and velocity as functions of time?

2.14 Suppose that during the preliminary design of a car, you assume its maximum acceleration is approximately constant. What constant acceleration is necessary if you want the car to be able to accelerate from rest to a velocity of 88 km/hr in 10 s? What distance would the car travel during that time?

2.15 An entomologist estimates that a flea 1 mm in length attains a velocity of 1.3 m/s in a distance of one body length when jumping. What constant acceleration is necessary to achieve that velocity?

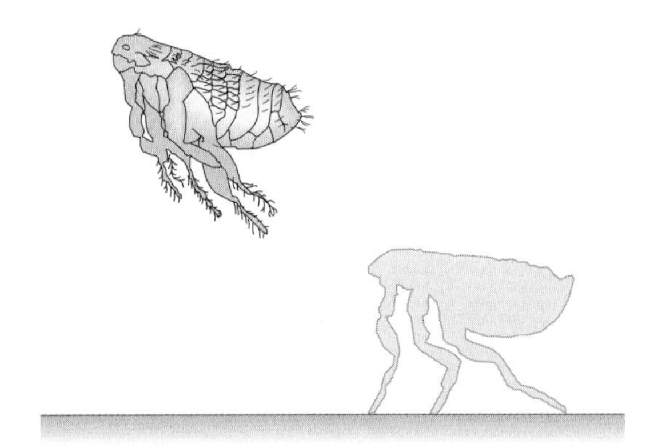

P2.15

2.16 Missiles designed for defence against ballistic missiles achieve accelerations in excess of 100 g's or one hundred times the acceleration of gravity. If a missile has a constant acceleration of 100 g's, how long does it take to go from rest to 96 km/hr? What is its displacement during that time?

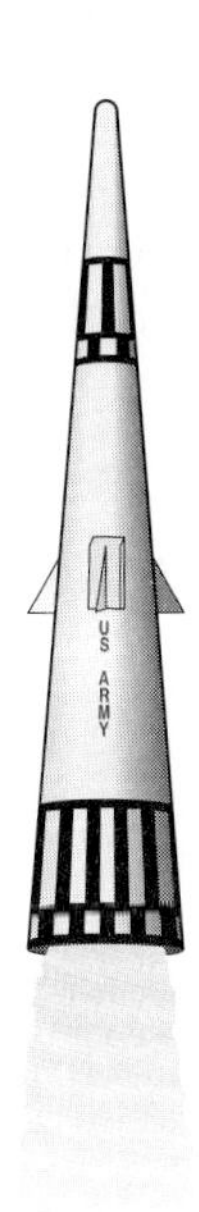

P2.16

2.17 Suppose you want to throw some keys to a friend standing on a first-floor balcony. If you release the keys at 1.5 m above the ground, what vertical velocity is necessary for them to reach your friend's hand 6 m above the ground?

2.18 The Lunar Module descends toward the surface of the moon at 1 m/s when its landing probes, which extend 2 m below the landing gear, touch the surface, automatically shutting off the engines. Determine the velocity with which the landing gear contacts the surface. (The acceleration due to gravity at the surface of the moon is 1.62 m/s^2.)

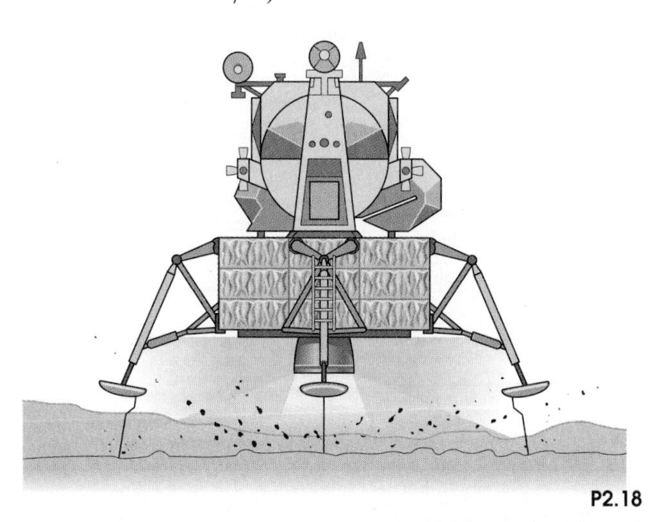

P2.18

2.19 In 1960 R. C. Owens of the Baltimore Colts blocked a Washington Redskins field goal attempt by jumping and knocking the ball away in front of the cross bar at a point 3.35 m above the field. If he was 1.90 m tall and could reach 0.36 m above his head, what was his vertical velocity as he left the ground?

2.20 The velocity of a bobsled is $v = 3t$ m/s. When $t = 2$ s, its position is $s = 7.5$ m. What is its position when $t = 10$ s?

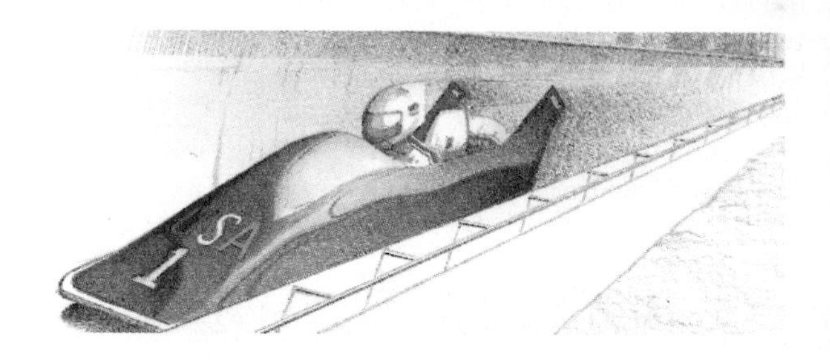

P2.20

2.21 The acceleration of an object is $a = (10 - 2t)$ m/s^2. When $t = 0$, $s = 0$ and $v = 0$. What is its maximum velocity during the interval of time from $t = 0$ to $t = 10$ s?

2.22 The velocity of an object is $v = (200 - 2t^2)$ m/s. When $t = 3$ s, its position is $s = 600$ m. What are the position and acceleration of the object at $t = 6$ s?

2.23 The acceleration of a part undergoing a machining operation is measured and determined to be $a = (12 - 6t)$ mm/s^2. When $t = 0$, $v = 0$. For the interval of time from $t = 0$ to $t = 4$ s, determine: (a) the maximum velocity; (b) the displacement.

2.24 The missile shown in Problem 2.16 starts from rest and accelerates straight up for 3 s at 100 g's. After 3 s, its weight and aerodynamic drag cause it to have a constant deceleration of 4 g's. How long does it take the missile to go from the ground to an altitude of 15 240 m?

2.25 A car is traveling at 48 km/hr when a traffic light 90 m ahead turns amber. The light will remain amber for 5 s before turning red.

(a) What constant acceleration will cause the car to reach the light at the instant it turns red, and what will the velocity of the car be when it reaches the light?

(b) If the driver decides not to try to make the light, what constant rate of acceleration will cause the car to come to a stop just as it reaches the light?

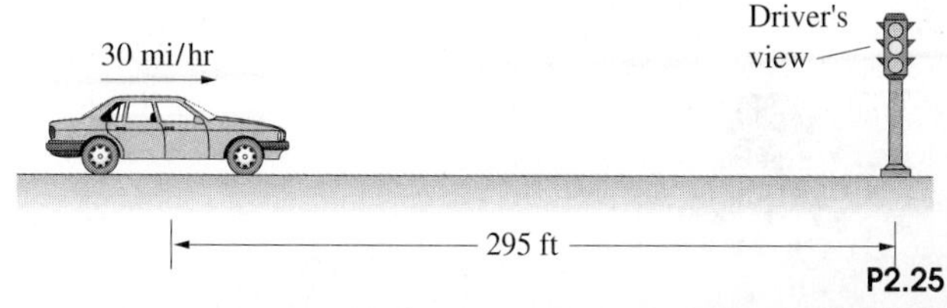

P2.25

2.26 At $t = 0$, a motorist travelling at 100 km/hr sees a deer standing in the road 100 m ahead. After a reaction time of 0.3 s, he applies the brakes and decelerates at a constant rate of $4\ \mathrm{m/s^2}$. If the deer takes 5 s from $t = 0$ to react and leave the road, does the motorist miss him?

2.27 A high-speed rail transportation system has a top speed of 100 m/s. For the comfort of the passengers, the magnitude of the acceleration and deceleration is limited to $2\ \mathrm{m/s^2}$. Determine the minimum time required for a trip of 100 km.

Strategy: A graphical approach can help you solve this problem. Recall that the change in the position from an initial time t_0 to a time t is equal to the area defined by the graph of the velocity as a function of time from t_0 to t.

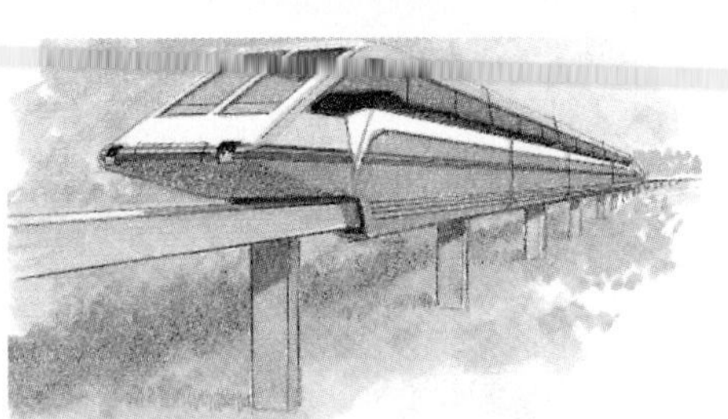

P2.27

2.28 The nearest star, Proxima Centauri, is 4.22 light years from the earth. Ignoring relative motion between the solar system and Proxima Centauri, suppose that a spacecraft accelerates from the vicinity of the earth at $0.01\,g$ (0.01 times the acceleration due to gravity at sea level) until it reaches one-tenth the speed of light, coasts until time to decelerate, then decelerates at $0.01\,g$ until it comes to rest in the vicinity of Proxima Centauri. How long does the trip take? (Light travels at $3 \times 10^8\ \mathrm{m/s}$. A solar year is 365.2422 solar days.)

2.29 A racing car starts from rest and accelerates at $a = (1.5 + 0.6t)\ \mathrm{m/s^2}$ for 10 s. The brakes are then applied, and the car has a constant acceleration $a = -9\ \mathrm{m/s^2}$ until it comes to rest. Determine: (a) the maximum velocity; (b) the total distance travelled; (c) the total time of travel.

2.30 When $t = 0$, the position of a point is $s = 6$ m and its velocity is $v = 2$ m/s. From $t = 0$ to $t = 6$ s, its acceleration is $a = (2 + 2t^2)\ \mathrm{m/s^2}$. From $t = 6$ s until it comes to rest, its acceleration is $a = -4\ \mathrm{m/s^2}$.

(a) What is the total time of travel?

(b) What total distance does it move?

2.31 Zoologists studying the ecology of the Serengeti Plain estimate that the average adult cheetah can run 100 km/hr and the average springbok can run 65 km/hr. If the animals run along the same straight line, start at the same time, and are each assumed to have constant acceleration and reach top speed in 4 s, how close must a cheetah be when the chase begins to catch a springbok in 15 s?

2.32 Suppose that a person unwisely drives at 120 km/hr in an 80 km/hr zone and passes a police car going at 80 km/hr in the same direction. If the police officers begin constant acceleration at the instant they are passed and increase their velocity to 130 km/hr in 4 s, how long does it take them to be level with the pursued car?

2.33 If $\theta = 1$ rad and $d\theta/dt = 1$ rad/s, what is the velocity of P relative to O?

Strategy: You can write the position of P relative to O as

$$s = (2\ \mathrm{m})\cos\theta + (2\ \mathrm{m})\cos\theta,$$

then take the derivative of this expression with respect to time to determine the velocity.

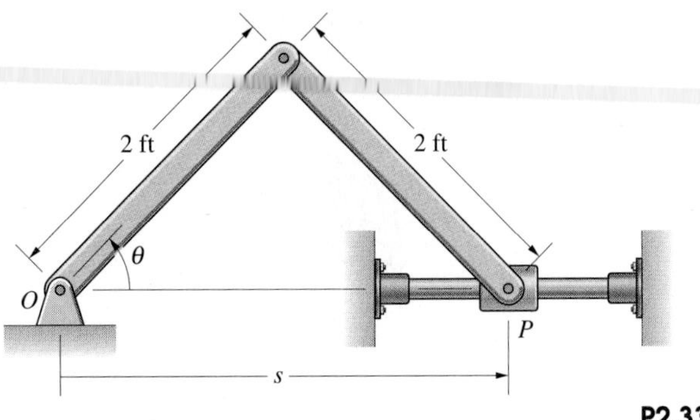

P2.33

2.34 In Problem 2.33, if $\theta = 1$ rad, $d\theta/dt = -2$ rad/s and $d^2\theta/dt^2 = 0$, what are the velocity and acceleration of P relative to O?

2.35 If $\theta = 1$ rad and $d\theta/dt = 1$ rad/s, what is the velocity of P relative to O?

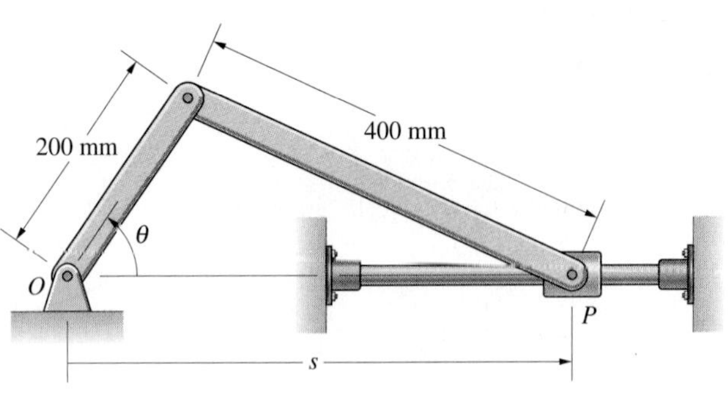

P2.35

Acceleration Specified as a Function of Velocity Aerodynamic and hydrodynamic forces can cause an object's acceleration to depend on its velocity (Figure 2.11). Suppose that the acceleration is a known function of velocity $a(v)$:

$$\frac{dv}{dt} = a(v) \tag{2.14}$$

Figure 2.11
Aerodynamic and hydrodynamic forces depend on an object's velocity. The faster the object moves relative to the fluid, the greater is the force resisting its motion.

We cannot integrate this equation with respect to time to determine the velocity, because $a(v)$ is not known as a function of time. But we can **separate variables**, putting terms involving v on one side of the equation and terms involving t on the other side:

$$\frac{dv}{a(v)} = dt \tag{2.15}$$

We can now integrate

$$\int_{v_0}^{v} \frac{dv}{a(v)} = \int_{t_0}^{t} dt \tag{2.16}$$

where v_0 is the velocity at time t_0. In principle, we can solve this equation for the velocity as a function of time, then integrate the relation

$$\frac{ds}{dt} = v$$

to determine the position as a function of time.

By using the chain rule, we can also determine the velocity as a function of the position. Writing the acceleration as

$$\frac{dv}{dt} = \frac{dv}{ds}\frac{ds}{dt} = \frac{dv}{ds}v$$

and substituting it into Equation (2.14), we obtain

$$\frac{dv}{ds}v = a(v)$$

Separating variables,

$$\frac{v\,dv}{a(v)} = ds$$

and integrating,

$$\int_{v_0}^{v} \frac{v\,dv}{a(v)} = \int_{s_0}^{s} ds$$

we can obtain a relation between the velocity and the position.

Acceleration Specified as a Function of Position Gravitational forces and forces exerted by springs can cause an object's acceleration to depend on its position. If the acceleration is a known function of position,

$$\frac{dv}{dt} = a(s) \tag{2.17}$$

we cannot integrate with respect to time to determine the velocity because s is not known as a function of time. Moreover, we cannot separate variables, because the equation contains three variables, v, t and s. However, by using the chain rule,

$$\frac{dv}{dt} = \frac{dv}{ds}\frac{ds}{dt} = \frac{dv}{ds}v$$

we can write Equation (2.17) as

$$\frac{dv}{ds}v = a(s)$$

Now we can separate variables,

$$v\,dv = a(s)\,ds \tag{2.18}$$

and integrate:

$$\int_{v_0}^{v} v\,dv = \int_{s_0}^{s} a(s)\,ds \tag{2.19}$$

In principle, we can solve this equation for the velocity as a function of the position:

$$v = \frac{ds}{dt} - v(s) \tag{2.20}$$

Then we can separate variables in this equation and integrate to determine the position as a function of time:

$$\int_{s_0}^{s} \frac{ds}{v(s)} = \int_{t_0}^{t} dt$$

The next two examples show how you can analyse the motion of an object when its acceleration is a function of velocity or position. The initial steps are summarized in Table 2.1.

Table 2.1. Determining the velocity when you know the acceleration as a function of velocity or position.

If you know $a = a(v)$:	Separate variables, $\frac{dv}{dt} = a(v)$ $\frac{dv}{a(v)} = dt$ or apply the chain rule, $\frac{dv}{dt} = \frac{dv}{ds}\frac{ds}{dt} = \frac{dv}{ds}v = a(v)$ then separate variables, $\frac{v\,dv}{a(v)} = ds$
If you know $a = a(s)$:	Apply the chain rule, $\frac{dv}{dt} = \frac{dv}{ds}\frac{ds}{dt} = \frac{dv}{ds}v = a(s)$ then separate variables, $v\,dv = a(s)\,ds$

Example 2.4

After deploying its drag parachute, the aeroplane in Figure 2.12 has an acceleration $a = -0.004v^2$ m/s.
(a) Determine the time required for the velocity to decrease from 80 m/s to 10 m/s.
(b) What distance does the plane cover during that time?

Figure 2.12

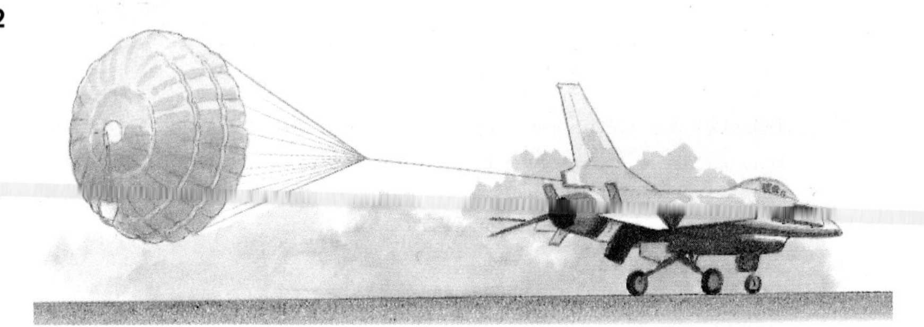

STRATEGY

In part (b), we will use the chain rule to express the acceleration in terms of a derivative with respect to position and integrate to obtain a relation between the velocity and the position.

SOLUTION

(a) The acceleration is

$$a = \frac{dv}{dt} = -0.004v^2$$

We separate variables,

$$\frac{dv}{v^2} = -0.004\,dt$$

and integrate, defining $t = 0$ to be the time at which $v = 80$ m/s:

$$\int_{80}^{v} \frac{dv}{v^2} = \int_{0}^{t} -0.004\,dt$$

Evaluating the integrals and solving for t, we obtain

$$t = 250\left(\frac{1}{v} - \frac{1}{80}\right)$$

The time required for the plane to slow to $v = 10\,\text{m/s}$ is 21.9 s. We show the velocity of the aeroplane as a function of time in Figure 2.13.

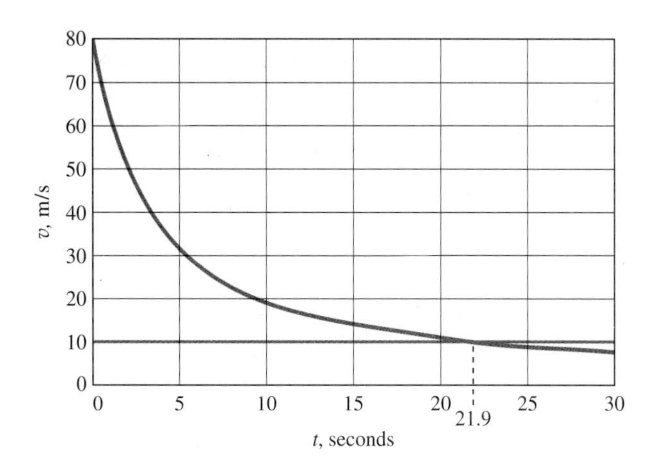

Figure 2.13
Graph of the aeroplane's velocity as a function of time.

(b) We write the acceleration as

$$a = \frac{dv}{dt} = \frac{dv}{ds}\frac{ds}{dt} = \frac{dv}{ds}v = -0.004v^2$$

separate variables,

$$\frac{dv}{v} = -0.004ds$$

and integrate, defining $s = 0$ to be the position at which $v = 80\,\text{m/s}$:

$$\int_{80}^{v} \frac{dv}{v} = \int_{0}^{s} -0.004ds$$

Evaluating the integrals and solving for s, we obtain

$$s = 250\ \ln\left(\frac{80}{v}\right)$$

The distance required for the plane to slow to $v = 10\,\text{m/s}$ is 519.9 m.

DISCUSSION

Notice that our results predict that the time elapsed and distance travelled continue to increase without bound as the aeroplane's velocity decreases. The reason is that the modelling is incomplete. The equation for the acceleration includes only aerodynamic drag and does not account for other forces, such as friction in the aeroplane's wheels.

Example 2.5

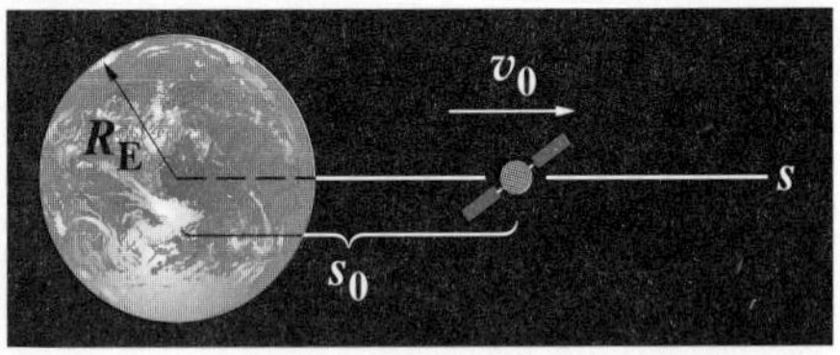

Figure 2.14

In terms of distance s from the centre of the earth, the magnitude of the acceleration due to gravity is gR_E^2/s^2, where R_E is the radius of the earth. (See the discussion of gravity in Section 1.3.) If a spacecraft is a distance s_0 from the centre of the earth (Figure 2.14), what outward velocity v_0 must it be given to reach a specified distance h from the centre of the earth?

SOLUTION

The acceleration due to gravity is *towards* the centre of the earth:

$$a = -\frac{gR_E^2}{s^2}$$

Applying the chain rule,

$$a = \frac{dv}{dt} = \frac{dv}{ds}\frac{ds}{dt} = \frac{dv}{ds}v = -\frac{gR_E^2}{s^2}$$

and separating variables, we obtain

$$v\,dv = -\frac{gR_E^2}{s^2}\,ds$$

We integrate this equation using the initial condition, $v = v_0$ when $s = s_0$, as the lower limits, and the final condition, $v = 0$ when $s = h$, as the upper limits:

$$\int_v^0 v\,dv = -\int_{s_0}^h \frac{gR_E^2}{s^2}\,ds$$

Evaluating the integrals and solving for v_0, we obtain the initial velocity v_0 necessary for the spacecraft to reach a distance h:

$$v_0 = \sqrt{2gR_E^2\left(\frac{1}{s_0} - \frac{1}{h}\right)}$$

DISCUSSION

We can make an interesting and important observation from the result of this example. Notice that as the distance h increases, the necessary initial velocity v_0 approaches a finite limit. This limit,

$$v_{esc} = \lim_{h \to \infty} v_0 = \sqrt{\frac{2gR_E^2}{s_0}}$$

is called the **escape velocity**. In the absence of other effects, an object with this initial velocity will continue moving outwards indefinitely. The existence of an escape velocity makes it feasible to send probes and persons to other planets. Once escape velocity is attained, it isn't necessary to expend additional fuel to keep going.

Problems

2.36 The acceleration of an object is $a = -2v \text{ m/s}^2$. When $t = 0$, $s = 0$ and $v = 2 \text{ m/s}$. Determine the object's velocity as a function of time.

2.37 In Problem 2.36, determine the object's position as a function of time.

2.38 The boat is moving at 20 m/s when its engine is shut down. Due to hydrodynamic drag, its acceleration is $a = -0.1v^2 \text{ m/s}^2$. What is the boat's velocity 2 s later?

P2.38

2.39 In Problem 2.38, what distance does the boat move in the 2 s following the shutdown of its engine?

2.40 A steel ball is released from rest in a container of oil. Its downward acceleration is $a = 0.9g - cv$, where g is the acceleration due to gravity at sea level and c is a constant. What is the velocity of the ball as a function of time?

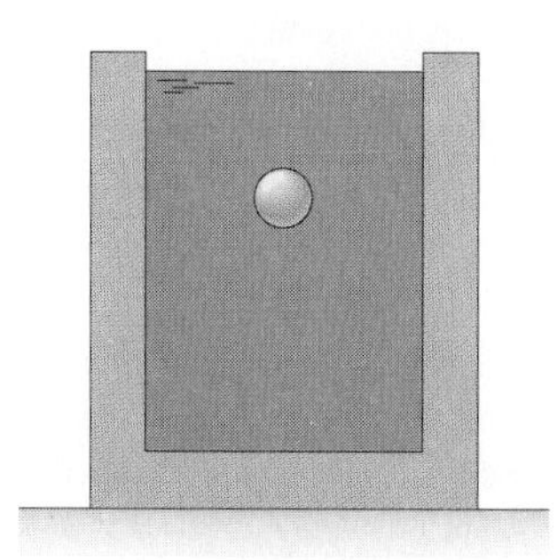

P2.40

2.41 In Problem 2.40, determine the position of the ball relative to its initial position as a function of time.

2.42 The greatest ocean depth yet discovered is in the Marianas Trench in the western Pacific Ocean. A steel ball released at the surface requires 64 min to reach the bottom. The ball's downward acceleration is $a = 0.9g - cv$, where g is the acceleration due to gravity at sea level and the constant $c = 3.02 \text{ s}^{-1}$. What is the depth of the Marianas Trench in kilometres?

2.43 To study the effects of meteor impacts on satellites, engineers use a rail gun to accelerate a plastic pellet to a high velocity. They determine that when the pellet has travelled 1 m from the gun, its velocity is 2.25 km/s, and when it has travelled 2 m from the gun, its velocity is 1.00 km/s. Assume that the acceleration of the pellet after it leaves the gun is given by $a = -cv^2$, where c is a constant.

(a) What is the value of c, and what are its SI units?

(b) What was the velocity of the pellet as it left the rail gun?

P2.43

2.44 If aerodynamic drag is taken into account, the acceleration of a falling object can be approximated by $a = g - cv^2$, where g is the acceleration due to gravity at sea level and c is a constant.

(a) If an object is released from rest, what is its velocity as a function of the distance s from the point of release?

(b) Determine the limit of your answer to part (a) as $c \to 0$, and show that it agrees with the solution you obtain by assuming that the acceleration $a = g$.

2.45 A sky diver jumps from a helicopter and is falling straight down at 30 m/s when her parachute opens. From then on, her downward acceleration is approximately $a = g - cv^2$, where $g = 9.81\ \text{m/s}^2$ and c is a constant. After an initial 'transient' period, she descends at a nearly constant velocity of 5 m/s.
(a) What is the value of c, and what are its SI units?
(b) What maximum deceleration is she subjected to?
(c) What is her downward velocity when she has fallen 2 m from the point where her parachute opens?

P2.45

2.46 A rocket sled starts from rest and accelerates at $a = 3t^2\ \text{m/s}^2$ until its velocity is 1000 m/s. It then hits a water brake, and its acceleration is $a = -0.001v^2\ \text{m/s}$ until its velocity decreases to 500 m/s. What total distance does the sled travel?

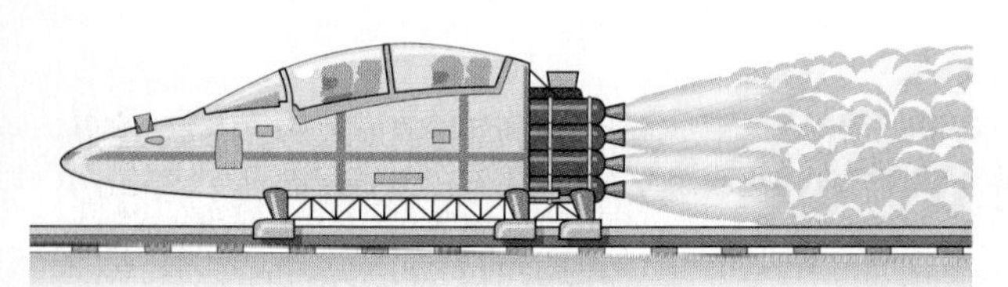

P2.46

2.47 The velocity of a point is given by the equation

$$v = (24 - 2s^2)^{1/2}\ \text{m/s}$$

What is its acceleration when $s = 2$ m?

2.48 The velocity of an object subjected to the earth's gravitational field is

$$v = \left[v_0^2 + 2gR_E^2\left(\frac{1}{s} - \frac{1}{s_0}\right)\right]^{1/2}$$

where v_0 is the velocity at position s_0 and R_E is the radius of the earth. Using this equation, show that the object's acceleration is $a = -gR_E^2/s^2$.

2.49 Engineers analysing the motion of a linkage determine that the velocity of an attachment point is given by $v = (A + 4s^2)\ \text{m/s}$, where A is a constant. When $s = 2$ m, its acceleration is measured and determined to be $a = 320\ \text{m/s}^2$. What is its velocity when $s = 2$ m?

2.50 The acceleration of an object is given by the function $a = 2v\ \text{m/s}^2$. When $t = 0$, $v = 1$ m/s. What is the velocity when the object has moved 2 m from its initial position?

2.51 The acceleration of an object is given by $a = 3s^2\ \text{m/s}^2$. At $s = 0$, its velocity is $v = 10$ m/s. What is its velocity when $s = 4$ m?

2.52 The velocity of an object is given by $v^2 = k/s$, where k is a constant. If $v = 4$ m/s and $s = 4$ m at $t = 0$, determine the constant k and the velocity at $t = 2$ s.

2.53 A spring-mass oscillator consists of a mass and a spring connected as shown. The coordinate s measures the displacement of the mass relative to its position when the spring is unstretched. If the spring is linear, the mass is subjected to a deceleration proportional to s. Suppose that $a = -4s\ \text{m/s}^2$ and that you give the mass a velocity $v = 1$ m/s in the position $s = 0$.
(a) How far will the mass move to the right before the spring brings it to a stop?
(b) What will be the velocity of the mass when it has returned to the position $s = 0$?

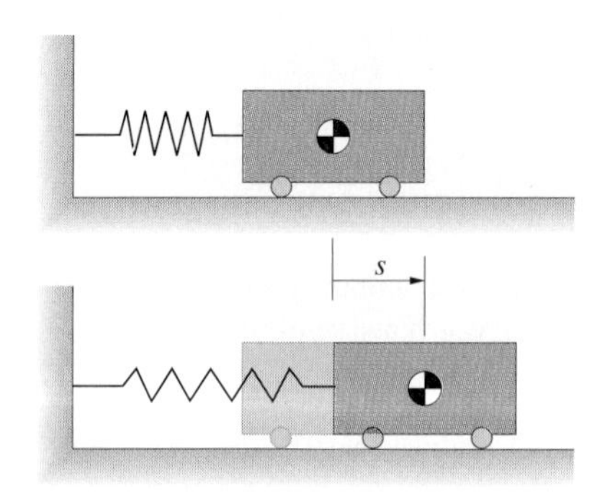

P2.53

2.54 In Prblem 2.53, suppose that at $t = 0$ you release the mass from rest in the position $s = 1$ m. Determine the velocity of the mass as a function of s as it moves from the initial position to $s = 0$.

2.55 In Problem 2.53, suppose that at $t = 0$ you release the mass from rest in the position $s = 1$ m. Determine the position of the mass as a function of time as it moves from its initial position to $s = 0$.

2.56 If a spacecraft is 160 km above the surface of the earth, what initial velocity v_0 straight away from the earth would be required for it to reach the moon's orbit 383 000 km from the centre of the earth? The radius of the earth is 6370 km. Neglect the effect of the moon's gravity. (See Example 2.5.)

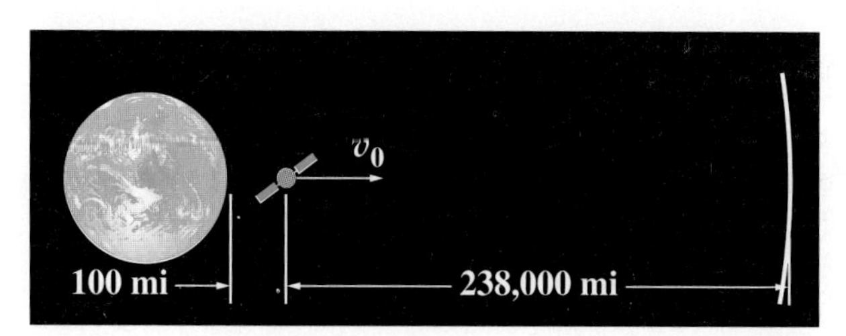

P2.56

2.57 The radius of the moon is $R_M = 1738$ km. The acceleration of gravity at its surface is $1.62\ \text{m/s}^2$. If an object is released from rest 1738 km above the surface of the moon, what is the magnitude of its velocity just before it impacts the surface?

2.58 Using the data in Problem 2.57, determine the escape velocity from the surface of the moon. (See Example 2.5.)

2.59 Suppose that a tunnel could be drilled straight through the earth from the North Pole to the South Pole and the air evacuated. An object dropped from the surface would fall with acceleration $a = -gs/R_E$, where g is the acceleration of gravity at sea level, R_E is the radius of the earth, and s is the distance of the object from the centre of the earth. (Gravitational acceleration is equal to zero at the centre of the earth and increases linearly with distance from the centre.) What is the magnitude of the velocity of the dropped object when it reaches the centre of the earth?

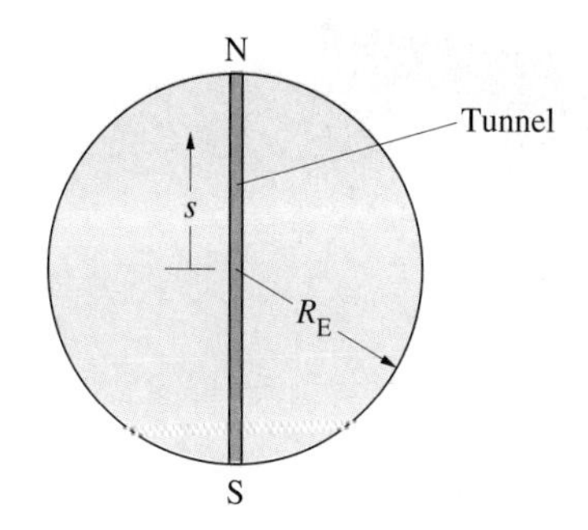

P2.59

2.60 The acceleration of gravity of a hypothetical two-dimensional planet would depend upon the distance s from the centre of the planet according to the relation $a = -k/s$, where k is a constant. Let the radius of the planet be R_T and let the magnitude of the acceleration due to gravity at its surface be g_T.
(a) If an object is given an initial outward velocity v_0 at a distance s_0 from the centre of the planet, determine its velocity as a function of s.
(b) Show that there is no escape velocity from a two-dimensional planet, thereby explaining why we have never been visited by any two-dimensional beings.

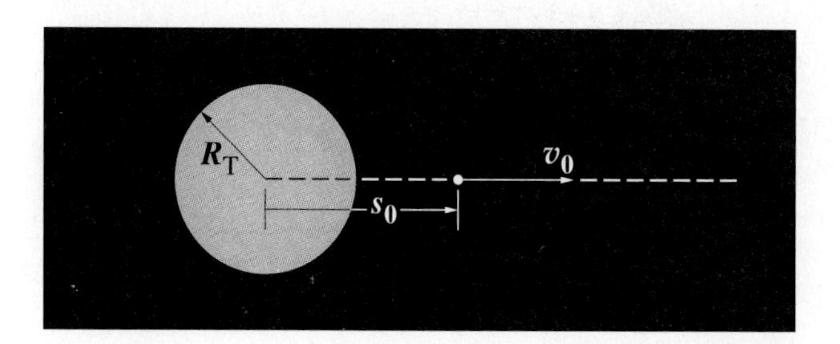

P2.60

2.3 Curvilinear Motion

If the motion of a point is confined to a straight line, its position vector **r**, velocity vector **v** and acceleration vector **a** are described completely by the scalars, s, v and a, respectively. We know the directions of these vectors because they are parallel to the straight line. But if a point describes a **curvilinear** path, we must specify both the magnitudes and directions of these vectors, and we require a coordinate system to express them in terms of scalar components. Although the directions and magnitudes of the position, velocity and acceleration vectors do not depend on the coordinate system used to express them, we will show that the *representations* of these vectors are different in different coordinate systems. Many problems can be expressed in terms of cartesian coordinates, but some situations, including the motions of satellites and rotating machines, can be expressed more naturally using other coordinate systems. In the following sections we show how curvilinear motions of points are analysed in terms of various coordinate systems.

Cartesian Coordinates

Let **r** be the position vector of a point P relative to a reference point O. To express the motion of P in terms of a cartesian coordinate system, we place the origin at O (Figure 2.15), so that the components of **r** are the x, y, z coordinates of P:

$$\mathbf{r} = x\,\mathbf{i} + y\,\mathbf{j} + z\,\mathbf{k}$$

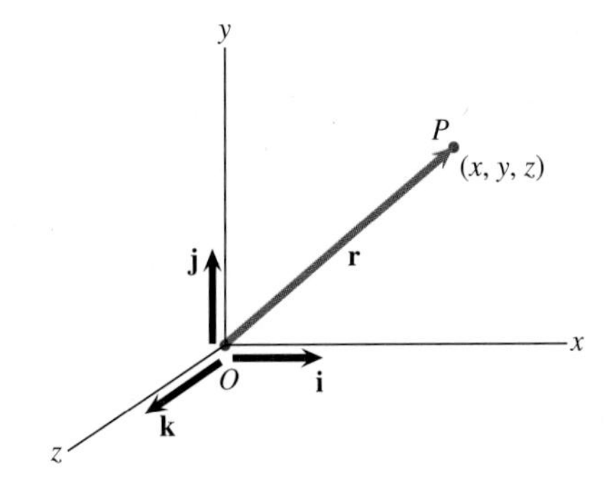

Figure 2.15
A cartesian coordinate system with its origin at the reference point O.

Assuming that the coordinate system does not rotate, the unit vectors **i**, **j** and **k** are constants. (We will discuss rotating coordinate systems in Chapter 6.) Thus the velocity of P is

$$\mathbf{v} = \frac{d\mathbf{r}}{dt} = \frac{dx}{dt}\mathbf{i} + \frac{dy}{dt}\mathbf{j} + \frac{dz}{dt}\mathbf{k} \qquad (2.21)$$

Expressing the velocity in terms of scalar components,

$$\mathbf{v} = v_x\,\mathbf{i} + v_y\,\mathbf{j} + v_z\,\mathbf{k} \qquad (2.22)$$

we obtain scalar equations relating the components of the velocity to the coordinates of P:

$$v_x = \frac{dx}{dt} \qquad v_y = \frac{dy}{dt} \qquad v_z = \frac{dz}{dt} \tag{2.23}$$

The acceleration of P is

$$\mathbf{a} = \frac{d\mathbf{v}}{dt} = \frac{dv_x}{dt}\mathbf{i} + \frac{dv_y}{dt}\mathbf{j} + \frac{dv_z}{dt}\mathbf{k}$$

and by expressing the acceleration in terms of scalar components,

$$\mathbf{a} = a_x\mathbf{i} + a_j\mathbf{j} + a_2\mathbf{k} \tag{2.24}$$

we obtain the scalar equations

$$a_x = \frac{dv_x}{dt} \qquad a_y = \frac{dv_y}{dt} \qquad a_z = \frac{dv_z}{dt} \tag{2.25}$$

Equations (2.23) and (2.25) describe the motion of a point relative to a cartesian coordinate system. Notice that the equations describing the motion in each coordinate direction are identical in form to the equations that describe the motion of a point along a straight line. As a consequence, you can often analyse the motion in each coordinate direction using the methods you applied to straight-line motion.

The **projectile problem** is the classic example of this kind. If an object is thrown through the air and aerodynamic drag is negligible, it accelerates downwards with the acceleration due to gravity. In terms of a cartesian coordinate system with its y axis upwards, the acceleration is $a_x = 0$, $a_y = -g$, $a_z = 0$. Suppose that $t = 0$, the projectile is located at the origin and has velocity v_0 in the x–y plane at an angle θ_0 above the horizontal (Figure 2.16(a)).

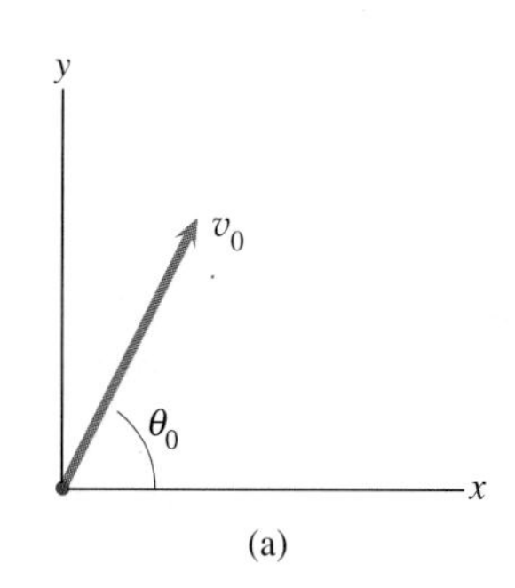

Figure 2.16
(a) Initial conditions for a projectile problem.

At $t = 0, x = 0$ and $v_x = v_0 \cos\theta_0$. The acceleration in the x direction is zero,

$$a_x = \frac{dv_x}{dt} = 0$$

so v_x is constant and remains equal to its initial value:

$$v_x = \frac{dx}{dt} = v_0 \cos\theta_0 \tag{2.26}$$

(This result may seem unrealistic; the reason is that your intuition, based upon everyday experience, accounts for drag, whereas this analysis does not.) Integrating this equation,

$$\int_0^x dx = \int_0^t v_0 \cos\theta_0 \, dt$$

we obtain the x coordinate of the object as a function of time:

$$x = v_0 \cos\theta_0 t \tag{2.27}$$

Thus we have determined the position and velocity in the x direction as functions of time without considering the motion in the y or z directions.

At $t = 0, y = 0$ and $v_u = v_0 \sin\theta_0$. The acceleration in the y direction is

$$a_y = \frac{dv_y}{dt} = -g$$

Integrating with respect to time,

$$\int_{v_0 \sin\theta_0}^{v_y} dv_y = \int_0^t -g \, dt$$

we obtain

$$v_y = \frac{dy}{dt} = v_0 \sin\theta_0 - gt \tag{2.28}$$

Integrating this equation,

$$\int_0^y dy = \int_0^t (v_0 \sin\theta_0 - gt) \, dt$$

we find that the y coordinate as a function of time is

$$y = v_0 \sin\theta_0 t - \frac{1}{2} g t^2 \tag{2.29}$$

You can see from this analysis that the same vertical velocity and position are obtained by throwing the projectile straight up with initial velocity $v_0 \sin\theta_0$ (Figures 2.16(b), 2.16(c)). The vertical motion is completely independent of the horizontal motion.

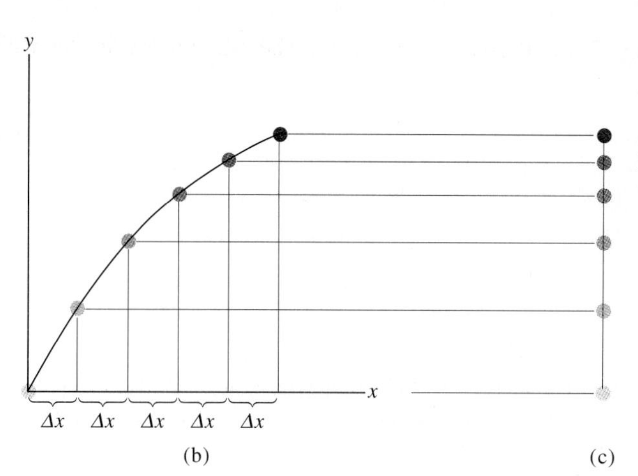

Figure 2.16
(b) Positions of the projectile at equal time intervals Δt. The distance $\Delta x = v_0 \cos\theta_0 \Delta t$. (c) Positions at equal time intervals Δt of a projectile given an initial vertical velocity equal to $v_0 \sin\theta_0$.

By solving Equation (2.27) for t and substituting the result into Equation (2.29), we obtain an equation describing the parabolic trajectory of the projectile:

$$y = \tan\theta_0 x - \frac{g}{2v_0^2 \cos^2\theta_0} x^2 \qquad (2.30)$$

In the following example we discuss a situation in which you can use Equations (2.23) and (2.25) to determine the motion of an object by analysing each coordinate direction independently.

Example 2.6

During a test flight in which a helicopter starts from rest at $t = 0$ (Figure 2.17), accelerometers mounted on board indicate that its components of acceleration from $t = 0$ to $t = 10$ s are closely approximated by

$$a_x = 0.6t \text{ m/s}^2$$
$$a_y = (1.8 - 0.36t) \text{ m/s}^2$$
$$a_z = 0$$

Determine the helicopter's velocity and position as functions of time.

Figure 2.17

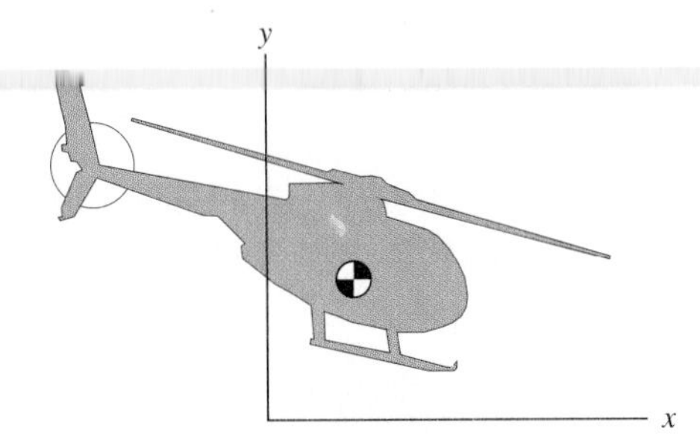

STRATEGY

We can analyse the motion in each coordinate direction independently, integrating the acceleration to determine the velocity and then integrating the velocity to determine the position.

SOLUTION

The velocity is zero at $t = 0$, and we assume that $x = y = z = 0$ at $t = 0$. The acceleration in the x direction is

$$a_x = \frac{dv_x}{dt} = 0.6t \text{ m/s}^2$$

Integrating with respect to time,

$$\int_0^{v_x} dv_x = \int_0^t 0.6t \, dt$$

we obtain the velocity component v_x as a function of time:

$$v_x = \frac{dx}{dt} = 0.3t^2 \text{ m/s}$$

Integrating again,

$$\int_0^x dx = \int_0^t 0.3t^2 \, dt$$

we obtain x as a function of time:

$$x = 0.1t^3 \text{ m}$$

Now we analyse the motion in the y direction in the same way. The acceleration is

$$a_y = \frac{dv_y}{dt} = (1.8 - 0.36t) \text{ m/s}^2$$

Integrating,

$$\int_0^{v_y} dv_y = \int_0^t (1.8 - 0.36t)\, dt$$

we obtain the velocity,

$$v_y = \frac{dy}{dt} = (1.8t - 0.18t^2) \text{ m/s}$$

Integrating again,

$$\int_0^y dy = \int_0^t (1.8t - 0.18t^2)\, dt$$

we determine the position:

$$y = (0.9t^2 - 0.06t^3) \text{ m}$$

You can easily show that the z components of the velocity and position are $v_z = 0$ and $z = 0$. We show the position of the helicopter as a function of time in Figure (a).

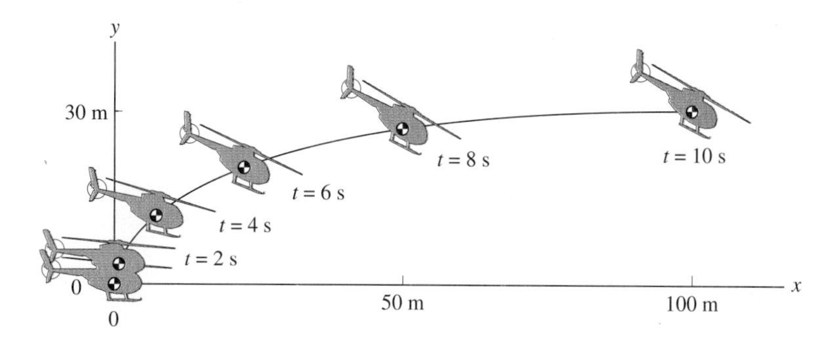

(a) Position of the helicopter at 2 s intervals.

DISCUSSION

This example demonstrates how the inertial navigation systems used in commercial aeroplanes and ships work. They contain accelerometers that measure the x, y and z components of acceleration. (Gyroscopes maintain the alignments of the accelerometers.) By integrating the acceleration components twice with respect to time, the systems compute changes in the x, y and z coordinates of aeroplane or ship.

Problems

2.61 The cartesian coordinates of a point (in metres) are $x = 2t + 4$, $y = t^3 - 2t$, $z = 4t^2 - 4$, where t is in seconds. What are its velocity and acceleration at $t = 4$ s?

Strategy: Since the cartesian coordinates are given as functions of time, you can use Equations (2.23) to determine the components of the velocity as functions of time and then use Equations (2.25) to determine the components of the acceleration as functions of time.

2.62 The velocity of a point is $\mathbf{v} = (2\,\mathbf{i} + 3t^2\,\mathbf{j})$ m/s. At $t = 0$ its position is $\mathbf{r} = -\mathbf{i} + 2\,\mathbf{j}$ (m). What is its position at $t = 2$ s?

2.63 The acceleration components of a point (in m/s^2) are $a_x = 3t^2$, $a_y = 6t$ and $a_z = 0$. At $t = 0$, $x = 5$ m, $v_x = 3$ m/s, $y = 1$ m, $v_y = -2$ m/s, $z = 0$ and $v_z = 0$. What are its position vector and velocity vector at $t = 3$ s?

2.64 The acceleration components of an object (in m/s^2) are $a_x = 2t$, $a_y = 4t^2 - 2$ and $a_z = -6$. At $t = 0$ the position of the object is $\mathbf{r} = (10\,\mathbf{j} - 10\,\mathbf{k})$ m and its velocity is $\mathbf{v} = (2\,\mathbf{i} - 4\,\mathbf{j})$ m/s. Determine its position when $t = 4$ s.

2.65 Suppose you are designing a mortar to send a rescue line from a Coast Guard boat to ships in distress. The line is attached to a weight that is fired by the mortar. The mortar is to be mounted so that it fires at 45° above the horizontal. If you neglect aerodynamic drag and the weight of the line for your preliminary design and assume a muzzle velocity of 30 m/s at $t = 0$, what are the x and y coordinates of the weight as functions of time?

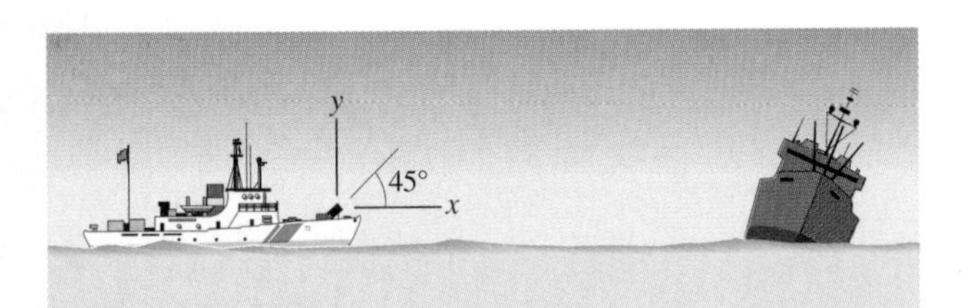

P2.65

2.66 In Problem 2.65, what must the mortar's muzzle velocity be to reach ships 300 m away?

2.67 If a stone is thrown horizontally from the top of a 30 m tall building at 15 m/s, at what horizontal distance from the point at which it is thrown does it hit the ground? (Assume level ground.) What is the magnitude of its velocity just before it hits?

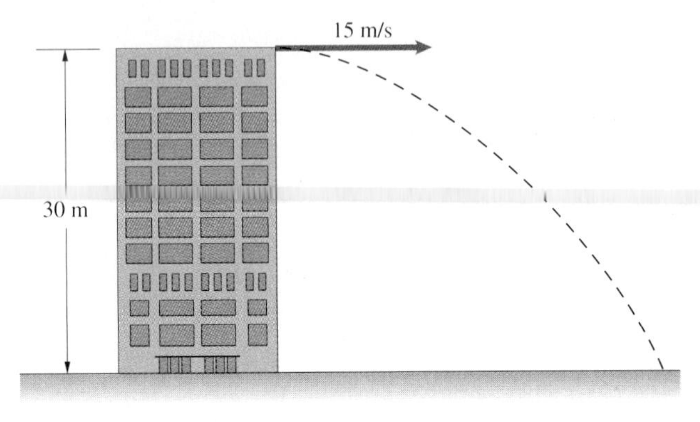

P2.67

2.68 A projectile is launched from ground level with an initial velocity v_0. What initial angle θ_0 above the horizontal causes the range R to be a maximum, and what is the maximum range?

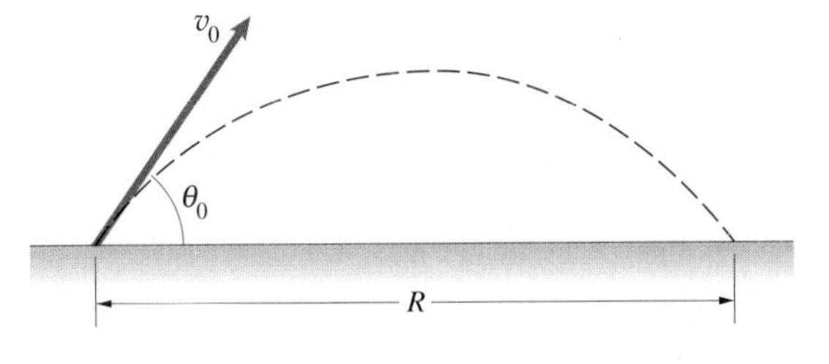

P2.68

2.69 A pilot wants to drop supplies to remote locations in the Australian outback. He intends to fly horizontally and release the packages with no vertical velocity. Derive an equation for the horizontal distance d at which he should release the package in terms of the aeroplane's velocity v_0 and altitude h.

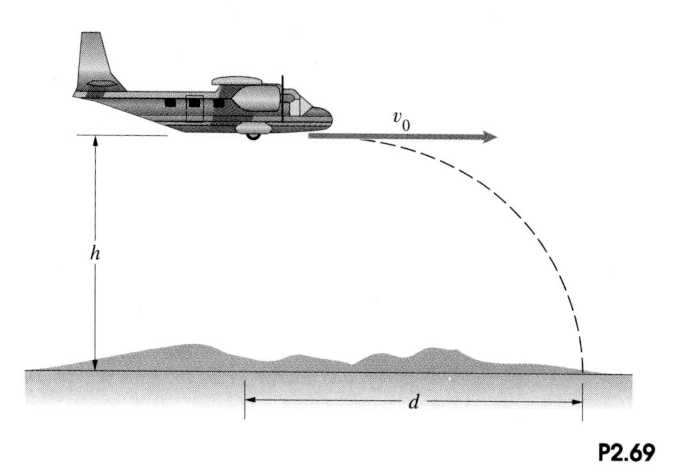

P2.69

2.70 A batter strikes a baseball at 1 m above home plate and pops it up at an angle of 60° above the horizontal. The second baseman catches it at 2 m above second base. What was the ball's initial velocity?

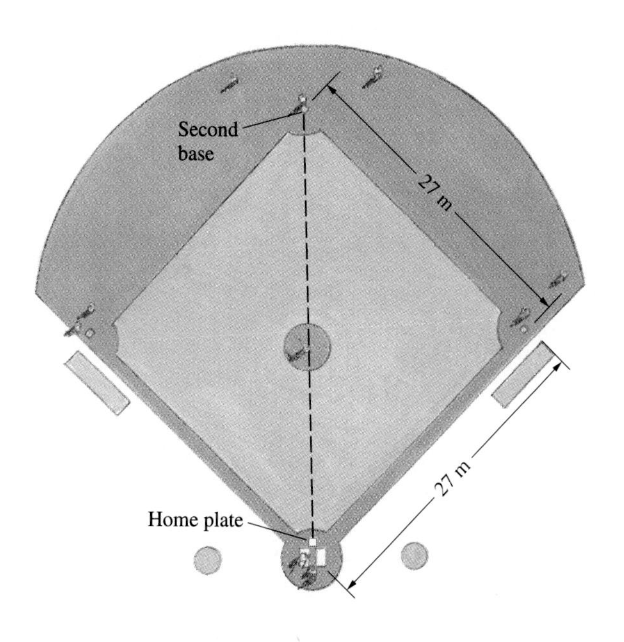

P2.70

2.71 In Problem 2.70, how high above the field did the ball go?

2.72 A baseball pitcher releases a fastball with an initial velocity $v_0 = 145\,\text{km/hr}$. Let θ be the initial angle of the ball's velocity vector above the horizontal. When it is released, the ball is 1.83 m above the ground and 17.68 m from the batter's plate. The batter's strike zone (between his knees and shoulders) extends from 0.56 m above the ground to 1.37 m above the ground. Neglecting aerodynamic effects, determine whether the ball will hit the strike zone: (a) if $\theta = 1°$; (b) if $\theta = 2°$.

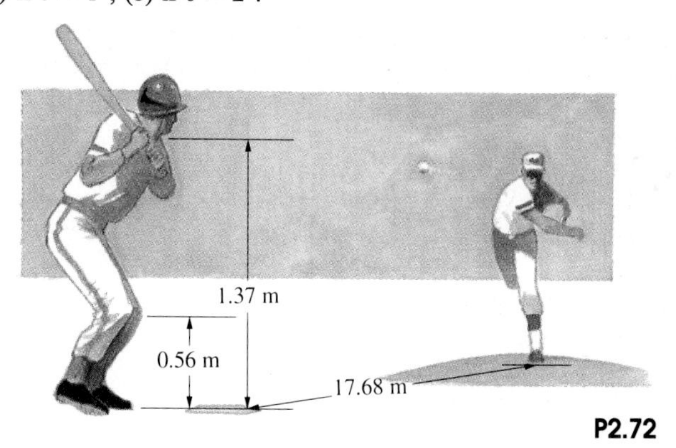

P2.72

2.73 In Problem 2.72, assume that the pitcher releases the ball at an angle $\theta = 1°$ above the horizontal and determine the range of velocities v_0 (in m/s) within which he must release the ball to hit the strike zone.

2.74 A zoology graduate student is armed with a bow and an arrow tipped with a syringe of tranquillizer and assigned to measure the temperature of a black rhinoceros (*Diceros bicornis*). The *maximum* range of his bow is 100 m. If a truculent rhino charges straight towards him at 30 km/hr and he aims his bow 20° above the horizontal, how far away should the rhino be when he releases the arrow?

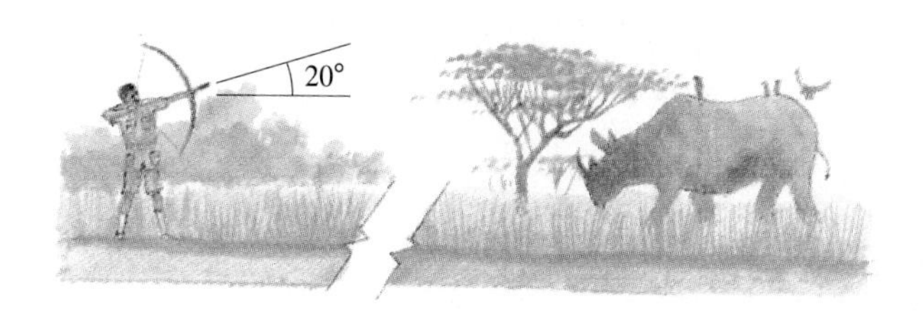

P2.74

2.75 The cliff divers of Acapulco, Mexico, must time their dives so that they enter the water at the crest (high point) of a wave. The crests of the waves are 0.6 m above the mean water depth $h = 3.6\,\text{m}$ and the horizontal velocity of the waves is $\sqrt{gh}$. The diver's aiming point is 2 m out from the base of the cliff. Assume that his velocity is horizontal when he begins the dive.
(a) What is the magnitude of his velocity in kilometres per hour when he enters the water?
(b) How far from his aiming point should a wave crest be when he dives in order for him to enter the water at the crest?

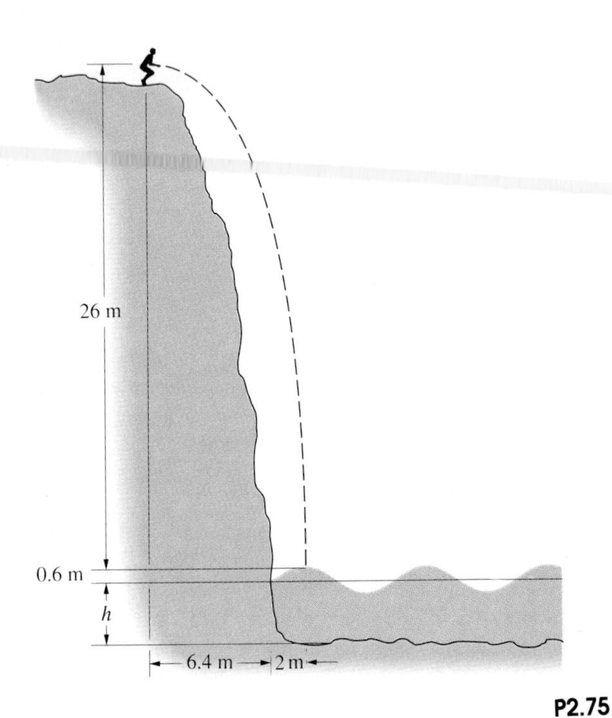

P2.75

2.76 A projectile is launched at 10 m/s from a sloping surface. Determine the range R.

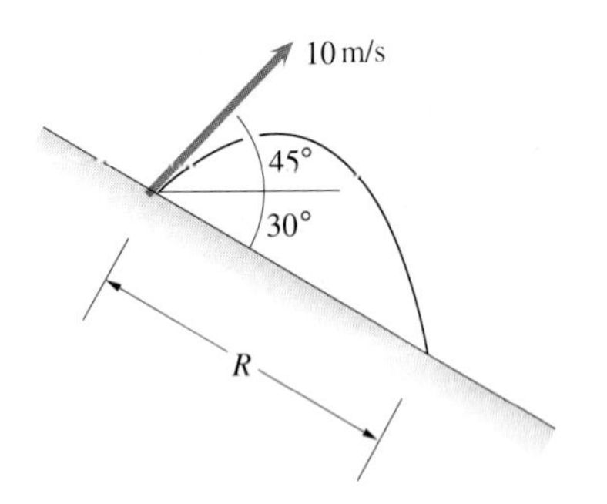

P2.76

2.77 A skier leaves a 20° slope at 15 m/s.
(a) Determine the distance d to the point where he lands.
(b) Determine his components of velocity parallel and perpendicular to the 45° slope when he lands.

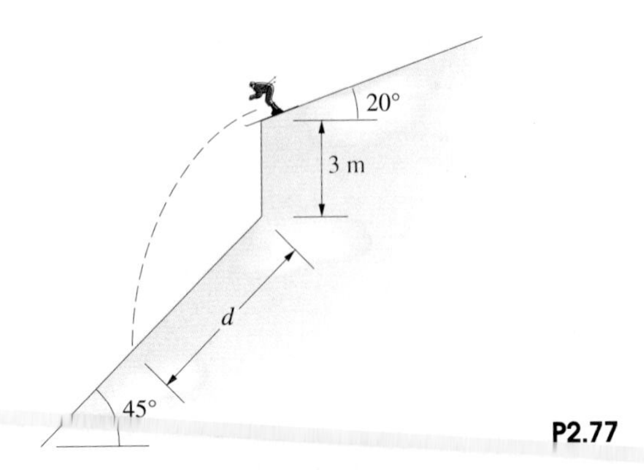

P2.77

2.78 At $t = 0$, a steel ball in a tank of oil is given a horizontal velocity $\mathbf{v} = 2\,\mathbf{i}\,\text{m/s}$. The component of its acceleration in m/s^2 are $a_x = -1.2v_x$, $a_y = -8 - 1.2v_y$, $a_z = -1.2v_z$. What is the velocity of the ball at $t = 1\,\text{s}$?

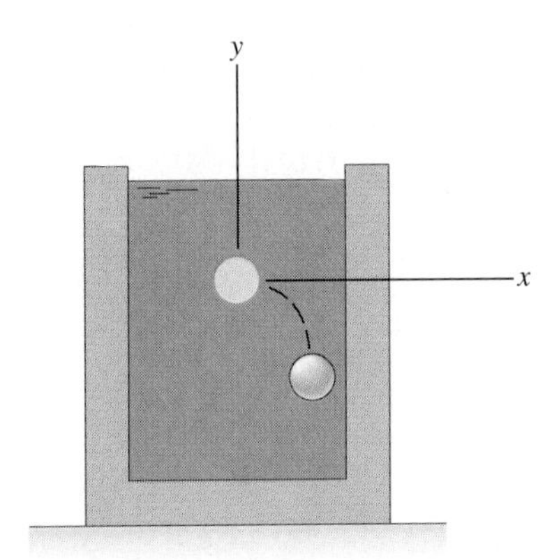

P2.78

2.79 In Problems 2.78, what is the position of the ball at $t = 1\,\text{s}$ relative to its position at $t = 0$?

2.80 You must design a device for an assembly line that launches small parts through the air into a bin. The launch point is $x = 200\,\text{mm}$, $y = -50\,\text{mm}$, $z = -100\,\text{mm}$. (The y axis is vertical and positive upwards.) To land in the bin, the parts must pass through the point $x = 600\,\text{mm}$, $y = 200\,\text{mm}$, $z = 100\,\text{mm}$ *moving horizontally*. Determine the components of velocity the launcher must give the parts.

2.81 If $y = 150$ mm, $dy/dt = 300$ mm/s, and $d^2y/dt^2 = 0$, what are the magnitudes of the velocity and acceleration of point P?

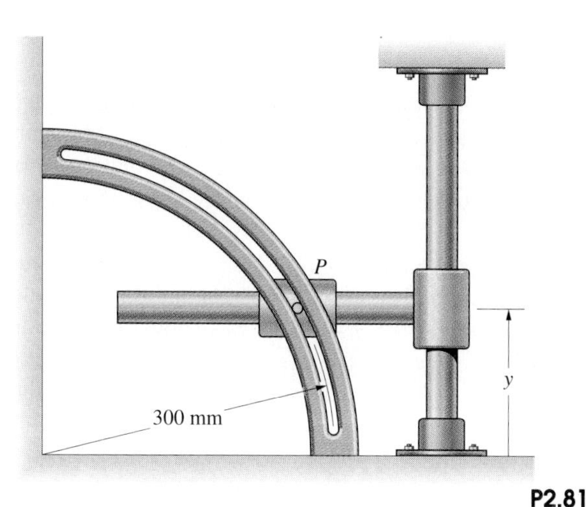

P2.81

2.82 A car travels at a constant speed of 100 km/hr on a straight road of increasing grade whose vertical profile can be approximated by the equation shown. When the car's horizontal coordinate is $x = 400$ m, what is its acceleration?

P2.82

2.83 Suppose that a projectile has the initial conditions shown in Figure 2.16(a). Show that in terms of the $x'y'$ coordinate system with its origin at the highest point of the trajectory, the equation describing the trajectory is

$$y' \frac{g}{2v_0^2 \cos^2\theta_0}(x')^2$$

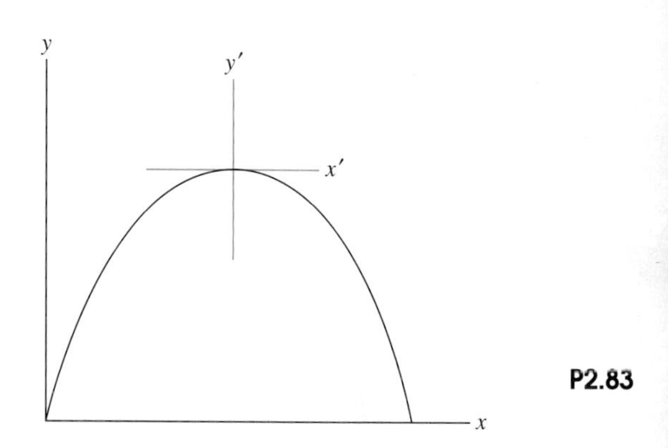

P2.83

2.84 The acceleration components of a point are $a_x = -4\cos 2t$, $a_y = -4\sin 2t$, $a_z = 0$. At $t = 0$ its position and velocity are $\mathbf{r} = \mathbf{i}$, $\mathbf{v} = 2\,\mathbf{j}$. Show that: (a) the magnitude of the velocity is constant; (b) the velocity and acceleration vectors are perpendicular; (c) the magnitude of the acceleration is constant and points towards the origin; (d) the trajectory of the point is a circle with its centre at the origin.

Angular Motion

We have seen that in some cases the curvilinear motion of a point can be analysed using cartesian coordinates. In the following sections we describe problems that can be analysed more simply in terms of other coordinate systems. To help you understand our discussion of these alternative coordinate systems, we introduce two preliminary topics in this section: the angular motion of a line in a plane and the time derivative of a unit vector rotating in a plane.

Angular Motion of a Line We can specify the angular position of a line L in a particular plane relative to a reference line L_0 in the plane by an angle θ (Figure 2.18). The **angular velocity** of L relative to L_0 is defined by

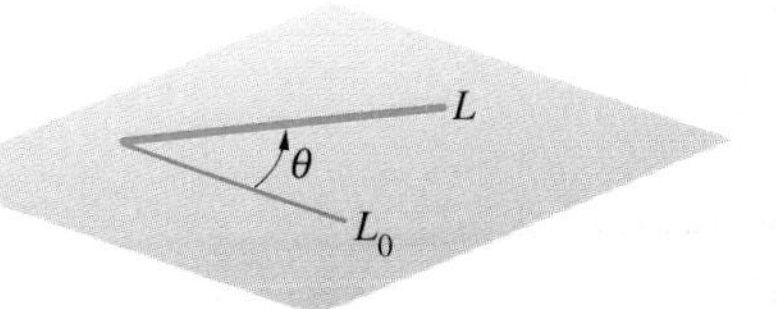

$$\omega = \frac{d\theta}{dt} \qquad (2.31)$$

and the **angular acceleration** of L relative to L_0 is defined by

$$\alpha = \frac{d\omega}{dt} = \frac{d^2\theta}{dt^2} \tag{2.32}$$

The dimensions of the angular position, angular velocity and angular acceleration are radians (rad), rad/s and rad/s^2, respectively. Although these quantities are often expressed in terms of degrees or revolutions instead of radians, you should convert them into radians before using them in calculations.

Notice the analogy between Equations (2.31) and (2.32) and the equations relating the position, velocity and acceleration of a point along a straight line (Table 2.2). In each case the position is specified by a single scalar coordinate, which can be positive or negative. (In Figure 2.18 the counterclockwise direction is positive.) Because the equations are identical in form, you can analyse problems involving angular motion of a line by the same methods you applied to straight-line motion.

Table 2.2. The equations governing straight-line motion and the equations governing the angular motion of a line are identical in form.

Straight-line motion	Angular motion
$v = \frac{ds}{dt}$	$\omega = \frac{d\theta}{dt}$
$a = \frac{dv}{dt} = \frac{d^2s}{dt^2}$	$\alpha = \frac{d\omega}{dt} = \frac{d^2\theta}{dt^2}$

Rotating Unit Vector We have seen that the cartesian unit vectors **i**, **j** and **k** are constants provided the coordinate system does not rotate. However, in other coordinate systems the unit vectors used to describe the motion of a point rotate as the point moves. To obtain expressions for the velocity and acceleration in such coordinate systems, we must know the time derivative of a rotating unit vector.

We can describe the angular motion of a unit vector **e** in a plane just as we described the angular motion of a line. The direction of **e** relative to a reference line L_0 is specified by the angle θ in Figure 2.19(a), and the rate of rotation of **e** relative to L_0 is specified by the angular velocity

$$\omega = \frac{d\theta}{dt}$$

The time derivative of **e** is defined by

$$\frac{d\mathbf{e}}{dt} = \lim_{\Delta t \to 0} \frac{\mathbf{e}(t + \Delta t) - \mathbf{e}(t)}{\Delta t}$$

Figure 2.19(b) shows the vector **e** at time t and at time $t + \Delta t$. The change in **e** during this interval is $\Delta\mathbf{e} = \mathbf{e}(t + \Delta t) - \mathbf{e}(t)$, and the angle through which **e** rotates is $\Delta\theta = \theta(t + \Delta t) - \theta(t)$. The triangle in Figure 2.19(b) is isosceles, so the magnitude of $\Delta\mathbf{e}$ is

$$|\Delta\mathbf{e}| = 2|\mathbf{e}|\sin(\Delta\theta/2) = 2\sin(\Delta\theta/2)$$

To write the vector $\Delta\mathbf{e}$ in terms of this expression, we introduce a unit vector $\mathbf{n}$ that points in the direction of $\Delta\mathbf{e}$ (Figure 2.19(b)):

$$\Delta\mathbf{e} = |\Delta\mathbf{e}|\mathbf{n} = 2\sin(\Delta\theta/2)\,\mathbf{n}$$

In terms of this expression, the time derivative of $\mathbf{e}$ is

$$\frac{d\mathbf{e}}{dt} = \lim_{\Delta t\to 0}\frac{\Delta\mathbf{e}}{\Delta t} = \lim_{\Delta t\to 0}\frac{2\sin(\Delta\theta/2)\,\mathbf{n}}{\Delta t}$$

To evaluate this limit, we write it in the form

$$\frac{d\mathbf{e}}{dt} = \lim_{\Delta t\to 0}\frac{\sin(\Delta\theta/2)}{\Delta\theta/2}\frac{\Delta\theta}{\Delta t}\mathbf{n}$$

In the limit as Δt approaches zero, $\sin(\Delta\theta/2)/(\Delta\theta/2)$ equals 1, $\Delta\theta/\Delta t$ equals $d\theta/dt$, and the unit vector $\mathbf{n}$ is perpendicular to $\mathbf{e}(t)$ (Figure 2.19(c)). Therefore the time derivative of $\mathbf{e}$ is

$$\frac{d\mathbf{e}}{dt} = \frac{d\theta}{dt}\mathbf{n} = \omega\mathbf{n} \tag{2.33}$$

where $\mathbf{n}$ is a unit vector that is perpendicular to $\mathbf{e}$ and points in the positive θ direction (Figure 2.19(d)). In the following sections we use this result in deriving expressions for the velocity and acceleration of a point in different coordinate systems.

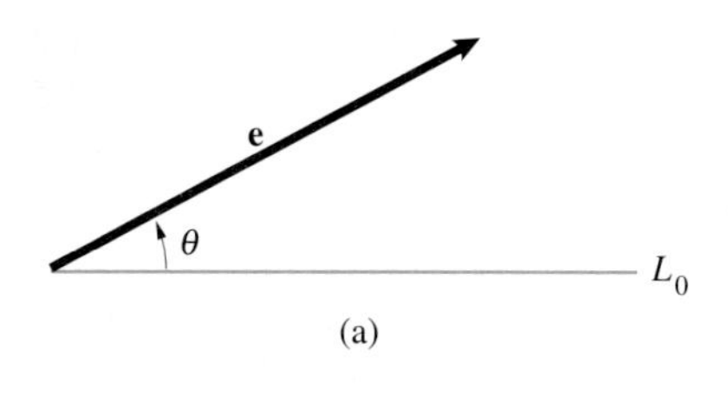

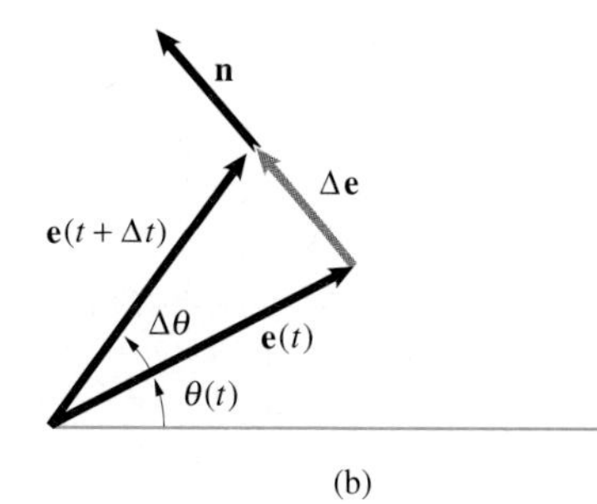

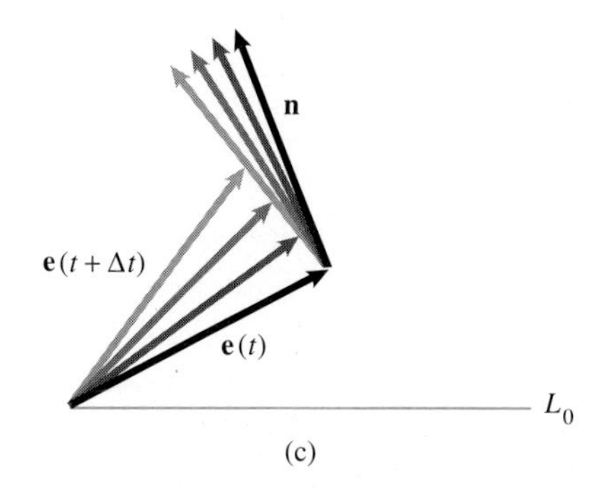

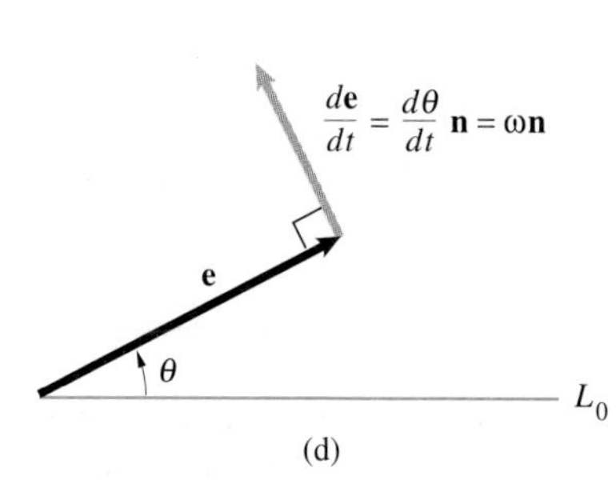

Figure 2.19
(a) A unit vector $\mathbf{e}$ and reference line L_0.
(b) The change $\Delta\mathbf{e}$ in $\mathbf{e}$ from t to $t + \Delta t$.
(c) As Δt goes to zero, $\mathbf{n}$ becomes perpendicular to $\mathbf{e}(t)$.
(d) The time derivative of $\mathbf{e}$.

Example 2.7

The rotor of a jet engine is rotating at 10 000 rpm (revolutions per minute) when the fuel is shut off. The ensuing angular acceleration is $\alpha = -0.02\omega$, where ω is the angular velocity in rad/s.

(a) How long does it take the rotor to slow to 1000 rpm?

(b) How many revolutions does the rotor turn while decelerating to 1000 rpm?

STRATEGY

To analyse the angular motion of the rotor, we define a line L that is fixed to the rotor and perpendicular to its axis (Figure 2.20). Then we examine the motion of L relative to the reference line L_0. The angular position, velocity and acceleration of L define the angular motion of the rotor.

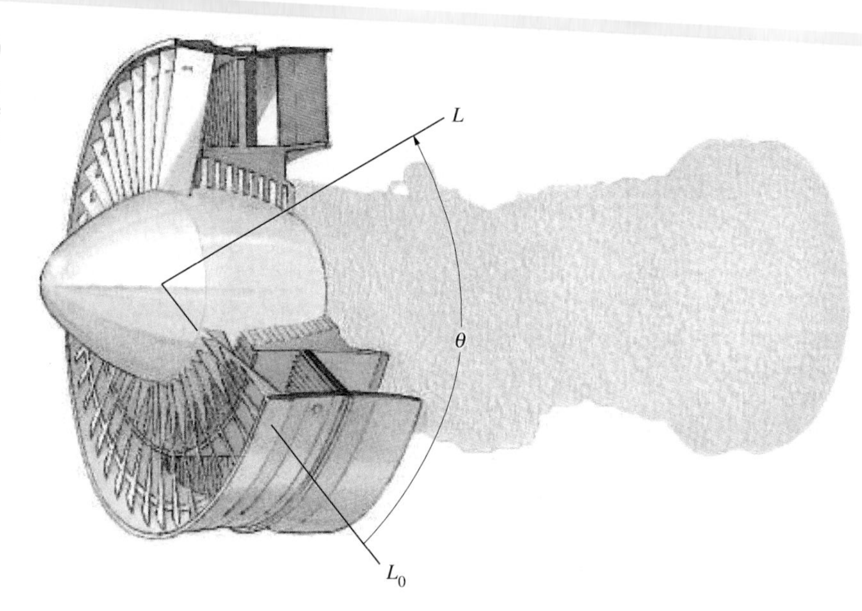

Figure 2.20
Introducing a line L and reference line L_0 to specify the angular position of the rotor.

SOLUTION

The conversion from rpm to rad/s is

$$1 \text{ rpm} = 1 \text{ revolution/min} \times \left(\frac{2\pi \text{ rad}}{1 \text{ revolution}}\right) \times \left(\frac{1 \text{ min}}{60 \ s}\right)$$
$$= (\pi/30) \text{ rad/s}$$

(a) The angular acceleration is

$$\alpha = \frac{d\omega}{dt} = -0.02\omega$$

We separate variables,

$$\frac{d\omega}{\omega} = -0.02\,dt$$

and integrate, defining $t = 0$ to be the time at which the fuel is turned off:

$$\int_{10\,000\pi/30}^{1000\pi/30} \frac{d\omega}{\omega} = \int_0^t -0.02\,dt$$

Evaluating the integrals and solving for t, we obtain

$$t = \left(\frac{1}{0.02}\right) \ln\left(\frac{10\,000\pi/30}{1000\pi/30}\right) = 115.1 \text{ s}$$

(b) We write the angular acceleration as

$$\alpha = \frac{d\omega}{dt} = \frac{d\omega}{d\theta}\frac{d\theta}{dt} = \frac{d\omega}{d\theta}\omega = -0.02\omega$$

separate variables,

$$d\omega = -0.02\,d\theta$$

and integrate, defining $\theta = 0$ to be the angular position at which the fuel is turned off:

$$\int_{10\,000\pi/30}^{1000\pi/30} d\omega = \int_0^\theta -0.02\,d\theta$$

Solving for θ, we obtain

$$\theta = \left(\frac{1}{0.02}\right)[(10\,000\pi/30) - (1000\pi/30)]$$
$$= 15\,000\pi \text{ rad} = 7500 \text{ revolutions}$$

Problems

2.85 What are the magnitudes of the angular velocities (in rad/s) of the minute hand and the hour hand of the clock?

P2.85

2.86 Let L be a line from the centre of the earth to a fixed point on the equator and let L_0 denote a fixed reference direction. The figure shows the earth seen from above the North Pole.
(a) Is $d\theta/dt$ positive or negative?
(b) What is the magnitudes of $d\theta/dt$ in rad/s?

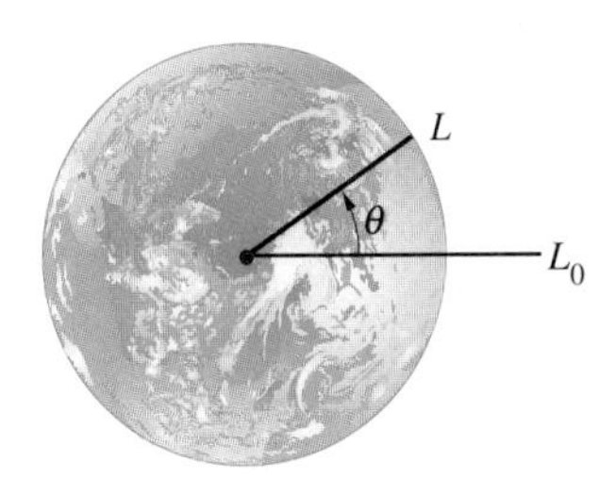

P2.86

2.87 The angle between a line L and a reference line L_0 is $\theta = 2t^2$ rad.
(a) What are the angular velocity and angular acceleration of L relative to L_0 at $t = 6$ s?
(b) How many revolutions does L rotate relative to L_0 during the interval of time from $t = 0$ to $t = 6$ s?

Strategy: Use Equations (2.31) and (2.32) to determine the angular velocity and angular acceleration as functions of time.

2.88 The angle θ between the bar and the horizontal line is $\theta = (t^3 - 2t^2 + 4)$ degrees. Determine the angular velocity and angular acceleration of the bar at $t = 10$ s.

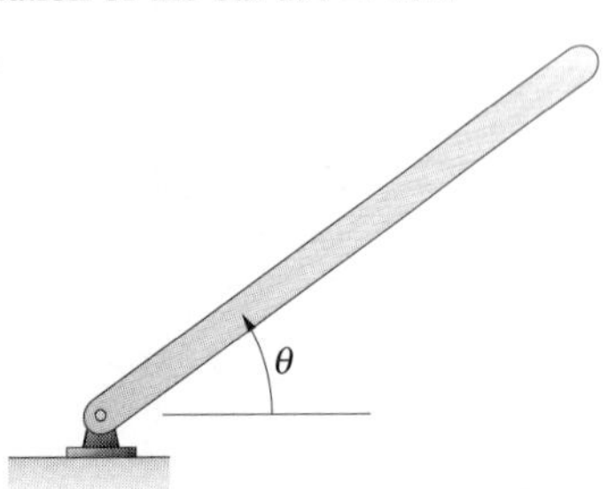

P2.88

2.89 The angular acceleration of a line L relative to a reference line L_0 is $\alpha = (30 - 6t)\,\text{rad/s}^2$. When $t = 0$, $\theta = 0$ and $\omega = 0$. What is the maximum angular velocity of L relative to L_0 during the interval of time from $t = 0$ to $t = 10$ s?

2.90 A gas turbine starts rotating from rest at $t = 0$ and has angular acceleration $\alpha = 6t\,\text{rad/s}^2$ for 3 s. It then slows down with constant angular deceleration $\alpha = -3\,\text{rad/s}^2$ until it stops.
(a) What maximum angular velocity does it attain?
(b) Through what total angle does it turn?

2.91 The rotor of an electric generator is rotating at 200 rpm (revolutions per minute) when the motor is turned off. Due to frictional effects, the angular deceleration of the rotor after it is turned off is $\alpha = -0.01\omega\,\text{rad/s}^2$, where ω is the angular velocity in rad/s. How many revolutions does the rotor turn after the motor is turned off?

2.92 A needle of a measuring instrument is connected to a torsional spring that subjects it to an angular acceleration $\alpha = -4\theta\,\text{rad/s}^2$, where θ is the needle's angular position in radians relative to a reference direction. If the needle is released from rest at $\theta = 1$ rad, what is its angular velocity at $\theta = 0$?

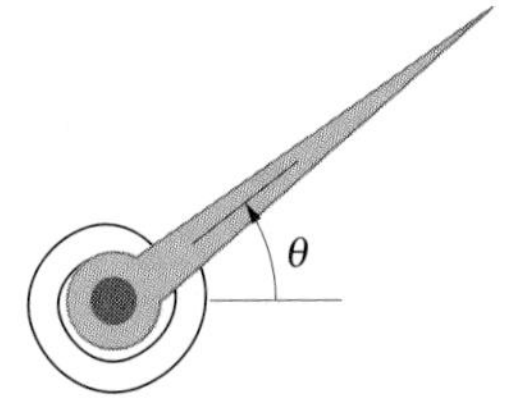

P2.92

2.93 The angle θ measures the direction of the unit vector **e** relative to the x axis. Given that $\omega = d\theta/dt = 2$ rad/s, determine the vector $d\mathbf{e}/dt$: (a) when $\theta = 0$; (b) when $\theta = 90°$; (c) when $\theta = 180°$.

Strategy: You can obtain these results either by using Equation (2.33) or by expressing **e** in terms of its x and y components and taking its time derivative.

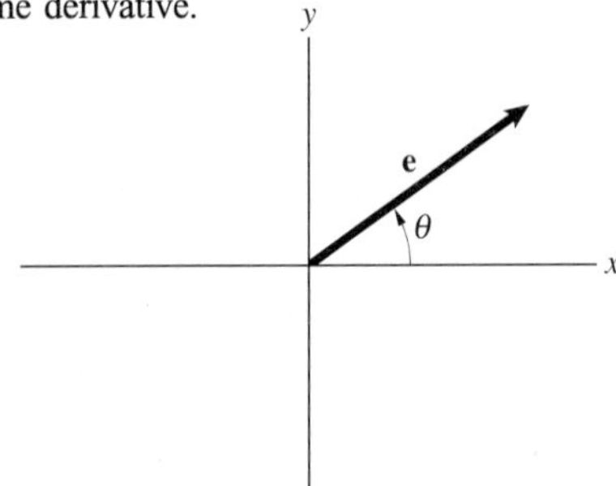

P2.93

2.94 In Problem 2.93, suppose that the angle $\theta = 2t^2$ rad. What is the vector $d\mathbf{e}/dt$ at $t = 4$ s?

2.95 The line OP is of constant length R. The angle $\theta = \omega_0 t$, where ω_0 is a constant.
(a) Use the relations

$$v_x = \frac{dx}{dt} \qquad v_y = \frac{dy}{dt}$$

to determine the velocity of point P relative to O.
(b) Use Equation (2.33) to determine the velocity of point P relative to O, and confirm that your result agrees with the result of part (a).

Strategy: In part (b), write the position vector of P relative to O as $\mathbf{r} = R\,\mathbf{e}$, where $\mathbf{e}$ is a unit vector that points from O towards P.

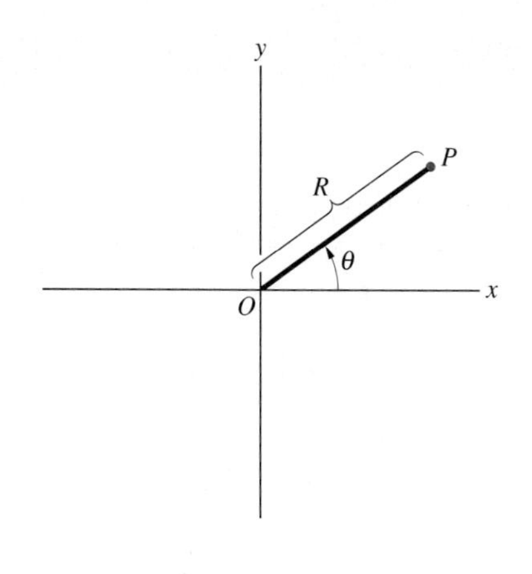

P2.95

Normal and Tangential Components

In this method of describing curvilinear motion, we specify the position of a point by its position measured *along its path*, and express the velocity and acceleration in terms of their components tangential and normal (perpendicular) to the path. Normal and tangential components are particularly useful when a point moves along a circular path. Furthermore, they provide unique insight into the character of the velocity and acceleration in curvilinear motion.

Consider a point P moving along a plane, curvilinear path (Figure 2.21(a)). The position vector $\mathbf{r}$ specifies the position of P relative to the reference point O, and the coordinate s measures the position of P along the path relative to a point O' on the path. The velocity of P relative to O is

$$\mathbf{v} = \frac{d\mathbf{r}}{dt} = \lim_{\Delta t \to 0} \frac{\mathbf{r}(t + \Delta t) - \mathbf{r}(t)}{\Delta t} = \lim_{\Delta t \to 0} \frac{\Delta \mathbf{r}}{\Delta t} \tag{2.34}$$

where $\Delta\mathbf{r} = \mathbf{r}(t + \Delta t) - \mathbf{r}(t)$ (Figure 2.21(b)). We denote the distance travelled along the path from t to $t + \Delta t$ by Δs. By introducing a unit vector $\mathbf{e}$ defined to point in the direction of $\Delta\mathbf{r}$, we can write Equation (2.34) as

$$\mathbf{v} = \lim_{\Delta t \to 0} \frac{\Delta s}{\Delta t}\mathbf{e} \tag{2.34}$$

As Δt approaches zero, $\Delta s/\Delta t$ becomes ds/dt and $\mathbf{e}$ becomes a unit vector tangent to the path at the position of P at time t, which we denote by $\mathbf{e}_t$ (Figure 2.21(c)):

$$\mathbf{v} = v\,\mathbf{e}_t = \frac{ds}{dt}\mathbf{e}_t \tag{2.35}$$

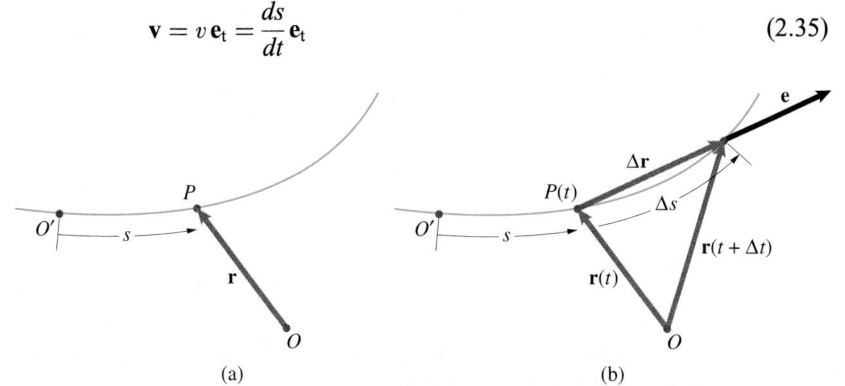

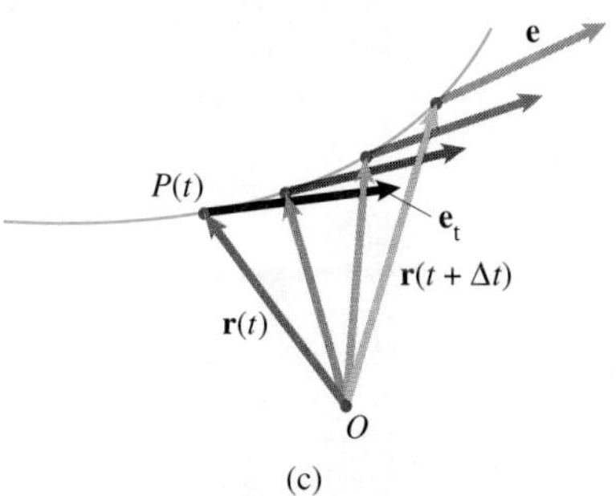

Figure 2.21
(a) The position of P along its paths is specified by the coordinate s.
(b) Position of P at time t and at time $t + \Delta t$.
(c) The limit of $\mathbf{e}$ as $\Delta t \to 0$ is a unit vector tangent to the path.

The velocity of a point in curvilinear motion is a vector whose magnitude equals the rate of distance travelled along the path and whose direction is tangent to the path.

To determine the acceleration of P, we take the time derivative of Equation (2.35):

$$\mathbf{a} = \frac{d\mathbf{v}}{dt} = \frac{dv}{dt}\mathbf{e}_t + v\frac{d\mathbf{e}_t}{dt} \tag{2.36}$$

If the path is not a straight line, the unit vector $\mathbf{e}_t$ rotates as P moves. As a consequence, the time derivative of $\mathbf{e}_t$ is not zero. In the previous section we derived an expression for the time derivative of a rotating unit vector in terms of the unit vector's angular velocity, Equation (2.33). To use that result, we define the **path angle** θ specifying the direction of $\mathbf{e}_t$ relative to a reference line (Figure 2.22). Then from Equation (2.33), the time derivative of $\mathbf{e}_t$ is

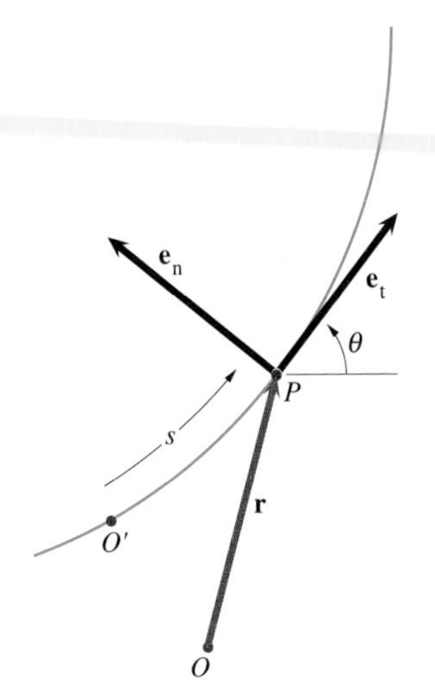

Figure 2.22
The path angle θ.

$$\frac{d\mathbf{e}_t}{dt} = \frac{d\theta}{dt}\mathbf{e}_n$$

where $\mathbf{e}_n$ is a unit vector that is normal to $\mathbf{e}_t$ and points in the positive θ direction if $d\theta/dt$ is positive (Figure 2.22). Substituting this expression into Equation (2.36), we obtain the acceleration of P:

$$\mathbf{a} = \frac{dv}{dt}\mathbf{e}_t + v\frac{d\theta}{dt}\mathbf{e}_n. \tag{2.37}$$

We can derive this result in another way that is less rigorous but gives additional insight into the meanings of the tangential and normal components of the acceleration. Figure 2.23(a) shows the velocity of P at times t and $t + \Delta t$. In Figures 2.23(b), you can see that the change in the velocity, $\mathbf{v}(t + \Delta t) - \mathbf{v}(t)$, consists of two components. The component Δv, which is tangent to the path at time t, is due to the change in the *magnitude* of the velocity. The component $v\Delta\theta$, which is perpendicular to the path at time t, is due to the change in the *direction* of the velocity vector. Thus the change in the velocity is (approximately)

$$\mathbf{v}(t + \Delta t) - \mathbf{v}(t) = \Delta v\,\mathbf{e}_t + v\Delta\theta\,\mathbf{e}_n$$

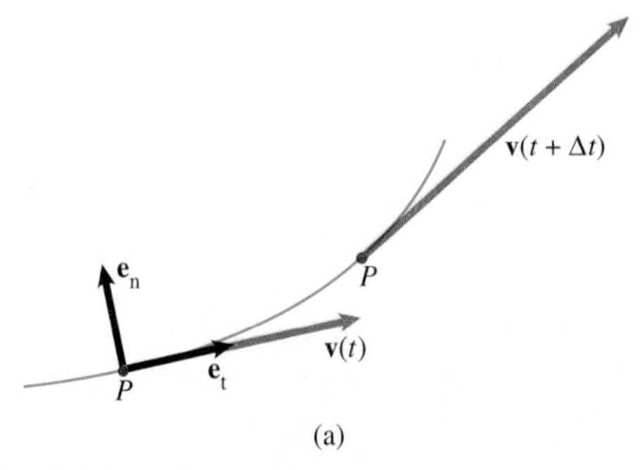

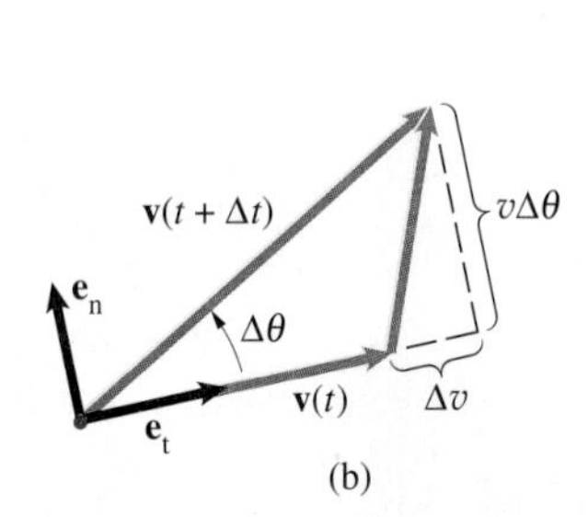

Figure 2.23
(a) Velocity of P at t and at $t + \Delta t$.
(b) The tangential and normal components of the change in the velocity.

To obtain the acceleration, we divide this expression by Δt and take the limit as $\Delta t \to 0$:

$$\mathbf{a} = \lim_{\Delta t \to 0} \frac{\Delta \mathbf{v}}{\Delta t} = \lim_{\Delta t \to 0} \left[\frac{\Delta v}{\Delta t} \mathbf{e}_{\mathrm{t}} + v \frac{\Delta \theta}{\Delta t} \mathbf{e}_{\mathrm{n}} \right]$$

$$= \frac{dv}{dt} \mathbf{e}_{\mathrm{t}} + v \frac{d\theta}{dt} \mathbf{e}_{\mathrm{n}}$$

Thus we again obtain Equation (2.37). However, this derivation clearly points out that the tangential component of the acceleration arises from the rate of change of the magnitude of the velocity, whereas the normal component arises from the rate of change in the direction of the velocity vector. Notice that if the path is a straight line at time t, the normal component of the acceleration equals zero, because in that case $d\theta/dt$ is zero.

We can express the acceleration in another form that is often more convenient to use. Figure 2.24 shows the positions on the path reached by P at times t and $t + dt$. If the path is curved, straight lines extended from these points perpendicular to the path will intersect as shown. The distance ρ from the path to the point where these two lines intersect is called the **instantaneous radius of curvature** of the path. (If the path is circular, ρ is simply the radius of the path.) The angle $d\theta$ is the change in the path angle, and ds is the distance travelled, from t to $t + dt$. You can see from the figure that ρ is related to ds by

$$ds = \rho d\theta$$

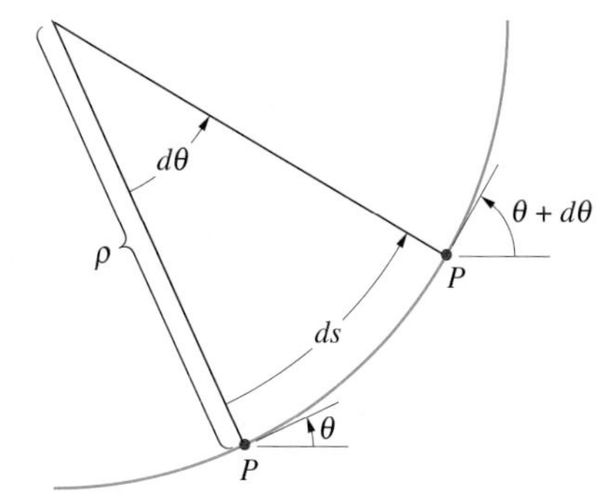

Figure 2.24
The instantaneous radius of curvature ρ.

Dividing by dt, we obtain

$$\frac{ds}{dt} = v = \rho \frac{d\theta}{dt}$$

Using this relation, we can write Equation (2.37) as

$$\mathbf{a} = \frac{dv}{dt} \mathbf{e}_{\mathrm{t}} + \frac{v^2}{\rho} \mathbf{e}_{\mathrm{n}}$$

For a given value of v, the normal component of the acceleration depends on the instantaneous radius of the curvature. The greater the curvature of the path, the greater the normal component of acceleration. When the acceleration is expressed in this way, the unit vector $\mathbf{e}_n$ must be defined to point towards the *concave* side of the path (Figure 2.25).

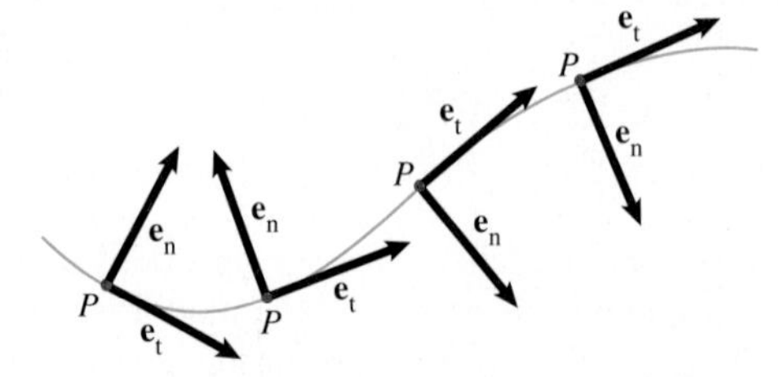

Figure 2.25
The unit vector normal to the path points towards the concave side of the path.

Thus the velocity and acceleration in terms of normal and tangential components are (Figure 2.26)

$$\mathbf{v} = v\,\mathbf{e}_t = \frac{ds}{dt}\mathbf{e}_t \tag{2.38}$$

$$\mathbf{a} = a_t\,\mathbf{e}_t + a_n\,\mathbf{e}_n \tag{2.39}$$

where

$$a_t = \frac{dv}{dt} \qquad a_n = v\frac{d\theta}{dt} = \frac{v^2}{\rho} \tag{2.40}$$

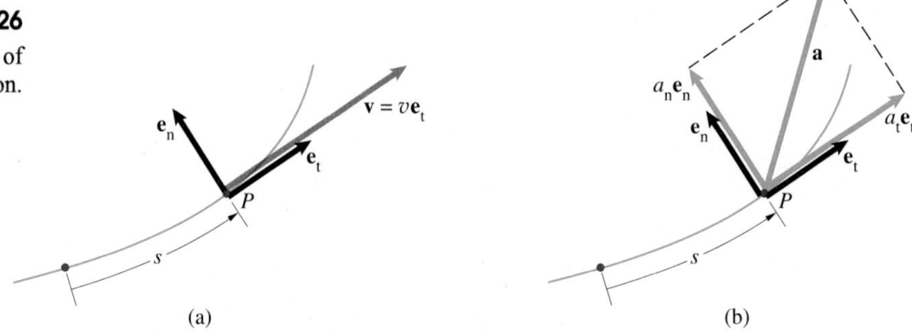

Figure 2.26
Normal and tangential components of (a) the velocity and (b) acceleration.

Circular Motion If a point P moves in a circular path of radius R (Figure 2.27), the distance s is related to the angle θ by

$$s = R\theta \qquad \textbf{Circular path}$$

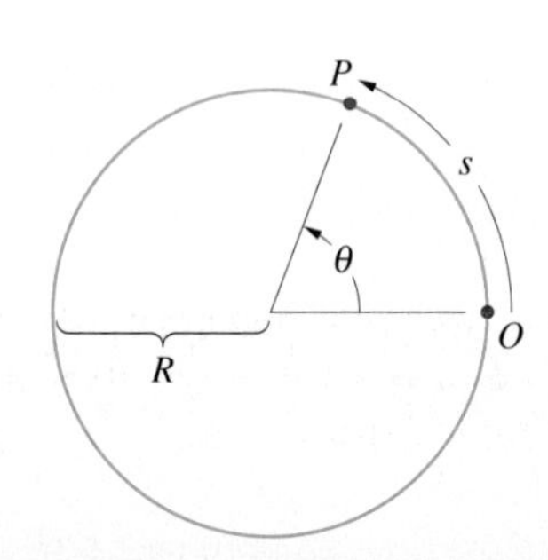

Figure 2.27
A point moving in a circular path.

This relation means we can specify the position of P along the circular path by either s or θ. Taking the time derivative of this equation, we obtain a relation between $v = ds/dt$ and the angular velocity of the line from the centre of the path to P:

$$v = R\frac{d\theta}{dt} = R\omega \qquad \textbf{Circular path} \tag{2.41}$$

Taking another time derivative, we obtain a relation between the tangential component of the acceleration $a_t = dv/dt$ and the angular acceleration:

$$a_t = R\frac{d\omega}{dt} = R\alpha \qquad \textbf{Circular path} \tag{2.42}$$

For the circular path, the instantaneous radius of curvature $\rho = R$, so the normal component of the acceleration is

$$a_n = \frac{v^2}{R} = R\omega^2 \qquad \textbf{Circular path} \tag{2.43}$$

Because problems involving circular motion of a point are so common, these relations are worth remembering. But you must be careful to use them *only* when the path is circular.

The following examples demonstrate the use of Equations (2.38) and (2.39) to analyse curvilinear motions of objects. Because the equations relating s, v and the tangential component of the acceleration,

$$v = \frac{ds}{dt}$$

$$a_t = \frac{dv}{dt}$$

are identical in form to the equations that govern the motion of a point along a straight line, in some cases you can solve them using the same methods you applied to straight-line motion.

Example 2.8

Figure 2.28

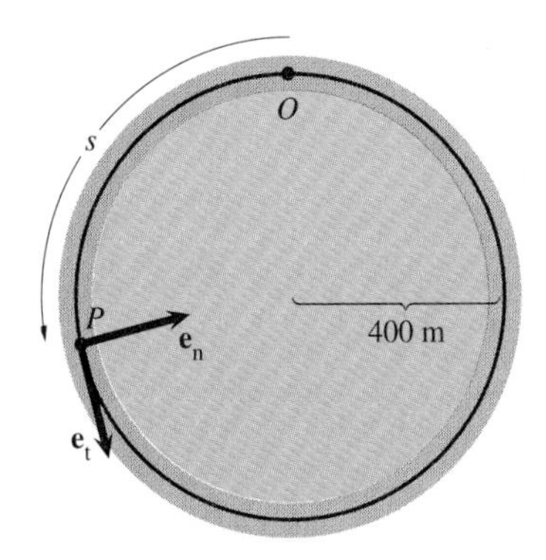

(a) The coordinate s measures the distance along the track.

The motorcycle in Figure 2.28 starts from rest at $t = 0$ on a circular track of 400 m radius. The tangential component of its acceleration is $a_t = (2 + 0.2t)\,\text{m/s}^2$. At $t = 10$ s, determine: (a) the distance it has moved along the track; (b) the magnitude of its acceleration.

STRATEGY

Let s be the distance along the track from the initial position O of the motorcycle to its position at time t (Figure (a)). Knowing the tangential acceleration as a function of time, we can integrate to determine v and s as functions of time.

SOLUTION

(a) The tangential acceleration is

$$a_t = \frac{dv}{dt} = (2 + 0.2t)\,\text{m/s}^2$$

Integrating,

$$\int_0^v dv = \int_0^t (2 + 0.2t)\,dt$$

we obtain v as a function of time:

$$v = \frac{ds}{dt} = (2t + 0.1t^2)\,\text{m/s}$$

Integrating this equation,

$$\int_0^s ds = \int_0^t (2t + 0.1t^2)\,dt$$

the coordinate s as a function of time is

$$s = \left(t^2 + \frac{0.1}{3}t^3\right)\,\text{m}$$

At $t = 10$ s, the distance moved along the track is

$$s = (10)^2 + \frac{0.1}{3}(10)^3 = 133.3\,\text{m}$$

(b) At $t = 10$ s, the tangential component of the acceleration is

$$a_t = 2 + 0.2(10) = 4\,\text{m/s}^2$$

We must also determine the normal component of acceleration. The instantaneous radius of curvature of the path is the radius of the circular track, $\rho = 400$ m. The magnitude of the velocity at $t = 10$ s is

$$v = 2(10) + 0.1(10)^2 = 30\,\text{m/s}$$

Therefore

$$a_n = \frac{v^2}{\rho} = \frac{(30)^2}{400} = 2.25\,\text{m/s}^2$$

The magnitude of the acceleration at $t = 10$ s is

$$|\mathbf{a}| = \sqrt{a_t^2 + a_n^2} = \sqrt{(4)^2 + (0.25)^2} = 4.59\,\text{m/s}^2$$

Example 2.9

A satellite is in a circular orbit of radius R around the earth. What is its velocity?

STRATEGY

The acceleration due to gravity at a distance R from the centre of the earth is gR_E^2/R^2, where R_E is the radius of the earth. By using this expression together with the equation for the acceleration in terms of normal and tangential components, we can obtain an equation for the satellite's velocity.

SOLUTION

In terms of normal and tangential components (Figure 2.29), the acceleration of the satellite is

$$\mathbf{a} = \frac{dv}{dt}\mathbf{e}_t + \frac{v^2}{R}\mathbf{e}_n$$

This expression must equal the acceleration due to gravity towards the centre of the earth:

$$\frac{dv}{dt}\mathbf{e}_t + \frac{v^2}{R}\mathbf{e}_n = \frac{gR_E^2}{R^2}\mathbf{e}_n$$

Because there is no $\mathbf{e}_t$ component on the right side, we conclude that the magnitude of the satellite's velocity is constant:

$$\frac{dv}{dt} = 0$$

Equating the $\mathbf{e}_n$ components and solving for v, we obtain

$$v = \sqrt{\frac{gR_E^2}{R}}$$

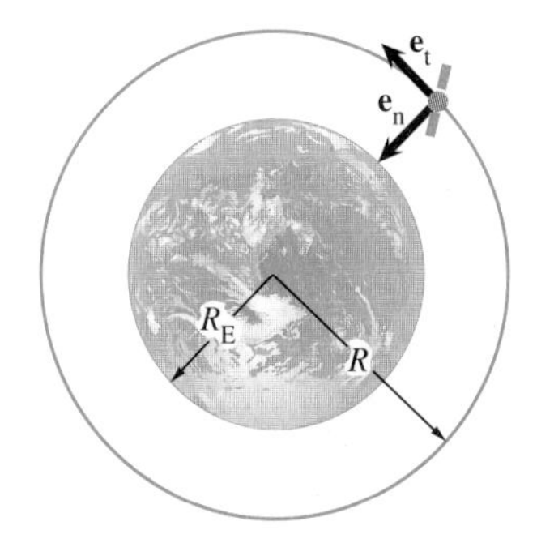

Figure 2.29
Describing the satellite's motion in terms of normal and tangential components.

DISCUSSION

In Example 2.5 we determined the escape velocity of an object travelling straight away from the earth in terms of its initial distance from the centre of the earth. The escape velocity for an object a distance R from the centre of the earth, $v_{esc} = \sqrt{2gR_E^2/R}$, is only $\sqrt{2}$ times the velocity of an object in a circular orbit of radius R. This explains why it was possible to begin launching probes to other planets not long after the first satellites were placed in orbit.

Example 2.10

During a flight in which a helicopter starts from rest at $t = 0$, the cartesian components of its acceleration are

$$a_x = 0.6t \text{ m/s}^2$$
$$a_y = (1.8 - 0.36t) \text{ m/s}^2$$

What are the normal and tangential components of its acceleration and the instantaneous radius of curvature of its path at $t = 4$ s?

STRATEGY

We can integrate the cartesian components of acceleration to determine the cartesian components of the velocity at $t = 4$ s. The velocity vector is tangent to the path, so knowing the cartesian components of the velocity allows us to determine the path angle.

SOLUTION

Integrating the components of acceleration with respect to time (see Example 2.6), the cartesian components of the velocity are

$$v_x = 0.3t^2 \text{ m/s}$$
$$v_y = (1.8t - 0.18t^2) \text{ m/s}$$

At $t = 4$ s, $v_x = 4.80$ m/s and $v_y = 4.32$ m/s. Therefore the path angle (Figure (a)) is

$$\theta = \arctan\left(\frac{4.32}{4.80}\right) = 42.0^\circ$$

The cartesian components of the acceleration at $t = 4$ s are

$$a_x = 0.6(4) = 2.4 \text{ m/s}^2$$
$$a_y = 1.8 - 0.36(4) = 0.36 \text{ m/s}^2$$

By calculating the components of these accelerations in the directions tangential and normal to the path (Figure (b)), we obtain a_t and a_n:

$$a_t = (2.4)\cos 42.0^\circ + (0.36)\sin 42.0^\circ = 2.02 \text{ m/s}^2$$
$$a_n = (2.4)\sin 42.0^\circ - (0.36)\cos 42.0^\circ = 1.34 \text{ m/s}^2$$

To determine the instantaneous radius of curvature of the path, we use the relation $a_n = v^2/\rho$. The magnitude of the velocity at $t = 4$ s is

$$v = \sqrt{v_x^2 + v_y^2} = \sqrt{(4.80)^2 + (4.32)^2} = 6.46 \text{ m/s}$$

so the value of $\rho = 4$ s is

$$\rho = \frac{v^2}{a_n} = \frac{(6.46)^2}{1.34} = 31.2 \text{ m}$$

Figure 2.30

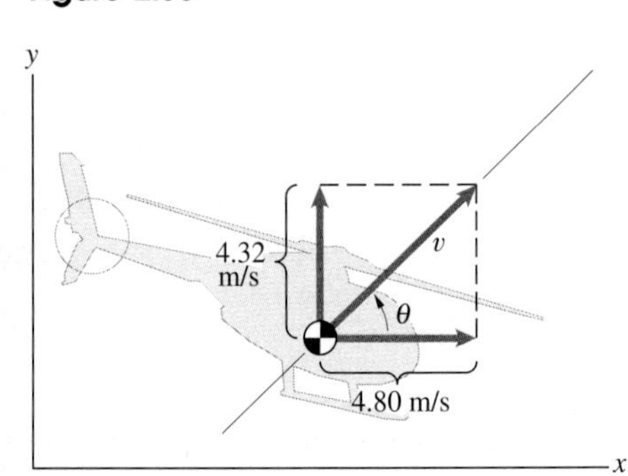

(a) Cartesian components of the velocity and the path angle θ.

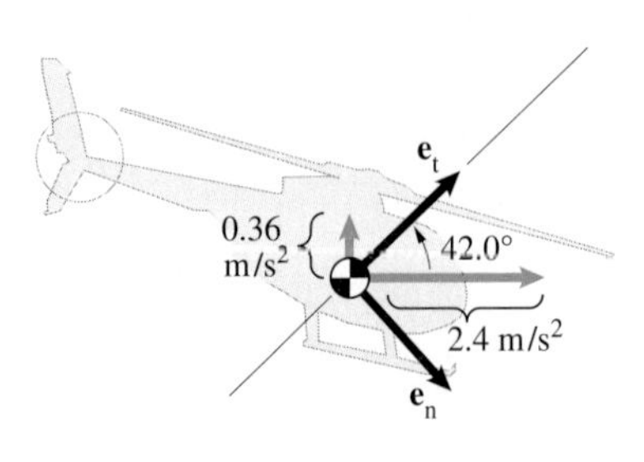

(b) Determining the tangential and normal components of the acceleration from the cartesian components.

Problems

2.96 The armature of an electric motor rotates at a constant rate. The magnitude of the velocity of point P relative to O is 4 m/s.
(a) What are the normal and tangential components of the acceleration of P relative to O?
(b) What is the angular velocity of the armature?

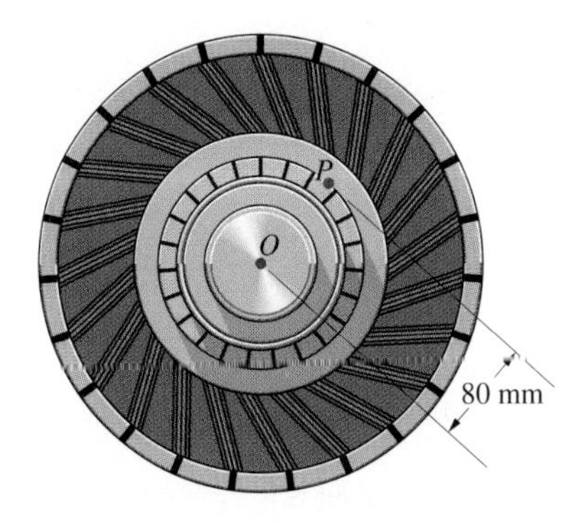

P2.96

2.97 The armature in Problem 2.96 starts from rest and has constant angular acceleration $\alpha = 10\,\text{rad/s}^2$. What are the velocity and acceleration of P relative to O in terms of normal and tangential components after 10 s?

2.98 Centrifuges are used in medical laboratories to increase the speed of precipitation (settling) of solid matter out of solutions. Suppose that you want to design a centrifuge to subject samples to accelerations of 1000 g's.
(a) If the distance from the centre of the centrifuge to the sample is 300 mm, what speed of rotation in rpm (revolutions per minute) is necessary?
(b) If you want the centrifuge to reach its design rpm in 1 min, what constant angular acceleration is necessary?

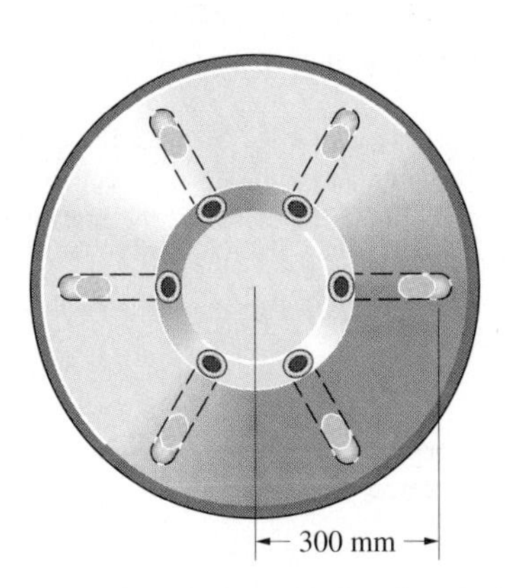

P2.98

2.99 A powerboat being tested for manoeuvrability is started from rest and driven in a circular path of 40 m radius. The magnitude of its velocity is increased at a constant rate of $2\,\text{m/s}^2$. In terms of normal and tangential components, determine: (a) the velocity as a function of time; (b) the acceleration as a function of time.

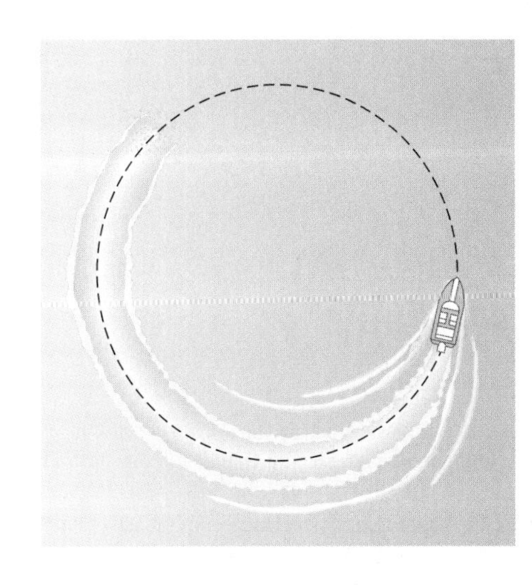

P2.99

2.100 The angle $\theta = 2t^2$ rad.
(a) What are the magnitudes of the velocity and acceleration of P relative to O at $t = 1$ s?
(b) What distance along the circular path does P move from $t = 0$ to $t = 1$ s?

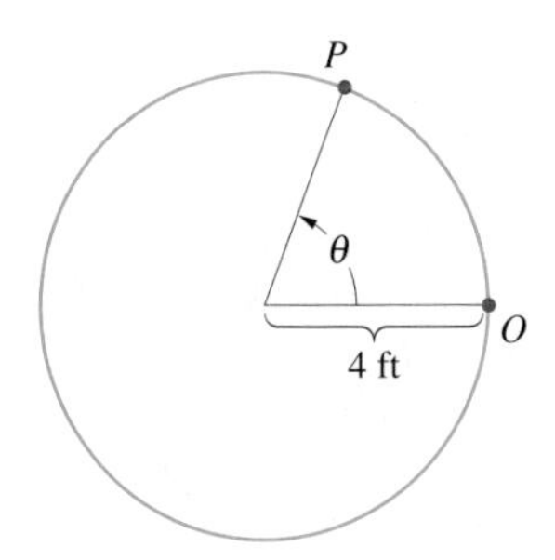

P2.100

2.101 In Problem 2.100, what are the magnitudes of the velocity and acceleration of P relative to O when P has gone one revolution around the circular path starting from $t = 0$?

2.102 The radius of the earth is 6370 km. If you are standing at the equator, what is the magnitude of your velocity relative to the centre of the earth?

2.103 The radius of the earth is 6370 km. If you are standing at the equator, what is the magnitude of your acceleration relative to the centre of the earth?

2.104 Suppose that you are standing at point P at 30° north latitude (that is, a point that is 30° north of the equator). The radius of the earth is $R_E = 6370$ km. What are the magnitudes of your velocity and acceleration relative to the centre of the earth?

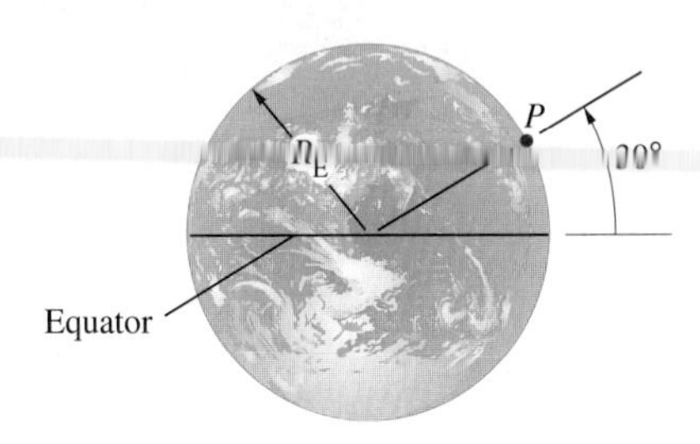

P2.104

2.105 The magnitude of the velocity of the aeroplane is constant and equal to 400 m/s. The rate of change of the path angle θ is constant and equal to 5°/s.
(a) What are the velocity and acceleration of the aeroplane in terms of normal and tangential components?
(b) What is the instantaneous radius of the curvature of the aeroplane's path?

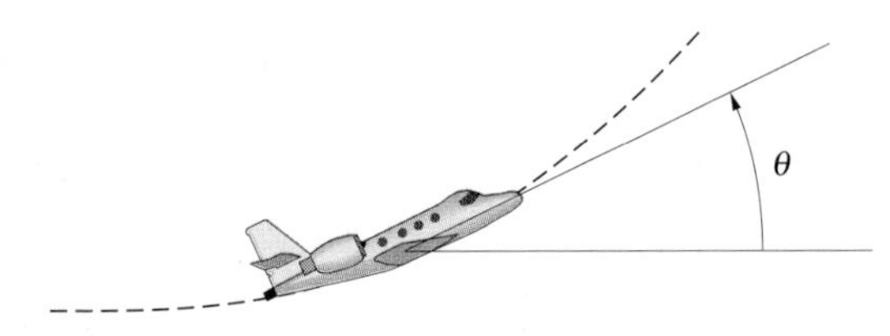

P2.105

2.106 At $t = 0$, a car starts from rest at point A. It moves towards the right, and the tangential component of its acceleration is $a_t = 0.4t$ m/s^2. What is the magnitude of the car's acceleration when it reaches point B?

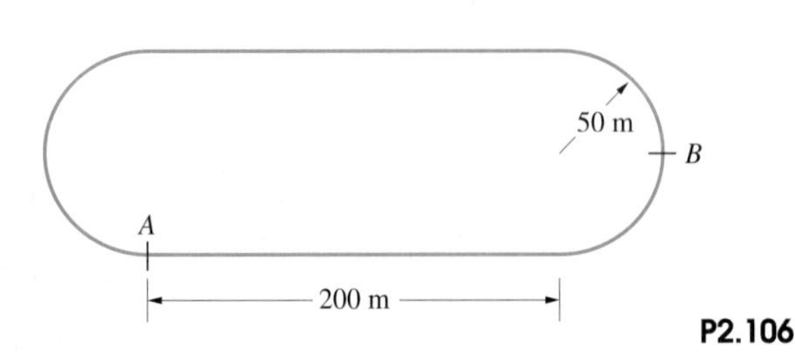

P2.106

2.107 A car increases its speed at a constant rate from 64 km/hr at A to 96 km/hr at B. What is the magnitude of its acceleration 2 s after it passes point A?

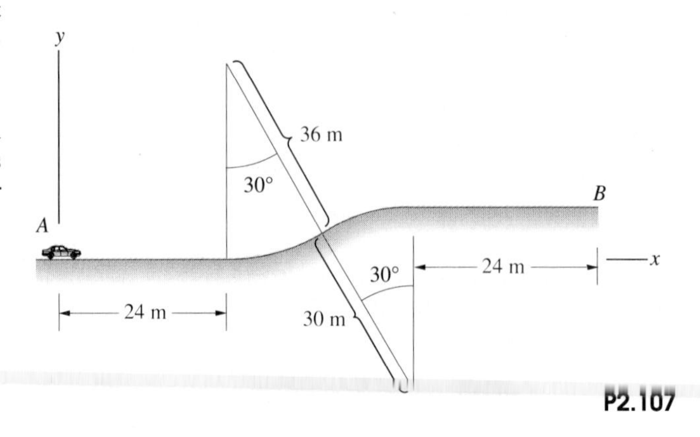

P2.107

2.108 Determine the magnitude of the acceleration of the car in Problem 2.107 when it has travelled along the road a distance (a) 36 m from A; (b) 48 m from A.

2.109 An astronaut candidate is to be tested in a centrifuge with a radius of 10 m. He will lose consciousness if his total horizontal acceleration reaches 14 g's. What is the maximum constant angular acceleration of the centrifuge, starting from rest, if he is not to lose consciousness within 1 min?

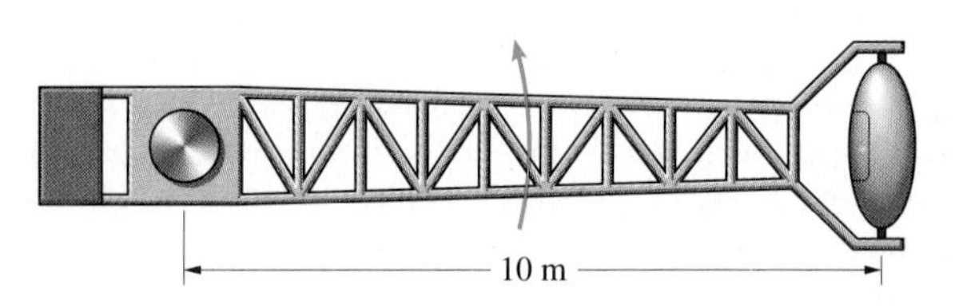

P2.109

2.110 A projectile has an initial velocity of 6 m/s at 30° above the horizontal.
(a) What are the velocity and acceleration of the projectile in terms of normal and tangential components when it is at the highest point of its trajectory?
(b) What is the instantaneous radius of curvature of the projectile's path when it is at the highest point of its trajectory?
Strategy: In part (b), you can determine the instantaneous radius of curvature from the relation $a_n = v^2/\rho$.

P2.110

2.111 In Problem 2.110, let $t = 0$ be the instant at which the projectile is launched.
(a) What are the velocity and acceleration in terms of normal and tangential components at $t = 0.2$ s?
(b) What is the instantaneous radius of curvature of the path at $t = 0$ s?

2.112 The cartesian coordinates of a point moving in the x-y plane are

$$x = (20 + 4t^2)\,\text{m} \quad y = (10 - t^3)\,\text{m}$$

What is the instantaneous radius of curvature of the path at $t = 3$ s?

2.113 A satellite is in a circular orbit 320 km above the surface of the earth. The radius of the earth is 6370 km.
(a) What is the magnitude v of the satellite's velocity relative to the centre of the earth?
(b) How long does it take for the satellite to complete one orbit?

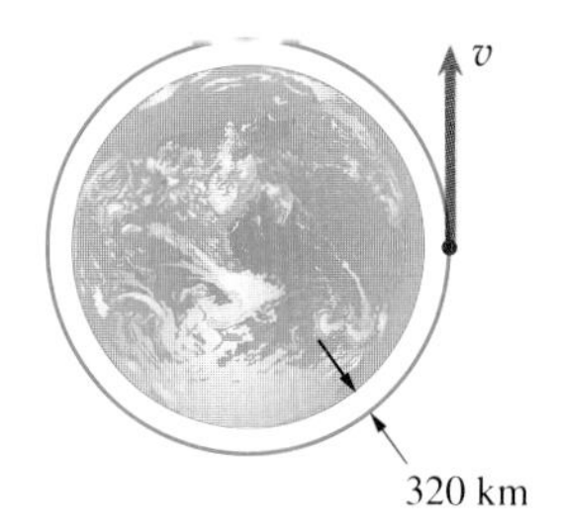

P2.113

2.114 For astronaut training, the aeroplane shown is to achieve 'weightlessness' for a short period of time by flying along a path such that its acceleration is $a_x = 0$, $a_y = -g$. If its velocity at O at time $t = 0$ is $\mathbf{v} = v_0\,\mathbf{i}$, show that the autopilot must fly the aeroplane so that its tangential component of acceleration as a function of time is

$$a_t = g\frac{(gt/v_0)}{\sqrt{1 + (gt/v_0)^2}}$$

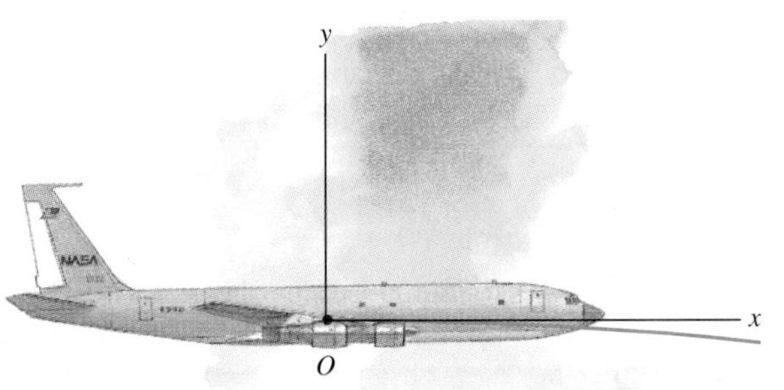

P2.115

2.115 In Problem 2.114, what is the aeroplane's normal component of acceleration as a function of time?

2.116 If $y = 100$ mm, $dy/dt = 200$ mm/s and $d^2y/dt^2 = 0$, what are the velocity and acceleration of P in terms of normal and tangential components?

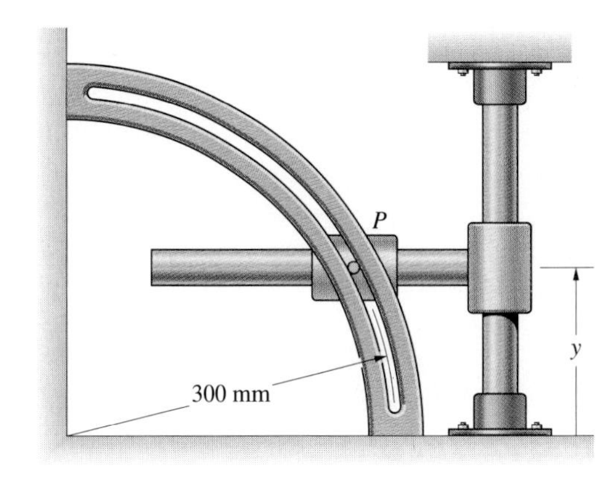

P2.116

2.117 Suppose that the point P in Problem 2.116 moves upwards in the slot with velocity $\mathbf{v} = 300\,\mathbf{e}_t$ mm/s. When $y = 150$ mm, what are dy/dt and d^2y/dt^2?

2.118 A car travels at 100 km/hr on a straight road of increasing gradient whose vertical profile can be approximated by the equation shown. When the car's horizontal coordinate is $x = 400$ m, what are the tangential and normal components of its acceleration?

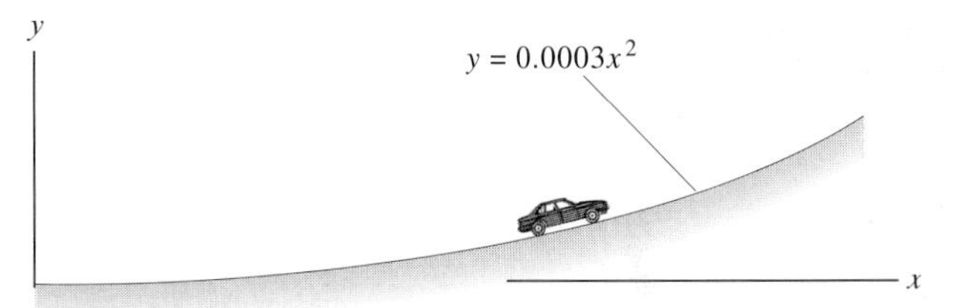

P2.118

2.119 A boy rides a skateboard on the concrete surface of an empty drainage canal described by the equation shown. He starts at $y = 6$ m and the magnitude of his velocity is approximated by $v = \sqrt{2(9.81)(6 - y)}$ m/s. What are the normal and tangential components of his acceleration when he reaches the bottom?

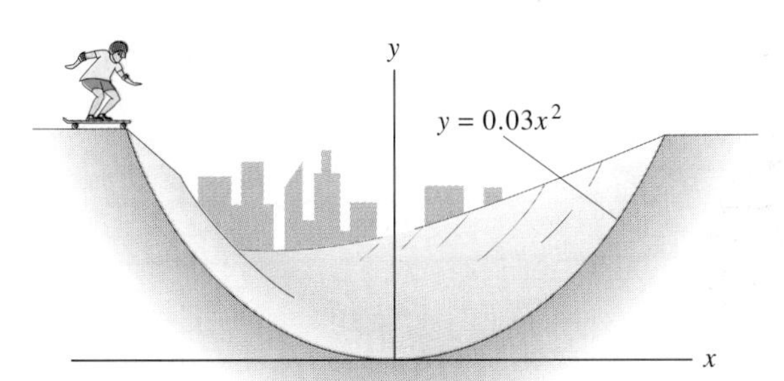

P2.119

2.120 In Problem 2.119, what are the normal and tangential components of the boy's acceleration when he has passed the bottom and reached $y = 3$ m?

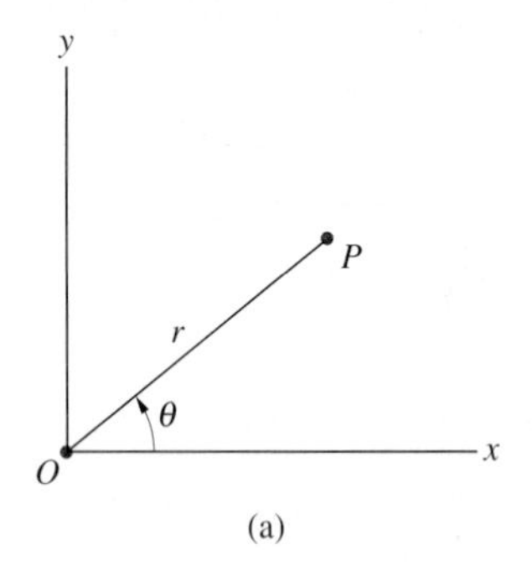

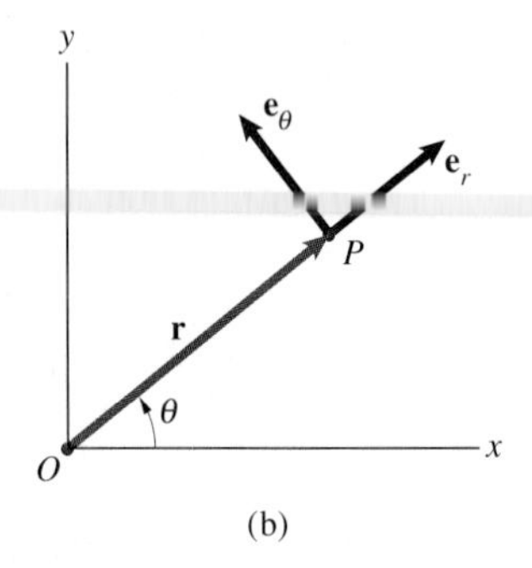

Figure 2.31
(a) The polar coordinates of P.
(b) The unit vector $\mathbf{e}_r$ and $\mathbf{e}_\theta$ and position vector $\mathbf{r}$.

Polar and Cylindrical Coordinates

Polar coordinates are often used to describe the curvilinear motion of a point. Circular motion, certain orbit problems and, more generally, **central force** problems, in which the acceleration of a point is directed towards a given point, can be expressed conveniently in polar coordinates.

Consider a point P in the x-y plane of a cartesian coordinate system. We can specify the position of P relative to the origin O either by its cartesian coordinates x, y or by its polar coordinates r, θ (Figure 2.31(a)). To express vectors in terms of polar coordinates, we define a unit vector $\mathbf{e}_r$ that points in the direction of the radial line from the origin to P and a unit vector $\mathbf{e}_\theta$ that is perpendicular to $\mathbf{e}_r$ and points in the direction of increasing θ (Figure 2.31(b)). In terms of these vectors, the position vector $\mathbf{r}$ from O to P is

$$\mathbf{r} = r\mathbf{e}_r \tag{2.44}$$

(Notice that $\mathbf{r}$ has no component in the direction of $\mathbf{e}_\theta$.)

We can determine the velocity of P in terms of polar coordinates by taking the time derivative of Equation (2.44):

$$\mathbf{v} = \frac{d\mathbf{r}}{dt} = \frac{dr}{dt}\mathbf{e}_r + r\frac{d\mathbf{e}_r}{dt} \tag{2.45}$$

As P moves along a curvilinear path, the unit vector $\mathbf{e}_r$ rotates with angular velocity $\omega = d\theta/dt$. Therefore, from Equation (2.33), we can express the time derivative of $\mathbf{e}_r$ in terms of $\mathbf{e}_\theta$ as

$$\frac{d\mathbf{e}_r}{dt} = \frac{d\theta}{dt}\mathbf{e}_\theta \tag{2.46}$$

Substituting this result into Equation (2.45), we obtain the velocity of P:

$$\mathbf{v} = \frac{dr}{dt}\mathbf{e}_r + r\frac{d\theta}{dt}\mathbf{e}_\theta = \frac{dr}{dt}\mathbf{e}_r + r\omega\,\mathbf{e}_\theta \tag{2.47}$$

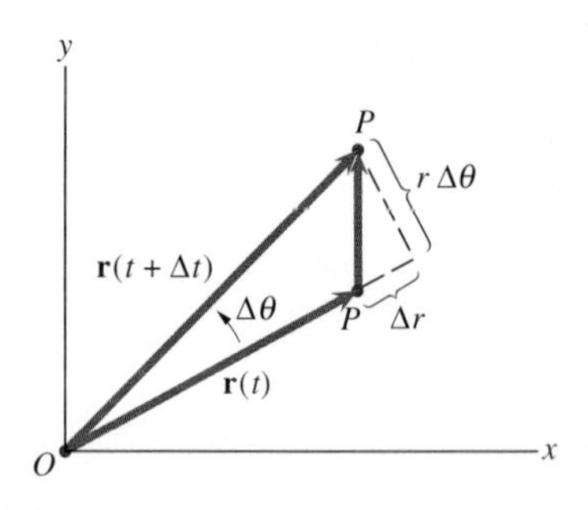

Figure 2.32
The position vector of P at t and at $t + \Delta t$.

We can obtain this result in another way that is less rigorous but more direct and intuitive. Figure 2.32 shows the position vector of P at times t and $t + \Delta t$. The change in the position vector, $\mathbf{r}(t + \Delta t) - \mathbf{r}(t)$, consists of two components. The component Δr is due to the change in the radial position r and is in the $\mathbf{e}_r$ direction. The component $r\,\Delta\theta$ is due to the change in θ and is in the $\mathbf{e}_\theta$ direction. Thus the change in the position of P is (approximately)

$$\mathbf{r}(t + \Delta t) - \mathbf{r}(t) = \Delta r\,\mathbf{e}_r + r\Delta\theta\,\mathbf{e}_\theta$$

Dividing this expression by Δt and taking the limit as $\Delta t \to 0$, we obtain the velocity of P:

$$\begin{aligned}\mathbf{v} &= \lim_{\Delta t\to 0}\left[\frac{\Delta r}{\Delta t}\mathbf{e}_r + r\frac{\Delta\theta}{\Delta t}\mathbf{e}_\theta\right]\\ &= \frac{dr}{dt}\mathbf{e}_r + r\omega\,\mathbf{e}_\theta\end{aligned}$$

One component of the velocity is in the radial direction and is equal to the rate of change of the radial position r. The other component is normal, or **transverse** to the radial direction, and is proportional to the radial distance and to the rate of change of θ.

We obtain the acceleration of P by taking the time derivative of Equation (2.47):

$$\mathbf{a} = \frac{d\mathbf{v}}{dt} = \frac{d^2r}{dt^2}\mathbf{e}_r + \frac{dr}{dt}\frac{d\mathbf{e}_r}{dt} + \frac{dr}{dt}\frac{d\theta}{dt}\mathbf{e}\theta + r\frac{d^2\theta}{dt^2}\mathbf{e}_\theta + r\frac{d\theta}{dt}\frac{d\mathbf{e}_\theta}{dt} \qquad (2.48)$$

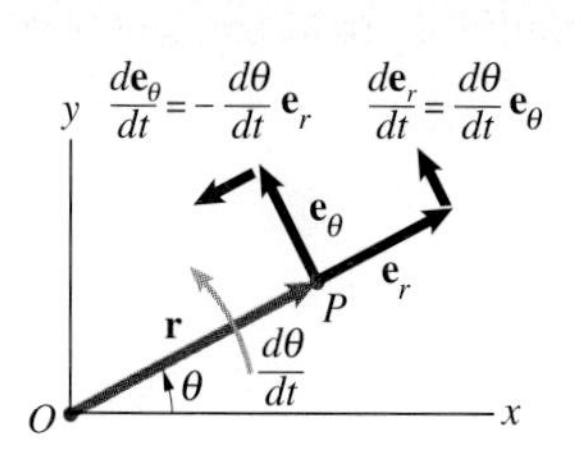

Figure 2.33
Time derivatives of $\mathbf{e}_r$ and $\mathbf{e}_\theta$.

The time derivative of the unit vector $\mathbf{e}_r$ due to the rate of change of θ is given by Equation (2.46). As P moves, $\mathbf{e}_\theta$ also rotates with angular velocity $d\theta/dt$ (Figure 2.33). You can see from this figure that the time derivative of $\mathbf{e}_\theta$ is in the $-\mathbf{e}_r$ direction if $d\theta/dt$ is positive:

$$\frac{d\mathbf{e}_\theta}{dt} = \frac{d\theta}{dt}\mathbf{e}_r$$

Substituting this expression and Equation (2.46) and Equation (2.48), we obtain the acceleration of P:

$$\mathbf{a} = \left[\frac{d^2r}{dt^2} - r\left(\frac{d\theta}{dt}\right)^2\right]\mathbf{e}_r + \left[r\frac{d^2\theta}{dt^2} + 2\frac{dr}{dt}\frac{d\theta}{dt}\right]\mathbf{e}_\theta$$

Thus the velocity and acceleration are (Figure 2.34)

$$\boxed{\mathbf{v} = v_r\,\mathbf{e}_r + v_\theta\,\mathbf{e}_\theta = \frac{dr}{dt}\mathbf{e}_r + r\omega\,\mathbf{e}_\theta} \qquad (2.49)$$

and

$$\boxed{\mathbf{a} = a_r\,\mathbf{e}_r + a_\theta\,\mathbf{e}_\theta} \qquad (2.50)$$

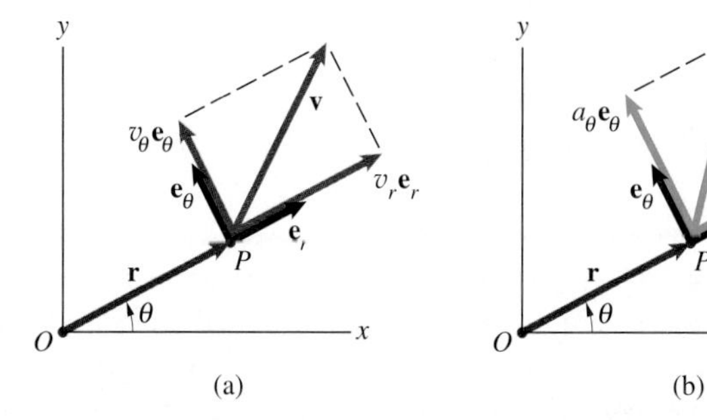

Figure 2.34
Radial and transverse components of (a) the velocity and (b) acceleration

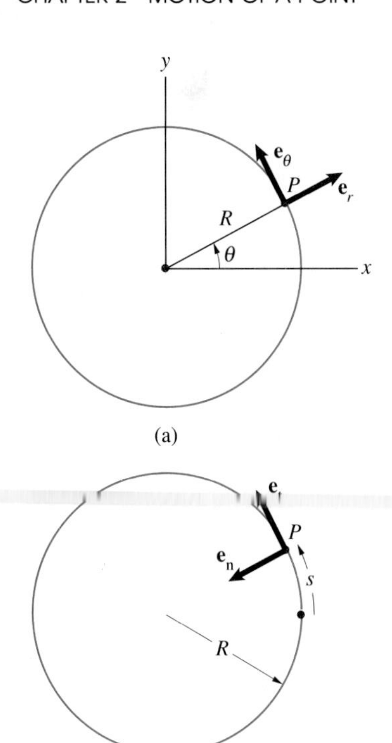

Figure 2.35
A point P moving in a circular path. (a) Polar coordinates. (b) Normal and tangential components.

where

$$
\boxed{\begin{aligned}
a_r &= \frac{d^2r}{dt^2} - r\left(\frac{d\theta}{dt}\right)^2 = \frac{d^2r}{dt^2} - r\omega^2 \\
a_\theta &= r\frac{d^2\theta}{dt^2} + 2\frac{dr}{dt}\frac{d\theta}{dt} = r\alpha + 2\frac{dr}{dt}\omega
\end{aligned}} \qquad (2.51)
$$

The term $-r\omega^2$ in the radial component of the acceleration is called the **centripetal acceleration**, and the terms $2(dr/dt)\omega$ in the transverse component is called the **Coriolis acceleration.**

Circular Motion Circular motion can be conveniently described using either radial and transverse or **normal and** tangential components. Let's compare these two methods of expressing the velocity and acceleration of a point P moving in a circular path of radius R (Figure 2.35). Because the polar coordinate $r = R$ is constant, Equation (2.49) for the velocity reduces to

$$\mathbf{v} = R\omega\,\mathbf{e}_\theta$$

In terms of normal and tangential components, the velocity is

$$\mathbf{v} = v\,\mathbf{e}_\mathrm{t}$$

Notice in Figure 2.35 that $\mathbf{e}_\theta = \mathbf{e}_\mathrm{t}$. Comparing these two expressions for the velocity, we obtain the relation between the velocity and the angular velocity in circular motion:

$$v = R\omega$$

From Equations (2.50) and (2.51), the acceleration in terms of polar coordinates for a circular path of radius R is

$$\mathbf{a} = -R\omega^2\,\mathbf{e}_r + R\alpha\,\mathbf{e}_\theta$$

and the acceleration in terms of normal and tangential components is

$$\mathbf{a} = \frac{dv}{dt}\mathbf{e}_\mathrm{t} + \frac{v^2}{R}\mathbf{e}_\mathrm{n}$$

The unit vector $\mathbf{e}_r = -\mathbf{e}_\mathrm{n}$. Because of the relation $v = R\omega$, the normal components of acceleration are equal: $v^2/R\omega^2$. Equating the transverse and tangential components, we obtain the relation

$$\frac{dv}{dt} = a_\mathrm{t} = R\alpha$$

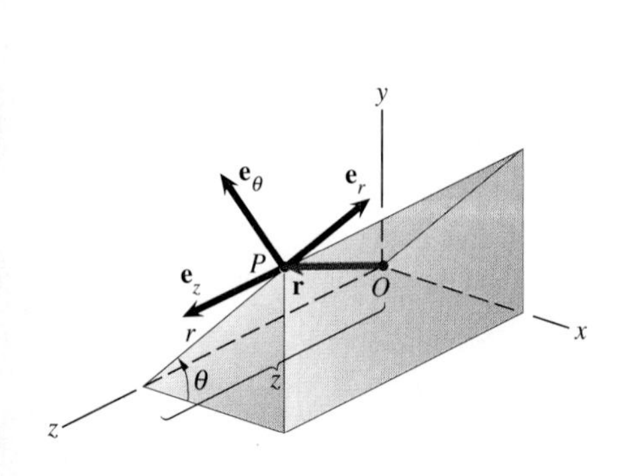

Figure 2.36
Cylindrical coordinates r, θ, z of point P and the unit vectors $\mathbf{e}_r$, $\mathbf{e}_\theta$, $\mathbf{e}_z$.

Cylindrical Coordinates Polar coordinates describe the motion of a point P in the x-y plane. We can describe three-dimensional motion by using **cylindrical coordinates** r, θ, z (Figure 2.36). The cylindrical coordinates r and θ are the polar coordinates of P measured in the plane parallel to the x-y

plane, and the definitions of the unit vectors $\mathbf{e}_r$ and $\mathbf{e}_\theta$ are unchanged. The position of P perpendicular to the x-y plane is measured by the coordinate z, and the unit vector $\mathbf{e}_z$ points in the positive z-axis direction.

The position vector $\mathbf{r}$ in terms of cylindrical coordinates is the sum of the expression for the position vector in polar coordinates and the z component:

$$\mathbf{r} = r\,\mathbf{e}_r + z\,\mathbf{e}_z \tag{2.52}$$

(The polar coordinate r does not equal the magnitude of $\mathbf{r}$ except when P lies in the x-y plane.) By taking time derivatives, we obtain the velocity,

$$\begin{aligned} \mathbf{v} &= \frac{d\mathbf{r}}{dt} = v_r\mathbf{e}_r + v_\theta\mathbf{e}_\theta + v_z\mathbf{e}_z \\ &= \frac{dr}{dt}\mathbf{e}_r + r\omega\mathbf{e}_\theta + \frac{dz}{dt}\mathbf{e}_z \end{aligned} \tag{2.53}$$

and the acceleration,

$$\mathbf{a} = \frac{d\mathbf{v}}{dt} = a_r\mathbf{e}_r + a_\theta\mathbf{e}_\theta + a_z\mathbf{e}_z \tag{2.54}$$

where

$$a_r = \frac{d^2r}{dt^2} - r\omega^2 \quad a_\theta = r\alpha + 2\frac{dr}{dt}\omega \quad a_z = \frac{d^2z}{dt^2} \tag{2.55}$$

Notice that Equations (2.53) and (2.54) reduce to the polar coordinate expressions for the velocity and acceleration, Equations (2.49) and (2.50), when P moves along a path in the x-y plane.

The next two examples demonstrate the use of Equations (2.49) and (2.50) to analyse curvilinear motions of objects in terms of polar coordinates.

Example 2.11

Suppose that you are standing on a large disc (a merry-go-round) rotating with constant angular velocity ω_0 and you start walking at constant speed v_0 along a straight radial line painted on the disc (Figure 2.37). What are your velocity and acceleration when you are a distance r from the centre of the disk?

Figure 2.37

STRATEGY

We can describe your motion in terms of polar coordinates (Figure (a)). By using the information given about your motion and the motion of the disc, we can evaluate the terms in the expressions for the velocity and acceleration in terms of polar coordinates.

SOLUTION

The speed with which you walk along the radial line is the rate of change of r, $dr/dt = v_0$, and the angular velocity of the disc is the rate of change of θ, $\omega = \omega_0$. Your velocity is

$$\mathbf{v} = \frac{dr}{dt}\mathbf{e}_r + r\omega\,\mathbf{e}_\theta = v_0\,\mathbf{e}_r + r\omega_0\,\mathbf{e}_\theta$$

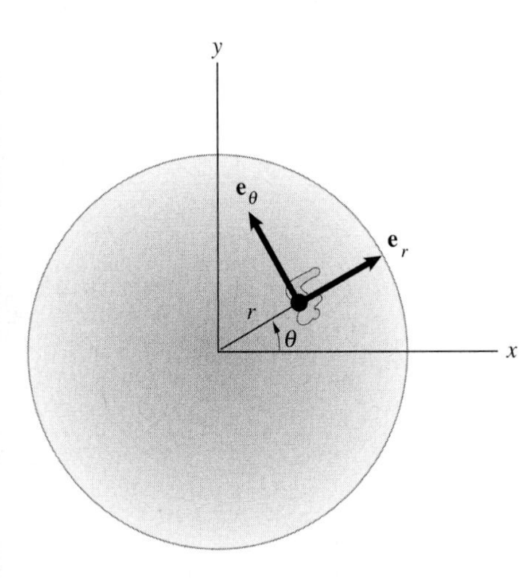

(a) Your position in terms of polar coordinates.

Your velocity consists of two components: the radial component due to the speed at which you are walking and a transverse component due to the disc's rate of rotation. The transverse component increases as your distance from the centre of the disc increases.

Your walking speed $v_0 = dr/dt$ is constant, so $d^2r/dt^2 = 0$. Also, the disc's angular velocity $\omega_0 = d\theta/dt$ is constant, so $d^2\theta/dt^2 = 0$. The radial component of your acceleration is

$$a_r = \frac{d^2r}{dt^2} - r\omega^2 = -r\omega_0^2$$

and the transverse component is

$$a_\theta = r\alpha + 2\frac{dr}{dt}\omega = 2v_0\omega_0$$

DISCUSSION

If you have ever tried walking on a merry-go-round, you know it is a difficult proposition. This example indicates why: subjectively, you are walking along a straight line with constant velocity, but you are actually experiencing the centripetal acceleration a_r and the Coriolis acceleration a_θ due to the disk's rotation.

Example 2.12

The robot arm in Figure 2.38 is programmed so that point P describes the path

$$r = (1 - 0.5\cos 2\pi t)\ \text{m}$$
$$\theta = (0.5 - 0.2\sin 2\pi t)\ \text{rad}$$

At $t = 0.8$ s, determine: (a) the velocity of P in terms of radial and transverse components; (b) the cartesian components of the velocity of P.

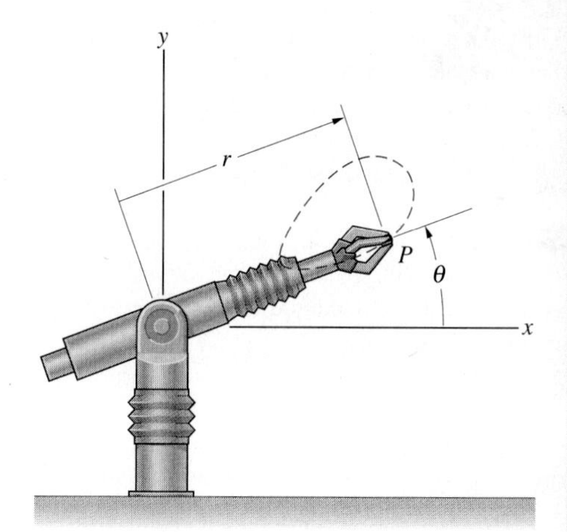

Figure 2.38

STRATEGY

(a) Since we are given r and θ as functions of time, we can calculate the derivatives in the expression for the velocity in terms of polar coordinates and obtain the velocity as a function of time.
(b) By determining the value of θ at $t = 0.8$ s, we can use trigonometry to determine the cartesian components in terms of the radial and transverse components.

SOLUTION

(a) From Equation (2.49), the velocity is

$$\mathbf{v} = \frac{dr}{dt}\mathbf{e}_r + r\frac{d\theta}{dt}\mathbf{e}_\theta$$
$$= (\pi\sin 2\pi t)\mathbf{e}_r + (1 - 0.5\cos 2\pi t)(-0.4\pi\cos 2\pi t)\mathbf{e}_\theta$$

At $t = 0.8$ s,

$$\mathbf{v} = (-2.99\,\mathbf{e}_r - 0.328\,\mathbf{e}_\theta)\ \text{m/s}$$

(b) At $t = 0.8$ s, $\theta = 0.690\ \text{rad} = 39.5^\circ$ (Figure (a)). The x component of the velocity of P is

$$v_x = v_r\cos 39.5^\circ - v_\theta\sin 39.5^\circ$$
$$= (-2.99)\cos 39.5^\circ - (-0.328)\sin 39.5^\circ = -2.09\ \text{m/s}$$

and the y component is

$$v_y = v_r\sin 39.5^\circ + v_\theta\cos 39.5^\circ$$
$$= (-2.99)\sin 39.5^\circ + (-0.328)\cos 39.5^\circ = -2.16\ \text{m/s}$$

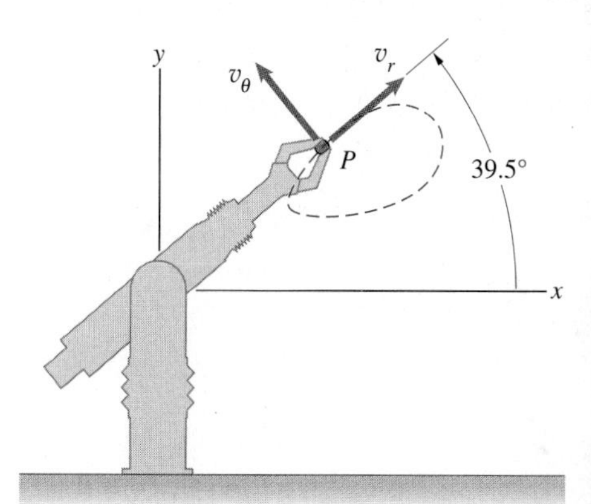

(a) Position at $t = 0.8$ s.

DISCUSSION

When you determine components of a vector in terms of different coordinate systems, you should always check them to make sure they give the same magnitude. In this example,

$$|\mathbf{v}| = \sqrt{(-2.99)^2 + (-0.328)^2} = \sqrt{(-2.09)^2 + (-2.16)^2} = 3.01\ \text{m/s}$$

Remember that although the components of the velocity are different in the two coordinate systems, those components describe the same velocity vector.

Problems

2.121 At a particular time, the polar coordinates of a point P moving in the x-y plane are $r = 4\,\text{m}$, $\theta = 0.5\,\text{rad}$, and their time derivatives are $dr/dt = 8\,\text{m/s}$ and $d\theta/dt = -2\,\text{rad/s}$.
(a) What is the magnitude of the velocity of P?
(b) What are the cartesian components of the velocity of P?

2.122 In Problem 2.121, suppose that $d^2r/dt^2 = 6\,\text{m/s}^2$ and $d^2\theta/dt^2 = 3\,\text{rad/s}^2$. At the instant described, determine: (a) the magnitude of the acceleration of P; (b) the instantaneous radius of curvature of the path.

2.123 The polar coordinates of a point P moving in the x-y plane are $r = (t^2 - 4t)\,\text{m}$, $\theta = (t^2 - t)\,\text{rad}$. Determine the velocity of P in terms of radial and transverse components at $t = 1\,\text{s}$.

2.124 In Problem 2.123, what is the acceleration of P in terms of radial and transverse components at $t = 1\,\text{s}$?

2.125 The radial rotates with a constant angular velocity of $2\,\text{rad/s}$. Point P moves along the line at a constant speed of $4\,\text{m/s}$. Determine the magnitudes of the velocity and acceleration of P when $r = 2\,\text{m}$.

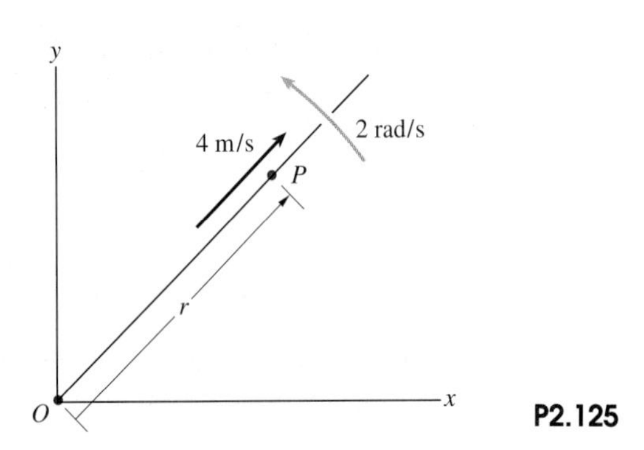

P2.125

2.126 A boat searching for underwater archaeology sites in the Aegean Sea moves at 4 knots and follows the path $r = 10\theta\,\text{m}$, where θ is in radians. (A knot is one nautical mile, or 1852 m, per hour.) When $\theta = 2\pi\,\text{rad}$, determine the boat's velocity (a) in terms of polar coordinates; (b) in terms of cartesian coordinates.

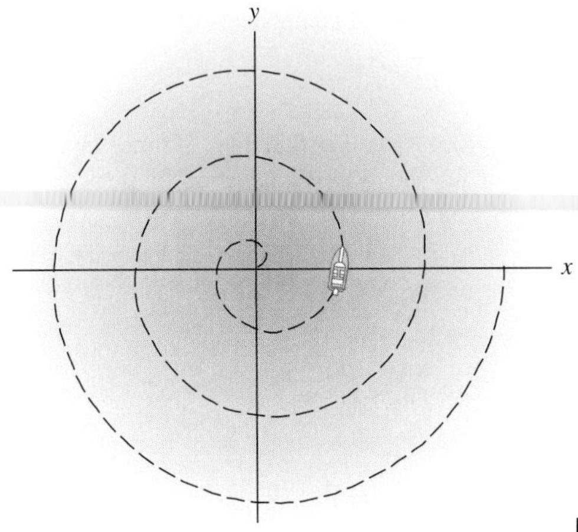

P2.126

2.127 In Problem 2.126, what is the boat's acceleration in terms of polar coordinates?

Strategy: The magnitude of the boat's velocity is constant, so you know that the tangential component of its acceleration equals zero.

2.128 A point P moves in the x-y plane along the path described by the equation $r = e^{\theta}$, where θ is in radians. The angular velocity $d\theta/dt = \omega_0 = \text{constant}$, and $\theta = 0$ at $t = 0$.
(a) Draw a polar graph of the path for values of θ from zero to 2π.
(b) Show that the velocity and acceleration as functions of time are $\mathbf{v} = \omega_0 e^{\omega_0 t}(\mathbf{e}_r + \mathbf{e}_\theta)$, $\mathbf{a} = 2\omega_0^2 e^{\omega_0 t}\,\mathbf{e}_\theta$.

2.129 In Problem 2.128, show that the instantaneous radius of curvature of the path as a function of time is $\rho = \sqrt{2}e^{\omega_0 t}$.

2.130 In Example 2.12, determine the acceleration of point P at $t = 0.8\,\text{s}$ (a) in terms of radial and transverse components; (b) in terms of cartesian components.

2.131 A bead slides along a wire that rotates in the x-y plane with constant angular velocity ω_0. The radial component of the bead's acceleration is zero. The radial component of its velocity is v_0 when $r = r_0$. Determine the radial and transverse components of the bead's velocity as a function of r.

Strategy: The radial component of the bead's velocity is

$$v_r = \frac{dr}{dt}$$

and the radial component of its acceleration is

$$a_r = \frac{d^2r}{dt^2} - r\left(\frac{d\theta}{dt}\right)^2 = \frac{dv_r}{dt} - r\omega_0^2$$

By using the chain rule,

$$\frac{dv_r}{dt} = \frac{dv_r}{dr}\frac{dr}{dt} = \frac{dv_r}{dr}v_r$$

you can express the radial component of the acceleration in the form

$$a_r = \frac{dv_r}{dr}v_r - r\omega_0^2$$

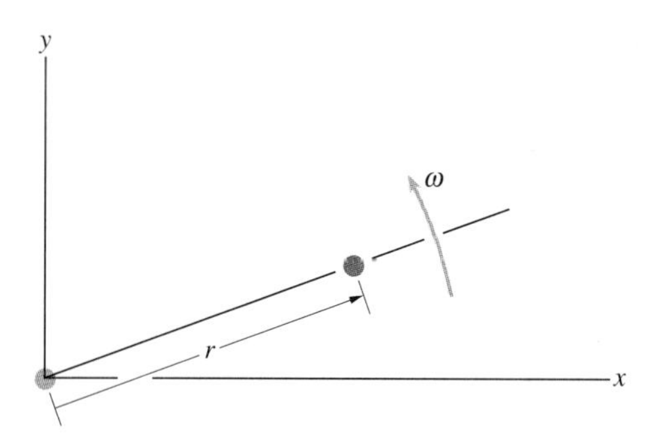

P2.131

2.132 The bar rotates in the x-y plane with constant angular velocity ω_0. The radial component of acceleration of the collar C is $a_r = -Kr$, where K is a constant. When $r = r_0$, the radial component of velocity of C is v_0. Determine the radial and transverse components of the velocity of C as a function of r.

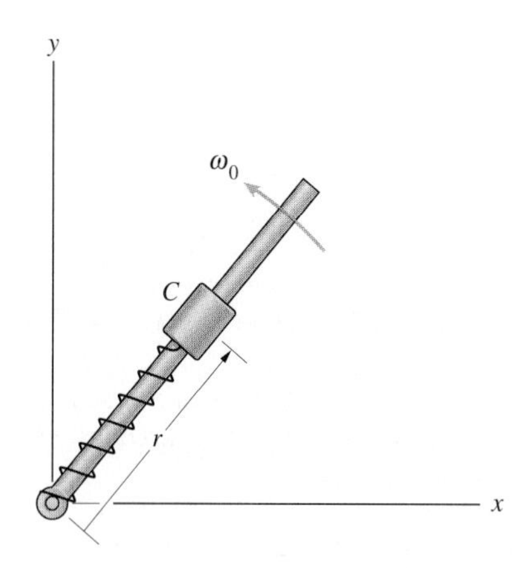

P2.132

2.133 The cartesian coordinates of a point P in the x-y plane are related to its polar coordinates by the relations $x = r\cos\theta$, $y = r\sin\theta$.

(a) Show that the unit vectors **i** and **j** are related to the unit vectors $\mathbf{e}_r$ and $\mathbf{e}_\theta$ by

$$\mathbf{i} = \mathbf{e}_r\cos\theta - \mathbf{e}_\theta\sin\theta$$
$$\mathbf{j} = \mathbf{e}_r\sin\theta + \mathbf{e}_\theta\cos\theta$$

(b) Beginning with the expression for the position vector of P in terms of cartesian coordinates, $\mathbf{r} = x\mathbf{i} + y\mathbf{j}$, derive Equation (2.44) for the position vector in terms of polar coordinates.

(c) By taking the time derivative of the position vector of point P expressed in terms of cartesian coordinates, derive Equation (2.47) for the velocity in terms of polar coordinates.

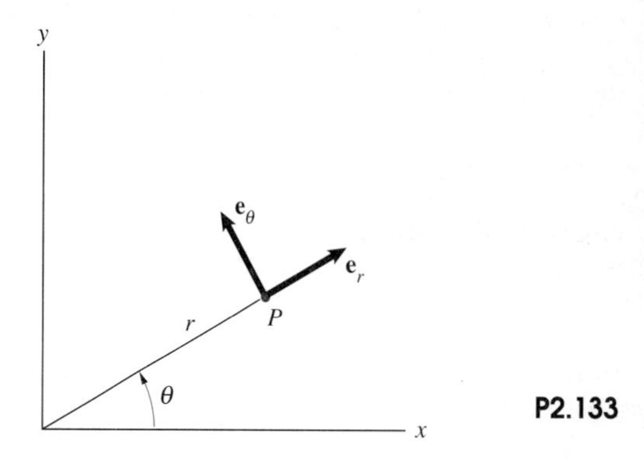

P2.133

2.134 The aeroplane flies in a straight line at 640 km/hr. The radius of its propeller is 1.5 m, and it turns at 2000 rpm (revolutions per minute) in the counterclockwise direction when seen from the front of the aeroplane. Determine the velocity and acceleration of a point on the tip of the propeller in terms of cylindrical coordinates. (Let the z axis be oriented as shown in the figure.)

P2.134

2.135 A charged particle P in a magnetic field moves along the spiral path described by $r = 1$ m, $\theta = 2z$ rad, where z is in metres. The particle moves along the path in the direction shown with constant speed $|\mathbf{v}| = 1$ km/s. What is the velocity of the particle in terms of cylindrical coordinates?

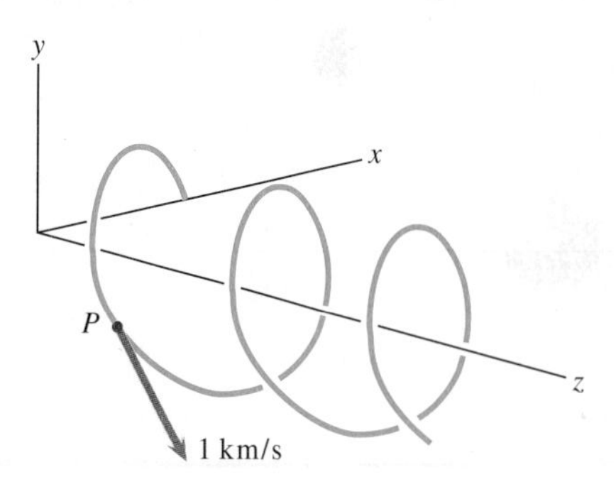

P2.135

2.136 What is the acceleration of the charged particle in Problem 2.135 in terms of cylindrical coordinates?

2.4 Orbital Mechanics

Newton's analytical determination of the elliptic orbits of the planets, which had been deduced from observational data by Johannes Kepler (1571–1630), was a triumph for Newtonian mechanics and confirmation of the inverse-square relation for gravitational acceleration. We can use the equations developed in this chapter to determine the orbit of an earth satellite or planet.

Suppose that at $t = 0$ a satellite has an initial velocity v_0 at a distance r_0 from the centre of the earth (Figure 2.39(a)). We assume that the initial velocity is perpendicular to the line from the centre of the earth to the satellite. The satellite's position during its subsequent motion is specified by its polar coordinates (r, θ), where θ is measured from its position at $t = 0$ (Figure 2.39(b)). Our objective is to determine r as a function of θ.

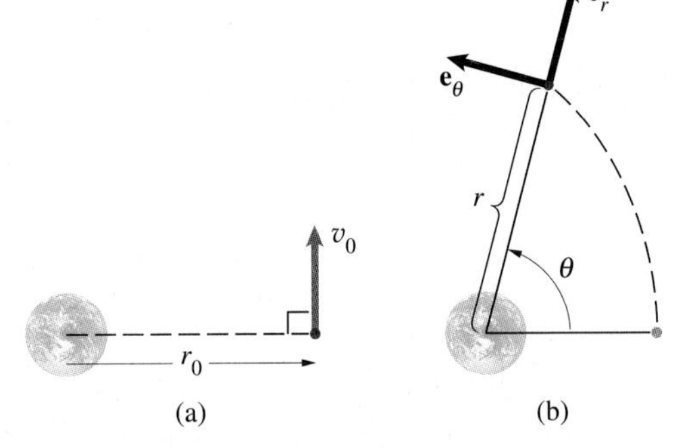

Figure 2.39
(a) Initial position and velocity of an earth satellite.
(b) Specifying the subsequent path in terms of polar coordinates.

If we model the earth as a homogeneous sphere, the acceleration due to gravity at a distance r from the centre is

$$\mathbf{a} = -\frac{gR_{\mathrm{E}}^2}{r^2}\mathbf{e}_r$$

where R_{E} is the earth's radius. By setting this expression equal to Equation (2.50) for the acceleration in terms of polar coordinates,

$$\left[\frac{d^2r}{dt^2} - r\left(\frac{d\theta}{dt}\right)^2\right]\mathbf{e}_r + \left[r\frac{d^2\theta}{dt^2} + 2\frac{dr}{dt}\frac{d\theta}{dt}\right]\mathbf{e}_\theta = -\frac{gR_{\mathrm{E}}^2}{r^2}\mathbf{e}_r$$

and equating $\mathbf{e}_r$ and $\mathbf{e}_\theta$ components, we obtain the equations

$$\frac{d^2r}{dt^2} - r\left(\frac{d\theta}{dt}\right)^2 = -\frac{gR_{\mathrm{E}}^2}{r^2} \tag{2.56}$$

$$r\frac{d^2\theta}{dt^2} + 2\frac{dr}{dt}\frac{d\theta}{dt} = 0 \tag{2.57}$$

We can write Equation (2.57) in the form

$$\frac{1}{r}\frac{d}{dt}\left(r^2\frac{d\theta}{dt}\right) = 0$$

which indicates that

$$r^2 \frac{d\theta}{dt} = r v_\theta = \text{constant} \tag{2.58}$$

At $t = 0$ the components of the velocity are $v_r = 0$, $v_\theta = v_0$, and the radial position is $r = r_0$. We can therefore write the constant in Equation (2.58) in terms of the initial conditions:

$$r^2 \frac{d\theta}{dt} = r v_\theta = r_0 v_0 \tag{2.59}$$

Using Equation (2.59) to eliminate $d\theta/dt$ from Equation (2.56), we obtain

$$\frac{d^2 r}{dt^2} - \frac{r_0^2 v_0^2}{r^3} = -\frac{g R_E^2}{r^2} \tag{2.60}$$

We can solve this differential equation by introducing the change of variable

$$u = \frac{1}{r} \tag{2.61}$$

In doing so, we also change the independent variable from t to θ because we want to determine r as a function of the angle θ instead of time. To express Equation (2.60) in terms of u, we must determine d^2r/dt^2 in terms of u. Using the chain rule, we write the derivative of r with respect to time as

$$\frac{dr}{dt} = \frac{d}{dt}\left(\frac{1}{u}\right) = -\frac{1}{u^2}\frac{du}{dt} = -\frac{1}{u^2}\frac{du}{d\theta}\frac{d\theta}{dt} \tag{2.62}$$

Notice from Equation (2.59) that

$$\frac{d\theta}{dt} = \frac{r_0 v_0}{r^2} = r_0 v_0 u^2 \tag{2.63}$$

Substituting this expression into Equation (2.62), we obtain

$$\frac{dr}{dt} = -r_0 v_0 \frac{du}{d\theta} \tag{2.64}$$

We differentiate this expression with respect to time and apply the chain rule again:

$$\frac{d^2 r}{dt^2} = \frac{d}{dt}\left(-r_0 v_0 \frac{du}{d\theta}\right) = -r_0 v_0 \frac{d\theta}{dt}\frac{d}{d\theta}\left(\frac{du}{d\theta}\right) = -r_0 v_0 \frac{d\theta}{dt}\frac{d^2 u}{d\theta^2}$$

Using Equation (2.63) to eliminate $d\theta/dt$ from this expression, we obtain the second time derivative of r in terms of u:

$$\frac{d^2 r}{dt^2} = -r_0^2 v_0^2 u^2 \frac{d^2 u}{d\theta^2}$$

Substituting this result into Equation (2.60) yields a linear differential equation for u as a function of θ:

$$\frac{d^2u}{d\theta^2} + u = \frac{gR_E^2}{r_0^2 v_0^2}$$

The general solution of this equation is

$$u = A\sin\theta + B\cos\theta + \frac{gR_E^2}{r_0^2 v_0^2} \tag{2.65}$$

where A and B are constants. We can use the initial conditions to determine A and B. When $\theta = 0$, $u = 1/r_0$. Also, when $\theta = 0$, the radial component of velocity $v_r = dr/dt = 0$, so from Equation (2.64) we see that $du/d\theta = 0$. From these two conditions, we obtain

$$A = 0 \quad B = \frac{1}{r_0} - \frac{gR_E^2}{r_0^2 v_0^2}$$

Substituting these results into Equation (2.65), we can write the resulting solution for $r = 1/u$ as

$$\frac{r}{r_0} = \frac{1+\varepsilon}{1+\varepsilon\cos\theta} \tag{2.66}$$

where

$$\varepsilon = \frac{r_0 v_0^2}{gR_E^2} - 1 \tag{2.67}$$

The curve called a **conic section** (Figure 2.40) has the property that the ratio of r to the perpendicular distance d to a straight line, called the **directrix**, is constant. This ratio, $r/d = r_0/d_0$, is called the **eccentricity** of the curve. From Figure 2.40 we see that

$$r\cos\theta + d = r_0 + d_0$$

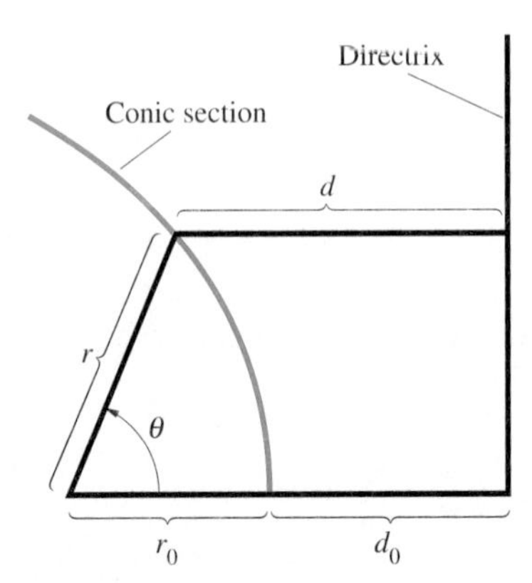

Figure 2.40
If the ratio r/d is constant, the curve describes a conic section.

which we can write in terms of the eccentricity as

$$\frac{r}{r_0} = \frac{1 + (r_0/d_0)}{1 + (r_0/d_0)\cos\theta}$$

Comparing this expression with Equation (2.66), we see that *the satellite's orbit describes a conic section with eccentricity* ε.

The value of the eccentricity determines the character of the orbit (Figure 2.41). If the initial velocity v_0 is chosen so that $\varepsilon = 0$, Equation (2.66) reduces to $r = r_0$ and the orbit is circular. Setting $\varepsilon = 0$ in Equation (2.67) and solving for v_0, we obtain

$$v_0 = \sqrt{\frac{gR_E^2}{r_0}} \tag{2.68}$$

which agrees with the velocity for a circular orbit we obtained in Example 2.9 by a different method.

If $0 < \varepsilon < 1$, the orbit is an ellipse. The maximum radius of the ellipse occurs when $\theta = 180°$. Setting θ equal to 180° in Equation (2.66), we obtain an expression for the maximum radius of the ellipse in terms of the initial radius and ε:

$$r_{max} = r_0\left(\frac{1+\varepsilon}{1-\varepsilon}\right) \tag{2.69}$$

Notice that the maximum radius of the ellipse increases without limit as $\varepsilon \to 1$. When $\varepsilon = 1$, the orbit is a parabola, which means that v_0 is the escape velocity. Setting $\varepsilon = 1$ in Equation (2.67) and solving for v_0, we obtain

$$v_0 = \sqrt{\frac{2gR_E^2}{r_0}}$$

which is the same value for the escape velocity we obtained in Example 2.5 for the case of motion straight away from the earth. If $\varepsilon > 1$, the orbit is a hyperbola.

The solution we have presented, based on the assumption that the earth is a homogeneous sphere, approximates the orbit of an earth satellite. Determining the orbit accurately requires taking into account the variations in the earth's gravitational field due to its actual mass distribution. Similarly, depending on the accuracy required, determining the orbit of a planet around the sun may require accounting for perturbations due to the gravitational attractions of the other planets.

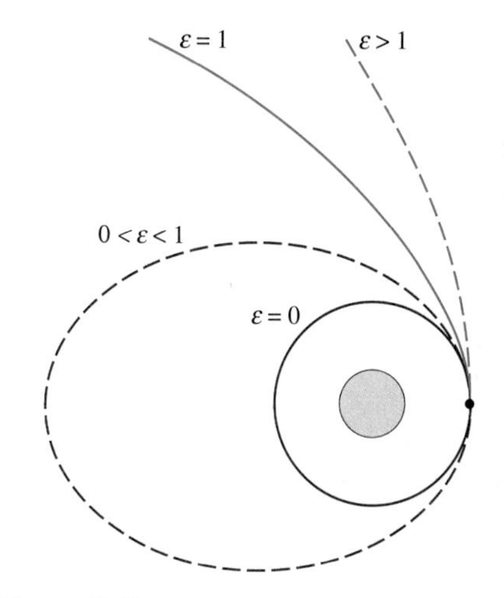

Figure 2.41
Orbits for different values of the eccentricity.

Example 2.13

An earth satellite is in an elliptic orbit with a minimum radius of 6700 km and a maximum radius of 16 090 km. The radius of the earth is 6370 km.
(a) Determine the satellite's velocity when it is at perigee (its minimum radius) and when it is at apogee (its maximum radius).
(b) Draw a graph of the orbit.

STRATEGY

We can regard the radius and velocity of the satellite at perigee as the initial conditions r_0 and v_0 used in obtaining Equation (2.66). Since we also know the maximum radius of the orbit, we can solve Equation (2.69) for the eccentricity of the orbit and then use Equation (2.67) to determine v_0. From Equation (2.58), the product of r and the transverse component of the velocity is constant. We can use this condition to determine the velocity at apogee.

SOLUTION

(a) Solving Equation (2.69) for ε, the eccentricity of the orbit is

$$\varepsilon = \frac{r_{max}/r_0 - 1}{r_{max}/r_0 + 1} = \frac{16\,090/6700 - 1}{16\,090/6700 + 1} = 0.412$$

Now from Equation (2.67), the velocity at perigee is

$$v_0 = \sqrt{\frac{(\varepsilon + 1)gR_E^2}{r_0}} = \frac{\sqrt{(0.412 + 1)(9.81)[(6370)(1000)]^2}}{(6700)(1000)}$$
$$= 9170 \text{ m/s}$$

At perigee and apogee, the velocity has only a transverse component. Therefore the velocity at apogee, v_a, is related to the velocity v_0 at perigee by

$$r_0 v_0 = r_{max} v_a$$

We solve this equation for the velocity at apogee:

$$v_a = \frac{r_0}{r_{max}} v_0 = \left(\frac{6700}{16\,090}\right)(9170) = 3819 \text{ m/s}$$

(b) By plotting Equation (2.66) with $\varepsilon = 0.412$, we obtain the graph of the orbit (Figure 2.42).

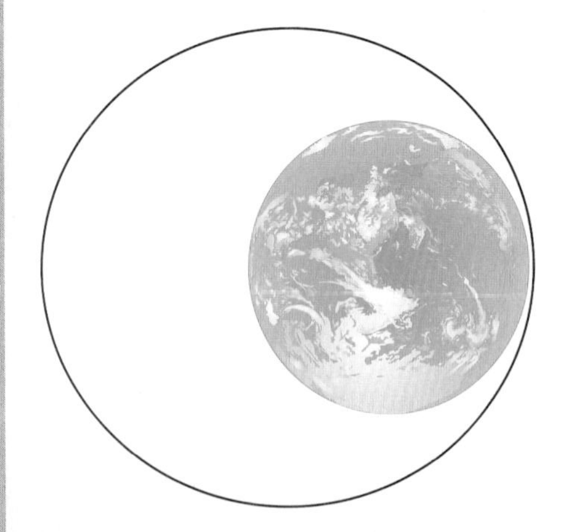

Figure 2.42
Orbit of an earth satellite with a perigee of 6700 km and an apogee of 16 090 km.

Example 2.14

Application to Engineering

Communication Satellites

A communication satellite is usually placed in geosynchronous orbit, a circular orbit above the equator in which the satellite remains above the same point on the earth as the earth rotates beneath it. The satellite is placed into geosynchronous orbit starting from a circular parking orbit nearer the earth by a procedure called a **Hohmann transfer** (Figure 2.43). Let v_1 be the velocity of the satellite in the circular parking orbit. The satellite is first boosted from v_1 to a velocity v_2 in the direction tangent to the parking orbit to put it into an elliptic orbit whose maximum radius equals the radius of the geosynchronous orbit. When the satellite reaches the geosynchronous orbit, its velocity has slowed from v_2 to a velocity v_3. It is then boosted to the velocity v_4 necessary for it to be in the circular geosynchronous orbit, completing the Hohmann transfer.
(a) Determine the radius r_g (in km) of the geosynchronous orbit.
(b) The radius of the earth is $R_E = 6370$ km. If the radius of the circular parking orbit is $r_p = 6670$ km, determine the velocities v_1, v_2, v_3 and v_4.

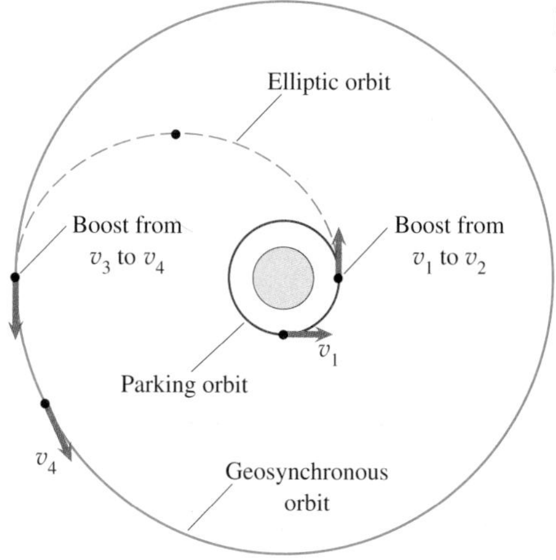

Figure 2.43
A Hohmann transfer.

STRATEGY

(a) To be in geosynchronous orbit, a satellite must complete one revolution in approximately 24 hr while the earth turns one revolution beneath it. This condition, together with Equation (2.68) for the velocity of a satellite in circular orbit, allows us to determine the radius of a geosynchronous orbit.
(b) Since the parking and geosynchronous orbits are circular, we can use Equation (2.68) to determine v_1 and v_4. The initial conditions for the elliptic orbit are $r_0 = r_p$, $v_0 = v_2$. We want the maximum radius of the elliptic orbit to be equal to the radius of the geosynchronous orbit: $r_{max} = r_g$. We can solve Equation (2.69) for the eccentricity of the elliptic orbit and then use Equation (2.67) to determine $v_0 = v_2$. From Equation (2.58), the product of r and the transverse component of the velocity is constant while the satellite is in the elliptic orbit, so we can determine the velocity v_3 from the relation $r_p v_2 = r_g v_3$.

SOLUTION

(a) Let $T = 24\,\text{hr} = (24)(3600)\,\text{s}$. In time T, a satellite in geosynchronous orbit must travel the distance $2\pi r_g$, so

$$2\pi r_g = v_4 T \tag{2.70}$$

From Equation (2.68), the velocity v_4 and radius r_g must also satisfy the equation

$$v_4 = \sqrt{\frac{gR_E^2}{r_g}}$$

Substituting this expression into Equation (2.70) and solving for r_g, we obtain

$$r_g = g^{1/3}\left(\frac{TR_E}{2\pi}\right)^{2/3} = (9.81)^{1/3}\left(\frac{(24)(3600)(6.37 \times 10^6)}{2\pi}\right)^{2/3}$$
$$= 4.22 \times 10^4\,\text{km}$$

(b) From Equation (2.68), the velocity of the satellite in the parking orbit is

$$v_1 = \sqrt{\frac{gR_E^2}{r_p}} = \sqrt{\frac{(9.81)(6.37 \times 10^6)^2}{6.67 \times 10^6}} = 7725\,\text{m/s}$$

and its velocity in the geosynchronous orbit is

$$v_4 = \sqrt{\frac{gR_E^2}{r_g}} = \sqrt{\frac{(9.81)(6.37 \times 10^6)^2}{4.22 \times 10^7}} = 3070\,\text{m/s}$$

From Equation (2.69), the maximum radius of the elliptic orbit is related to its eccentricity by

$$r_g = r_p \frac{1+\varepsilon}{1-\varepsilon}$$

Solving for ε, we obtain

$$\varepsilon = \frac{r_g/r_p - 1}{r_g/r_p + 1} = 0.727$$

Now we can solve Equation (2.67) for v_2:

$$v_2 = \sqrt{\frac{gR_E^2(\varepsilon + 1)}{r_p}} = \sqrt{\frac{(9.81)(6.37 \times 10^6)^2(0.727 + 1)}{6.67 \times 10^6}}$$
$$= 10\,153\,\text{m/s}$$

From the relation $r_p v_2 = r_g v_3$, the velocity v_3 is

$$v_3 = \left(\frac{r_p}{r_g}\right) v_2 = 1604\,\text{m/s}$$

DESIGN ISSUES

Communication satellites (Figure 2.44) have revolutionized communications, making possible the real-time transmission of audiovisual information to every part of the planet. Because satellites are placed in geosynchronous orbit, earth stations used to send signals to and receive signals from the satellites can use simple and relatively inexpensive fixed antennas. (The familiar 'dish' antennas used to receive television transmissions are aimed at satellites in geosynchronous orbit.)

Because the radius of a geosynchronous orbit is large in comparison to the earth's radius – $r_g = 4.22 \times 10^4$ km, which is approximately 26 200 mi – building communication satellites and launching them is a formidable problem in system design. In Example 2.14, the satellite in circular parking orbit must be boosted from v_1 to v_2, an increase in velocity of 2427 m/s, to initiate the elliptic orbit. It must then be boosted from v_3 to v_4, an increase in velocity of 1467 m/s, to achieve geosynchronous orbit. The satellite must be equipped with rocket engines capable of producing these substantial increases in velocity. In addition, it must have guidance and attitude (orientation) control systems that can align the satellite so that the necessary changes in velocity occur in the correct directions. Once in geosynchronous orbit, the satellite must be able to determine its orientation and aim its own antennas to receive and transmit signals.

Figure 2.44
Intelsat V Communication Satellite.

Problems

Use the value $R_E = 6370$ km for the radius of the earth.

2.137 A satellite is in a circular orbit 320 km above the earth's surface.
(a) What is the magnitude of its velocity?
(b) How long does it take to complete one revolution?

2.138 The moon is approximately 383 000 km from the earth. Assuming that the moon's orbit around the earth is circular with velocity given by Equation (2.68), determine the time required for the moon to make one revolution around the earth.

2.139 A satellite is given an initial velocity $v_0 = 6700$ m/s at a distance $r_0 = 2R_E$ from the centre of the earth, as shown in Figure 2.39(a).
(a) What is the maximum radius of the resulting elliptic orbit?
(b) What is the magnitude of the velocity of the satellite when it is at its maximum radius?

2.140 Draw a graph of the elliptic orbit described in Problem 2.139.

2.141 A satellite is given an initial velocity v_0 at a distance $r_0 = 6800$ km from the centre of the earth, as shown in Figure 2.39(a). The resulting elliptic orbit has a maximum radius of 20 000 km. What is v_0?

2.142 In Problem 2.141, what velocity v_0 would be necessary to put the satellite into a parabolic escape orbit?

2.143 From astronomical data, Kepler deduced that the line from the sun to a planet traces out equal areas in equal times (Figure (a)). Show that this result follows from the fact that the transverse component a_θ of the planet's acceleration is zero. (When r changes by an amount dr and θ changes by an amount $d\theta$ (Figure (b)), the resulting differential element of area is $dA = \frac{1}{2}\, r(r\, d\theta)$.)

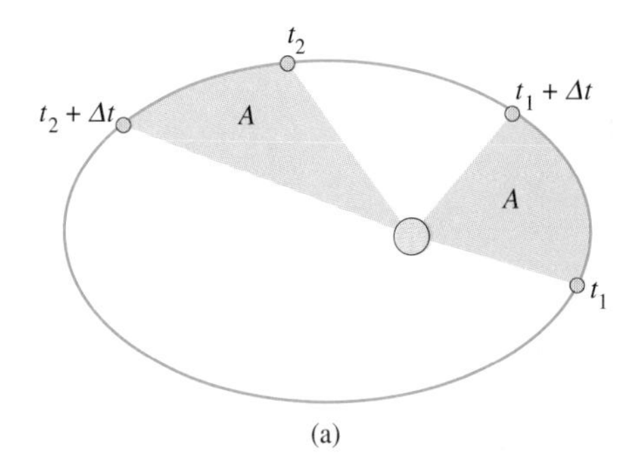

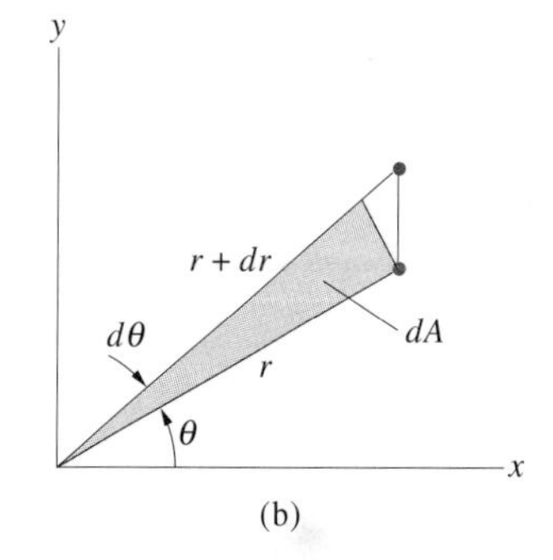

P2.143

2.144 At $t = 0$, an earth satellite is a distance r_0 from the centre of the earth and has an initial velocity v_0 in the direction shown. Show that the polar equation for the resulting orbit is

$$\frac{r}{r_0} = \frac{(\varepsilon + 1)\cos^2\beta}{[(\varepsilon + 1)\cos^2\beta - 1]\cos\theta - (\varepsilon + 1)\sin\beta\cos\beta\sin\theta + 1}$$

where $\varepsilon = (r_0 v_0^2 / g R_E^2) - 1$.

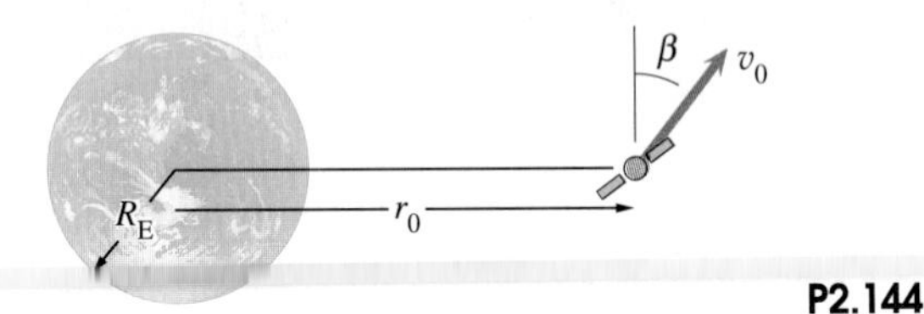

P2.144

2.145 Draw the graphs of the orbits given by the polar equation obtained in Problem 2.144 for $\varepsilon = 0$ and $\beta = 0$, 30° and 60°.

Problems 2.146–2.149 are related to Example 2.14.

2.146 An earth satellite is in a circular parking orbit of radius $r_p = 6800$ km. Determine the increase in velocity $v_2 - v_1$ necessary to put it into an elliptic orbit with maximum radius equal to the radius r_g of a geosynchronous orbit.

2.147 (a) Determine the velocity v_3 of the satellite in Problem 2.146 when it reaches the radius of geosynchronous orbit. (b) Determine the increase in velocity $v_4 - v_3$ necessary to place the satellite in geosynchronous orbit.

2.148 A satellite is in a circular parking orbit of radius $r_p = 7337$ km from the centre of the earth. Determine the velocity increases $v_2 - v_1$ and $v_4 - v_3$ necessary to perform a Hohmann transfer to a circular orbit with radius equal to the radius of the moon's orbit 383 000 km.

2.149 A satellite is in a circular parking orbit of radius $r_p = 3500$ km from the centre of Mars. The radius of Mars is 3394 km, the acceleration due to gravity at its surface is 3.73 m/s^2, and it turns on its polar axis once every 24 hr 37 min. Determine the velocity increases $v_2 - v_1$ and $v_4 - v_3$ necessary to place the satellite in a synchronous orbit around Mars.

2.5 *Relative Motion*

Our discussion so far has been limited to the motion of a single point. However, often it is not the motion of an individual point, but motions of two or more points *relative to each other* that we must consider. For example, if a pilot wants to land on an aircraft carrier (Figure 2.45(a)), the individual motions of the carrier and his plane relative to the earth concern him less than

(a)

Figure 2.45(a)
In many situations the relative motion of objects is of greater importance than their individual motions.

the motion of his plane *relative to the carrier*. Pairs skaters (Figure 2.45(b)) must carefully control both their individual motions relative to the ice and their motion *relative to each other* to successfully complete their moves. In this section we discuss the analysis of the relative motions of points.

Suppose that A and B are two points whose individual motions we measure relative to a reference point O, and let's consider how to describe the motion of A relative to B. Let $\mathbf{r}_A$ and $\mathbf{r}_B$ be the position vectors of points A and B relative to O (Figure 2.46). The vector $\mathbf{r}_{A/B}$ is the position vector of point A relative to point B. These vectors are related by

$$\boxed{\mathbf{r}_A = \mathbf{r}_B + \mathbf{r}_{A/B}} \tag{2.71}$$

Taking the time derivative of this relation, we obtain

$$\boxed{\mathbf{v}_A = \mathbf{v}_B + \mathbf{v}_{A/B}} \tag{2.72}$$

where $\mathbf{v}_A$ is the velocity of A relative to O, $\mathbf{v}_B$ is the velocity of B relative to O, and $\mathbf{v}_{A/B} = d\mathbf{r}_{A/B}/dt$ is the velocity of A relative to B.

(b)

Figure 2.45(b)

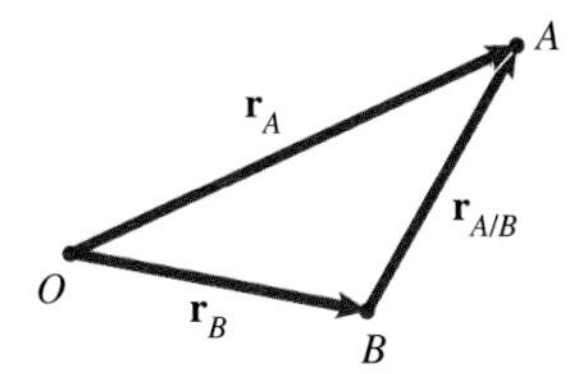

Figure 2.46
Two points A and B and a reference point O. The vectors $\mathbf{r}_A$ and $\mathbf{r}_B$ specify the positions of A and B relative to O, and $\mathbf{r}_{A/B}$ specifies the position of A relative to B.

In our example of an aeroplane approaching an aircraft carrier, the plane could be point A and the carrier point B. The individual motions of the carrier and the plane would be measured (for example, by using on-board inertial navigation systems) relative to a reference point O fixed relative to the earth. Knowing the velocities of the plane, $\mathbf{v}_A$, and the carrier, $\mathbf{v}_B$, the pilot could use Equation (2.72) to determine his velocity relative to the carrier.

Taking the time derivative of Equation (2.72), we obtain

$$\boxed{\mathbf{a}_A = \mathbf{A}_B + \mathbf{a}_{A/B}} \tag{2.73}$$

where $\mathbf{a}_A$ and $\mathbf{a}_B$ are the accelerations of A and B relative to O and $\mathbf{a}_{A/B} = d\mathbf{v}_{A/B}/dt$ is the acceleration of A relative to B. In deriving Equations (2.72) and (2.73), we have assumed that the position, velocity and acceleration vectors are expressed in terms of a reference frame that *does not rotate*. We discuss relative motion expressed in terms of a rotating reference frame in Chapter 6.

The following examples show how you can use Equations (2.71)–(2.73) to analyse relative motions of objects.

Example 2.15

An aircraft carrier travels north at 15 knots (nautical miles per hour) relative to the earth. With its radar, the carrier determines that the velocity of an aeroplane relative to the carrier is horizontal and of magnitude 300 knots towards the northeast. What are the magnitude and direction of the plane's velocity relative to the earth?

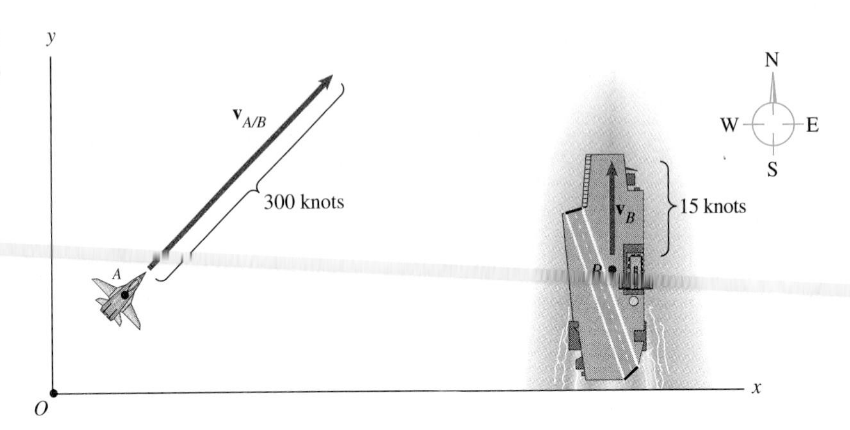

Figure 2.47
The aeroplane (A), carrier (B) and a point O fixed relative to the earth.

STRATEGY

Since we know the carrier's velocity relative to the earth and the velocity of the plane relative to the carrier, we can use Equation (2.72) to determine the plane's velocity relative to the earth.

SOLUTION

Let the aeroplane be point A and let the aircraft be point B (Figure 2.47). Point O and the xy coordinate system are fixed relative to the earth. The velocity of the carrier relative to the earth and the velocity of the plane *relative to the carrier* are shown. The velocity of the carrier is

$$\mathbf{v}_B = 15\mathbf{j} \text{ knots}$$

and the velocity of the plane relative to the carrier is

$$\mathbf{v}_{A/B} = (300 \cos 45^\circ \, \mathbf{i} + 300 \sin 45^\circ \, \mathbf{j}) \text{ knots}$$

Therefore the velocity of the plane relative to the earth is

$$\mathbf{v}_A = \mathbf{v}_B + \mathbf{v}_{A/B} = 300 \cos 45^\circ \, \mathbf{i} + (15 + 300 \sin 45^\circ) \, \mathbf{j}$$
$$= (212.1 \, \mathbf{i} + 227.1 \, \mathbf{j}) \text{ knots}$$

The magnitude of the aeroplane's velocity relative to the earth is

$$\sqrt{(212.1)^2 + (227.1)^2} = 310.8 \text{ knots}$$

and its direction is arctan $(212.1/227.1) = 43.0^\circ$ east of north.

Example 2.16

A ship moving at 5 m/s relative to the water is in a uniform current flowing east at 2 m/s. If the captain wants to sail northwest relative to the earth, what direction should she point the ship? What will the resulting magnitude of the ship's velocity relative to the earth?

STRATEGY

Let the ship be point A and let B be a point moving with the water (Figure (a)). Point O and the xy coordinate system are fixed relative to the earth. We know $\mathbf{v}_B$, the desired direction of $\mathbf{v}_A$ and the magnitude of $\mathbf{v}_{A/B}$. We can use Equation (2.72) to determine the magnitude of $\mathbf{v}_A$ and the direction of $\mathbf{v}_{A/B}$.

Figure 2.48

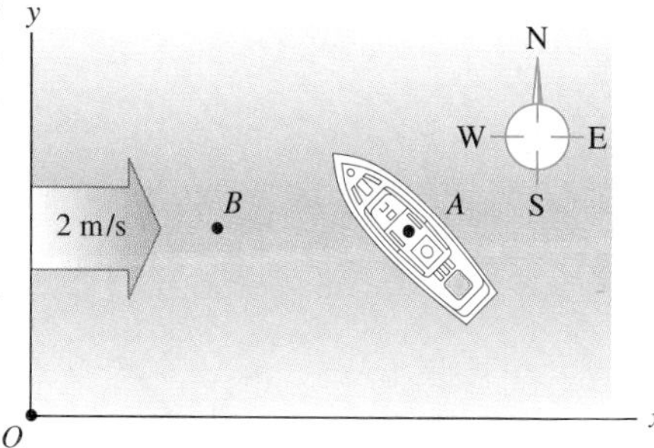

(a) The ship A and a point B moving with the water.

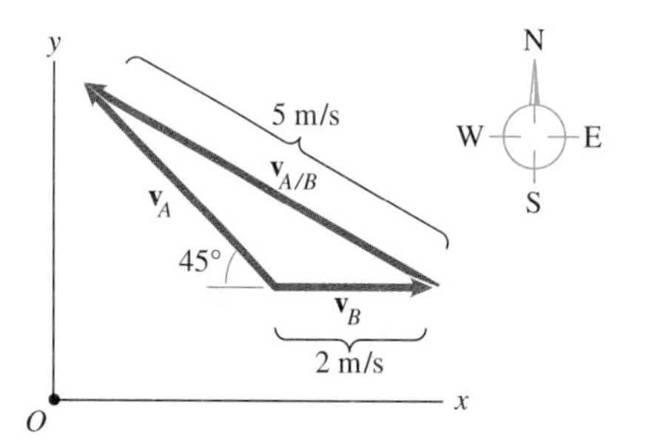

(b) Diagram of the velocity vectors.

SOLUTION

The velocity of the ship relative to the earth is equal to the velocity of the water relative to the earth plus the velocity of the ship relative to the water:

$$\mathbf{v}_A = \mathbf{v}_B + \mathbf{v}_{A/B}$$

In Figure (b) we show this relationship together with the information we know about these velocities: the velocity of the current is 2 m/s towards the east, the magnitude of the velocity of the ship relative to the water is 5 m/s, and the direction of the velocity of the ship relative to the earth is northwest. In terms of the coordinate system shown, the velocity of the current is $\mathbf{v}_B = 2\,\mathbf{i}\,\text{m/s}$. We don't know the magnitude of $\mathbf{v}_A$ but, because we know its direction, we can write it in terms of components as

$$\mathbf{v}_A = -|\mathbf{v}_A| \cos 45^\circ\, \mathbf{i} + |\mathbf{v}_A| \sin 45^\circ\, \mathbf{j}$$

The velocity of the ship relative to the water is

$$\mathbf{v}_{A/B} = \mathbf{v}_A - \mathbf{v}_B = -(|\mathbf{v}_A| \cos 45^\circ + 2)\,\mathbf{i} + |\mathbf{v}_A| \sin 45^\circ\, \mathbf{j}$$

The magnitude of this vector is

$$|\mathbf{v}_{A/B}| = \sqrt{(|\mathbf{v}_A \cos 45^\circ + 2)^2 + (|\mathbf{v}_A| \sin 45^\circ)^2} = 5\,\text{m/s}$$

Solving this equation, we obtain $|\mathbf{v}_A| = 3.38$ m/s, so the velocity of the ship relative to the water is

$$\mathbf{v}_{A/B} = (-4.39\,\mathbf{i} + 2.39\,\mathbf{j})\,\text{m/s}$$

The captain must point her ship at arctan $(4.39/2.39) = 61.4^\circ$ west of north to cause the ship to travel in the northwest direction relative to the earth.

DISCUSSION

The problem described in this example must be solved whenever a ship travels in a current or an aeroplane flies in a wind that is not parallel to its desired course.

Example 2.17

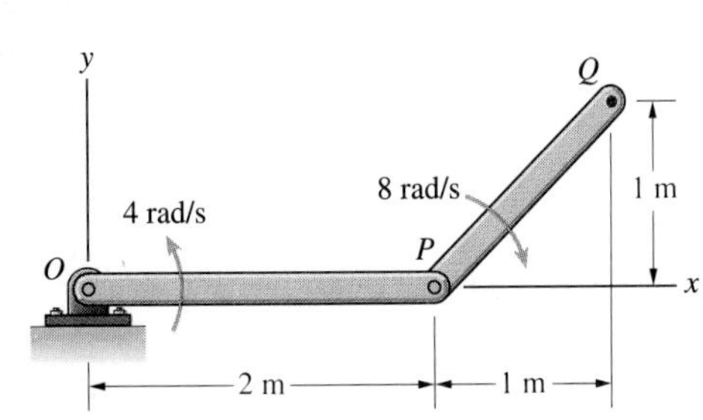

Figure 2.49

The bars OP and PQ in Figure 2.49 rotate in the x-y plane with constant angular velocities. In terms of the fixed coordinate system shown, what is the acceleration of point Q relative to the fixed point O?

STRATEGY

Relative to point P, point Q moves in a circular path around P with constant angular velocity. We can use polar coordinates to determine $\mathbf{a}_{Q/P}$ and then express it in terms of components in the xy coordinate system. Point P moves in a circular path about O with constant angular velocity, so we can also use polar coordinates to determine the acceleration $\mathbf{a}_P$ and then express it in terms of components in the xy coordinate system. Then the acceleration of Q relative to O is $\mathbf{a}_Q = \mathbf{a}_p + \mathbf{a}_{Q/P}$.

SOLUTION

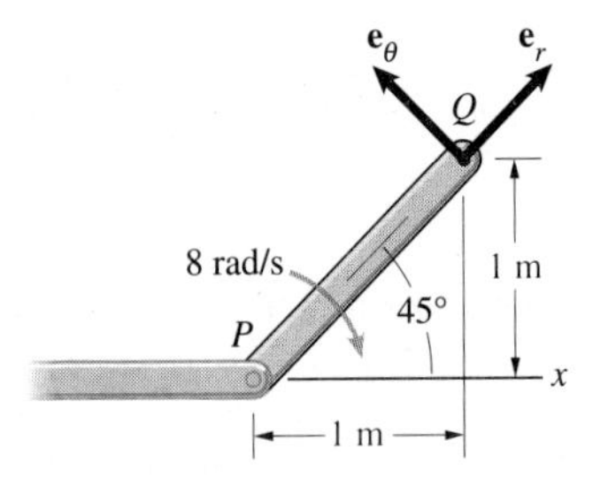

(a) Determining the acceleration of Q relative to P.

Expressing the motion of Q relative to P in terms of polar coordinates (Figure (a)), we obtain the radial component of the acceleration,

$$a_r = \frac{d^2r}{dt^2} - r\omega^2 = 0 - (\sqrt{2})(-8)^2 = -90.51\ \text{m/s}^2$$

and the transverse component,

$$a_\theta = r\alpha + 2\frac{dr}{dt}\omega = 0$$

Therefore, the acceleration of Q relative to P in terms of the xy coordinate system is

$$\mathbf{a}_{Q/P} = a_r \cos 45^\circ\,\mathbf{i} + a_r \sin 45^\circ\,\mathbf{j} = (-64\,\mathbf{i} - 64\,\mathbf{j})\ \text{m/s}^2$$

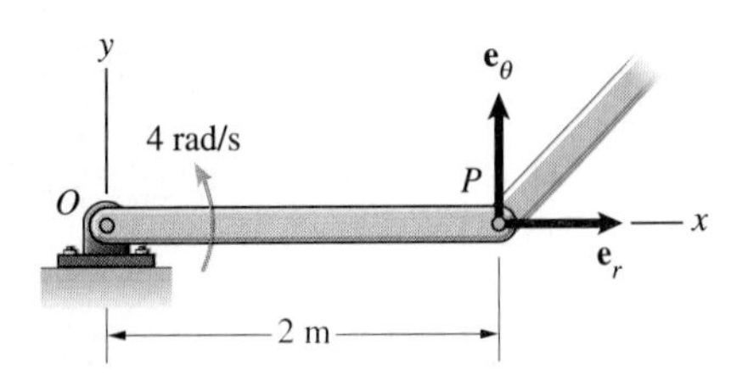

(b) Determining the acceleration of P relative to O.

We also express the acceleration of P relative to O in terms of polar coordinates (Figure (b)). The radial component is

$$a_r = \frac{d^2r}{dt^2} - r\omega^2 = 0 - (2)(4)^2 = -32\ \text{m/s}^2$$

and the transverse component is

$$a_\theta = r\alpha + 2\frac{dr}{dt}\omega = 0$$

The acceleration of P relative to O in terms of the xy coordinate system is

$$\mathbf{a}_P = a_r\,\mathbf{i} = -32\mathbf{i}\ (\text{m/s}^2)$$

Therefore the acceleration of Q relative to O is

$$\mathbf{a}_Q = \mathbf{a}_P + \mathbf{a}_{Q/P} = -32\,\mathbf{i} - 64\,\mathbf{i} - 64\,\mathbf{j} = (-96\,\mathbf{i} - 64\,\mathbf{j})\ \text{m/s}^2$$

DISCUSSION

By using polar coordinates in this example, do we violate our assumption that the vectors in Equations (2.71)–(2.73) are expressed in terms of a reference frame that does not rotate? We do not, because the expressions for the velocity and acceleration in polar coordinates account for the fact that the unit vectors rotate. They give the velocity and acceleration relative to the reference frame in which the polar coordinates (r, θ) are measured. For the same reason, you can also use normal and tangential components to evaluate the terms in Equations (2.71)–(2.73).

This example demonstrates an important use of the concept of relative motion. The motion of point Q relative to point O is quite complicated. But because the motion of Q relative to P and the motion of P relative to Q are comparatively simple, we can take advantage of the equations describing relative motion to obtain information about the motion of Q relative to O.

Problems

2.150 Two cars A and B approach an intersection. Car A is going 20 m/s and is decelerating at 2 m/s^2, and car B is going 10 m/s and is decelerating at 3 m/s^2. In terms of the earth-fixed coordinate system shown, determine the velocity of car A relative to car B and the velocity of car B relative to car A.

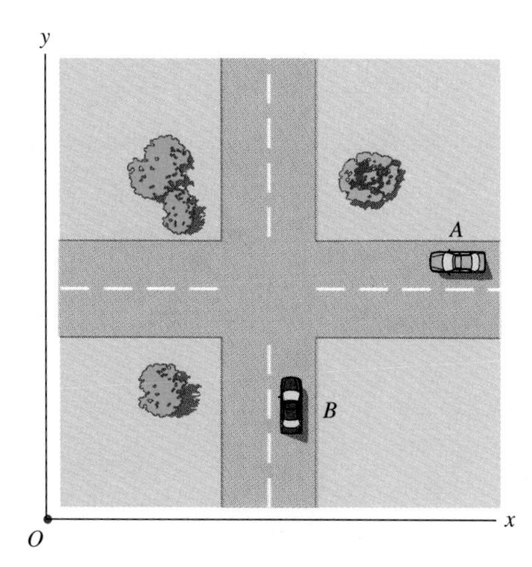

P2.150

2.151 In Problem 2.150, determine the acceleration of car A relative to car B and the acceleration of car B relative to car A.

2.152 Suppose that the two cars in Problem 2.150 approach the intersection with constant velocities. Prove that the cars will reach the intersection at the same time if the velocity of car A relative to car B points from car A towards car B.

2.153 Two sailing boats have constant velocities $\mathbf{v}_A$ and $\mathbf{v}_B$ relative to the earth. The skipper of boat A sights a point on the horizon behind boat B. Seeing that boat B appears stationary relative to that point, he knows he must change course to avoid a collision. Use Equation (2.72) to explain why.

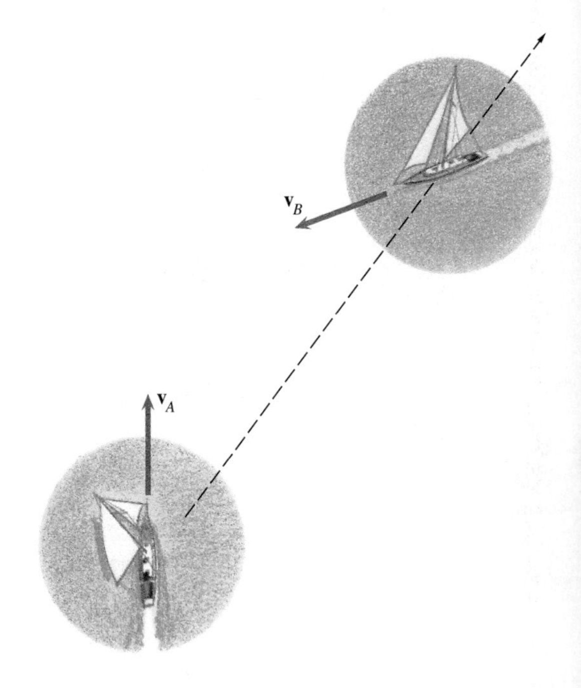

P2.153

2.154 Two projectiles A and B are launched from O at the same time with the initial velocities and elevation angles shown relative to the earth-fixed coordinate systems. At the instant B reaches its highest point, determine: (a) the acceleration of A relative to B; (b) the velocity of A relative to B; (c) the position vector of A relative to B.

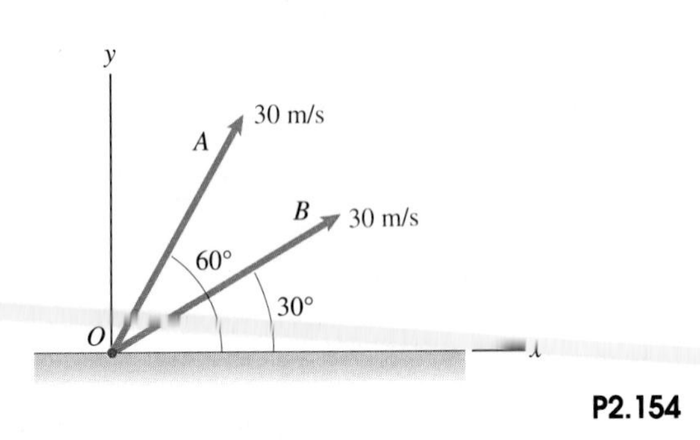

P2.154

2.155 In a machining process, the disk rotates about the fixed point O with a constant angular velocity of 10 rad/s. In terms of the non-rotating coordinate system shown, what is the magnitude of the velocity of A relative to B?

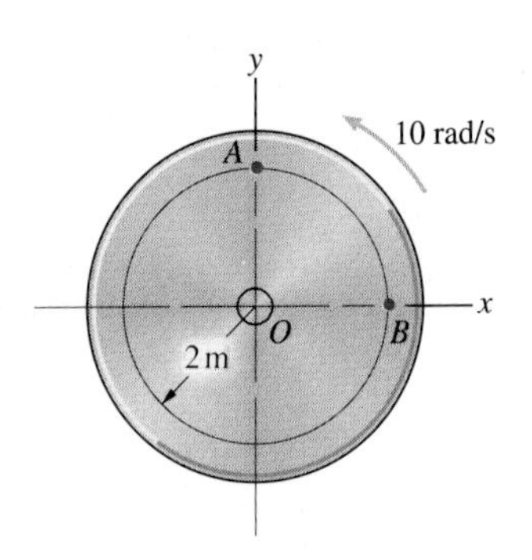

P2.155

2.156 In Problem 2.155, what is the magnitude of the acceleration of A relative to B?

2.157 The bar rotates about the fixed point O with a constant angular velocity of 2 rad/s. Point A moves outwards along the bar at a constant rate of 100 mm/s. Point B is a fixed point on the bar. At the instant shown, what is the magnitude of the velocity of point A relative to point B?

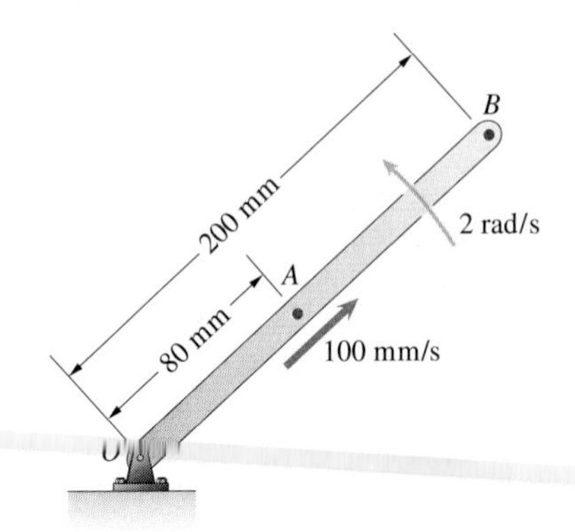

P2.157

2.158 In Problem 2.157, what is the magnitude of the acceleration of point B relative to point A at the instant shown?

2.159 The bars OA and AB are each 400 mm long and rotate in the x-y plane. OA has a counterclockwise angular velocity of 10 rad/s and a counterclockwise angular acceleration of 2 rad/s^2. AB has a constant counterclockwise angular velocity of 5 rad/s. What is the velocity of point B relative to point A in terms of the fixed coordinate system?

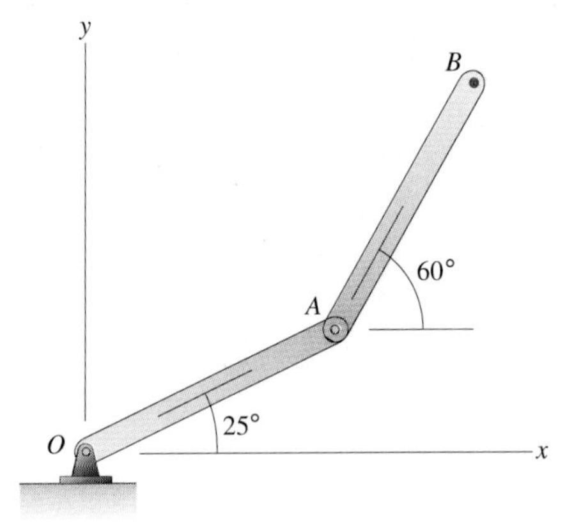

P2.159

2.160 In Problem 2.159, what is the acceleration of point B relative to point A?

2.161 In Problem 2.159, what is the velocity of point B relative to the fixed point O?

2.162 In Problem 2.159, what is the acceleration of point B relative to the fixed point O?

2.163 The train on the circular track is travelling at a constant speed of 15 m/s. The train on the straight track is travelling at 6 m/s and is increasing its speed at 0.6 m/s^2. In terms of the earth-fixed coordinate system shown, what is the velocity of passenger *A* relative to passenger *B*?

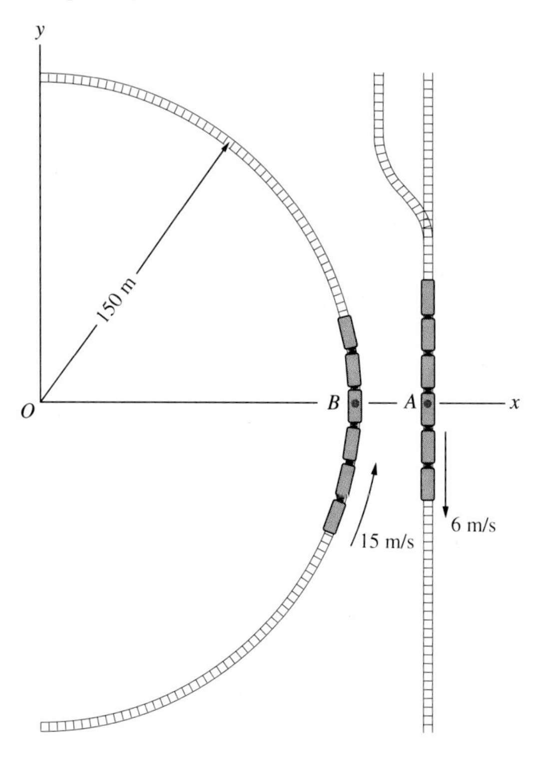

P2.163

2.164 In Problem 2.163, what is the acceleration of passenger *A* relative to passenger *B*?

2.165 The velocity of the boat relative to the earth-fixed coordinate system is 12 m/s and is constant. The length of the tow rope is 15 m. The angle θ is 30° and is increasing at a constant rate of 10°/s. What are the velocity and acceleration of the skier relative to the boat?

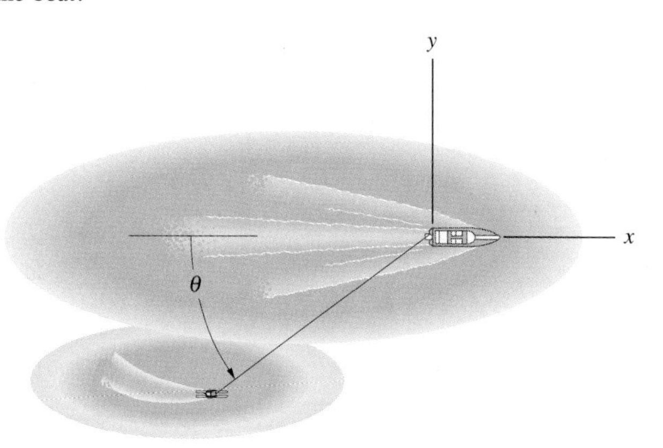

P2.165

2.166 In Problem 2.165, what are the velocity and acceleration of the skier relative to the earth?

2.167 The hockey player is skating with velocity components $v_x = 1.2$ m/s, $v_z = -6$ m/s when he hits a slap shot with a velocity of magnitude 30 m/s *relative to him*. The position of the puck when he hits it is $x = 3.6$ m, $z = 3.6$ m. If he hits the puck so that its velocity vector *relative to him* is directed towards the centre of the goal, where will the puck intersect the *x* axis? Will it enter the 2 m wide goal?

P2.167

2.168 In Problem 2.167 at what point on the *x* axis should the player aim the puck's velocity vector relative to him so that it enters the centre of the goal?

2.169 An aeroplane flies in a jet stream flowing east at 160 km/hr. The aeroplane's airspeed (its velocity relative to the air) is 800 km/hr towards the northwest. What are the magnitude and direction of the aeroplane's velocity relative to the earth?

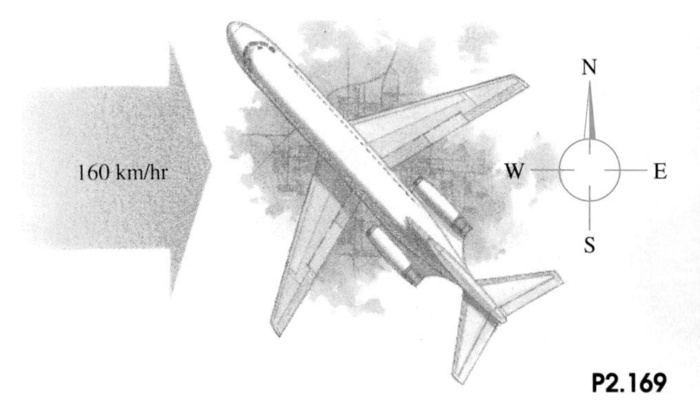

P2.169

2.170 In Problem 2.169, if the pilot wants to fly towards a city that is northwest of his current position, in which direction should he point the aeroplane, and what will be the magnitude of his velocity relative to the earth?

2.171 A river flows north at 3 m/s. (Assume that the current is uniform.) If you want to travel in a straight line from point C to point D in a boat that moves at a constant speed of 10 m/s relative to the water, in what direction do you point the boat? How long does it take to make the crossing?

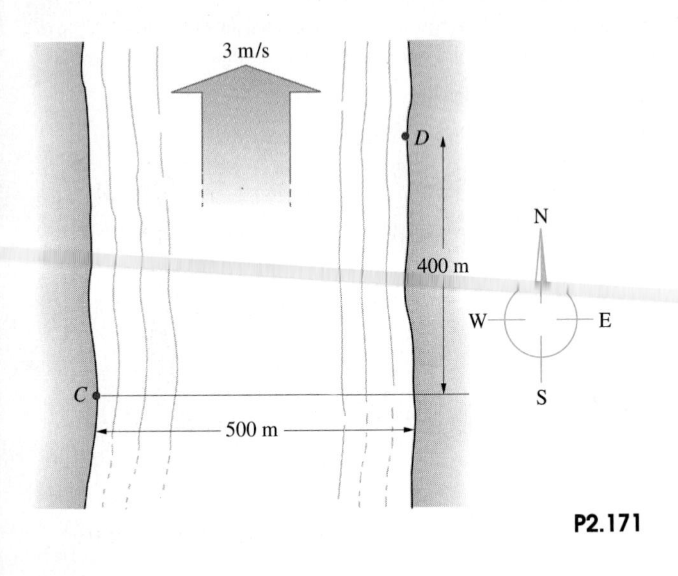

P2.171

2.172 In Problem 2.171, what is the minimum boat speed relative to the water necessary to make the trip from point C to point D?

2.173 Relative to the earth, a sailing boat sails north at velocity v_0 and then sails east at the same velocity. The velocity of the wind is uniform and constant. A 'tell-tale' on the boat points in the direction of the velocity of the wind relative to the boat. What are the direction and magnitude of the wind's velocity relative to the earth? (Your answer for the magnitude of the velocity of the wind will be in terms of v_0.)

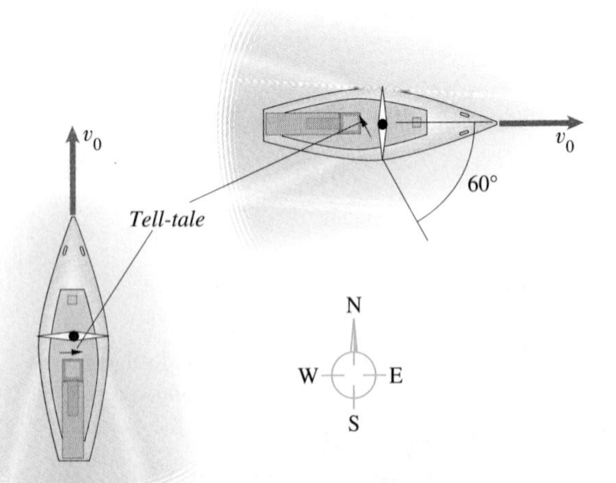

P2.173

2.174 The origin O of the non-rotating coordinate system is at the centre of the earth, and the y axis points north. The satellite A on the x axis is in a circular polar orbit of radius R, and its velocity is $v_A\,\mathbf{j}$. Let ω be the angular velocity of the earth. What is the satellite's velocity relative to the point B on the earth directly below the satellite?

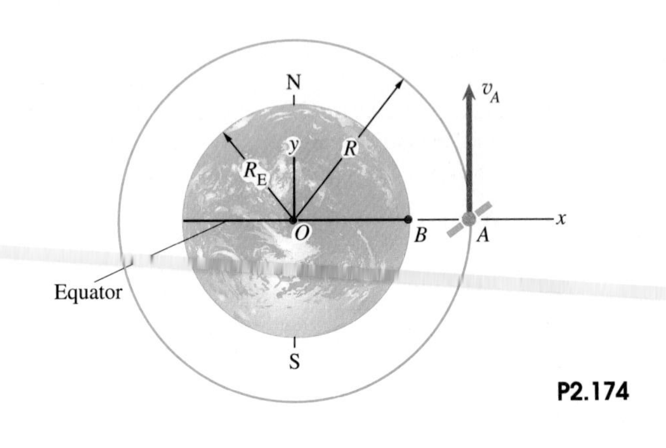

P2.174

2.175 In Problem 2.174, what is the satellite's acceleration relative to the point B on the earth directly below the satellite?

Computational Mechanics

The following example and problems are designed for the use of a programmable calculator or computer.

Example 2.18

With buoyancy accounted for, the downward acceleration of a steel ball falling in a particular liquid is $a = 0.9g - cv$, where c is a constant that is proportional to the viscosity of the liquid. To determine the viscosity, a rheologist releases the ball from rest at the top of a 2 m tank of the liquid. If the ball requires 2 s to fall to the bottom, what is the value of c?

Figure 2.50

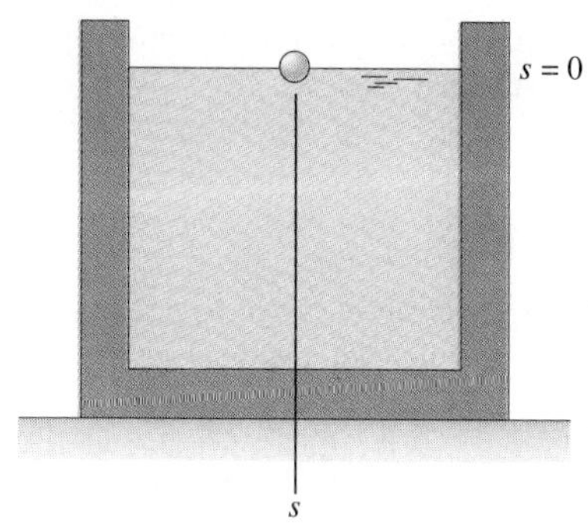

(a) The ball is released from rest at the surface.

STRATEGY

We can obtain an equation for c by determining the distance the ball falls as a function of time.

SOLUTION

We measure the ball's position s downward from the point of release (Figure (a)) and let $t = 0$ be the time of release.

The acceleration is

$$a = \frac{dv}{dt} = 0.9g - cv$$

Separating variables and integrating,

$$\int_0^v \frac{dv}{0.9g - cv} = \int_0^t dt$$

we obtain

$$v = \frac{ds}{dt} = \frac{0.9g}{c}(1 - e^{-ct})$$

Integrating this equation with respect to time, we obtain the distance the ball has fallen as a function of the time from its release:

$$s = \frac{0.9g}{c^2}(ct - 1 + e^{-ct})$$

We know that $s = 2$ m when $t = 2$ s, so determining c requires the equation

$$f(c) = \frac{(0.9)(9.81)}{c^2}(2c - 1 + e^{-2c}) - 2 = 0$$

We can't solve this transcendental equation in closed form to determine c. Problem-solving programs such as *Mathcad* and *TK! Solver* are designed to obtain roots of such equations. Another approach is to compute the value of $f(c)$ for a range of values of c and plot the results, as we have done in Figure 2.51. From the graph we estimate that $c = 8.3\ \mathrm{s}^{-1}$.

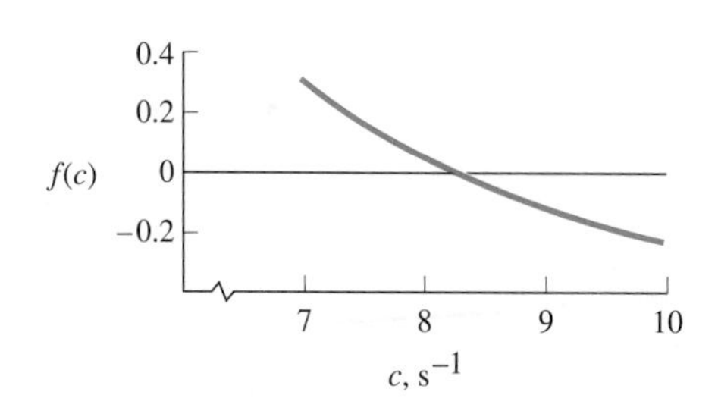

Figure 2.51
Graph of the function $f(c)$.

Problems

2.176 An engineer analysing a machining process determines that from $t = 0$ to $t = 4$ s the workpiece starts from rest and moves in a straight line with acceleration

$$a = (2 + t^{0.5} - t^{1.5})\,\text{m/s}^2$$

(a) Draw a graph of the position of the workpiece relative to its position at $t = 0$ for values of time from $t = 0$ to $t = 4$ s.
(b) Estimate the maximum velocity during this time interval and the time at which it occurs.

2.177 In Problem 2.72, determine the range of angles θ within which the pitcher must release the ball to hit the strike zone.

2.178 A catapault designed to throw a line to ships in distress throws a projectile with initial velocity $v_0(1 - 0.4 \sin\theta_0)$, where θ_0 is the angle above the horizontal. Determine the value of θ_0 for which the distance the projectile is thrown is a maximum, and show that the maximum distance is $0.559 v_0^2/g$.

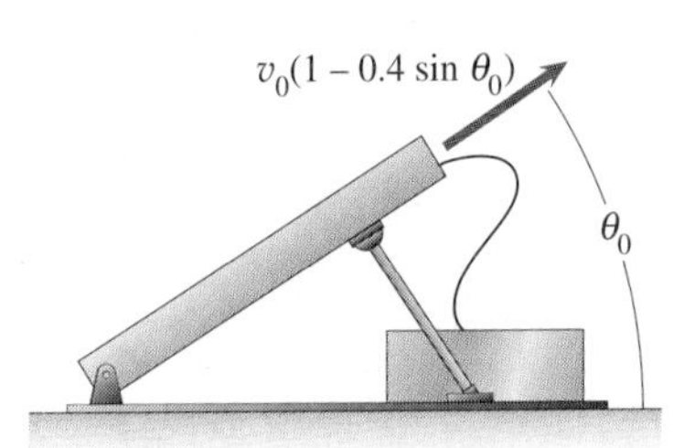

P2.178

2.179 At $t = 0$, a projectile is located at the origin and has a velocity of 20 m/s at 40° above the horizontal. The profile of the ground surface it strikes can be approximated by the equation $y = 0.4x - 0.006x^2$, where x and y are in metres. Determine the approximate coordinates of the point where it hits the ground.

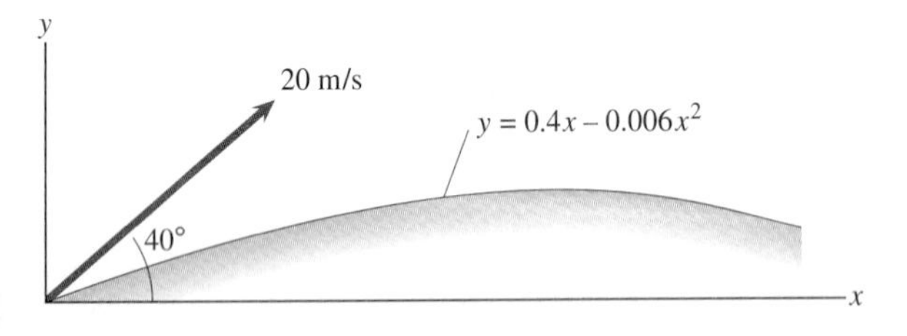

P2.179

2.180 A carpenter working on a house asks his apprentice to throw him an apple. The apple is thrown at 10 m/s. What two values of θ_0 will cause the apple to land in the carpenter's hand, 3.85 m horizontally and 3.85 m vertically from the point where it is thrown?

P2.180

2.181 A motorcycle starts from rest at $t = 0$ and moves along a circular track with 400 m radius. The tangential component of its acceleration is $a_t = (2 + 0.2t)\,\text{m/s}^2$. When the magnitude of its total acceleration reaches $6\,\text{m/s}^2$, friction can no longer keep it on the circular track and it spins out. How long after it starts does it spin out, and how fast is it going?

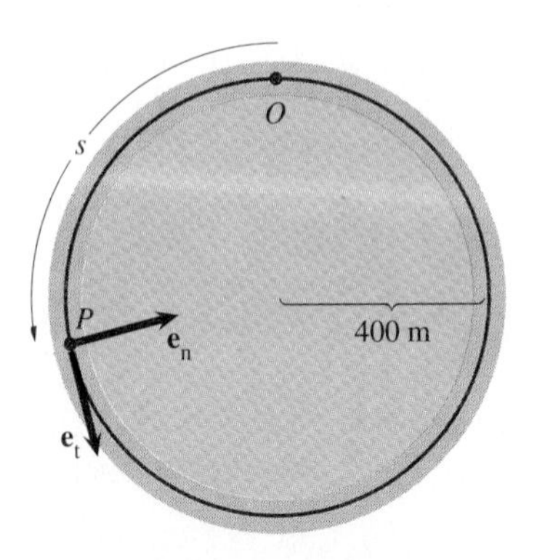

P2.181

2.182 At $t = 0$, a steel ball in a tank of oil is given a horizontal velocity $\mathbf{v} = 2\mathbf{i}$ m/s. The components of its acceleration are $a_x = -cv_x$, $a_y = -0.8g - cv_y$, $a_z = -cv_z$, where c is a constant. When the ball hits the bottom of the tank, its position relative to its position at $t = 0$ is $\mathbf{r} = (0.8\mathbf{i} - \mathbf{j})$ m. What is the value of c?

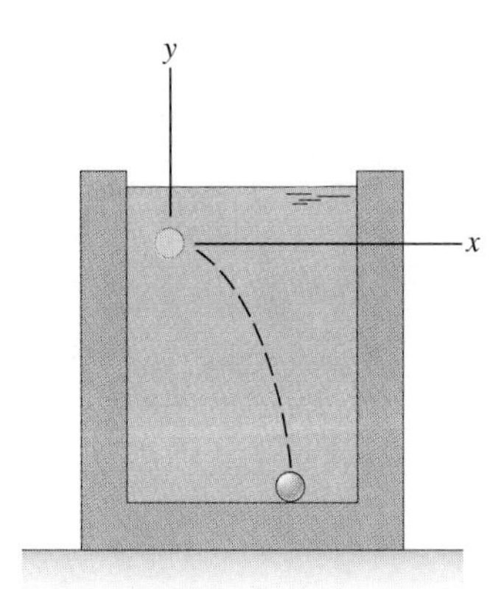

P2.182

2.183 The polar coordinates of a point P moving in the x-y plane are $r = (t^3 - 4t)$ m, $\theta = (t^2 - t)$ rad.
(a) Draw a graph of the magnitude of the velocity of P from $t = 0$ to $t = 2$ s.
(b) Estimate the minimum magnitude of the velocity and the time at which it occurs.

2.184 (a) Draw a graph of the magnitude of the acceleration of the point P in Problem 2.183 from $t = 0$ to $t = 2$ s.
(b) Estimate the minimum magnitude of the acceleration and the time at which it occurs.

2.185 The robot is programmed so that point P describes the path

$$r = (1 - 0.5\cos 2\pi t) \text{ m}$$
$$\theta = \{0.5 - 0.2\sin[2\pi(t - 0.1)]\} \text{ rad}$$

Determine the values of r and θ at which the magnitude of the velocity of P attains its maximum value.

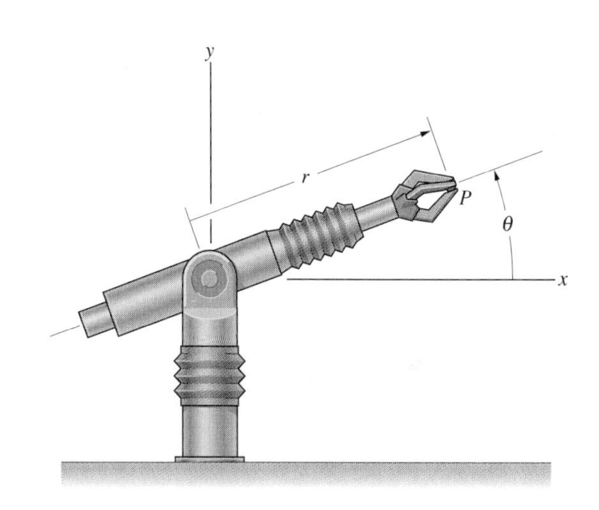

P2.185

2.186 In Problem 2.185, determine the values of r and θ at which the magnitude of the acceleration of P attains its maximum value.

Chapter Summary

The position of a point P relative to a reference point O is specified by the **position vector r** from O to P. The **velocity** of P relative to O is

$$\mathbf{v} = \frac{d\mathbf{r}}{dt} \quad \textbf{Equation (2.1)}$$

and the **acceleration** of P relative to O is

$$\mathbf{a} = \frac{d\mathbf{v}}{dt} \quad \textbf{Equation (2.2)}$$

Straight-Line Motion

The position of a point P on a straight line relative to a reference point O is specified by a coordinate s measured along the line from O to P. The coordinate s and the velocity and acceleration of P along the line are related by

$$v = \frac{ds}{dt} \qquad \textbf{Equation (2.3)}$$

$$a = \frac{dv}{dt} \qquad \textbf{Equation (2.4)}$$

If the acceleration is specified as a function of time, the velocity and position can be determined as functions of time by integration. If the acceleration is specified as a function of velocity, $dv/dt = a(v)$, the velocity can be determined as a function of time by separating variables:

$$\int_{v_0}^{v} \frac{dv}{a(v)} = \int_{t_0}^{t} dt \qquad \textbf{Equation (2.16)}$$

If the acceleration is specified as a function of position, $dv/dt = a(s)$, the chain rule can be used to express the acceleration in terms of a derivative with respect to position:

$$\frac{dv}{dt} = \frac{dv}{ds}\frac{ds}{dt} = \frac{dv}{ds}v = a(s)$$

Separating variables, the velocity can be determined as a function of position:

$$\int_{v_0}^{v} v\,dv = \int_{s_0}^{s} a(s)ds \qquad \textbf{Equation (2.19)}$$

Cartesian Coordinates

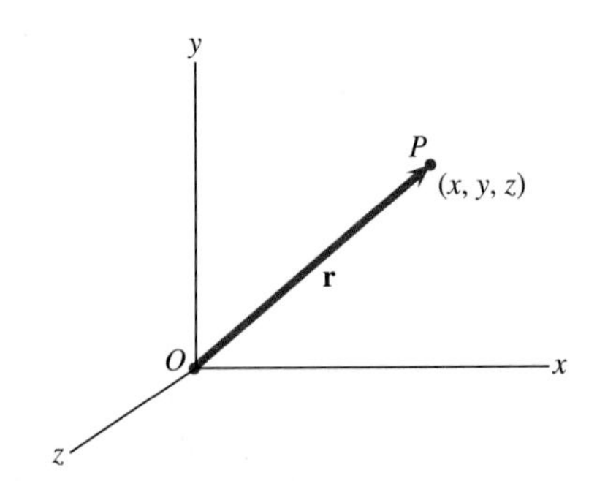

The position, velocity and acceleration are [Equations (2.21)–(2.25)]

$$\mathbf{r} = x\,\mathbf{i} + y\,\mathbf{j} + z\,\mathbf{k}$$

$$\mathbf{v} = v_x\,\mathbf{i} + v_y\,\mathbf{j} + v_z\,\mathbf{k} = \frac{dx}{dt}\mathbf{i} + \frac{dy}{dt}\mathbf{j} + \frac{dz}{dt}\mathbf{k}$$

$$\mathbf{a} = a_x\,\mathbf{i} + a_y\,\mathbf{j} + a_z\,\mathbf{k} = \frac{dv_x}{dt}\mathbf{i} + \frac{dv_y}{dt}\mathbf{j} + \frac{dv_z}{dt}\mathbf{k}$$

The equations describing the motion in each coordinate direction are identical in form to the equations that describe the motion of a point along a straight line.

Angular Motion

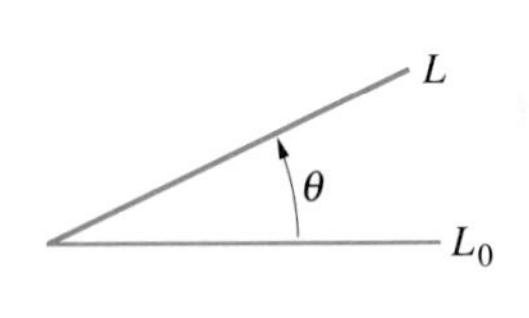

The angular velocity ω and angular acceleration α of L relative to L_0 are

$$\omega = \frac{d\theta}{dt} \qquad \textbf{Equation (2.31)}$$

$$\alpha = \frac{d\omega}{dt} = \frac{d^2\theta}{dt^2} \qquad \textbf{Equation (2.32)}$$

Normal and Tangential Components

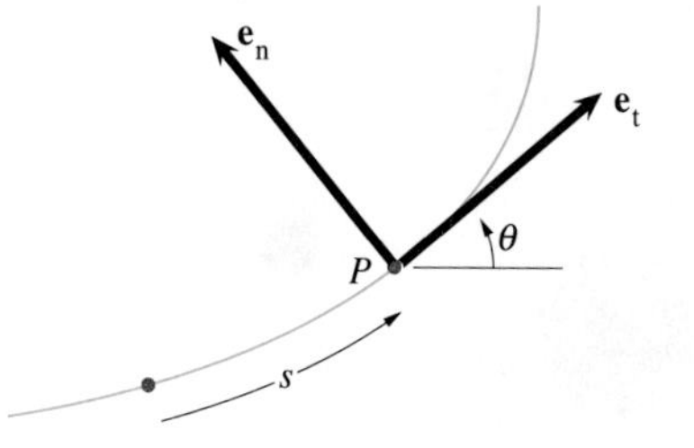

The velocity and acceleration are

$$\mathbf{v} = v\,\mathbf{e}_t = \frac{ds}{dt}\mathbf{e}_t \qquad \textbf{Equation (2.38)}$$

$$\mathbf{a} = a_t\,\mathbf{e}_t + a_n\,\mathbf{e}_n \qquad \textbf{Equation (2.39)}$$

where

$$a_t = \frac{dv}{dt} \qquad a_n = v\frac{d\theta}{dt} = \frac{v^2}{\rho} \qquad \textbf{Equation (2.40)}$$

The unit vector $\mathbf{e}_n$ points toward the concave side of the path. The term ρ is the instantaneous radius of curvature of the path.

Polar Coordinates

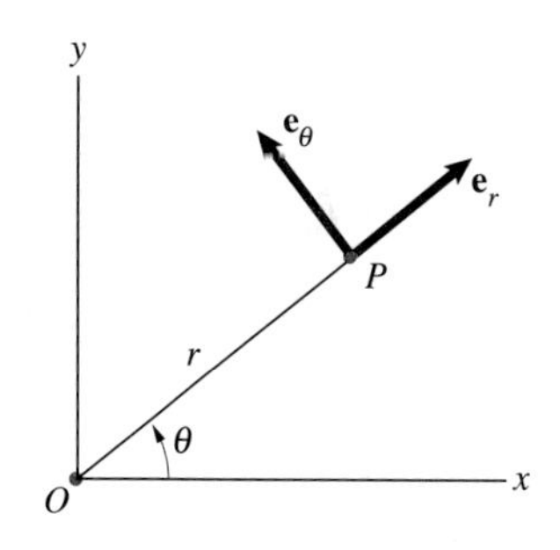

The position, velocity and acceleration are

$$\mathbf{r} = r\,\mathbf{e}_r \qquad \textbf{Equation (2.44)}$$

$$\mathbf{v} = v_r\,\mathbf{e}_r + v_\theta\,\mathbf{e}_\theta = \frac{dr}{dt}\mathbf{e}_r + r\omega\,\mathbf{e}_\theta \qquad \textbf{Equation (2.49)}$$

$$\mathbf{a} = a_r\,\mathbf{e}_r + a_\theta\,\mathbf{e}_\theta \qquad \textbf{Equation (2.50)}$$

where

$$a_r = \frac{d^2r}{dt^2} - r\left(\frac{d\theta}{dt}\right)^2 = \frac{d^2r}{dt^2} - r\omega^2$$

$$a_\theta = r\frac{d^2\theta}{dt^2} + 2\frac{dr}{dt}\frac{d\theta}{dt} = r\alpha = 2\frac{dr}{dt}\omega$$

Equation (2.51)

Relative Motion

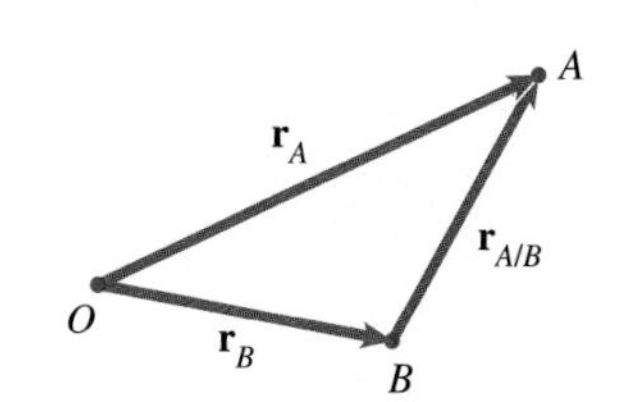

The vectors $\mathbf{r}_A$ and $\mathbf{r}_B$ specify the positions of A and B relative to O, and $\mathbf{r}_{A/B}$ specifies the position of A relative to B:

$$\mathbf{r}_A = \mathbf{r}_B + \mathbf{r}_{A/B} \qquad \textbf{Equation (2.71)}$$

Taking time derivatives of this equation gives the relations

$$\mathbf{v}_A = \mathbf{v}_B + \mathbf{v}_{A/B} \qquad \textbf{Equation (2.72)}$$

where $\mathbf{v}_A$ and $\mathbf{v}_B$ are the velocities of A and B relative to O and $\mathbf{v}_{A/B} = d\mathbf{r}_{A/B}/dt$ is the velocity of A relative to B, and

$$\mathbf{a}_A = \mathbf{a}_B + \mathbf{a}_{A/B} \qquad \textbf{Equation (2.73)}$$

where $\mathbf{a}_A$ and $\mathbf{a}_B$ are the accelerations of A and B relative to O and $\mathbf{a}_{A/B} = d\mathbf{v}_{A/B}/dt$ is the acceleration of A relative to B.

Review Problems

2.187 Suppose that you must determine the duration of the amber light at a highway intersection. Assume that cars will be approaching the intersection travelling as fast as 105 km/hr, that drivers' reaction times are as long as 0.5 s, and that cars can safely achieve a deceleration of at least $0.4g$.
(a) How long must the light remain amber to allow drivers to come to a stop safely before the light turns red?
(b) What is the minimum distance cars must be from the intersection when the light turns amber to come to a stop safely at the intersection?

2.188 The acceleration of a point moving along a straight line is $a = (4t + 2)\ \text{m/s}^2$. When $t = 2$ s, its position is $s = 36$ m, and when $t = 4$ s, its position is $s = 90$ m. What is its velocity when $t = 4$ s?

2.189 A model rocket takes off straight up. Its acceleration during the 2 s its motor burns is $25\ \text{m/s}^2$. Neglect aerodynamic drag. Determine: (a) the maximum velocity during the flight; (b) the maximum altitude reached.

P2.189

2.190 In Problem 2.189, if the rocket's parachute fails to open, what is the total time of flight from take-off until the rocket hits the ground?

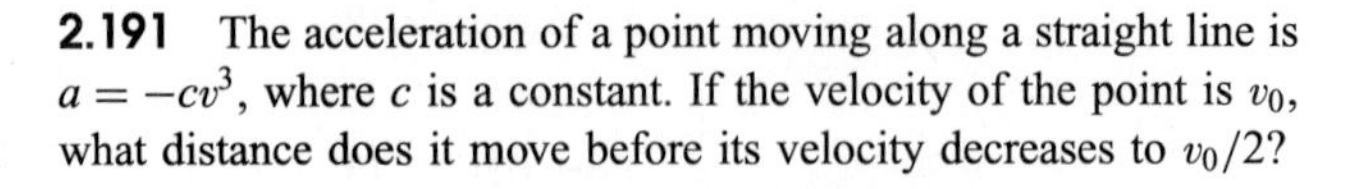
2.191 The acceleration of a point moving along a straight line is $a = -cv^3$, where c is a constant. If the velocity of the point is v_0, what distance does it move before its velocity decreases to $v_0/2$?

2.192 Water leaves the nozzle at 20° above the horizontal and strikes the wall at the point indicated. What was the velocity of the water as it left the nozzle?

Strategy: Determine the motion of the water by treating each particle of water as a projectile.

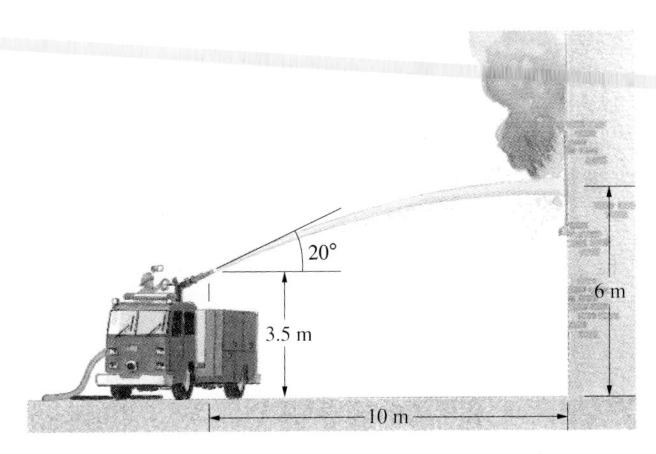

P2.192

2.193 In practice, a quarterback throws the football with velocity v_0 at 45° above the horizontal. At the same instant, the receiver standing 6 m in front of him starts running straight downfield at 3 m/s and catches the ball. Assume that the ball is thrown and caught at the same height above the ground. What is the velocity v_0?

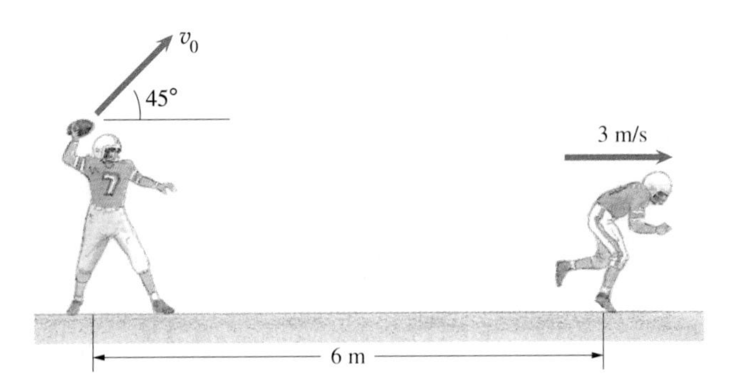

P2.193

2.194 The constant velocity $v = 2\,\text{m/s}$. What are the magnitudes of the velocity and acceleration of point P when $x = 0.25\,\text{m}$?

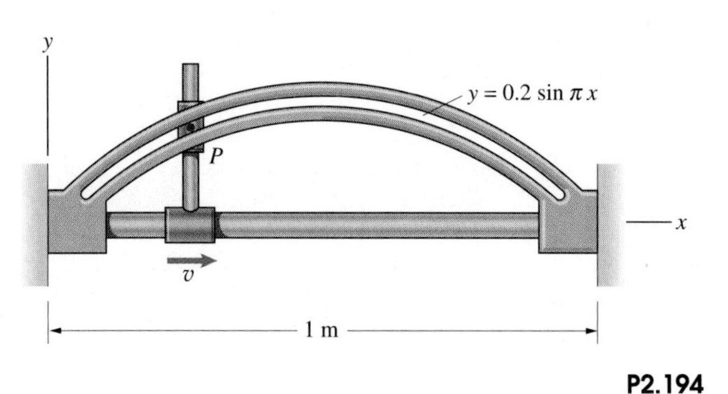

P2.194

2.195 In Problem 2.194, what is the acceleration of point P in terms of normal and tangential components when $x = 0.25\,\text{m}$? What is the instantaneous radius of curvature of the path?

2.196 In Problem 2.194, what is the acceleration of point P in terms of radial and transverse components (polar coordinates) when $x = 0.25\,\text{m}$?

2.197 A point P moves along the spiral path $r = (0.1)\theta\,\text{m}$, where θ is in radians. The angular position $\theta = 2t\,\text{rad}$, where t is in seconds, and $r = 0$ at $t = 0$. Determine the magnitudes of the velocity and acceleration of P at $t = 1\,\text{s}$.

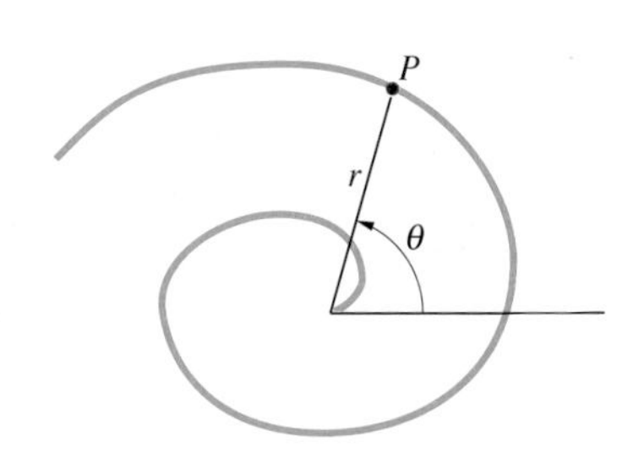

P2.197

2.198 A manned vehicle (M) attempts to rendezvous with a satellite (S) to repair it. (They are not shown to scale.) The magnitude of the satellite's velocity is $|\mathbf{v}_S| = 6\,\text{km/s}$, and a sighting determines that the angle $\beta = 40^\circ$. If you assume that their velocities remain constant and that the vehicles move along the straight lines shown, what should be the magnitude of $\mathbf{v}_M$ to achieve rendezvous?

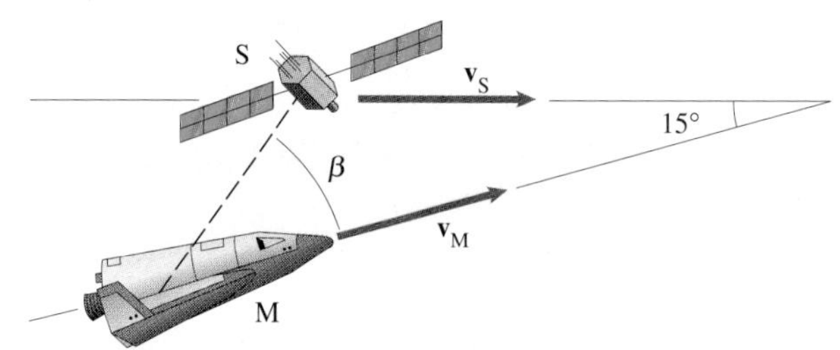

P2.198

2.199 In Problem 2.198, what is the magnitude of the velocity of the manned vehicle relative to the spacecraft once the magnitude of $\mathbf{v}_M$ has been adjusted to achieve rendezvous?

2.200 The three 1 m bars rotate in the x-y plane with constant angular velocity ω. If $\omega = 20\,\text{rad/s}$, what is the magnitude of the velocity of point C relative to point A in terms of the fixed coordinate system?

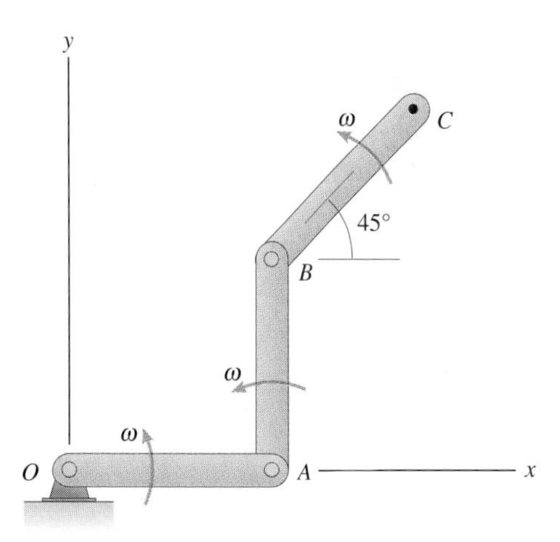

P2.200

2.201 In Problem 2.200, what is the velocity of point C relative to the fixed point O?

2.202 In Problem 2.200, accelerometers mounted at C indicate that the acceleration of point C relative to the fixed point O is $\mathbf{a}_C = (-1500\,\mathbf{i} - 1500\,\mathbf{j})\,\text{m/s}^2$. What is the angular velocity ω? Can you determine from this information whether ω is counterclockwise or clockwise?

A racing motorcycle can accelerate from rest to 60 mi/hr (96.6 km/hr) in 3 s. Its acceleration is related by Newton's second law to the combined mass of the motorcycle and rider and the external forces acting on them. In this chapter we will use free-body diagrams and Newton's second law to determine the motions that result from the forces acting on objects.

Chapter 3

Force, Mass and Acceleration

UNTIL now we have analysed motions of objects without considering the forces causing them. Here we relate cause and effect: by drawing the free-body diagram of an object to identify the forces acting on it, we can use Newton's second law to determine its acceleration. Once the acceleration is known, we can determine its velocity and position by the methods developed in Chapter 2.

3.1 *Newton's Second Law*

Newton stated that the force on a particle is equal to the rate of change of its **linear momentum**, which is the product of its mass and velocity:

$$\mathbf{f} = \frac{d}{dt}(m\mathbf{v})$$

If the particle's mass is constant, the force equals the product of its mass and acceleration:

$$\boxed{\mathbf{f} = m\frac{d\mathbf{v}}{dt} = m\mathbf{a}} \tag{3.1}$$

We pointed out in Chapter 1 that the second law gives precise meanings to the terms **force** and **mass**. Once a unit of mass is chosen, a unit of force is defined to be the force necessary to give one unit of mass an acceleration of unit magnitude. For example, the unit of force in SI units, the newton, is the force necessary to give a mass of one kilogram an acceleration of one metre per second squared. In principle, the second law then gives the value of any force and the mass of any object. By subjecting a one-kilogram mass to an arbitrary force and measuring the acceleration, we can solve the second law for the direction of the force and its magnitude in newtons. By subjecting an arbitrary mass to a one-newton force and measuring the acceleration, we can solve the second law for the value of the mass in kilograms.

If you know a particle's mass and the force acting on it, you can use Newton's second law to determine its acceleration. In Chapter 2 you learned how to determine the velocity, position and path, or trajectory, of a point when you know its acceleration. Therefore, *with the second law you can determine a particle's motion when you know the force acting on it.*

3.2 *Inertial Reference Frames*

When we discussed the motion of a point in Chapter 2, we specified its position, velocity and acceleration relative to an arbitrary reference point O. But Newton's second law cannot be expressed in terms of just any reference point. Suppose that no force acts on a particle, and we measure the particle's motion relative to a particular reference point O and determine that its acceleration is zero. In terms of this reference point, Newton's second law agrees with our observation. But if we then measure the particle's motion relative to a reference point O' that is accelerating relative to O, we determine that its acceleration is not zero. Relative to O', Newton's second law, at least in the form given by Equation (3.1), does not predict the correct result. Equation (3.1) also will not predict the correct result if we use a coordinate system, or reference frame, that is rotating.

Newton stated that the second law should be expressed in terms of a non-rotating reference frame that does not accelerate relative to the 'fixed stars'.

Even if the stars were fixed, that would not be practical advice because virtually every convenient reference frame accelerates, rotates, or both due to the earth's motion. Newton's second law *can* be applied rigorously using reference frames that accelerate and rotate, by properly accounting for the acceleration and rotation. We explain how in Chapter 6. But for now, we need to indicate when you can apply Newton's second law and when you cannot.

Fortunately, in nearly all 'down to earth' situations, you can express Newton's second law in the form given by Equation (3.1) in terms of a reference frame that is fixed relative to the earth and obtain sufficiently accurate answers. For example, if you throw a piece of chalk across a room, you can use a coordinate system that is fixed relative to the room to predict the chalk's motion. While the chalk is in motion, the earth rotates, and therefore the coordinate system rotates. But *because the chalk's flight is brief*, the effect on your prediction is very small. (The earth rotates slowly – its angular velocity is one-half that of a clock's hour hand.) You can also obtain accurate answers in most situations using a reference frame that translates with constant velocity relative to the earth. For example, if you and a friend play tennis on the deck of a cruise ship moving with constant velocity, you can apply Equation (3.1) in terms of a coordinate system fixed relative to the ship to analyse the ball's motion. But you cannot if the ship is turning or changing its speed. In situations that are not 'down to earth', such as the motions of earth satellites and spacecraft near the earth, you can apply Equation (3.1) by using a non-rotating coordinate system with its origin at the centre of the earth.

If a reference frame can be used to apply Equation (3.1), we say that it is **inertial**. We discuss inertial reference frames in greater detail in Chapter 6. For now, you should assume that examples and problems are expressed in terms of inertial reference frames.

3.3 *Equation of Motion for the Centre of Mass*

Newton's second law is postulated for a particle, or small element of matter, but an equation of precisely the same form describes the motion of the **centre of mass** of an **arbitrary** object. We can show that the total external force on an arbitrary object is equal to the product of its mass and the acceleration of its centre of mass.

To do so, we *conceptually* divide an arbitrary object into N particles. Let m_i be the mass of the ith particle, and let $\mathbf{r}_i$ be its position vector (Figure 3.1(a)). The object's mass m is the sum of the masses of the particles,

$$m = \sum_i m_i$$

where the summation sign with subscript i means 'the sum over i from 1 to N'. The position of the object's centre of mass is

$$\mathbf{r} = \frac{\sum_i m_i \mathbf{r}_i}{m}$$

By taking two time derivatives of this expression, we obtain

$$\sum_i m_i \frac{d^2\mathbf{r}_i}{dt^2} = m\frac{d^2\mathbf{r}}{dt^2} = m\mathbf{a} \tag{3.2}$$

where $\mathbf{a}$ is the acceleration of the object's centre of mass.

The ith particle of the object may be subjected to forces by the other particles of the object. Let $\mathbf{f}_{ij}$ be the force exerted on the ith particle by the jth particle. Newton's third law states that the ith particle exerts a force on the jth particle of equal magnitude and opposite direction: $\mathbf{f}_{ji} = -\mathbf{f}_{ij}$. Denoting the external force on the ith particle (that is, the total force exerted on the ith particle by objects other than the object we are considering) by $\mathbf{f}_i^{\mathrm{E}}$, Newton's second law for the ith particle is (Figure 3.1(b))

$$\sum_j \mathbf{f}_{ij} + \mathbf{f}_i^{\mathrm{E}} = m_i \frac{d^2\mathbf{r}_i}{dt^2}$$

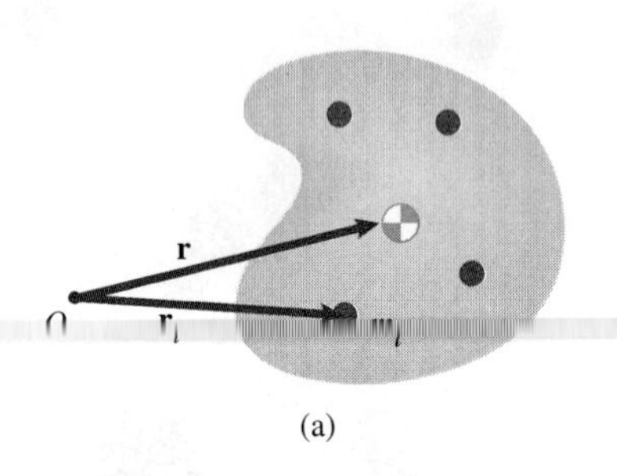

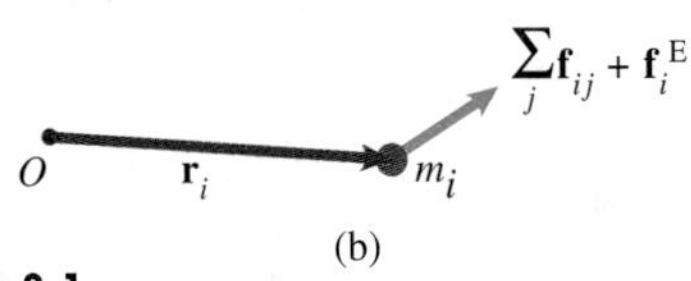

Figure 3.1
(a) Dividing an object into particles. The vector $\mathbf{r}_i$ is the position vector of the ith particle and $\mathbf{r}$ is the position vector of the object's centre of mass.
(b) Forces on the ith particle.

We can write this equation for each particle of the object. Summing the resulting equations from $i = 1$ to N, we obtain

$$\sum_i \sum_j \mathbf{f}_{ij} + \sum_i \mathbf{f}_i^{\mathrm{E}} = m\mathbf{a} \tag{3.3}$$

where we have used Equation (3.2). The first term on the left side, the sum of the internal forces on the object, is zero due to Newton's third law:

$$\sum_i \sum_j \mathbf{f}_{ij} = \mathbf{f}_{12} + \mathbf{f}_{21} + \mathbf{f}_{13} + \mathbf{f}_{31} + \cdots = \mathbf{0}$$

The second term on the left side of Equation (3.3) is the sum of the external forces on the object. Denoting it by $\Sigma\mathbf{F}$, we conclude that the sum of the external forces equals the product of the mass and the acceleration of the centre of mass:

$$\boxed{\Sigma\mathbf{F} = m\mathbf{a}} \tag{3.4}$$

Because this equation is identical in form to Newton's postulate for a particle, for convenience we also refer to it as Newton's second law. Notice that we made no assumptions restricting the nature of the 'object' or its state of motion in obtaining this result. The sum of the external forces on any object or collection of objects – solid, liquid or gas – equals the product of the total mass and the acceleration of the centre of mass. For example, suppose that the space shuttle is in orbit and has fuel remaining in its tanks. If its engines are turned on, the fuel sloshes in a complicated manner, affecting the shuttle's motion due to internal forces between the fuel and the shuttle. Nevertheless, we can use Equation (3.4) to determine the *exact* acceleration of the centre of mass of the shuttle, including the fuel it contains, and thereby determine the velocity, position and trajectory of the centre of mass.

3.4 Applications

To apply Newton's second law in a particular situation, you must choose a coordinate system. Often you will find that you can resolve the forces into components most conveniently in terms of a particular coordinate system, or your choice may be determined by the object's path. In the following sections we use different types of coordinate systems to determine the motions of objects.

Cartesian Coordinates and Straight-Line Motion

Expressing the total force and the acceleration in Newton's second law in terms of their components in a cartesian coordinate system,

$$(\Sigma F_x \mathbf{i} + \Sigma F_y \mathbf{j} + \Sigma F_z \mathbf{k}) = m(a_x \mathbf{i} + a_y \mathbf{j} + a_z \mathbf{k})$$

we obtain three scalar equations of motion:

$$\boxed{\Sigma F_x = ma_x \quad \Sigma F_y = ma_y \quad \Sigma F_z = ma_z} \qquad (3.5)$$

The total force in each coordinate direction equals the product of the mass and the component of the acceleration in that direction (Figure 3.2(a)).

If the motion is confined to the x-y plane, $a_z = 0$, so the sum of the forces in the z direction is zero. Thus when the motion of an object is confined to a fixed plane, the component of the total force normal to the plane equals zero. For straight-line motion along the x axis (Figure 3.2(b)), Equations (3.5) are

$$\Sigma F_x = ma_x \quad \Sigma F_y = 0 \quad \Sigma F_z = 0$$

In straight-line motion, the components of the total force perpendicular to the line equal zero and the component of the total force tangent to the line equals the product of the mass and the acceleration of the object along the line.

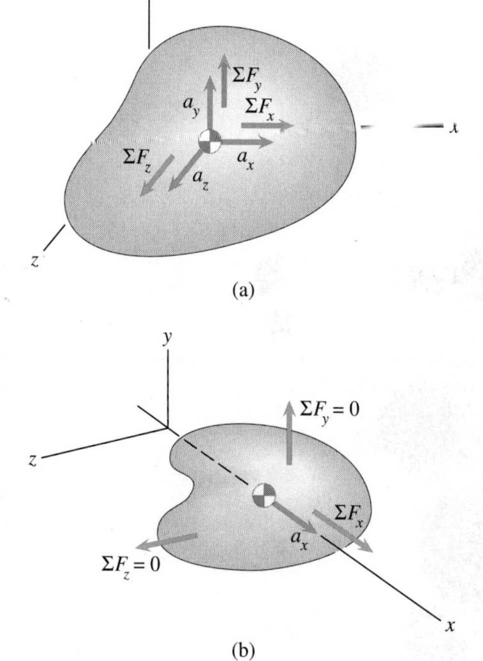

Figure 3.2
(a) Force and acceleration components in cartesian coordinates.
(b) An object in straight-line motion along the x axis.

The following examples demonstrate the use of Newton's second law to analyse motions of objects. By drawing the free-body diagram of an object, you can identify the external forces acting on it and use Newton's second law to determine its acceleration. Conversely, if you know the motion of an object, you can use Newton's second law to determine the external forces acting on it. In particular, if you know that an object's acceleration in a specific direction is zero, the sum of the external forces in that direction must also equal zero.

Example 3.1

The aeroplane in Figure 3.3 touches down on the aircraft carrier with a horizontal velocity of 50 m/s relative to the carrier. The horizontal component of the force exerted on the arresting gear is of magnitude $10\,000v$ N, where v is the plane's velocity in metres per second. The plane's mass is 6500 kg.
(a) What maximum horizontal force does the arresting gear exert on the plane?
(b) If other horizontal forces can be neglected, what distance does the plane travel before coming to rest?

Figure 3.3

STRATEGY

(a) Since the plane begins to decelerate when it contacts the arresting gear, the maximum force occurs at first contact when $v = 50\,\text{m/s}$.
(b) The horizontal force exerted by the arresting gear equals the product of the plane's mass and its acceleration. Once we know the acceleration, we can integrate to determine the distance required for the plane to come to rest.

SOLUTION

(a) The magnitude of the maximum force is

$$10\,000v = (10\,000)(50) = 500\,000\,\text{N}$$

(b) Using the coordinate system shown in Figure (a), we obtain the equation of motion:

$$\Sigma F_x = ma_x:$$
$$-10\,000v_x = ma_x$$

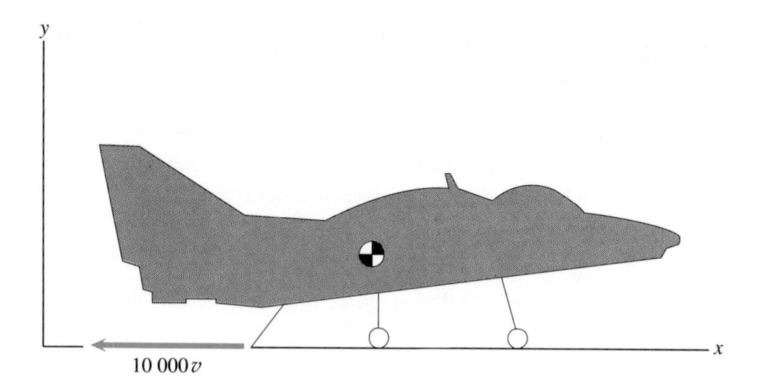

(a) The x axis is aligned with the plane's horizontal motion.

The aeroplane's acceleration is a function of its velocity. We use the chain rule to express the acceleration in terms of a derivative with respect to x:

$$ma_x = m\frac{dv_x}{dt} = m\frac{dv_x}{dx}\frac{dx}{dt} = m\frac{dv_x}{dx}v_x = 10\,000v_x$$

Now we integrate, defining $x = 0$ to be the position at which the plane contacts the arresting gear:

$$\int_{50}^{0} m dv_x = -\int_{0}^{x} 10\,000\,dx$$

Solving for x, we obtain

$$x = \frac{50m}{10\,000} = \frac{(50)(6500)}{10\,000} = 32.5 \text{ m}$$

DISCUSSION

As we demonstrate in this example, once you have used Newton's second law to determine the acceleration, you can apply the methods developed in Chapter 2 to determine the position and velocity.

Example 3.2

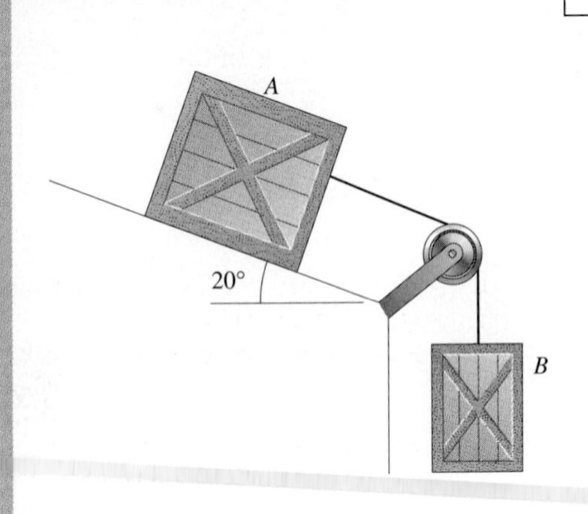

Figure 3.4

The two crates in Figure 3.4 are released from rest. Their masses are $m_A = 40\,\text{kg}$ and $m_B = 30\,\text{kg}$, and the coefficients of friction between crate A and the inclined surface are $\mu_s = 0.2$, $\mu_k = 0.15$. What is their acceleration?

STRATEGY

We must first determine whether A slips. We will assume the crates remain stationary and see whether the friction force necessary for equilibrium exceeds the maximum friction force. If slip occurs, we can determine the resulting acceleration by drawing free-body diagrams of the crates and applying Newton's second law to them individually.

SOLUTION

We draw the free-body diagram of crate A in Figure (a). If we assume it does not slip, the equilibrium equations apply,

$$\Sigma F_x = T + m_A g \sin 20^\circ - f = 0$$
$$\Sigma F_y = N - m_A g \cos 20^\circ = 0$$

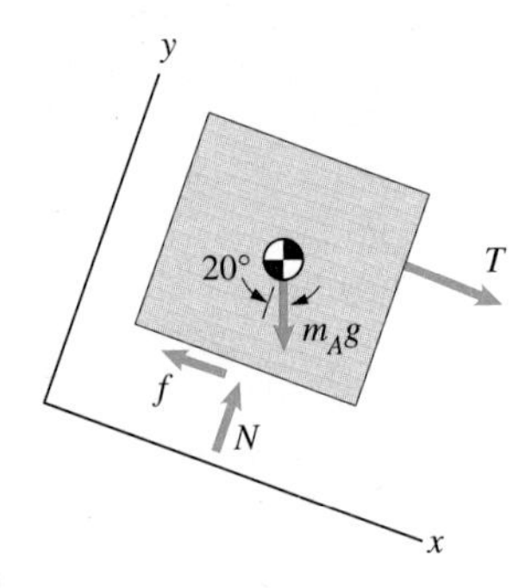

(a) Free-body diagram of crate A.

and the tension T equals the weight of crate B. Therefore the friction force necessary for equilibrium is

$$f = m_B g + m_A g \sin 20^\circ = (30 + 40 \sin 20^\circ)(9.81) = 428.5\,\text{N}$$

The normal force $N = m_A g \cos 20^\circ$, so the maximum friction force the surface will support is

$$f_{\text{max}} = \mu_s N = (0.2)[(40)(9.81)\cos 20^\circ] = 73.7\,\text{N}$$

Crate A will therefore slip, and the friction force is $f = \mu_k N$. Applying Newton's second law,

$$\Sigma F_x = T + m_A g \sin 20^\circ - \mu_k N = m_A a_x$$
$$\Sigma F_y = N - m_A g \cos 20^\circ = 0$$

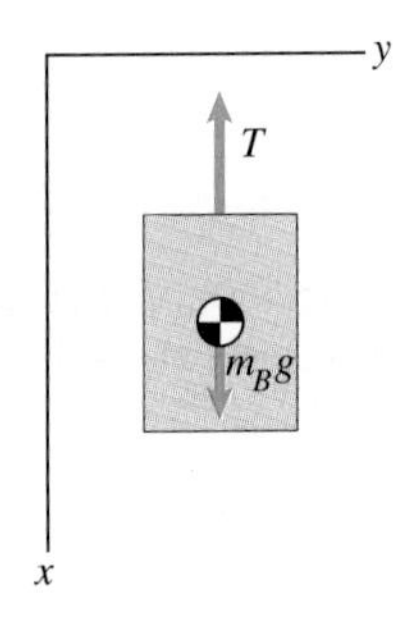

(b) Free-body diagram of crate B.

Crate A has no acceleration normal to the surface, so the sum of the forces in the y direction equals zero. In this case *we do not know* the tension T because crate B is not in equilibrium. From the free-body diagram of crate B (Figure (b)) we obtain the equation of motion

$$\Sigma F_x = m_B g - T = m_B a_x$$

(Notice that in terms of the two coordinate systems we use, the two crates have the same acceleration a_x.) By applying Newton's second law to both crates, we have obtained three equations in terms of the unknowns T, N and a_x. Solving for a_x, we obtain $a_x = 5.33\,\text{m/s}^2$.

DISCUSSION

Notice that we assumed the tension in the cable to be the same on each side of the pulley (Figure (c)). In fact, the tensions must be different because a moment is necessary to cause angular acceleration of the pulley. For now, our only recourse is to assume that the pulley is light enough that the moment necessary to accelerate it is negligible. In Chapter 7, we include the analysis of the angular motion of the pulley in problems of this type and obtain more realistic solutions.

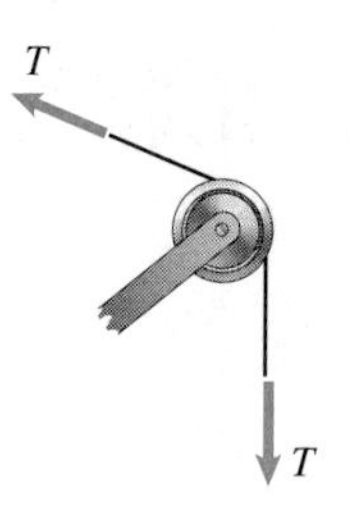

(c) The tension is assumed to be the same on both sides of the pulley.

Problems

3.1 The total external force on a 10 kg object is (90 **i** − 60 **j** + 20 **k**) N. What is the magnitude of its acceleration relative to an inertial reference frame?

3.2 The total external force acting on a 20 kg object is (10 **i** + 20 **j**) N. When $t = 0$, its position vector relative to an inertial reference frame is $\mathbf{r} = 0$ and its velocity is $\mathbf{v} = (2\,\mathbf{i} - \mathbf{j})$ m/s. Determine the position and velocity of the object when $t = 2$ s.

3.3 The total external force on an object is $(10t\,\mathbf{i} + 60\,\mathbf{j})$ N. When $t = 0$, its position vector relative to an inertial reference frame is $\mathbf{r} = 0$ and its velocity is $\mathbf{v} = 0.2\,\mathbf{j}$ m/s. When $t = 5$ s, the magnitude of its position vector is measured and determined to be 8.75 m. What is the mass of the object?

3.4 The position of a 10 kg object relative to an inertial reference frame is $\mathbf{r} = (\frac{1}{3}t^3\,\mathbf{i} + 4t\,\mathbf{j} - 30t^2\,\mathbf{k})$ m. What are the components of the total external force acting on the object at $t = 10$ s?

3.5 If the 7000 kg helicopter starts from rest and its rotor exerts a constant 90 kN vertical force, how high does it rise in 2 s?

P3.5

3.6 The 1 kg collar A is initially at rest in the position shown on the smooth horizontal bar. At $t = 0$, a force $\mathbf{F} = (\frac{1}{20}t^2\,\mathbf{i} + \frac{1}{10}t\,\mathbf{j} - \frac{1}{30}t^3\,\mathbf{k})$ N is applied to the collar, causing it to slide along the bar. What is the velocity of the collar when it reaches the right end of the bar?

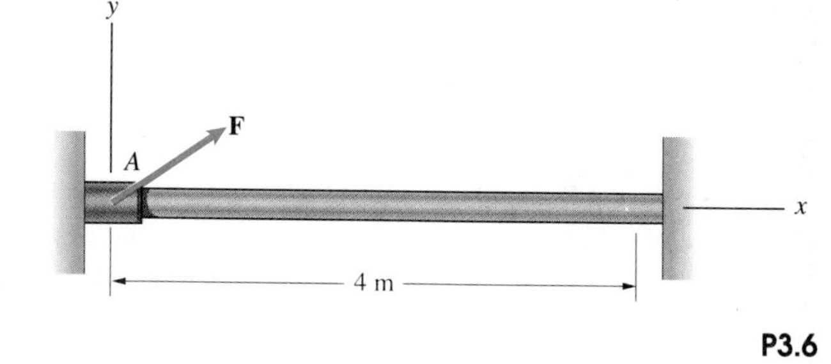

P3.6

3.7 Suppose you are in an elevator and standing on a set of scales. When the elevator is stationary, the scales read your weight, W.

(a) What is the acceleration of the elevator if the scales read 1.01 W?

(b) What is its acceleration if the scales read 0.99W?

Strategy: Draw your free-body diagram. The upward force exerted on you by the scales equals the force you exert on the scales.

3.8 A cart partially filled with water is initially stationary (Figure (a)). The total mass of the cart and water is m. The cart is subjected to a time-dependent force (Figure (b)). If the horizontal forces exerted on the wheels by the floor are negligible and no water sloshes out, what is the x coordinate of the centre of the cart after the motion of the water has subsided?

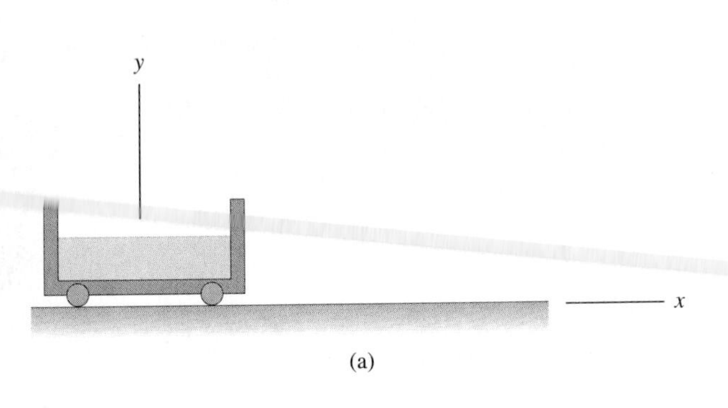

(a)

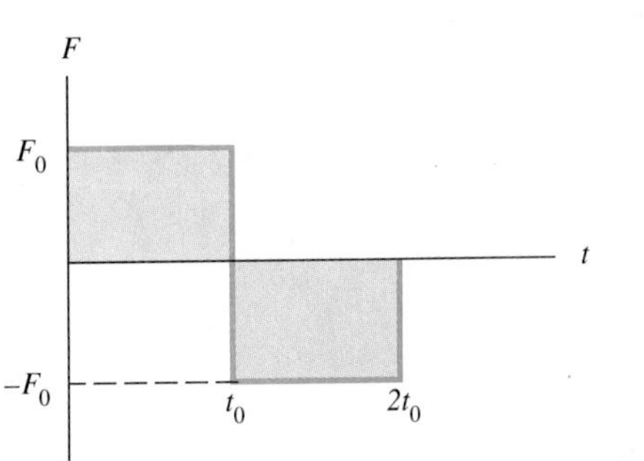

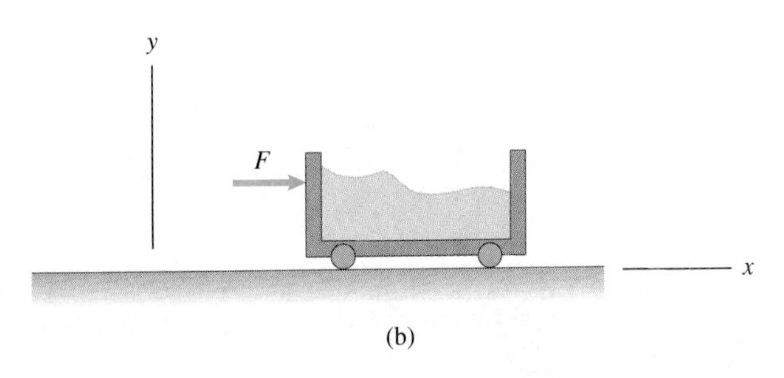

(b)

P3.8

3.9 The rocket travels straight up at low altitude. Its weight at the present time is 890 kN and the thrust of its engine is 1200 kN. An on-board accelerometer indicates that its acceleration is $3\,m/s^2$ upwards. What is the magnitude of the aerodynamic drag force on the rocket?

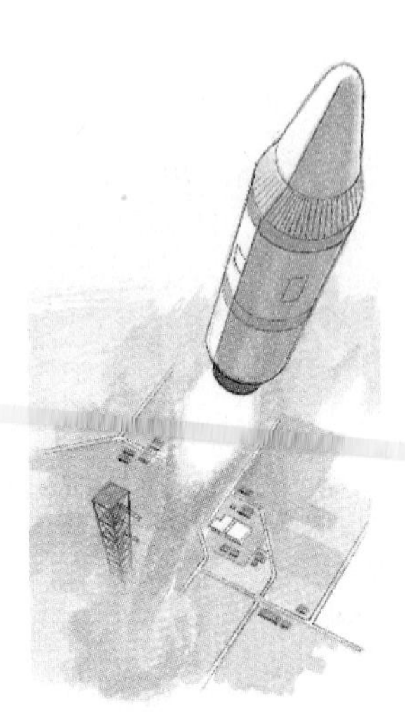

P3.9

3.10 The aeroplane weights 90 kN. At the instant shown, the pilot increases the thrust T of the engine by 22.5 kN. The horizontal component of the aeroplane's acceleration the instant before the thrust is increased is $6\,m/s^2$. What is the horizontal component of the aeroplane's acceleration the instant after the thrust is increased?

P3.10

3.11 The combined weight of the motorcycle and rider is 1600 N. The coefficient of kinetic friction between the motorcycle's tyres and the road is $\mu_k = 0.8$. If he spins the rear (drive) wheel, the normal force between the rear wheel and the road is 1100 N, and the horizontal force exerted on the front wheel by the road is negligible, what is the resulting horizontal acceleration?

P3.11

3.12 The bucket B weighs 1800 N and the acceleration of its centre of mass is $\mathbf{a} = (-10\,\mathbf{i} - 3\,\mathbf{j})\,\text{m/s}^2$. Determine the x and y components of the total force exerted on the bucket by its supports.

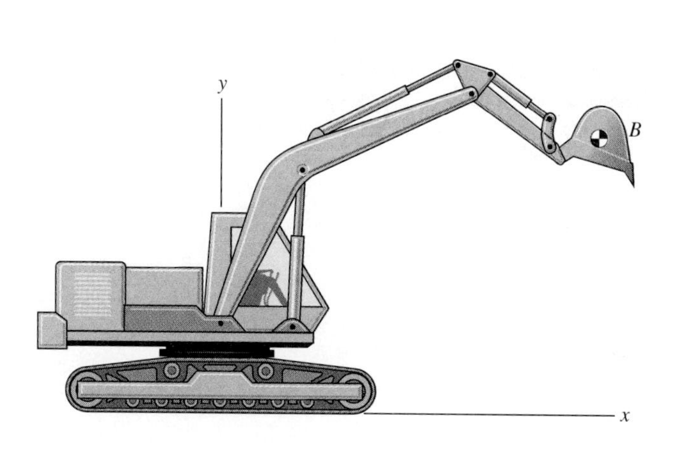

P3.12

3.13 During a test flight in which a 9000 kg helicopter starts from rest at $t = 0$, the acceleration of its centre of mass from $t = 0$ to $t = 10$ s is

$$\mathbf{a} = [(0.6t)\mathbf{i} + (1.8 - 0.36t)\,\mathbf{j}]\ \text{m/s}^2$$

What is the magnitude of the total external force on the helicopter (including its weight) at $t = 6$ s?

3.14 The engineers conducting the test described in Problem 3.13 want to express the total force on the helicopter at $t = 6$ s in terms of three forces: the weight W, a component T tangent to the path, and a component L normal to the path. What are the values of W, T and L?

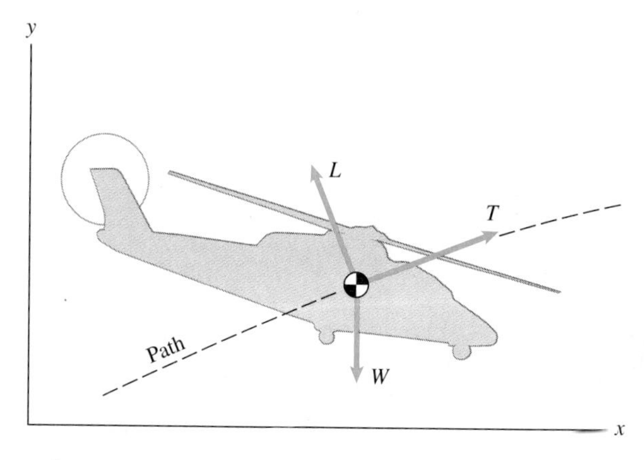

P3.14

3.15 The robot manipulator is programmed so that $x = (100 + 25t^2)$ mm, $y = 6x^2$ mm, $z = 0$ during the interval of time from $t = 0$ to $t = 4$ s. What are the x and y components of the total force exerted by the jaws of the manipulator on the 5 kg widget A at $t = 2$ s?

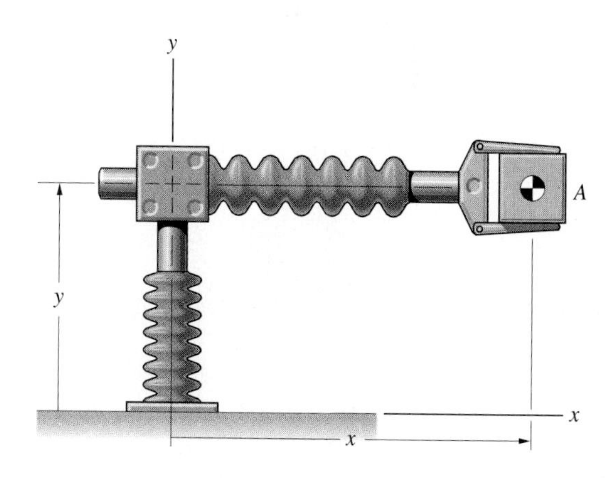

P3.15

3.16 The robot manipulator in Problem 3.15 is stationary at $t = 0$ and is programmed so that $a_x = (50 - 0.4v_x)\,\text{mm/s}^2$, $a_y = (25 - 0.2v_y)\,\text{mm/s}^2$, $a_z = 0$ during the interval of time from $t = 0$ to $t = 4$ s. What are the x and y components of the total force exerted by the jaws of the manipulator on the 5 kg widget A at $t = 2$ s?

3.17 In the sport of curling, the object is to slide a 'stone' weighing 200 N onto the centre of a target located 28 m from the point of release. If $\mu_k = 0.01$ and the stone is thrown directly towards the target, what initial velocity would result in a perfect shot?

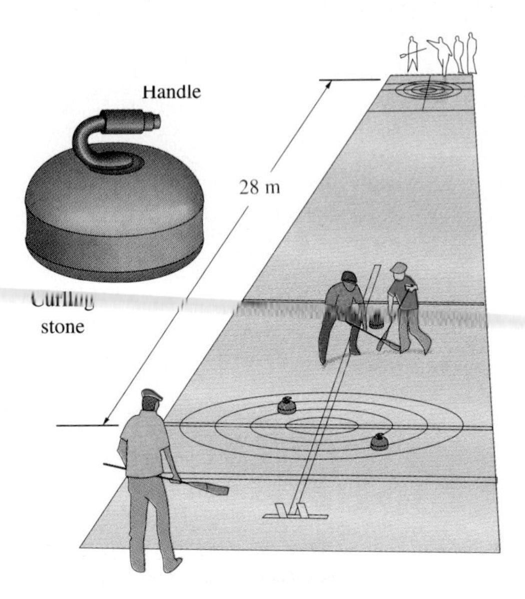

P3.17

3.18 The two weights are released from rest. How far does the 50 N weight fall in one-half second?

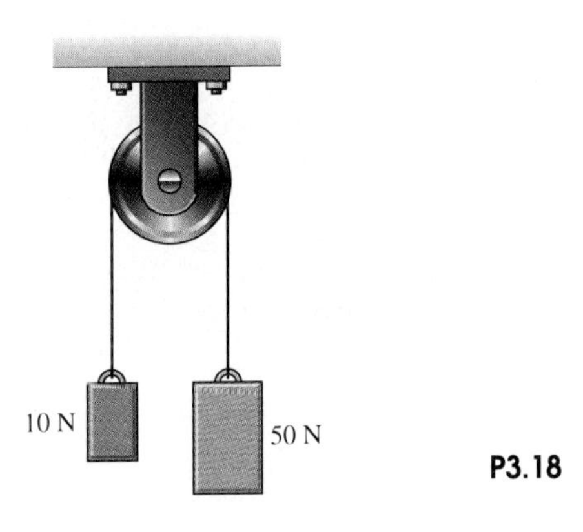

P3.18

3.19 In Example 3.2, what is the ratio of the tension in the cable to the weight of crate B after the crates are released from rest?

3.20 Each box weighs 200 N and friction can be neglected. If the boxes start from rest at $t = 0$, determine the magnitude of their velocity and the distance they have moved from their initial position at $t = 1$ s.

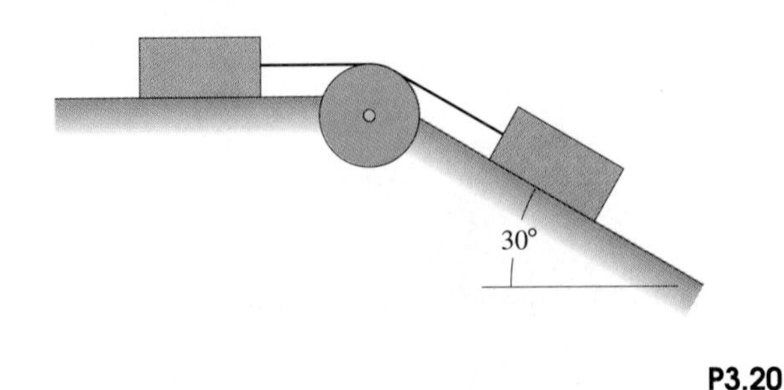

P3.20

3.21 In Problem 3.20, determine the magnitude of the velocity of the boxes and the distance they have moved from their initial position at $t = 1$ s if the coefficient of kinetic friction between the boxes and the surface is $\mu_k = 0.15$.

3.22 The masses $m_A = 15$ kg, $m_B = 30$ kg, and the coefficients of friction between all of the surfaces are $\mu_s = 0.4$, $\mu_k = 0.35$. What is the largest force F that can be applied without causing A to slip relative to B? What is the resulting acceleration?

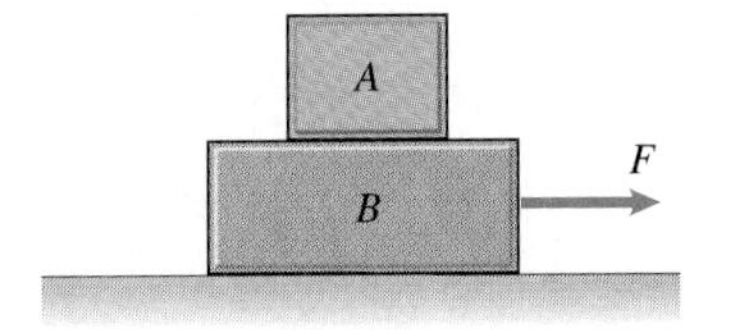

P3.22

3.23 The crane's trolley at A moves to the right with constant acceleration, and the 800 kg load moves without swinging.
(a) What is the acceleration of the trolley and load?
(b) What is the sum of the tensions in the parallel cables supporting the load?

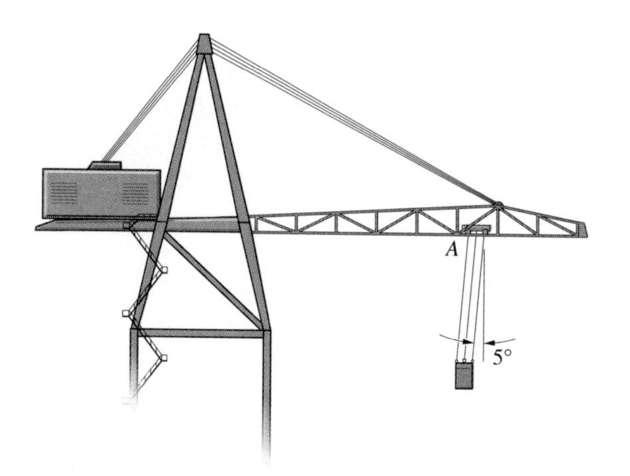

P3.23

3.24 The 50 kg crate is initially stationary. The coefficients of friction between the crate and the inclined surface are $\mu_s = 0.2$, $\mu_k = 0.16$. Determine how far the crate moves from its initial position in 2 s if the horizontal force $f = 500$ N.

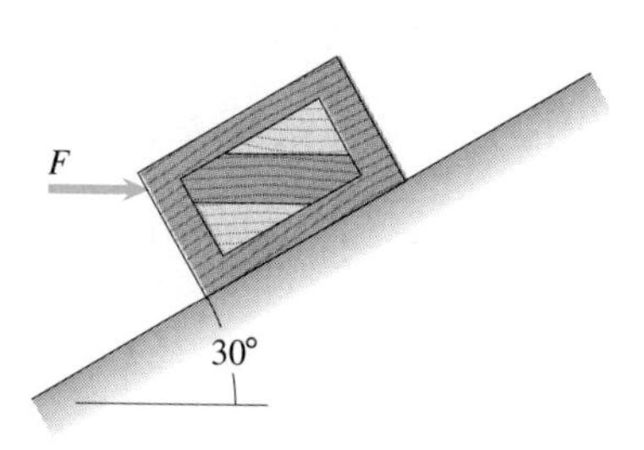

P3.24

3.25 In Problem 3.24, determine how far the crate moves from its initial position in 2 s if the horizontal force $F = 150$ N.

3.26 The crate has a mass of 120 kg and the coefficients of friction between it and the sloping dock are $\mu_s = 0.6$, $\mu_k = 0.5$.
(a) What tension must the winch exert on the cable to start the stationary crate sliding up the dock?
(b) If the tension is maintained at the value determined in part (a), what is the magnitude of the crate's velocity when it has moved 10 m up the dock?

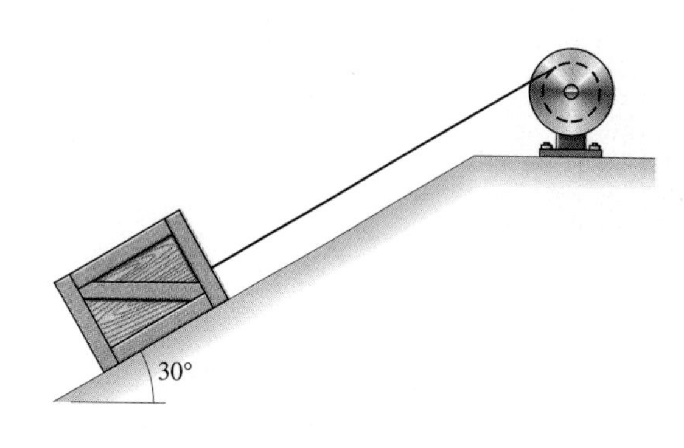

P3.26

3.27 The utility vehicle is moving forwards at 3 m/s. The coefficients of friction between its load A and the bed of the vehicle are $\mu_s = 0.5$, $\mu_k = 0.45$. If $\alpha = 0$, determine the shortest distance in which the vehicle can be brought to a stop without causing the load to slide on the bed.

P3.27

3.28 In Problem 3.27, determine the shortest distance if the angle α is (a) 15°; (b) −15°.

3.29 In an assembly-line process, the 20 kg package A starts from rest and slides down the smooth ramp. Suppose that you want to design the hydraulic device B to exert a constant force of magnitude F on the package and bring it to rest in a distance of 100 mm. What is the required force F?

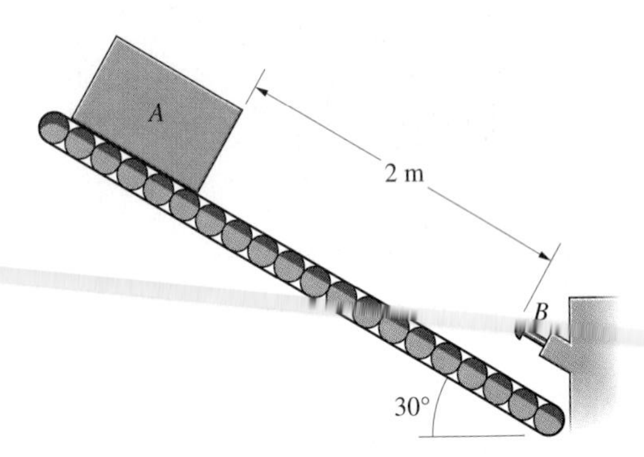

P3.29

3.30 The force exerted on the 10 kg mass by the linear spring is $F = -ks$, where k is the spring constant and s is the displacement of the mass from its position when the spring is unstretched. The value of k is 50 N/m. The mass is released from rest in the position $s = 1$ m.
(a) What is the acceleration of the mass at the instant it is released?
(b) What is the velocity of the mass when it reaches the position $s = 0$?

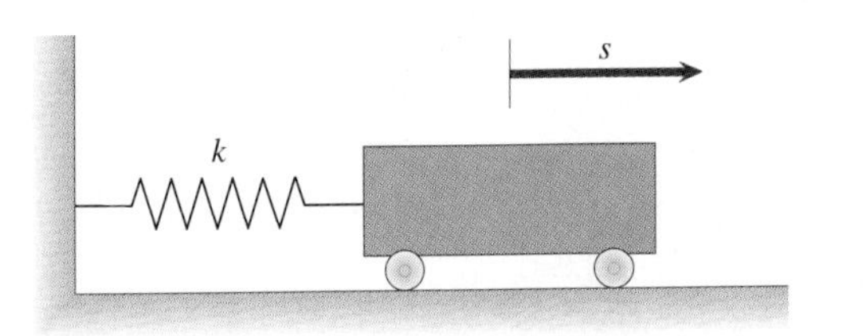

P3.30

3.31 A sky diver and his parachute weight 900 N. He is falling vertically at 30 m/s when his parachute opens. With the parachute open, the magnitude of the drag force is $0.5v^2$.
(a) What is the magnitude of his acceleration at the instant the parachute opens?
(b) What is the magnitude of his velocity when he has descended 6 m from the point where his parachute opens?

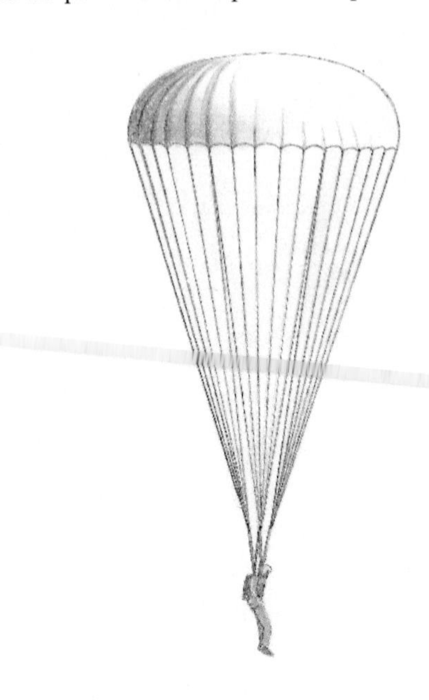

P3.31

3.32 A 100 kg 'bungee jumper' jumps from a bridge 40 m above a river. The bungee cord has an unstretched length of 18 m and has a spring constant $k = 200$ N/m.
(a) How far above the river is he when the cord brings him to a stop?
(b) What maximum force does the cord exert on him?

P3.32

3.33 In Problem 3.32, what maximum velocity does the jumper reach, and at what height above the river does it occur?

3.34 In a cathode-ray tube, an electron (mass $= 9.11 \times 10^{-31}$ kg) is projected at O with velocity $[\mathbf{v} + (2.2 \times 10^7)\mathbf{i}]$ m/s. While it is between the charged plates, the electric field generated by the plates subjects it to a force $\mathbf{F} = -eE\,\mathbf{j}$, where the charge of the electron $e = 1.6 \times 10^{-19}$ C (coulombs) and the electric field strength $E = 15$ kN/C. External forces on the electron are negligible when it is not between the plates. Where does it strike the screen?

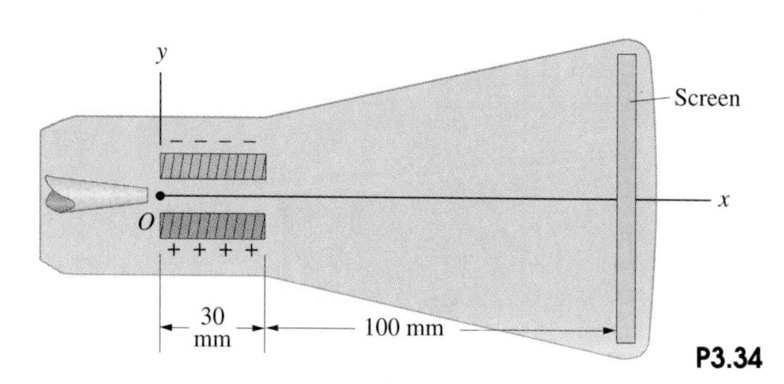

P3.34

3.35 In Problem 3.34, determine where the electron strikes the screen if the electric field strength is $E = 15\sin(\omega t)$ kN/C, where the circular frequency $\omega = 2 \times 10^9\ \mathrm{s}^{-1}$.

3.36 An astronaut wants to travel from a space station to a satellite that needs repair. He departs the space station at O. A spring-loaded launching device gives his manoeuvring unit an initial velocity of 1 m/s (relative to the space station) in the y direction. At that instant, the position of the satellite is $x = 70$ m, $y = 50$ m, $z = 0$, and it is drifting at 2 m/s (relative to the station) in the x direction. The astronaut intercepts the satellite by applying a constant thrust parallel to the x axis. The total mass of the astronaut and his manoeuvring unit is 300 kg.

(a) How long does it take him to reach the satellite?

(b) What is the magnitude of the thrust he must apply to make the intercept?

(c) What is his velocity *relative to the satellite* when he reaches it?

P3.36

3.37 What is the acceleration of the 8 kg collar A relative to the smooth bar?

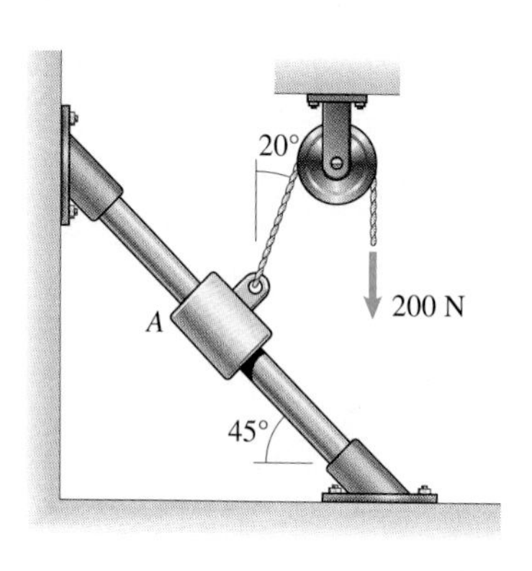

P3.37

3.38 In Problem 3.37, determine the acceleration of the collar A relative to the bar if the coefficient of kinetic friction between the collar and the bar is $\mu_k = 0.1$.

3.39 The acceleration of the 10 kg collar A is $(2\,\mathbf{i} + 3\,\mathbf{j} - 3\,\mathbf{k})$ m/s^2. What is the force F?

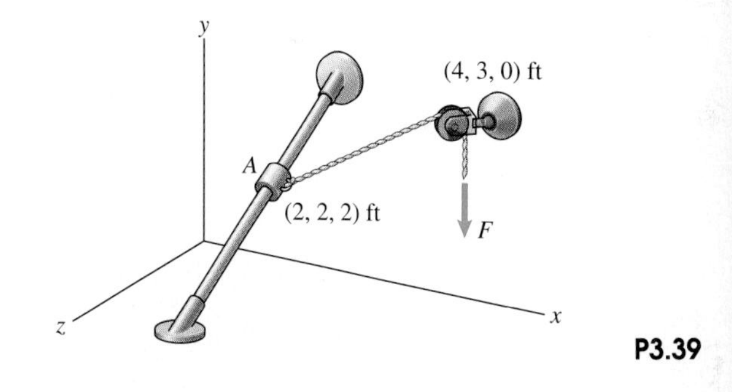

P3.39

3.40 In Problem 3.39, determine the force F if the coefficient of kinetic friction between the collar and the bar is $\mu_k = 0.1$.

3.41 The crate is drawn across the floor by a winch that retracts the cable at a constant rate of 0.2 m/s. The crate's mass is 120 kg, and the coefficient of kinetic friction between the crate and the floor is $\mu_k = 0.24$.
(a) At the instant shown, what is the tension in the cable?
(b) Obtain a 'quasi-static' solution for the tension in the cable by ignoring the crate's acceleration, and compare it with your result in part (a).

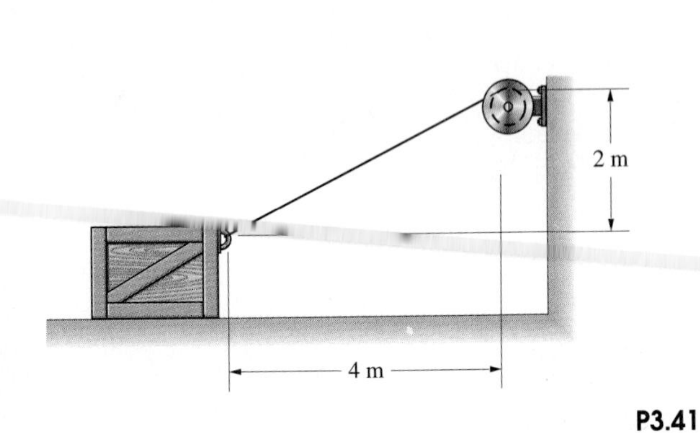

P3.41

3.42 If $y = 100$ mm, $dy/dt = 600$ mm/s, and $d^2y/dt^2 = -200$ mm/s^2, what horizontal force is exerted on the 0.4 kg slider A by the smooth circular slot?

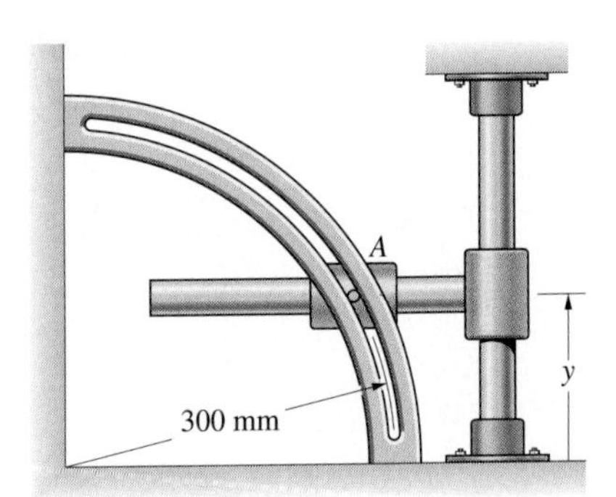

P3.42

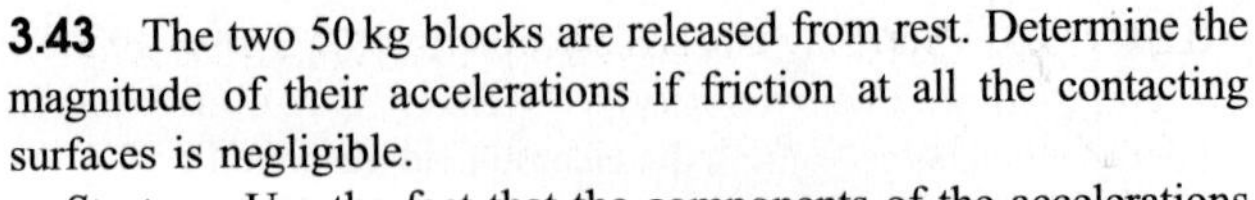

3.43 The two 50 kg blocks are released from rest. Determine the magnitude of their accelerations if friction at all the contacting surfaces is negligible.

Strategy: Use the fact that the components of the accelerations of the blocks perpendicular to their mutual interface must be equal.

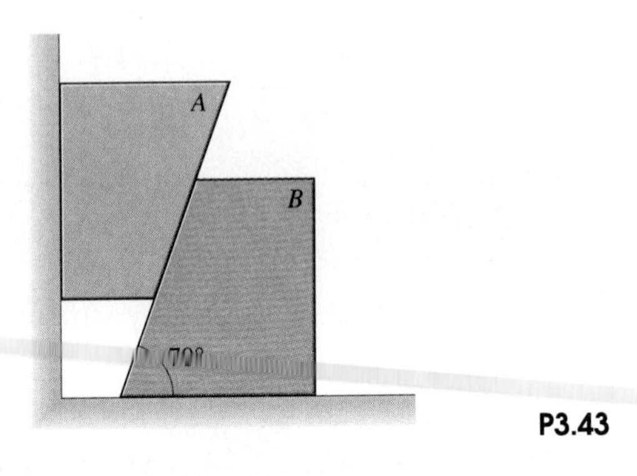

P3.43

3.44 In Problem 3.43, determine how long it takes block A to fall 1 m if $\mu_k = 0.1$ at all the contacting surfaces.

Normal and Tangential Components

When an object moves in a plane curved path, we can express Newton's second law in terms of normal and tangential components:

$$(\Sigma F_t \mathbf{e}_t + \Sigma F_n \mathbf{e}_n) = m(a_t \, \mathbf{e}_t + a_n \, \mathbf{e}_n) \tag{3.6}$$

where

$$a_t = \frac{dv}{dt} \qquad a_n = \frac{v^2}{\rho}$$

Equating the normal and tangential components in Equation (3.6), we obtain two scalar equations of motion:

$$\boxed{\Sigma F_t = m\frac{dv}{dt} \qquad \Sigma F_n = m\frac{v^2}{\rho}} \tag{3.7}$$

The sum of the forces in the tangential direction equals the product of the mass and the rate of change of the magnitude of the velocity, and the sum of the forces in the normal direction equals the product of the mass and the normal component of acceleration (Figure 3.5). The sum of the forces perpendicular to the plane curved path must equal zero.

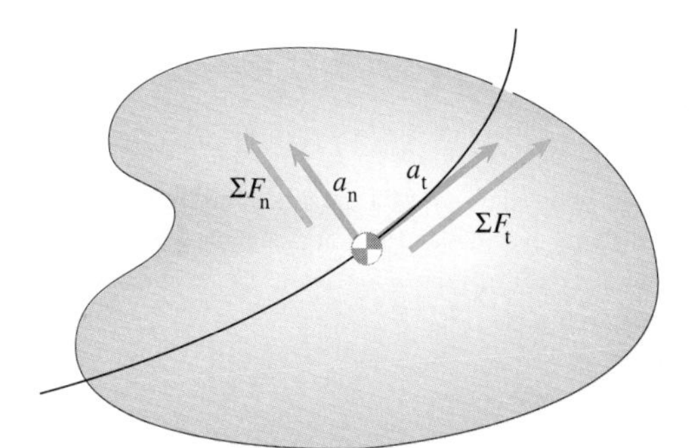

Figure 3.5
Normal and tangential components of $\Sigma \mathbf{F}$ and $\mathbf{a}$.

In the following examples we use Newton's second law expressed in terms of normal and tangential components to analyse motions of objects. By drawing the free-body diagram of an object, you can identify the components of the forces acting on it and use Newton's second law to determine the components of its acceleration. Or, if you know the components of the acceleration, you can use Newton's second law to determine the external forces. When an object follows a circular path, normal and tangential components are usually the simplest choice for analysing its motion.

Example 3.3

Future space stations may be designed to rotate in order to provide simulated gravity for their inhabitants (Figure 3.6). If the distance from the axis of rotation of the station to the occupied outer ring is $R = 100$ m, what rotation rate is necessary to simulate one-half of earth's gravity?

Figure 3.6

STRATEGY

By drawing the free-body diagram of a person in equilibrium and expressing Newton's second law in terms of normal and tangential components, we can relate the force exerted on the person by the floor to the angular velocity of the station. The person exerts an equal and opposite force on the floor, which is his effective weight.

SOLUTION

We draw the free-body diagram of a person standing in the outer ring in Figure (a), where N is the force exerted by the floor. Relative to the centre of the station, he moves in a circular path of radius R. Newton's second law in terms of normal and tangential components is

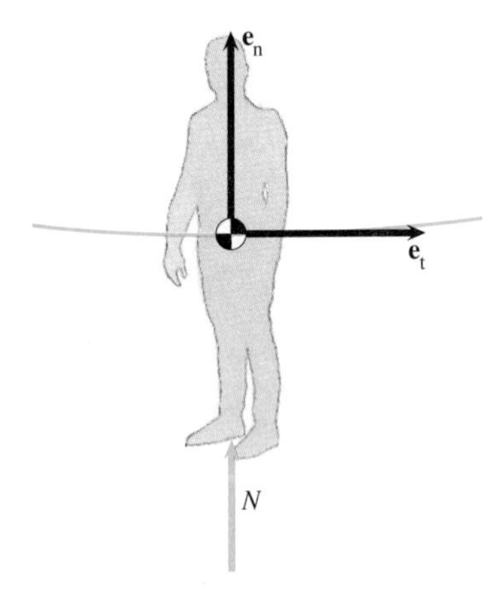

(a) Free-body diagram of a person standing in the occupied ring.

$$\Sigma \mathbf{F} = m\mathbf{a}:$$

$$N\,\mathbf{e}_n = m\left(\frac{dv}{dt}\mathbf{e}_t + \frac{v^2}{R}\mathbf{e}_n\right)$$

Therefore $n = mv^2/R$. The magnitude of his velocity is $v = R\omega$, where ω is the angular velocity of the station. If one-half of earth's gravity is simulated, $N = \frac{1}{2}mg$. Therefore

$$\frac{1}{2}mg = m\frac{(R\omega)^2}{R}$$

Solving for ω, we obtain the necessary angular velocity of the station,

$$\omega = \sqrt{\frac{g}{2R}} = \sqrt{\frac{9.81}{(2)(100)}} = 0.221 \text{ rad/s}$$

which is one revolution every 28.4 s.

Example 3.4

The experimental magnetically levitated train in Figure 3.7 is supported by magnetic repulsion forces exerted normal to the tracks. Motion of the train transverse to the tracks is prevented by lateral supports. The 20 Mg (megagram) train is travelling at 30 m/s on a circular segment of track of radius $R = 150$ m, and the bank angle of the track is 40°. What force must the magnetic levitation system exert to support the train, and what total force is exerted by the lateral supports?

Figure 3.7

STRATEGY

We know the train's velocity and the radius of its circular path, so we can determine its normal component of acceleration. By expressing Newton's second law in terms of normal and tangential components, we can determine the components of force normal and transverse to the track.

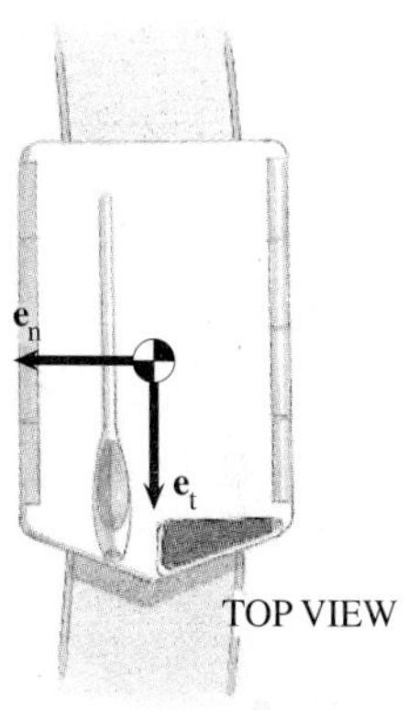

(a) The train's circular path viewed from above.

SOLUTION

The train's path viewed from above is circular (Figure (a)). The unit vector $\mathbf{e}_n$ is horizontal and points towards the centre of the circular path. In Figure (b) we draw the free-body diagram of the train seen from the front, where M is the magnetic force normal to the tracks and S is the transverse force. The sum of the forces in the vertical direction (perpendicular to the train's path) must equal zero:

$$M \cos 40^\circ + S \sin 40^\circ - mg = 0$$

The sum of the forces in the $\mathbf{e}_n$ direction equals the product of the mass and the normal component of the acceleration:

$$\Sigma F_n = m\frac{v^2}{\rho}:$$

$$M \sin 40^\circ - S \cos 40^\circ = m\frac{v^2}{R}$$

Solving these two equations for M and S, we obtain $M = 227.4$ kN, $S = 34.2$ kN.

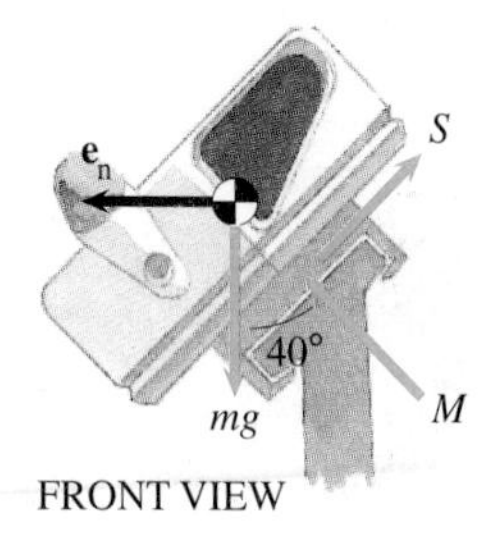

(b) Free-body diagram of the train.

Example 3.5

Application to Engineering

Motor Vehicle Dynamics

A civil engineer's preliminary design for a freeway off-ramp is circular with radius $R = 100\,\mathrm{m}$ (Figure 3.8). If she assumes that the coefficient of static friction between tyres and road is at least $\mu_s = 0.4$, what is the maximum speed at which vehicles can enter the ramp without losing traction?

Figure 3.8

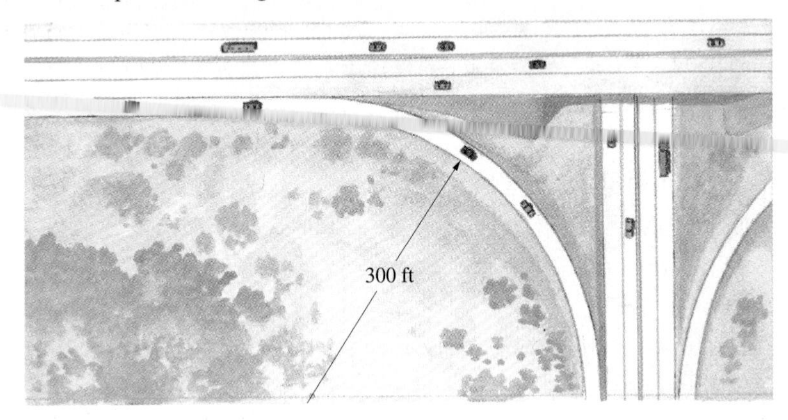

STRATEGY

Since a vehicle on the off-ramp moves in a circular path, it has a normal component of acceleration that depends on its velocity. The necessary normal component of force is exerted by friction between the tyres and the road, and the friction force cannot be greater than the product of μ_s and the normal force. By assuming that the friction force is equal to this value, we can determine the maximum velocity for which slipping will not occur.

SOLUTION

We view the free-body diagram of a car on the off-ramp from above the car in Figure (a) and from the front of the car in Figure (b). The friction force f must equal the

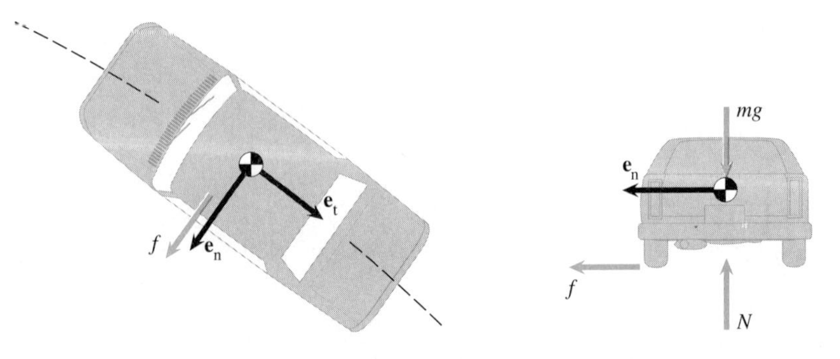

(a) Top view of the free-body diagram.

(b) Front view.

product of the car's mass m and its normal component of acceleration:

$$f = m\frac{v^2}{R}$$

The required friction force increases as v increases. The maximum friction force the surfaces will support is $f_{max} = \mu_s N = \mu_s mg$. Therefore the maximum velocity for which slipping does not occur is

$$v = \sqrt{\mu_s gR} = \sqrt{(0.4)(9.81)(100)} = 19.8 \text{ m/s}$$

or 71.3 km/hr.

DESIGN ISSUES

Automotive engineers, civil engineers who design highways, and engineers who study traffic accidents and their prevention must analyse and measure the motions of vehicles under different conditions. By using the methods discussed in this chapter, they can relate the forces acting on vehicles to their motions and study, for example, the factors influencing the distance necessary for a car to be brought to a stop in an emergency, or the effects of banking and curvature on the velocity at which a car can safely be driven on a curved road (Figure 3.9).

In example 3.5, the analysis indicates that vehicles will lose traction if they enter the freeway off-ramp at speeds greater than 71.3 km/hr. This result can be used as an indication of the speed limit that must be posted in order for vehicles to enter the ramp safely, or the off-ramp could be designed for a greater speed by increasing the radius of curvature. Or, if a larger safe speed is desired but space limitations forbid a larger radius of curvature, the off-ramp could be designed to incorporate banking (see Problem 3.65).

Figure 3.9

Tests of the capabilities of vehicles to negotiate curves influence the design of both vehicles and highways.

Problems

3.45 If you choose the velocity of the train in Example 3.4 properly, the lateral force S exerted on it as it travels along the circular track is zero.
(a) What is the necessary velocity?
(b) Explain why this would be the most desirable velocity from the passengers' point of view.

3.46 An earth satellite with a mass of 4000 kg is in a circular orbit of radius $R = 8000$ km. Its velocity relative to the centre of the earth is 7038 m/s.
(a) Use the given information to determine the gravitational force acting on the satellite, and compare it with the satellite's weight at sea level.
(b) The acceleration due to gravity at a distance R from the centre of the earth is gR_E^2/R^2, where the radius of the earth is $R_E = 6370$ km. Use this expression to confirm your answer to part (a).

3.47 The 2 kg slider A starts from rest and slides *in the horizontal plane* along the smooth circular bar under the action of a tangential force $F_t = 4t$ N. At $t = 4$ s, determine (a) the magnitude of the velocity of the slider; (b) the magnitude of the horizontal force exerted on the slider by the bar.

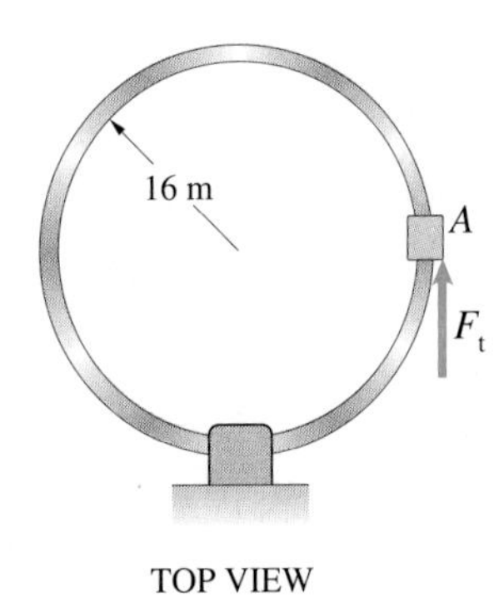

P3.47

3.48 Small parts on a conveyer belt moving with constant velocity v are allowed to drop into a bin. Show that the angle α at which the parts start sliding on the belt satisfies the equation

$$\cos\alpha - \frac{1}{\mu_s}\sin\alpha = \frac{v^2}{gR}$$

where μ_s is the coefficient of static friction between the parts and the belt.

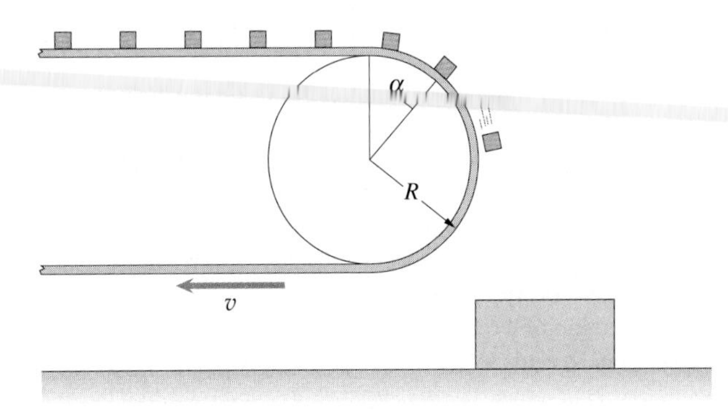

P3.48

3.49 The mass m rotates around the vertical pole in a horizontal circular path. Determine the magnitude of its velocity in terms of θ and L.

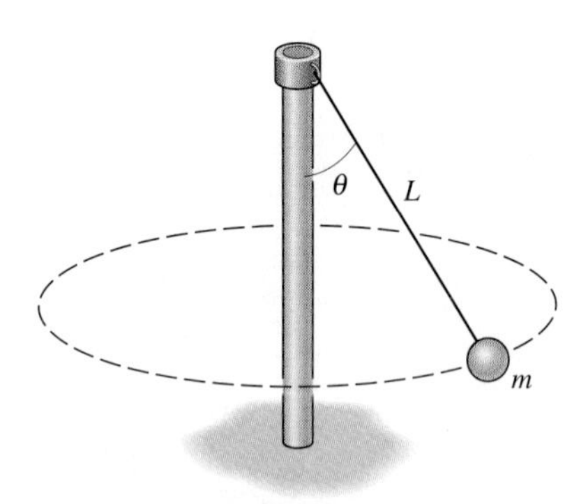

P3.49

3.50 In Problem 3.49, if $m = 15$ kg, $L = 1$ m and the mass is moving in its circular path at $v = 5$ m/s, what is the tension in the string?

3.51 The 10 kg mass m rotates around the vertical pole in a horizontal circular path of radius $R = 1$ m. If the magnitude of its velocity is $v = 3$ m/s, what are the tensions in the strings A and B?

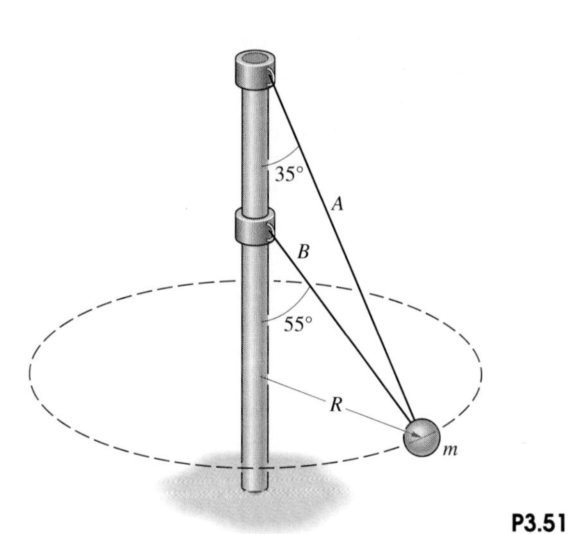

P3.51

3.52 In Problem 3.51, what is the range of values of v for which the mass will remain in the circular path described?

3.53 Suppose you are designing a monorail transportation system that will travel at 50 m/s, and you decide that the angle α that the cars swing out from the vertical when they go through a turn must not be larger than 20°. If the turns in the track consist of circular arcs of constant radius R, what is the minimum allowable value of R?

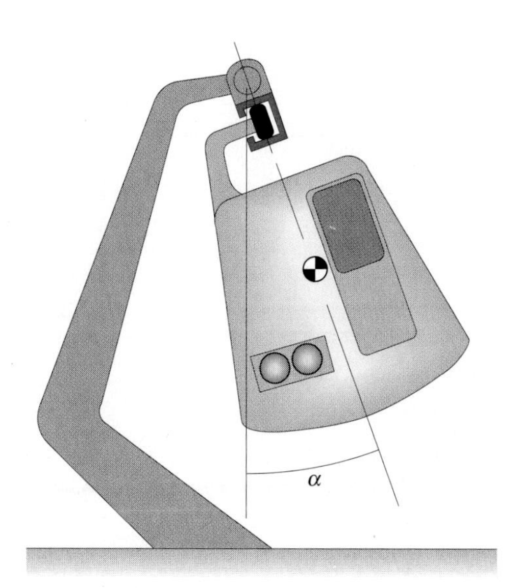

P3.53

3.54 An aeroplane of weight $W = 900$ kN makes a turn at constant altitude and at constant velocity $v = 180$ m/s. The bank angle is 15°.

(a) Determine the lift force L.

(b) What is the radius of curvature of the plane's path?

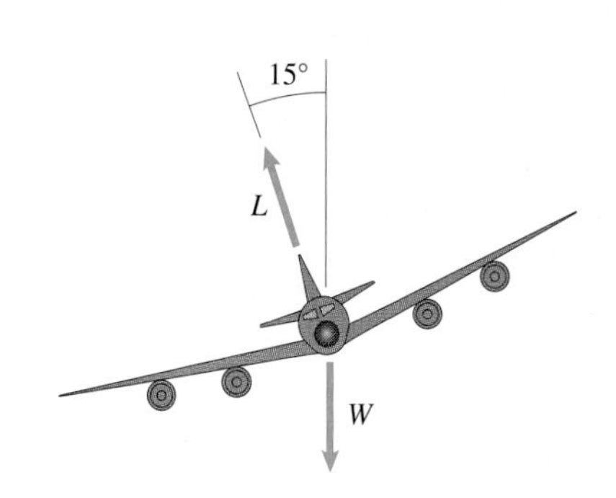

P3.54

3.55 The suspended mass m is stationary.

(a) What are the tensions in the strings?

(b) If string A is cut, what is the tension in string B immediately afterwards?

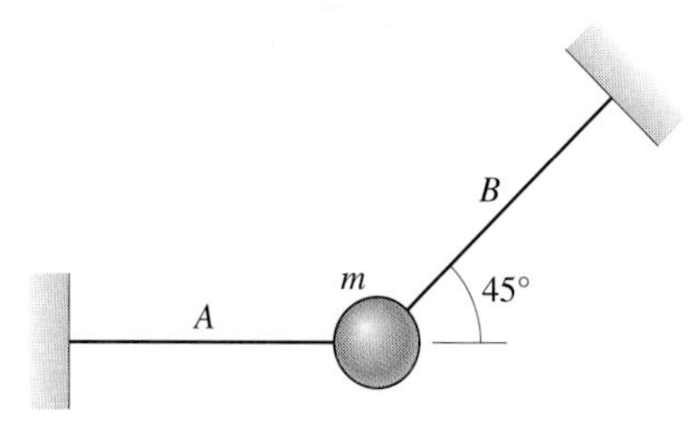

P3.55

3.56 An aeroplane flies with constant velocity v along a circular path in the vertical plane. The radius of its circular path is 1500 m. The pilot weighs 660 N.
(a) The pilot will experience 'weightlessness' at the top of the circular path if the aeroplane exerts no net force on him at that point. Draw a free-body diagram of the pilot, and use it to determine the velocity v necessary to achieve this condition.
(b) Determine the force exerted on the pilot by the aeroplane at the top of the circular path if the aeroplane is travelling at twice the velocity determined in part (a).

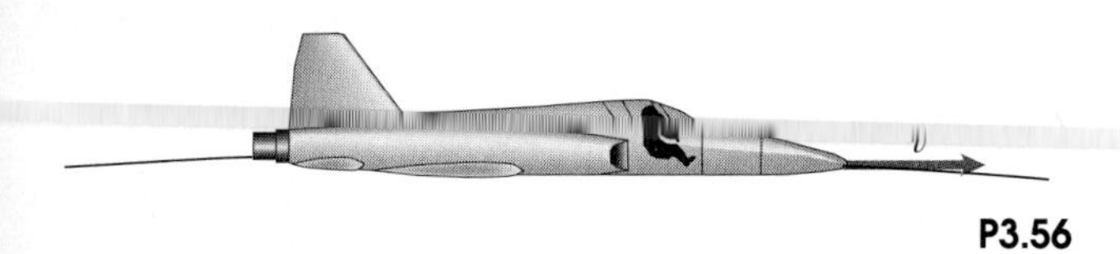

P3.56

3.57 The smooth circular bar rotates with constant angular velocity ω_0 about the vertical axis AB. Determine the angle β at which the slider of mass m will remain stationary relative to the circular bar.

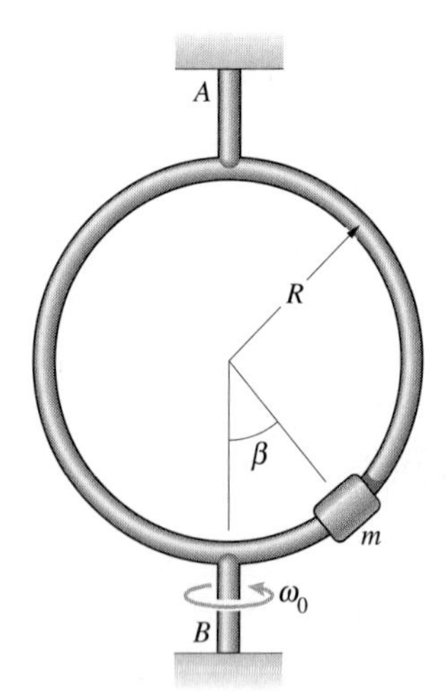

P3.57

3.58 The force exerted on a charged particle by a magnetic field is

$$\mathbf{F} = q\mathbf{v} \times \mathbf{B}$$

where q and $\mathbf{v}$ are the charge and velocity vector of the particle and $\mathbf{B}$ is the magnetic field vector. A particle of mass m and positive charge q is projected at O with velocity $\mathbf{v} = v_0\mathbf{i}$ into a uniform magnetic field $\mathbf{B} = B_0\mathbf{k}$. Using normal and tangential components, show that: (a) the magnitude of the particle's velocity is constant; (b) the particle's path is a circle with radius mv_0/qB_0.

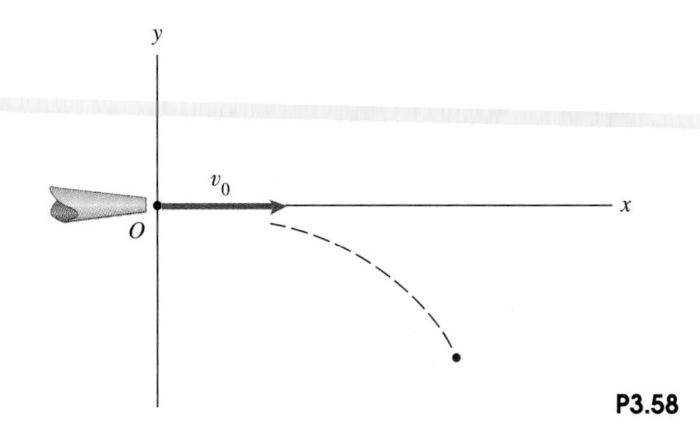

P3.58

3.59 A mass m is attached to a string that is wrapped around a fixed post of radius R. At $t = 0$, the object is given a velocity v_0 as shown. Neglect external forces on m other than the force exerted by the string. Determine the tension in the string as a function of the angle θ.

Strategy: The velocity vector of the mass is perpendicular to the string. Express Newton's second law in terms of normal and tangential components.

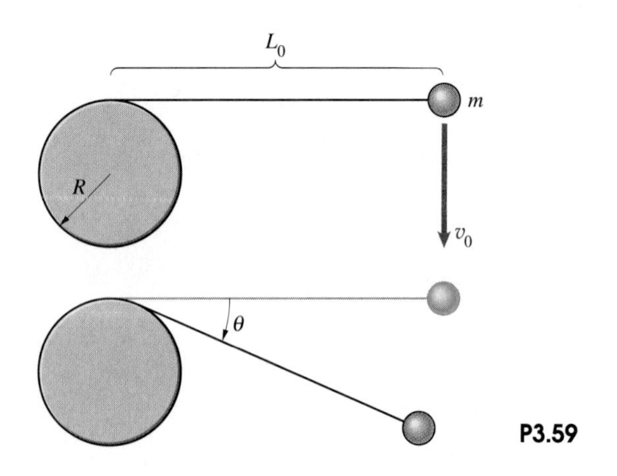

P3.59

3.60 In Problem 3.59, determine the angle θ as a function of time.

Problems 3.61–3.65 are related to Example 3.5.

3.61 A car is travelling on a straight, level road when the driver perceives a hazard ahead. After a reaction time of 0.5 s, he applies the brakes, locking the wheels. The coefficient of kinetic friction between the tyres and the road is $\mu_k = 0.6$. Determine the total distance the car travels before coming to rest, including the distance travelled before the brakes are applied, if it is travelling at (a) 88 km/hr; (b) 105 km/hr.

3.62 If the car in Problem 3.61 is travelling at 105 km/hr and rain decreases the value of μ_k to 0.4, what total distance does the car travel before coming to rest?

3.63 A car travelling at 30 m/s is at the top of a hill. The coefficient of kinetic friction between the tyres and the road is $\mu_k = 0.8$ and the instantaneous radius of curvature of the car's path is 200 m. If the driver applies the brakes and the car's wheels lock, what is the resulting deceleration of the car in the direction tangent to its path?

P3.63

3.64 Suppose that the car in Problem 3.63 is at the bottom of a depression whose radius of curvature is 200 m when the driver applies the brakes. What is the resulting deceleration of the car in the direction tangent to its path?

P3.64

3.65 A freeway off-ramp is circular with radius R (Figure (a)), and the roadway is banked at an angle β (Figure (b)). Show that the maximum constant velocity at which a car can travel the off-ramp without losing traction is

$$v = \sqrt{gR\left(\frac{\sin\beta + \mu_s \cos\beta}{\cos\beta - \mu_s \sin\beta}\right)}$$

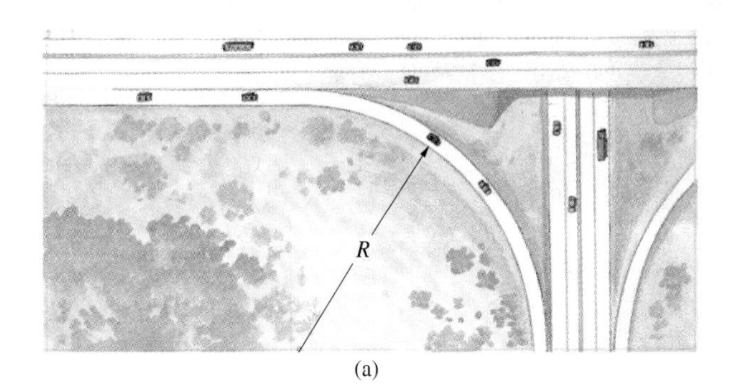

(a)

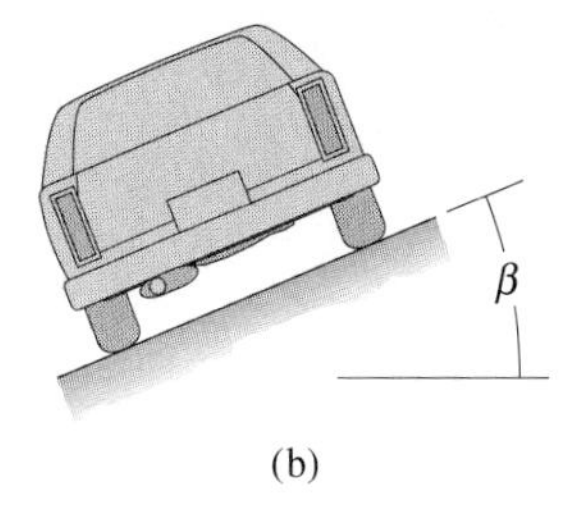

(b)

P3.65

Polar Coordinates

In terms of polar coordinates, Newton's second law for an object moving in the x-y plane is

$$(\Sigma F_r \mathbf{e}_r + \Sigma F_\theta \, \mathbf{e}_\theta) = m(a_r \, \mathbf{e}_r + a_\theta \, \mathbf{e}_\theta) \tag{3.8}$$

where

$$a_r = \frac{\mathrm{d}^2 r}{\mathrm{d}t^2} - r\left(\frac{\mathrm{d}\theta}{\mathrm{d}t}\right)^2 = \frac{\mathrm{d}^2 r}{\mathrm{d}t^2} - r\omega^2$$

$$a_\theta = r\frac{\mathrm{d}^2\theta}{\mathrm{d}t^2} + 2\frac{\mathrm{d}r}{dt}\frac{d\theta}{\mathrm{d}t} = r\alpha + 2\frac{\mathrm{d}r}{\mathrm{d}t}\omega$$

Equating the $\mathbf{e}_r$ and $\mathbf{e}_\theta$ components in Equation (3.8), we obtain the scalar equations

$$\Sigma F_r = ma_r = m\left(\frac{d^2 r}{dt^2} - r\omega^2\right) \tag{3.9}$$

$$\Sigma F_\theta = ma_\theta = m\left(r\alpha + 2\frac{dr}{dt}\omega\right) \tag{3.10}$$

The sum of the forces in the radial direction equals the product of the mass and the radial component of the acceleration, and the sum of the forces in the transverse direction equals the product of the mass and the transverse component of the acceleration (Figure 3.10).

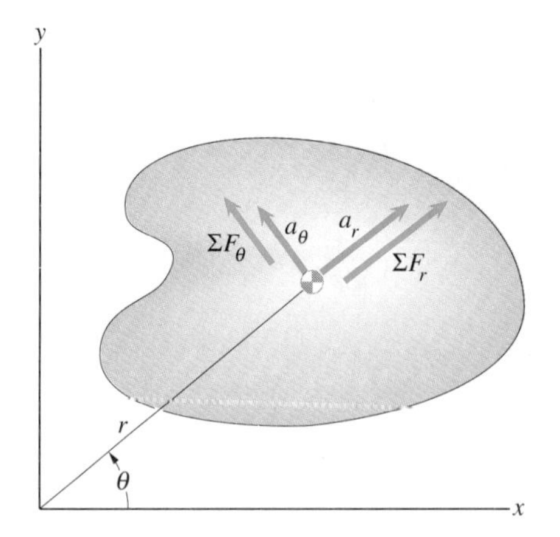

Figure 3.10
Radial and transverse components of $\Sigma \mathbf{F}$ and $\mathbf{a}$.

In the following example we use Newton's second law expressed in terms of polar coordinates, or radial and transverse components, to analyse the motions of an object. By drawing the free-body diagram of an object, you can identify the components of the forces acting on it and use Newton's second law to determine the components of its acceleration. Or, if you know the components of the acceleration, you can use Newton's second law to determine the external forces.

Example 3.6

The smooth bar in Figure 3.11 rotates *in the horizontal plane* with constant angular velocity ω_0. The unstretched length of the linear spring is r_0. The collar A has mass m and is released at $r = r_0$ with no radial velocity.
(a) Determine the radial velocity of the collar as a function of r.
(b) Determine the horizontal force exerted on the collar by the bar as a function of r.

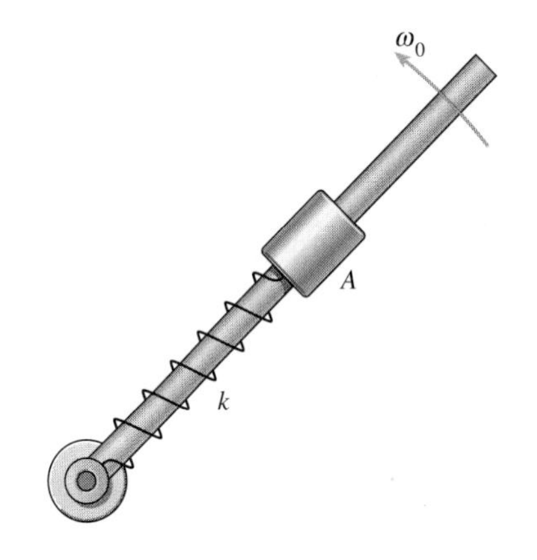

Figure 3.11

STRATEGY

(a) The only force on the collar in the radial direction is the spring force, which we can express in polar coordinates in terms of r. By integrating Equation (3.9), we can determine the radial velocity v_r as a function of r.
(b) Once $v_r = dr/dt$ is known in terms of r, we can use Equation (3.10) to determine the transverse force exerted on the collar by the bar.

SOLUTION

(a) The spring exerts a radial force $k(r - r_0)$ in the negative r direction (Figure (a)). Since the bar is smooth, it exerts no radial force on A, but may exert a transverse force F_θ. From Equation (3.9),

$$\Sigma F_r = -k(r - r_0) = m\left(\frac{d^2r}{dt^2} - r\omega^2\right) = m\left(\frac{dv_r}{dt} - r\omega_0^2\right)$$

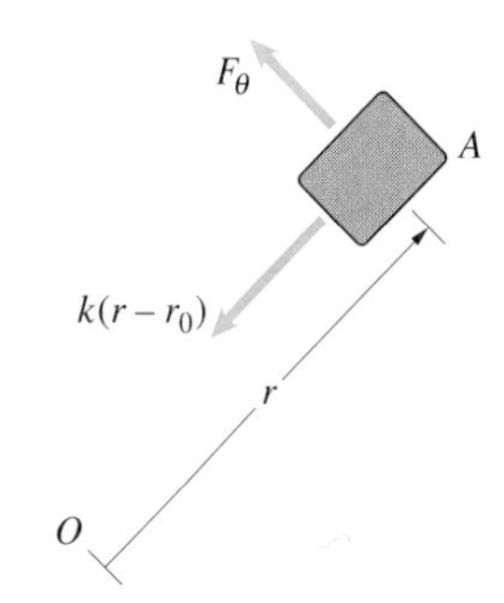

(a) Radial and transverse forces on A.

By using the chain rule to express the time derivative of v_r in terms of a derivative with respect to r,

$$\frac{dv_r}{dt} = \frac{dv_r}{dr}\frac{dr}{dt} = \frac{dv_r}{dr}v_r$$

we obtain

$$v_r\,dv_r = \left[\left(\omega_0^2 - \frac{k}{m}\right)r + \frac{k}{m}r_0\right]dr$$

Integrating,

$$\int_0^{v_r} v_r\,dv_r = \int_{r_0}^{r}\left[\left(\omega_0^2 - \frac{k}{m}\right)r + \frac{k}{m}r_0\right]dr$$

we obtain the radial velocity as a function of r:

$$v_r = \sqrt{\left(\omega_0^2 - \frac{k}{m}\right)(r^2 - r_0^2) + \frac{2k}{m}r_0(r - r_0)}$$

(b) From Equation (3.10), the transverse force exerted on A by the bar is

$$F_\theta = m\left(r\alpha + 2\frac{dr}{dt}\omega\right) = 2m\omega_0 v_r$$

Substituting our expression for v_r as a function of r, we obtain the horizontal force exerted by the bar as a function of r:

$$F_\theta = 2m\omega_0\sqrt{\left(\omega_0^2 - \frac{k}{m}\right)(r^2 - r_0^2) + \frac{2k}{m}r_0(r - r_0)}$$

Problems

3.66 The polar coordinates of an object are $r = (t^2 + 2)$ m, $\theta = 2t^3 - t^2$ rad, and its mass is 44 kg. What are the radial and transverse components of the total external force on the object at $t = 1s$?

3.67 The polar coordinates of an object are $r = (2t^3 + 4t)$ m, $\theta = (t^2 - t)$ rad, and its mass is 20 kg. What are the radial and transverse components of the total external force on the object at $t = 1$ s?

3.68 The robot is programmed so that the 0.4 kg part A describes the path

$$r = (1 - 0.5\cos 2\pi t)\ \text{m}$$
$$\theta = (0.5 - 0.2\sin 2\pi t)\ \text{rad}$$

At $t = 2$ s, determine the radial and transverse components of force exerted on A by the robot's jaws.

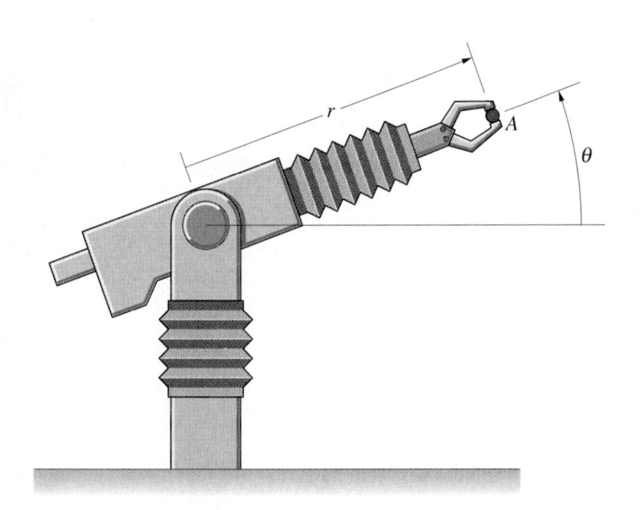

P3.68

3.69 In Example 3.6, what is the maximum radial distance reached by the collar A?

3.70 The smooth bar rotates *in the horizontal plane* with constant angular velocity $\omega_0 = 60$ rpm (revolutions per minute). If the 2 kg collar A is released at $r = 1$ m with no radial velocity, what is the magnitude of its velocity when it reaches the end of the bar?

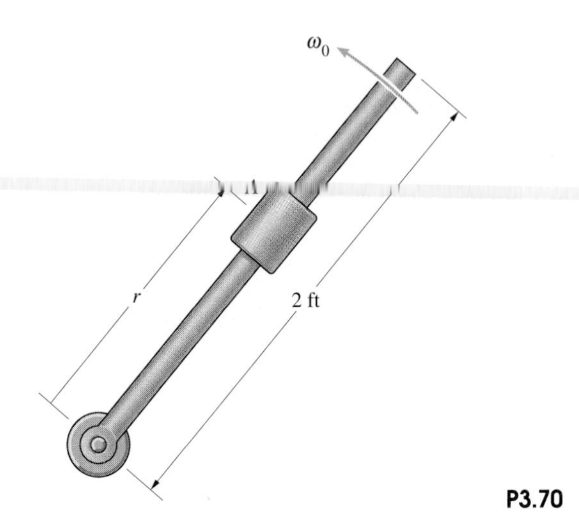

P3.70

3.71 In Problem 3.70, what is the maximum horizontal force exerted on the collar by the bar?

3.72 The mass m is released from rest with the string horizontal. By using Newton's second law in terms of polar coordinates, determine the magnitude of the velocity of the mass and the tension in the string as functions of θ.

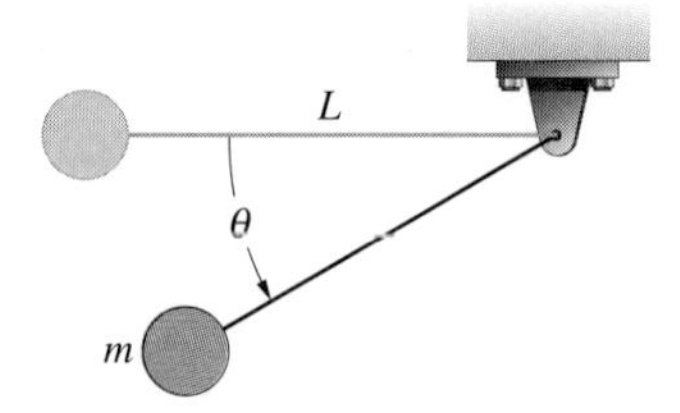

P3.72

3.73 The skier passes point A 17 m/s. From A to B, the radius of his circular path is 6 m. By using Newton's second law in terms of polar coordinates, determine the magnitude of his velocity as he leaves the jump at B. Neglect transverse forces other than the transverse component of his weight.

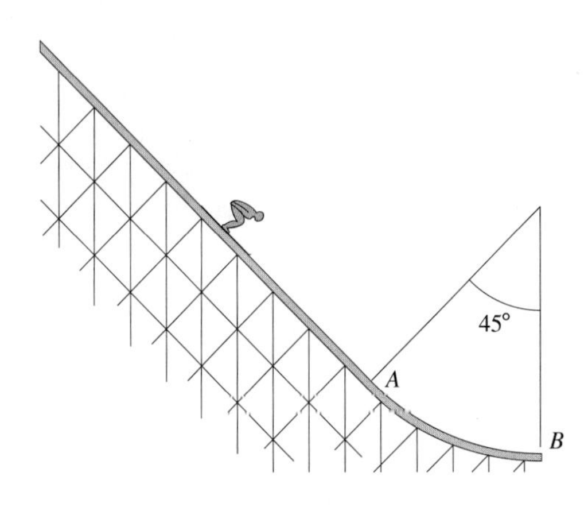

P3.73

3.74 A 2 kg mass rests on a flat horizontal bar. The bar begins rotating *in the vertical plane* about O with a constant angular acceleration of 1 rad/s^2. The mass is observed to slip relative to the bar when the bar is 30° above the horizontal. What is the static coefficient of friction between the mass and the bar? Does the mass slip towards or away from O?

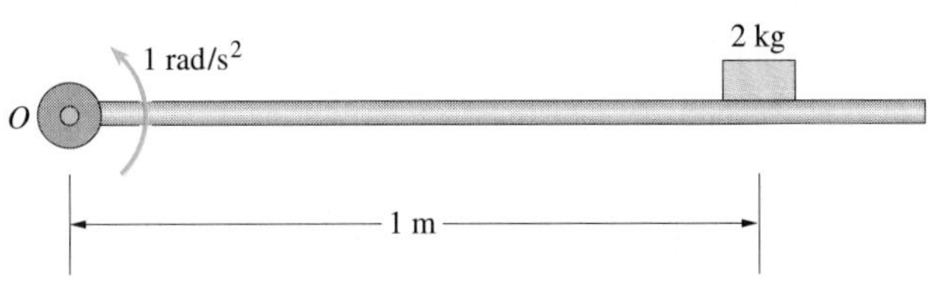

P3.74

3.75 The 0.25 kg slider A is pushed along the circular bar by the slotted bar. The circular bar lies *in the horizontal plane*. The angular position of the slotted bar is $\theta = 10t^2$ rad. Determine the radial and transverse components of the total external force exerted on the slider at $t = 0.2$ s.

P3.75

3.76 In Problem 3.75, supposed that the circular bar lies *in the vertical plane*. Determine the radial and transverse components of the total force exerted on the slider by the circular and slotted bars at $t = 0.25$ s.

3.77 The slotted bar rotates *in the horizontal plane* with constant angular velocity ω_0. The mass m has a pin that fits in the slot of the bar. A spring holds the pin against the surface of the fixed cam. The surface of the cam is described by $r = r_0(2 - \cos\theta)$. Determine the radial and transverse components of the total external force exerted on the pin as functions of θ.

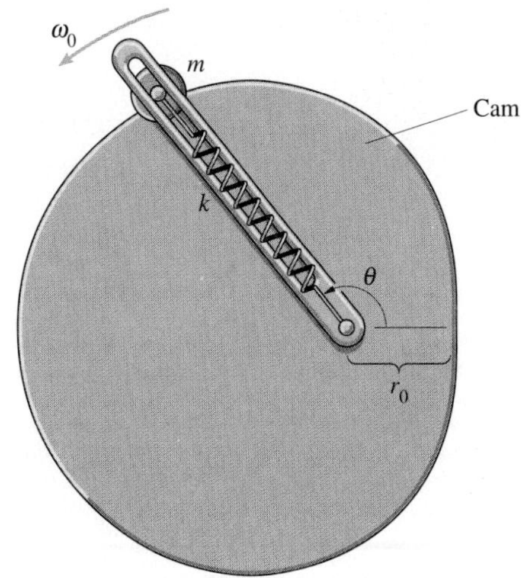

P3.77

3.78 In Problem 3.77, suppose that the unstretched length of the spring is r_0. Determine the smallest value of the spring constant k for which the pin will remain on the surface of the cam.

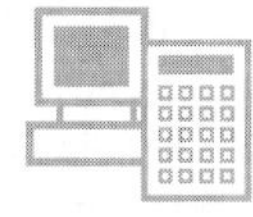

Computational Mechanics

The material in this section is designed for the use of a programmable calculator or computer.

So far in this chapter you have seen many situations in which you were able to determine the motion of an object by a simple procedure: after using Newton's second law to determine the acceleration, you integrated to obtain analytical, or **closed-form**, expressions for the object's velocity and position. These examples are very valuable – they teach you to use free-body diagrams and express problems in different coordinate systems, and they develop your intuitive understanding of forces and motions. But you would be misled if we presented examples of this kind only, because most problems that must be dealt with in engineering cannot be solved in this way. The functions describing the forces, and therefore the acceleration, are often too complicated for you to integrate and obtain closed-form solutions. In other situations, you will not know the forces in terms of functions but instead will know them in terms of data, either as a continuous recording of force as a function of time (analogue data) or as values of force measured at discrete times (digital data).

You can obtain approximate solutions to such problems by using numerical integration. Let's consider an object of mass m in straight-line motion along the x axis (Figure 3.12) and assume that the x component of the total force may depend on the time, position and velocity:

$$\Sigma F_x = \Sigma F_x(t, x, v_x) \tag{3.11}$$

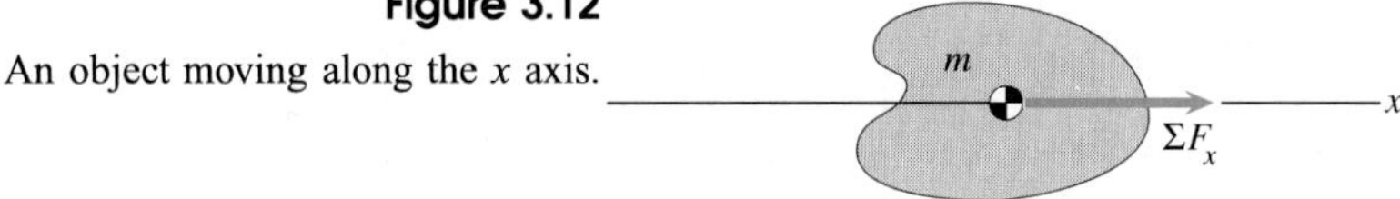

Figure 3.12
An object moving along the x axis.

Suppose that at a particular time t_0, we know the position $x(t_0)$ and velocity $v_x(t_0)$. The acceleration of the object at t_0 is

$$\frac{dv_x}{dt}(t_0) = \frac{1}{m}\Sigma F_x[t_0, x(t_0), v_x(t_0)] \tag{3.12}$$

The definition of the time derivative of v_x at t_0 is

$$\frac{dv_x}{dt}(t_0) = \lim_{\Delta t \to 0} \frac{v_x(t_0 + \Delta t) - v_x(t_0)}{\Delta t}$$

By choosing a sufficiently small value of Δt, we can approximate this derivative by

$$\frac{dv_x}{dt}(t_0) = \frac{v_x(t_0 + \Delta t) - v_x(t_0)}{\Delta t}$$

and substitute it into Equation (3.12) to obtain an approximate expression for the velocity at $t_0 + \Delta t$:

$$v_x(t_0 + \Delta t) = v_x(t_0) + \frac{1}{m}\Sigma F_x[t_0, x(t_0), v_x(t_0)]\,\Delta t \tag{3.13}$$

The relation between the velocity and position at t_0 is

$$\frac{dx}{dt}(t_0) = v_x(t_0)$$

Approximating this derivative by

$$\frac{dx}{dt}(t_0) = \frac{x(t_0 + \Delta t) - x(t_0)}{\Delta t}$$

we obtain an approximate expression for the position at $t_0 + \Delta t$:

$$x(t_0 + \Delta t) = x(t_0) + v_x(t_0)\,\Delta t \tag{3.14}$$

Thus, if we know the position and velocity at a time t_0, we can approximate their values at $t_0 + \Delta t$ by using Equations (3.13) and (3.14). We can then repeat the procedure, using $x(t_0 + \Delta t)$ and $v_x(t_0 + \Delta t)$ as initial conditions to determine the approximate position and velocity at $t_0 + 2\Delta t$. By continuing in this way, we obtain approximate solutions for the position and velocity in terms of time. This procedure is easy to carry out using a calculator or computer. It is called a **finite-difference method** because it determines changes in the dependent variables over finite intervals of time. The particular method we describe, due to Leonhard Euler (1707–83), is called **forward differencing**: the value of the derivative of a function at t_0 is approximated by using its value at t_0 and its value forward in time, at $t_0 + \Delta t$. Although more elaborate finite-difference methods exist that result in smaller errors in each time step, Euler's method is adequate to introduce you to numerical solutions of problems in dynamics. Notice that Equation (3.11) does not need to be a functional expression to carry out this process. The values of the total force must be known at times $t_0, t_0 + \Delta t, \ldots,$ and can be determined either from a function or from analogue or digital data.

You can determine the velocity and position of an object in curviliner motion by the same approach. Suppose that an object moves in the x-y plane and that the components of force may depend on the time, position and velocity:

$$\Sigma F_x = \Sigma F_x(t, x, y, v_x, v_y) \quad \Sigma F_y = \Sigma F_y(t, x, y, v_x, v_y)$$

If the position and velocity are known at a time t_0, we can use the same steps leading to Equations (3.13) and (3.14) to obtain approximate expressions for the components of position and velocity at $t_0 + \Delta t$:

$$\begin{aligned}
x(t_0 + \Delta t) &= x(t_0) + v_x(t_0)\,\Delta t \\
y(t_0 + \Delta t) &= y(t_0) + v_y(t_0)\,\Delta t \\
v_x(t_0 + \Delta t) &= v_x(t_0) + \frac{1}{m}\Sigma F_x[t_0, x(t_0), y(t_0), v_x(t_0), v_y(t_0)]\,\Delta t \\
v_y(t_0 + \Delta t) &= v_y(t_0) + \frac{1}{m}\Sigma F_y[t_0, x(t_0), y(t_0), v_x(t_0), v_y(t_0)]\,\Delta t
\end{aligned} \tag{3.15}$$

The material in this section is designed for the use of a programmable calculator or computer.

Example 3.7

A 1450 kg projectile is launched from $x = 0, y = 0$ with initial velocity $v_x = 120\,\text{m/s}$, $v_y = 120\,\text{m}$. (The y axis is positive upwards.) The aerodynamic drag force is of magnitude $C|\mathbf{v}|^2$, where C is a constant. Determine the trajectory for values of C of 0.1, 0.2 and 0.3.

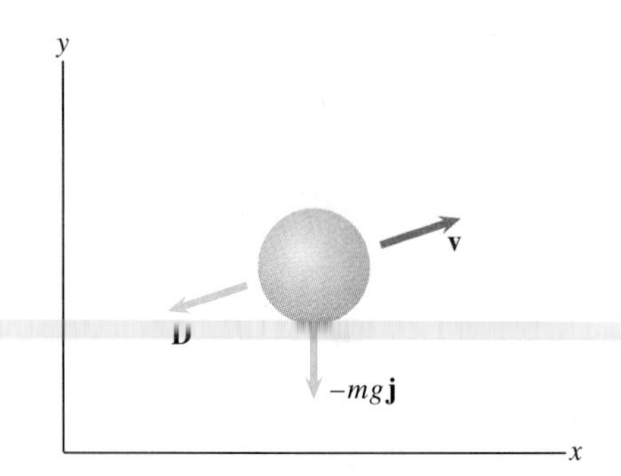

Figure 3.13
The forces on the projectile are its weight and the drag force **D**.

SOLUTION

To apply Equations (3.15), we must determine the x and y components of the total force on the projectile. Let **D** be the drag force (Figure 3.13). Since $\mathbf{v}/|\mathbf{v}|$ is a unit vector in the direction of **v**, we can write **D** as

$$\mathbf{D} = -C|\mathbf{v}|^2 \frac{\mathbf{v}}{|\mathbf{v}|} = -C|\mathbf{v}|\mathbf{v}$$

The external forces on the projectile are its weight and the drag,

$$\Sigma \mathbf{F} = -mg\,\mathbf{j} - C|\mathbf{v}|\mathbf{v}$$

so the components of the total force are

$$\Sigma F_x = -C\sqrt{v_x^2 + v_y^2}\, v_x \qquad \Sigma F_y = -mg - C\sqrt{v_x^2 + v_y^2}\, v_y \tag{3.16}$$

Consider the case $C = 0.1$, and let $\Delta t = 0.1$. At the initial time $t_0 = 0$, $x(t_0)$ and $y(t_0)$ are zero, $v_x(t_0) = 120\,\text{m/s}$ and $v_y(t_0) = 120\,\text{m/s}$. The components of the position and velocity after the first time step are

$$\begin{aligned} x(t_0 + \Delta t) &= x(t_0) + v_x(t_0)\,\Delta t : \\ x(0.1) &= x(0) + v_x(0)\,\Delta t \\ &= 0 + (120)(0.1) = 12\ \text{m} \end{aligned}$$

$$\begin{aligned} y(t_0 + \Delta t) &= y(t_0) + v_y(t_0)\,\Delta t : \\ y(0.1) &= y(0) + v_y(0)\,\Delta t \\ &= 0 + (120)(0.1) = 12\ \text{m} \end{aligned}$$

$$\begin{aligned} v_x(t_0 + \Delta t) &= v_x(t_0) + \frac{1}{m}\Sigma F_x[t_0, x(t_0), y(t_0), v_x(t_0), v_y(t_0)]\,\Delta t : \\ v_x(0.1) &= v_x(0) + \left\{-\frac{C}{m}\sqrt{[v_x(0)]^2 + [v_y(0)]^2}\, v_x(0)\right\}\Delta t \\ &= 120 + \left[-\frac{0.1}{1450}\sqrt{(120)^2 + (120)^2}\,(120)\right](0.1) \\ &= 119.86\,\text{m/s} \end{aligned}$$

$$\begin{aligned} v_y(t_0 + \Delta t) &= v_y(t_0) + \frac{1}{m}\Sigma F_y[t_0, x(t_0), y(t_0), v_x(t_0), v_y(t_0)]\,\Delta t : \\ v_y(0.1) &= v_y(0) + \left\{-g - \frac{C}{m}\sqrt{[v_x(0)]^2 + [v_y(0)]^2}\, v_y(0)\right\}\Delta t \\ &= 120 + \left[-9.81 - \frac{0.2}{1450}\sqrt{(120)^2 + (120)^2}\,(120)\right](0.1) \\ &= 118.88\ \text{m/s} \end{aligned}$$

Continuing in this way, we obtain the following results for the first five time steps:

Time, s	x, m	y, m	v_x, m/s	v_y, m/s
0.0	0.00	0.00	120.00	120.00
0.1	12.00	12.00	119.86	118.88
0.2	23.99	23.89	119.72	117.76
0.3	35.96	35.66	119.58	116.64
0.4	47.92	47.33	119.44	115.33
0.5	59.86	58.88	119.31	114.41

When there is no drag ($C = 0$), we can obtain the closed-form solution for the trajectory and compare it with numerical solutions. In Figure 3.14, we present this comparison using $\Delta t = 3.5$ s, 1.0 s and 0.1 s. Notice that the numerical solution with $\Delta t = 0.1$ s closely approximates the closed-form solution.

In Figure 3.15, we show the numerical solutions for the various values of C obtained using $\Delta t = 0.1$ s. As expected, the range of the projectile decreases as C increases. Also, when drag is present, the shape of the trajectory is changed. The projectile descends at an angle steeper than that when it ascends.

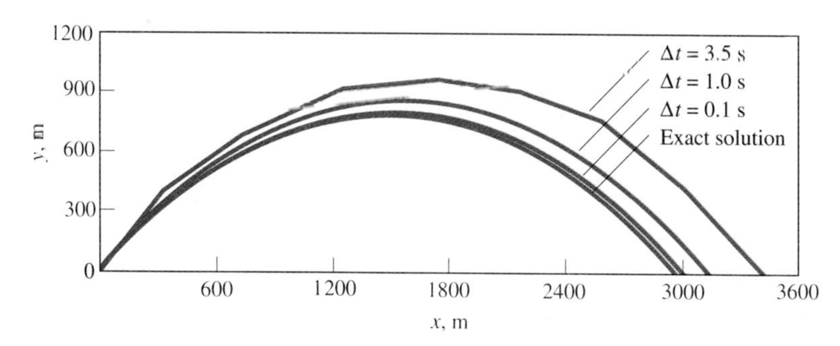

Figure 3.14
The closed-form solution for the trajectory when $C = 0$ compared with numerical solutions.

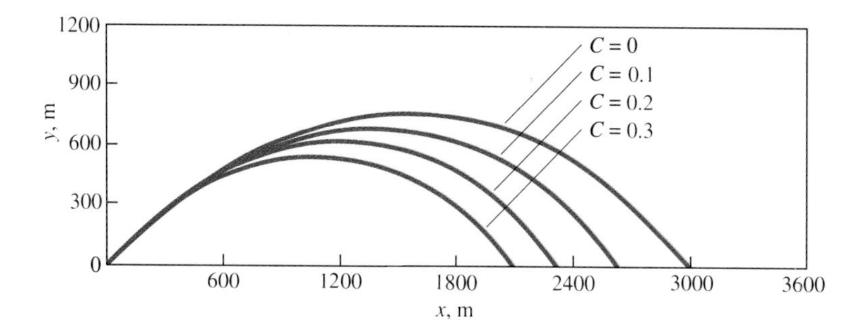

Figure 3.15
Trajectories for various values of C.

DISCUSSION

The development of the first completely electronic digital computer, the ENIAC (Electronic Numerical Integrator and Computer), built at the University of Pennsylvania between 1943 and 1945, was motivated in part by the need to calculate trajectories of projectiles. A room-size machine with 18 000 vacuum tubes, it had 20 bytes of random-access memory and 450 bytes of read-only memory.

Problems

3.79 A 1 kg object moves along the x axis under the action of the force $F_x = 6t$ N. At $t = 0$, its position and velocity are $x = 0$ and $v_x = 10$ m/s. Using numerical integration with $\Delta t = 0.1$ s, determine the position and velocity of the object for the first five time steps.

Strategy: At the initial time $t_0 = 0, x(t_0) = 0$ and $v_x(t_0) = 10$ m/s. You can use Equations (3.13) and (3.14) to determine the velocity and position at time $t_0 + \Delta t = 0.1$ s. The position is

$$x(t_0 + \Delta t) = x(t_0) + v_x(t_0)\,\Delta t:$$

$$x(0.1) = x(0) + v_x(0)\,\Delta t$$

$$= 0 + (10)(0.1) = 1 \text{ m}$$

and the velocity is

$$v_x(t_0 + \Delta t) = v_x(t_0) + \frac{1}{m} F_x(t_0)\,\Delta t:$$

$$v_x(0.1) = 10 + \frac{1}{(1)} 6(0)(0.1) = 10 \text{ m/s}$$

Use these values of the position and velocity as the initial conditions for the next time step.

3.80 For the 1 kg object described in Problem 3.79, draw a graph comparing the exact solution from $t = 0$ to $t = 10$ s with the solutions obtained using numerical integration with $\Delta t = 2$ s, $\Delta t = 0.5$ s and $\Delta t = 0.1$ s.

3.81 At $t = 0$, an object released from rest falls with constant acceleration $g = 9.81$ m/s^2.
(a) Using the closed-form solution, determine the velocity of the object and the distance it has fallen at $t = 2$ s.
(b) Approximate the answers to part (a) by using numerical integration with $\Delta t = 0.2$ s.

3.82 In Problem 3.81, draw a graph of the distance the object falls as a function of time from $t = 0$ to $t = 4$ s, comparing the closed-form solution, the numerical solution using $\Delta t = 0.5$ s, and the numerical solution using $\Delta t = 0.05$ s.

3.83 A 1000 kg rocket starts from rest and travels straight up. The total force exerted on it is $F = (100\,000 + 10\,000t - v_2)$ N. Using numerical integration with $\Delta t = 0.1$ s, determine the rocket's height and velocity for the first time steps. (Assume that the change in the rocket's mass is negligible over this time interval.)

P3.83

3.84 The force exerted on the 50 kg mass by the linear spring is $F = -kx$, where x is the displacement of the mass from its position when the spring is unstretched. The spring constant k is 50 N/m. The mass is released from rest in the position $x = 1$ m. Use numerical integration with $\Delta t = 0.01$ s to determine the position and velocity of the mass for the first five time steps.

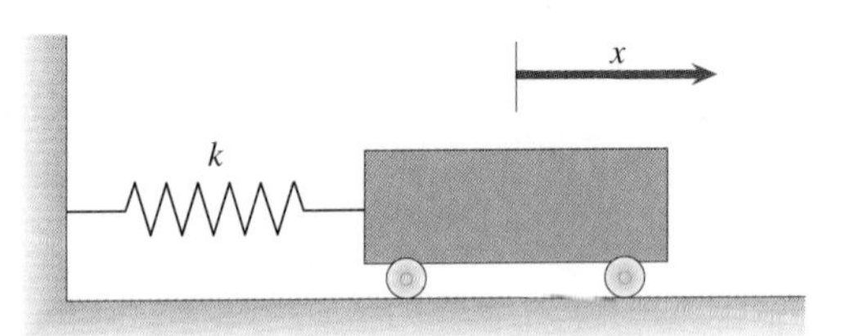

P3.84

3.85 In Problem 3.84, use numerical integration with $\Delta t = 0.01$ s to determine the position and velocity of the mass in terms of time from $t = 0$ to $t = 10$ s. Draw graphs of your results.

3.86 At $t = 0$, the velocity of a 50 kg machine element that moves along the x axis is $v_x = 7\,\text{m/s}$. Measurements of the total force ΣF_x acting on the element at 0.1 s intervals from $t = 0$ to $t = 0.9\,\text{s}$ give the following values:

Time, s	Force, N	Time, s	Force, N
0.0	50.0	0.5	58.8
0.1	51.1	0.6	57.6
0.2	56.0	0.7	55.4
0.3	57.2	0.8	52.1
0.4	58.5	0.0	49.9

Determine approximately how far the element moves from $t = 0$ to $t = 1\,\text{s}$ and its approximate velocity at $t = 1\,\text{s}$.

3.87 The lateral supports of a 100 kg structural element exert the horizontal force components

$$F_x = -2000x \qquad F_y = -2000y$$

where x and y are the coordinates of the centre of mass in metres. At $t = 0$, the coordinates and component of velocity of the centre of mass are $x = 0.1\,\text{m}$, $y = 0$, $v_x = 0$ and $v_y = 1\,\text{m/s}$. Using $\Delta t = 0.1\,\text{s}$, determine the approximate position and velocity of the centre of mass for the first five time steps.

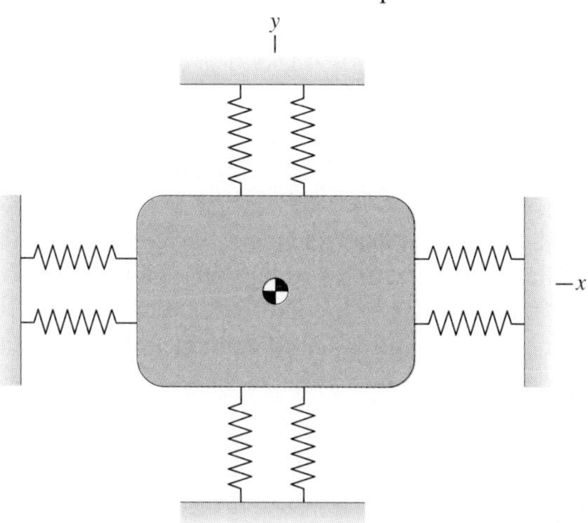

P3.87

3.88 In Problem 3.87, use numerical integration with $\Delta t = 0.001\,\text{s}$ to determine the elliptical path described by the centre of mass, and draw a graph of the path.

3.89 A car starts from rest at $t = 0$. Its acceleration is

$$a = (10 + 2t - 0.0185t^3)\,\text{m/s}^2$$

(a) Using the closed-form solution, determine the distance the car has travelled and its velocity at $t = 6\,\text{s}$.
(b) Use numerical integration with $\Delta t = 0.1\,\text{s}$ to approximate the answers obtained in part (a).
(c) Use numerical integration with $\Delta t = 0.01\,\text{s}$ to approximate the answers obtained in part (a).

3.90 A 20 kg projectile is launched from the ground with velocity components $v_x = 100\,\text{m/s}$, $v_y = 49\,\text{m/s}$. The magnitude of the aerodynamic drag force is $C|\mathbf{v}|^2$, where C is a constant. If the range of the projectile is 600 m, what is the constant C? (Use numerical integration with $\Delta t = 0.01\,\text{s}$ to compute the trajectory.)

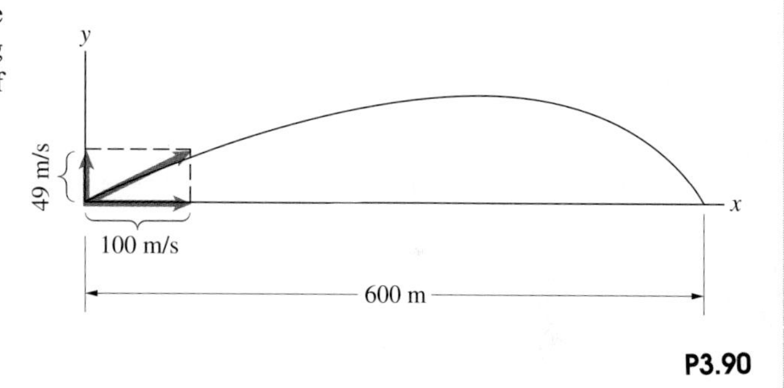

P3.90

Chapter Summary

The total external force on an object is equal to the product of its mass and the acceleration *of its centre of mass* relative to an inertial reference frame:

$$\Sigma \mathbf{F} = m\mathbf{a} \qquad \textbf{Equation (3.4)}$$

A reference frame is said to be inertial if it is one in which the second law can be applied in this form. A reference frame translating at constant velocity relative to an inertial reference frame is also inertial.

Expressing Newton's second law in terms of a coordinate system yields scalar equations of motion:

Cartesian Coordinates

$$\Sigma F_x = ma_x \qquad \Sigma F_y = ma_y \qquad \Sigma F_z = ma_z \qquad \textbf{Equation (3.5)}$$

Normal and Tangential Components

$$\Sigma F_{\mathrm{t}} = m\frac{dv}{dt} \qquad \Sigma F_{\mathrm{n}} = m\frac{v^2}{\rho} \qquad \textbf{Equation (3.7)}$$

Polar Coordinates

$$\Sigma F_r = m\left(\frac{d^2r}{dt^2} - r\omega^2\right) \qquad \textbf{Equation (3.9)}$$

$$\Sigma F_\theta = m\left(r\alpha + 2\frac{dr}{dt}\omega\right) \qquad \textbf{Equation (3.10)}$$

If the motion of an object is confined to a fixed plane, the component of the total force normal to the plane equals zero. In straight-line motion, the components of the total force perpendicular to the line equal zero and the component of the total force tangent to the line equals the product of the mass and the acceleration of the object along the line.

Review Problems

3.91 In a future mission, a spacecraft approaches the surface of an asteroid passing near the earth. Just before it touches down, the spacecraft is moving downwards at constant velocity relative to the surface of the asteroid and its downward thrust is 0.01 N. The computer decreases the downward thrust to 0.005 N, and an on-board laser interferometer determines that the acceleration of the spacecraft relative to the surface becomes 5×10^{-6} m/s^2 downwards. What is the gravitational acceleration of the asteroid near its surface?

P3.91

3.92 A 'cog' engine hauls three cars of sightseers to a mountaintop in Bavaria. The mass of each car including its passengers is 10 Mg and the friction forces exerted by the wheels of the cars are negligible. Determine the forces in the couplings 1, 2 and 3 if (a) the engine is moving at constant velocity; (b) the engine is accelerating up the mountain at $1.2\,\text{m/s}^2$.

P3.92

3.93 The car drives at constant velocity up the straight segment of road on the left. If the car's tyres continue to exert the same tangential force on the road after the car has gone over the crest of the hill and is on the straight segment of road on the right, what will be the car's acceleration?

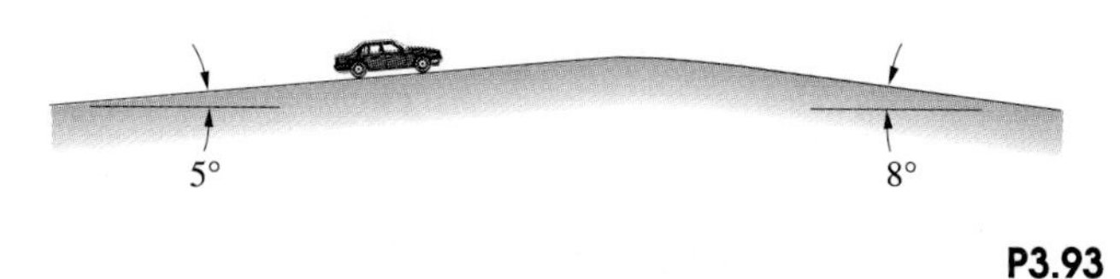

P3.93

3.94 The aircraft carrier *Nimitz* weighs 810 MN. Suppose that it is travelling at its top speed of approximately 30 knots (a knot is 1852 m/hr) when its engines are shut down. If the water exerts a drag force of magnitude $292v$ kN, were v is the carrier's velocity in metres per second, what distance does the carrier move before coming to rest?

3.95 If $m_A = 10\,\text{kg}$, $m_B = 40\,\text{kg}$ and the coefficients of kinetic friction between all surfaces is $\mu_k = 0.11$, what is the acceleration of B down the inclined surface?

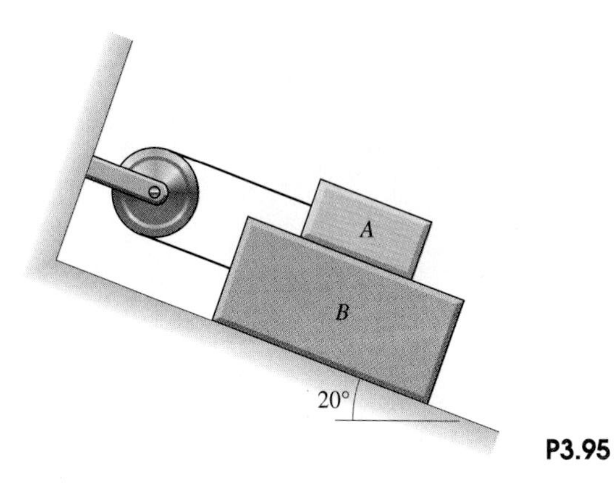

P3.95

3.96 In Problem 3.95, if A weighs 300 N, B weighs 1500 N, and the coefficient of kinetic friction between all surfaces is $\mu_k = 0.15$, what is the tension in the cord as B slides down the inclined surface?

3.97 A gas gun is used to accelerate projectiles to high velocities for research on material properties. The projectile is held in place while gas is pumped into the tube to a high pressure p_0 on the left and the tube is evacuated on the right. The projectile is then released and is accelerated by the expanding gas. Assume that the pressure p of the gas is related to the volume V it occupies by $pV^\gamma = \text{constant}$, where γ is a constant. If friction can be neglected, show that the velocity of the projectile at this position x is

$$v = \sqrt{\frac{2p_0 A x_0^\gamma}{m(\gamma - 1)}\left(\frac{1}{x_0^{\gamma-1}} - \frac{1}{x^{\gamma-1}}\right)}$$

where m is the mass of the projectile and A is the cross-sectional area of the tube.

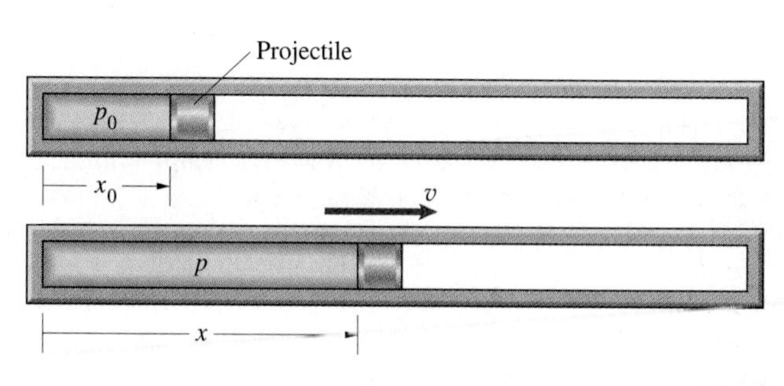

P3.97

3.98 The weights of the blocks are $W_A = 1800\,\text{N}$ and $W_B = 300\,\text{N}$ and the surfaces are smooth. Determine the acceleration of block A and the tension in the cord.

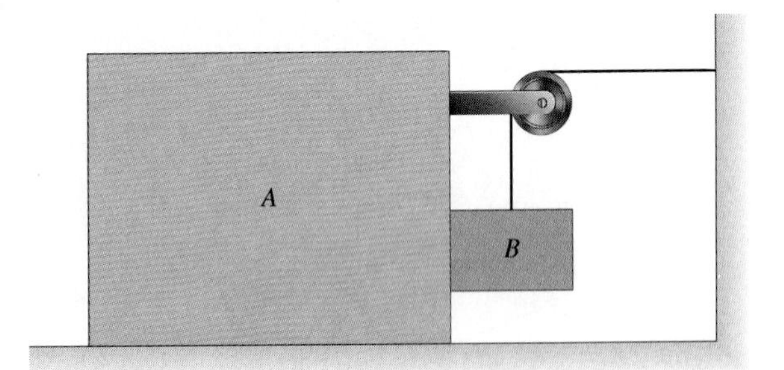

P3.98

3.99 The 100 Mg space shuttle is in orbit when its engines are turned on, exerting a thrust force $\mathbf{T} = (10\,\mathbf{i} - 20\,\mathbf{j} + 10\,\mathbf{k})\,\text{kN}$ for 2 s. Neglect the resulting change in its mass. At the end of the 2 s burn, fuel is still sloshing back and forth in the shuttle's tanks. What is the change in the velocity of the centre of mass of the shuttle (including the fuel it contains) due to the 2 s burn?

3.100 The water skier contacts the ramp with a velocity of 40 km/hr parallel to the surface of the ramp. Neglecting friction and assuming that the tow rope exerts no force on him once he touches the ramp, estimate the horizontal length of his jump from the end of the ramp.

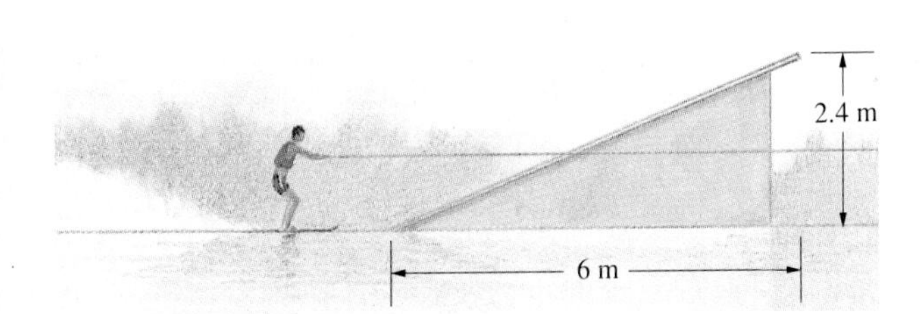

P3.100

3.101 Suppose you are designing a roller coaster track that will take the cars through a vertical loop of 12 m radius. If you decide that, for safety, the downward force exerted on a passenger by his seat at the top of the loop should be at least one-half his weight, what is the minimum safe velocity of the cars at the top of the loop?

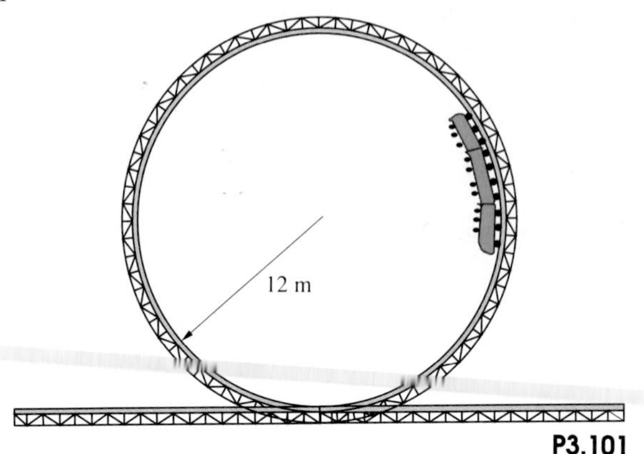

P3.101

3.102 If you want to design the cars of a train to tilt as the train goes around curves to achieve maximum passenger comfort, what is the relationship between the desired tilt angle α, the velocity v of the train, and the instantaneous radius of curvature ρ of the track?

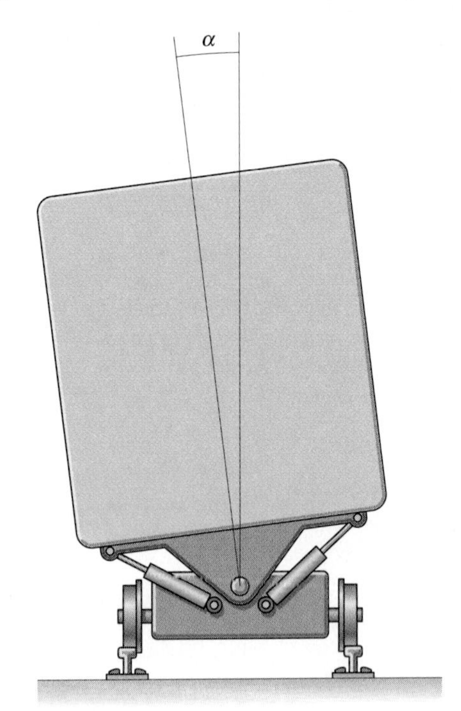

P3.102

3.103 If a car is travelling at 48 km/hr on a straight road and the coefficient of static friction between its tyres and the road is $\mu_s = 0.8$, what is the largest deceleration the driver can achieve by applying the brakes?

3.104 If the car in Problem 3.103 is travelling on an unbanked, circular curve of 30 m radius, what is the largest tangential deceleration the driver can achieve by applying the brakes?

3.105 To determine the coefficient of static friction between two materials, an engineer places a small sample of one material on a horizontal disc surfaced with the other one, then rotates the disc from rest with a constant angular acceleration of $0.4\,\mathrm{rad/s^2}$. If she determines that the small sample slips on the disc after 9.903 s, what is the coefficient of friction?

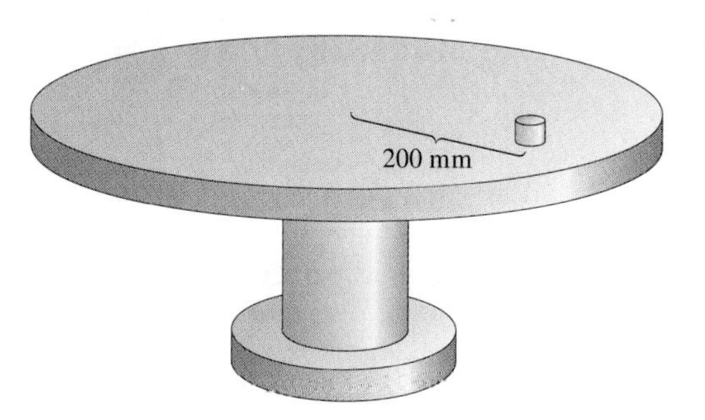

P3.105

3.106 As the smooth bar rotates *in the horizontal plane*, the string winds up on the fixed cylinder and draws the 1 kg collar A inwards. The bar starts from rest at $t = 0$ in the position shown and rotates with constant angular acceleration. What is the tension in the string at $t = 1$ s?

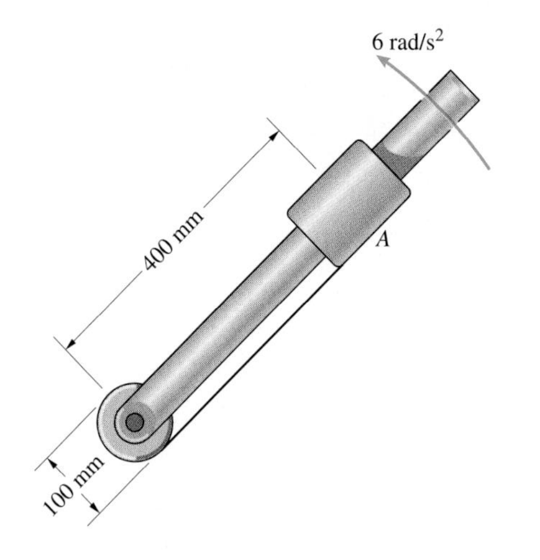

P3.106

3.107 In Problem 3.106, suppose that the coefficient of kinetic friction between the collar and the bar is $\mu_k = 0.2$. What is the tension in the string at $t = 1$ s?

3.108 The 1 kg slider A is pushed along the curved bar by the slotted bar. The curved bar lies *in the horizontal plane*, and its profile is described by $r = 2(\theta/2\pi + 1)$ m, where θ is in radians. The angular position of the slotted bar is $\theta = 2t$ rad. Determine the radial and transverse components of the total external force exerted on the slider when $\theta = 120°$.

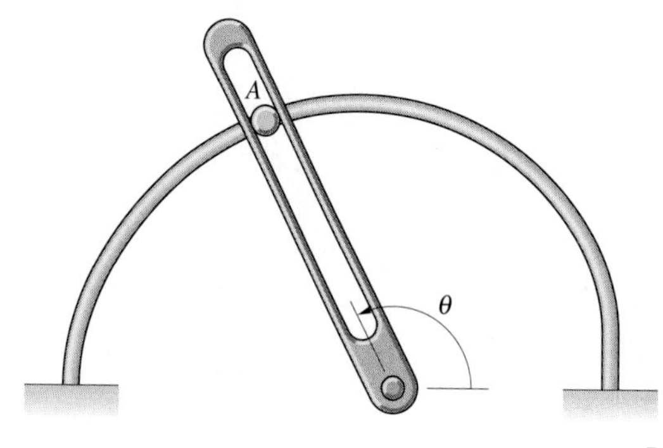

P3.108

3.109 In Problem 3.108, suppose that the curved bar lies *in the vertical plane*. Determine the radial and transverse components of the total force exerted on A by the curved and slotted bars at $t = 0.5$ s.

3.110 The ski boat moves relative to the water with a constant velocity of magnitude $|\mathbf{v}_B| = 10\,\mathrm{m/s}$. The magnitude of the 80 kg skier's velocity relative to the boat is $|\mathbf{v}_{S/B}| = 3\,\mathrm{m/s}$. The tension in the 11 m tow rope is 180 N, and the horizontal force exerted on the skier by the water is perpendicular to the direction of his motion relative to the water. If you can neglect other horizontal forces, what is the skier's acceleration in the direction of his motion relative to the water?

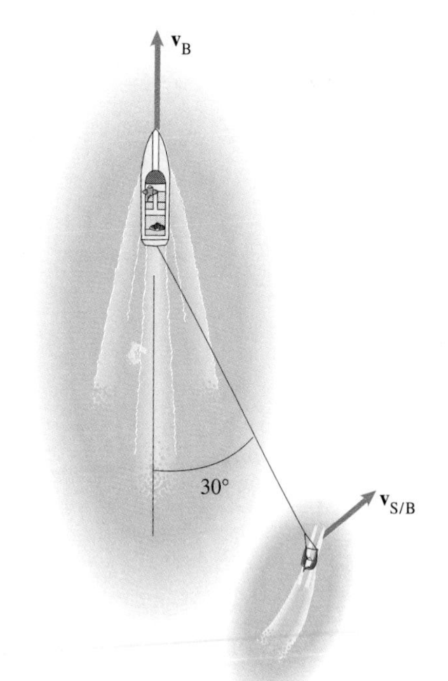

P3.110

3.111 In Problem 3.110, what is the magnitude of the horizontal force exerted on the skier by the water?

The ski lift performs work on the skiers, increasing their gravitational potential energy. Going down the hill, the skiers trade their gravitational potential energy for kinetic energy. To avoid going too fast, they must ski so that the snow performs negative work on them decreasing their kinetic energy. In this chapter we use the concepts of work and energy to analyse motions of objects.

Chapter 4

Energy Methods

ENERGY methods are used in nearly every area of science and engineering. Changes in energy must be considered in the design of any device that moves, including ski lifts as well as skis. The concepts of energy and conservation of energy originated in large part from the study of classical mechanics. A simple transformation of Newton's second law results in an equation that motivates the definitions of work, kinetic energy (energy due to an object's motion) and potential energy (energy due to an object's position). This equation relates the work done by the external forces acting on an object to the change in magnitude of its velocity. This relationship can greatly simplify the solution of problems involving forces that depend on an object's position, such as gravitational forces or forces exerted by springs. In addition, studying the derivation and applications in this chapter will develop your intuition concerning energy and its transformations and give you insight into applications of these ideas in other fields.

Work and Kinetic Energy

4.1 Principle of Work and Energy

You have used Newton's second law to relate the acceleration of an object's centre of mass to its mass and the external forces acting on it. We will now show how this vector equation can be mathematically transformed into a scalar form that is extremely useful in certain circumstances. We begin with Newton's second law in the form

$$\Sigma \mathbf{F} = m\frac{d\mathbf{v}}{dt} \tag{4.1}$$

and take the dot product of both sides with the velocity:

$$\Sigma \mathbf{F}\cdot\mathbf{v} = m\frac{d\mathbf{v}}{dt}\cdot\mathbf{v} \tag{4.2}$$

By expressing the velocity on the left side of this equation as $d\mathbf{r}/dt$ and observing that

$$\frac{d}{dt}(\mathbf{v}\cdot\mathbf{v}) = \frac{d\mathbf{v}}{dt}\cdot\mathbf{v} + \mathbf{v}\cdot\frac{d\mathbf{v}}{dt} = 2\frac{d\mathbf{v}}{dt}\cdot\mathbf{v}$$

we can write Equation (4.2) as

$$\Sigma \mathbf{F}\cdot d\mathbf{r} = \frac{1}{2}m\,d(v^2) \tag{4.3}$$

where $v^2 = \mathbf{v}\cdot\mathbf{v}$ is the square of the magnitude of $\mathbf{v}$. The term on the left is the **work** expressed in terms of the total external force acting on the object and the infinitesimal displacement $d\mathbf{r}$. We integrate this equation,

$$\int_{\mathbf{r}_1}^{\mathbf{r}_2} \Sigma \mathbf{F}\cdot d\mathbf{r} = \int_{v_1^2}^{v_2^2} \frac{1}{2}m\,d(v^2) \tag{4.4}$$

where v_1 and v_2 are the magnitudes of the velocity at the positions $\mathbf{r}_1$ and $\mathbf{r}_2$. Evaluating the integral on the right side, we obtain

$$\boxed{U = \frac{1}{2}mv_2^2 - \frac{1}{2}mv_1^2} \tag{4.5}$$

where

$$\boxed{U = \int_{\mathbf{r}_1}^{\mathbf{r}_2} \Sigma \mathbf{F}\cdot d\mathbf{r}} \tag{4.6}$$

is the work done as the centre of mass of the object moves from position $\mathbf{r}_1$ to position $\mathbf{r}_2$. The term $\frac{1}{2}mv^2$ is called the **kinetic energy**. The dimensions of the work, and therefore the dimensions of the kinetic energy, are (force) × (length). In US Customary units, work is expressed in foot-pounds. In SI units, work is expressed in newton-metres, or joules (J).

Equation (4.5) states that the work done on an object as it moves from a position $\mathbf{r}_1$ to a position $\mathbf{r}_2$ is equal to the change in its kinetic energy. This is called the **principle of work and energy**. If you can evaluate the work, this principle allows you to determine the change in the magnitude of an object's velocity as it moves from one position to another. You can also equate the total work done by external forces on a system of objects to the change in the total kinetic energy of the system *if no net work is done by internal forces*. Internal friction forces can do net work on a system (See Example 4.3.)

Although the principle of work and energy relates changes in position to changes in velocity, you cannot use it to obtain other information about the motion, such as the time required to move from one position to another. Furthermore, since the work is an integral with respect to position, you can usually evaluate it only when the forces doing work are known as functions of position. Despite these limitations, this principle is extremely useful for certain problems because the work can be determined very easily.

4.2 Work and Power

In this section we discuss how to determine the work done on an object, both in general and in several common and important special cases. We also define the power done by the forces acting on an object and show how it is calculated.

Evaluating the Work

Let's consider an object in curvilinear motion (Figure 4.1(a)) and specify its position by the coordinate s measured along its path from a reference point O. In terms of the tangential unit vector $\mathbf{e}_t$, the object's velocity is

$$\mathbf{v} = \frac{ds}{dt}\mathbf{e}_t$$

Because $\mathbf{v} = d\mathbf{r}/dt$, we can multiply the velocity by dt to obtain an expression for the vector $d\mathbf{r}$ describing an infinitesimal displacement along the path (Figure 4.1(b)):

$$d\mathbf{r} = \mathbf{v}\,dt = ds\,\mathbf{e}_t$$

The work done by the external forces acting on the object as a result of the displacement $d\mathbf{r}$ is

$$\Sigma\mathbf{F}\cdot d\mathbf{r} = (\Sigma\mathbf{F}\cdot\mathbf{e}_t)ds = \Sigma F_t\,ds$$

where ΣF_t is the tangential component of the total force. Therefore, as the

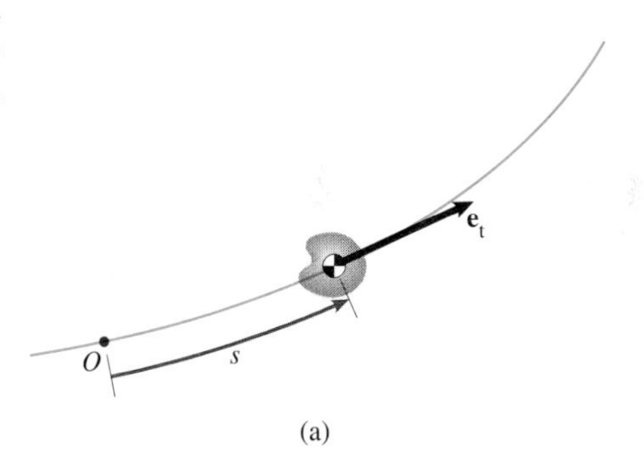

(a)

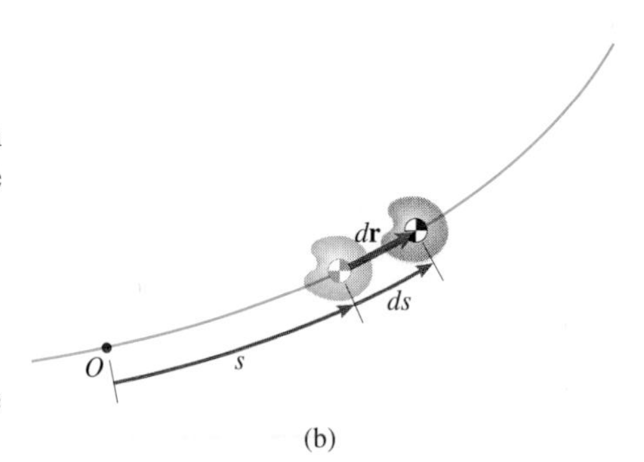

(b)

Figure 4.1

(a) The coordinate s and tantential unit vector.
(b) An infinitesimal displacement $d\mathbf{r}$.

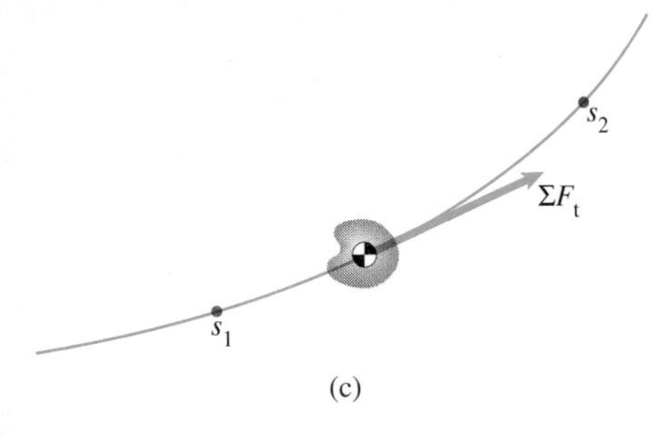

(c) The work done from s_1 and s_2 is determined by the tangential component of the external forces

object moves from a position s_1 to a position s_2 (Figure 4.1(c)), the work is

$$U = \int_{s_1}^{s_2} \Sigma F_t \, ds \tag{4.7}$$

The work is equal to the integral of the tangential component of the total force with respect to distance along the path. Thus the work done is equal to the area defined by the graph of the tangential force from s_1 to s_2 (Figure 4.2(a)). *Components of force perpendicular to the path do no work.* Notice that if ΣF_t is opposite to the direction of motion over some part of the path, which means the object is decelerating, the work is negative (Figure 4.2(b)). If ΣF_t is constant between s_1 and s_2, the work is simply the product of the total tangential force and the displacement (Figure 4.2(c)):

$$U = \Sigma F_t(s_2 - s_1) \qquad \textbf{Constant tangential force} \tag{4.8}$$

Figure 4.2

(a) The work equals the area defined by the graph of the tangential force as a function of distance along the path.
(b) Negative work is done if the tangential force is opposite to the direction of motion.
(c) The work done by a constant tangential force equals the product of the force and the distance.

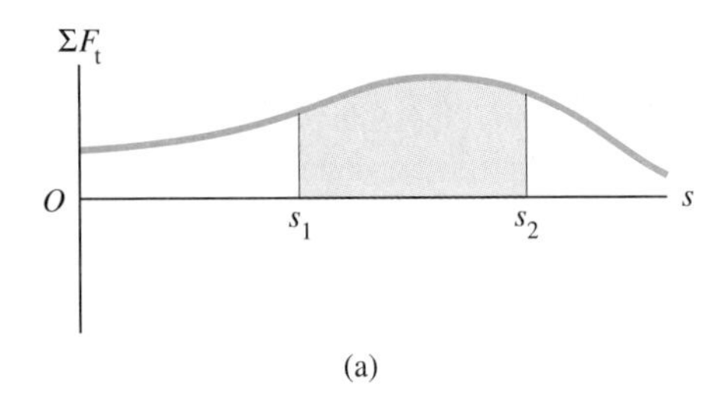

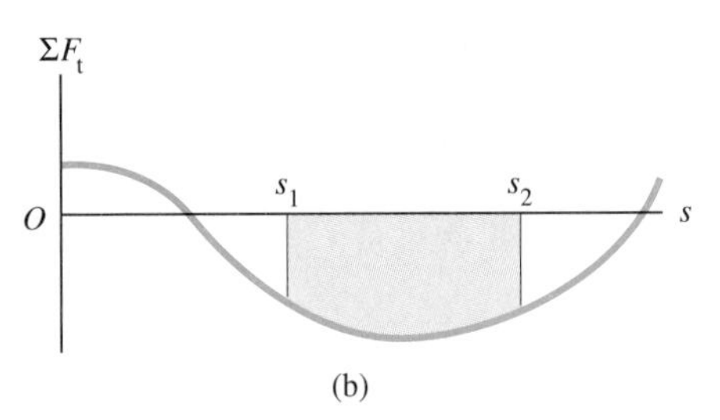

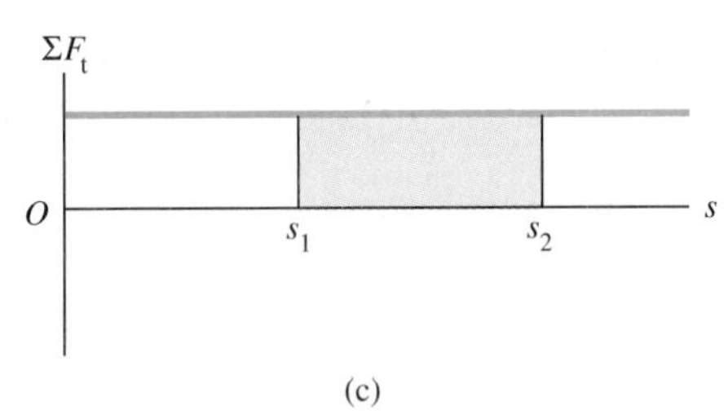

In the following examples we apply the principle of work and energy and use Equations (4.7) and (4.8) to evaluate the work. You should consider using work and energy when you want to relate the change in velocity of an object to a change in its position. This typically involves two steps:

(1) Identify the forces that do work – *By drawing a free-body diagram, you must determine which external forces do work on the object.*

(2) Apply work and energy – *Equate the total work done during a change in position to the change in the object's kinetic energy.*

Example 4.1

The 180 kg container A in Figure 4.3 starts from rest at position $s = 0$ and is subjected to a horizontal force $F = 700 - 45s$ N by the hydraulic cylinder. The coefficient of kinetic friction between the container and the floor is $\mu_k = 0.26$. What is the velocity of the container when it has reached the position $s = 1.2$ m?

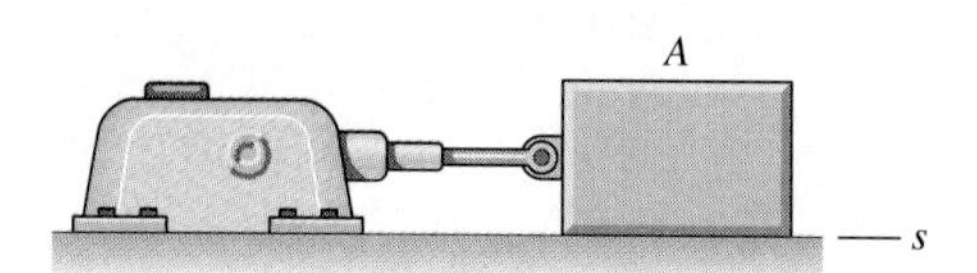

Figure 4.3

SOLUTION

Identify the Forces That Do Work We draw the free-body diagram of the container in Figure (a). The forces teengent to its path are the force exerted by the hydraulic cylinder and the friction force. The container's acceleration in the vertical direction is zero, so $N = 1766$ N.

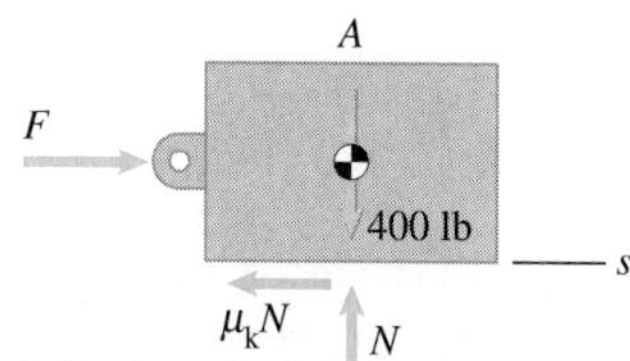

(a) Free-body diagram of the container.

Apply Work and Energy Let v be the magnitude of the container's velocity at $s = 1.2$ m. Using Equation (4.7) to evaluate the work, we obtain

$$\int_{s_1}^{s_2} \Sigma F_t \, ds = \frac{1}{2} m v_2^2 = \frac{1}{2} m_1^2 :$$

$$\int_0^{1.2} (F - \mu_k N) ds = \frac{1}{2} m v^2 - 0$$

$$\int_0^{1.2} [(700 - 45\,s) - (0.26)(1766)] ds = \frac{1}{2} \left(\frac{180}{9.81} \right) v^2$$

Evaluating the integral and solving for v, we obtain $v = 1.69$ m/s.

Example 4.2

The two crates in Figure 4.4 are released from rest. Their masses are $m_A = 40\,\text{kg}$ and $m_B = 30\,\text{kg}$, and the kinetic coefficient of friction between crate A and the inclined surface is $\mu_k = 0.15$. What is their velocity when they have moved 400 mm?

Figure 4.4

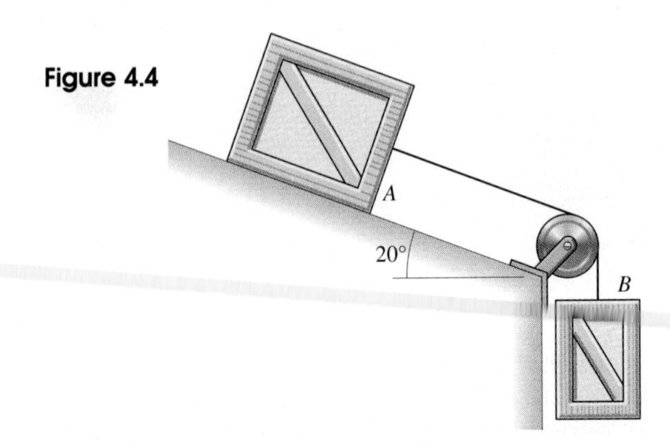

STRATEGY

We will determine the velocity in two ways.

First method By drawing free-body diagrams of each of the crates and applying the principle of work and energy to them individually, we can obtain two equations in terms of the magnitude of the velocity and the tension in the cable.

Second method We can draw a single free-body diagram of the two crates, the cable and the pulley and apply the principle of work and energy to the entire system.

SOLUTION

First Method We draw the free-body diagram of crate A in Figure (a). The forces that do work as the crate moves down the plane are the forces tangential to its path: the tension T, the tangential component of the weight $m_A g \sin 20°$, and the friction force $\mu_k N$. Because the acceleration of the crate normal to the surface is zero, $N = m_A g \cos 20°$. Let v be the magnitude of the crate's velocity when it has moved

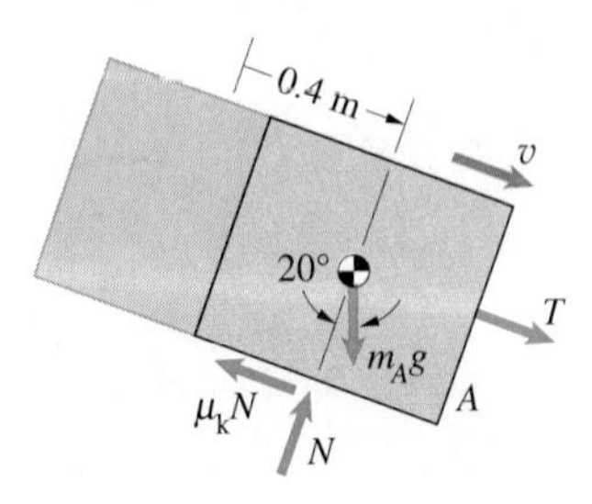

(a) Free-body diagram of A.

400 mm. Using Equation (4.7) to determine the work, we equate the work done on A to the change in its kinetic energy:

$$\int_{s_1}^{s_2} \Sigma F_t \, ds = \frac{1}{2}mv_2^2 - \frac{1}{2}mv_1^2 :$$

$$\int_0^{0.4} [T + m_A g \sin 20^\circ - \mu_k(m_A g \cos 20^\circ)]ds = \frac{1}{2}m_A v^2 - 0 \qquad (4.9)$$

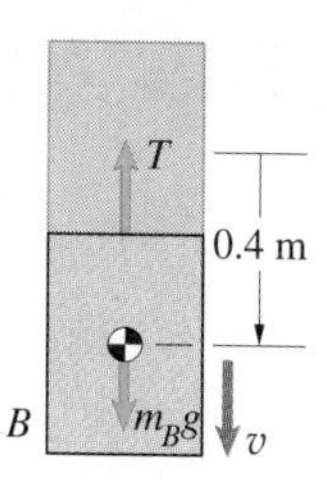

(b) Free-body diagram of B.

The forces that do work on crate B are its weight $m_B g$ and the tension T (Figure (b)). The magnitude of its velocity is the same as that of crate A. The work done on B equals the change in its kinetic energy:

$$\int_{s_1}^{s_2} \Sigma F_t \, ds = \frac{1}{2}mv_2^2 - \frac{1}{2}mv_1^2 :$$

$$\int_0^{0.4} (m_B g - T)\, ds = \frac{1}{2}m_B v^2 - 0 \qquad (4.10)$$

By summing Equations (4.9) and (4.10), we eliminate T, obtaining

$$\int_0^{0.4} (m_A g \sin 20^\circ - \mu_k m_A g \cos 20^\circ + m_B g)\, ds = \frac{1}{2}(m_A + m_B)v^2 :$$

$$[40 \sin 20^\circ - (0.15)(40) \cos 20^\circ + 30](9.81)(0.4) = \frac{1}{2}(40 + 30)v^2$$

Solving for v, the velocity of the boxes is $v = 2.07$ m/s.

Second Method We draw the free-body diagram of the system consisting of the crates, cable and pulley in Figure (c). Notice that the cable tension does not appear in this free-body diagram. The reactions at the pin support of the pulley do no work, because the support does not move. The total work done by external forces on the system as the boxes move 400 mm is equal to the change in the total kinetic energy of the system:

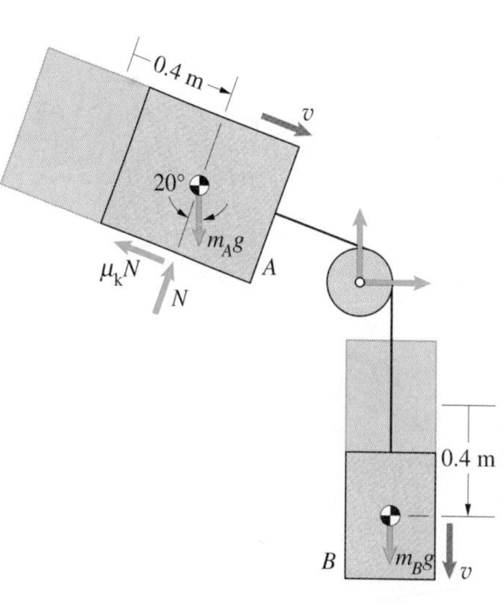

(c) Free-body diagram of the system.

$$\int_0^{0.4} [m_A g \sin 20^\circ - \mu_k(m_A g \cos 20^\circ)]\, ds + \int_0^{0.4} m_B g \, ds$$

$$= \left(\frac{1}{2}m_A v^2 + \frac{1}{2}m_B v^2\right) - 0 :$$

$$[40 \sin 20^\circ - (0.15)(40) \cos 20^\circ + 30](9.81)(0.4) = \frac{1}{2}(40 + 30)v^2$$

This equation is identical to that we obtained by applying the principle of work and energy to the individual crates.

DISCUSSION

You will often find it simpler to apply the principle of work and energy to an entire system instead of its separate parts. However, as we demonstrate in the next example, you need to be aware that internal forces in a system can do net work.

Example 4.3

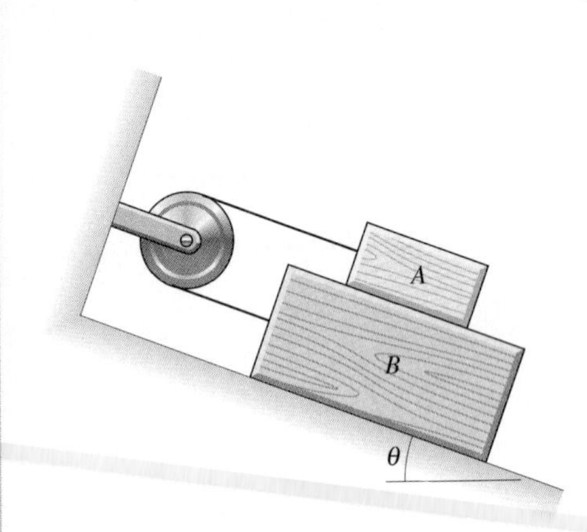

Figure 4.5

Crates A and B in Figure 4.5 are released from rest. The coefficient of kinetic friction between A and B is μ_k, and friction between B and the inclined surface can be neglected. What is their velocity when they have moved a distance b?

STRATEGY

By applying the principle of work and energy to each crate, we can obtain two equations in terms of the tension in the cable and the velocity.

SOLUTION

We draw the free-body diagrams of the crates in Figures (a) and (b). The acceleration of A normal to the inclined surface is zero, so $N = m_A g \cos\theta$. Let v be the magnitude of the velocity when the crates have moved a distance b. The work done on A equals the change in its kinetic energy,

$$\int_0^b (T - m_A g \sin\theta - \mu_k m_A g \cos\theta)\, ds = \frac{1}{2} m_A v^2 \tag{4.11}$$

and the work done on B equals the change in its kinetic energy,

$$\int_0^b (-T + m_B g \sin\theta - \mu_k m_A g \cos\theta)\, ds = \frac{1}{2} m_B v^2 \tag{4.12}$$

Summing these equations to eliminate T and solving for v, we obtain

$$v = \sqrt{2bg[(m_B - m_A)\sin\theta - 2\mu_k m_A \cos\theta]/(m_A + m_B)}.$$

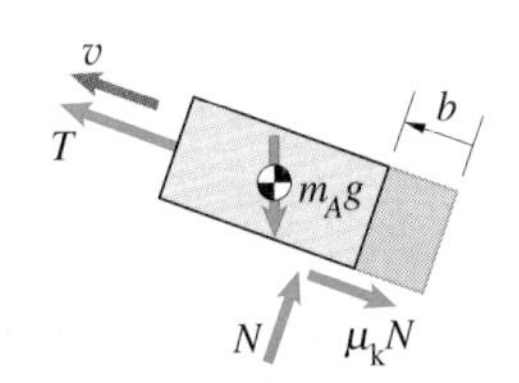

(a) Free-body diagram of A.

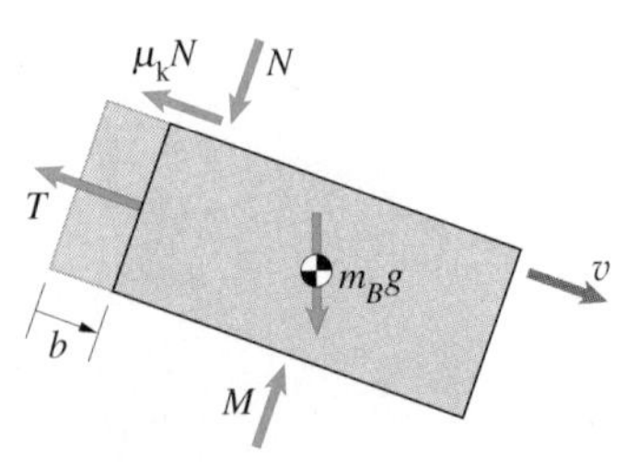

(b) Free-body diagram of B.

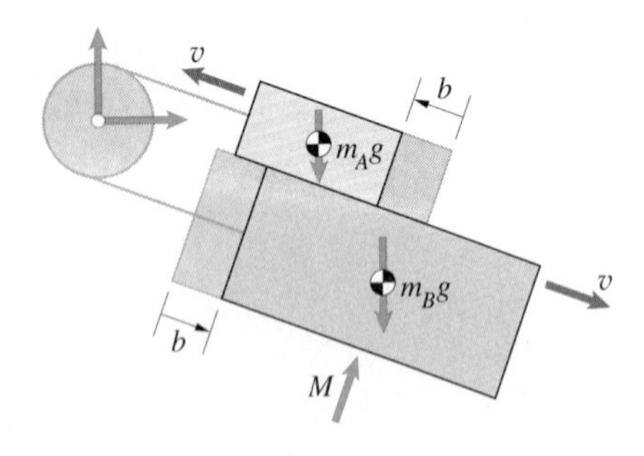

(c) Free-body diagram of the system.

DISCUSSION

If we attempt to solve this example by applying the principle of work and energy to the system consisting of the crates, the cable and the pulley (Figure (c)), we obtain an incorrect result. Equating the work done by external forces to the change in the total kinetic energy of the system, we obtain

$$\int_0^b m_B g \sin\theta\, ds - \int_0^b m_A g \sin\theta\, ds = \frac{1}{2} m_A v^2 + \frac{1}{2} m_B v^2 :$$

$$(m_B g \sin\theta) b - (m_A g \sin\theta) b = \frac{1}{2} m_A v^2 + \frac{1}{2} m_B v^2$$

But if we sum our work and energy equations for the individual crates – Equations (4.11) and (4.12) – we obtain the correct equation:

$$\underbrace{[(m_B g \sin\theta) b - (m_A g \sin\theta) b]}_{\substack{\text{Work by} \\ \text{external forces}}} + \underbrace{[-(2\mu_k m_A g \cos\theta) b]}_{\substack{\text{Work by} \\ \text{internal forces}}} = \frac{1}{2} m_A v^2 + \frac{1}{2} m_B v^2$$

The internal friction forces the crates exert on each other do net work on the system. We did not account for this work in applying the principle of work and energy to the free-body diagram of the system.

Work Done by Various Forces

You have seen that if the tangential component of the total external force on an object is known as a function of distance along the object's path, you can use the principle of work and energy to relate a change in position to the change in the object's velocity. For certain types of forces, however, not only can you determine the work without knowing the tangential component of the force as a function of distance along the path, you don't even need to know the path. Two important examples are weight and the force exerted by a spring.

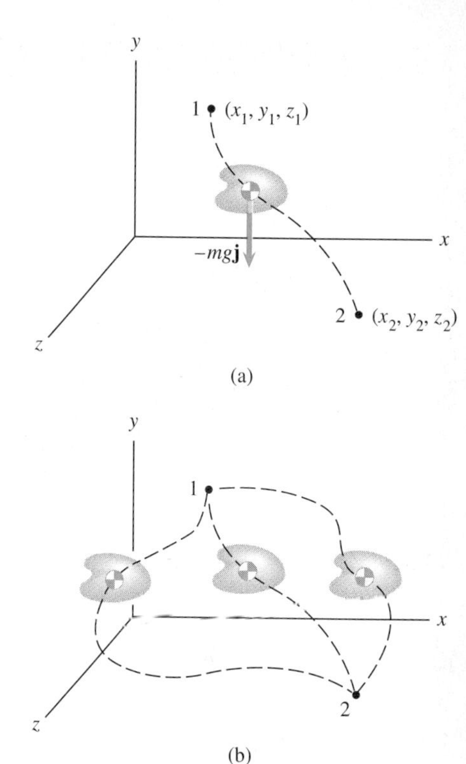

Figure 4.6

(a) An object moving between two positions (b) the work done by the weight is the same for any path.

Weight To evaluate the work done by an object's weight, we orient a cartesian coordinate system with the y axis upwards and suppose that the object moves from position 1 with coordinates (x_1, y_1, z_1) to position 2 with coordinates (x_2, y_2, z_2) (Figure 4.6(a)). The force exerted by its weight if $\mathbf{F} = -mg\,\mathbf{j}$. (Other forces may act on the object, but we are concerned only with the work done by its weight.) Because $\mathbf{v} = d\mathbf{r}/dt$, we can multiply the velocity, expressed in cartesian coordinates, by dt to obtain an expression for the vector $d\mathbf{r}$:

$$d\mathbf{r} = \left(\frac{dx}{dt}\,\mathbf{i} + \frac{dy}{dt}\,\mathbf{j} + \frac{dz}{dt}\,\mathbf{k}\right) dt = dx\,\mathbf{i} + dy\,\mathbf{j} + dz\,\mathbf{k}$$

Taking the dot product of $\mathbf{F}$ and $d\mathbf{r}$,

$$\mathbf{F}\cdot d\mathbf{r} = (-mg\,\mathbf{j})\cdot(dx\,\mathbf{i} + dy\,\mathbf{j} + dz\,\mathbf{k}) = -mg\;dy$$

the work done as the object moves from position 1 to position 2 reduces to an integral with respect to y:

$$U = \int_{\mathbf{r}_1}^{\mathbf{r}_2} \mathbf{F}\cdot d\mathbf{r} = \int_{y_1}^{y_2} -mg\;dy$$

Evaluating the integral, we obtain the work done by the weight of an object as it moves between two positions:

$$\boxed{U = -mg(y_2 - y_1)} \qquad (4.13)$$

The work is simply the product of the weight and the change in the object's height. The work done is negative if the height increases and positive if it decreases. Notice that *the work done is the same no matter what path the object follows from position* 1 *to position* 2 (Figure 4.6(b)). You don't need to know the path to determine the work done by an object's weight – you only need to know the relative heights of the two positions.

What work is done by an object's weight if we account for its variation with distance from the centre of the earth? In terms of polar coordinates, we can write the weight of an object at a distance r from the centre of the earth as (Figure 4.7)

$$\mathbf{F} = -\frac{mgR_{\mathrm{E}}^2}{r^2}\,\mathbf{e}_r$$

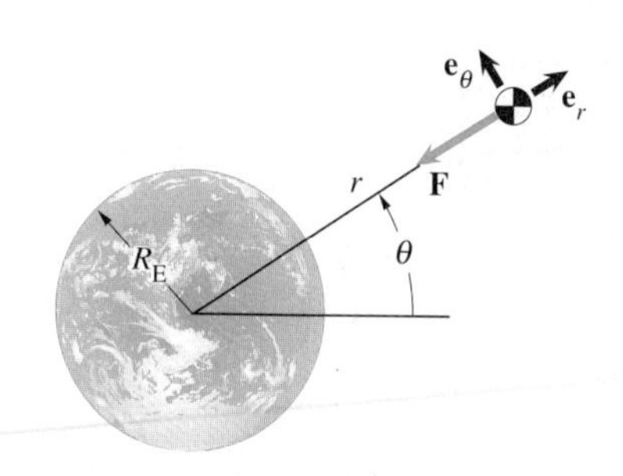

Figure 4.7

Expressing an object's weight in polar coordinates.

Using the expression for the velocity in polar coordinates, the vector $d\mathbf{r} = \mathbf{v}\,dt$ is

$$d\mathbf{r} = \left(\frac{dr}{dt}\,\mathbf{e}_r + r\,\frac{d\theta}{dt}\,\mathbf{e}_\theta\right) dt = dr\,\mathbf{e}_r + r\,d\theta\,\mathbf{e}_\theta \tag{4.14}$$

The dot product of $\mathbf{F}$ and $d\mathbf{r}$ is

$$\mathbf{F}\cdot d\mathbf{r} = \left(-\frac{mgR_{\mathrm{E}}^2}{r^2}\,\mathbf{e}_r\right)\cdot(dr\,\mathbf{e}_r + r\,d\theta\,\mathbf{e}_\theta) = -\frac{mgR_{\mathrm{E}}^2}{r^2}\,dr$$

so the work reduces to an integral with respect to r:

$$U = \int_{\mathbf{r}_1}^{\mathbf{r}_2} \mathbf{F}\cdot d\mathbf{r} = \int_{r_1}^{r_2} -\frac{mgR_{\mathrm{E}}^2}{r^2}\,dr$$

Evaluating the integral, we obtain the work done by an object's weight accounting for the variation of the weight with height:

$$\boxed{U = mgR_{\mathrm{E}}^2\left(\frac{1}{r_2} - \frac{1}{r_1}\right)} \tag{4.15}$$

Again, the work is independent of the path from position 1 to position 2. To evaluate it, you only need to know the object's radial distance from the centre of the earth at the two positions.

Springs Suppose that a linear spring connects an object to a fixed support. In terms of polar coordinates (Figure 4.8), the force exerted on the object is

$$\mathbf{F} = -k(r - r_0)\,\mathbf{e}_r$$

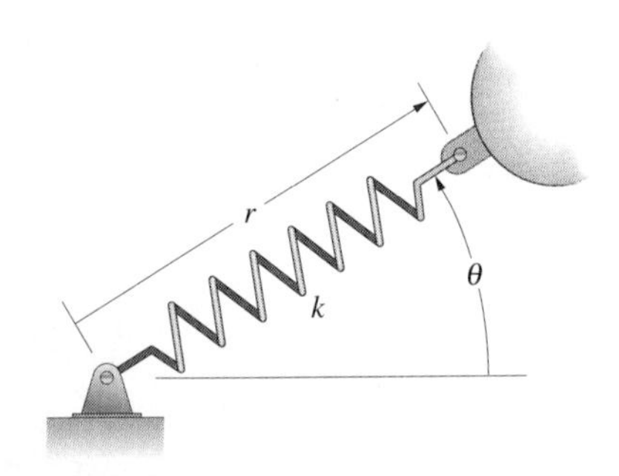

Figure 4.8

Expressing the force exerted by a linear spring in polar coordinates.

where k is the spring constant and r_0 is the unstretched length of the spring. Using Equation (4.14), the dot product of $\mathbf{F}$ and $d\mathbf{r}$ is

$$\mathbf{F}\cdot d\mathbf{r} = [-k(r - r_0)\,\mathbf{e}_r]\cdot(dr\,\mathbf{e}_r + r d\theta\,\mathbf{e}_\theta) = -k(r - r_0)\,dr$$

It is convenient to express the work done by a spring in terms of its **extension** defined by $S = r - r_0$. (Although the word *extension* usually means an increase in length, we use this term more generally to denote the change in length of the spring. A negative extension is a decrease in length.) In terms of this variable, $\mathbf{F}\cdot d\mathbf{r} = -kS\,dS$, and the work is

$$U = \int_{\mathbf{r}_1}^{\mathbf{r}_2} \mathbf{F}\cdot d\mathbf{r} = \int_{S_1}^{S_2} -kS\,dS$$

The work done on an object by a spring attached to a fixed support is

$$\boxed{U = -\frac{1}{2}k(S_2^2 - S_1^2)} \tag{4.16}$$

where S_1 and S_2 are the values of the extension at the initial and final positions.

You don't need to know the object's path to determine the work done by the spring. You must remember that Equation (4.16) applies to a *linear* spring. In Figure 4.9 we determine the work done in stretching a linear spring by calculating the area defined by the graph of kS as a function of S.

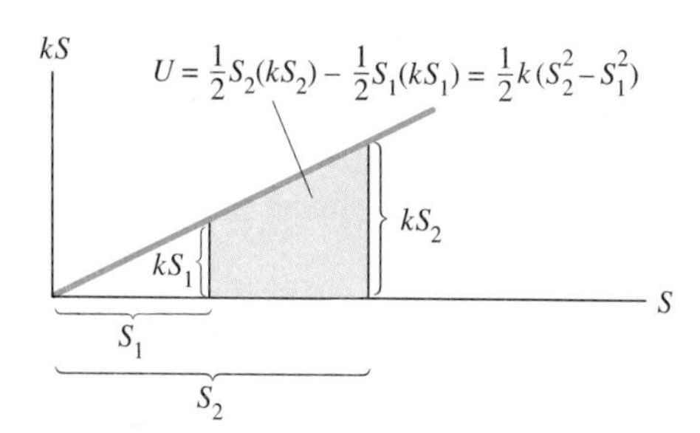

Figure 4.9

Work done in stretching a linear spring from S_1 to S_2. (If $S_2 > S_1$, the work done *on* the spring is positive, so the work done *by* the spring is negative.)

Power

Power is the rate at which work is done. The work done by the external forces acting on an object during an infinitesimal displacement $d\mathbf{r}$ is

$$\Sigma \mathbf{F} \cdot d\mathbf{r}$$

We obtain the power P by dividing this expression by the interval of time dt during which the displacement takes place:

$$P = \Sigma \mathbf{F} \cdot \mathbf{v} \tag{4.17}$$

This is the power transferred to or from the object, depending on whether P is positive or negative. In SI units, power is expressed in newton-metres per second, which is joules per second (J/s) or watts (W). In US Customary units, power is expressed in foot-pounds per second or in the anachronistic horse-power (hp), which is 746 W or 550 ft-lbs/s.

Notice from Equation (4.3) that the power equals the rate of change of the kinetic energy of the object:

$$P = \frac{d}{dt}\left(\frac{1}{2}mv^2\right)$$

Transferring power to or from an object causes its kinetic energy to increase or decrease. Using this relation, we can write the average with respect to time of the power during an interval of time from t_1 to t_2 as

$$P_{\text{av}} = \frac{1}{t_2 - t_1}\int_{t_1}^{t_2} P\, dt = \frac{1}{t_2 - t_1}\int_{v_2^2}^{v_2^2} \frac{1}{2}m\, d(v^2)$$

This result states that the average power transferred to or from an object during an interval of time is equal to the change in its kinetic energy, or the work done, divided by the interval of time:

$$P_{\text{av}} = \frac{\frac{1}{2}mv_2^2 - \frac{1}{2}mv_1^2}{t_2 - t_1} = \frac{U}{t_2 - t_1} \tag{4.18}$$

Example 4.4

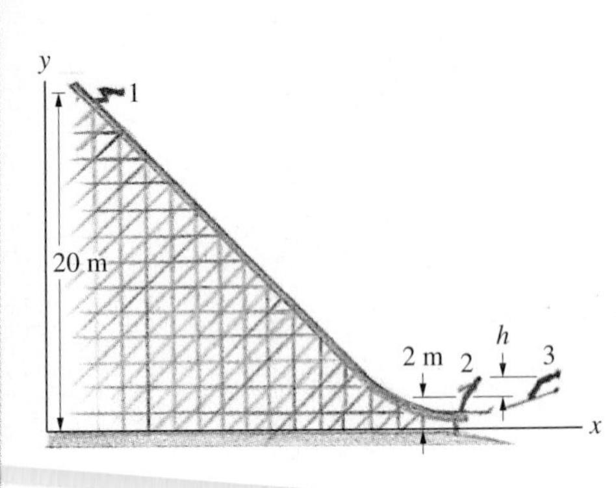

Figure 4.10

The skier in Figure 4.10 is travelling at 15 m/s at position 1. When he reaches the level end of the ramp at position 2, he jumps upwards, achieving a vertical component of velocity of 3 m/s. (Disregard the change in the vertical position of his centre of mass due to his jump.) Neglect aerodynamic drag and the friction forces on his skis.

(a) What is the magnitude of his velocity as he leaves the ramp at position 2?

(b) At the highest point of his jump, position 3, what is his height h above position 2?

STRATEGY

(a) If we neglect aerodynamic and friction forces, the only force doing work from position 1 to position 2 is the skier's weight, so we can apply the principle of work and energy to determine his velocity at position 2 before he jumps.

(b) From the time he leaves the ramp at position 2 until he reaches position 3, the only force is his weight, so $a_x = 0$ and the horizontal component of his velocity is constant. That means that we know the magnitude of his velocity at position 3, because he is moving horizontally at that point. Therefore we can apply the principle of work and energy to his motion from position 2 to position 3 to determine h.

SOLUTION

(a) Using Equation (4.13) to evaluate the work done by his weight from position 1 to position 2, the principle of work and energy is

$$-mg(y_2 - y_1) = \frac{1}{2}mv_2^2 - \frac{1}{2}mv_1^2 :$$

$$-m(9.81)(2 - 20) = \frac{1}{2}mv_2^2 - \frac{1}{2}m(15)^2$$

Solving for v_2, the magnitude of his velocity at position 2 before he jumps upwards is 24.04 m/s. After he jumps upwards the magnitude of his velocity at position 2 is $v_2' = \sqrt{(24.04)^2 + (3)^2} = 24.23$ m/s.

(b) The magnitude of his velocity at position 3 is equal to the horizontal component of his velocity at position 2: $v_3 = 24.04$ m/s. Applying work and energy to his motion from position 2 to position 3,

$$-mg(y_3 - y_2) = \frac{1}{2}mv_3^2 - \frac{1}{2}m(v_2')^2 :$$

$$-m(9.81)h = \frac{1}{2}m(24.04)^2 - \frac{1}{2}m(24.23)^2$$

we obtain $h = 0.459$ m.

DISCUSSION

Although we neglected aerodynamic effects, a ski jumper is actually subjected to substantial aerodynamic forces, both parallel to his path (drag) and perpendicular to it (lift).

Example 4.5

In the forgoing device shown in Figure 4.11, the 40 kg hammer is lifted to position 1 and released from rest. It falls and strikes a workpiece when it is in position 2. The spring constant $k = 1500$ N/m, and the tension in each spring is 150 N when the hammer is in position 2. Neglect friction.
(a) What is the velocity of the hammer just before it strikes the workpiece?
(b) Assuming that all of the hammer's kinetic energy is transferred to the workpiece, what average power is transferred if the duration of the impact is 0.02 s?

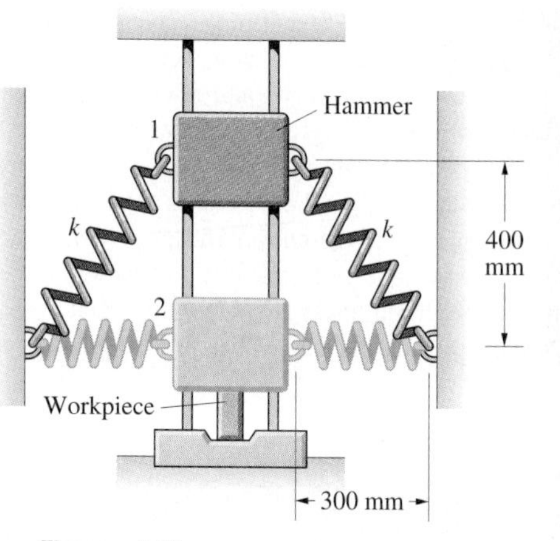

Figure 4.11

STRATEGY

Work is done on the hammer by its weight and by the two springs. We can apply the principle of work and energy to the motion of the hammer from position 1 to position 2 to determine its velocity at position 2.

SOLUTION

(a) Let r_0 be the unstretched length of one of the springs. In position 2, the tension in the spring is 150 N and its length is 0.3 m. From the relation between the tension in a linear spring and its extension,

$$150 - k(0.3 - r_0) = (1500)(0.3 - r_0)$$

we obtain $r_0 = 0.2$ m. The values of the extension of each spring in positions 1 and 2 are $S_1 = \sqrt{(0.4)^2 + (0.3)^2} - 0.2 = 0.3$ m and $S_2 = 0.3 - 0.2 = 0.1$ m. From Equation (4.16), the total work done on the manner by the two springs from position 1 to position 2 is

$$U_{\text{springs}} = 2\left[-\frac{1}{2}k(S_2^2 - S_1^2)\right] = -(1500)[(0.1)^2 - (0.3)^2] = 120\,\text{N.m}$$

The work done by the weight from position 1 to position 2 is positive and equal to the product of the weight and the change in height:

$$U_{\text{weight}} = mg(0.4\,\text{m}) = (40)(9.81)(0.4) = 156.96\,\text{N.m}$$

From the principle of work and energy,

$$U_{\text{springs}} + U_{\text{weight}} = \frac{1}{2}mv_2^2 - \frac{1}{2}mv_1^2 :$$

$$120 + 156.96 = \frac{1}{2}(40)v_2^2 - 0$$

we obtain $v_2 = 3.72$ m/s.
(b) All of the hammer's kinetic energy is transferred to the workpiece, so Equation (4.18) indicates that the average power equals the kinetic energy of the hammer divided by the duration of the impact:

$$P_{\text{av}} = \frac{(1/2)(40\,\text{kg})(3.72\,\text{m/s})^2}{0.02\,\text{s}} = 13.8\,\text{kW (kilowatts)}$$

Problems

4.1 The fictional starship *Enterprise* obtains its power by combining matter and antimatter, achieving complete conversion of mass into energy. The amount of energy contained in an amount of matter of mass m is given by Einstein's equation $E = mc^2$, where c is the speed of light (3×10^8 m/s).
(a) The mass of the *Enterprise* is approximately 5×10^9 kg. How much mass must be converted into kinetic energy to accelerate it from rest to one-tenth the speed of light?
(b) How much mass must be converted into kinetic energy to accelerate a 90 000 kg airliner from rest to 965 km/hr?

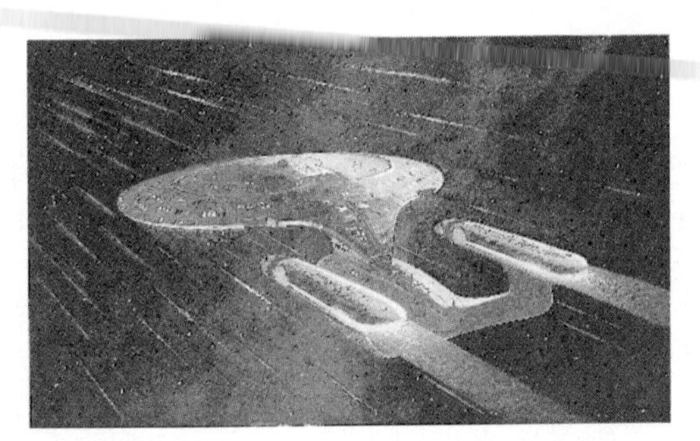

P4.1

4.2 The meteor crater near Winslow, Arizona, is 1200 m in diameter. An explosion of energy E near ground level causes a crater whose diameter is roughly proportional to $E^{1/3}$. Tests indicate that an explosion of 1 tonne of TNT, with an energy of 4.6×10^9 N.m, causes a crater approximately 10 m in diameter.
(a) How many tonnes of TNT would be equivalent to the energy due to the impact of the meteor?
(b) If the meteor was moving at 7600 m/s when it struck the ground and you assume as a first approximation that all of its kinetic energy went into creating the crater, what was the meteor's mass?

P4.2

4.3 The force exerted on a charged particle by a magnetic field is

$$\mathbf{F} = q\mathbf{v} \times \mathbf{B}$$

where q and $\mathbf{v}$ are the charge and velocity vector of the particle and $\mathbf{B}$ is the magnetic field vector. If other forces on the particle are negligible, use the principle of work and energy to show that the magnitude of the particle's velocity is constant.

4.4 A 1 tonne drag racer can accelerate from rest to 480 km/hr in 400 m.
(a) How much work is done on the car?
(b) If you assume as a first approximation that the tangential force exerted on the car is constant, what is the magnitude of the force?

P4.4

4.5 Assume that all of the weight of the drag racer in Problem 4.4 acts on its rear (drive) wheels and that the coefficients of friction between the wheels and the road are $\mu_s = \mu_k = 0.9$. Use the principle of work and energy to determine the maximum velocity in kilometres per hour the car can theoretically reach in 400 m. What do you think might account for the discrepancy between your answer and the car's actual velocity of 480 km/hr?

4.6 Assuming as a first approximation that the tangential force exerted on the drag racer in Problem 4.4 is constant, what is the maximum power transferred to the car as it accelerates from rest to 480 km/hr?

4.7 A 10 Mg (megagram) aeroplane must reach a velocity of 60 m/s to take off. If the horizontal force exerted by its engine is 60 kN and you neglect other horizontal forces, what length runway is needed?

4.8 Suppose you want to design an auxiliary rocket unit that will allow the aeroplane in Problem 4.7 to reach its takeoff speed using only 100 m of runway. For your preliminary design calculation, you can assume that the combined mass of the rocket and aeroplane is constant and equal to 10.5 Mg. What horizontal component of thrust must the rocket unit provide?

P4.8

4.9 The force exerted on a car by a prototype crash barrier as the barrier crushes is $F = -(3000 + 150\,000s)$ N, where s is the distance in metres from the initial contact. Suppose you want to design the barrier so that it can stop a 2200 kg car travelling at 130 km/hr. What is the necessary effective length of the barrier? That is, what is the distance required for the barrier to bring the car to a stop?

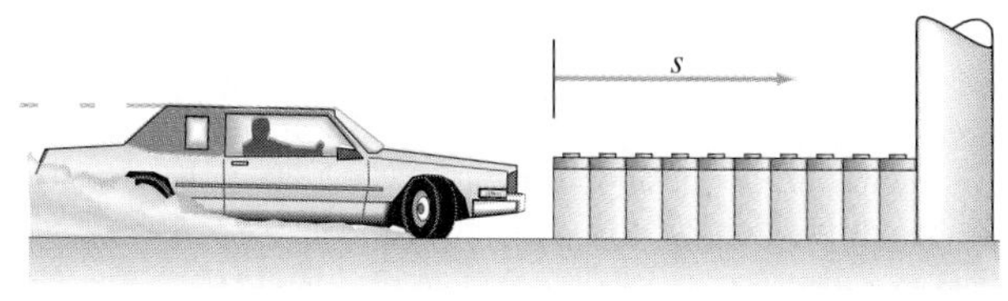

P4.9

4.10 The component of the total external force tangent to a 1 kg object's path is $\Sigma F_t = (60s - 50s^2)$ N, where s is its position measured along the path in metres. At $s = 0$, the object's velocity is $v = 3$ m/s.

(a) How much work is done on the object as it moves from $s = 0$ to $s = 1.2$ m?

(b) What is its velocity when it reaches $s = 1.2$ m?

4.11 The component of the total external force tangent to a 10 kg object's path is $\Sigma F_t = (100–20t)$ N, where t is in seconds. When $t = 0$, its velocity is $v = 4$ m/s. How much work is done on the object from $t = 2$ to $t = 4$ s?

4.12 The component of the total external force tangent to the path of an object of mass m is $\Sigma F_t = -cv$, where v is the magnitude of the object's velocity and c is a constant. When the position $s = 0$, its velocity is $v = v_0$. How much work is done on the object as it moves from $s = 0$ to a position $s = s_f$?

4.13 The 200 mm diameter tube is evacuated on the right of the 8 kg piston. On the left of the piston the tube contains gas with pressure $p_0 = 1 \times 10^5$ Pa (N/m^2). The force F is slowly increased, moving the piston 0.5 m to the left from the position shown. The force is then removed and the piston accelerates to the right. If you neglect friction and assume that the pressure of the gas is related to its volume by pV = constant, what is the velocity of the piston when it has returned to its original position?

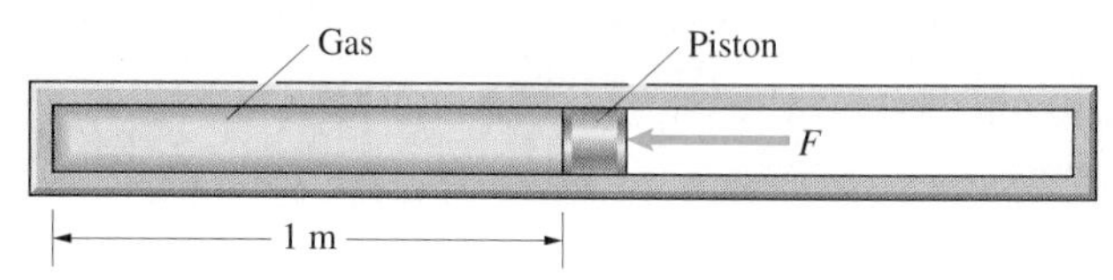

P4.13

4.14 In Problem 4.13, if you assume that the pressure of the gas is related to its volume by pV = constant while it is compressed (an isothermal process) and by $pV^{1.4}$ = constant while it is expanding (an isentropic process), what is the velocity of the piston when it has returned to its original position?

4.15 The system is released from rest. By applying the principle of work and energy to each weight, determine the magnitude of the velocity of the weights when they have moved 1 m.

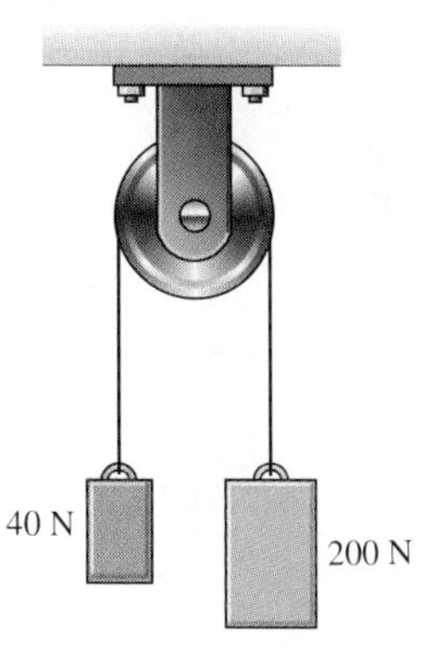

P4.15

4.16 In Problem 4.15, what is the tension in the cable during the motion of the system?

4.17 Solve Problem 4.15 by applying the principle of work and energy to the system consisting of the two weights, the cable and the pulley.

4.18 Suppose that you want to design a 'bumper' that will bring a 25 kg package moving at 3 m/s to rest 150 mm from the point of contact. If friction is neglible, what is the necessary spring constant k?

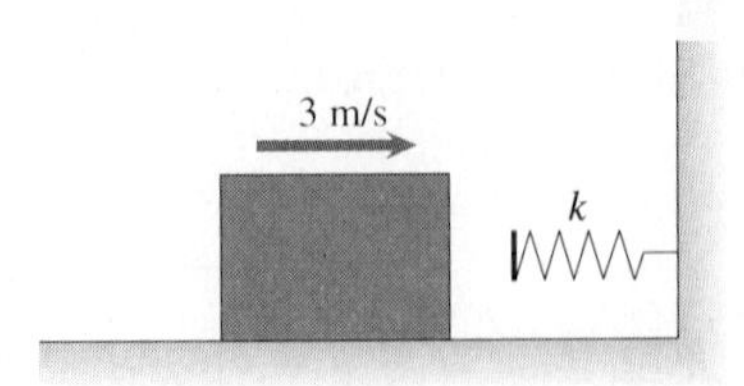

P4.18

4.19 In Problem 4.18, what spring constant is necessary if the coefficient of kinetic friction between the package and the floor is $\mu_k = 0.3$ and the package contacts the spring moving at 3 m/s?

4.20 The system is released from rest with the spring unstretched. If the spring constant is $k = 440$ N/m, what maximum velocity do the weights attain?

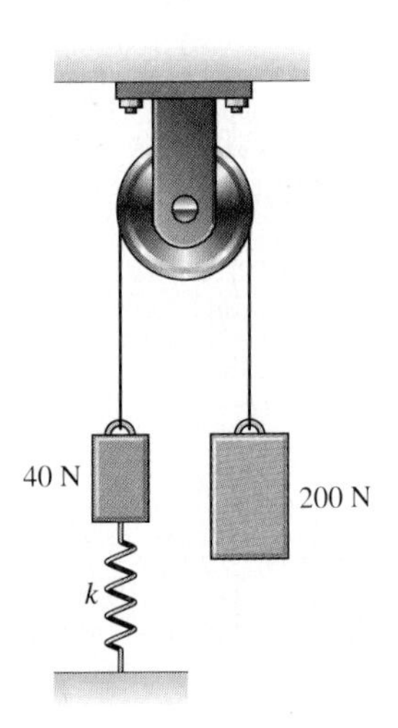

P4.20

4.21 Suppose you don't know the spring constant k of the system in Problem 4.20. If you release the system from rest with the spring unstretched and you observe that the 200 N weight falls 0.6 m before rebounding, what is k?

4.22 In Example 4.5, suppose that the unstretched length of each spring is 200 mm and you want to design the device so that the hammer strikes the workpiece at 5 m/s. Determine the necessary spring constant k.

4.23 The 20 kg crate is released from rest with the spring unstretched. The spring constant $k = 100$ N/m. Neglect friction.
(a) How far down the inclined surface does the crate slide before it stops?
(b) What maximum velocity does it attain on the way down?

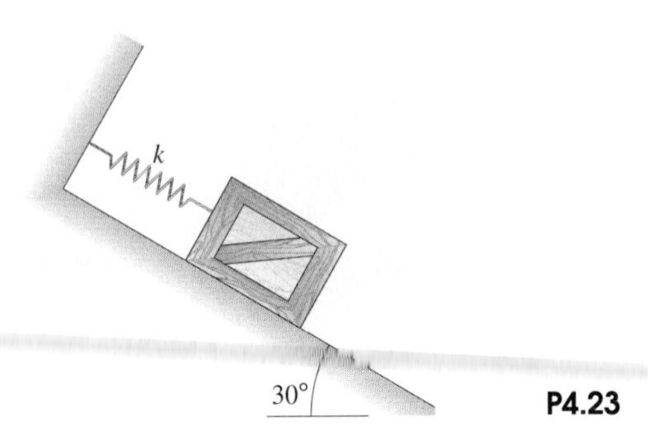

P4.23

4.24 Solve Problem 4.23 if the coefficient of kinetic friction between the crate and the surface is $\mu_k = 0.12$.

4.25 Solve Problem 4.23 if the coefficient of kinetic friction between the crate and the surface is $\mu_k = 0.16$ and the tension in the spring when the crate is released is 20 N.

4.26 The 30 kg box starts from rest at position 1. Neglect friction. For cases (a) and (b), determine the work done on the box from position 1 to position 2 and the magnitude of the velocity of the box at position 2.

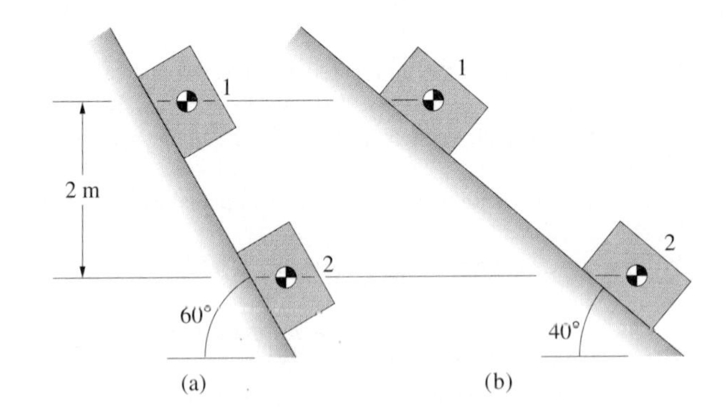

P4.26

4.27 Solve Problem 4.26 if the coefficient of kinetic friction between the box and the inclined surface is $\mu_k = 0.2$.

4.28 The masses of the three blocks are $m_A = 40$ kg, $m_B = 16$ kg, and $m_C = 12$ kg. Neglect the mass of the bar holding C in place. Friction is negligible. By applying the principle of work and energy to A and B individually, determine the magnitude of their velocity when they have moved 500 mm.

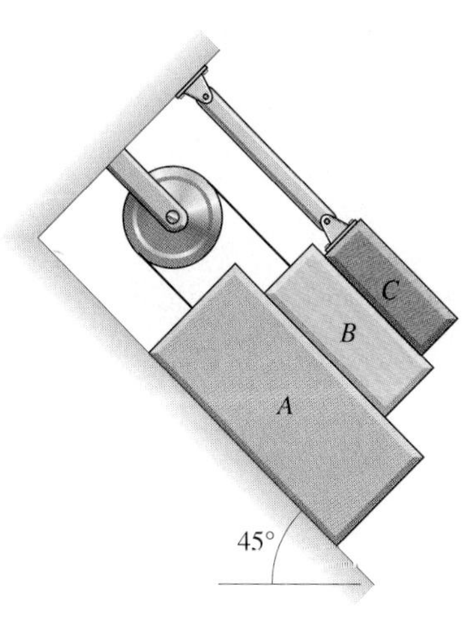

P4.28

4.29 Solve Problem 4.28 by applying the principle of work and energy to the system consisting of A, B, the cable connecting them, and the pulley.

4.30 In Problem 4.28, determine the magnitude of the velocity of A and B when they have moved 500 mm if the coefficient of kinetic friction between all surfaces is $\mu_k = 0.1$.

Strategy: The simplest approach is to apply the principle of work and energy to A and B individually. If you treat them as a single system, you must account for the work done by internal friction forces. See Example 4.3.

4.31 The 2 kg collar starts from rest at position 1 and slides down the smooth rigid wire. The y axis points upwards. What is the collar's velocity when it reaches position 2?

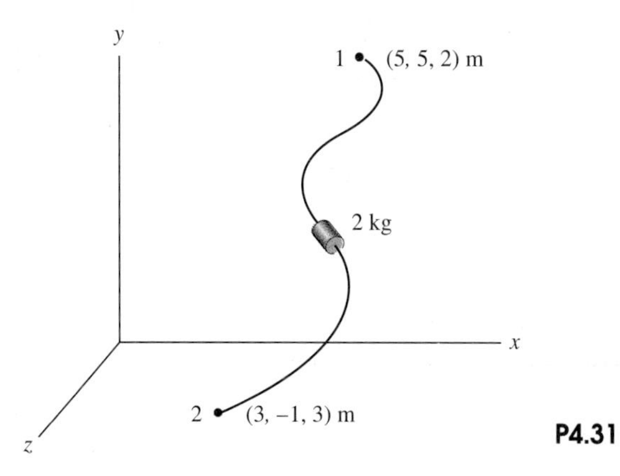

P4.31

4.32 The coefficients of friction between the 160 kg crate and the ramp are $\mu_s = 0.3$ and $\mu_k = 0.28$.

(a) What tension T_0 must the winch exert to start the crate moving up the ramp?

(b) If the tension remains at the value T_0 after the crate starts sliding, what total work is done on the crate as it slides a distance $s = 3$ m up the ramp, and what is the resulting velocity of the crate?

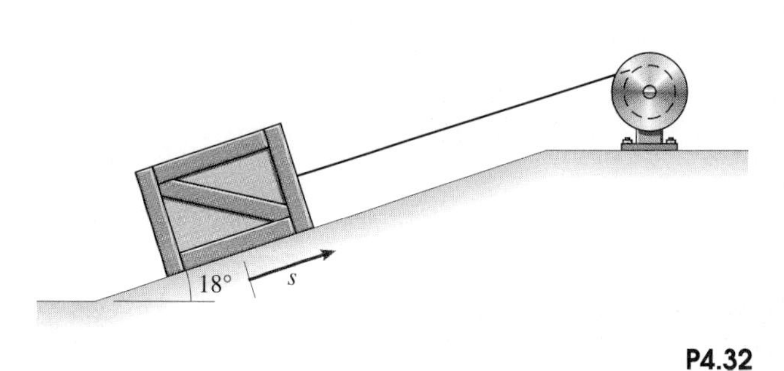

P4.32

4.33 In Problem 4.32, if the winch exerts a tension $T = T_0(1 + 0.1s)$ after the crate starts sliding, what total work is done on the crate as it slides a distance $s = 3$ m up the ramp, and what is the resulting velocity of the crate?

4.34 The mass of the rocket is 250 kg, and it has a constant thrust of 6000 N. The total length of the launching ramp is 10 m. Neglecting friction, drag, and the change in mass of the rocket, determine the magnitude of its velocity when it reaches the end of the ramp.

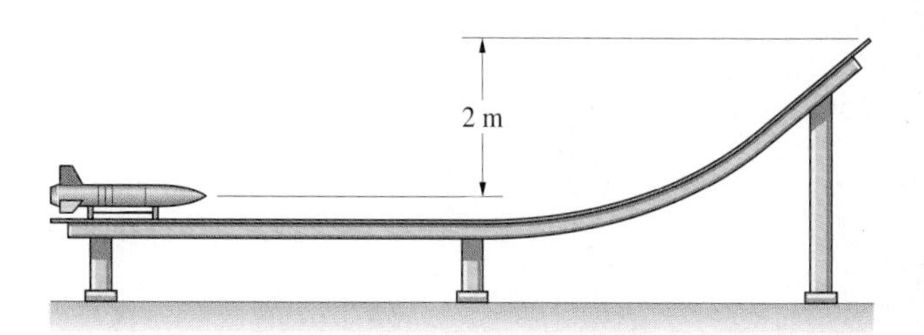

P4.34

4.35 The 1100 kg car is travelling at 64 km/hr at position 1. If the combined effect of the aerodynamic drag on the car and the tangential force exerted on the road by its wheels is that they exert no net tangential force on the car, what is its velocity at position 2?

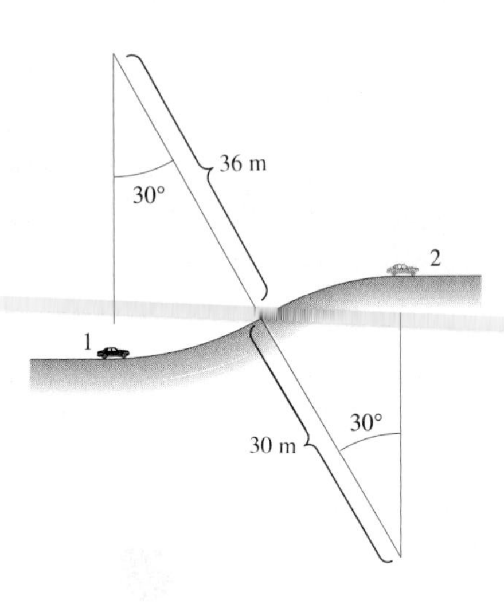

P4.35

4.36 In Problem 4.35, if the combined effect of the aerodynamic drag on the car and the tangential force exerted on the road by its wheels is that they exert a constant 1800 N tangential force on the car in the direction of its motion, what is its velocity at position 2?

4.37 The ball of mass m is released from rest in position 1. Determine the work done on the ball as it swings to position 2 (a) by its weight; (b) by the force exerted on it by the string. (c) What is the magnitude of its velocity at position 2?

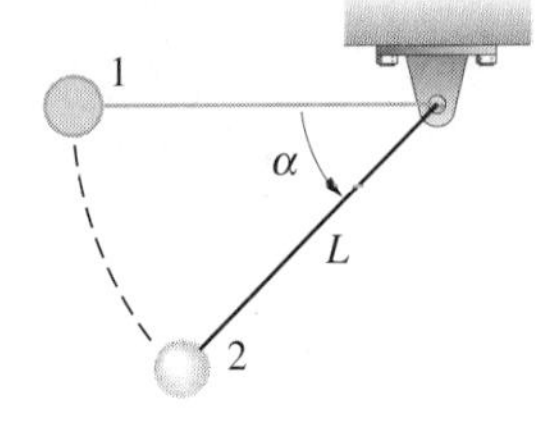

P4.37

4.38 In Problem 4.37, what is the tension in the string in position 2?

4.39 The 200 kg wrecker's ball hangs from a 6 m cable. If it is stationary at position 1, what is the magnitude of its velocity just before it hits the wall at position 2?

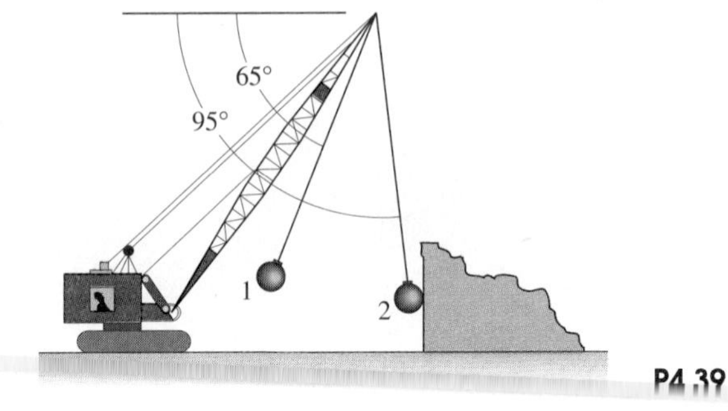

P4.39

4.40 In Problem 4.39, what is the maximum tension in the cable during the motion of the ball from position 1 to position 2?

4.41 A stunt driver wants to drive a car through a circular loop of radius R and hires you as a consultant to tell him the necessary velocity v_0 at which the car must enter the loop so that it can coast through without losing contact with the track.
(a) What is v_0 if you neglect friction and aerodynamic drag for your first rough estimate?
(b) What is the resulting velocity of the car at the top of the loop?

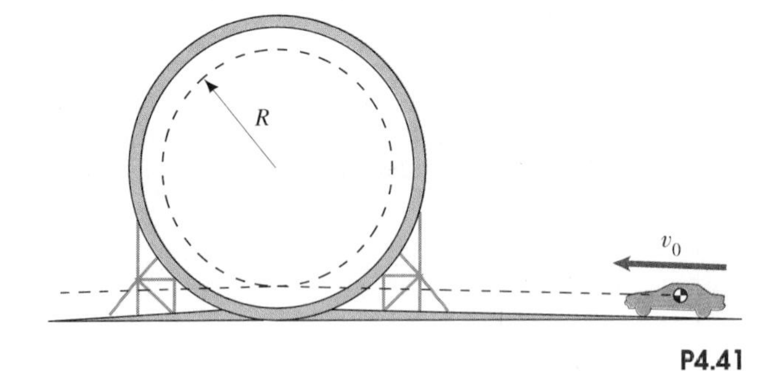

P4.41

4.42 Suppose that you throw rocks from the top of a 200 m cliff with a velocity of 10 m/s in the three directions shown. Neglecting aerodynamic drag, use the principle of work and energy to determine the magnitude of the velocity of the rock just before it hits the gound in each case.

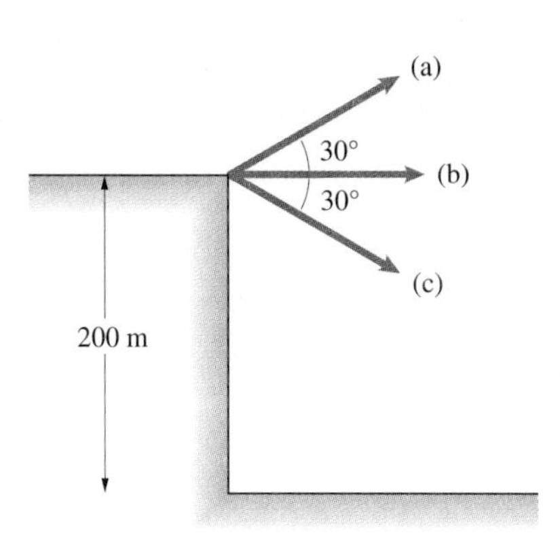

P4.42

4.43 A small pellet of mass m starts from rest at position 1 and slides down the smooth surface of the cylinder.
(a) What work is done on the pellet as it slides from position 1 to position 2?
(b) What is the magnitude of the pellet's velocity at position 2?

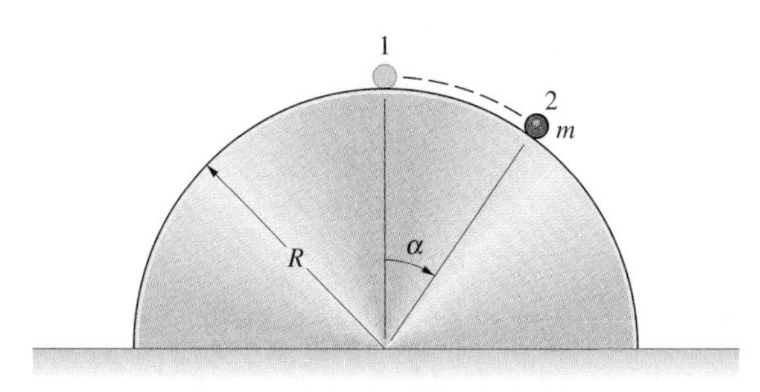

P4.43

4.44 In Problem 4.43, what is the value of the angle α at which the pellet leaves the surface of the cylinder?

4.45 In Problem 4.43, at what distance from the centre of the cylinder does the pellet strike the floor?

4.46 The 10 kg collar starts from rest at position 1 with the spring unstretched. The spring constant is $k = 600$ N/m. Neglect friction. How far does the collar fall relative to position 1?

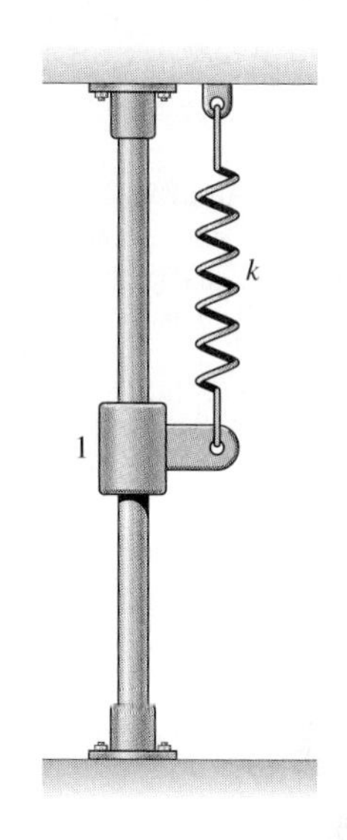

P4.46

4.47 In Problem 4.46, what maximum velocity does the collar attain?

4.48 What is the solution of Problem 4.46 if the tension in the spring in position 1 is 18 N?

4.49 The 4 kg collar is released from rest at position 1. Neglect friction. If the spring constant is $k = 6$ kN/m and the spring is unstretched in position 2, what is the velocity of the collar when it has fallen to position 2?

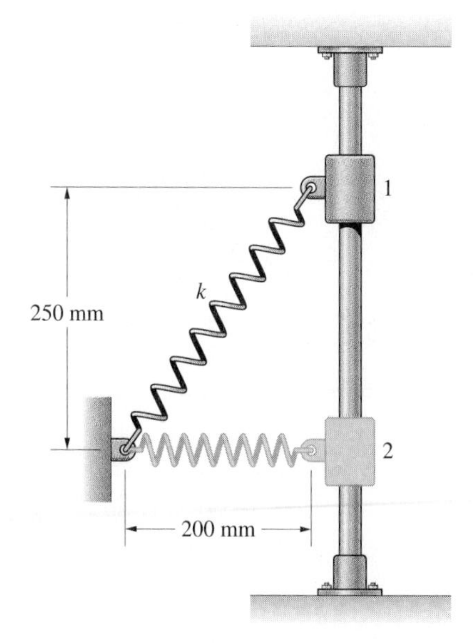

P4.49

4.50 In Problem 4.49, if the spring constant is $k = 4$ kN/m and the tension in the spring in position 2 is 500 N, what is the velocity of the collar when it has fallen to position 2?

4.51 In Problem 4.49, suppose that you don't know the spring constant k. If the spring is unstretched in position 2 and the velocity of the collar when it has fallen to position 2 is 4 m/s, what is k?

4.52 The 10 kg collar starts from rest at position 1 and slides along the smooth bar. The y axis points upwards. The spring constant is $k = 100$ N/m and the unstretched length of the spring is 2 m. What is the velocity of the collar when it reaches position 2?

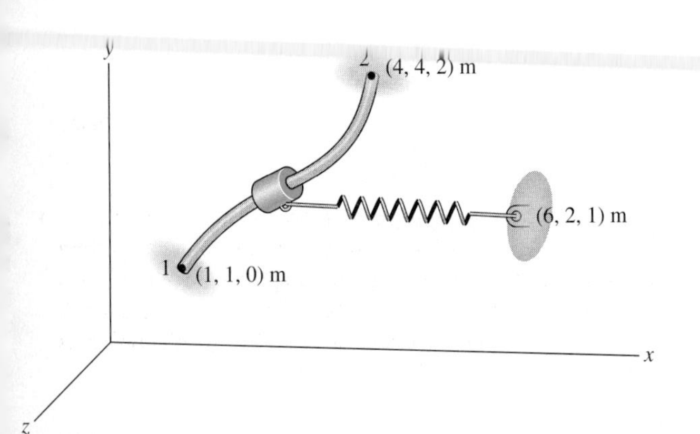

P4.52

4.53 Suppose an object has a string or cable with *constant* tension T attached as shown. The force exerted on the object can be expressed in terms of polar coordinates as $\mathbf{F} = -T\mathbf{e}_r$. Show that the work done on the object as it moves along an *arbitrary* plane path from a radial position r_1 to a radial position r_2 is $U = -T(r_2 - r_1)$.

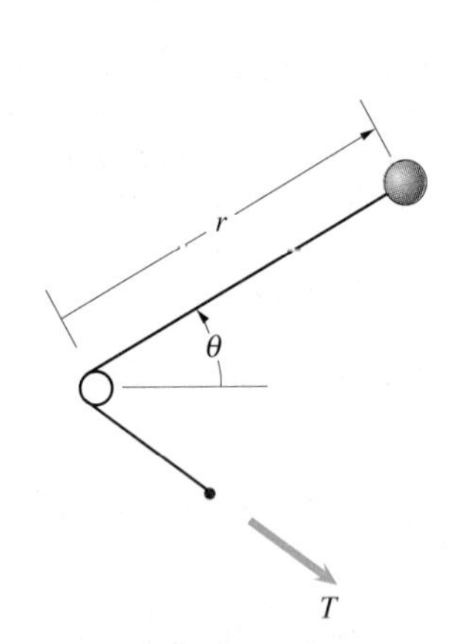

P4.53

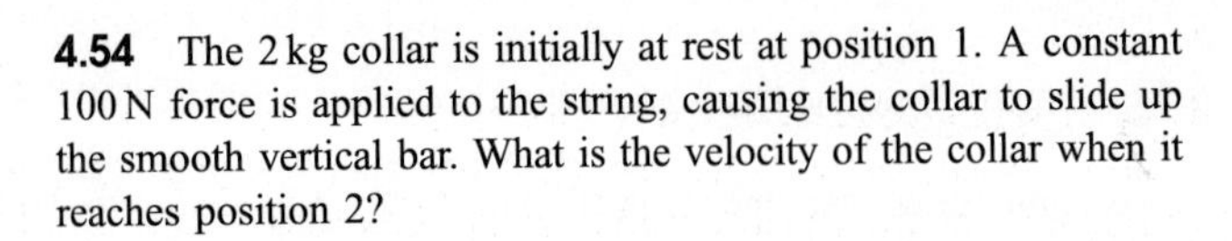

4.54 The 2 kg collar is initially at rest at position 1. A constant 100 N force is applied to the string, causing the collar to slide up the smooth vertical bar. What is the velocity of the collar when it reaches position 2?

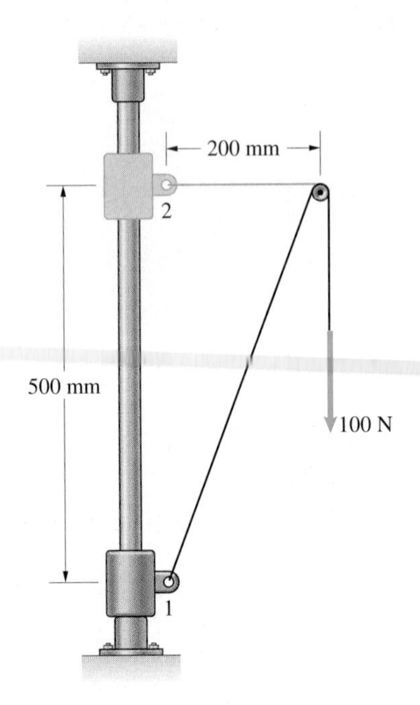

P4.54

4.55 The 10 kg collar starts from rest at position 1. The tension in the string is 200 N, and the y axis points upwards. If friction is negligible, what is the magnitude of the collar's velocity when it reaches position 2?

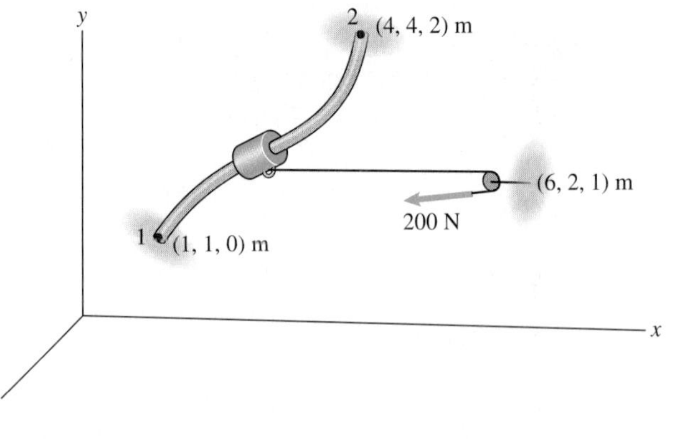

P4.55

4.56 A spring-powered mortar is used to launch 5 kg packages of fireworks into the air. The package starts from rest with the spring compressed to a length of 150 mm; the unstretched length of the spring is 750 mm. If the spring constant is $k = 19$ kN/m what is the magnitude of the velocity of the package as it leaves the mortar?

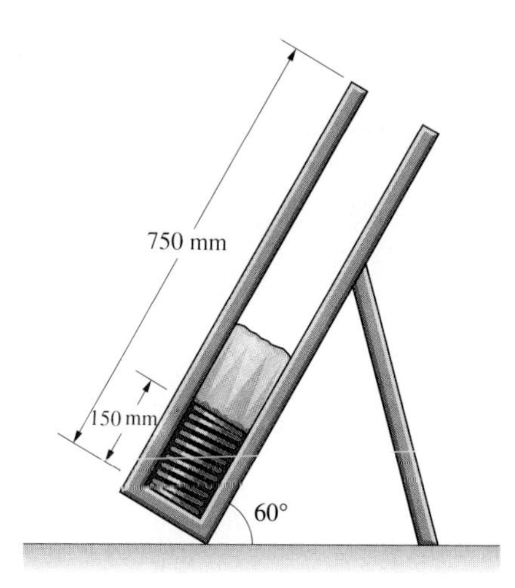

P4.56

4.57 Suppose you want to design the mortar in Problem 4.56 to throw the package to a height of 45 m above its initial position. Neglecting friction and drag, determine the necessary spring constant.

4.58 The system is released from rest in the position shown. The weights are $W_A = 180$ N and $W_B = 1350$ N. Neglect friction. What is the magnitude of the velocity of A when it has risen 1.2 m?

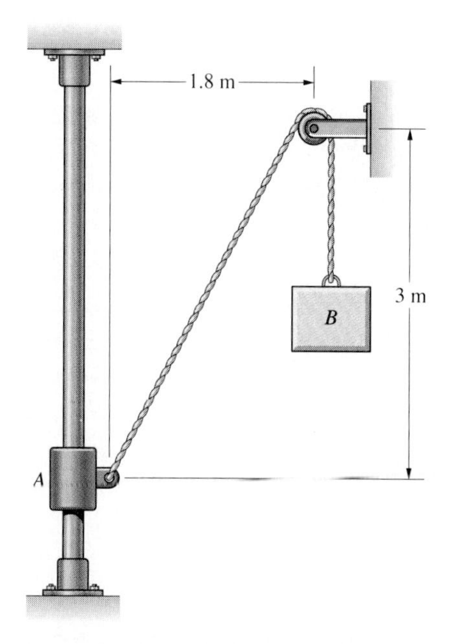

P4.58

4.59 In Problem 4.58, suppose that the system is released from rest with the weight A level with the pulley. What is the magnitude of the velocity of A when it has fallen 0.3 m?

4.60 A spacecraft 320 km above the surface of the earth has escape velocity $v_{esc} = \sqrt{2gR_E^2/r}$, where r is its distance from the centre of the earth and $R_E = 6370$ km is the radius of the earth. What is the magnitude of the spacecraft's velocity when it reaches the moon's orbit 383 000 km from the centre of the earth?

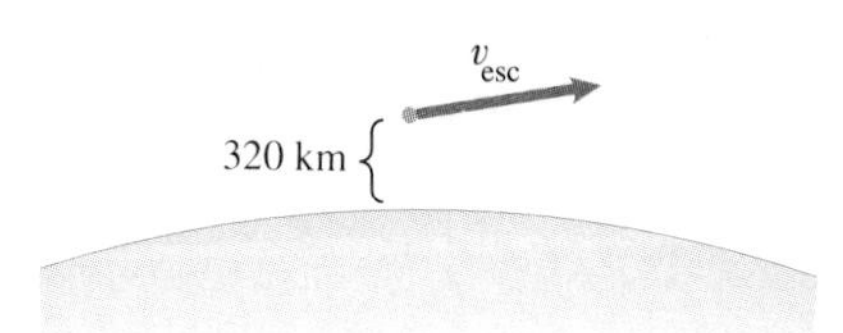

P4.60

4.61 A piece of ejecta thrown up by the impact of a meteor on the moon has a velocity of 200 m/s magnitude relative to the centre of the moon when it is 1000 km above the moon's surface. What is the magnitude of its velocity just before it strikes the moon's surface? (The acceleration due to gravity at the moon's surface is 1.62 m/s^2 and the moon's radius is 1738 km.)

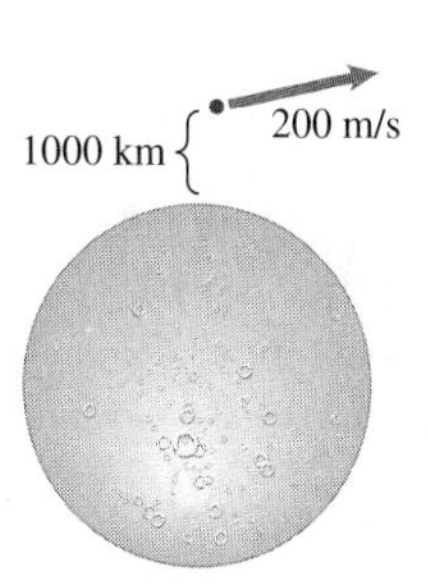

P4.61

4.62 A satellite in a circular orbit of radius r around the earth has velocity $v = \sqrt{gR_E^2/r}$, where $R_E = 6370$ km is the radius of the earth. Suppose you are designing a rocket to transfer a 900 kg communication satellite from a parking orbit with 6700 km radius to a geosynchronous orbit with 42 222 km radius. How much work must the rocket do on the satellite?

4.63 A 900 kg drag racer can accelerate from zero to 480 km/hr in 6 s. What average power is transferred to the car?

4.64 In Problem 4.9, what power is transferred from the car when it first contacts the barrier?

4.65 In Problem 4.32, what maximum power must the winch provide while pulling the crate up the ramp?

4.66 In Problem 4.39, if the wrecker's ball is brought to rest in 0.1 s as a result of hitting the wall, what average power does it transmit to the wall?

4.67 A Boeing 737 weighing 554 kN can accelerate to a takeoff speed of 55 m/s in 30 s.
(a) What average power is transferred to the plane?
(b) If you assume the tangential force exerted on the plane is constant, what is the maximum power transferred to the plane during its takeoff run?

P4.67

4.68 The Winter Park ski area in Colorado has a vertical drop of 700 m. Four skiers get on a chair lift to the top every 8 s, the chair moves at 1.2 m/s and the ride to the top takes 18 min. If the average skier with equipment weighs 700 N, approximately how much power is necessary to operate the chair lift?

P4.68

Potential Energy

4.3 Conservation of Energy

The work done on an object by some forces can be expressed as the change of a function of the object's position, called the **potential energy**. When all the forces that do work on a system have this property, we can state the principle of work and energy as a conservation law: the sum of the kinetic and potential energies is constant.

When we derived the principle of work and energy by integrating Newton's second law, we were able to evaluate the integral on one side of the equation, obtaining the change in the kinetic energy:

$$U = \int_{\mathbf{r}_1}^{\mathbf{r}_2} \Sigma \mathbf{F} \cdot d\mathbf{r} = \frac{1}{2} m v_2^2 - \frac{1}{2} m v_1^2 \tag{4.19}$$

Suppose we could determine a scalar function of position V such that

$$dV = -\Sigma \mathbf{f} \cdot d\mathbf{r} \tag{4.20}$$

Then we could also evaluate the integral defining the work:

$$U = \int_{\mathbf{r}_1}^{\mathbf{r}_2} \Sigma \mathbf{F} \cdot d\mathbf{r} = \int_{V_1}^{V_2} -dV = V_1 - V_2 \tag{4.21}$$

where V_1 and V_2 are the values of V at the positions $\mathbf{r}_1$ and $\mathbf{r}_2$. The principle of work and energy would then have the simple form

$$\frac{1}{2}mv_1^2 + V_1 = \frac{1}{2}mv_2^2 + V_2 \tag{4.22}$$

which means that the sum of the kinetic energy and the function V is constant:

$$\boxed{\frac{1}{2}mv^2 + V = \text{constant}} \tag{4.23}$$

If the kinetic energy increases, V must decrease, and vice versa, as if V represents a reservoir of 'potential' kinetic energy. For this reason, V is called the **potential energy**.

If a potential energy exists for a given force **F**, which means that a function of position V exists such that $dV = -\mathbf{F} \cdot d\mathbf{r}$, then **F** is said to be **conservative**. If all the forces that do work on a system are conservative, the total energy – the sum of the kinetic energy and the potential energies of the forces – is constant, or conserved. In that case, the system is said to be conservative, and you can use conservation of energy instead of the principle of work and energy to relate a change in its position to the change in its kinetic energy. The two approaches are equivalent, and you obtain the same quantitative information. But you gain greater insight by using conservation of energy, because you can interpret the motion of the object or system in terms of transformations between potential and kinetic energies.

4.4 Conservative Forces

You can apply conservation of energy only if the forces doing work on an object or system are conservative and you know (or can determine) their potential energies. In this section, we determine the potential energies of some conservative forces and use the results to demonstrate applications of conservation of energy. But before discussing forces that are conservative, we demonstrate with a simple example that friction forces are not.

The work done by a conservative force as an object moves from a position 1 to a position 2 is independent of the object's path. This result follows from Equation (4.21), which states that the work depends only on the values of the potential energy at positions 1 and 2. It also implies that if the object moves along a closed path, returning to position 1, the work done by a conservative force is zero. Suppose that a book of mass m rests on a table and you push it horizontally so that it slides along a path of length L. The magnitude of the friction force is $\mu_k mg$, and it points opposite to the direction of the book's motion (Figure 4.12). The work done is

$$U = \int_0^L -\mu_k mg\, ds = -\mu_k mgL$$

The work is proportional to the length of the path and therefore is not independent of the object's path. Friction forces are not conservative.

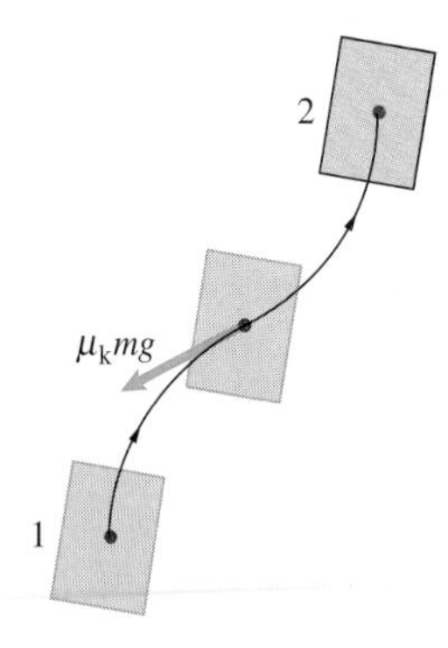

Figure 4.12

The book's path from position 1 to position 2. The friction force points opposite to the direction of the motion.

Potential Energies of Various Forces

The weight of an object and the force exerted by a spring attached to a fixed support are conservative forces. Using them as examples, we demonstrate how you can determine the potential energies of other conservative forces. We also use the potential energies of these forces in examples of the use of conservation of energy to analyse the motions of conservative systems.

Weight To determine the potential energy associated with an object's weight, we use a cartesian coordinate system with its y axis upwards (Figure 4.13). The weight is $\mathbf{F} = -mg\,\mathbf{j}$, and its dot product with the vector $d\mathbf{r}$ is

$$\mathbf{F} \cdot d\mathbf{r} = (-mg\,\mathbf{j}) \cdot (dx\,\mathbf{i} + dy\,\mathbf{j} + dz\,\mathbf{k}) = -mg\;dy$$

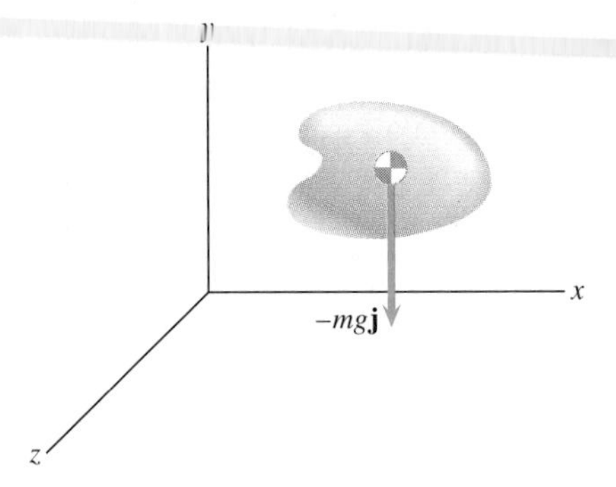

Figure 4.13
Weight of an object expressed in terms of a coordinate system with the y axis upwards.

From Equation (4.20), the potential energy V must satisfy the relation

$$dV = -\mathbf{F} \cdot d\mathbf{r} = mg\;dy \tag{4.24}$$

which we can write as

$$\frac{dV}{dy} = mg$$

Integrating this equation, we obtain

$$V = mgy + C$$

where C is an integration constant. The constant C is arbitrary, because this expression satisfies Equation (4.24) for any value of C. Another way of understanding why C is arbitrary is to notice in Equation (4.22) that it is the *difference* in the potential energy between two positions that determines the change in the kinetic energy. We will let $C = 0$ and write the potential energy of the weight of an object as

$$\boxed{V = mgy} \tag{4.25}$$

The potential energy is the product of the object's weight and height. The height can be measured from any convenient reference level, or **datum**. Since the difference in potential energy determines the change in the kinetic energy, it is the difference in height that matters, not the level from which the height is measured.

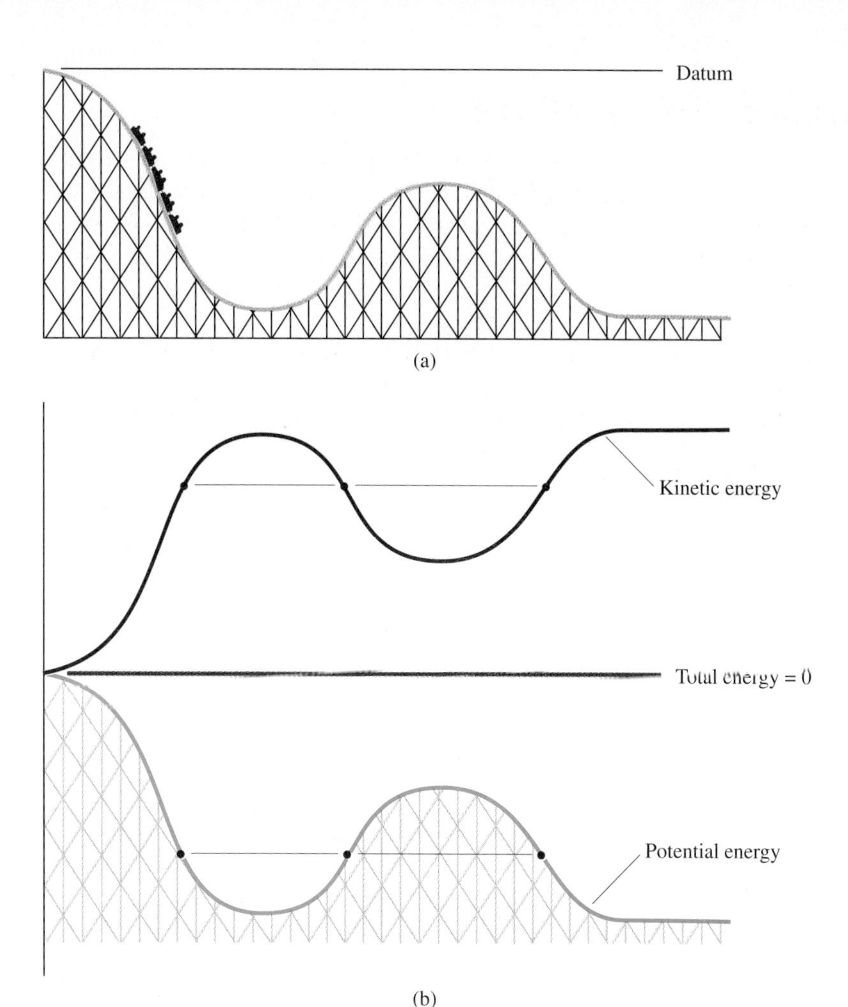

Figure 4.14

(a) Roller coaster and a reference level, or datum
(b) The sum of the potential and kinetic energies is constant.

The roller coaster (Figure 4.14(a)) is a classic example of conservation of energy. If aerodynamic and friction forces are neglected, the weight is the only force doing work and the system is conservative. The potential energy of the roller coaster is proportional to the height of the track relative to a datum. In Figure 4.14(b), we assume the roller coaster started from rest at the datum level. The sum of the kinetic and potential energies is constant, so the kinetic energy 'mirrors' the potential energy. At points of the track that have equal heights, the magnitudes of the velocities are equal.

To account for the variation of the weight with distance from the centre of the earth, we can express the weight in polar coordinates as

$$\mathbf{F} = -\frac{mgR_{\mathrm{E}}^2}{r^2}\mathbf{e}_r$$

where r is the distance from the centre of the earth (Figure 4.15). From Equation (4.14), the vector $d\mathbf{r}$ in terms of polar coordinates is

$$d\mathbf{r} = dr\,\mathbf{e}_r + r\,d\theta\,\mathbf{e}_\theta \tag{4.26}$$

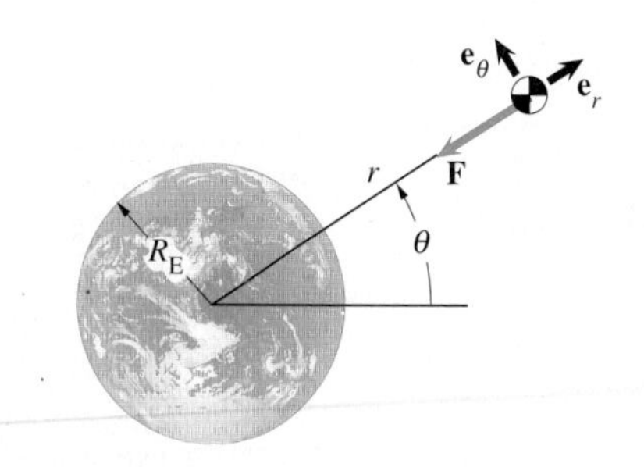

Figure 4.15

Expressing the weight in terms of polar coordinates.

The potential energy must satisfy

$$dV = -\mathbf{F}\cdot d\mathbf{r} = \frac{mgR_E^2}{r^2}\,dr$$

or

$$\frac{dV}{dr} = \frac{mgR_E^2}{r^2}$$

We integrate this equation and let the constant of integration be zero, obtaining the potential energy

$$\boxed{V = -\frac{mgR_E^2}{r}} \qquad (4.27)$$

Springs In terms of polar coordinates, the force exerted on an object by a linear spring is

$$\mathbf{F} = -k(r - r_0)\,\mathbf{e}_r$$

where r_0 is the unstretched length of the spring (Figure 4.16). Using Equation (4.26), the potential energy must satisfy

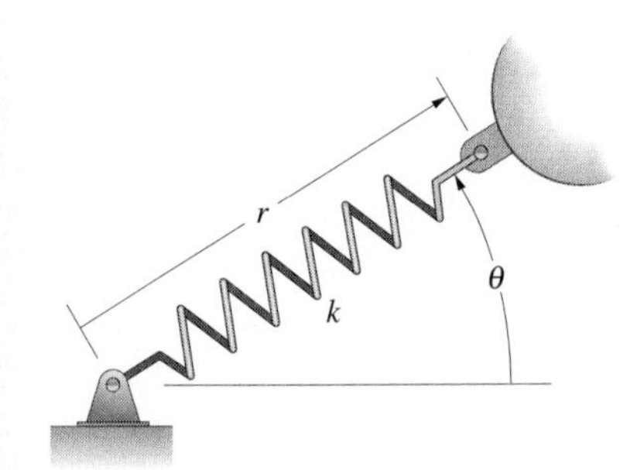

Figure 4.16
Expressing the force exerted by a linear spring in polar coordinates.

$$dV = -\mathbf{F}\cdot d\mathbf{r} = k(r - r_0)\,dr$$

Expressed in terms of the extension of the spring $S = r - r_0$, this equation is $dV = kS\,dS$, or

$$\frac{dV}{dS} = kS$$

Integrating this equation, we obtain the potential energy of a linear spring:

$$\boxed{V = \frac{1}{2}kS^2} \qquad (4.28)$$

In the following examples we use conservation of energy to relate changes in the positions of conservative systems to changes in their kinetic energies. This typically involves two steps:

(1) Determine the potential energy – *You must identify the conservative forces that do work and evaluate their potential energies in terms of the position of the system.*

(2) Apply conservation of energy – *By equating the sum of the kinetic and potential energies of the system at two positions, you can obtain an expression for the change in the kinetic energy.*

Example 4.6

In Example 4.5, the 40 kg hammer is lifted into position 1 and released from rest. Its weight and the two springs ($k = 1500$ N/m) accelerate the hammer downwards to position 2, where it strikes a workpiece. Use conservation of energy to determine the hammer's velocity when it reaches position 2.

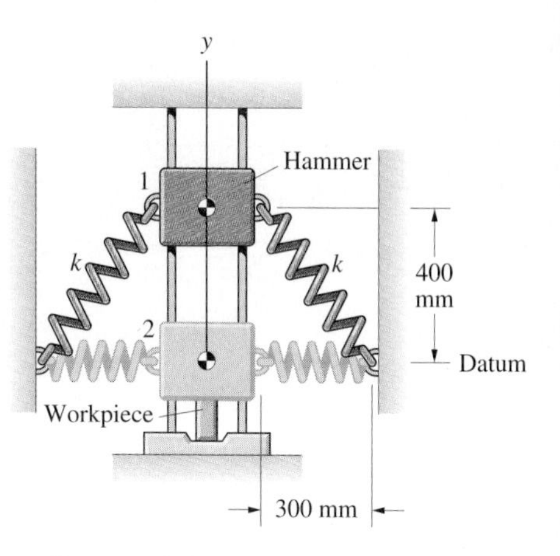

Figure 4.17

(a) Measuring the height of the hammer relative to position 2.

STRATEGY

Work is done on the hammer by its weight and the two springs, so the system is conservative. By equating the sums of the potential and kinetic energies at positions 1 and 2, we can obtain an equation for the velocity of the hammer at position 2.

SOLUTION

Determine the Potential Energy The potential energy of each spring is $\frac{1}{2}kS^2$, where S is the extension, so the total potential energy of the two springs is

$$V_{\text{springs}} = 2\left(\frac{1}{2}kS^2\right)$$

In Example 4.5 the extensions in positions 1 and 2 were determined to be $S_1 = 0.3$ m, $S_2 = 0.1$ m. The potential energy associated with the weight is

$$V_{\text{weight}} = mgy$$

where y is the height relative to a convenient datum (Figure (a)).

Apply Conservation of Energy The sums of the potential and kinetic energies at positions 1 and 2 must be equal:

$$2\left(\frac{1}{2}kS_1^2\right) + mgy_1 + \frac{1}{2}mv_1^2 = 2\left(\frac{1}{2}kS_2^2\right) + mgy_2 + \frac{1}{2}mv_2^2:$$

$$(1500)(0.3)^2 + (40)(9.81)(0.4) + 0 = (1500)(0.1)^2 + 0 + \frac{1}{2}(4)v_2^2$$

Solving this equation, we obtain $v_2 = 3.72$ m/s.

DISCUSSION

From the graphs of the total potential energy associated with the springs and the weight and the kinetic energy of the hammer as functions of y (Figure 4.18), you can see the transformation of the potential energy into kinetic energy as the hammer falls. Notice that the total energy of the conservative system remains constant.

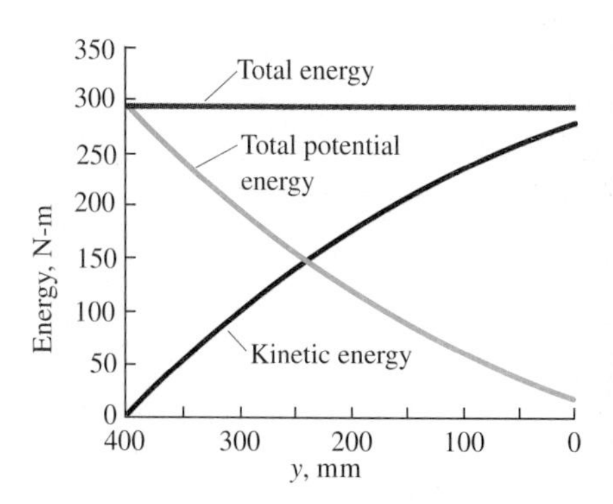

Figure 4.18

The potential and kinetic energies as functions of the y coordinate of the hammer.

Example 4.7

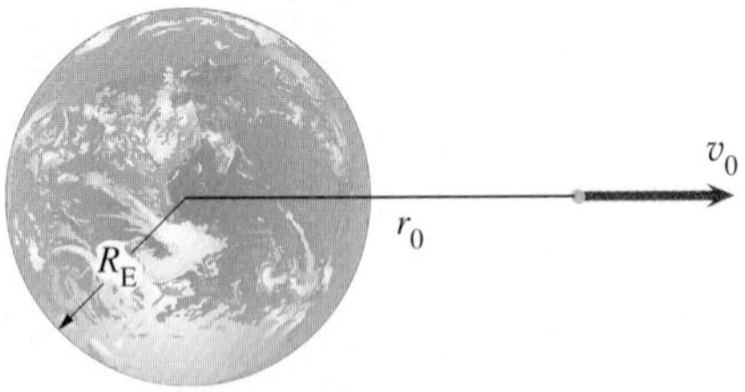

A spacecraft at a distance $r_0 = 2R_E$ from the centre of the earth is moving outwards with initial velocity $v_0 = \sqrt{2gR_E/3}$ (Figure 4.19). Determine its velocity as a function of its distance from the centre of the earth.

Figure 4.19

SOLUTION

Determine the Potential Energy The potential energy associated with the spacecraft's weight is given in terms of its distance r from the centre of the earth by Equation (4.27):

$$V = -\frac{mgR_E^2}{r}$$

Apply Conservation of Energy Let v be the magnitude of the spacecraft's velocity at an arbitrary distance r. The sums of the potential and kinetic energies at r_0 and at r must be equal:

$$-\frac{mgR_E^2}{r_0} + \frac{1}{2}mv_0^2 = -\frac{mgR_E^2}{r} + \frac{1}{2}mv^2 :$$

$$-\frac{mgR_E^2}{2R_E} + \frac{1}{2}m\left(\frac{2}{3}g\,R_E\right) = -\frac{mgR_E^2}{r} + \frac{1}{2}mv^2$$

Solving for v, the spacecraft's velocity as a function of r is

$$v = \sqrt{gR_E\left(\frac{2R_E}{r} - \frac{1}{3}\right)}$$

DISCUSSION

We show graphs of the kinetic energy, potential energy and total energy as functions of r/R_E in Figure 4.20. The kinetic energy decreases and the potential energy increases as the spacecraft moves outwards until its velocity decreases to zero at $r = 6R_E$.

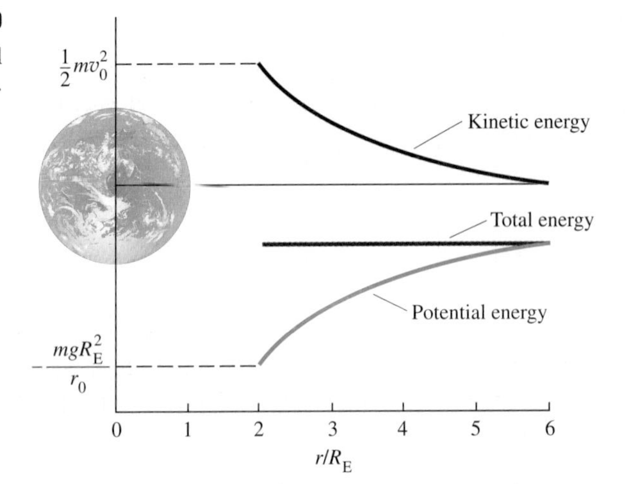

Figure 4.20
Energies as functions of the radial coordinate.

Relationships Between Force and Potential Energy

Here we consider two questions: (1) Given a potential energy, how can you determine the corresponding force? (2) Given a force, how can you determine whether it is conservative? That is, how can you tell whether an associated potential energy exists?

The potential energy V of a force $\mathbf{F}$ is a function of position that satisfies the relation

$$dV = -\mathbf{F}\cdot d\mathbf{r} \tag{4.29}$$

If we express V in terms of a cartesian coordinate system,

$$V = V(x, y, z)$$

its differential dV is

$$dV = \frac{\partial V}{\partial x}\,dx + \frac{\partial V}{\partial y}\,dy + \frac{\partial V}{\partial z}\,dz \tag{4.30}$$

Expressing $\mathbf{F}$ and $d\mathbf{r}$ in terms of their cartesian components, their dot product is

$$\begin{aligned}\mathbf{F}\cdot d\mathbf{r} &= (F_x\,\mathbf{i} + F_y\,\mathbf{j} + F_z\,\mathbf{k})\cdot(dx\,\mathbf{i} + dy\,\mathbf{j} + dz\,\mathbf{k})\\ &= F_x\,dx + F_y\,dy + F_z\,dz\end{aligned}$$

Substituting this expression and Equation (4.30) into Equation (4.29), we obtain

$$\frac{\partial V}{\partial x}\,dx + \frac{\partial V}{\partial y}\,dy + \frac{\partial V}{\partial z}\,dz = -(F_x dx + F_y dy + F_z dz)$$

which implies that

$$F_x = -\frac{\partial V}{\partial x} \qquad F_y = -\frac{\partial V}{\partial y} \qquad F_z = -\frac{\partial V}{\partial z} \tag{4.31}$$

Given a potential energy V expressed in cartesian coordinates, you can use these relations to determine the corresponding force. The force $\mathbf{F}$ is

$$\mathbf{F} = -\left(\frac{\partial V}{\partial x}\,\mathbf{i} + \frac{\partial V}{\partial y}\,\mathbf{j} + \frac{\partial V}{\partial z}\,\mathbf{k}\right) = -\nabla V \tag{4.32}$$

where ∇V is the **gradient** of V. By using expressions for the gradient in terms of other coordinate systems, you can determine the force $\mathbf{F}$ when you know the potential energy in terms of those coordinate systems. For example, in terms of cylindrical coordinates,

$$\mathbf{F} = -\left(\frac{\partial V}{\partial r}\,\mathbf{e}_r + \frac{1}{r}\frac{\partial V}{\partial \theta}\,\mathbf{e}_\theta + \frac{\partial V}{\partial z}\,\mathbf{e}_z\right) \tag{4.33}$$

If a force $\mathbf{F}$ is conservative, its **curl** $\nabla \times \mathbf{F}$ is zero. The expression for the curl of $\mathbf{F}$ in cartesian coordinates is

$$\nabla \times \mathbf{F} = \begin{vmatrix} \mathbf{i} & \mathbf{j} & \mathbf{k} \\ \dfrac{\partial}{\partial x} & \dfrac{\partial}{\partial y} & \dfrac{\partial}{\partial z} \\ F_x & F_y & F_z \end{vmatrix} \tag{4.34}$$

Substituting Equations (4.31) into this expression confirms that $\nabla \times \mathbf{F} = 0$ when $\mathbf{F}$ is conservative. *The converse is also true:* a force $\mathbf{F}$ is conservative if its curl is zero. You can use this condition to determine whether a given force is conservative. In terms of cylindrical coordinates, the curl of $\mathbf{F}$ is

$$\nabla \times \mathbf{F} = \frac{1}{r} \begin{vmatrix} \mathbf{e}_r & r\mathbf{e}_\theta & \mathbf{e}_z \\ \dfrac{\partial}{\partial r} & \dfrac{\partial}{\partial \theta} & \dfrac{\partial}{\partial z} \\ F_r & rF_\theta & F_z \end{vmatrix} \tag{4.35}$$

Example 4.8

From Equation (4.27), the potential energy associated with the weight of an object of mass m at a distance r from the centre of the earth is (in polar coordinates)

$$V = -\frac{mgR_{\mathrm{E}}^2}{r}$$

where R_{E} is the radius of the earth. Use this expression to determine the force exerted on the object by its weight.

STRATEGY

The force $\mathbf{F} = -\nabla V$. The potential energy is expressed in terms of polar coordinates, so we can use Equation (4.33) to determine the force.

SOLUTION

The partial derivatives of V with respect to r, θ and z are

$$\frac{\partial V}{\partial r} = \frac{mgR_{\mathrm{E}}^2}{r^2} \qquad \frac{\partial V}{\partial \theta} = 0 \qquad \frac{\partial V}{\partial z} = 0$$

From Equation (4.33), the force is

$$\mathbf{F} = -\nabla V = -\frac{mgR_{\mathrm{E}}^2}{r^2}\,\mathbf{e}_r$$

DISCUSSION

We already know that the force is conservative, because we know its potential energy, but we can use Equation (4.35) to confirm that its curl is zero:

$$\nabla \times \mathbf{F} = \frac{1}{r} \begin{vmatrix} \mathbf{e}_r & r\mathbf{e}_\theta & \mathbf{e}_z \\ \dfrac{\partial}{\partial r} & \dfrac{\partial}{\partial \theta} & \dfrac{\partial}{\partial z} \\ -\dfrac{mgR_E^2}{r^2} & 0 & 0 \end{vmatrix} = 0$$

Although we used cylindrical coordinates in determining **F** and in evaluating the cross product, the expression for V and our resulting expression for **F** are valid only if the object remains in the plane $z = 0$.

Problems

4.69 Suppose that you kick a soccer ball straight up. When it leaves your foot, it is 1 m above the ground and moving at 12 m/s. Neglecting drag, use conservation of energy to determine how high above the ground the ball goes and how fast it will be going just before it hits the ground. Obtain the answers by expressing the potential energy in terms of a datum (a) at the level of the ball's initial position; (b) at ground level.

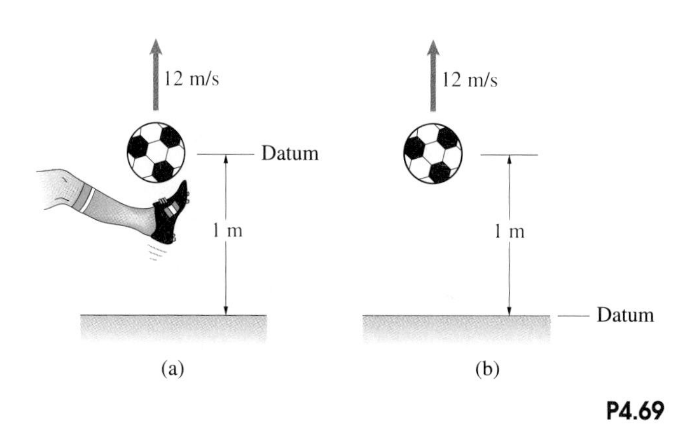

P4.69

4.70 The Lunar Module could make a safe landing if its vertical velocity at impact was 5 m/s or less. Suppose that you want to determine the greatest height h at which the pilot could shut off the engine if the velocity of the lander relative to the surface was (a) zero; (b) 2 m/s downwards; (c) 2 m/s upwards. Use conservation of energy to determine h in each case. The acceleration due to gravity at the surface of the moon is $1.62\ \text{m/s}^2$.

P4.70

4.71 The ball is released from rest in position 1.
(a) Use conservation of energy to determine the magnitude of its velocity at position 2.
(b) Draw graphs of the kinetic energy, the potential energy and the total energy for values of α from zero to 180°.

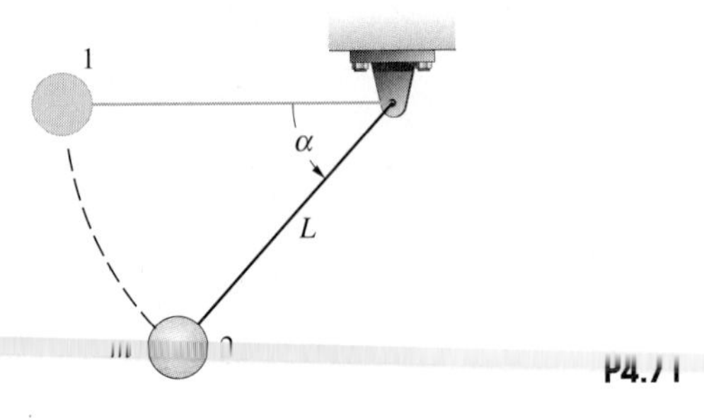

P4.71

4.72 If a ball is released from rest in position 1, use conservation of energy to determine the initial angle α necessary for it to swing to position 2.

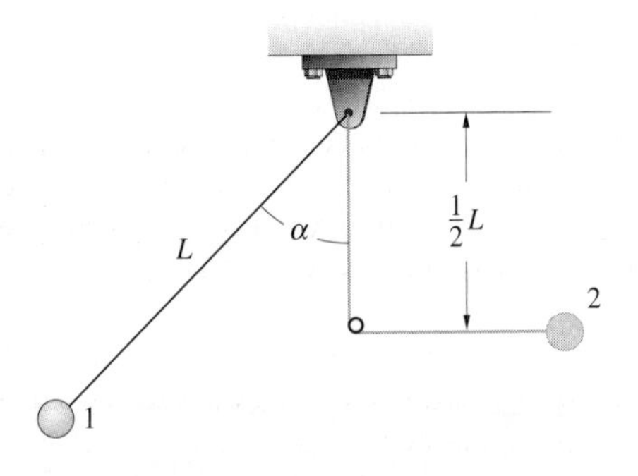

P4.72

4.73 The bar is smooth. Use conservation of energy to determine the minimum velocity the 10 kg slider must have at A (a) to reach C; (b) to reach D.

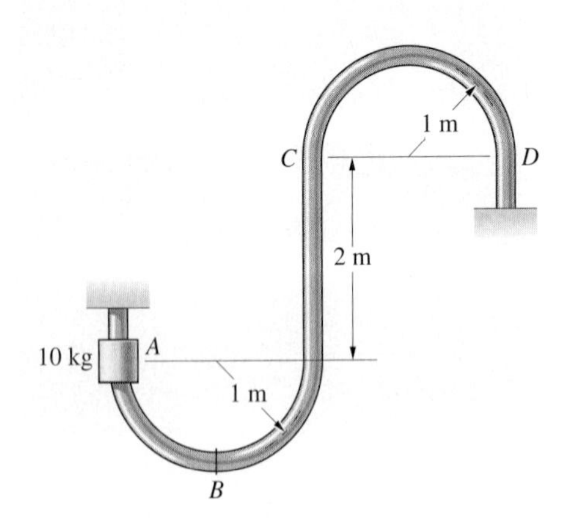

P4.73

4.74 In Problem 4.73, what normal force does the bar exert on the slider at B in cases (a) and (b)?

4.75 The 10 kg collar starts from rest at position 1 and slides along the bar. The y axis points upwards. The spring constant is $k = 100$ N/m, and the unstretched length of the spring is 2 m. Use conservation of energy to determine the collar's velocity when it reaches position 2.

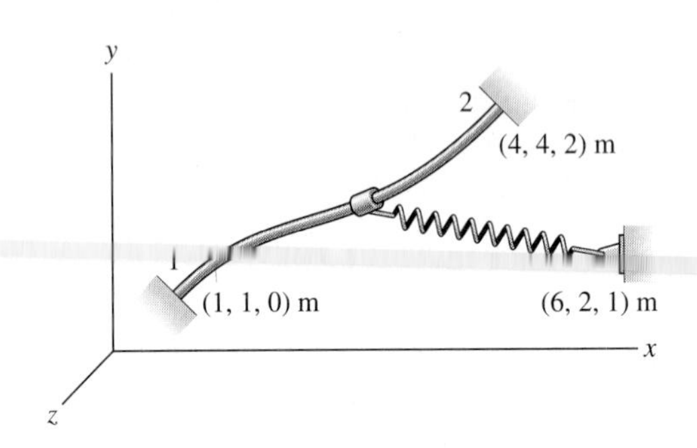

P4.75

4.76 A rock climber of weight W has a rope attached a distance h below him for protection. Suppose that he falls, and assume that the rope behaves like a linear spring with unstretched length h and spring constant $k = C/h$, where C is a constant. Use conservation of energy to determine the maximum force exerted on him by the rope. (Notice that the maximum force is independent of h, which is a reassuring result for climbers – the maximum force resulting from a long fall is the same as that resulting from a very short one.)

P4.76

4.77 The 5 kg collar starts from rest at A and slides along the semicircular bar. The spring constant is $k = 3200$ N/m and the unstretched length of the spring is 1 m. Use conservation of energy to determine the velocity of the collar at B.

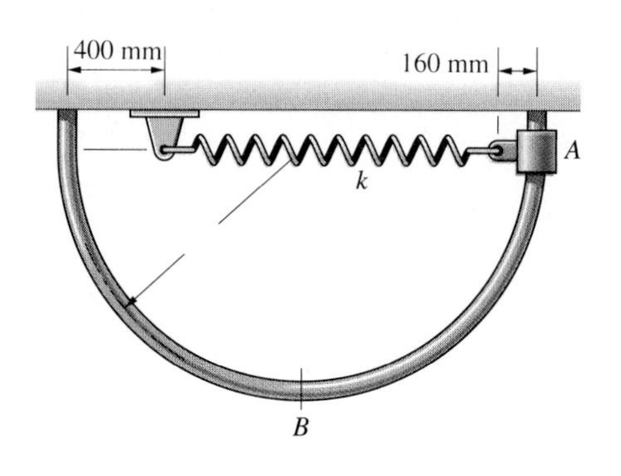

P4.77

4.78 The force exerted on an object by a *nonlinear* spring is

$$\mathbf{F} = -[k(r - r_0) + q(r - r_0)^3]\mathbf{e}_r$$

where k and q are constants and r_0 is the unstretched length. Determine the potential energy of the spring in terms of its extension $S = r - r_0$.

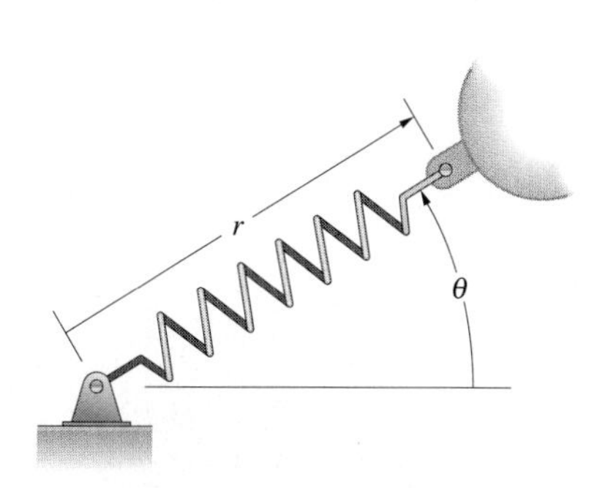

P4.78

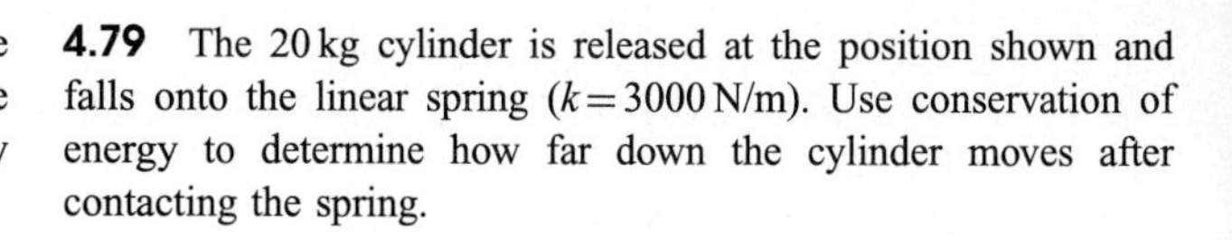

4.79 The 20 kg cylinder is released at the position shown and falls onto the linear spring ($k = 3000$ N/m). Use conservation of energy to determine how far down the cylinder moves after contacting the spring.

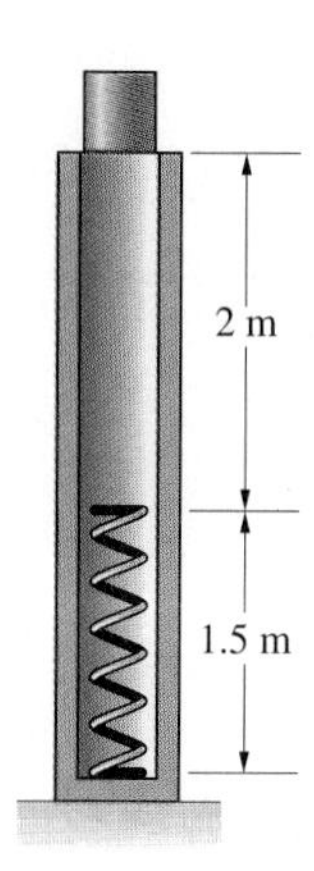

P4.79

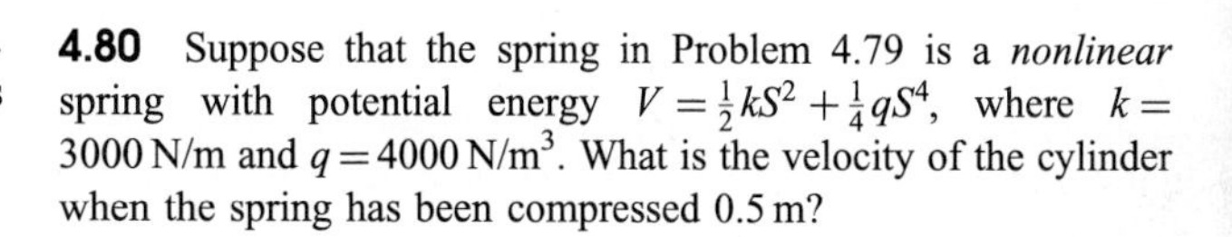

4.80 Suppose that the spring in Problem 4.79 is a *nonlinear* spring with potential energy $V = \frac{1}{2}kS^2 + \frac{1}{4}qS^4$, where $k = 3000$ N/m and $q = 4000$ N/m^3. What is the velocity of the cylinder when the spring has been compressed 0.5 m?

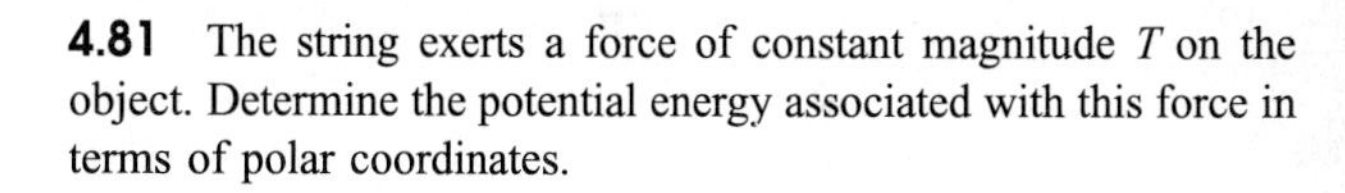

4.81 The string exerts a force of constant magnitude T on the object. Determine the potential energy associated with this force in terms of polar coordinates.

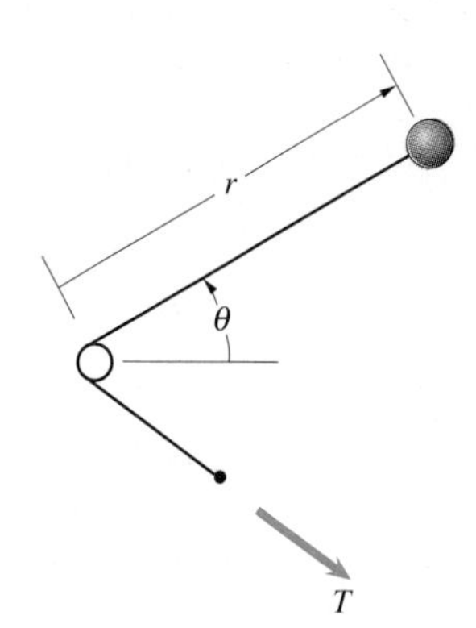

P4.81

4.82 The system is at rest in the position shown, with the 5 kg collar A resting on the spring ($k = 300$ N/m), when a constant 150 N force is applied to the cable. What is the velocity of the collar when it has risen 0.3 m?

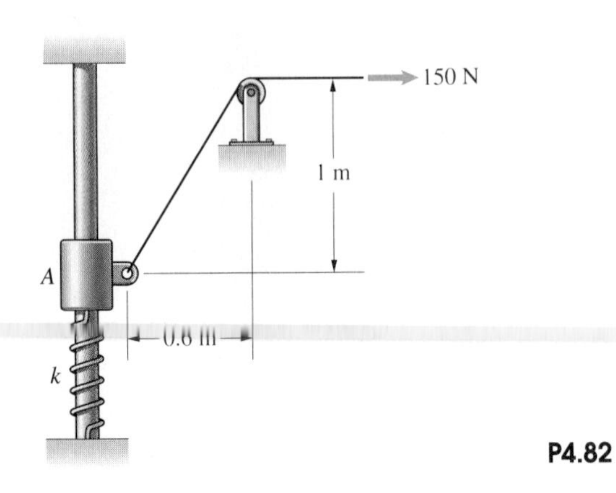

P4.82

4.83 The tube (cross-sectional area A) is evacuated on the right of the piston of mass m and on the left it contains gas at pressure p. Let the value of the pressure when $s = s_0$ be p_0, and assume that the pressure of the gas is related to its volume V by $pV = \text{constant}$.
(a) Determine the potential energy associated with the force exerted on the piston in terms of s.
(b) If the piston starts from rest at $s = s_0$ and friction is negligible, what is its velocity as a function of s?

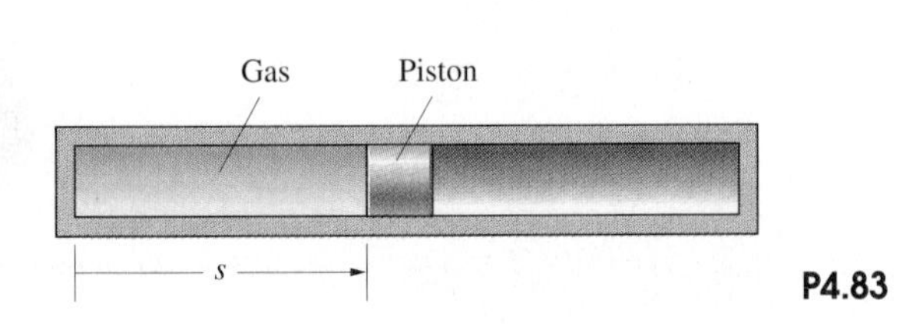

P4.83

4.84 Solve Problem 4.83, assuming that the pressure of the gas is related to its volume by $pV^{\gamma} = \text{constant}$, where γ is a constant.

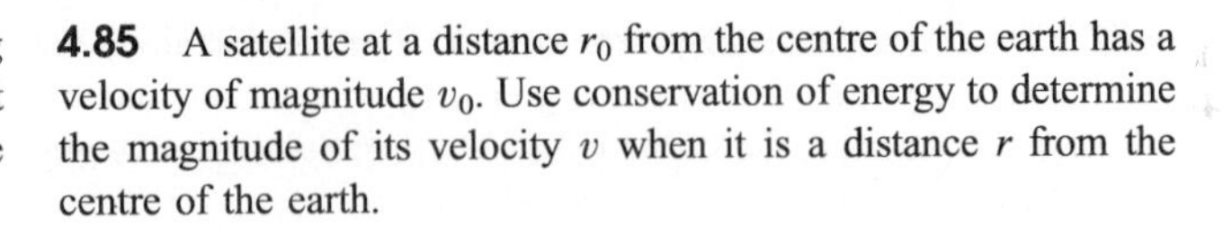

4.85 A satellite at a distance r_0 from the centre of the earth has a velocity of magnitude v_0. Use conservation of energy to determine the magnitude of its velocity v when it is a distance r from the centre of the earth.

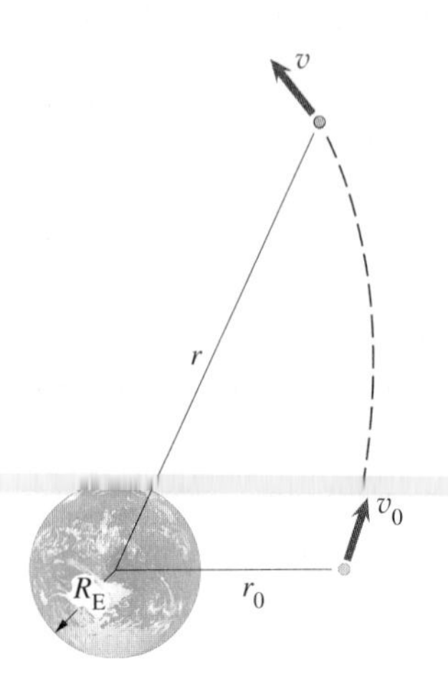

P4.85

4.86 Astronomers detect an asteroid 100 000 km from the earth moving at 2 km/s relative to the centre of the earth. If it should strike the earth, use conservation of energy to determine the magnitude of its velocity as it enters the atmosphere (You can neglect the thickness of the atmosphere in comparison to the earth's 6370 km radius.)

4.87 A satellite is in an elliptic orbit around the earth. Its velocity at the perigee A is 8620 m/s. Use conservation of energy to determine its velocity at B. The radius of the earth is 6370 km.

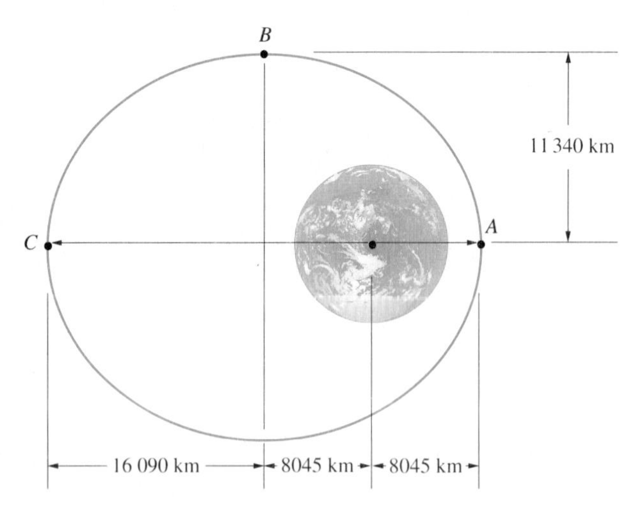

P4.87

4.88 For the satellite orbit in Problem 4.87, use conservation of energy to determine the velocity at the apogee C. Using your result, confirm numerically that the velocities at perigee and apogee satisfy the relation $r_A v_A = r_C v_C$.

4.89 The component of the total external force tangential to the path of a 10 kg object moving along the x axis is $\Sigma F_x = 3x^2\mathbf{i}$ N, where x is in metres. At $x = 2$ m, the object's velocity is $v_x = 4$ m/s.
(a) Use the principle of work and energy to determine its velocity at $x = 6$ m.
(b) Determine the potential energy associated with the force ΣF_x and use conservation of energy to determine its velocity at $x = 6$ m.

4.90 The potential energy associated with a force **F** acting on an object is $V = 2x^2 - y$ N.m, where x and y are in metres.
(a) Determine **F**.
(b) If the object moves from position 1 to position 2 along the paths A and B, determine the work done by **F** along each path.

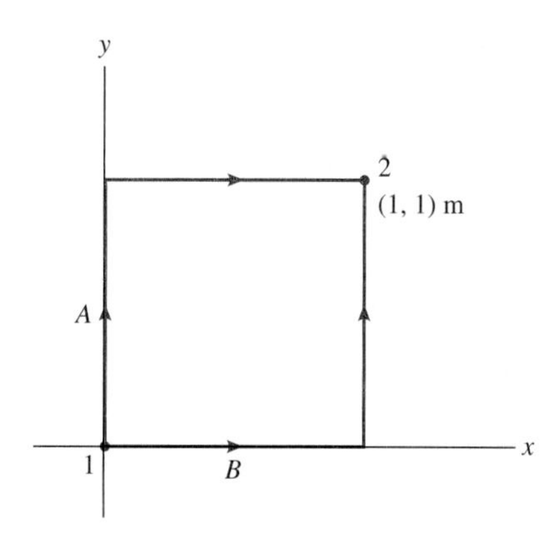

P4.90

4.91 An object is subjected to the force $\mathbf{F} = y\mathbf{i} - x\mathbf{j}$ N, where x and y are in metres.
(a) Show that **F** is *not* conservative.
(b) If the object moves from point 1 to point 2 along the paths A and B shown in Problem 4.90, determine the work done by **F** along each path.

4.92 In terms of polar coordinates, the potential energy associated with the force **F** exerted on an object by a *nonlinear* spring is

$$V = \frac{1}{2}k(r - r_0)^2 + \frac{1}{4}q(r - r_0)^2$$

where k and q are constants and r_0 is the unstretched length. Determine **F** in terms of polar coordinates.

4.93 In terms of polar coordinates, the force exerted on an object by a *nonlinear* spring is

$$\mathbf{F} = -[k(r - r_0) + q(r - r_0)^3]\,\mathbf{e}_r$$

where k and q are constants and r_0 is the unstretched length. Use Equation (4.35) to show that **F** is conservative.

4.94 The potential energy associated with a force **F** acting on an object is $V = -r\sin\theta + r^2\cos^2\theta$ N.m, where r is in metres.
(a) Determine **F**.
(b) If the object moves from point 1 to point 2 along the circular path, how much work is done by **F**?

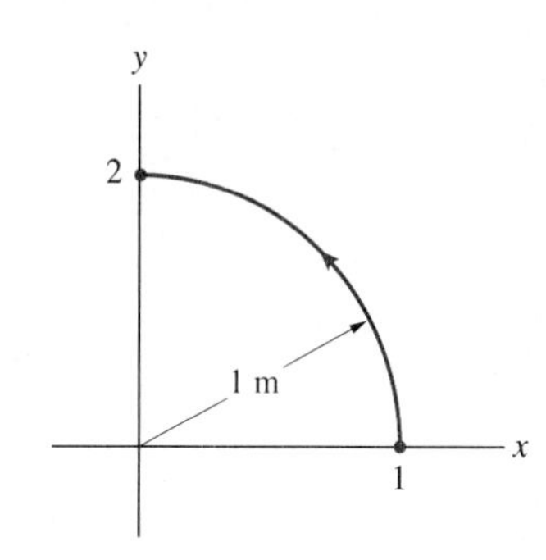

P4.94

4.95 In terms of polar coordinates, the force exerted on an object of mass m by the gravity of a hypothetical two-dimensional planet is $\mathbf{F} = -(mg_T R_T/r)\,\mathbf{e}_r$, where g_T is the acceleration due to gravity at the surface, R_T is the radius of the planet, and r is the distance from the centre of the planet.
(a) Determine the potential energy associated with this gravitational force.
(b) If the object is given a velocity v_0 at a distance r_0, what is its velocity v as a function r?

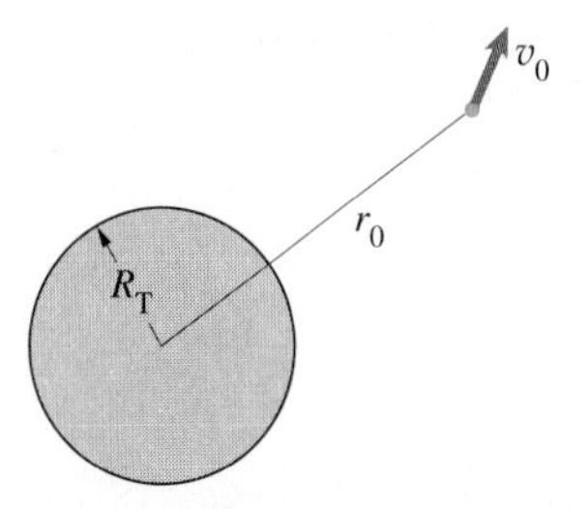

P4.95

4.96 By substituting Equations (4.31) into Equation (4.34), confirm that $\nabla \times \mathbf{F} = 0$ if **F** is conservative.

4.97 Determine which of the following forces are conservative:
(a) $\mathbf{F} = (3x^2 - 2xy)\mathbf{i} - x^2\mathbf{j}$;
(b) $\mathbf{F} = (x - xy^2)\mathbf{i} + x^2y\mathbf{j}$;
(c) $\mathbf{F} = (2xy^2 + y^3)\mathbf{i} + (2x^2y - 3xy^2)\mathbf{j}$.

4.98 Determine which of the following forces are conservative:
(a) $\mathbf{F} = 3r^2\sin^2\theta\,\mathbf{e}_r + 2r^2\sin\theta\cos\theta\,\mathbf{e}_\theta$;
(b) $\mathbf{F} = (2r\sin\theta - \cos\theta)\,\mathbf{e}_r + (r\cos\theta - \sin\theta)\,\mathbf{e}_\theta$;
(c) $\mathbf{F} = (\sin\theta + r\cos^2\theta)\,\mathbf{e}_r + (\cos\theta - r\sin\theta\cos\theta)\,\mathbf{e}_\theta$.

Computational Mechanics

The following example and problems are designed for the use of a programmable calculator or computer.

Example 4.9

In the mechanical delay switch shown in Figure 4.21, an electromagnet releases the 1 kg slider at position 1. Under the actions of gravity and the linear spring, the slider moves along the smooth bar from position 1 to position 2, closing the switch. The constant of the spring is $k = 40$ N/m, and its unstretched length is $r_0 = 50$ mm. The dimensions are $R = 200$ mm and $h = 100$ mm. What is the magnitude of the slider's maximum velocity, and where does it occur?

Figure 4.21

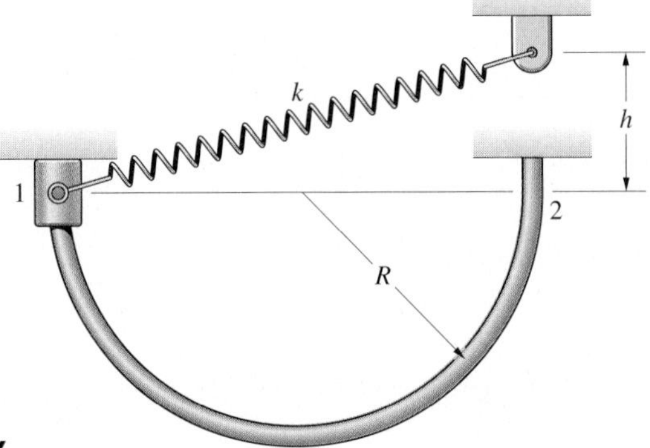

STRATEGY

We can use conservation of energy to obtain an equation relating the slider's velocity to its position. By drawing a graph of the velocity as a function of the position, we can estimate the maximum velocity and the position where it occurs.

SOLUTION

We can specify the slider's position by the angle θ through which it has moved relative to position 1 (Figure (a)). In position 1, the extension of the spring equals its

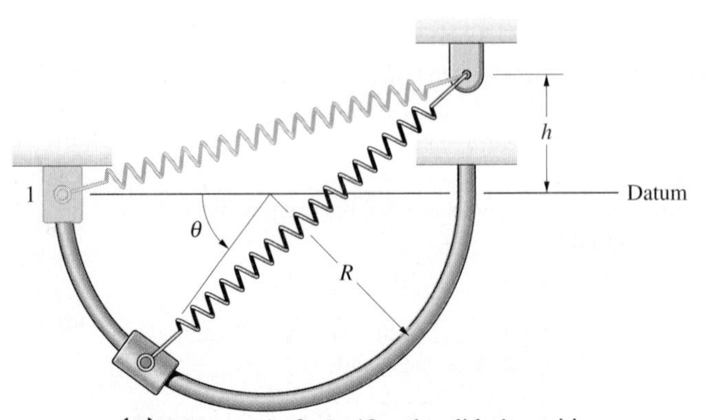

(a) The angle θ specifies the slider's position.

length in position 1 minus its unstretched length:

$$S_1 = \sqrt{(2R)^2 + h^2} - r_0$$

When the slider has moved through the angle θ, the extension of the spring is

$$S = \sqrt{(R + R\cos\theta)^2 + (h + R\sin\theta)^2} - r_0$$

We express the potential energy of the slider's weight using the datum shown in Figure (a). The sum of the potential and kinetic energies at position 1 must equal the sum of the potential and kinetic energies when the slider has moved through the angle θ:

$$\frac{1}{2}kS_1^2 + mgy_1 + \frac{1}{2}mv_1^2 = \frac{1}{2}kS^2 + mgy + \frac{1}{2}mv^2 :$$

$$\frac{1}{2}k\left[\sqrt{(2R)^2 + h^2} - r_0\right]^2 + 0 + 0$$

$$= \frac{1}{2}k\left[\sqrt{(R + R\cos\theta)^2 + (h + R\sin\theta)^2} - r_0\right]^2$$

$$- mgR\sin\theta + \frac{1}{2}mv^2$$

Solving for v, we obtain

$$v = \left\{(k/m)\left[\sqrt{(2R)^2 + h^2} - r_0\right]^2\right.$$

$$\left. -(k/m)\left[\sqrt{(R + R\cos\theta)^2 + (h + R\sin\theta)^2} - r_0\right]^2 + 2gR\sin\theta\right\}^{1/2}$$

Computing the values of this expression as a function of θ, we obtain the graph shown in Figure 4.22. The velocity is a maximum at approximately $\theta = 135°$. By examining the computed results near 135°,

θ	m/s
132°	2.5393
133°	2.5397
134°	2.5399
135°	2.5398
136°	2.5394
137°	2.5389
138°	2.5380

we estimate that a maximum velocity of 2.54 m/s occurs at $\theta = 134°$

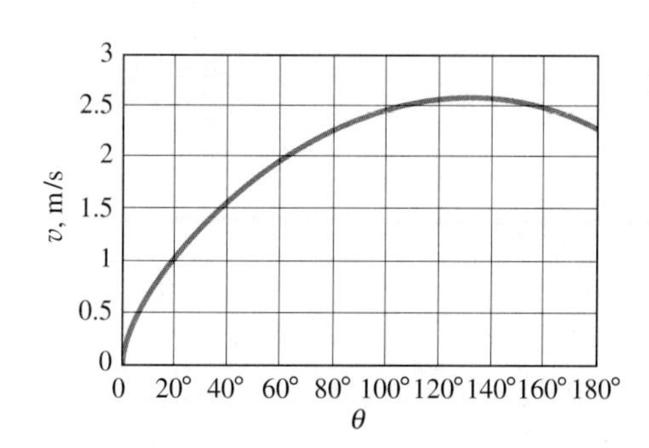

Figure 4.22

Magnitude of the velocity as a function of θ.

Problems

4.99 The component of the total external force tangential to a 4 kg object's path is $\Sigma F_t = (200 + 2s^2 - 0.2s^3)$ N, where s is its position measured along the path in metres. At $s = 0$, the object's velocity is $v = 10$ m/s. What distance along its path has the object travelled when its velocity reaches 30 m/s?

4.100 The 6 kg collar is released from rest in the position shown. If the spring constant is $k = 4$ kN/m and the unstretched length of the spring is 150 mm, how far does the mass fall from its initial position before rebounding?

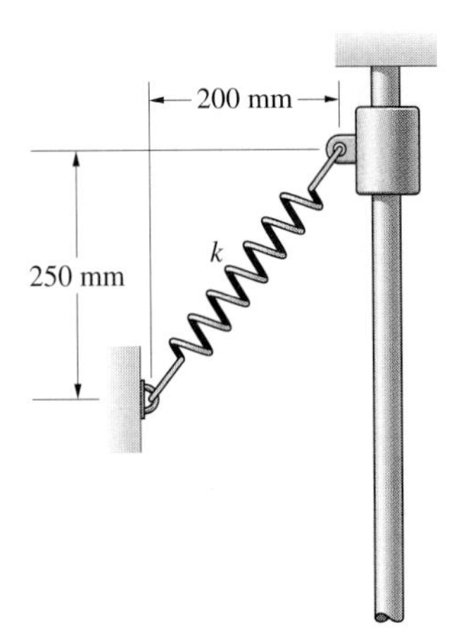

P4.100

4.101 How far below its initial position does the collar in Problem 4.100 reach its maximum velocity, and what is the maximum velocity?

4.102 How far below its initial position does the power being transferred to the collar in Problem 4.100 reach its maximum, and what is the maximum power?

4.103 The system is released from rest in the position shown. The weights are $W_A = 900$ N and $W_B = 1350$ N. Neglect friction. Determine the maximum velocity attained by A as it rises.

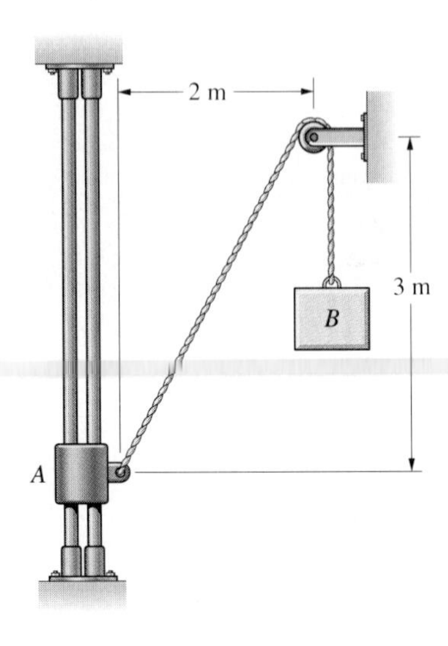

P4.103

4.104 In Problem 4.103, what maximum height is reached by A relative to its initial position?

4.105 The 16 kg cylinder is released at the position shown and falls onto a nonlinear spring with potential energy $V = \frac{1}{2}kS^2 + \frac{1}{4}qS^4$, where $k = 2400$ N/m and $q = 3000$ N/m^3. Determine how far down the cylinder moves after contacting the spring.

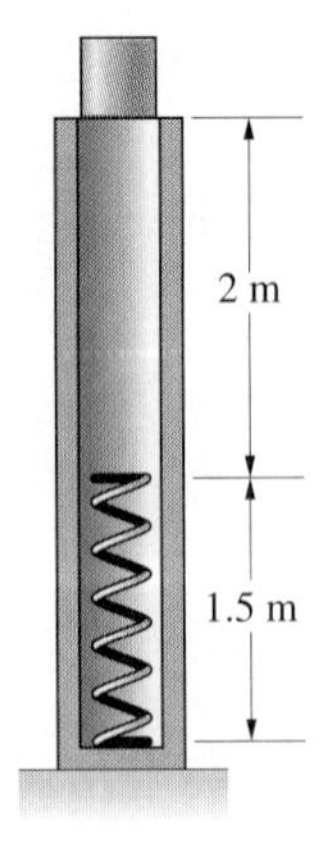

P4.105

4.106 In problem 4.105, what is the maximum velocity attained by the cylinder?

4.107 In Problem 4.82, how high does the collar A rise relative to its initial position?

4.108 In Problem 4.9, what is the maximum power transferred from the car by the barrier, and what distance has the car travelled from its initial contact when it occurs?

4.109 A student runs at 4.5 m/s, grabs a rope, and swings out over a lake. Determine the angle θ at which he should release the rope to maximize the horizontal distance b. What is the resulting value of b?

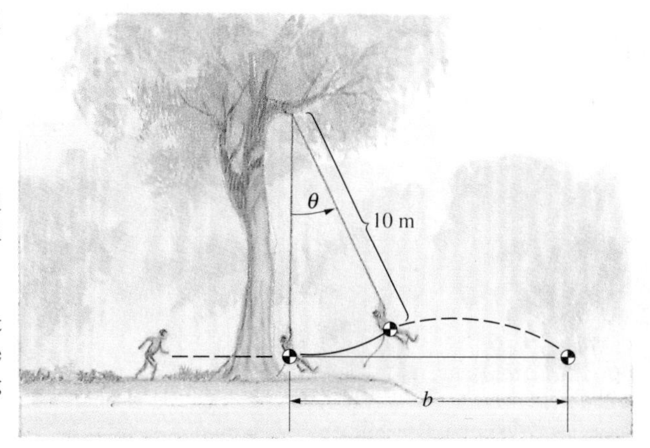

P4.109

Chapter Summary

Principle of Work and Energy

The **principle of work and energy** states that the work U done on an object as it moves from a position $\mathbf{r}_1$ to a position $\mathbf{r}_2$ is equal to the change in its kinetic energy,

$$U = \frac{1}{2}mv_2^2 - \frac{1}{2}mv_1^2 \qquad \textbf{Equation (4.5)}$$

where

$$U = \int_{\mathbf{r}_1}^{\mathbf{r}_2} \Sigma \mathbf{F} \cdot d\mathbf{r} \qquad \textbf{Equation (4.6)}$$

The total work done by external forces on a system of objects equals the change in the total kinetic energy of the system if no net work is done by internal forces.

Evaluating the Work

Let s be the position of an object's centre of mass along its path. The work done on the object from a position s_1 to a position s_2 is

$$U = \int_{s_1}^{s_2} \Sigma F_t \, ds \qquad \textbf{Equation (4.7)}$$

where ΣF_t is the tangential component of the total external force on the object. *Components of force perpendicular to the path do no work.*

Weight In terms of a coordinate system with the positive y axis upwards, the work done by an object's weight as its centre of mass moves from position 1 to position 2 is

$$U = -mg(y_2 - y_1) \qquad \textbf{Equation (4.13)}$$

The work is the product of the weight and the change in the height of the centre of mass. The work is negative if the height increases and positive if it decreases.

When the variation of an object's weight with distance r from the centre of the earth is accounted for, the work done by its weight is

$$U = mgR_E^2\left(\frac{1}{r_2} - \frac{1}{r_1}\right) \quad \textbf{Equation (4.15)}$$

where R_E is the radius of the earth.

Springs The work done on an object by a spring attached to a fixed support is

$$U = -\frac{1}{2}k(S_2^2 - S_1^2) \quad \textbf{Equation (4.16)}$$

where S_1 and S_2 are the values of the extension at the initial and final positions.

Power

The **power** is the rate at which work is done. The power transferred to an object by the external forces acting on it is

$$P = \Sigma \mathbf{F} \cdot \mathbf{v} \quad \textbf{Equation (4.17)}$$

The power equals the rate of change of the object's kinetic energy. The average with respect to time of the power during an interval of time from t_1 to t_2 is equal to the change in its kinetic energy, or the work done, divided by the interval of time:

$$P_{av} = \frac{\frac{1}{2}mv_2^2 = \frac{1}{2}mv_1^2}{t_2 - t_1} = \frac{U}{t_2 - t_1} \quad \textbf{Equation (4.18)}$$

Potential Energy

For a given force $\mathbf{F}$ acting on an object, if a function V of the object's position exists such that

$$dV = -\mathbf{F} \cdot d\mathbf{r}$$

then $\mathbf{F}$ is said to be **conservative** and V is called the **potential energy** associated with $\mathbf{F}$. The work done by $\mathbf{F}$ from a position 1 to a position 2 is

$$U = V_1 - V_2 \quad \textbf{Equation (4.21)}$$

If all the forces that do work on a system are conservative, the total energy – the sum of the kinetic energy and the potential energies of the forces – is conserved:

$$\frac{1}{2}mv^2 + V = \text{constant} \quad \textbf{Equation (4.23)}$$

Weight In terms of a cartesian coordinate system with its y axis upwards, the potential energy of the weight of an object is

$$V = mgy \qquad \textbf{Equation (4.25)}$$

The potential energy is the product of the object's weight and the height of its centre of mass measured from any convenient reference level, or **datum**.

When the variation of an object's weight with distance r from the centre of the earth is accounted for, the potential energy of its weight is

$$V = -\frac{mgR_E^2}{r} \qquad \textbf{Equation (4.27)}$$

where R_E is the radius of the earth.

Springs The potential energy of the force exerted on an object by a linear spring is

$$V = \frac{1}{2}kS^2 \qquad \textbf{Equation (4.28)}$$

where S is the extension of the spring.

Relationships Between Force and Potential Energy

A force **F** is related to its associated potential energy by

$$\mathbf{F} = -\left(\frac{\partial V}{\partial x}\mathbf{i} + \frac{\partial V}{\partial y}\mathbf{j} + \frac{\partial V}{\partial z}\mathbf{k}\right) = -\nabla V \qquad \textbf{Equation (4.23)}$$

A force **F** is conservative if its **curl** is zero:

$$\nabla \times \mathbf{F} = \begin{vmatrix} \mathbf{i} & \mathbf{j} & \mathrm{k} \\ \dfrac{\partial}{\partial x} & \dfrac{\partial}{\partial y} & \dfrac{\partial}{\partial z} \\ F_x & F_y & F_z \end{vmatrix} = 0$$

Review Problems

4.110 The driver of a 1360 kg car moving at 64 km/hr applies an increasing force on the brake pedal. The magnitude of the resulting friction force exerted on the car by the road is $f = (1000 + 40\,s)$ N, where s is the car's horizontal position in metres relative to its position when the brakes were applied. Assuming that the car's tyres do not slip, determine the distance required for the car to stop (a) by using Newton's second law; (b) by using the principle of work and energy.

4.111 Suppose that the car in Problem 4.110 is on wet pavement and the coefficients of friction between the tyres and the road are $\mu_s = 0.4$, $\mu_k = 0.35$. Determine the distance required for the car to stop.

4.112 An astronaut in a small rocket vehicle (combined mass = 450 kg) is hovering 100 m above the surface of the Moon when he discovers he is nearly out of fuel and can only exert the thrust necessary to cause the vehicle to hover for 5 more seconds. He quickly considers two strategies for getting to the surface: (a) fall 20 m, turn on the thrust for 5 s, then fall the rest of the way; (b) fall 40 m, turn on the thrust for 5 s, then fall the rest of the way. Which strategy gives him the best change of surviving? How much work is done by the engine's thrust in each case? (g_{moon} = 1.62 m/s^2.)

4.113 The coefficients of friction between the 20 kg crate and the inclined surface are $\mu_s = 0.24$ and $\mu_k = 0.22$. If the crate starts from rest and the horizontal force $F = 200$ N, what is the magnitude of its velocity when it has moved 2 m?

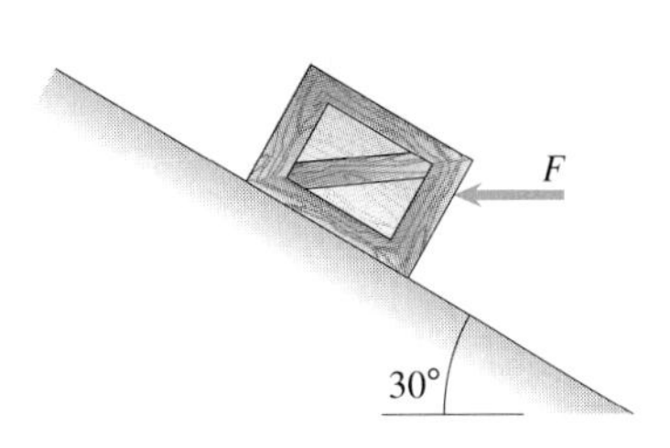

P4.113

4.114 In Problem 4.113, what is the magnitude of the crate's velocity when it has moved 2 m if the horizontal force $F = 40$ N?

4.115 The Union Pacific *Big Boy* locomotive weighs 5.29 million N, and the tractive effort (tangential force) of its drive wheels is 600 000 N. If you neglect other tangential forces, what distance is required for it to accelerate from zero to 100 km/hr?

P4.115

4.116 In Problem 4.115, suppose that the acceleration of the locomotive as it accelerates from zero to 100 km/hr is $(F_0/m)(1 - v/100)$, where $F_0 = 600\,000$ N, m is its mass, and v is its velocity in km/hr.
(a) How much work is done in accelerating it to 100 km/hr?
(b) Determine its velocity as a function of time.

4.117 If a car travelling at 105 km/hr hits the crash barrier described in Problem 4.9, determine the maximum decleration the passengers are subjected to if the car weighs (a) 11 120 N; (b) 22 240 N.

4.118 In a preliminary design for a mail sorting machine, parcels moving at 0.6 m/s slide down a smooth ramp and are brought to rest by a linear spring. What should the spring constant be if you don't want a 5 kg parcel to be subjected to a maximum deceleration greater than 10 g's?

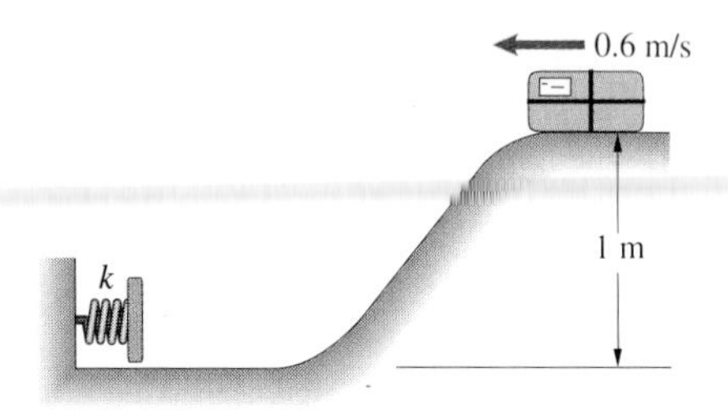

P4.118

4.119 When the 1 kg collar is in position 1, the tension in the spring is 50 N, and the unstretched length of the spring is 260 mm. If the collar is pulled to position 2 and released from rest, what is its velocity when it returns to 1?

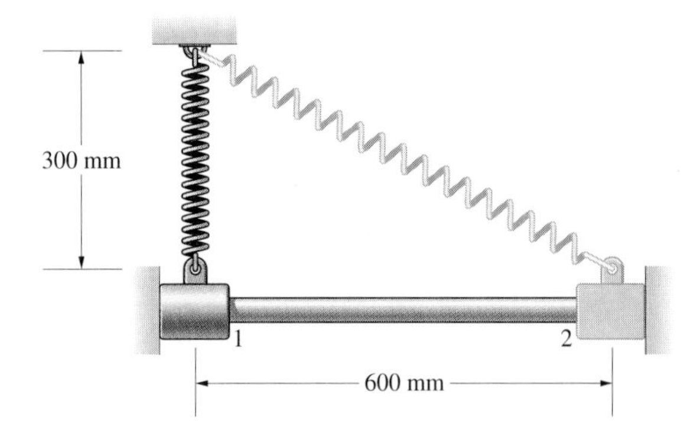

P4.119

4.120 In Problem 4.119, suppose that the tensions in the spring in positions 1 and 2 are 100 N and 400 N, respectively.
(a) What is the spring constant k?
(b) If the collar is given a velocity of 15 m/s at 1, what is its velocity when it reaches 2?

4.121 The 14 kg weight is released from rest with the two springs ($k_A = 440$ N/m, $k_B = 220$ N/m) unstretched.
(a) How far does the weight fall before rebounding?
(b) What maximum velocity does it attain?

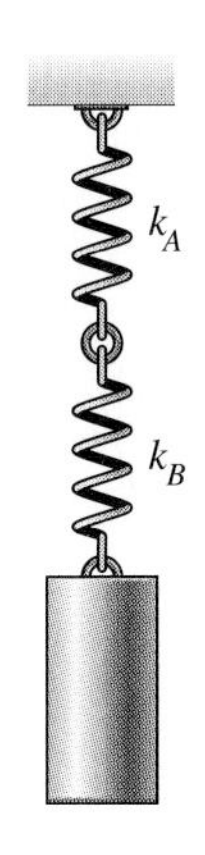

P4.121

4.122 The 12 kg collar A is at rest in the position shown at $t = 0$ and is subjected to the tangential force $F = (24 - 12t^2)$ N for 1.5 s. Neglecting friction, what maximum height h does it reach?

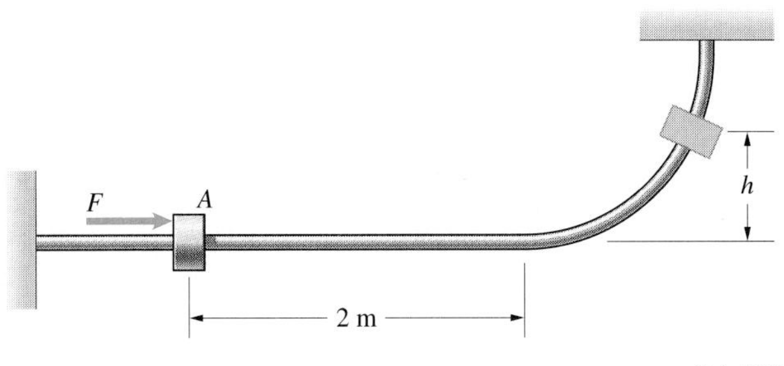

P4.122

4.123 When a 22 Mg rocket's engine burns out at an altitude of 2 km, its velocity is 3 km/s and it is travelling at an angle of 60° relative to the horizontal. Neglect the variation in the gravitational force with altitude.
(a) If you neglect aerodynamic forces, what is the magnitude of the rocket's velocity when it reaches an altitude of 6 km?
(b) If the rocket's actual velocity when it reaches an altitude of 6 km is 2.8 km/s, how much work is done by aerodynamic forces as the rocket moves from 2 km to 6 km altitude?

4.124 The piston and the load it supports are accelerated upwards by the gas in the cylinder. The total weight of the piston and load is 4450 N. The cylinder wall exerts a constant 220 N friction force on the piston as it rises. The net force exerted on the piston by pressure is $(p - p_{atm})A$, where p is the pressure of the gas, $p_{atm} = 101\,300$ Pa is atmospheric pressure, and $A = 0.1$ m^2 is the cross-sectional area of the piston. Assume that the product of p and the volume of the cylinder is constant. When $s = 0.3$ m the piston is stationary and $p = 239\,250$ Pa. What is the velocity of the piston when $s = 0.6$ m?

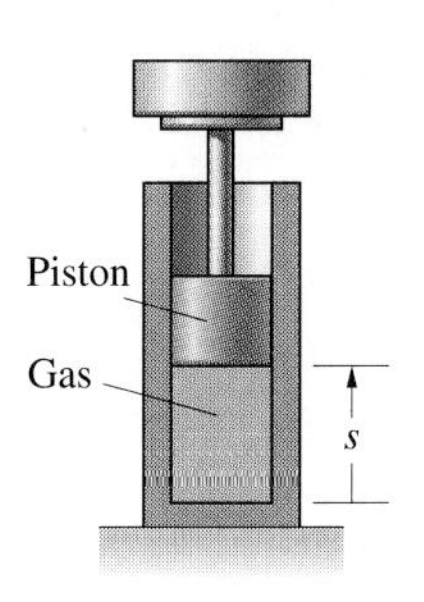

P4.124

4.125 Suppose that in designing a loop for a roller coaster's track, you establish as a safety criterion that at the top of the loop, the normal force exerted on a passenger by the roller coaster should equal 10 per cent of the passenger's weight. (That is, the passenger's 'effective weight' pressing him down into his seat is 10 per cent of his weight.) The roller coaster is moving at 20 m/s when it enters the loop. What is the necessary instantaneous radius of curvature ρ of the track at the top of the loop?

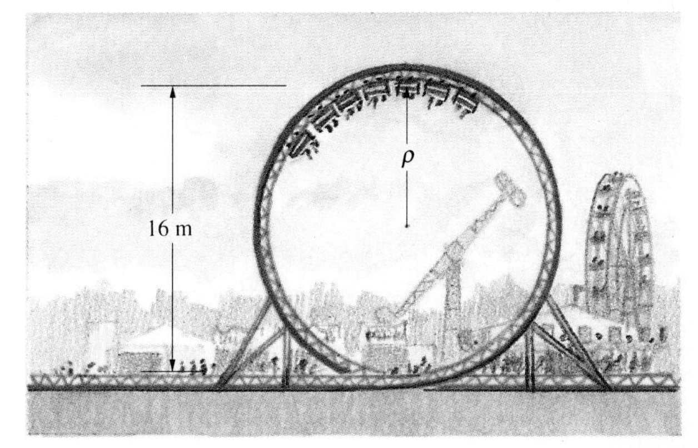

P4.125

4.126 An 80 kg student runs at 4.5 m/s, grabs a rope, and swings out over a lake. He releases the rope when his velocity is zero.
(a) What is the angle θ when he releases the rope?
(b) What is the tension in the rope just before he releases it?
(c) What is the maximum tension in the rope?

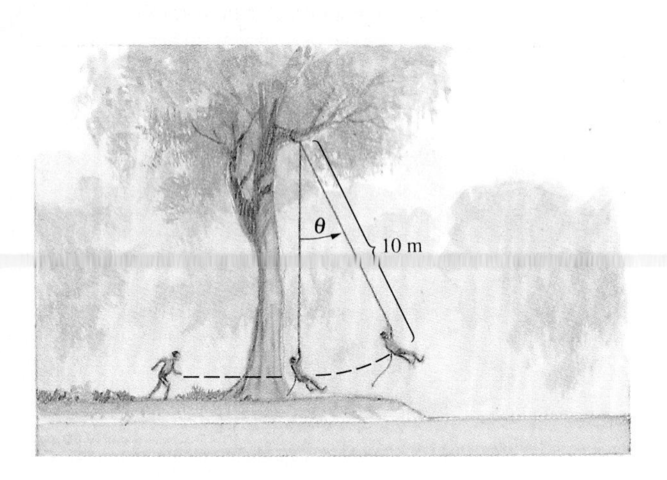

P4.126

4.127 If the student in Problem 4.126 releases the rope when $\theta = 25°$, what maximum height does he reach relative to his position when he grabs the rope?

4.128 A boy takes a running start and jumps on his sled at 1. He leaves the ground at 2 and lands in deep snow at a distance $b = 6$ m. How fast was he going at 1?

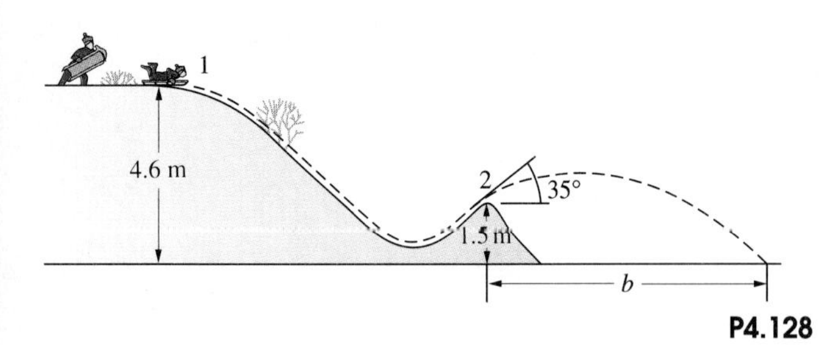

P4.128

4.129 In Problem 4.128, if the boy starts at 1 going at 4.5 m/s, what distance b does he travel through the air?

4.130 The 1 kg collar A is attached to the linear spring ($k = 500$ N/m) by a string. The collar starts from rest in the position shown, and the initial tension in the string is 100 N. What distance does the collar slide up the smooth bar?

P4.130

4.131 The y axis is vertical and the curved bar is smooth. If the magnitude of the velocity of the 4 kg slider is 6m/s at position 1, what is the magnitude of its velocity when it reaches position 2?

P4.131

4.132 In Problem 4.131, determine the magnitude of the slider's velocity when it reaches position 2 if it is subjected to the additional force $\mathbf{F} = (3x\,\mathbf{i} - 2\,\mathbf{j})$ N during its motion.

4.133 Suppose that an object of mass m is beneath the surface of the earth. In terms of a polar coordinate system with its origin at the earth's centre the gravitational force on the object is $-(mgr/R_E)\mathbf{e}_r$, where R_E is the radius of the earth. Show that the potential energy associated with the gravitational force is $V = mgr^2/2R_E$.

4.134 It has been pointed out that if tunnels could be drilled straight through the earth between points on the surface, trains could travel between those points using gravitational force for acceleration and deceleration. (The effects of friction and aerodynamic drag could be minimized by evacuating the tunnels and using magnetically levitated trains.) Suppose that such a train travels from the North Pole to a point on the equator. Determine the magnitude of the train's velocity (a) when it arrives at the equator; (b) when it is halfway from the North Pole to the equator. The radius of the earth is $R_E = 6370$ km.

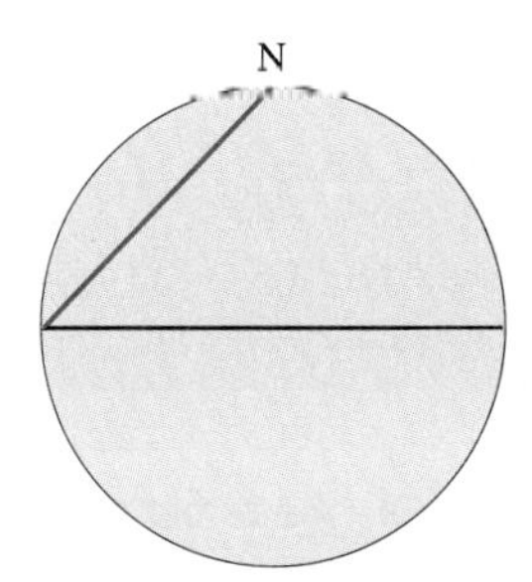

P4.134

4.135 In Problem 4.115, what is the maximum power transferred to the locomotive during its acceleration?

4.136 Just before it lifts off, the 10.5 Mg aeroplane is travelling at 60 m/s. The total horizontal force exerted by its engines is 189 kN, and the plane is accelerating at 15 m/s^2.
(a) How much power is being transferred to the plane by its engines?
(b) What is the total power being transferred to the plane?

P4.136

4.137 The 'Paris Gun', used by Germany in World War I, had a range of 120 km, a 37.5 m barrel, a muzzle velocity of 1550 m/s, and it fired a 120 kg shell.
(a) If you assume the shell's acceleration to be constant, what maximum power was transferred to it as it travelled along the barrel?
(b) What everage power was transferred to the shell?

P4.137

The total linear momentum of the vehicles is the same immediately after their collision. By analysing simulated traffic accidents, engineers obtain information useful in the design of the structures of vehicles, their steering and braking systems, and devices for protecting passengers. In this chapter you will use methods based on linear and angular momentum to analyse motions of objects.

Chapter 5

Momentum Methods

IN Chapter 4 we transformed Newton's second law to obtain the principle of work and energy. In this chapter we integrate Newton's second law with respect to time, obtaining a relation between the time integral of the forces acting on an object and the change in the object's linear momentum. With this result, called the principle of impulse and momentum, we can determine the change in an object's velocity when the external forces are known as functions of time.

By applying the principle of impulse and momentum to two or more objects, we obtain the principle of conservation of linear momentum. This conservation law allows us to analyse impacts between objects and evaluate forces exerted by continuous flows of mass, as in jet and rocket engines.

By another transformation of Newton's second law, we obtain a relation between the time integral of the moments exerted on an object and the change in a quantity called angular momentum. We show that in the circumstance called central-force motion, an object's angular momentum is conserved.

5.1 *Principle of Impulse and Momentum*

The principle of work and energy is a very useful tool in mechanics. We can derive another useful tool for the analysis of motion by integrating Newton's second law with respect to time. We express Newton's second law in the form

$$\Sigma \mathbf{F} = m\frac{d\mathbf{v}}{dt}$$

Then we integrate with respect to time to obtain

$$\int_{t_1}^{t_2} \Sigma \mathbf{F}\, dt = m\mathbf{v}_2 - m\mathbf{v}_1 \tag{5.1}$$

where $\mathbf{v}_1$ and $\mathbf{v}_2$ are the velocities of the centre of mass of the objects at the times t_1 and t_2. The term on the left is called the **linear impulse**, and $m\mathbf{v}$ is the **linear momentum**. This result is called the **principle of impulse and momentum**: the impulse applied to an object during an interval of time is equal to the change in its linear momentum (Figure 5.1). The dimensions of the linear impulse and linear momentum are (force) × (time).

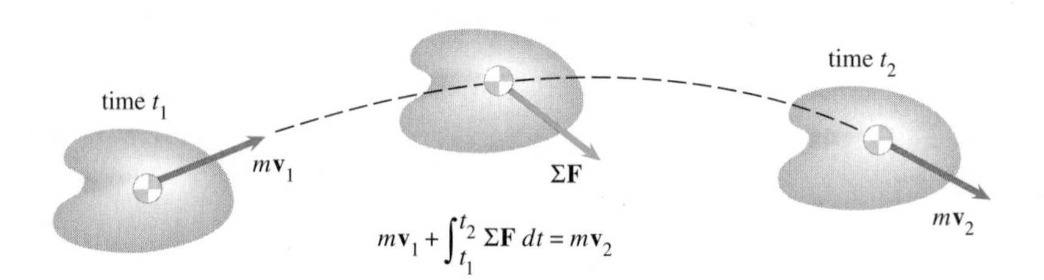

Figure 5.1
Principle of impulse and momentum.

Notice that Equation (5.1) and the principle of work and energy, Equation (4.5), are quite similar. They both relate an integral of the external forces to the change in an object's velocity. Equation (5.1) is a vector equation that tells you the change in both the magnitude and direction of the velocity, whereas the principle of work and energy, a scalar equation, tells you only the change in the magnitude of the velocity. There is a greater difference between the two methods, however: in the case of impulse and momentum, there is no class of forces equivalent to the conservation forces that make work and energy so easy to apply.

When you know the external forces acting on an object as functions of time, the principle of impulse and momentum allows you to determine the change in its velocity during an interval of time. Although this is an important result, it is not new. When you used Newton's second law in Chapter 3 to determine an object's acceleration and then integrated the acceleration with respect to time to determine the velocity, you were effectively applying the principle of impulse and momentum. However, in the rest of this chapter we show that this principle can be extended to new and interesting applications.

The average with respect to time of the total force acting on an object from t_1 to t_2 is

$$\Sigma \mathbf{F}_{\text{av}} = \frac{1}{t_2 - t_1} \int_{t_1}^{t_2} \Sigma \mathbf{F}\, dt$$

so we can write Equation (5.1) as

$$(t_2 - t_1)\Sigma \mathbf{F}_{\text{av}} = m\mathbf{v}_2 - m\mathbf{v}_1 \qquad (5.2)$$

With this equation you can determine the average value of the total force acting on an object during a given interval of time if you know the change in its velocity.

A force of relatively large magnitude that acts over a small interval of time is called an **impulsive force** (Figure 5.2). Determining the actual time history of such a force is usually impractical, but its average value can often be determined. For example, a golf ball struck by a club is subjected to an impulsive force. By making high-speed motion pictures, we can determine the duration of the impact. Also, the ball's velocity can be measured from motion pictures of its motion following the impact. Knowing the duration and the ball's linear momentum following the impact, we can use Equation (5.2) to determine the average force exerted on the ball by the club. (See Example 5.2.)

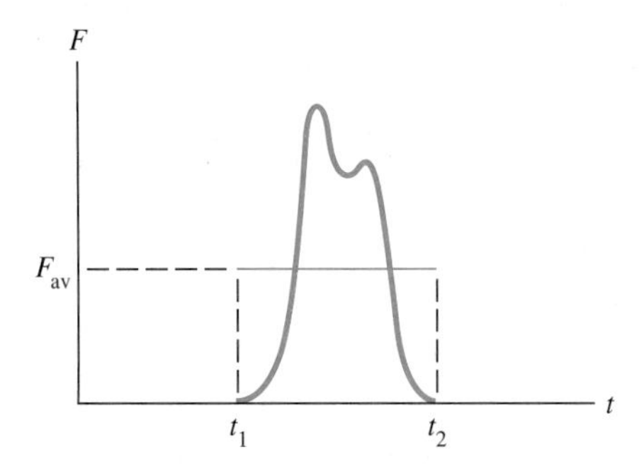

Figure 5.2

An impulsive force and its average value.

Example 5.1

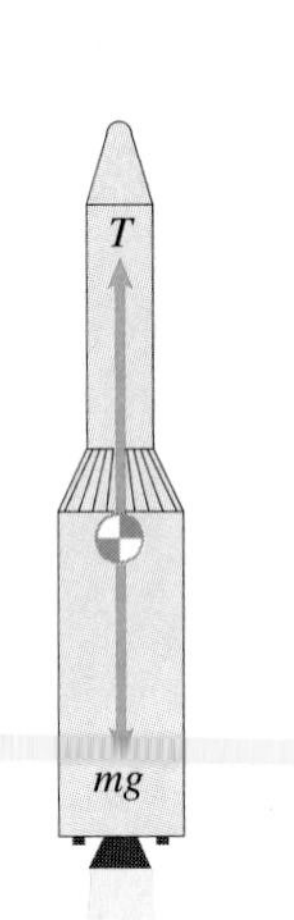

Figure 5.3

The rocket booster in Figure 5.3 is travelling straight up when it suddenly starts rotating counterclockwise at 0.25 rev/s. The range safety officer destroys it 2 s later. The booster's mass is $m = 90\,\text{Mg}$, its thrust is $T = 1.0\,\text{MN}$, and it is moving upwards at 10 m/s when it starts rotating. If aerodynamic forces are neglected, what is the booster's velocity at the time it is destroyed?

STRATEGY

Because we know the angular velocity, we can determine the direction of the booster's thrust as a function of time and calculate the impulse during the 2 s period.

SOLUTION

The booster's angular velocity is $\pi/2$ rad/s. Letting $t = 0$ be the time at which it starts rotating, the angle between its axis and the vertical is $(\pi/2)t$ (Figure (a)). The total force on the booster is

$$\Sigma \mathbf{F} = \left(-T \sin\frac{\pi}{2}t\right)\mathbf{i} + \left(T\cos\frac{\pi}{2}t - mg\right)\mathbf{j}$$

so the impulse from $t = 0$ to $t = 2$ s is

$$\begin{aligned}\int_0^2 \Sigma \mathbf{F}\,dt &= \int_0^2 \left[\left(-T \sin\frac{\pi}{2}t\right)\mathbf{i} + \left(T\cos\frac{\pi}{2}t - mg\right)\mathbf{j}\right]dt \\ &= \left[\left(T\frac{2}{\pi}\cos\frac{\pi}{2}t\right)\mathbf{i} + \left(T\frac{2}{\pi}\sin\frac{\pi}{2}t - mgt\right)\mathbf{j}\right]_0^2 \\ &= -\frac{4}{\pi}T\,\mathbf{i} - 2mg\,\mathbf{j}\end{aligned}$$

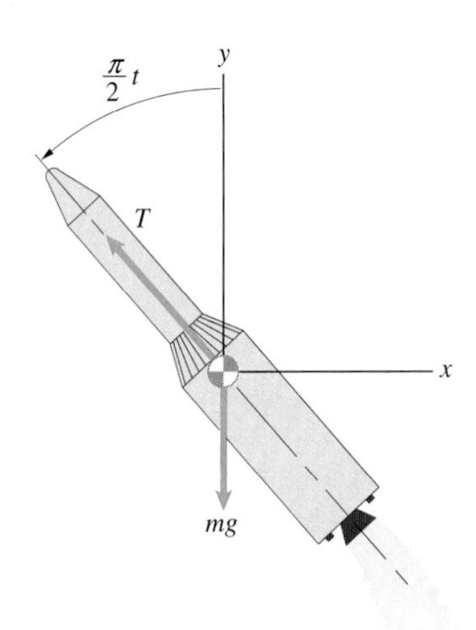

(a) The rotating booster.

From the principle of impulse and momentum,

$$\int_0^2 \Sigma \mathbf{F}\,dt = m\mathbf{v}_2 - m\mathbf{v}_1:$$

$$-\frac{4}{\pi}(1\times 10^6)\mathbf{i} - 2(90\times 10^3)(9.81)\mathbf{j} = (90\times 10^3)(\mathbf{v}_2 - 10\,\mathbf{j})$$

Solving for $\mathbf{v}_2$, we obtain $\mathbf{v}_2 = (-14.15\,\mathbf{i} - 9.62\,\mathbf{j})$ m/s.

DISCUSSION

Notice that the rocket's thrust has no net effect on its y component of velocity during the 2 s interval. The effect of the positive y component of the thrust during the first quarter revolution is cancelled by the effect of the negative y component during the second quarter revolution. The change in the y component of velocity is caused entirely by the rocket's weight. The thrust has a negative x component during the entire 2 s interval, giving the rocket its negative x component of velocity at the time it is destroyed.

Example 5.2

A golf ball in flight is photographed at intervals of 0.001 s (Figure 5.4). The 0.046 kg ball is 43 mm in diameter. If the club was in contact with the ball for 0.0006 s, estimate the average value of the impulsive force exerted by the club.

Figure 5.4

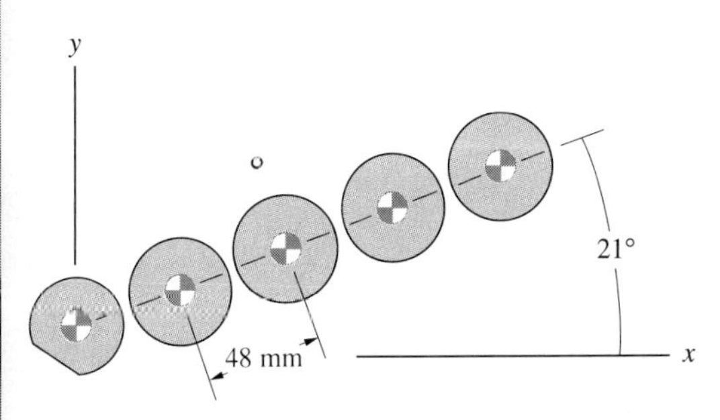

(a) Estimating the distance travelled during one 0.001 s interval.

STRATEGY

By measuring the distance travelled by the ball in one of the 0.001 s intervals, we can estimate its velocity after being struck, then use Equation (5.2) to determine the average total force on the ball.

SOLUTION

By comparing the distance moved during one of the 0.001 s intervals with the known diameter of the ball, we estimate that the ball travelled 48 mm and that its direction is 21° above the horizontal (Figure (a)). The magnitude of the ball's velocity is

$$\frac{0.048 \text{ m}}{0.001 \text{ s}} = 48 \text{ m/s}$$

The weight of the ball is (0.046)(9.81) = 0.451 N, and its mass is 0.046 kg. From Equation (5.2),

$$(t_2 - t_1)\Sigma \mathbf{F}_{\text{av}} = m\mathbf{v}_2 - m\mathbf{v}_1$$
$$(0.0006)\Sigma \mathbf{F}_{\text{av}} = (0.046)(48)(\cos 21^\circ \,\mathbf{i} + \sin 21^\circ \,\mathbf{j}) - 0$$

we obtain

$$\Sigma \mathbf{F}_{\text{av}} = (3436\,\mathbf{i} + 1319\,\mathbf{j}) \text{ N}$$

DISCUSSION

The average force during the time the club is in contact with the ball includes both the impulsive force exerted by the club and the ball's weight. In comparison with the large average impulsive force exerted by the club, the weight ($-0.45\,\mathbf{j}$ N) is negligible.

Problems

5.1 The aircraft carrier *Nimitz* weighs 810 MN. Suppose that its engines and hydrodynamic drag exert a constant 4.45 MN decelerating force on it.
(a) Use the principle of impulse and momentum to determine how long it requires the ship to come to rest from its top speed of approximately 30 knots. (A knot is approximately 1.85 km/hr).
(b) Use the principle of work and energy to determine the distance the ship travels during the time it takes to come to rest.

P5.1

5.2 The 900 kg drag racer accelerates from rest to 480 km/hr in 6 s.
(a) What impulse is applied to the car during the 6 s?
(b) If you assume as a first approximation that the tangential force exerted on the car is constant, what is the magnitude of the force?

P5.2

5.3 The 21 900 kg Gloster Saro Protector, designed for rapid response to airport emergencies, accelerates from rest to 80 km/hr in 35 s.
(a) What impulse is applied to the vehicle during the 35 s?
(b) If you assume as a first approximation that the tangential force exerted on the vehicle is constant, what is the magnitude of the force?
(c) What average power is transferred to the vehicle?

P5.3

5.4 The combined weight of the motorcycle and rider is 1350 N. The coefficient of kinetic friction between the motorcycle's tyres and the road is $\mu_k = 0.8$. Suppose that the rider starts from rest and spins the rear (drive) wheel. The normal force between the rear wheel and the road is 1100 N.
(a) What impulse does the friction force on the rear wheel exert in 5 s?
(b) If you neglect other horizontal forces, what velocity is attained in 5 s?

P5.4

5.5 An astronaut drifts towards a space station at 8 m/s. He carries a manoeuvring unit (a small hydrogen peroxide rocket) that has an impulse rating of 720 N.s. The total mass of the astronaut, his suit and the manoeuvring unit is 120 kg. If he uses all of the impulse to slow himself down, what will be his velocity relative to the station?

P5.5

5.6 The total external force on a 10 kg object is constant and equal to $(90\mathbf{i} - 60\mathbf{j} + 20\mathbf{k})$ N. At $t = 2$ s, the object's velocity is $(-8\mathbf{i} + 6\mathbf{j})$ m/s.
(a) What impulse is applied to the object from $t = 2$ s to $t = 4$ s?
(b) What is the object's velocity at $t = 4$ s?

5.7 The total external force on an object is $\mathbf{F} = (10t\mathbf{i} + 60\mathbf{j})$ N. At $t = 0$, its velocity is $\mathbf{v} = 20\mathbf{j}$ m/s. At $t = 12$ s, the x component of its velocity is 48 m/s.
(a) What impulse is applied to the object from $t = 0$ to $t = 6$ s?
(b) What is its velocity at $t = 6$ s?

5.8 During the first 5 s of the 15 000 kg aeroplane's takeoff roll, the pilot increases the engine's thrust at a constant rate from 25 kN to its full thrust of 125 kN.
(a) What impulse does the thrust exert on the aeroplane during the 5 s?
(b) If you neglect other forces, what total time is required for the aeroplane to reach its takeoff speed of 50 m/s?

P5.8

5.9 The 45 kg box starts from rest and is subjected to the force shown. If you neglect friction, what is the box's velocity at $t = 8$ s?

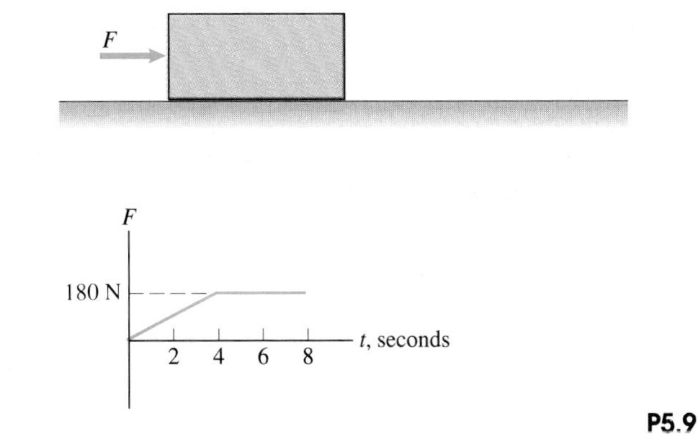

P5.9

5.10 Solve Problem 5.9 if the coefficients of friction between the box and the floor are $\mu_s = \mu_k = 0.2$.

5.11 The crate has a mass of 120 kg and the coefficients of friction between it and the sloping dock are $\mu_s = 0.6$, $\mu_k = 0.5$. The crate starts from rest, and the winch exerts a tension $T = 1220$ N.
(a) What impulse is applied to the crate during the first second of motion?
(b) What is the crate's velocity after 1 s?

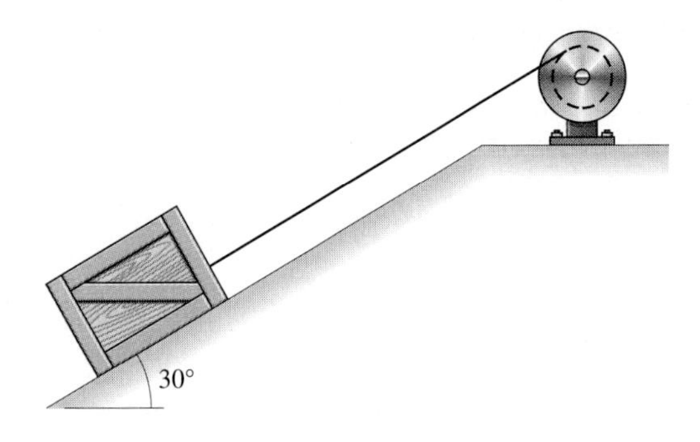

P5.11

5.12 Solve Problem 5.11 if the crate starts from rest at $t = 0$ and the winch exerts a tension $T = (1220 + 200t)$ N.

5.13 In an assembly-line process the 20 kg package A starts from rest and slides down the smooth ramp. Suppose that you want to design the hydraulic device B to exert a constant force of magnitude F on the package and bring it to rest in 0.2 s. What is the required force F?

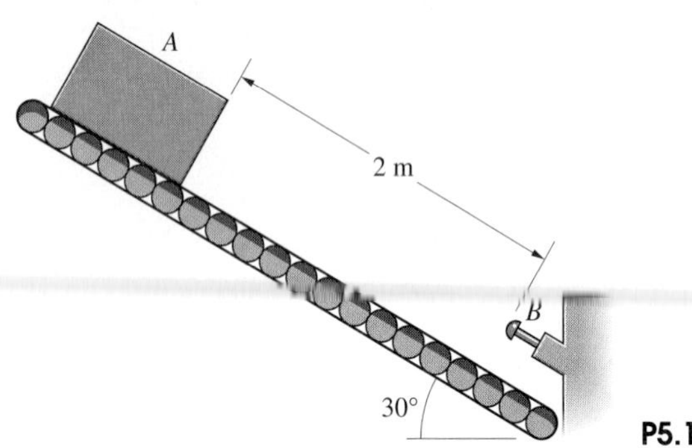

P5.13

5.14 In Problem 5.13, if the hydraulic device B exerts a force of magnitude $F = 540(1 + 0.4t)$ N on the package, where t is in seconds measured from the time of first contact, what time is required to bring the package to rest?

5.15 In a cathode-ray tube, an electron (mass $= 9.11 \times 10^{-31}$ kg). is projected at O with velocity $\mathbf{v} = (2.2 \times 10^7)\,\mathbf{i}$ (m/s). While it is between the charged plates, the electric field generated by the plates subjects it to a force $\mathbf{F} = -eE\,\mathbf{j}$. The charge of the electron is $e = 1.6 \times 10^{-19}$C (coulombs), and the electric field strength is $E = 15\sin(\omega t)$ kN/C, where the frequency $\omega = 2 \times 10^9\ \text{s}^{-1}$.
(a) What impulse does the electric field exert on the electron while it is between the plates?
(b) What is the velocity of the electron as it leaves the region between the plates?

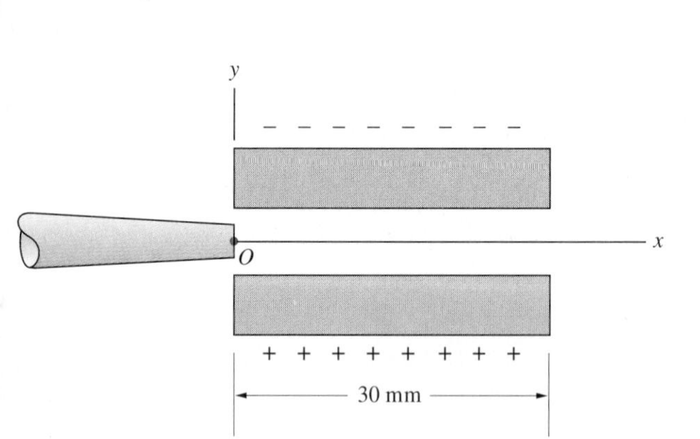

P5.15

5.16 The two weights are released from rest. What is the magnitude of their velocity after one-half second?

Strategy: Apply the principle of impulse and momentum to each weight individually.

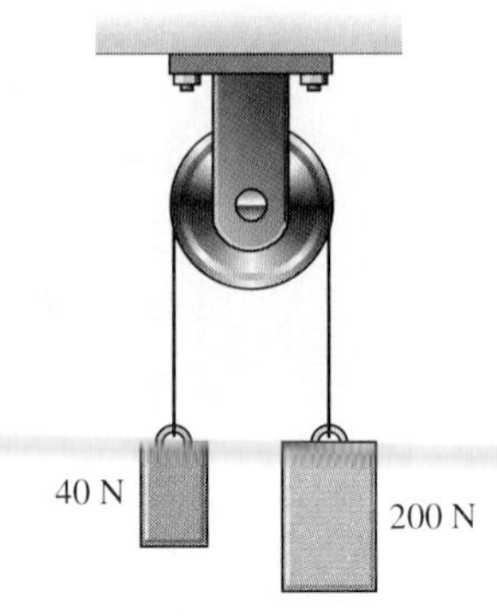

P5.16

5.17 The two crates are released from rest. Their masses are $m_A = 40$ kg and $m_B = 30$ kg, and the coefficient of kinetic friction between crate A and the inclined surface is $\mu_k = 0.15$. What is the magnitude of their velocity after 1 s?

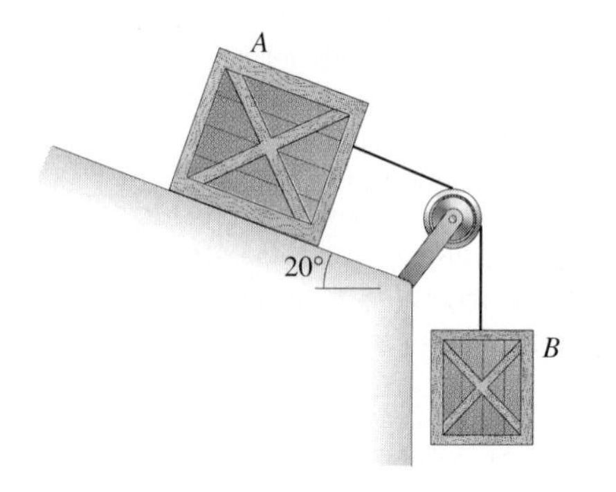

P5.17

5.18 In Example 5.1, if the range safety officer destroys the booster 1 s after it starts rotating, what is its velocity at the time it is destroyed?

5.19 An object of mass m slides with constant velocity v_0 on a horizontal table (seen from above in the figure). The object is attached by a string to the fixed point O and is in the position shown, with the string parallel to the x axis, at $t = 0$.
(a) Determine the x and y components of the force exerted on the mass by the string as functions of time.
(b) Use your results from part (a) and the principle of impulse and momentum to determine the velocity vector of the mass when it has travelled one-quarter of a revolution about point O.

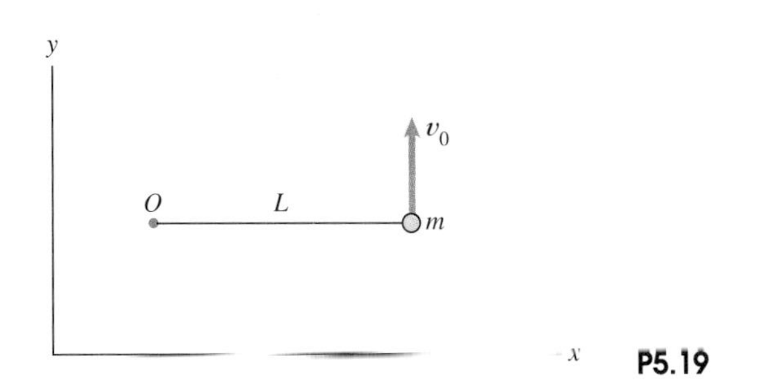

P5.19

5.20 At $t = 0$, a 25 kg projectile is given an initial velocity of 12 m/s at 60° above the horizontal. Neglect drag.
(a) What impulse is applied to the projectile from $t = 0$ to $t = 2$ s?
(b) What is the projectile's velocity at $t = 2$ s?

5.21 A rail gun, which uses an electromagnetic field to accelerate an object, accelerates a 30 g projectile to 5 km/s in 0.0005 s. What average force is exerted on the projectile?

5.22 The powerboat is going at 80 km/hr when its motor is turned off. In 5 s its velocity decreases to 48 km/hr. The boat and its passengers weigh 8 kN. Determine the magnitude of the average force exerted on the boat by hydrodynamic and aerodynamic drag during the 5 s.

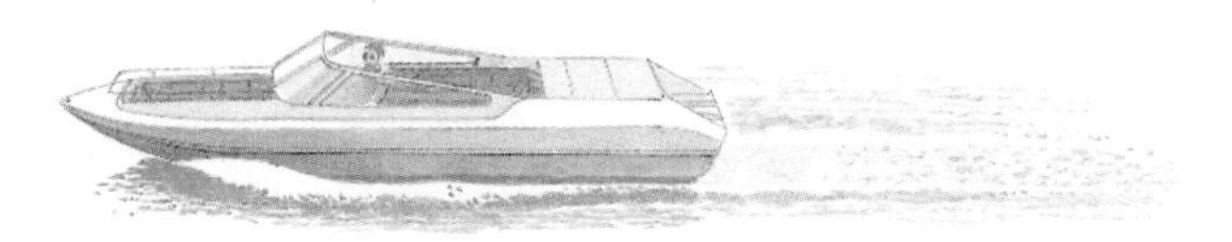
P5.22

5.23 The 77 kg skier is travelling at 10 m/s at 1, and he goes from 1 to 2 in 0.7 s.
(a) If you neglect friction and aerodynamic drag, what is the time average of the tangential component of force exerted on him as he moves from 1 to 2?
(b) If his actual velocity is measured at 2 and determined to be 13.1 m/s, what is the time average of the tangential component of force exerted on him as he moves from 1 to 2?

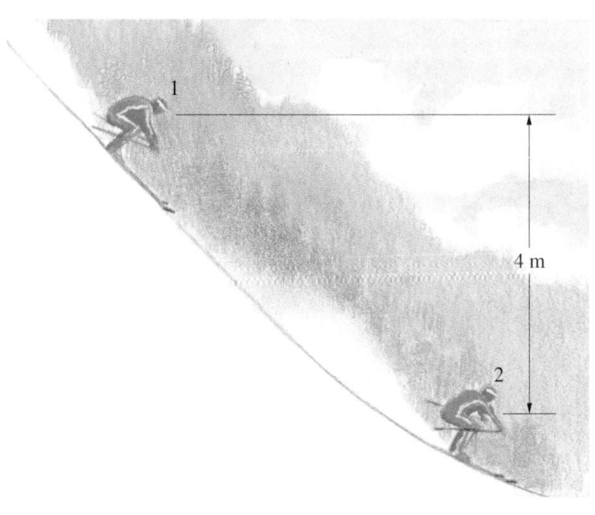

P5.23

5.24 In a test of an energy-absorbing bumper, a 1270 kg car is driven into a barrier at 8 km/hr. The duration of the impact is 0.4 s, and the car bounces back from the barrier at 1.6 km/hr.
(a) What is the magnitude of the average horizontal force exerted on the car during the impact?
(b) What is the average deceleration of the car during the impact?

P5.24

5.25 A bioengineer, using an instrumented dummy to test a protective mask for a hockey goalie, launches the 170 g puck so that it strikes the mask moving horizontally at 40 m/s. From photographs of the impact, she estimates its duration to be 0.02 s and observes that the puck rebounds at 5 m/s.
(a) What linear impulse does the puck exert?
(b) What is the average value of the impulsive force exerted on the mask by the puck?

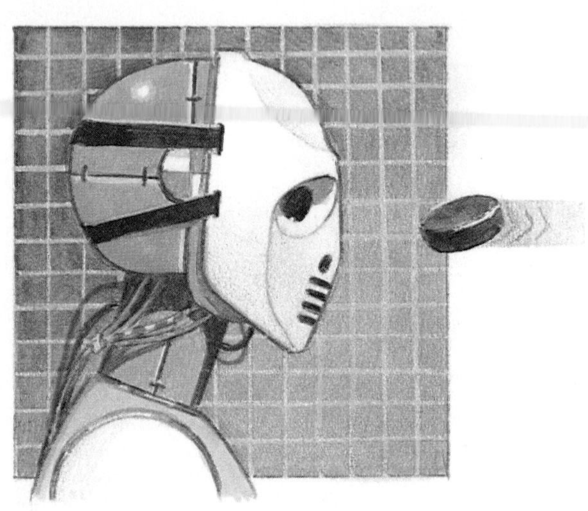

P5.25

5.26 A fragile object dropped onto a hard surface breaks because it is subjected to a large impulsive force. If you drop a 0.057 kg watch from 1.2 m above the floor, the duration of the impact is 0.001 s, and the watch bounces 50 mm above the floor, what is the average value of the impulsive force?

5.27 A 25 kg projectile is subjected to an impulsive force with a duration of 0.01 s that accelerates it from rest to a velocity of 12 m/s at 60° above the horizontal. What is the average value of the impulsive force?

Strategy: Use Equation (5.2) to determine the average total force on the projectile. To determine the average value of the impulsive force, you must subtract the projectile's weight.

5.28 An entomologist measures the motion of a 3 g locust during its jump and determines that it accelerates from rest to 3.4 m/s in 25 ms (milliseconds). The angle of takeoff is 55° above the horizontal. What are the horizontal and vertical components of the average impulsive force exerted by the insect's hind legs during the jump?

5.29 A 0.14 kg baseball is 1 m above the ground when it is struck by a bat. The horizontal distance to the point where the ball strikes the ground is 55 m. Photographic studies indicate that the ball was moving approximately horizontally at 30 m/s before it was struck, the duration of the impact was 0.015 s, and the ball was travelling at 30° above the horizontal after it was struck. What was the magnitude of the average impulsive force exerted on the ball by the bat?

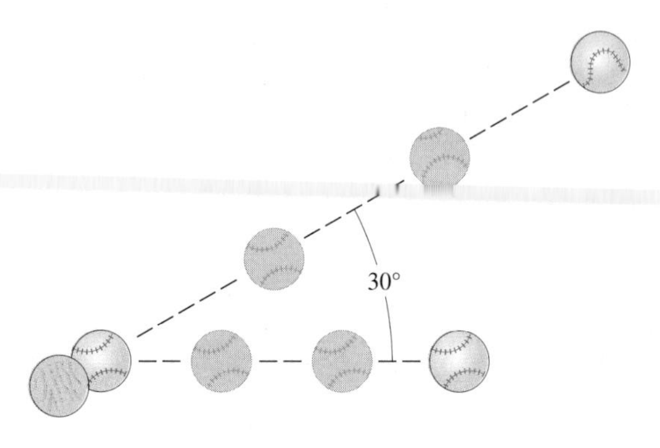

P5.29

5.30 The 1 kg ball is given a horizontal velocity of 1.2 m/s at A. Photographic measurements indicate that $b = 1.2$ m, $h = 1.3$ m, and the duration of the bounce at B is 0.1 s. What are the components of the average impulsive force exerted on the ball by the floor at B?

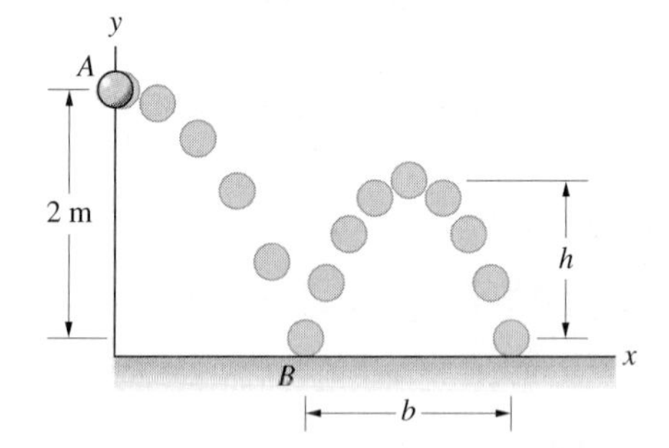

P5.30

5.2 Conservation of Linear Momentum

In this section we consider the motions of several objects and show that if the effects of external forces can be neglected, total linear momentum is conserved. This result provides you with a powerful tool for analysing interactions between objects, such as collisions, and also permits you to determine forces exerted on objects as a result of gaining or losing mass.

Consider the objects A and B in Figure 5.5. $\mathbf{F}_{AB}$ is the force exerted on A by B and $\mathbf{F}_{BA}$ is the force exerted on B by A. These forces could result from the two objects being in contact, or could be exerted by a spring connecting them. As a consequence of Newton's third law, these forces are equal and opposite:

$$\mathbf{F}_{AB} + \mathbf{F}_{BA} = 0 \tag{5.3}$$

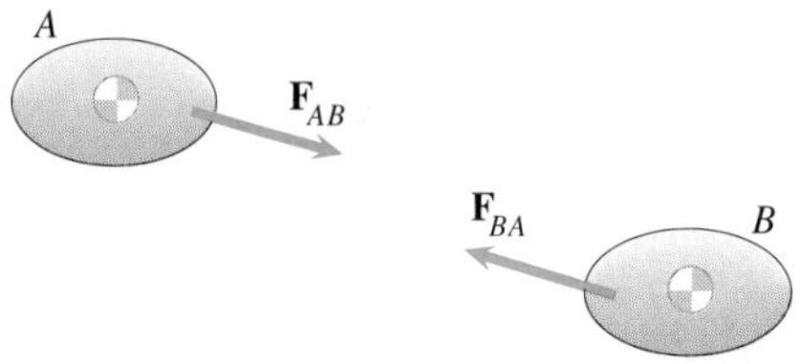

Figure 5.5

Two objects and the forces they exert on each other.

Suppose that no other external forces act on A and B, or that other external forces are negligible in comparison with the forces that A and B exert on each other. We can apply the principle of impulse and momentum to each object for arbitrary times t_1 and t_2:

$$\int_{t_1}^{t_2} \mathbf{F}_{AB}\, dt = m_A\mathbf{v}_{A2} - m_A\mathbf{v}_{A1}$$

$$\int_{t_1}^{t_2} \mathbf{F}_{BA}\, dt = m_B\mathbf{v}_{B2} - m_B\mathbf{v}_{B1}$$

If we sum these equations, the terms on the left cancel and we obtain

$$m_A\mathbf{v}_{A1} + m_B\mathbf{v}_{B1} = m_A\mathbf{v}_{A2} + m_B\mathbf{v}_{B2}$$

which means that the total linear momentum of A and B is conserved:

$$\boxed{m_A\mathbf{v}_A + m_B\mathbf{v}_B = \text{constant}} \tag{5.4}$$

We can show that the velocity of the combined centre of mass of the objects A and B (that is, the centre of mass A and B regarded as a single object) is also constant. Let $\mathbf{r}_A$ and $\mathbf{r}_B$ be the position vectors of their individual centres of mass (Figure 5.6). The position of the combined centre of mass is

$$\mathbf{r} = \frac{m_A\mathbf{r}_A + m_B\mathbf{r}_B}{m_A + m_B}$$

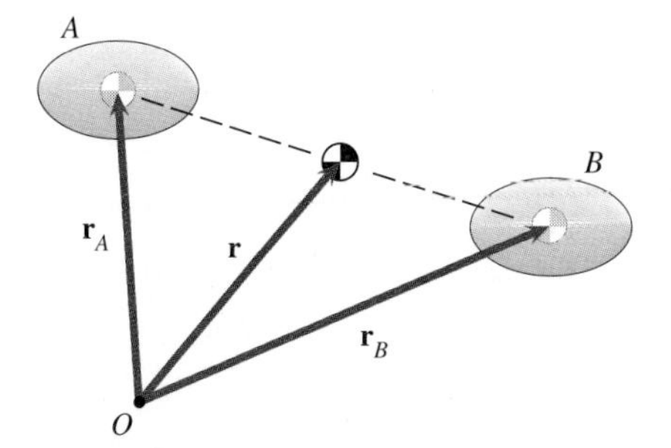

Figure 5.6

Position vector $\mathbf{r}$ of the centre of mass of A and B.

By taking the time derivative of this equation, we obtain

$$(m_A + m_B)\mathbf{v} = m_A\mathbf{v}_A + m_B\mathbf{v}_B = \text{constant} \tag{5.5}$$

where $\mathbf{v} = d\mathbf{r}/dt$ is the velocity of the combined centre of mass. Although your goal will usually be to determine the individual motions of the objects, knowing that the velocity of the combined centre of mass is constant can contribute to your understanding of a problem, and in some instances the motion of the combined centre of mass may be the only information you can obtain.

Even when significant external forces act on A and B, if the external forces are negligible in a particular direction, Equations (5.4) and (5.5) apply in that direction. These equations also apply to an arbitrary number of objects: if the external forces acting on any collection of objects are negligible, the total linear momentum of the objects is conserved and the velocity of their centre of mass is constant.

In the following example we demonstrate the use of Equations (5.4) and (5.5) to analyse motions of objects. When you know initial positions and velocities of objects and you can neglect external forces, these equations relate their positions and velocities at any subsequent time.

Example 5.3

A person of mass m_P stands at the centre of a stationary barge of mass m_B (Figure 5.7). Neglect horizontal forces exerted on the barge by the water.
(a) If the person starts running to the right with velocity v_P relative to the water, what is the resulting veocity of the barge relative to the water?
(b) If the person stops when he reaches the right end of the barge, what are his position and the barge's position relative to their original positions?

Figure 5.7

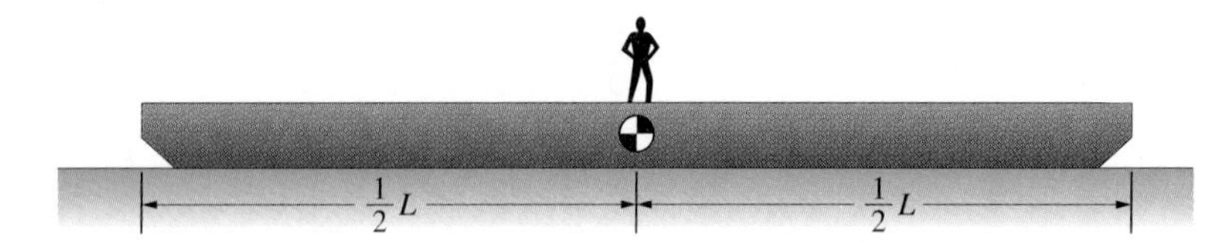

STRATEGY

(a) The only horizontal forces exerted on the person and the barge are the forces they exert on each other. Therefore their total linear momentum *in the horizontal direction* is conserved and we can use Equation (5.4) to determine the barge's velocity while the person is running.
(b) The combined centre of mass of the person and the barge is initially stationary, so it must remain stationary. Knowing the position of the combined centre of mass, we can determine the positions of the person and the barge when the person is at the right end of the barge.

SOLUTION

(a) Before the person starts running, the total linear momentum of the person and the barge in the horizontal direction is zero, so it must be zero after he starts running. Letting v_B be the value of the barge's velocity *to the left* while the person is running (Figure (a)), we obtain

$$m_P v_P + m_B(-v_B) = 0$$

so the velocity of the barge while he runs is

$$v_B = \left(\frac{m_P}{m_B}\right) v_P$$

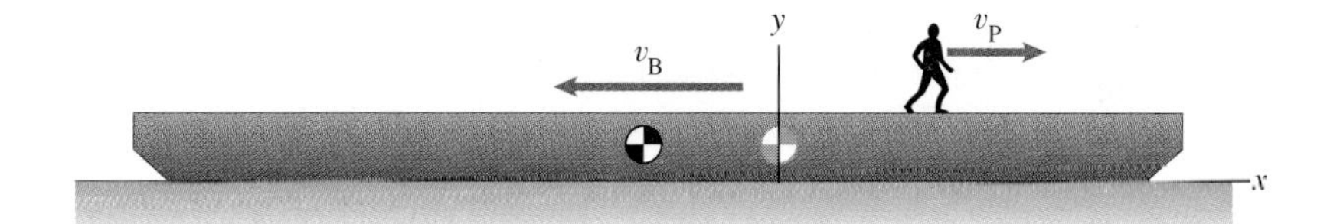

(a) Velocities of the person and barge.

(b) Let the origin of the coordinate system in Figure (b) be the original horizontal position of the centres of mass of the barge and the person, and let x_B be the position of the barge's centre of mass *to the left of the origin*. When the person has stopped at the right end of the barge, the combined centre of mass must still be at $x = 0$:

$$\frac{x_P m_P + (-x_B) m_B}{m_P + m_B} = 0$$

Solving this equation together with the relation $x_P + x_B = L/2$, we obtain

$$x_P = \frac{m_B L}{2(m_P + m_B)} \qquad x_B = \frac{m_P L}{2(m_P + m_B)}$$

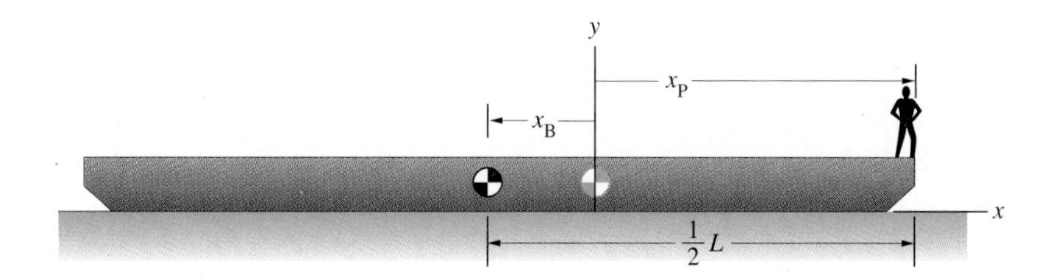

(b) Positions after the person has stopped.

DISCUSSION

This example is a well-known illustration of the power of momentum methods. Notice that we were able to determine the velocity of the barge and the final positions of the person and barge even though we did not know the complicated time dependence of the horizontal forces they exert on each other.

5.3 *Impacts*

In machines that perform stamping or forging operations, dies impact against workpieces. Mechanical printers create images by impacting metal elements against the paper and platen. Vehicles impact each other intentionally, as when railway cars are rolled against each other to couple them, and unintentionally in accidents. Impacts occur in many situations of concern in engineering. In this section we consider a basic question: if you know the velocities of two objects before they collide, how do you determine their velocities afterwards? In other words, what is the effect of the impact on their motions?

If colliding objects are not subjected to external forces, their total linear momentum must be the same before and after the impact. Even when they are subjected to external forces, the force of the impact is often so large, and its duration so brief, that the effect of external forces on their motions during the impact is negligible. Suppose that objects A and B with velocities $\mathbf{v}_A$ and $\mathbf{v}_B$ collide, and let $\mathbf{v}'_A$ and $\mathbf{v}'_B$ be their velocities after the impact (Figure 5.8(a)). If the effects of external forces are negligible, their total linear momentum is conserved:

$$m_A\mathbf{v}_A + m_B\mathbf{v}_B = m_A\mathbf{v}'_A + m_B\mathbf{v}'_B \tag{5.6}$$

Furthermore, the velocity $\mathbf{v}$ of their centre of mass is the same before and after the impact. From Equation (5.5),

$$\mathbf{v} = \frac{m_A\mathbf{v}_A + m_B\mathbf{v}_B}{m_A + m_B} \tag{5.7}$$

If A and B adhere and remain together after they collide, they are said to undergo a **perfectly plastic impact**. Equation (5.7) gives the velocity of the centre of mass of the object they form after the impact (Figure 5.8(b)). A remarkable feature of this result is that you can determine the velocity following the impact *without considering the physical nature of the impact.*

If A and B do not adhere, linear momentum conservation alone is not sufficient to determine their velocities after the impact. We first consider the case in which they travel along the same straight line before and after they collide.

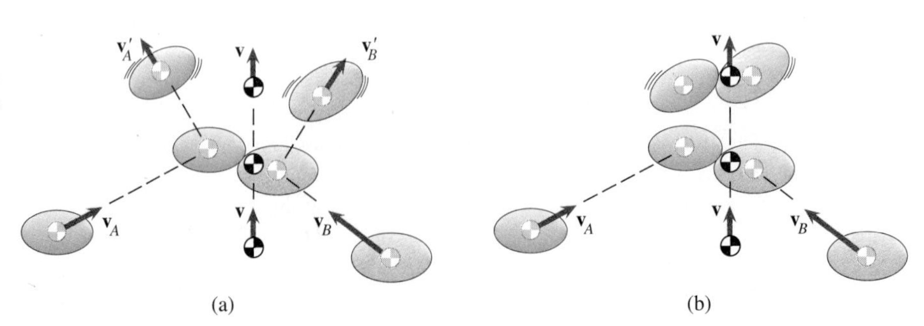

Figure 5.8
(a) Velocities of A and B before and after the impact and the velocity $\mathbf{v}$ of their centre of mass.
(b) A perfectly plastic impact.

Direct Central Impacts

Suppose that the centres of mass of A and B travel along the same straight line with velocities v_A and v_B before their impact (Figure 5.9(a)). Let R be the magnitude of the force they exert on each other during the impact (Figure 5.9(b)). We assume that the contacting surfaces are oriented so that R is parallel to the line along which they travel and directed towards their centres of mass. This condition, called **direct central impact**, means that they continue to travel along the same straight line after their impact (Figure 5.9(c)). If the effects of external forces during the impact are negligible, their total linear momentum is conserved:

$$m_A v_A + m_B v_B = m_A v'_A + m_B v'_B \tag{5.8}$$

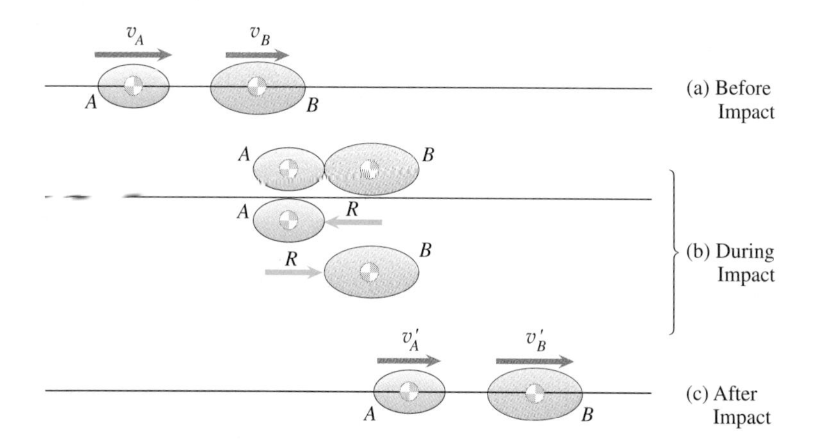

Figure 5.9

(a) Objects A and B travelling along the same straight line.
(b) During the impact, they exert a force R on each other.
(c) They travel along the same straight line after the central impact.

However, we need another question to determine the velocities v'_A and v'_B. To obtain it, we must consider the impact in more detail.

Let t_1 be the time at which A and B first come into contact (Figure 5.10(a)). As a result of the impact, they will deform and their centres of mass will continue to approach each other. At a time t_C, their centres of mass will have reached their nearest proximity (Figure 5.10(b)). At this time the relative velocity of the two centres of mass is zero, so they have the same velocity. We denote it by v_C. The objects then begin to move apart and separate at a time t_2 (Figure 5.10(c)). We apply the principle of impulse and momentum to A during the intervals of time from t_1 to the time of closest approach t_C and also from t_C to t_2:

$$\int_{t_1}^{t_C} -R\,dt = m_A v_C - m_A v_A \tag{5.9}$$

$$\int_{t_C}^{t_2} -R\,dt = m_A v'_A - m_A v_C \tag{5.10}$$

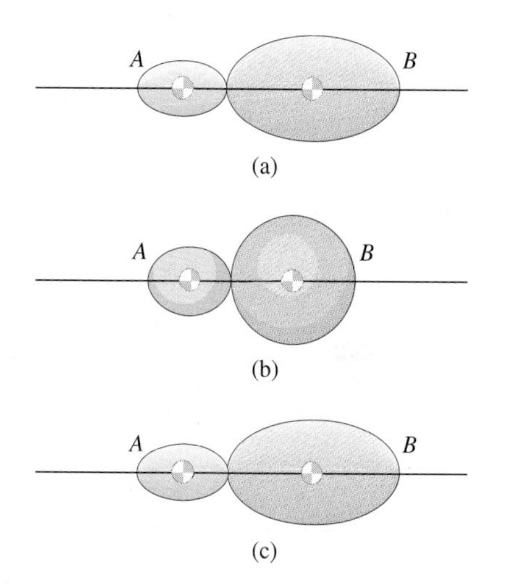

Figure 5.10

(a) First contact, $t = t_1$.
(b) Closest approach, $t = t_C$.
(c) End of contact, $t = t_2$.

Then we apply this principle to B for the same intervals of time:

$$\int_{t_1}^{t_C} R\,dt = m_B v_C - m_B v_B \tag{5.11}$$

$$\int_{t_C}^{t_C} R\,dt = m_B v'_B - m_B v_C \tag{5.12}$$

As a result of the impact, part of the objects' kinetic energy can be lost due to a variety of mechanisms, including permanent deformation and generation of heat and sound. As a consequence, the impulse they impart to each other during the 'restitution' phase of the impact from t_C to t_2 is in general smaller than the impulse they impart from t_1 to t_C. The ratio of these impulses is called the **coefficient of restitution**:

$$e = \frac{\int_{t_C}^{t_2} R\,dt}{\int_{t_1}^{t_C} R\,dt} \tag{5.13}$$

Its value depends on the properties of the objects as well as their velocities and orientations when they collide, and it can be determined only by experiment or by a detailed analysis of the deformations of the objects during the impact.

If we divide Equation (5.10) by Equation (5.9) and divide Equation (5.12) by Equation (5.11), we can express the resulting equations in the forms

$$(v_C - v_A)e = v'_A - v_C$$

$$(v_C - v_B)e = v'_B - v_C$$

Subtracting the first equation from the second one, we obtain

$$\boxed{e = \frac{v'_B - v'_A}{v_A - v_B}} \tag{5.14}$$

Thus the coefficient of restitution is related in a simple way to the relative velocities of the objects before and after the impact. If e is known, you can use Equation (5.14) together with the equation of conservation of linear momentum, Equation (5.8), to determine v'_A and v'_B.

If $e = 0$, Equation (5.14) indicates that $v'_B = v'_A$. The objects remain together after the impact, and the impact is perfectly plastic. If $e = 1$, it can be shown that the total kinetic energy is the same before and after the impact:

$$\frac{1}{2}m_A v_A^2 + \frac{1}{2}m_B v_B^2 = \frac{1}{2}m_A(v'_A)^2 + \frac{1}{2}m_B(v'_B)^2 \quad \text{(When } e = 1\text{)}$$

An impact in which kinetic energy is conserved is called **perfectly elastic**. Although this is sometimes a useful approximation, energy is lost in any impact in which material objects come into contact. If you can hear a collision, kinetic energy has been converted into sound. Permanent deformations and vibrations of the colliding objects after the impact also represent losses of kinetic energy.

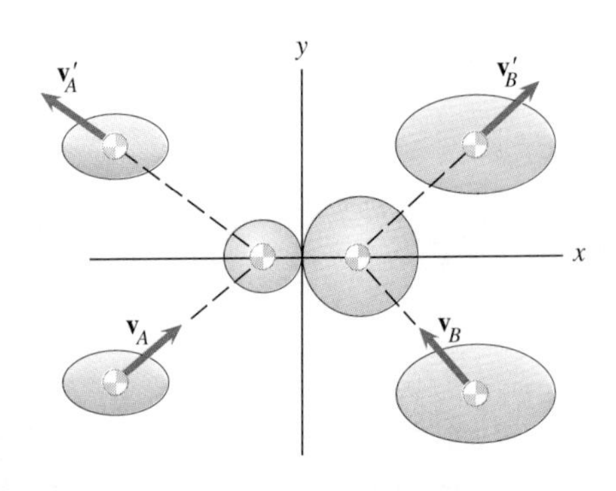

Figure 5.11

An oblique central impact.

Oblique Central Impacts

We can extend the procedure used to analyse direct central impacts to the case in which the objects approach each other at an oblique angle. Suppose that A and B approach with arbitrary velocities $\mathbf{v}_A$ and $\mathbf{v}_B$ (Figure 5.11) and that the forces they exert on each other during their impact are parallel to the x axis and point towards their centre of mass. No forces are exerted on them in the y or z directions, so their velocities in those directions are unchanged by the impact:

$$(\mathbf{v}'_A)_y = (\mathbf{v}_A)_y \qquad (\mathbf{v}'_B)_y = (\mathbf{v}_B)_y$$
$$(\mathbf{v}'_A)_z = (\mathbf{v}_A)_z \qquad (\mathbf{v}'_B)_z = (\mathbf{v}_B)_z \tag{5.15}$$

In the x direction, linear momentum is conserved,

$$m_A(\mathbf{v}_A)_x + m_B(\mathbf{v}_B)_x = m_A(\mathbf{v}'_A)_x + m_B(\mathbf{v}'_B)_x \tag{5.16}$$

and by the same analysis we used to arrive at Equation (5.14), the x components of velocity satisfy the relation

$$e = \frac{(\mathbf{v}'_B)_x - (\mathbf{v}'_A)_x}{(\mathbf{v}_A)_x - (\mathbf{v}_B)_x} \tag{5.17}$$

We can analyse an impact in which A hits a stationary object like a wall (Figure 5.12) as an oblique central impact if friction is negligible. The y and z components of A's velocity are unchanged, and the x component after the impact is given by Equation (5.17) with B's velocity equal to zero:

$$(\mathbf{v}'_A)_x = -e(\mathbf{v}_A)_x$$

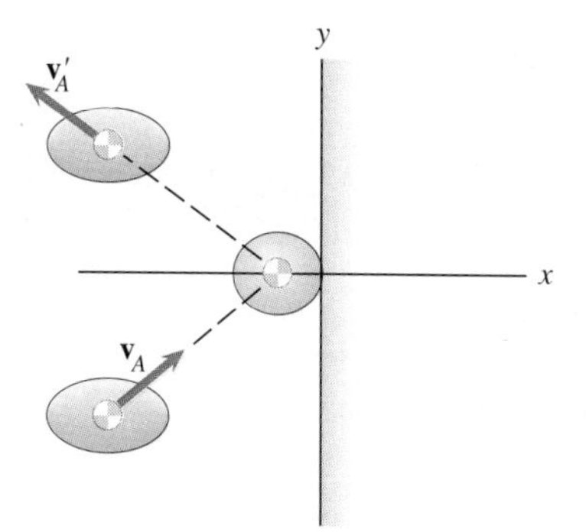

Figure 5.12
Impact with a stationary object.

In the following example we analyse the impact of two objects. If an impact is perfectly plastic, which means the objects adhere and remain together, you can determine from Equation (5.7) the velocity of their centre of mass after the impact. In a direct central impact, in terms of the coordinate system shown in Figure 5.11, the y and z components of the velocities of the objects are unchanged and you can solve Equations (5.16) and (5.17) for the x components of the velocities after the impact.

Example 5.4

The *Apollo* CSM (A) attempts to dock with the *Soyuz* capsule (B), 15 July 1975 (Figure 5.13). Their masses are $m_A = 18\,\text{Mg}$ and $m_B = 6.6\,\text{Mg}$. The *Soyuz* is stationary relative to the reference frame shown, and the CSM approaches with velocity $\mathbf{v}_A = (0.2\,\mathbf{i} + 0.03\,\mathbf{j} - 0.02\,\mathbf{k})\,\text{m/s}$.

(a) If the first attempt at docking is successful, what is the velocity of the centre of mass of the combined vehicles afterwards?

(b) If the first attempt is unsuccessful and the coefficient of restitution of the resulting impact is $e = 0.95$, what are the velocities of the two spacecraft after the impact?

Figure 5.13

STRATEGY

(a) If the docking is successful, the impact is perfectly plastic and we can use Equation (5.7) to determine the velocity of the centre of mass of the combined object after the impact.

(b) By assuming an oblique central impact with the forces exerted by the docking collars parallel to the x axis, we can use Equations (5.16) and (5.17) to determine the velocities of both spacecraft after the impact.

SOLUTION

(a) From Equation (5.7), the velocity of the centre of mass of the combined vehicle is

$$\mathbf{v} = \frac{m_A\mathbf{v}_A + m_B\mathbf{v}_B}{m_A + m_B}$$

$$= \frac{(18)(0.2\,\mathbf{i} + 0.03\,\mathbf{j} - 0.02\,\mathbf{k}) + 0}{18 + 6.6}$$

$$= (0.146\,\mathbf{i} + 0.022\,\mathbf{j} - 0.015\,\mathbf{k})\ \text{m/s}$$

(b) The y and z components of the velocities of both spacecraft are unchanged. To determine the x components, we use conservation of linear momentum, Equation (5.16),

$$m_A(\mathbf{v}_A)_x + m_B(\mathbf{v}_B)_x = m_A(\mathbf{v}'_A)_x + m_B(\mathbf{v}'_B)_x$$

$$(18)(0.2) = (18)(\mathbf{v}'_A)_x + (6.6)(\mathbf{v}'_B)_x$$

and the coefficient of restitution, Equation (5.17),

$$e = \frac{(\mathbf{v}'_B)_x - (\mathbf{v}'_A)_x}{(\mathbf{v}_A)_x - (\mathbf{v}_B)_x}$$

$$0.95 = \frac{(\mathbf{v}'_B)_x - (\mathbf{v}'_A)_x}{0.2 - 0}$$

Solving these two equations, we obtain $(\mathbf{v}'_A)_x = 0.095\,\mathrm{m/s}$ and $(\mathbf{v}'_B)_x = 0.285\,\mathrm{m/s}$, so the velocities of the spacecraft after the impact are

$$\mathbf{v}'_A = (0.095\,\mathbf{i} + 0.03\,\mathbf{j} - 0.02\,\mathbf{k})\ \mathrm{m/s}$$

$$\mathbf{v}'_B = 0.285\,\mathbf{i}\ \mathrm{m/s}$$

Problems

5.31 A girl weighing 440 N stands at rest on a barge weighing 2200 N. She starts running at 3 m/s *relative to the barge* and runs off the end. Neglect the horizontal force exerted on the barge by the water.

(a) Just before she hits the water, what is the horizontal component of her velocity relative to the water?

(b) What is the velocity of the barge relative to the water while she runs?

P5.31

5.32 A 60 kg astronaut aboard the space shuttle kicks off towards the centre of mass of the 105 Mg shuttle at 1 m/s relative to the shuttle. He travels 6 m relative to the shuttle before coming to rest at the opposite wall.

(a) What is the magnitude of the change in the velocity of the shuttle while he is in motion?

(b) What is the magnitude of the displacement of the centre of mass of the shuttle due to his 'flight'?

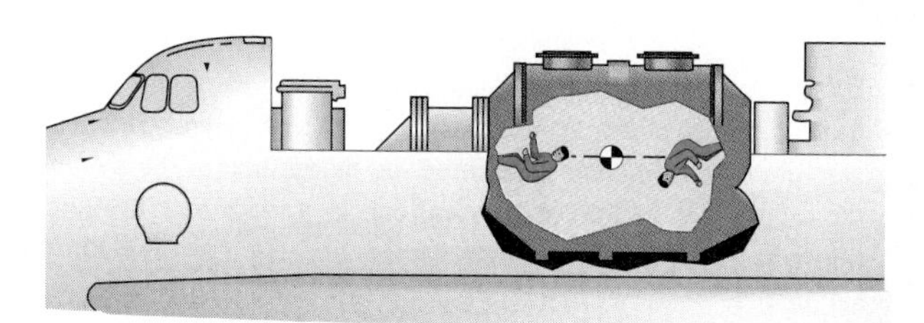

P5.32

5.33 A 36 kg boy sitting in a stationary 9 kg wagon wants to simulate rocket propulsion by throwing bricks out of the wagon. Neglect horizontal forces on the wagon's wheels. If he has three bricks weighing 40 N each and throws them with a horizontal velocity of 3 m/s relative to the wagon, determine the velocity he attains (a) if he throws the bricks one at a time; (b) if he throws them all at once.

P5.33

5.34 Two railroad cars ($m_A = 1.7m_B$) collide and become coupled. Car A is full and car B is half-full of carbolic acid. When the cars impact, the acid in B sloshes back and forth violently.
(a) Immediately after the impact, what is the velocity of the common centre of mass of the two cars?
(b) A few seconds later, when the sloshing has subsided, what is the velocity of the two cars?

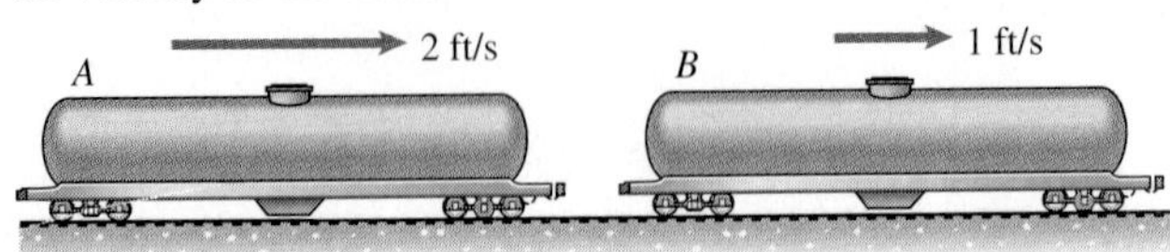

P5.34

5.35 In Problems 5.34, if the track slopes one-half degree upwards to the right and the cars are initially 3 m apart, what is the velocity of their common centre of mass immediately after the impact?

5.36 A 400 kg satellite S travelling at 7 km/s is hit by a 1 kg meteor M travelling at 12 km/s. The meteor is embedded in the satellite by the impact. Determine the magnitude of the velocity of their common centre of mass after the impact and the angle β between the path of the centre of mass and the original path of the satellite.

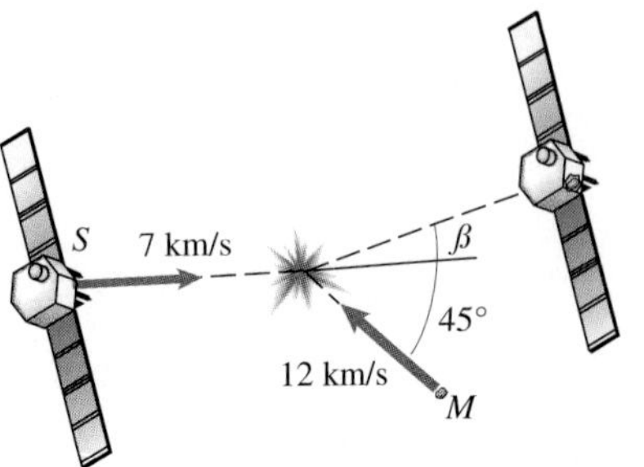

P5.36

5.37 The cannon weighed 1800 N, fired a cannonball weighing 45 N, and had a muzzle velocity of 50 m/s. For the 10° elevation angle shown, determine (a) the velocity of the cannon after it was fired; (b) the distance the cannonball travelled. (Neglect drag.)

P5.37

5.38 A bullet (mass m) hits a stationary block of wood (mass m_B) and becomes embedded in it. The coefficient of kinetic friction between the block and the floor is μ_k. As a result of the impact the block slides a distance D before stopping. What was the velocity v of the bullet?

Strategy: First solve the impact problem to determine the velocity of the block and the embedded bullet after the impact in terms of v, then relate the initial velocity of the block and the embedded bullet to the distance D that the block slides.

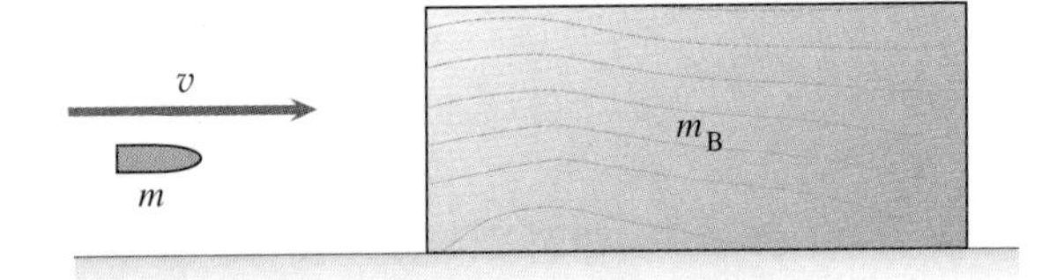

P5.38

5.39 The overhead conveyor drops the 12 kg package A into the 1.6 kg carton B. The package is 'tacky' and sticks to the bottom of the carton. If the coefficient of friction between the carton and the horizontal conveyor is $\mu_k = 0.2$, what distance does the carton slide after the impact?

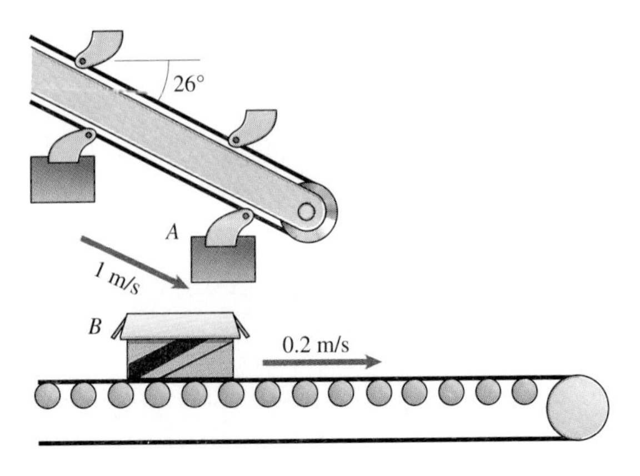

P5.39

5.40 A 0.028 kg bullet moving horizontally hits a suspended 50 kg block of wood and becomes embedded in it. If you measure the angle through which the wires supporting the block swings as a result of the impact and determine it to be 7°, what was the bullet's velocity?

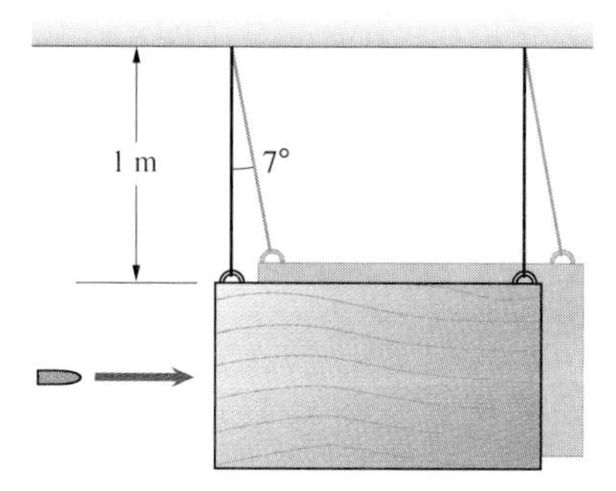

P5.40

5.41 Suppose you investigate an accident in which a 1360 kg car with velocity $\mathbf{v}_C = 32\,\mathbf{j}$ km/hr collided with a 5440 kg bus with velocity $\mathbf{v}_B = 16\,\mathbf{i}$ km/hr. The vehicles became entangled and remained together after the collision.

(a) What was the velocity of the common centre of mass of the two vehicles after the collision?

(b) If you estimate the coefficient of friction between the sliding vehicles and the road after the collision to be $\mu_k = 0.4$, what is the approximate final position of the common centre of mass relative to its position when the impact occurs?

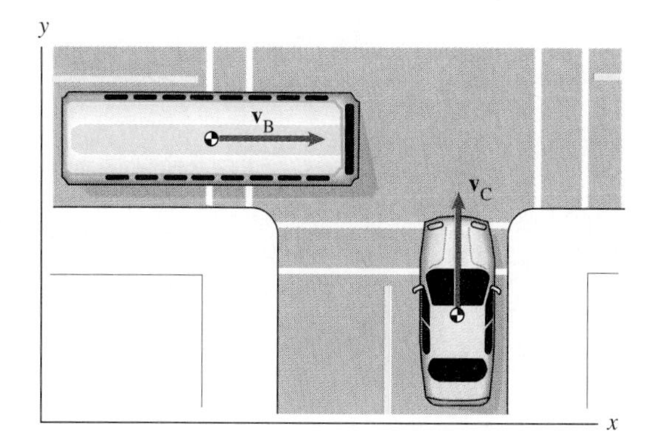

P5.41

5.42 The velocity of the 100 kg astronaut A relative to the space station is $(40\,\mathbf{i} + 30\,\mathbf{j})$ mm/s. The velocity of the 200 kg structural member B relative to the station is $(-20\,\mathbf{i} + 30\,\mathbf{j})$mm/s. When they approach each other, the astronaut grasps and clings to the structural member.

(a) Determine the velocity of their common centre of mass when they arrive at the station.

(b) Determine the approximate position at which they contact the station.

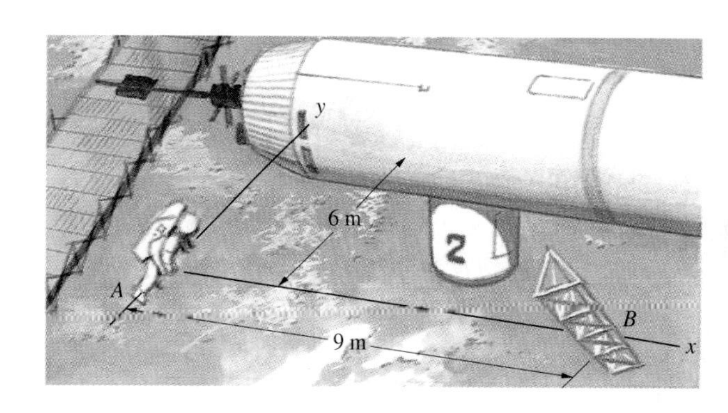

P5.42

5.43 Objects A and B with the same mass m undergo a direct central impact. The velocity of A before the impact is v_A, and B is stationary. Determine the velocities of A and B after the impact if it is (a) perfectly plastic ($e = 0$); (b) perfectly elastic ($e = 1$).

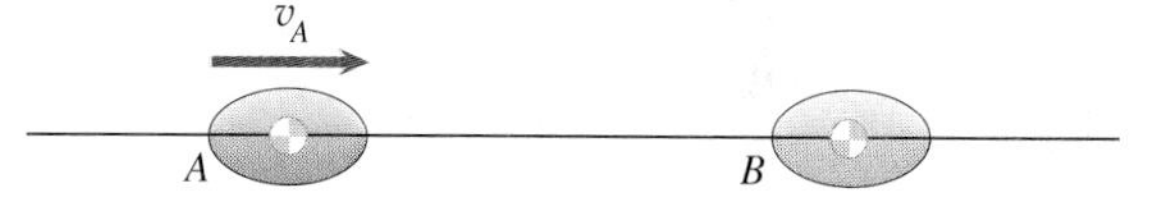

P5.43

5.44 In Problem 5.43, if the velocity of B after the impact is $0.6v_A$, determine the coefficient of restitution e and the velocity of A after the impact.

5.45 Objects A and B with masses m_A and m_B undergo a direct central impact.

(a) If $e = 1$, show that the total kinetic energy after the impact is equal to the total kinetic energy before the impact.

(b) If $e = 0$, how much kinetic energy is lost as a result of the collision?

P5.45

5.46 The two 5 kg weights slide on the smooth horizontal bar. Determine their velocities after they collide if the weights are coated with Velcro and stick together.

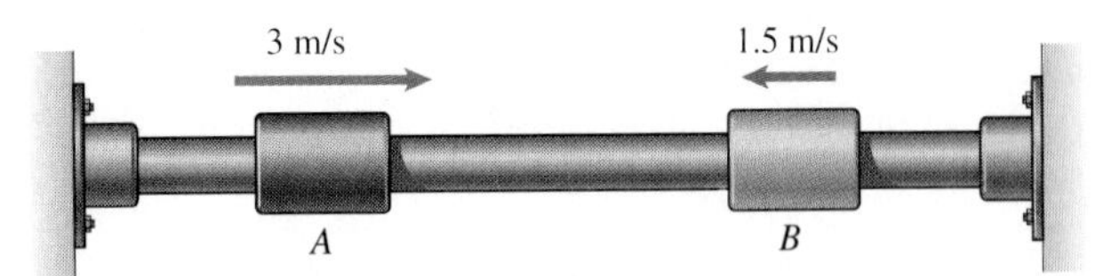

P5.46

5.47 Determine the velocities of the weights in Problem 5.46 after their impact if you assume it to be perfectly elastic.

5.48 Determine the velocities of the weights in Problem 5.46 after their impact if the coefficient of restitution is $e = 0.8$.

5.49 Two cars with energy-absorbing bumpers collide with speeds $v_A = v_B = 8$ km/hr. Their weights are $W_A = 12$ kN and $W_B = 20$ kN. If the coefficient of restitution of the collision is $e = 0.2$, what are the velocities of the cars after the collision?

P5.49

5.50 In Problem 5.49, if the duration of the collision is 0.1 s, what are the magnitudes of the average acceleration to which the occupants of the two cars are subjected?

5.51 The 10 kg mass A is moving at 5 m/s when it is 1 m from the stationary 10 kg mass B. The coefficient of kinetic friction between the floor and the two masses is $\mu_k = 0.6$, and the coefficient of restitution of the impact is $e = 0.5$. Determine how far B moves from its initial position as a result of the impact.

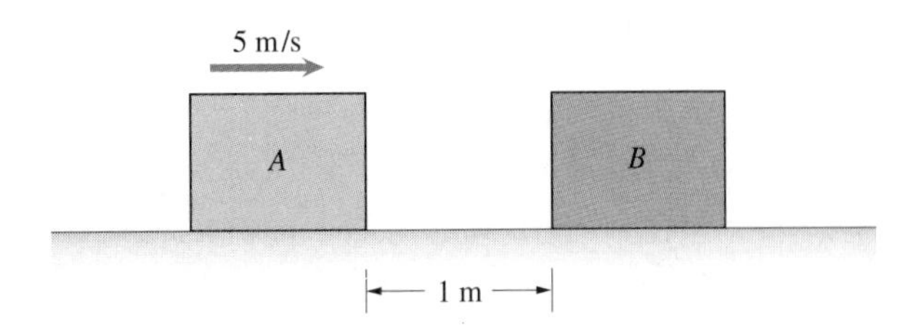

P5.51

5.52 Suppose you investigate an accident in which a 1300 kg car A struck a parked 1200 kg car B. All four of B's wheels were locked, and skid marks indicate that it slid 2 m after the impact. If you estimate the coefficient of friction between B's tyres and the road to be $\mu_k = 0.8$ and the coefficient of restitution of the impact to be $e = 0.4$, what was A's velocity just before the impact? (Assume that only one impact occurred.)

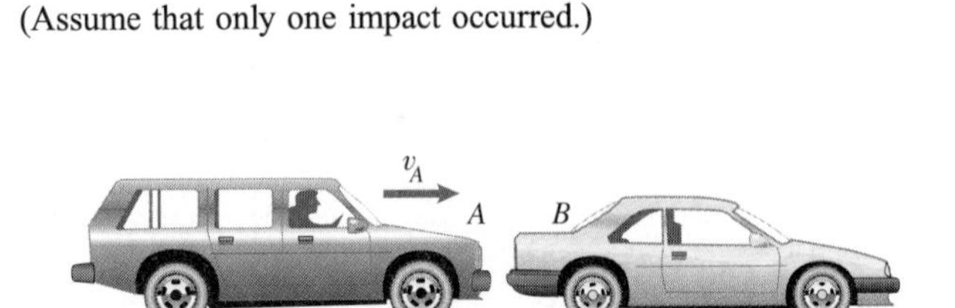

P5.52

5.53 Suppose you drop a basketball 1.5 m above the floor and it bounces to a height of 1.2 m. If you then throw the ball downwards, releasing it 1 m above the floor moving at 10 m/s, how high does it bounce?

5.54 By making measurements directly from the photograph of the bouncing golf ball, estimate the coefficients of restitution.

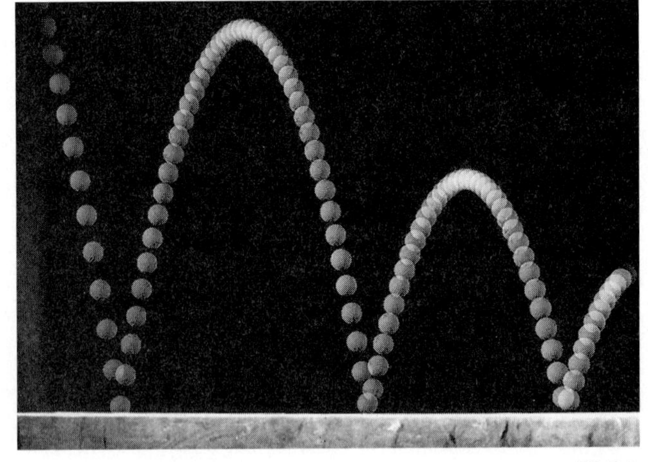

P5.54

5.55 If you throw the golf ball in Problem 5.54 horizontally at 0.6 m/s and release it 1.2 m above the surface, what is the distance between the first two bounces?

5.56 A bioengineer studying helmet design strikes a 2.4 kg helmet containing a 2 kg simulated human head against a rigid surface at 6 m/s. The head, being suspended within the helmet, is not immediately affected by the impact of the helmet with the surface and continues to move to the right at 6 m/s, so it then undergoes an impact with the helmet. If the coefficient of restitution of the helmet's impact with the surface is 0.8 and the coefficient of restitution of the following impact of the head and helmet is 0.2, what are the velocities of the helmet and head after their initial interaction?

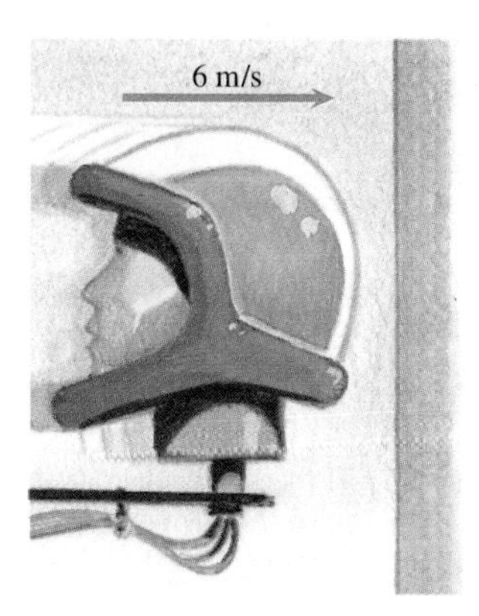

P5.56

5.57 (a) In Problem 5.56, if the duration of the impact of the head with the helmet is 0.008 s, what average force is the head subjected to?
(b) Suppose that the simulated head alone strikes the surface at 6 m/s, the coefficient of restitution is 0.3, and the duration of the impact is 0.002 s. What average force is the head subjected to?

5.58 Two small balls, each of mass m, hang from strings of length L. The left ball is released from rest in the position shown. As a result of the first collision, the right ball swings through an angle β. Determine the coefficient of restitution.

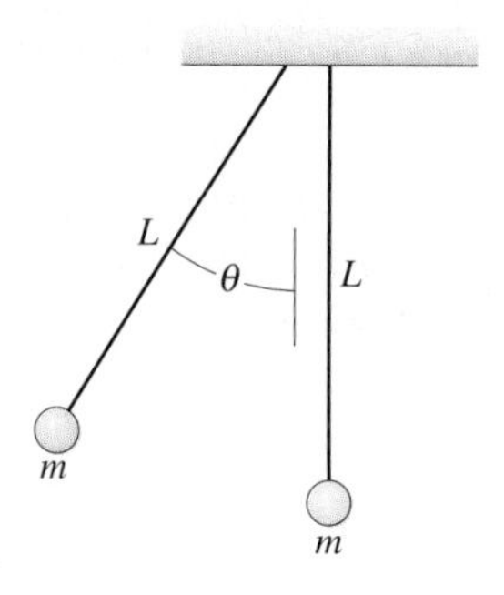

P5.58

5.59 A 15 kg object A and a 30 kg object B undergo an oblique central impact. The coefficient of restitution is $e = 0.8$. Before the impact, $\mathbf{v}_B = -10\,\mathbf{i}$ m/s, and after the impact, $\mathbf{v}'_A = (-15\,\mathbf{i} + 4\,\mathbf{j} + 2\,\mathbf{k})$ m/s. Determine the velocity of A before the impact and the velocity of B after the impact.

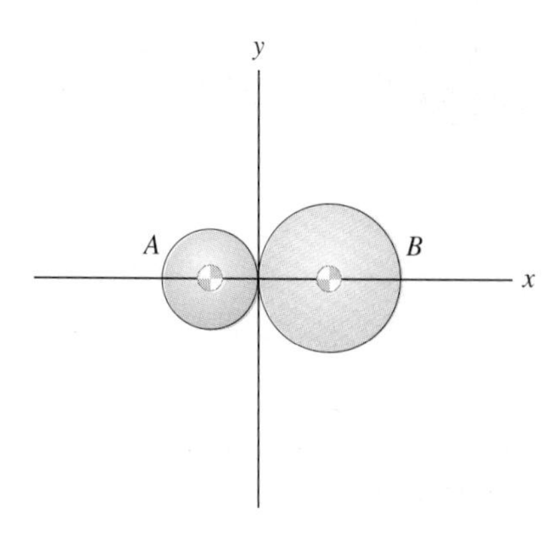

P5.59

5.60 The cue gives the cue ball A a velocity parallel to the y axis. It hits the 8-ball B and knocks it straight into the corner pocket. If the magnitude of the velocity of the cue ball just before the impact is 2 m/s and the coefficient of restitution is $e = 1$, what are the velocity vectors of the two balls just after the impact? (The balls are of equal mass.)

P5.60

5.61 In Problem 5.60, what are the velocity vectors of the two balls just after the impact if the coefficient of restitution is $e = 0.9$?

5.62 If the coefficient of restitution is the same for both impacts, show that the cue ball's path after two banks is parallel to its original path.

P5.62

5.63 The cue gives the cue ball A a velocity of magnitude 3 m/s. The angle $\beta = 0$ and the coefficient of restitution of the impact of the cue ball and the 8-ball B is $e = 1$. If the magnitude of the 8-ball's velocity after the impact is 0.9 m/s, what was the coefficient of restitution of the cue ball's impact with the cushion? (The balls are of equal mass.)

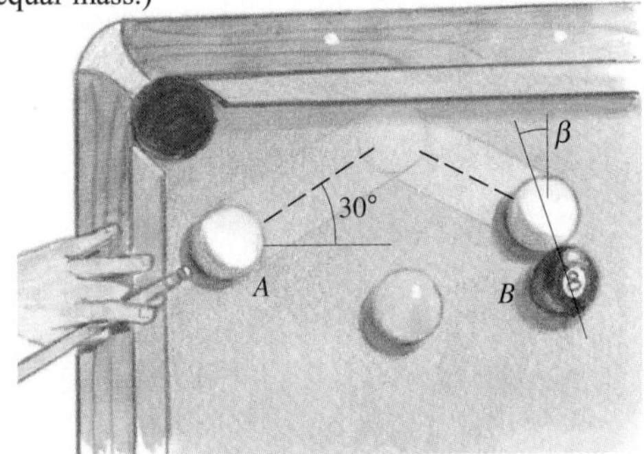

P5.63

5.64 What is the solution of Problem 5.63 if the angle $\beta = 10°$?

5.65 What is the solution of Problem 5.63 if the angle $\beta = 10°$ and the coefficient of restitution of the impact between the two balls is $e = 0.9$?

5.66 A ball is given a horizontal velocity of 3 m/s at 2 m above the smooth floor. Determine the distance D between its first and second bounces if the coefficient of restitution is $e = 0.6$.

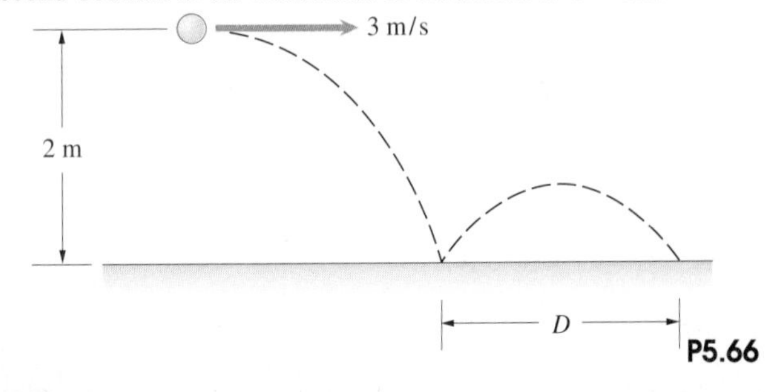

P5.66

5.67 The velocity of the 170 g hockey puck is $\mathbf{v}_P = (10\mathbf{i} - 4\mathbf{j})$ m/s. If you neglect the change in the velocity $\mathbf{v}_S = v_S\,\mathbf{j}$ of the stick resulting from the impact and the coefficient of restitution is $e = 0.6$, what should v_S be to send the puck towards the goal?

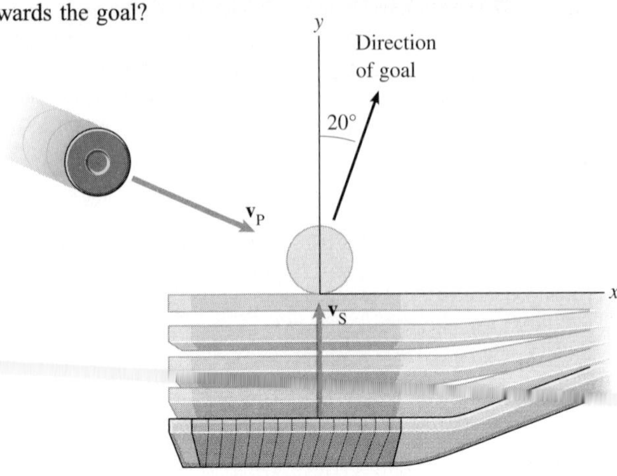

P5.67

5.68 In Problem 5.67, if the stick responds to the impact like an object with the same mass as the puck and the coefficient of restitution is $e = 0.6$, what should v_S be to send the puck towards the goal?

5.69 In a forging operation, the 500 N weight is lifted into position 1 and released from rest. It falls and strikes a workpiece in position 2. If the weight is moving at 5 m/s immediately before the impact and the coefficient of restitution is $e = 0.3$, what is its velocity immediately after the impact?

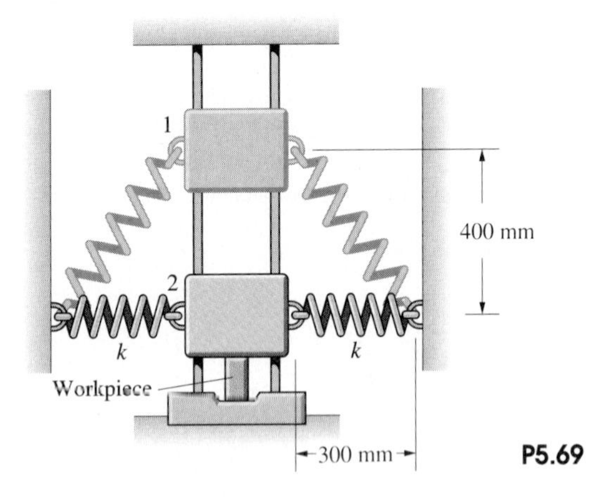

P5.69

5.70 In Problem 5.69, suppose that the spring constant is $k = 1750$ N/m, the springs are unstretched in position 2, and the coefficient of restitution is $e = 0.2$. Determine the velocity of the weight immediately after the impact.

5.71 In Problem 5.69, suppose that the spring constant is $k = 2400$ N/m, the springs are unstretched in position 2, and the weight bounces 75 mm after impact. Find the coefficient of restitution.

5.4 Angular Momentum

Here we derive a result, analogous to the principle of impulse and momentum, that relates the time integral of a moment to the change in a quantity called the angular momentum. We also obtain a useful conservation law: when the total moment due to the external forces acting on an object is zero, angular momentum is conserved.

Principle of Angular Impulse and Momentum

We describe the position of an object by the position vector $\mathbf{r}$ of its centre of mass relative to a reference point O (Figure 5.14(a)). Recall that we obtained the very useful principle of work and energy by taking the dot product of Newton's second law with the velocity. Here we obtain another useful result by taking the cross product of Newton's second law with the position vector. This procedure gives us a relation between the moment of the external forces about O and the object's motion.

The cross product of Newton's second law with $\mathbf{r}$ is

$$\mathbf{r} \times \Sigma \mathbf{F} = \mathbf{r} \times m\mathbf{a} = \mathbf{r} \times m\frac{d\mathbf{v}}{dt} \tag{5.18}$$

Notice that the time derivative of the quantity $\mathbf{r} \times m\mathbf{v}$ is

$$\frac{d}{dt}(\mathbf{r} \times m\mathbf{v}) = \underbrace{\left(\frac{d\mathbf{r}}{dt} \times m\mathbf{v}\right)}_{=0} + \left(\mathbf{r} \times m\frac{d\mathbf{v}}{dt}\right)$$

(The first term on the right is zero because $d\mathbf{r}/dt = \mathbf{v}$, and the cross product of parallel vectors is zero.) Using this result, we can write Equation (5.18) as

$$\mathbf{r} \times \Sigma \mathbf{F} = \frac{d\mathbf{H}_0}{dt} \tag{5.19}$$

where the vector

$$\mathbf{H}_0 = \mathbf{r} \times m\mathbf{v} \tag{5.20}$$

is called the **angular momentum** about O (Figure 5.14(b)). If we interpret the angular momentum as the moment of the linear momentum of the object about O, this equation states that the moment $\mathbf{r} \times \Sigma \mathbf{F}$ equals the rate of change of the moment of momentum about O. If the moment is zero during an interval of time, $\mathbf{H}_0$ is constant.

Integrating Equation (5.19) with respect to time, we obtain

$$\boxed{\int_{t_1}^{t_2} (\mathbf{r} \times \Sigma \mathbf{F})\, dt = (\mathbf{H}_0)_2 - (\mathbf{H}_0)_1} \tag{5.21}$$

The integral on the left is called the **angular impulse**, and this equation is

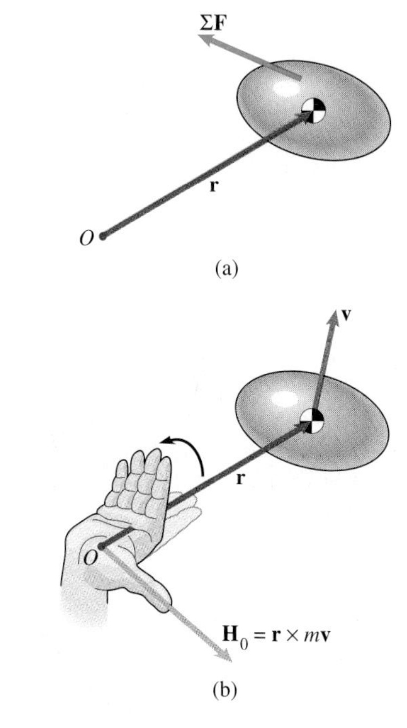

Figure 5.14
(a) The position vector and the total external force on an object.
(b) The angular momentum vector and the right-hand rule for determining its direction.

called the **principle of angular impulse and momentum**: the angular impulse applied to an object during an interval of time is equal to the change in its angular momentum. If you know the moment $\mathbf{r} \times \Sigma \mathbf{F}$ as a function of time, you can determine the change in the angular momentum.

Central-Force Motion

If the total force acting on an object remains directed towards a point that is fixed relative to an inertial reference frame, the object is said to be in **central-force motion**. The fixed point is called the **centre of the motion**. Orbit problems are the most familiar instances of central-force motion. For example, the gravitational force on an earth satellite remains directed towards the centre of the earth.

If we place the reference point O at the centre of the motion (Figure 5.15(a)), the position vector $\mathbf{r}$ is parallel to the total force, so $\mathbf{r} \times \Sigma \mathbf{F}$ equals zero. Therefore Equation (5.21) indicates that in central-force motion, an object's angular momentum is conserved:

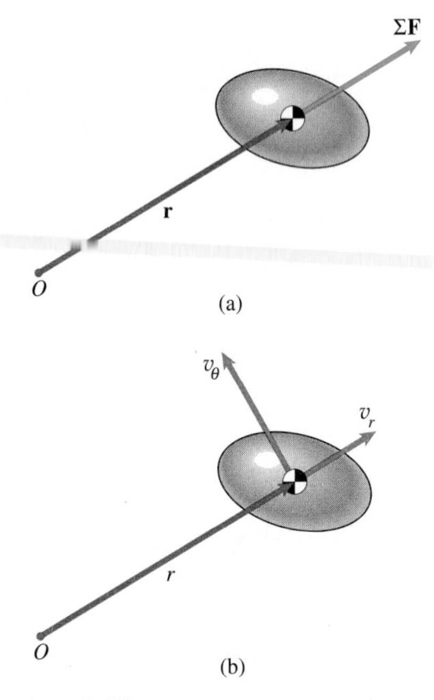

Figure 5.15
(a) Central-force motion
(b) Expressing the position and velocity in cylindrical coordinates.

$$\mathbf{H}_0 = \text{constant} \tag{5.22}$$

In plane central-force motion, we can express $\mathbf{r}$ and $\mathbf{v}$ in cylindrical coordinates (Figure 5.15(b)):

$$\mathbf{r} = r\,\mathbf{e}_r \qquad \mathbf{v} = v_r\,\mathbf{e}_r + v_\theta\,\mathbf{e}_\theta$$

Substituting these expressions into Equation (5.20), we obtain the angular momentum:

$$\mathbf{H}_0 = (r\,\mathbf{e}_r) \times m(v_r\,\mathbf{e}_r + v_\theta\,\mathbf{e}_\theta) = mrv_\theta\,\mathbf{e}_z$$

From this expression, we see that in plane central-force motion, *the product of the radial distance from the centre of the motion and the transverse component of the velocity is constant:*

$$rv_\theta = \text{constant} \tag{5.23}$$

In the following examples we show how you can use the principle of angular impulse and momentum and conservation of angular momentum to analyse motions of objects. If you know the moment $r \times \Sigma F$ during an interval of time, you can calculate the angular impulse and determine the change in an object's angular momentum. In central-force motion – the total force acting on an object points towards a point O – you know that the angular momentum about O is conserved.

Example 5.5

A disc of mass m attached to a string slides on a smooth horizontal table under the action of a constant transverse force F (Figure 5.16). The string is drawn through a hole in the table at O at constant velocity v_0. At $t = 0, r = r_0$ and the transverse velocity of the disc is zero. What is the disc's velocity as a function of time?

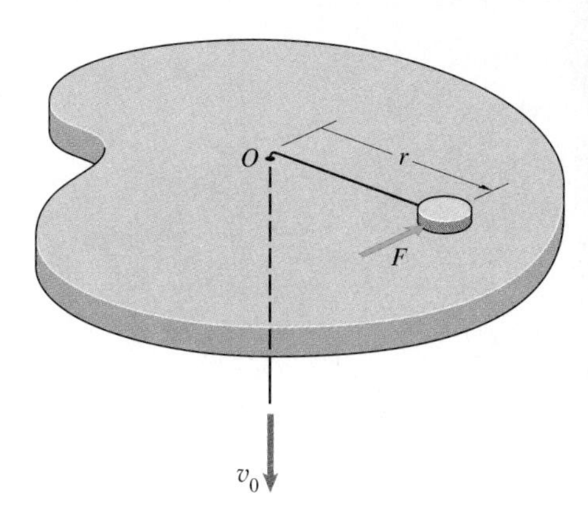

Figure 5.16

STRATEGY

By expressing r as a function of time, we can determine the moment of the forces on the disc about O as a function of time. The disc's angular momentum depends on its velocity, so we can apply the principle of angular impulse and momentum to obtain information about its velocity as a function of time.

SOLUTION

The radial position as a function of time is $r = r_0 - v_0t$. In terms of polar coordinates (Figure (a)), the moment about O of the forces on the disk is

$$\mathbf{r} \times \Sigma \mathbf{F} = r\,\mathbf{e}_r \times (-T\,\mathbf{e}_r + F\,\mathbf{e}_\theta) = F(r_0 \quad v_0t)\,\mathbf{e}_z$$

where T is the tension in the string. The angular momentum at time t is

$$\mathbf{H}_0 = \mathbf{r} \times m\mathbf{v} = r\,\mathbf{e}_r \times m(v_r\,\mathbf{e}_r + v_\theta\,\mathbf{e}\theta)$$

$$= mv_\theta(r_0 - v_0t)\,\mathbf{e}_z$$

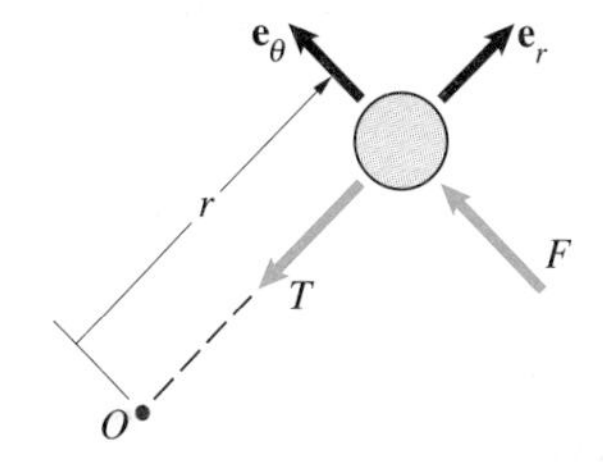

(a) Expressing the moment in terms of polar coordinates.

Substituting these expressions into the principle of angular impulse and momentum, we obtain

$$\int_{t-1}^{t_2} (\mathbf{r} \times \Sigma \mathbf{F})\,dt = (\mathbf{H}_0)_2 - (\mathbf{H}_0)_1 :$$

$$\int_0^t F(r_0 - v_0t)\,\mathbf{e}_z\,dt = mv_\theta(r_0 - v_0t)\,\mathbf{e}_z - \mathbf{0}$$

Evaluating the integral, we obtain the transverse component of velocity as a function of time:

$$v_\theta = \frac{[r_0t - (1/2)v_0t^2]F}{(r_0 - v_0t)m}$$

The disc's velocity as a function of time is

$$\mathbf{v} = v_0\,\mathbf{e}_r + \frac{[r_0t - (1/2)v0t^2]F}{(r_0 - v_0t)m}\,\mathbf{e}_\theta$$

Example 5.6

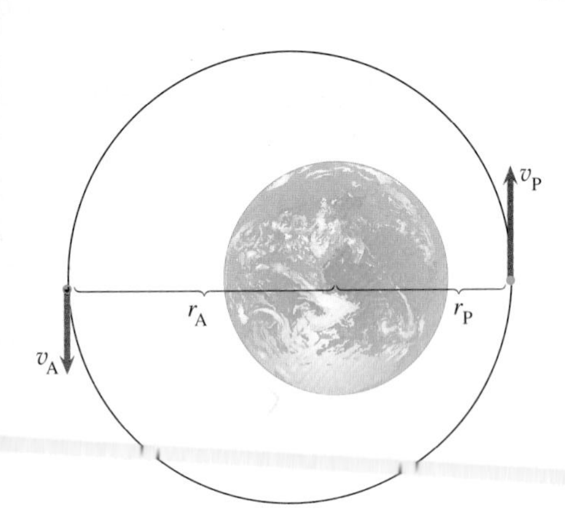

Figure 5.17

When an earth satellite is at perigee (the point at which it is nearest to the earth), the magnitude of its velocity is $v_P = 7000\,\text{m/s}$ and its distance from the centre of the earth is $r_P = 10\,000\,\text{km}$ (Figure 5.17). What are the magnitude of its velocity v_A and its distance r_A from the earth at apogee (the point at which it is farthest from the earth)? The radius of the earth is $R_E = 6370\,\text{km}$.

STRATEGY

Because this is central-force motion about the centre of the earth, we know that the product of the distance from the centre of the earth and the transverse component of the satellite's velocity is constant. This gives us one equation relating v_A and r_A. We can obtain a second equation relating v_A and r_A by using conservation of energy.

SOLUTION

From Equation (5.23), conservation of angular momentum requires that

$$r_A v_A = r_P v_P$$

From Equation (4.27), the satellite's potential energy in terms of distance from the centre of the earth is

$$V = -\frac{mgR_E^2}{r}$$

The sum of the kinetic and potential energies at apogee and perigee must be equal:

$$\frac{1}{2}mv_A^2 - \frac{mgR_E^2}{r_A} = \frac{1}{2}mv_P^2 - \frac{mgR_E^2}{r_P}$$

Substituting $r_A = r_P v_P / v_A$ into this equation and rearranging, we obtain

$$(v_A - v_P)\left(v_A + v_P - \frac{2gR_E^2}{r_P v_P}\right) = 0$$

This equation yields the trivial solution $v_A = v_P$ and also the solution for the velocity at apogee:

$$v_A = \frac{2gR_E^2}{r_P v_P} - v_P$$

Substituting the values of g, R_E, r_P and v_P, we obtain $v_A = 4373\,\text{m/s}$ and $r_A = 16\,007\,\text{km}$.

Problems

5.72 An object located at $\mathbf{r} = 12\mathbf{i} + 4\mathbf{j} - 3\mathbf{k}$ (m) relative to a point O is moving 130 m/s, and its angular momentum about O is zero. What is its velocity vector?

5.73 The total external force on a 2 kg object is $\Sigma\mathbf{F} = 2t\mathbf{i} + 4\mathbf{j}$ (N), where t is time in seconds. At time $t_1 = 0$, its position and velocity are $\mathbf{r} = 0$, $\mathbf{v} = 0$.
(a) Use Newton's second law to determine the object's position $\mathbf{r}$ and velocity $\mathbf{v}$ as functions of time.
(b) By integrating $\mathbf{r} \times \Sigma\mathbf{F}$ with respect to time, determine the angular impulse from $t_1 = 0$ to $t_2 = 6$ s.
(c) Use your results from part (a) to determine the change in the object's angular momentum from $t_1 = 0$ to $t_2 = 6$ s.

5.74 An astronaut moves in the x-y plane at the end of a 10 m tether attached to a large space station at O. The total mass of the astronaut and his equipment is 120 kg.
(a) What is his angular momentum about O before the tether becomes taut?
(b) What is the magnitude of the component of his velocity perpendicular to the tether immediately after the tether becomes taut?

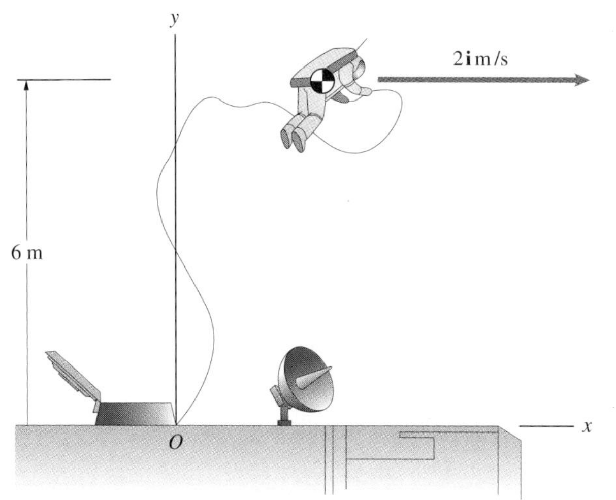

P5.74

5.75 In Problem 5.74, if the coefficient of restitution of the 'impact' that occurs when the astronaut reaches the end of the tether is $e = 0.8$, what are the x and y components of his velocity immediately after the tether becomes taut?

5.76 In Example 5.5, determine the disk's velocity as a function of time if the force is $F = Ct$, where C is a constant.

5.77 A 2 kg disc slides on a smooth horizontal table and is connected to an elastic cord whose tension is $T = 6r$ N, where r is the radial position of the disc in metres. If the disk is at $r = 1$ m and is given an initial velocity of 4 m/s in the transverse direction, what are the magnitudes of the radial and transverse components of its velocity when $r = 2$ m?

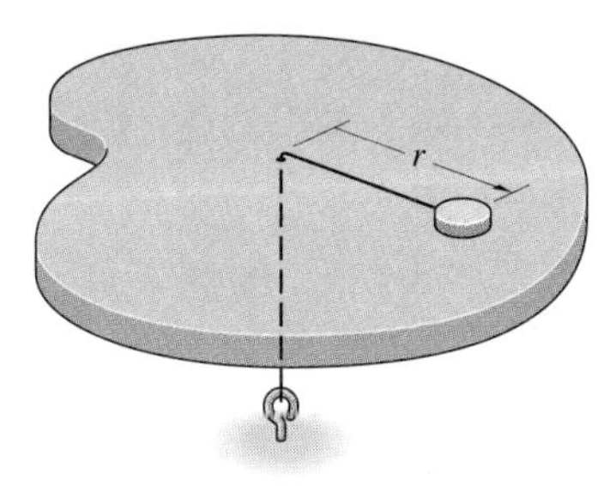

P5.77

5.78 In Problem 5.77, determine the maximum value of r reached by the disc.

5.79 A disc of mass m slides on a smooth horizontal table and is attached to a string that passes through a hole in the table.
(a) If the mass moves in a circular path of radius r_0 with transverse velocity v_0, what is the tension T?
(b) Starting from the initial condition described in part (a), the tension is increased in such a way that the string is pulled through the hole at a constant rate until $r = \frac{1}{2}r_0$. Determine T as a function of r while this is taking place.
(c) How much work is done on the mass in pulling the string through the hole as described in part (b)?

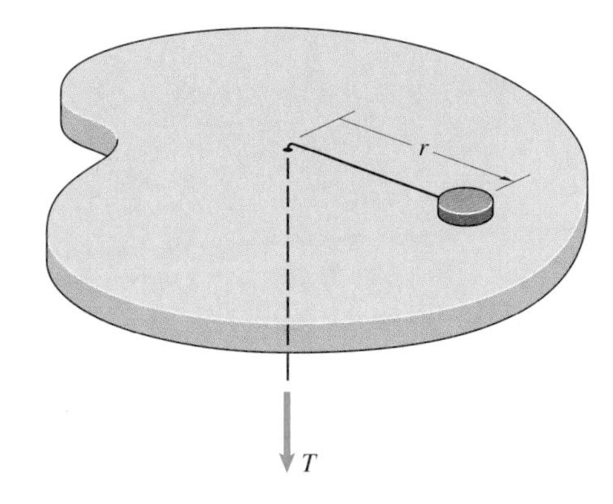

P5.79

5.80 Two gravity research satellites ($m_A = 250\,\text{kg}$, $m_B = 50\,\text{kg}$) are tethered by a cable. The satellites and cable rotate with angular velocity $\omega_0 = 0.25$ revolution per minute. Ground controllers order satellite A to slowly unreel 6 m of additional cable. What is the angular velocity afterwards?

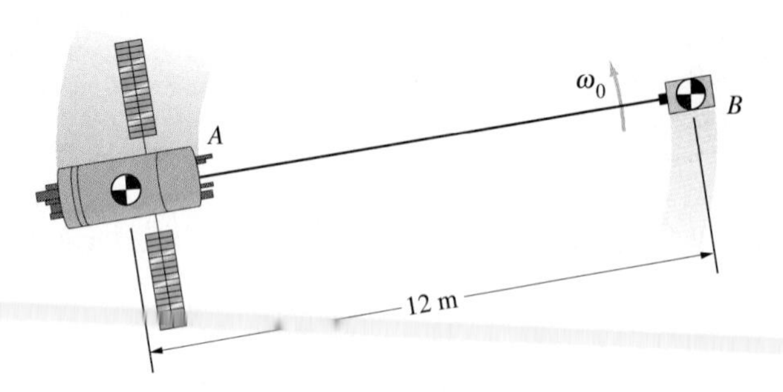

P5.80

5.81 A satellite at $r_0 = 16\ 000\,\text{km}$ from the centre of the earth is given an initial velocity $v_0 = 6000\,\text{m/s}$ in the direction shown. Determine the magnitude of its transverse component of velocity when $r = 32\ 000\,\text{km}$. The radius of the earth is 6370 km.

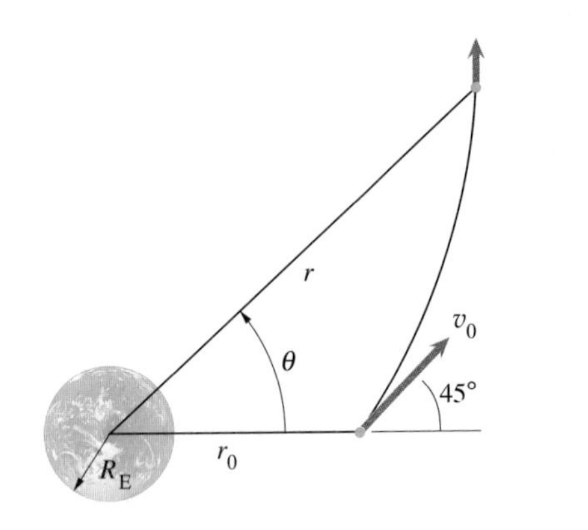

P5.81

5.82 In Problem 5.81, determine the magnitudes of the radial and transverse components of the satellite's velocity when $r = 24\ 000\,\text{km}$.

5.83 In Problem 5.81, determine the maximum distance r reached by the satellite.

5.84 A ball suspended from a string that goes through a hole in the ceiling at O moves with velocity v_A in a horizontal circular path of radius r_A. The string is then drawn through the hole until the ball moves with velocity v_B in a horizontal circular path of radius r_B. Use the principle of angular impulse and momentum to show that $r_A v_A = r_B v_B$.

Strategy: Let $\mathbf{e}$ be a unit vector that is perpendicular to the ceiling. Although this is not a central-force problem – the ball's weight does not point towards O – you can show that $\mathbf{e} \cdot (\mathbf{r} \times \Sigma \mathbf{F}) = 0$, so that $\mathbf{e} \cdot \mathbf{H}_O$ is conserved.

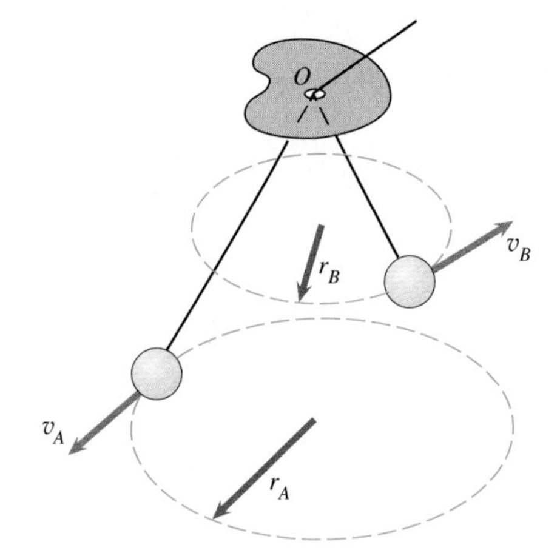

P5.84

5.5 *Mass Flows*

In this section we use conservation of linear momentum to determine the force exerted on an object as a result of emitting or absorbing a continuous flow of mass. The resulting equation applies to a variety of situations including determining the thrust of a rocket and calculating the forces exerted on objects by flows of liquids or granular materials.

Suppose that an object of mass m and velocity $\mathbf{v}$ is subjected to no external forces (Figure 5.18(a)) and it emits an element of mass Δm_f with velocity $\mathbf{v}_\mathrm{f}$ *relative to the object* (Figure 5.18(b)). We denote the new velocity of the object by $\mathbf{v} + \Delta\mathbf{v}$. The linear momentum of the object before the element of mass is emitted equals the total linear momentum of the object and the element afterwards.

$$m\mathbf{v} = (m - \Delta m_\mathrm{f})(\mathbf{v} + \Delta\mathbf{v}) + \Delta m_\mathrm{f}(\mathbf{v} + \mathbf{v}_\mathrm{f})$$

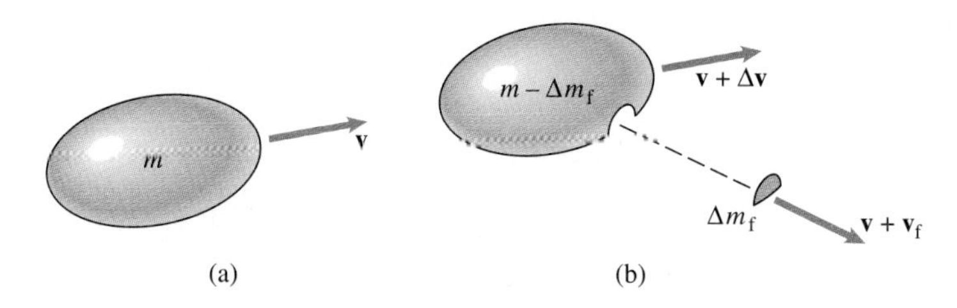

Figure 5.18
An object's mass and velocity (a) before and (b) after emitting an element of mass.

Evaluating the products and simplifying, we obtain

$$m\Delta\mathbf{v} + \Delta m_\mathrm{f}\mathbf{v}_\mathrm{f} - \Delta m_\mathrm{f}\mathbf{v} = 0 \tag{5.24}$$

Now we assume that, instead of a discrete element of mass, the object emits a continuous flow of mass and that Δm_f is the amount emitted in an interval of time Δt. We divide Equation (5.24) by Δt and write the resulting equation as

$$m\frac{\Delta\mathbf{v}}{\Delta t} + \frac{\Delta m_\mathrm{f}}{\Delta t}\mathbf{v}_\mathrm{f} - \frac{\Delta m_\mathrm{f}}{\Delta t}\frac{\Delta\mathbf{v}}{\Delta t}\Delta t = 0$$

Taking the limit of this equation as $\Delta t \to 0$, we obtain

$$-\frac{dm_\mathrm{f}}{dt}\mathbf{v}_\mathrm{f} = m\mathbf{a}$$

where $\mathbf{a}$ is the acceleration of the object's centre of mass. The term dm_f/dt is the **mass flow rate**, the rate at which mass flows from the object. Comparing this equation with Newton's second law, we conclude that a flow of mass *from* an object exerts a force

$$\boxed{\mathbf{F}_\mathrm{f} = -\frac{dm_\mathrm{f}}{dt}\mathbf{v}_\mathrm{f}} \tag{5.25}$$

on the object. The force is proportional to the mass flow rate and to the magnitude of the *relative* velocity of the flow, and its direction is *opposite* to the direction of the relative velocity. Conversely, a flow of mass *to* an object exerts a force in the same direction as the relative velocity.

The classic example of a force created by a mass flow is the rocket (Figure 5.19). Suppose that it has a uniform exhaust velocity v_f parallel to the x axis and the mass flow rate from the exhaust is dm_f/dt. In terms of the coordinate system shown, the velocity vector of the exhaust is $\mathbf{v}_f = -v_f\,\mathbf{i}$, so from Equation (5.25) the force exerted on the rocket is

$$\mathbf{F}_f = -\frac{dm_f}{dt}\mathbf{v}_f = \frac{dm_f}{dt}v_f\,\mathbf{i}$$

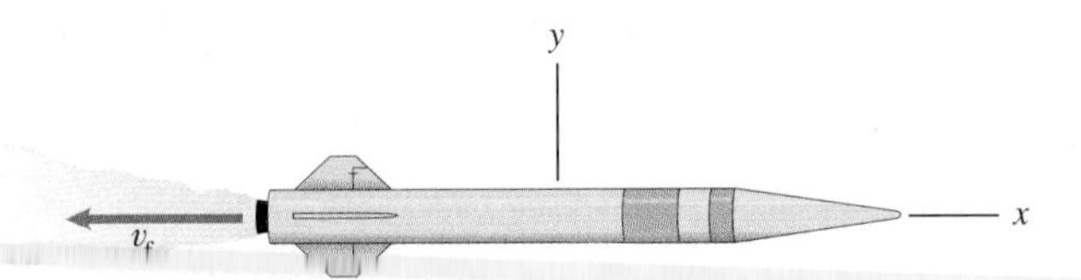

Figure 5.19
A rocket with its exhaust aligned with the x axis.

The force exerted on the rocket by its exhaust is towards the right, opposite to the direction of the flow of its exhaust. If we assume that no external forces act on the rocket, Newton's second law is

$$\frac{dm_f}{dt}v_f = m\frac{dv}{dt} \tag{5.26}$$

The mass flow rate of fuel is the rate at which the rocket's mass is being consumed. Therefore the rate of change of the mass of the rocket is

$$\frac{dm}{dt} = \frac{dm_f}{dt}$$

Using this expression, we can write Equation (5.26) as

$$dv = -v_f\frac{dm}{m}$$

Suppose that the rocket starts from rest with initial mass m_0. If the exhaust velocity is constant, we can integrate this equation to determine the velocity of the rocket as a function of its mass:

$$\int_0^v dv = \int_{m_0}^{m} -v_f\frac{dm}{m}$$

The result is

$$v = v_f \ln\left(\frac{m_0}{m}\right) \tag{5.27}$$

The rocket can gain more velocity by expending more mass, but notice that increasing the ratio m_0/m from 10 to 100 only increases the velocity attained by a factor of 2. In contrast, increasing the exhaust velocity results in a proportional increase in the rocket's velocity.

Example 5.7

A horizontal stream of water with velocity v_0 and mass flow rate dm_f/dt hits a plate that deflects the water in the horizontal plane through an angle θ (Figure 5.20). Assume that the magnitude of the velocity of the water when it leaves the plate is approximately equal to v_0. What force is exerted on the plate by the water?

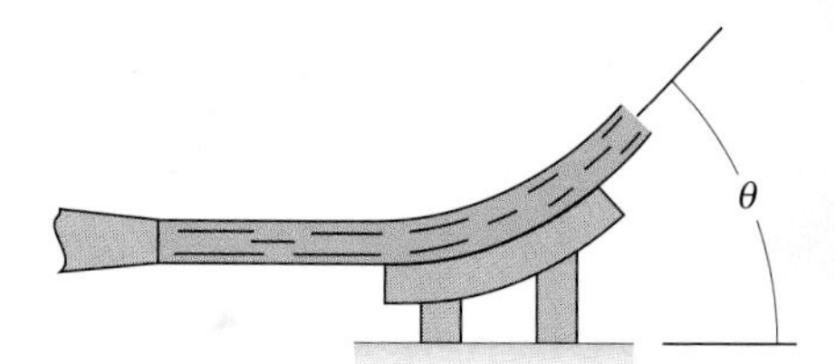

Figure 5.20

STRATEGY

We can determine the force exerted on the plate by treating the part of the stream in contact with the plate as an 'object' with mass flows entering and leaving it.

SOLUTION

In Figure (a) we draw the free-body diagram of the stream in contact with the plate. Streams of mass with velocity v_0 enter and leave this 'object', and $\mathbf{F}_P$ is the force exerted on the stream by the plate. It is the force $-\mathbf{F}_P$ exerted on the plate by the stream that we wish to determine. First we consider the departing stream of water. The mass flow rate of water leaving the free-body diagram must be equal to the mass flow rate entering. In terms of the coordinate system shown, the velocity of the departing stream is

$$\mathbf{v}_f = v_0 \cos\theta\,\mathbf{i} + v_0 \sin\theta\,\mathbf{j}$$

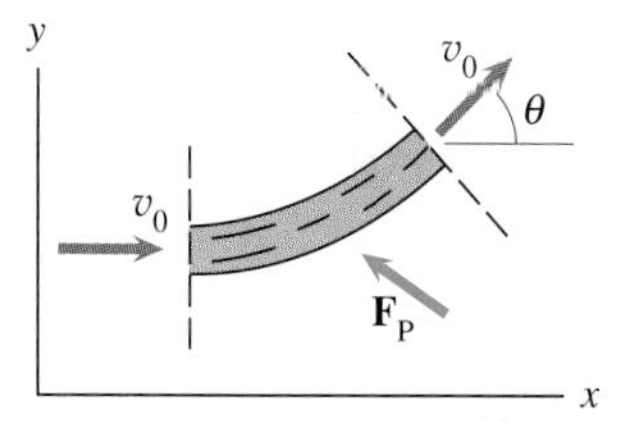

(a) Free-body diagram of the stream.

Let $\mathbf{F}_D$ be the force exerted on the object by the departing stream. From Equation (5.25),

$$\mathbf{F}_D = -\frac{dm_f}{dt}\mathbf{v}_r = -\frac{dm_f}{dt}(v_0 \cos\theta\,\mathbf{i} + v_0 \sin\theta\,\mathbf{j})$$

The velocity of the entering stream is $\mathbf{v}_f = v_0\,\mathbf{i}$. Since this flow is entering the object rather than leaving it, the resulting force $\mathbf{F}_E$ is in the same direction as the relative velocity:

$$\mathbf{F}_E = \frac{dm_f}{dt}\mathbf{v}_r = \frac{dm_f}{dt}v_0\,\mathbf{i}$$

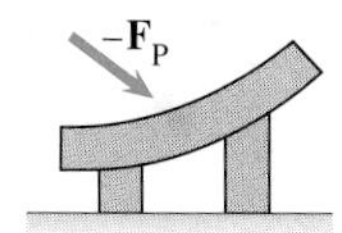

(b) Force exerted on the plate.

The sum of the forces on the free-body diagram must equal zero,

$$\mathbf{F}_D + \mathbf{F}_E + \mathbf{F}_P = 0$$

so the force exerted on the plate by the water is (Figure (b))

$$-\mathbf{F}_P = \mathbf{F}_D + \mathbf{F}_E = \frac{dm_f}{dt}v_0[(1-\cos\theta)\,\mathbf{i} - \sin\theta\,\mathbf{j}]$$

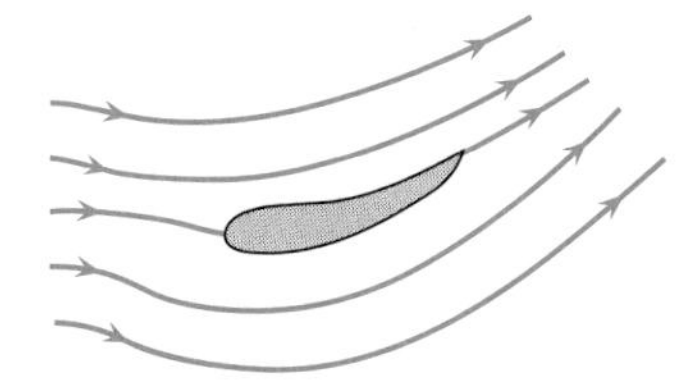

(c) Pattern of moving fluid around an aeroplane wing.

DISCUSSION

This simple example gives you insight into how turbine blades and aeroplane wings can create forces by deflecting streams of liquid or gas (Figure (c)).

Example 5.8

Application to Engineering

Jet Engines

In a turbulent engine (Figure 5.21), a mass flow rate dm_c/dt of inlet air enters the compressor with velocity v_i. The air is mixed with fuel and ignited in the combustion chamber. The mixture then flows through the turbine, which powers the compressor. The exhaust, with a mass flow rate equal to that of the air plus the mass flow rate of the fuel, $dm_c/dt + dm_f/dt$, exits at a high exhaust velocity v_e, exerting a large force on the engine. Suppose that $dm_c/dt = 13.5\ \mathrm{kg/s}$ and $dm_f/dt = 0.13\ \mathrm{kg/s}$. The inlet air velocity is $v_i = 120\ \mathrm{m/s}$, and the exhaust velocity is $v_e = 490\ \mathrm{m/s}$. What is the engine's thrust?

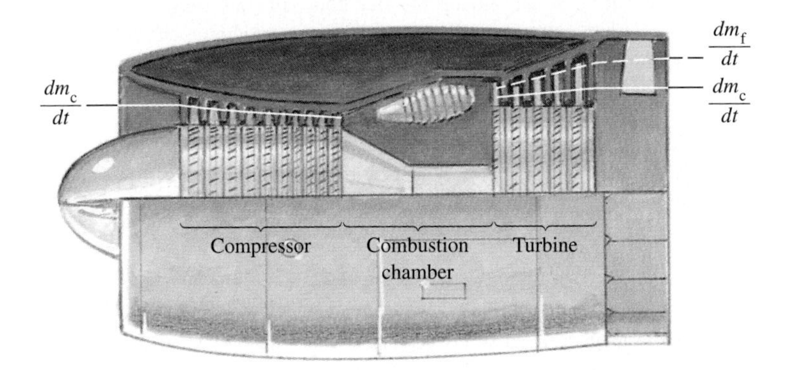

Figure 5.21

STRATEGY

We can determine the engine's thrust by using Equation (5.25). We must include both the force exerted by the engine's exhaust and the force exerted by the mass flow of air entering the compressor to determine the net thrust.

SOLUTION

The engine's exhaust exerts a force to the left equal to the product of the mass flow rate of the fuel–air mixture and the exhaust velocity. The inlet air exerts a force to the right equal to the product of the mass flow rate of the inlet air and the inlet velocity. The engine's thrust (the net force to the left) is

$$T = \left(\frac{dm_c}{dt} + \frac{dm_f}{dt}\right)v_e - \frac{dm_c}{dt}v_i$$

$$= (13.5 + 0.13)(490) - (13.5)(120)$$

$$= 5059\ \mathrm{N}$$

DESIGN ISSUES

The jet engine was developed in Europe in the years just prior to World War II. Although the turbojet engine in Figure 5.21 was a very successful design that dominated both military and commercial aviation for many years, it has the drawback of relatively large fuel consumption.

During the past 30 years, the fan-jet engine, shown in Figure 5.22, has become the most commonly used design, particularly for commercial aeroplanes. Part of its thrust is provided by air that is accelerated by the fan. The ratio of the mass flow rate of air entering the fan, dm_b/dt, to the mass flow rate of air entering the compressor, dm_c/dt, is called the *bypass ratio*.

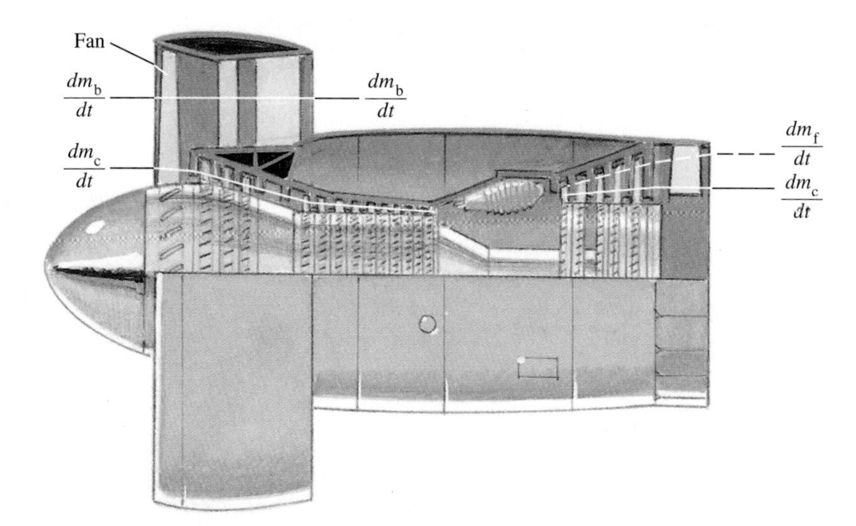

Figure 5.22

A fan-jet engine. Part of the entering mass flow of air is accelerated by the fan and does not enter the compressor.

The force exerted by a jet engine's exhaust equals the product of the mass flow rate and the exhaust velocity. In the fan-jet engine, the air passing through the fan is not heated by the combustion of fuel and therefore has a higher density than the exhaust of the turbojet engine. As a result, the fan-jet engine can provide a given thrust with a lower average exhaust velocity. Since the work that must be expended to create the thrust depends on the kinetic energy of the exhaust, the fan-jet engine creates thrust more efficiently.

Problems

5.85 The Cheverton fire-fighting and rescue boat can pump 3.8 kg/s of water from each of its two pumps at a velocity of 44 m/s. If both pumps point in the same direction, what total force do they exert on the boat?

P5.85

5.86 A nozzle mounted on a fire truck emits a stream of water at 24 m/s with a mass flow rate of 50 kg/s. Determine the moment about A due to the force exerted by the steam of water.

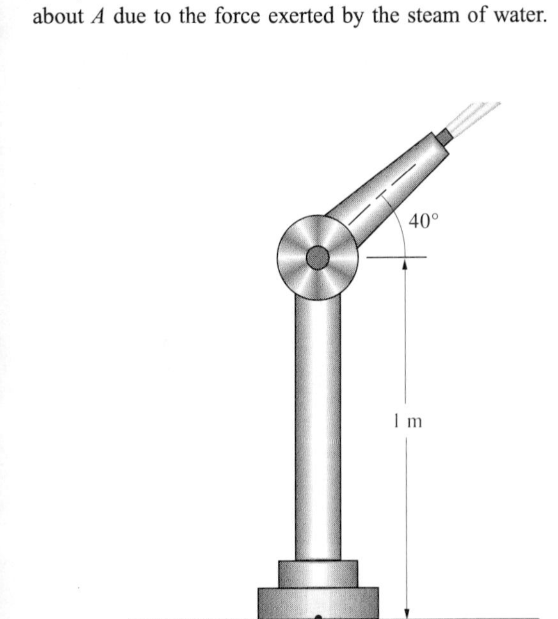

P5.86

5.87 A front-end loader moves at 3 km/hr and scoops up 30 000 kg of iron ore in 3 s. What horizontal force must its tyres exert?

P5.87

5.88 The snowblower moves at 1 m/s and scoops up 750 kg/s of snow. Determine the force exerted by the entering flow of snow.

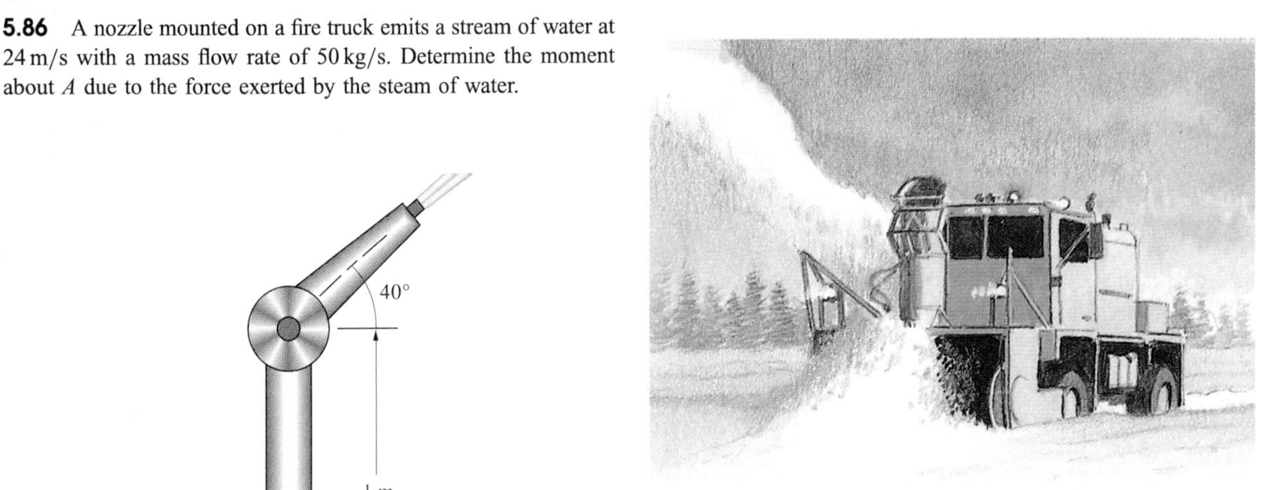

P5.88

5.89 If you design the snowblower in Problem 5.88 so that it blows snow out at 45° above the horizontal from a port 2 m above the ground and the snow lands 20 m away, what horizontal force is exerted on the blower by the departing flow of snow?

5.90 A nozzle ejects a stream of water horizontally at 40 m/s with a mass flow rate of 30 kg/s, and the stream is deflected in the horizontal plane by a plate. Determine the force exerted on the plate by the stream in cases (a), (b), (c).

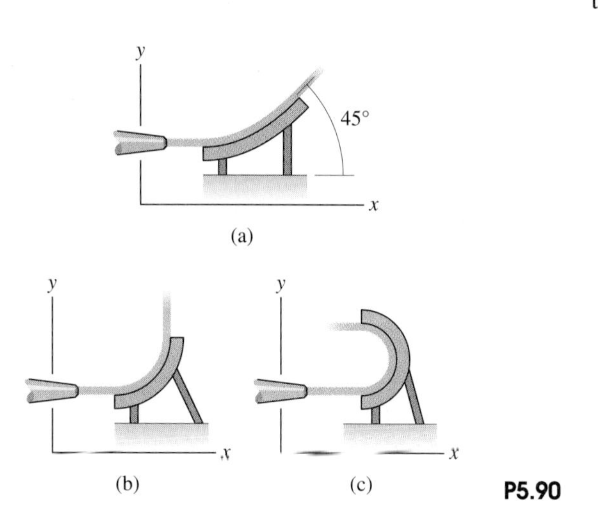

P5.90

5.91 A stream of water with velocity $80\,\mathbf{i}$ m/s and a mass flow rate of 6 kg/s strikes a turbine blade moving with constant velocity $20\,\mathbf{i}$ m/s.

(a) What force is exerted on the blade by the water?

(b) What is the magnitude of the velocity of the water as it leaves the blade?

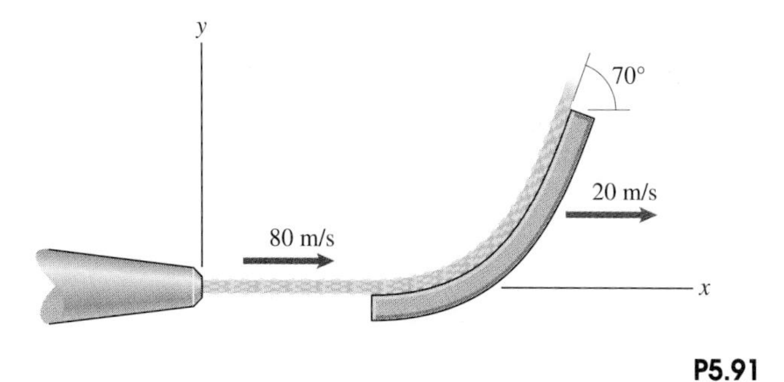

P5.91

5.92 The nozzle A of the lawn sprinkler is located at (175, −12, 12) mm. Water exits each nozzle at 8 m/s with a mass flow rate of 0.25 kg/s. The direction cosines of the flow direction from A are $(\frac{1}{\sqrt{3}} - \frac{1}{\sqrt{3}}, \frac{1}{\sqrt{3}})$. What is the total moment about the z axis exerted on the sprinkler by the flows from all four nozzles?

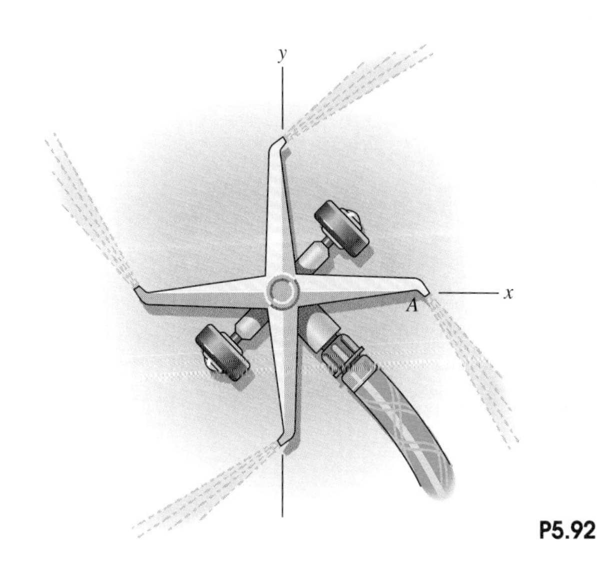

P5.92

5.93 A 45 kg/s flow of gravel exits the chute at 2 m/s and falls onto a conveyer moving at 0.3 m/s. Determine the components of the force exerted on the conveyer by the flow of gravel if $\theta = 0$.

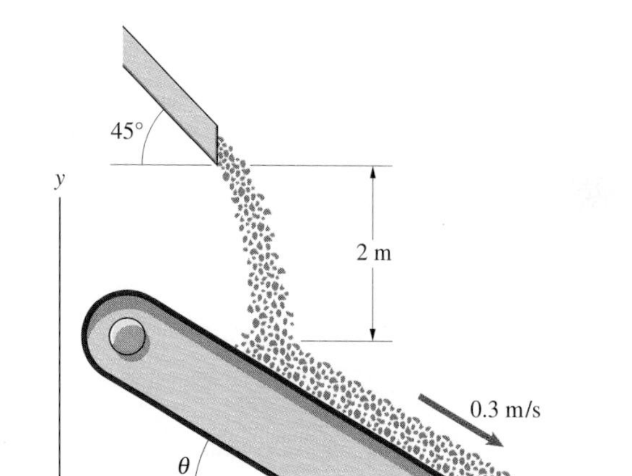

P5.93

5.94 Solve Problem 5.93 if $\theta = 30°$.

5.95 A toy car is propelled by water that squirts from an internal tank at 3 m/s relative to the car. If the mass of the empty car is 1 kg, it holds 2 kg of water, and you neglect other tangential forces, what is its top speed?

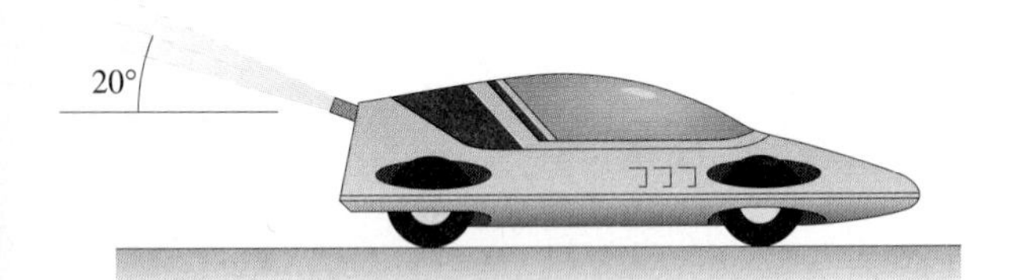

P5.95

5.96 A rocket consists of a 2 Mg payload and a 40 Mg booster. Eighty per cent of the booster's mass is fuel, and its exhaust velocity is 1 km/s. If the rocket starts from rest and you neglect external forces, what velocity will it reach?

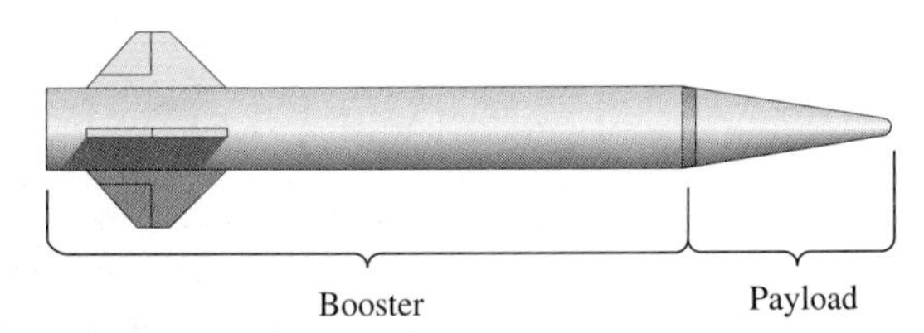

P5.96

5.97 A rocket consists of a 2 Mg payload and a booster. The booster has two stages whose total mass is 40 Mg. Eighty per cent of the mass of each stage is fuel. When the fuel of stage 1 is expended, it is discarded and the motor in stage 2 is ignited. The exhaust velocity of both stages is 1 km/s. Assume that the rocket starts from rest and neglect external forces. Determine the velocity reached by the rocket if the two stages are of equal mass and compare your results to the answer to Problem 5.96.

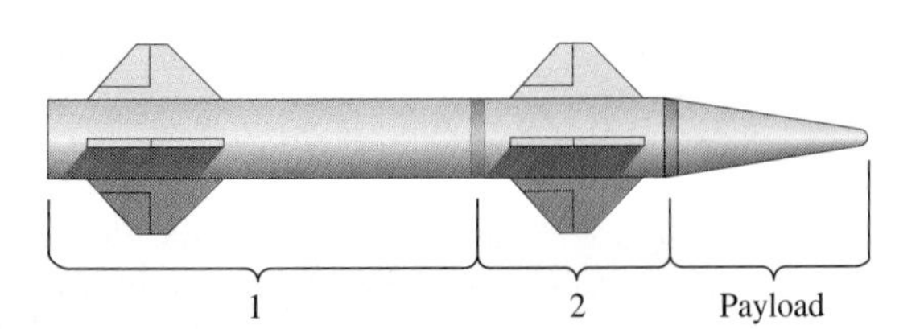

P5.97

5.98 In Problem 5.97, determine the velocity reached by the rocket for three sets of values of the masses of the two stages: (a) $m_1 = 25$ Mg, $m_2 = 15$ Mg; (b) $m_1 = 35$ Mg, $m_2 = 5$ Mg; (c) $m_1 = 38$ Mg, $m_2 = 2$ Mg.

5.99 After its rocket motor burns out, a rocket sled is slowed by a water brake. A tube extends into a trough of water so that water flows through the tube at the velocity of the sled and flows out in a direction perpendicular to the motion of the sled. The mass flow rate through the tube is $\rho v A$, where $\rho = 1000\,\text{kg/m}^3$ is the mass density of the water, v is the flow velocity, and $A = 0.01\,\text{m}^2$ is the cross-sectional area of the tube. The mass of the sled is 50 kg. Neglecting friction and aerodynamic drag, determine the time and the distance required for the sled to decelerate from 300 m/s to 30 m/s.

P5.99

5.100 Suppose that you grasp the end of a chain that weighs 45 N/m and lift it straight up off the floor at a constant speed of 0.6 m/s.

(a) Determine the upward force F you must exert as a function of the height s.

(b) How much work do you do in lifting the top of the chain to $s = 1.2$ m?

Strategy: Treat the part of the chain you have lifted as an object that is gaining mass.

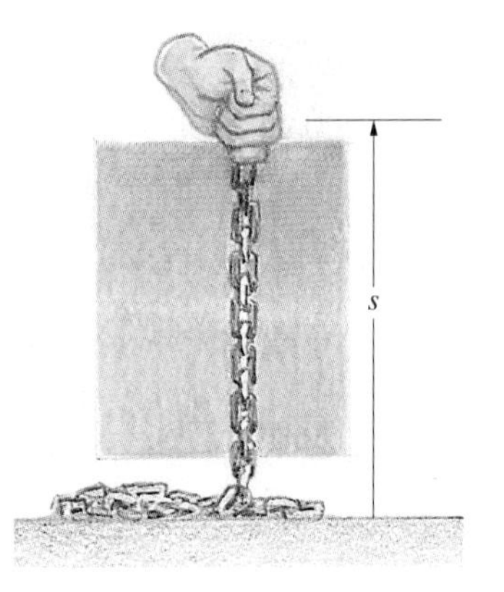

P5.100

5.101 Solve Problem 5.100, assuming that you lift the end of the chain straight up off the floor with a constant acceleration of 0.6 m/s.

5.102 It has been suggested that a heavy chain could be used to gradually stop an aeroplane that rolls past the end of the runway. A hook attached to the end of the chain engages the plane's nose wheel, and the plane drags an increasing length of the chain as it rolls. Let m be the aeroplane's mass and v_0 its initial velocity, and let ρ_L be the mass per unit length of the chain. If you neglect friction and aerodynamic drag, what is the aeroplane's velocity as a function of s?

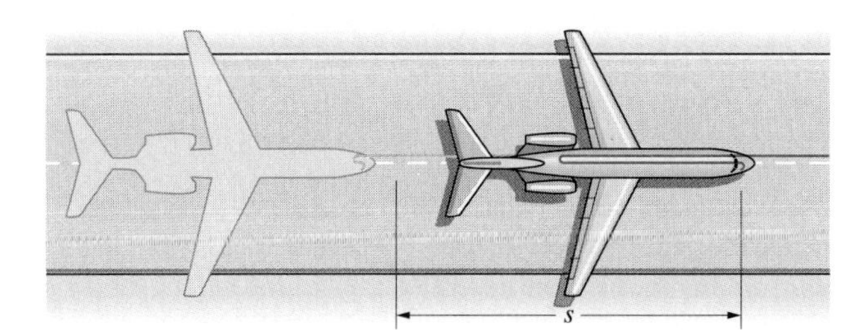

P5.102

5.103 In Problem 5.102, the friction force exerted on the chain by the ground would actually dominate other forces as the distance s increases. If the coefficient of kinetic friction between the chain and the ground is μ_k and you neglect all forces except the friction force, what is the aeroplane's velocity as a function of s?

Problems 5.104–5.108 are related to Example 5.8.

5.104 The turbojet engine in Figure 5.21 is being operated on a test stand. The mass flow rate of air entering the compressor is 13.5 kg/s, and the mass flow rate of fuel is 0.13 kg/s. The effective velocity of the air entering the compressor is zero, and the exhaust velocity is 500 m/s. What is the thrust of the engine?

5.105 Suppose that the engine described in Problem 5.104 is in an aeroplane flying at 400 km/hr. The effective velocity of the air entering the inlet is equal to the aeroplane's velocity. What is the thrust of the engine?

5.106 A turbojet engine's thrust reverser causes the exhaust to exit the engine at 20° from the engine centreline. The mass flow rate of air entering the compressor is 45 kg/s and it enters at 60 m/s. The mass flow rate of fuel is 1.5 kg/s, and the exhaust velocity is 360 m/s. What braking force does the engine exert on the aeroplane?

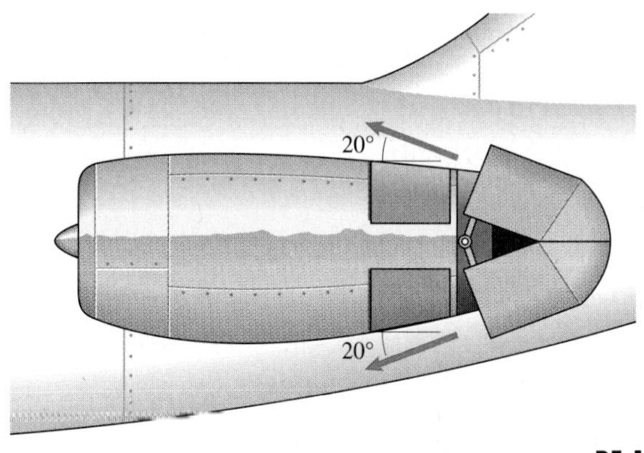

P5.106

5.107 The 13.6 Mg aeroplane is moving at 400 km/hr. The total mass flow rate of air entering the compressors of its turbojet engines is 280 kg/s, and the total mass flow rate of fuel is 2.6 kg/s. The effective velocity of the air entering the compressors is equal to the aeroplane's velocity, and the exhaust velocity is 480 m/s. The ratio of the lift force L to the drag force D is 6, and the z component of the aeroplane's acceleration is zero. What is the x component of its acceleration?

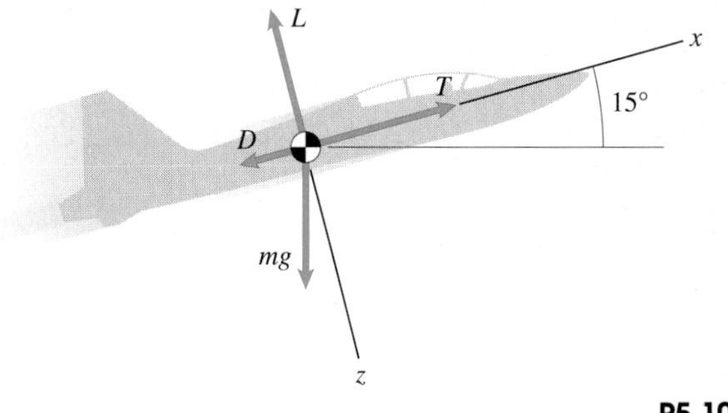

P5.107

5.108 The fan-jet engine in Figure 5.22 is similar to the Pratt and Whitney JT9D-3A engine used on early models of the Boeing 747. When the aeroplane begins its takeoff run, the velocity of the air entering the compressor and fan is negligible. A mass flow rate of 560 kg/s enters the fan and is accelerated to 270 m/s. A mass flow rate of 112 kg/s enters the compressor. The mass flow rate of fuel is 3.36 kg/s and the exhaust velocity is 363 m/s (a) What is the bypass ratio? (b) What is the thrust of the engine? (c) If the aeroplane weighs 2.22 MN, what is its initial acceleration? (It has four engines.)

Chapter Summary

Principle of Impulse and Momentum

The linear impulse applied to an object during an interval of time is equal to the change in its linear momentum:

$$\int_{t-1}^{t_2} \Sigma \mathbf{F}\, dt = m\mathbf{v}_2 - m\mathbf{v}_1 \qquad \textbf{Equation (5.1)}$$

This result can also be expressed in terms of the average with respect to time of the total force:

$$(t_2 - t_1)\Sigma \mathbf{F}_{\text{av}} = m\mathbf{v}_2 - m\mathbf{v}_1 \qquad \textbf{Equation (5.2)}$$

Conservation of Linear Momentum

If objects A and B are not subjected to external forces other than the forces they exert on each other (or if the effects of other external forces are negligible), their total linear momentum is conserved,

$$m_A\mathbf{v}_A + m_B\mathbf{v}_B = \text{constant} \qquad \textbf{Equation (5.4)}$$

and the velocity of their common centre of mass is constant.

Impacts

If colliding objects are not subjected to external forces, their total linear momentum must be the same before and after the impact. Even when they are subjected to external forces, the force of the impact is often so large, and its duration so brief, that the effect of external forces on their motions during the impact is negligible.

If objects A and B adhere and remain together after they collide, they are said to undergo a **perfectly plastic impact**. The velocity of their common centre of mass before and after the impact is given by

$$\mathbf{v} = \frac{m_A\mathbf{v}_A + m_B\mathbf{v}_B}{m_A + m_B} \qquad \textbf{Equation (5.7)}$$

Central Impacts

In a **direct central impact** (Figure (a)), linear momentum is conserved,

$$m_Av_A + m_Bv_B = m_Av'_A + m_Bv'_B \qquad \textbf{Equation (5.8)}$$

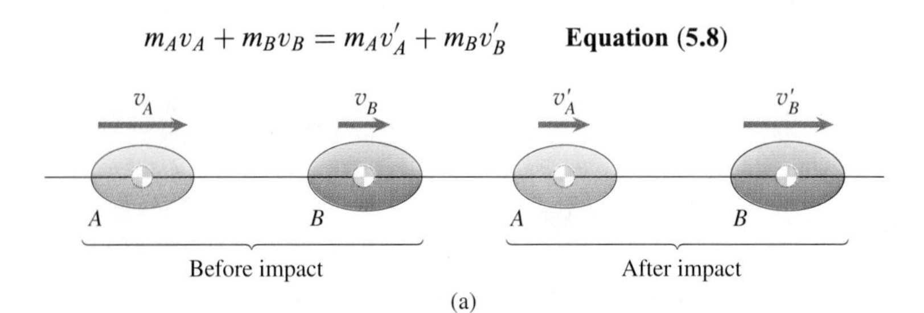

(a)

and the velocities are related by the **coefficient of restitution**:

$$e = \frac{v'_B - v'_A}{v_A - v_B} \qquad \textbf{Equation (5.14)}$$

If $e = 0$, the impact is perfectly plastic. If $e = 1$, the total kinetic energy is conserved and the impact is called **perfectly elastic**.

In an **oblique central impact** (Figure (b)), the components of velocity in the y and z directions are unchanged by the impact:

$$\begin{aligned} (\mathbf{v}'_A)_y &= (\mathbf{v}_A)_y \quad (\mathbf{v}'_B)_y = (\mathbf{v}_B)_y \\ (\mathbf{v}'_A)_z &= (\mathbf{v}_A)_z \quad (\mathbf{v}'_B)_z = (\mathbf{v}_B)_z \end{aligned} \qquad \textbf{Equation (5.5)}$$

In the x direction, linear momentum is conserved,

$$m_A(\mathbf{v}_A)_x + m_B(\mathbf{v}_B)_x = m_A(\mathbf{v}'_A)_x + m_B(\mathbf{v}'_B)_x \qquad \textbf{Equation (5.16)}$$

and the velocity components are related by the coefficient of restitution:

$$e = \frac{(\mathbf{v}'_B)_x - (\mathbf{v}'_A)_x}{(\mathbf{v}_A)_x - (\mathbf{v}_B)_x} \qquad \textbf{Equation (5.17)}$$

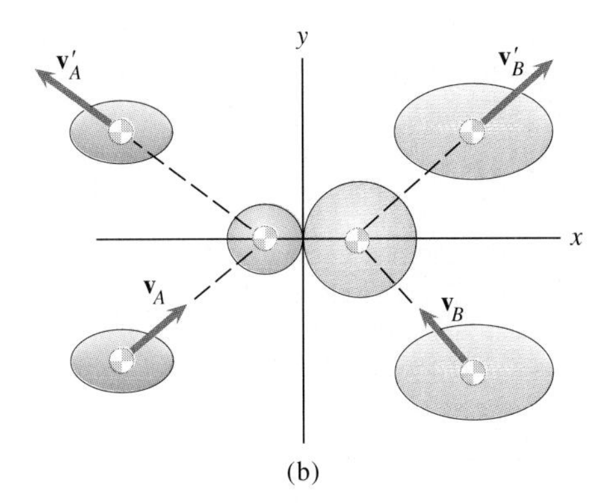

(b)

Principle of Angular Impulse and Momentum

The angular impulse about a point O applied to an object during an interval of time is equal to the change in its angular momentum about O:

$$\int_{t_1}^{t_2} (\mathbf{r} \times \Sigma\,\mathbf{F})\,dt = (\mathbf{H}_O)_2 - (\mathbf{H}_O)_1 \qquad \textbf{Equation (5.21)}$$

where the angular momentum is

$$\mathbf{H}_O = \mathbf{r} \times m\mathbf{v} \qquad \textbf{Equation (5.20)}$$

Central-Force Motion

If the total force acting on an object remains directed towards a fixed point, the object is said to be in **central-force motion**, and its angular momentum about the fixed point is conserved:

$$\mathbf{H}_0 = \text{constant} \qquad \textbf{Equation (5.22)}$$

In plane central-force motion, the product of the radial distance and the transverse component of the velocity is constant:

$$rv_\theta = \text{constant} \qquad \textbf{Equation (5.23)}$$

Mass Flows

A flow of mass *from* an object with velocity $\mathbf{v}_\mathrm{f}$ *relative to the object* exerts a force

$$\mathbf{F}_\mathrm{f} = -\frac{dm_\mathrm{f}}{dt}\mathbf{v}_\mathrm{f} \qquad \textbf{Equation (5.25)}$$

on the object, where dm_f/dt is the **mass flow rate**. The direction of the force is opposite to the direction of the relative velocity. A flow of mass *to* an object exerts a force in the same direction as the relative velocity.

Review Problems

5.109 An aircraft arresting system is used to stop aeroplanes whose braking systems fail. The system stops a 47.5 Mg aeroplane moving at 80 m/s in 9.15 s.
(a) What impulse is applied to the aeroplane during the 9.15 s?
(b) What is the average deceleration to which the passengers are subjected?

P5.109

5.110 The 1895 Austrian 150 mm howitzer had a 1.94 m long barrel, a muzzle velocity of 300 m/s, and fired a 38 kg shell. If the shell took 0.013 s to travel the length of the barrel, what average force was exerted on the shell?

5.111 A spacecraft is in an elliptic orbit around a large asteroid. The acceleration due to gravity of the asteroid is unknown. When the spacecraft is at its closest approach, its distance from the centre of the asteroid is $r_\mathrm{P} = 2$ km and its velocity is $v_\mathrm{P} = 1$ m/s. When it is at its farthest point from the asteroid, its distance is $r_\mathrm{A} = 6$ km. What is the velocity v_A?

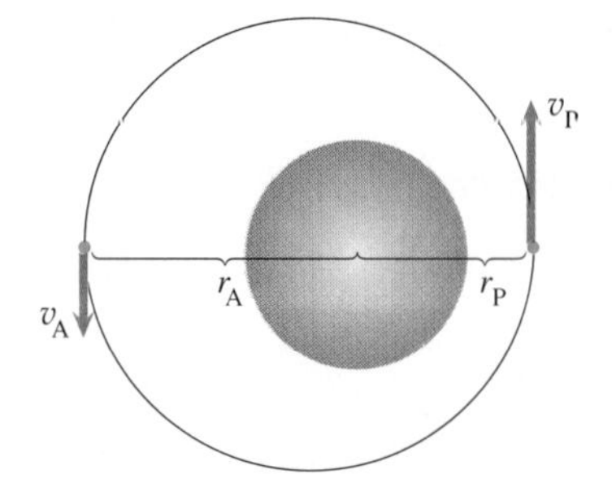

P5.111

5.112 In Problem 5.111, what is the asteroid's mass? If you assume that the asteroid is approximately spherical with an average mass density of 7000 kg/m^3, what is its radius?

Strategy: Use conservation of energy and express the gravitational potential energy in the form $V = -Gmm_A/r$, where $G = 6.67 \times 10^{-11}$ N.m^2/kg^2 is the universal gravitational constant and mA is the mass of the asteroid.

5.113 An athlete throws a shot put weighing 72 N. When he releases it, the shot put is 2.1 m above the ground and its components of velocity are $v_x = 9.5$ m/s, $v_y = 8$ m/s.

(a) If he accelerates the shot put from rest in 0.8 s and you assume as a first approximation that the force **F** he exerts on it is constant, use the principle of impulse and momentum to determine the x and y components of **F**.

(b) What is the horizontal distance from the point where he releases the shot put to the point where it strikes the ground?

P5.113

5.114 The 3000 kg pickup truck A moving at 12 m/s collides with the 2000 kg car B moving at 10 m/s.

(a) What is the magnitude of the velocity of their common centre of mass after the impact?

(b) If you treat the collision as a perfectly plastic impact, how much kinetic energy is lost?

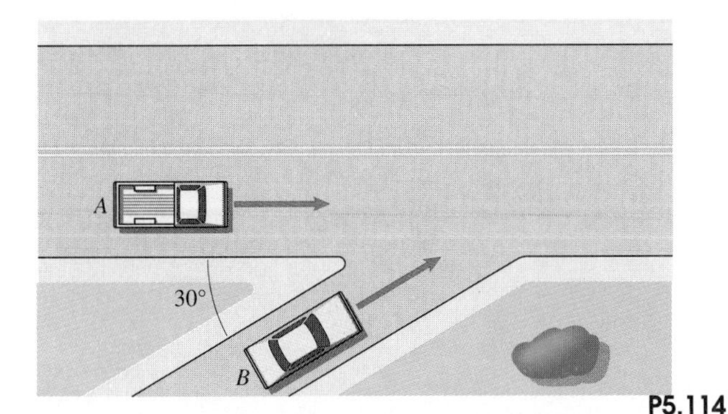

P5.114

5.115 Two hockey players ($m_A = 80$ kg, $m_B = 90$ kg) converging on the puck at $x = 0$, $y = 0$ become entangled and fall. Before the collision, $v_A = (9\,\mathbf{i} + 4\,\mathbf{j})$ m/s and $v_B = (-3\,\mathbf{i} + 6\,\mathbf{j})$ m/s. If the coefficient of kinetic friction between the players and the ice is $\mu_k = 0.1$, what is their approximate position when they stop sliding?

P5.115

5.116 An acceptable handball will bounce to a height between 1.07 m and 1.22 m when it is dropped onto a hardwood floor from a height of 1.78 m. What is the acceptable range of coefficients of restitution for handballs?

5.117 A 1 kg ball moving horizontal at 12 m/s strikes a 10 kg block. The coefficient of restitution of the impact is $e = 0.6$, and the coefficient of kinetic friction between the block and the inclined surface is $\mu_k = 0.4$. What distance does the block slide before stopping?

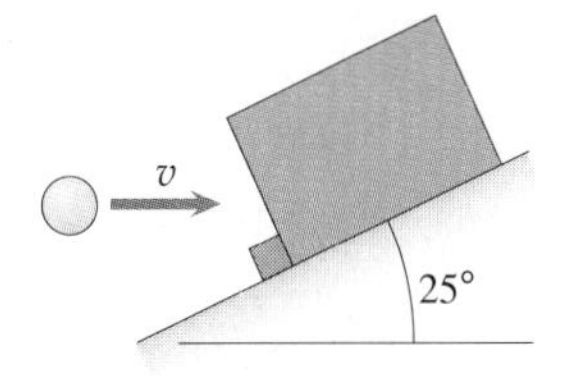

P5.117

5.118 A Peace Corps volunteer designs the simple device shown for drilling water wells in remote areas. A 70 kg 'hammer', such as a section of log or a steel drum partially filled with concrete, is hoisted to $h = 1$ m and allowed to drop onto a protective cap on the section of pipe being pushed into the ground. The mass of the cap and section of pipe is 20 kg. Assume the coefficient of restitution is nearly zero.
(a) What is the velocity of the cap and pipe immediately after the impact?
(b) If the pipe moves 30 mm downwards when the hammer is dropped, what resistive force was exerted on the pipe by the ground? (Assume the resistive force is constant during the motion of the pipe.)

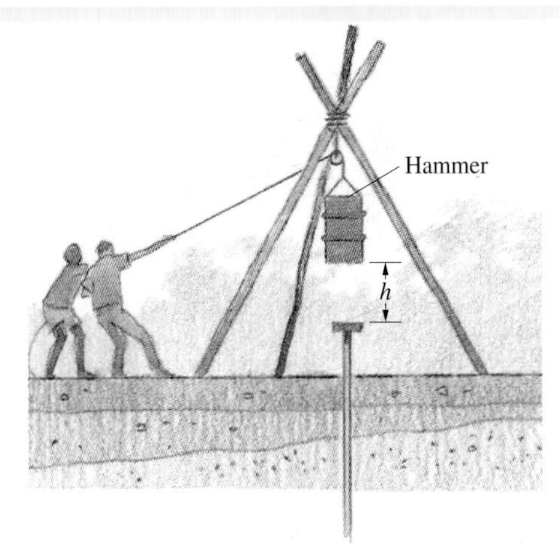

P5.118

5.119 A tugboat (mass = 40 Mg) and a barge (mass = 160 Mg) are stationary with a slack hawser connecting them. The tugboat accelerates to 2 knots (one knot = 1852 m/hr) before the hawser becomes taut. Determine the velocities of the tugboat and the barge just after the hawser becomes taut (a) if the 'impact' is perfectly plastic ($e = 0$); (b) if the 'impact' is perfectly elastic ($e = 1$). Neglect the forces exerted by the water and the tugboat's engines.

P5.119

5.120 In Problems 5.119, determine the magnitude of the impulsive force exerted on the tugboat in the two cases if the duration of the 'impact' is 4 s. Neglect the forces exerted by the water and the tugboat's engines during this period.

5.121 The balls are of equal mass m. Balls B and C are connected by an unstretched linear spring and are stationary. Ball A moves towards ball B with velocity v_A. The impact of A and B is perfectly elastic ($e = 1$). Neglect external forces.
(a) What is the velocity of the common centre of mass of the balls B and C immediately after the impact?
(b) What is the velocity of the common centre of mass of the balls B and C at time t after the impact?

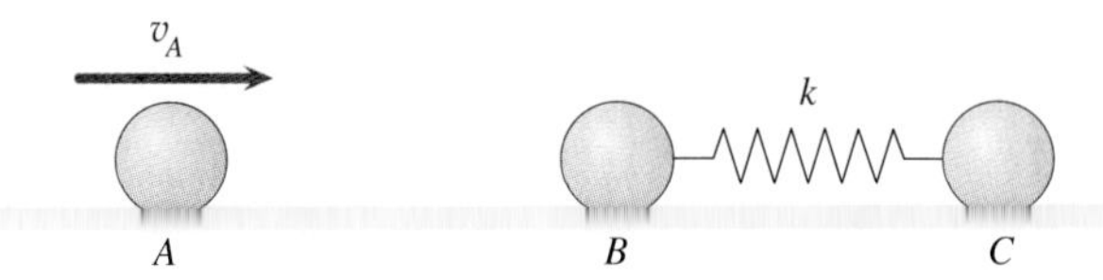

P5.121

5.122 In Problem 5.121, what is the maximum compressive force in the spring as a result of the impact?

5.123 Suppose you interpret Problem 5.121 as an impact between the ball A and an 'object' D consisting of the connected balls B and C.
(a) What is the coefficient of restitution of the impact between A and D?
(b) If you consider the total energy after the impact to be the sum of the kinetic energies $\frac{1}{2}m(v_A')^2 + \frac{1}{2}(2m)(v_D')^2$, where v_D' is the velocity of the centre of mass of D after the impact, how much energy is 'lost' as a result of the impact?
(c) How much energy is actually lost as a result of the impact? (This problem is an interesting model for one of the mechanisms of energy loss in impacts between objects. The energy 'loss' calculated in part (b) is transformed into 'internal energy' – the vibrational motions of B and C relative to their common centre of mass.)

5.124 A small object starts from rest at A and slides down the smooth ramp. The coefficient of restitution of its impact with the floor is $e = 0.8$. At what height above the floor does it hit the wall?

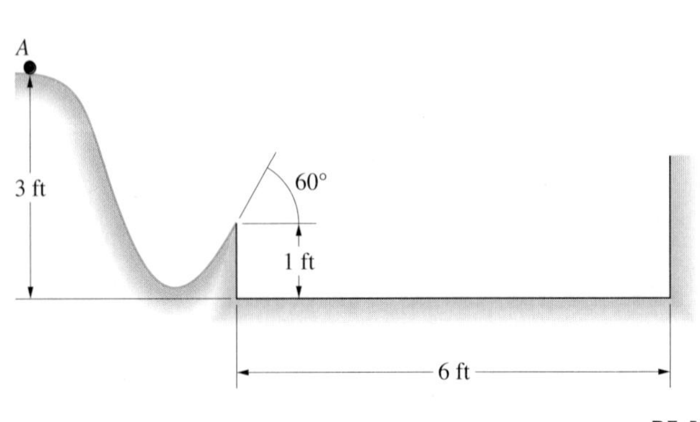

P5.124

5.125 A basketball dropped on the floor from a height of 1.2 m rebounds to a height of 0.9 m. In the 'lay-up' shot shown, the magnitude of the ball's velocity is 1.5 m/s and the angles between its velocity vector and the positive coordinate axes are $\theta_x = 42°$, $\theta_y = 68°$ and $\theta_z = 124°$ just before it hits the backboard. What are the magnitude of its velocity and the angles between its velocity vector and the positive coordinate axes just after it hits the backboard?

P5.125

5.126 In Problem 5.125, the basketball's diameter is 240 mm, the coordinates of the centre of the basket rim are $x = 0, y = 0, z = 300$ mm, and the backboard lies in the *x-y* plane. Determine the *x* and *y* coordinates of the point where the ball must hit the backboard so that the centre of the ball passes through the centre of the basket rim.

5.127 The snow is 0.6 m deep and weighs 3150 N/m^3, the snowplough is 2.4 m wide, and the truck travels at 8 km/hr. What force does the snow exert on the truck?

P5.127

5.128 An empty 25 kg drum, 1 m in diameter, stands on a set of scales. Water begins pouring into the drum at 550 kg/min from 2.4 m above the bottom of the drum. The density of water is approximately 1000 kg/m^3. What do the scales read 40 s after the water starts pouring?

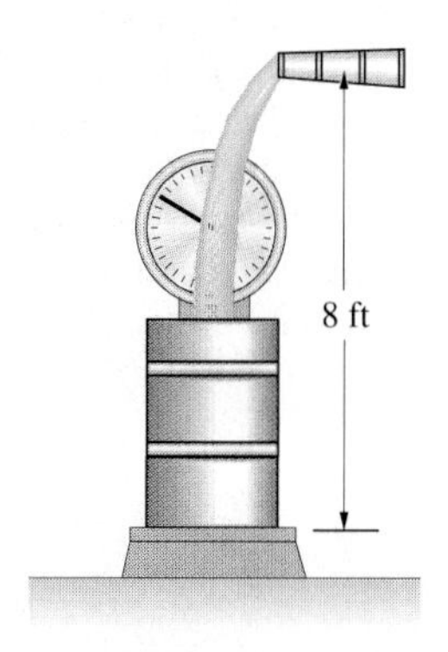

P5.128

5.129 The ski boat's jet propulsive system draws water in at *A* and expels it at *B* at 25 m/s relative to the boat. Assume that the water drawn in enters with no horizontal velocity relative to the surrounding water. The maximum mass flow rate of water through the engine is 36 kg/s. Hydrodynamic drag exerts a force on the boat of magnitude $24v$ N, where v is the boat's velocity in metres per second. If you neglect aerodynamic drag, what is the ski boat's maximum velocity?

P5.129

5.130 The ski boat in Problem 5.129 weighs 12.5 kN. The mass flow rate of water through its engine is 36 kg/s, and it starts from rest at $t = 0$. Determine the boat's velocity (a) at $t = 20$ s; (b) at $t = 60$ s.

5.131 A crate of mass m slides across the smooth floor pulling chain from a stationary pile. The mass per unit length of the chain is ρ_L. If the velocity of the crate is v_0 when $s = 0$, what is its velocity as a function of s?

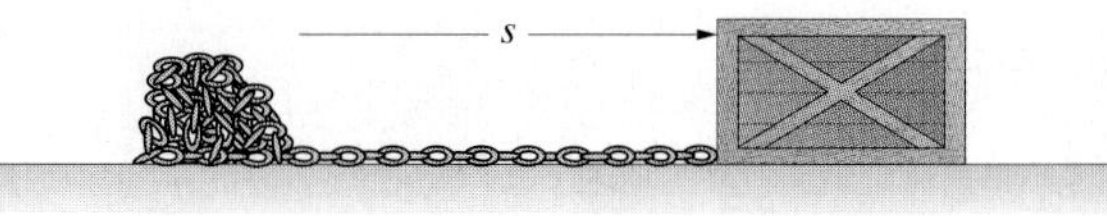

P5.131

Project 5.1 By making measurements, determine the coefficient of restitution of a tennis ball bouncing on a rigid surface. Try to determine whether your result is independent of the velocity with which the ball strikes the surface. Write a brief report describing your procedure and commenting on possible sources of error.

The gear that is engaged determines the ratio of the angular velocity of the pedals and sprocket to that of the bicycle's rear wheel. The ratio of the sprocket's radius to that of the gear equals the ratio of the wheel's angular velocity to that of the sprocket. In this chapter you will obtain results of this kind by modelling objects as rigid bodies.

Chapter 6

Planar Kinematics of Rigid Bodies

UNTIL now, we have considered situations in which you could determine the motion of an object's centre of mass by using Newton's second law alone. But you must often determine an object's rotational motion as well, even when your only objective is to determine the motion of its centre of mass. Moreover, the rotational motion itself can be of interest or even central to the situation you are considering, as in the motions of gears, wheels, generators, turbines and gyroscopes.

In this chapter we discuss the **kinematics** of objects, the description and analysis of the motions of objects without consideration of the forces and couples that cause them. In particular, we show how the motions of individual points of an object are related to the object's angular motion.

6.1 *Rigid Bodies and Types of Motion*

If you throw a brick (Figure 6.1(a)), you can determine the motion of its centre of mass without having to be concerned about its rotational motion. The only significant force is its weight, and Newton's second law determines the acceleration of its centre of mass. But suppose that the brick is standing on the floor, you tip it over (Figure 6.1(b)), and you want to determine the motion of its centre of mass as it falls. In this case, the brick is subjected to its weight and also a force exerted by the floor. You cannot determine the force exerted by the floor, and the motion of the brick's centre of mass, without also analysing its rotational motion.

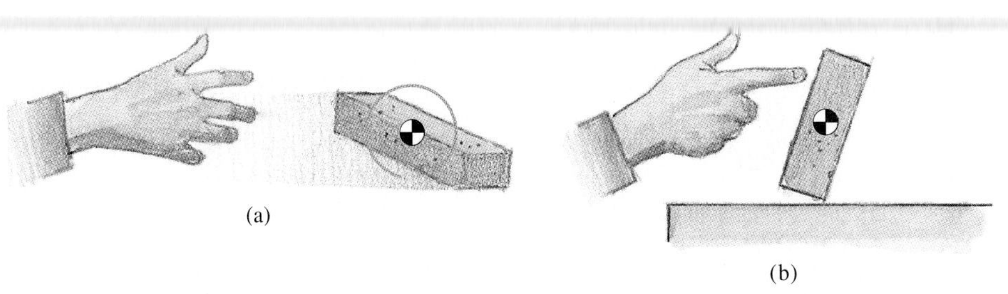

Figure 6.1
(a) A thrown brick – its rotation doesn't affect the motion of its centre of mass.
(b) A tipped brick – the rotation and the motion of the centre of mass are interrelated.

Before we can analyse such motions, we must consider how to describe them. A brick is an example of an object whose motion can be described by treating it as a rigid body. A **rigid body** is an idealized model of an object that does not deform, or change shape. The precise definition is that the distance between every pair of points of a rigid body remains constant. Although any object does deform as it moves, if its deformation is small *you can approximate its motion by modelling it as a rigid body*. For example, you can model a twirler's baton in normal use as a rigid body (Figure 6.2(a)), but not a fly-casting rod (Figure 6.2(b)).

To describe a rigid body's motion, it is sufficient to describe the motion of a single point, such as the centre of mass, and the rigid body's rotational motion about that point. Some particular types of motion occur frequently in applications. To help you visualize them, we use a coordinate system that is fixed relative to the rigid body and so moves with it. Such a coordinate system is said to be **body-fixed**.

Figure 6.2
(a) A baton can be modelled as a rigid body.
(b) A fishing rod is too flexible under normal use to model as a rigid body.

Translation If a rigid body in motion does not rotate, it is said to be in **translation**. Every point of a rigid body in translation has the same velocity and acceleration, so you completely describe the motion of the rigid body if you describe the motion of a single point. The point may move in a straight line, or it may undergo curvilinear motion. The directions of the axes of a body-fixed coordinate system remain constant (Figure 6.3(a)). For example, the child's swing in Figure 6.3(b) is designed to translate so that it will be easier to ride. Each point of the swing moves in a circular path, but the swing does not rotate – it remains level.

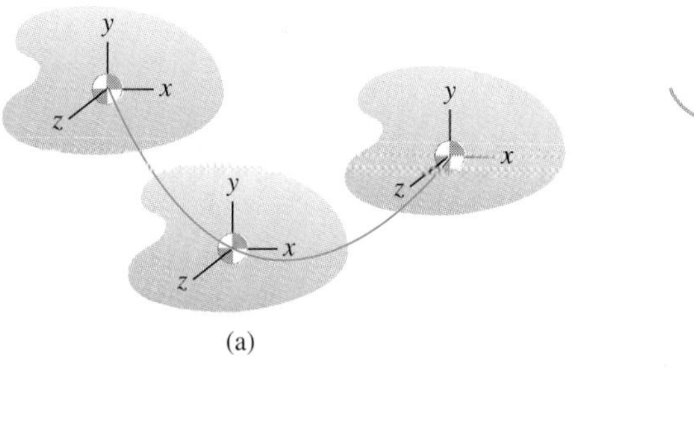

Figure 6.3

(a) An object in translation does not rotate.

(b) The translating swing remains level.

Rotation About a Fixed Axis After translation, the simplest type of rigid-body motion is rotation about a fixed axis. For example, in Figure 6.4(a) the z axis of the body-fixed coordinate system remains fixed and the x and y axes rotate about the z axis. Each point of the rigid body not on the axis moves in a circular path about the axis. A disc in a compact disc player and the rotor of an electric motor (Figure 6.4(b)) are examples of objects rotating about a fixed axis. The motion of a ship's propeller *relative to the ship* is also rotation about a fixed axis. We discuss this type of motion in more detail in the next section.

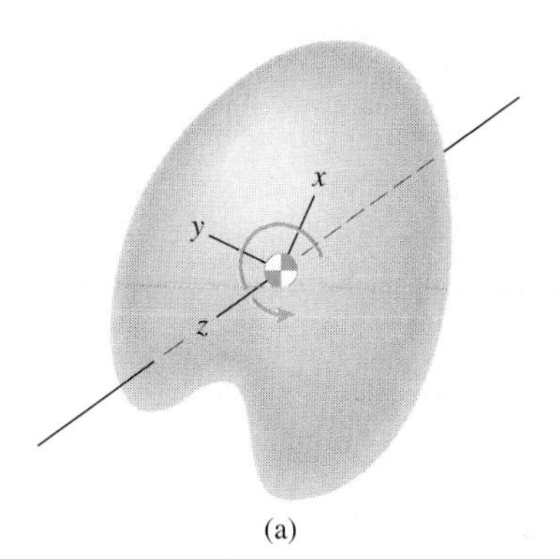

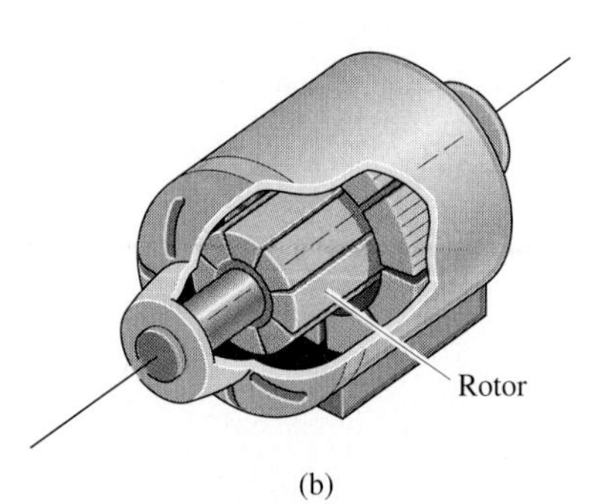

Figure 6.4

(a) A rigid body rotating about the z axis.

(b) If the motor's frame is stationary, its rotor rotates about a fixed axis.

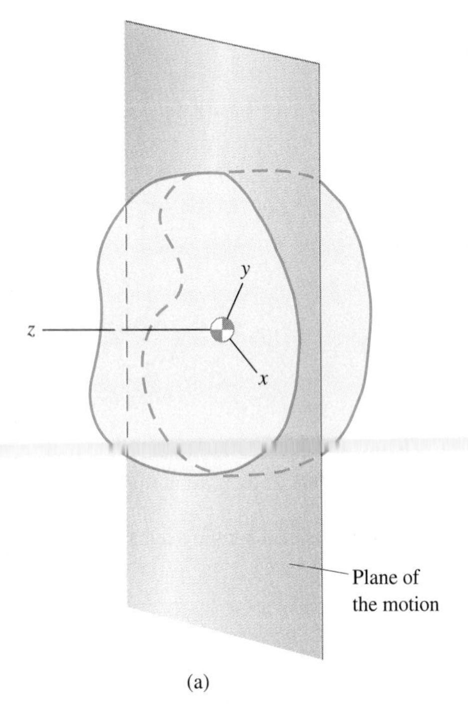

(a)

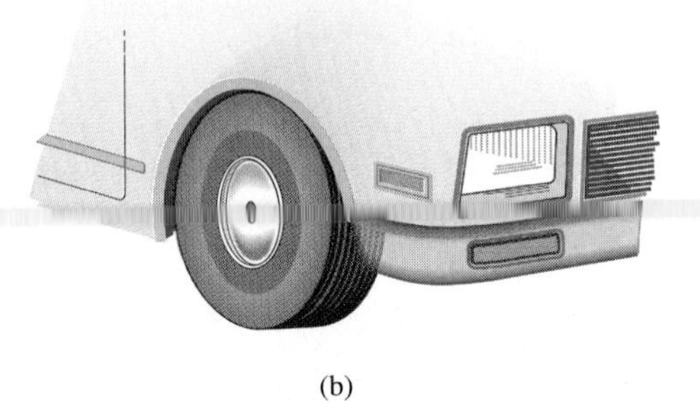

(b)

Figure 6.5

(a) Two-dimensional, or planar, motion.
(b) A wheel in planar motion.

Two-Dimensional Motion In this chapter we are concerned with two-dimensional motions of rigid bodies. A rigid body is said to undergo **two-dimensional**, or **planar**, motion if its centre of mass moves in a fixed plane and an axis of a body-fixed coordinate system remains perpendicular to the plane (Figure 6.5(a)). We refer to the fixed plane as the **plane of the motion**. Rotation of a rigid body about a fixed axis is a special case of two-dimensional motion. When a car moves in a straight path, its wheels are in two-dimensional motion (Figure 6.5(b)).

The components of an internal combustion engine running on a test stand illustrate these three types of motion (Figure 6.6). The pistons translate within the cylinders. The connecting rods are in two-dimensional motion, and the crankshaft rotates about a fixed axis.

We begin our analysis of rigid-body motion in the next section with a discussion of rotation about a fixed axis. In this type of motion, points of the rigid body move in circular paths about the fixed axis. We can therefore use results developed in Chapter 2 for the motion of a point in a circular path. Using normal and tangential components, we express the velocity and acceleration of a point of the rigid body in terms of the rigid body's angular velocity and angular acceleration. In the following sections we consider general two-dimensional motion and obtain expressions relating the relative velocity and acceleration of points of a rigid body to its angular velocity and angular acceleration. With these relations we analyse particular examples of general two-dimensional motion, such as rolling, and also analyse motions of connected rigid bodies.

Figure 6.6

Translation, rotation about a fixed axis, and planar motion in an automobile engine.

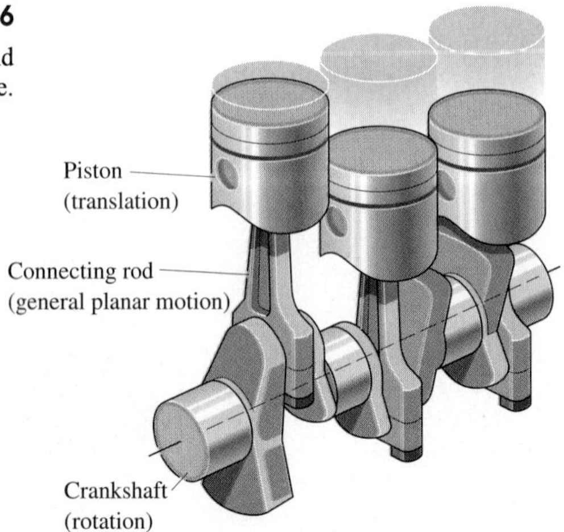

6.2 Rotation About a Fixed Axis

We can introduce some of the concepts involved in describing the motion of a rigid body by first considering an object rotating about a fixed axis. Consider a body-fixed straight line within such an object that is perpendicular to the fixed axis. To describe the object's position, or **orientation** about the fixed axis, we specify the angle θ between this body-fixed line and a reference direction (Figure 6.7). The object's **angular velocity** ω, or rate of rotation, and its **angular acceleration** α are

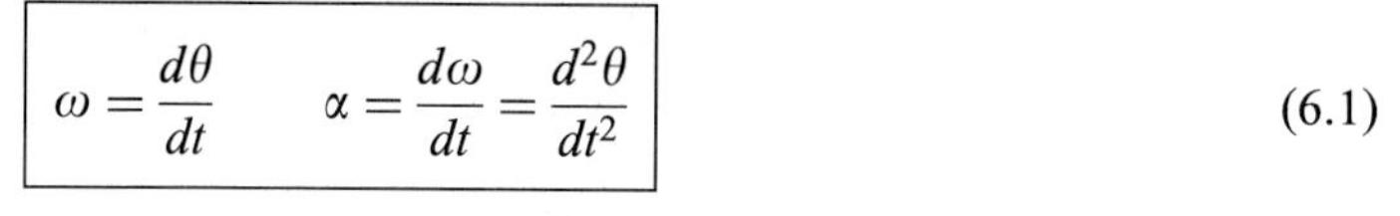

$$\omega = \frac{d\theta}{dt} \qquad \alpha = \frac{d\omega}{dt} = \frac{d^2\theta}{dt^2} \tag{6.1}$$

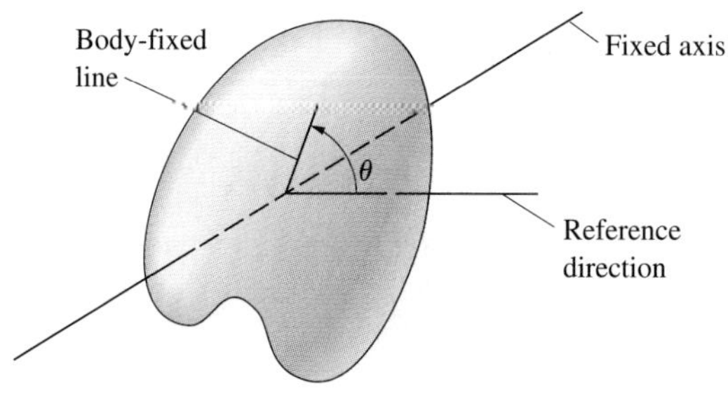

Figure 6.7
Specifying the orientation of an object rotating about a fixed axis.

Each point of the object not on the fixed axis moves in a circular path about the axis. Using our knowledge of a point in a circular path, we can relate the velocity and acceleration of the motion of a point to the object's angular velocity and angular acceleration. In Figure 6.8, we view the object in the direction parallel to the fixed axis. The velocity of a point at a distance r from the fixed axis is tangent to its circular path (Figure 6.8(a)) and is given in terms of the angular velocity of the object by

$$v = r\omega \tag{6.2}$$

A point has components of acceleration tangential and normal to its circular path (Figure 6.8(b)). In terms of the angular velocity and angular acceleration of the object, the components are given by

$$a_{\mathrm{t}} = r\alpha \qquad a_{\mathrm{n}} = \frac{v^2}{r} = r\omega^2 \tag{6.3}$$

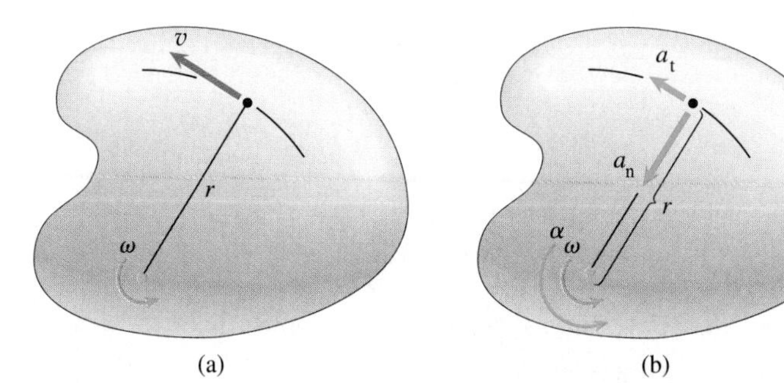

Figure 6.8
(a) Velocity and (b) acceleration of a point of a rigid body rotating about a fixed axis.

With these relations we can analyse problems involving objects rotating about fixed axes. For example, suppose that we know the angular velocity ω_A and angular acceleration α_A of the left gear in Figure 6.9, and we want to determine ω_B and α_B. Because the velocities of the gears must be equal at P (there is no relative motion between them at P),

$$r_A \omega_A = r_B \omega_B$$

so $\omega_B = (r_A/r_B)\omega_A$. Then, either by taking the time derivative of this equation or by equating the tangential components of acceleration at P, we obtain $\alpha_B = (r_A/r_B)\alpha_A$.

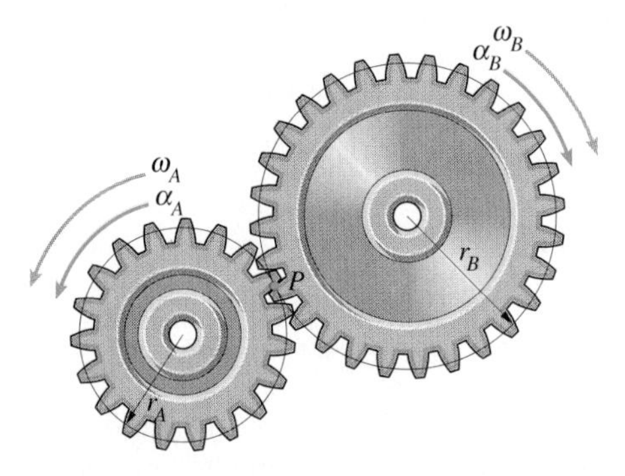

Figure 6.9
Relating the angular velocities and angular accelerations of meshing gears.

In the following example we demonstrate the analysis of motions of objects rotating about fixed axes. You can use Equations (6.1) to analyse the angular motions and use Equations (6.2) and (6.3) to determine the velocities and accelerations of points.

Example 6.1

Gear A of the winch in Figure 6.10 turns gear B, raising the hook H. If the gear A starts from rest at $t = 0$ and its clockwise angular acceleration is $\alpha_A = 0.2t \text{ rad/s}^2$, what vertical distance has the hook H risen and what is its velocity at $t = 10 \text{ s}$?

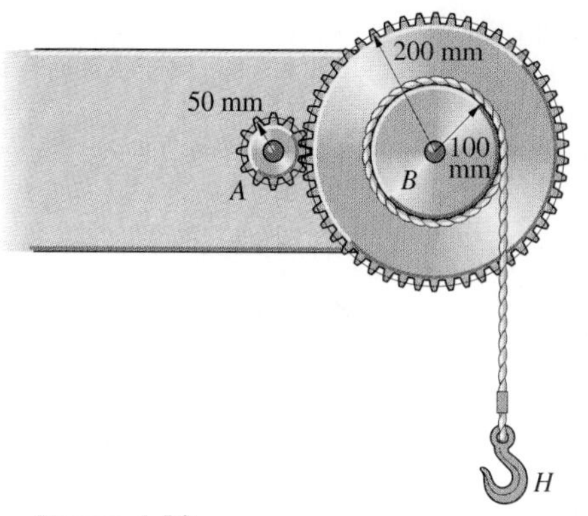

Figure 6.10

STRATEGY

By equating the tangential components of acceleration of gears A and B at their point of contact, we can determine the angular acceleration of gear B. Then we can integrate to obtain the angular velocity of gear B and the angle through which it has turned at $t = 10$ s.

SOLUTION

The tangential acceleration of the point of contact of the two gears (Figure (a)) is

$$a_t = (0.05 \text{ m})(0.2t \text{ rad/s}^2) = (0.2 \text{ m})(\alpha_B)$$

Therefore the angular acceleration of gear B is

$$\alpha_B = \frac{d\omega_B}{dt} = \frac{(0.05 \text{ m})(0.2t \text{ rad/s}^2)}{(0.2 \text{ m})} = 0.05t \text{ rad/s}^2$$

Integrating this equation,

$$\int_0^{\omega_B} d\omega_B = \int_0^t 0.05t\, dt$$

we obtain the angular velocity of gear B:

$$\omega_B = \frac{d\theta_B}{dt} = 0.025t^2 \text{ rad/s}$$

Integrating again, we obtain the angle through which gear B has turned:

$$\theta_B = 0.00833t^3 \text{ rad}$$

At $t = 10$ s, $\theta_B = 8.33$ rad. The amount of cable wound around the drum, which is the distance the hook H has risen, is the product of θ_B and the radius of the drum: $(8.33 \text{ rad})(0.1 \text{ m}) = 0.833 \text{ m}$.

At $t = 10$ s, $\omega_B = 2.5$ rad/s. The velocity of a point on the rim, which equals the velocity of the hook H (Figure (b)), is

$$v_H = (0.1 \text{ m})(2.5 \text{ rad/s}) = 0.25 \text{ m/s}$$

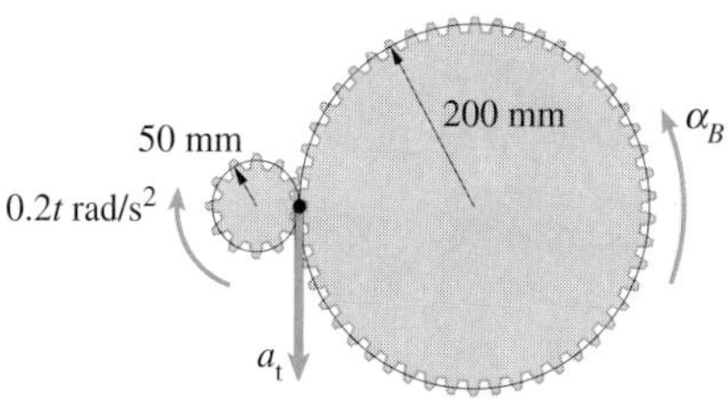

(a) The tangential accelerations of the gears are equal at their point of contact.

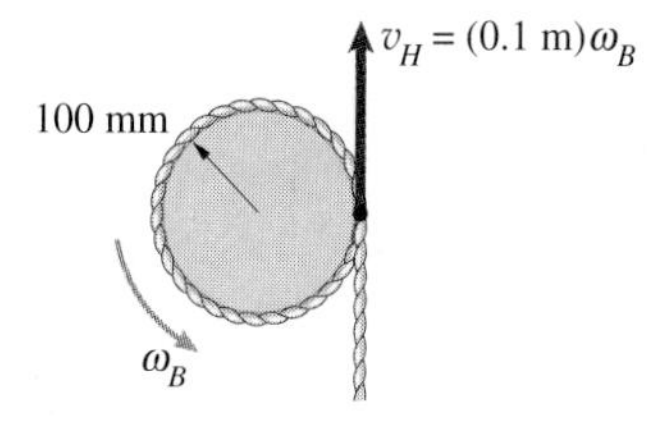

(b) Determining the hook's velocity.

Problems

6.1 The disc rotates about the fixed shaft O. It starts from rest at $t = 0$ and has constant counterclockwise angular acceleration $\alpha = 4\ \mathrm{rad/s^2}$. At $t = 5$ s, determine (a) its angular velocity and the number of revolutions it has turned; (b) the magnitudes of the velocity and acceleration of point A.

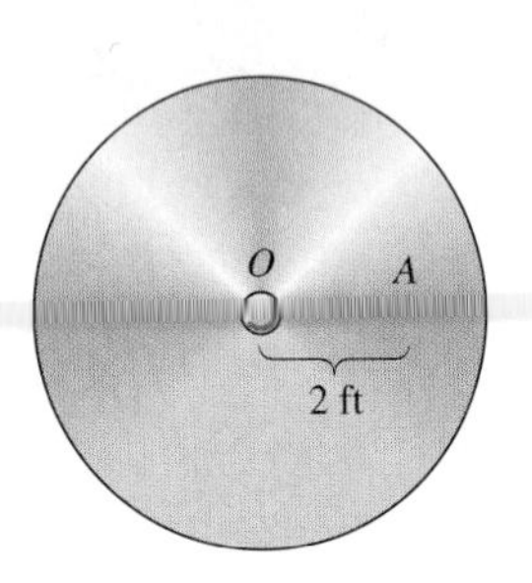

P6.1

6.2 The weight A starts from rest at $t = 0$ and falls with a constant acceleration of $2\ \mathrm{m/s^2}$, causing the disc to turn.
(a) What is the angular acceleration of the disc?
(b) How many revolutions has the disc turned at $t = 1$ s?

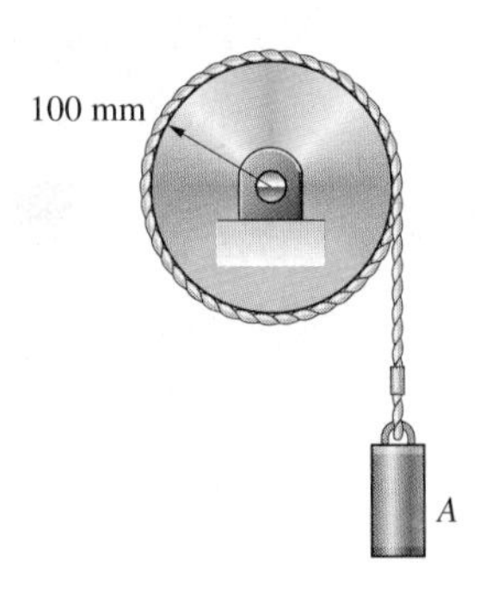

P6.2

6.3 Determine ω_B/ω_A and ω_C/ω_A.

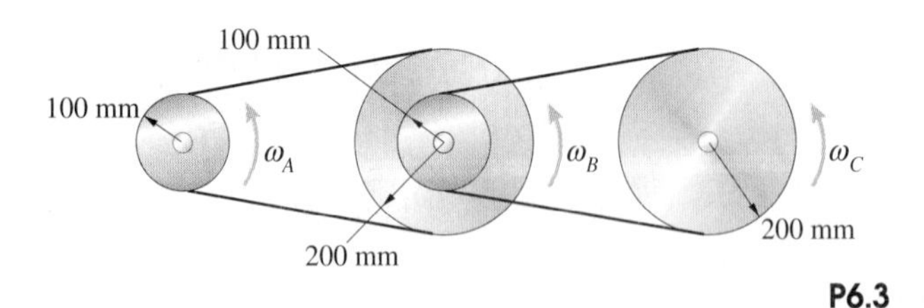

P6.3

6.4 The bicycle's 120 mm sprocket wheel turns at 3 rad/s. What is the angular velocity of the 45 mm gear?

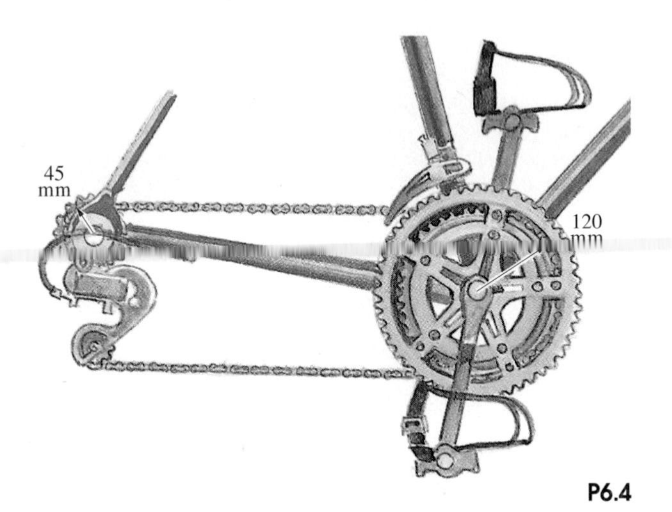

P6.4

6.5 The rear wheel of the bicycle in Problem 6.4 has a 330 mm radius and is rigidly attached to the 45 mm gear. If the rider turns the pedals, which are rigidly attached to the 120 mm sprocket wheel, at one revolution per second, what is the bicycle's velocity?

6.6 The disc rotates with a constant counterclockwise angular velocity of 10 rad/s. What are the velocity and acceleration of point A in terms of the coordinate system shown?

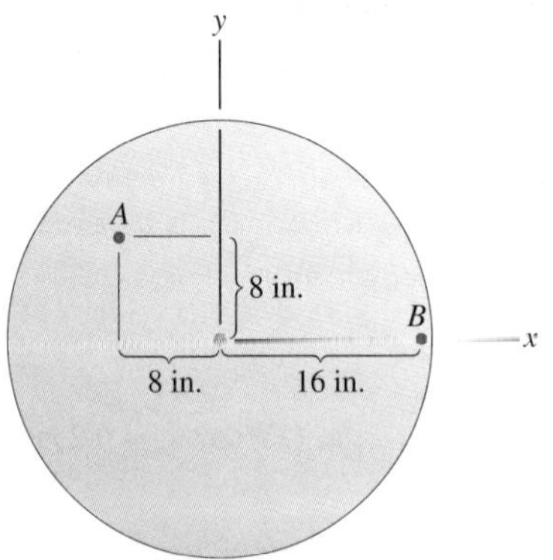

P6.6

6.7 In Problem 6.6, what are the velocity and acceleration of point A relative to point B?

6.8 In Problem 6.6, suppose that the disc starts from rest in the position shown at $t = 0$ and is subjected to a constant counterclockwise angular acceleration of $6\ \mathrm{rad/s^2}$. Determine the velocity of point B in terms of the coordinate system shown at $t = 1$ s if the coordinate system (a) is body-fixed; (b) remains oriented with the axes horizontal and vertical as shown.

6.9 The bracket rotates around the fixed shaft at O. If it has a counterclockwise angular velocity of 20 rad/s and a clockwise angular acceleration of 200 rad/s^2, what are the magnitudes of the accelerations of points A and B?

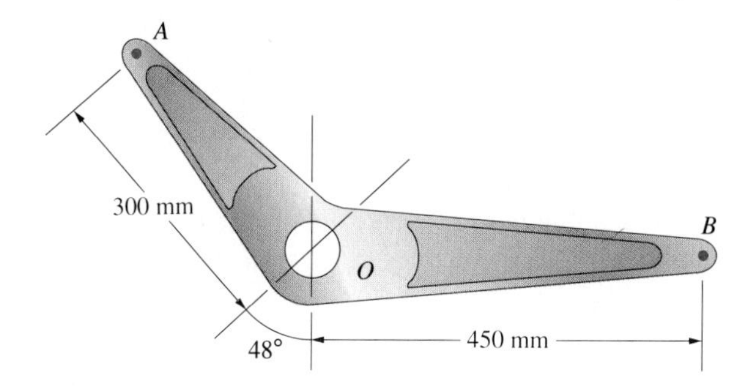

P6.9

6.10 Consider the bracket in Problem 6.9. If $|\mathbf{v}_A| = 3$ m/s $|\mathbf{a}_A| = 60$ m/s^2, what are $|\mathbf{v}_B|$ and $|\mathbf{a}_B|$?

6.11 Consider the bracket in Problem 6.9. If $|\mathbf{v}_A| = 0.9$ m/s and $|\mathbf{a}_B| = 15$ m/s^2, what are $|\mathbf{v}_B|$ and $|\mathbf{a}_A|$?

6.3 *General Motions: Velocities*

Each point of a rigid body in translation undergoes the same motion. Each point of a rigid body rotating about a fixed axis undergoes circular motion about the axis. To analyse more complicated motions that combine translation and rotation, we must develop equations that relate the relative motions of points of a rigid body to its angular motion.

Relative Velocities

In Figure 6.11(a) we view a rigid body perpendicular to the plane of its motion. Points A and B are points of the rigid body contained in that plane, and O is a reference point. We can show that the velocity of A relative to B is related in a simple way to the rigid body's angular velocity. The position of A relative to B, $\mathbf{r}_{A/B}$, is related to the positions of the points relative to O by

$$\mathbf{r}_A = \mathbf{r}_B + \mathbf{r}_{A/B}$$

Taking the time derivative of this equation, we obtain

$$\mathbf{v}_A = \mathbf{v}_B + \mathbf{v}_{A/B} \tag{6.4}$$

where $\mathbf{v}_{A/B} = d\mathbf{r}_{A/B}/dt$ is the velocity of A relative to B. Since A and B are points of the rigid body, the distance between them, $|\mathbf{r}_{A/B}|$, is constant. That means that *A moves in a circular path relative to B as the rigid body rotates* (Figure 6.11(b)). The velocity of A relative to B is tangent to the circular path, and its value equals the product of $|\mathbf{r}_{A/B}|$ and the angular velocity ω of the rigid body. You can use this result to relate velocities of points of a rigid body in two-dimensional motion when you know its angular velocity.

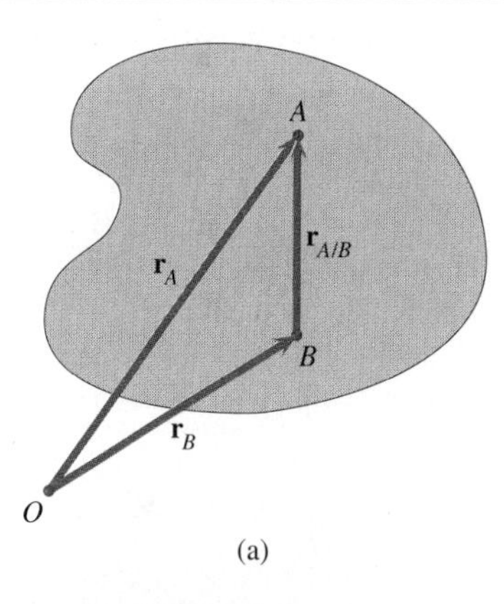

(a)

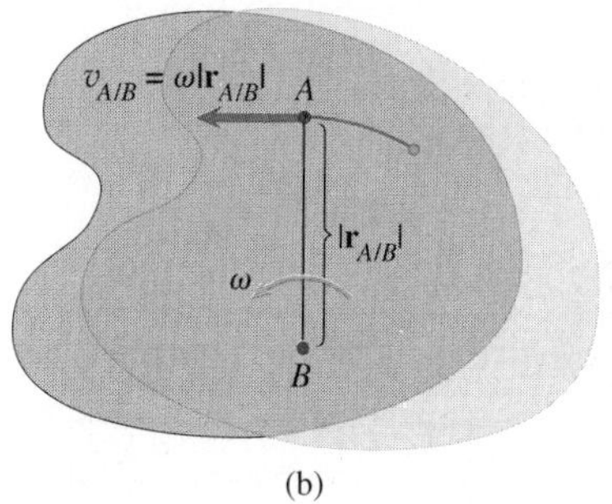

(b)

Figure 6.11
(a) A rigid body in two-dimensional motion.
(b) The motion viewed by an 'observer' stationary with respect to B.

For example, let's consider a circular disc of radius R rolling on a stationary plane surface with counterclockwise angular velocity ω (Figure 6.12(a)). By **rolling**, we mean that the velocity of the disc relative to the surface is zero at their point of contact C. Let B be the centre of the disc. *Relative to C*, point B moves in a circular path of radius R (Figure 6.12(b)). In terms of the coordinate system shown, the velocity of B relative to C is $\mathbf{v}_{B/C} = -R\omega\,\mathbf{i}$. Since the velocity of C is zero, the velocity of B is

$$\mathbf{v}_B = \mathbf{v}_C + \mathbf{v}_{B/C} = -R\omega\,\mathbf{i}$$

This result is worth remembering: *the magnitude of the velocity of the centre of a round object rolling on a stationary surface is the product of the radius and the magnitude of the angular velocity.*

We can determine the velocity of any other point of the disc in the same way. Consider the point A in Figure 6.12(c). Relative to the centre B, point A moves in a circular path of radius R, resulting in the relative velocity $\mathbf{v}_{A/B} = R\omega\,\mathbf{j}$ (Figure 6.12(d)). Therefore the velocity of A is

$$\mathbf{v}_A = \mathbf{v}_B + \mathbf{v}_{A/B} = -R\omega\,\mathbf{i} + R\omega\,\mathbf{j}$$

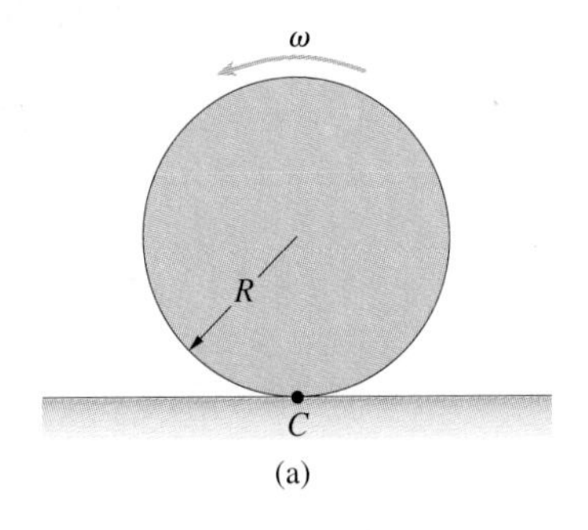

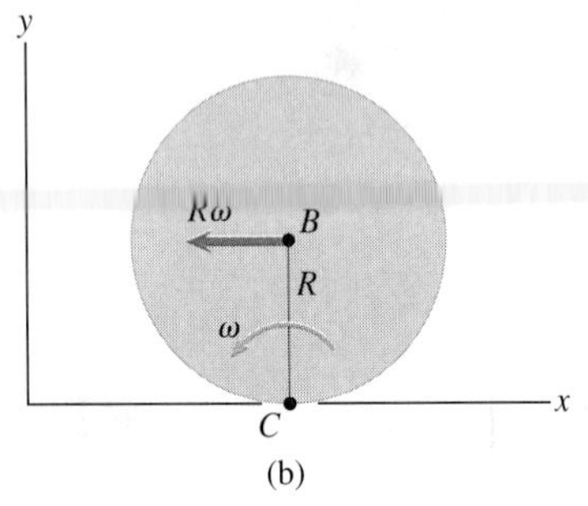

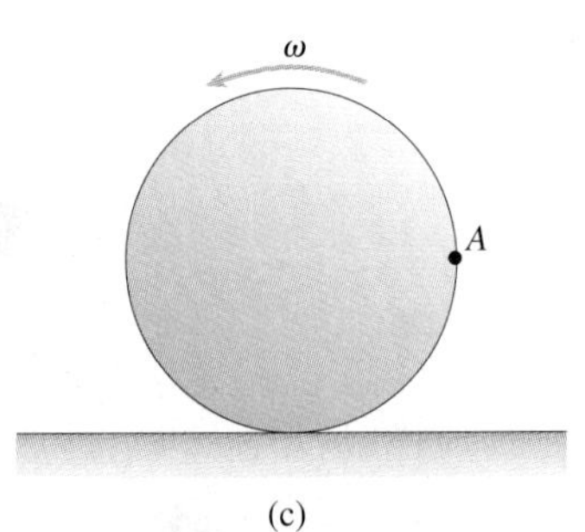

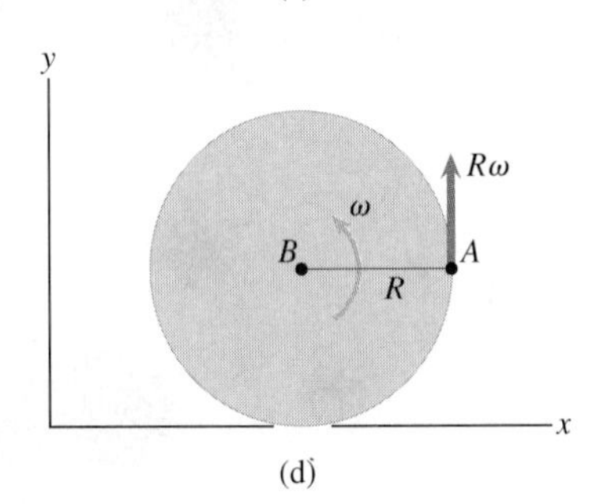

Figure 6.12
(a) A disk rolling with angular velocity ω.
(b) Determining the velocity of the centre B relative to C.
(c) A point A on the rim of the disk.
(d) Determining the velocity of A relative to B.

The Angular Velocity Vector

We can express the rate of rotation of a rigid body as a vector. **Euler's theorem** states that a rigid body constrained to rotate about a fixed point B can move between any two positions by a single rotation about some axis through B. Suppose that we choose an arbitrary point B of a rigid body that is undergoing an *arbitrary* motion at a time t. Euler's theorem allows us to express the rigid body's change in position relative to B during an interval of time from t to $t + dt$ as a single rotation through an angle $d\theta$ about some axis. At time t the rigid body's rate of rotation about the axis is its angular velocity $\omega = d\theta/dt$, and the axis about which it rotates is called the **instantaneous axis of rotation**.

The **angular velocity vector**, denoted by $\boldsymbol{\omega}$, specifies both the direction of the instantaneous axis of rotation and the angular velocity. It is defined to be parallel to the instantaneous axis of rotation (Figure 6.13(a)), and its magnitude is the rate of rotation, the absolute value of ω. Its direction is related to the direction of the rigid body's rotation through a right-hand rule: if you point the thumb of your right hand in the direction of $\boldsymbol{\omega}$, the fingers curl around $\boldsymbol{\omega}$ in the direction of the rotation (Figure 6.13(b)).

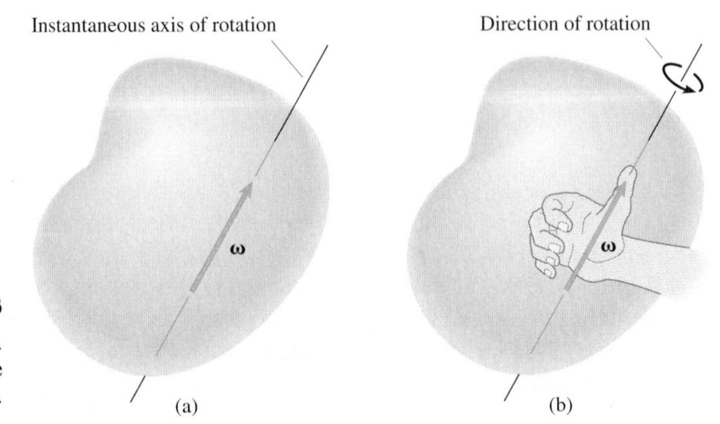

Figure 6.13
(a) An angular velocity vector.
(b) Right-hand rule for the direction of the vector.

For example, the axis of rotation of the rolling disc in Figure 6.12 is parallel to the z axis, so its angular velocity vector is parallel to the z axis and its magnitude is ω. If you curl the fingers of your right hand around the z axis in the direction of the rotation, your thumb points in the positive z direction (Figure 6.14). The angular velocity vector of the disc is $\boldsymbol{\omega} = \omega\,\mathbf{k}$.

The angular velocity vector allows us to express the results of the previous section in a very convenient form. Let A and B be points of a rigid body with angular velocity $\boldsymbol{\omega}$ (Figure 6.15(a)). We can show that the velocity of A relative to B is

$$\mathbf{v}_{A/B} = \frac{d\mathbf{r}_{A/B}}{dt} = \boldsymbol{\omega} \times \mathbf{r}_{A/B} \tag{6.5}$$

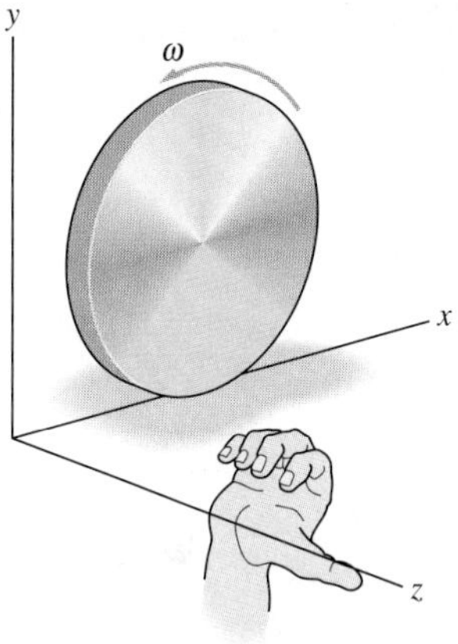

Figure 6.14
Determining the direction of the angular velocity vector of a rolling disk.

Relative to B, point A is moving at the present instant in a circular path of radius $|\mathbf{r}_{A/B}| \sin\beta$, where β is the angle between the vectors $\mathbf{r}_{A/B}$ and $\boldsymbol{\omega}$ (Figure 6.15(b)). The magnitude of the velocity of A relative to B is equal to the product of the radius of the circular path and the angular velocity of the rigid body, $|\mathbf{v}_{A/B}| = (|\mathbf{r}_{A/B}| \sin\beta)|\boldsymbol{\omega}|$, which is the magnitude of the cross product of $\mathbf{r}_{A/B}$ and $\boldsymbol{\omega}$. In addition, $\mathbf{v}_{A/B}$ is perpendicular to $\boldsymbol{\omega}$ and perpendicular to $\mathbf{r}_{A/B}$. But is $\mathbf{v}_{A/B}$ equal to $\boldsymbol{\omega} \times \mathbf{r}_{A/B}$ or $\mathbf{r}_{A/B} \times \boldsymbol{\omega}$? Notice in Figure 6.15(b) that, pointing the fingers of the right hand in the direction of $\boldsymbol{\omega}$ and closing them towards $\mathbf{r}_{A/B}$, the thumb points in the direction of the velocity of A relative to B, so $\mathbf{v}_{A/B} = \boldsymbol{\omega} \times \mathbf{r}_{A/B}$. Substituting Equation (6.5) into Equation (6.4), we obtain an equation for the relation between the velocities of two points of a rigid body in terms of its angular velocity:

$$\mathbf{v}_A = \mathbf{v}_B + \underbrace{\boldsymbol{\omega} \times \mathbf{r}_{A/B}}_{\mathbf{v}_{A/B}} \tag{6.6}$$

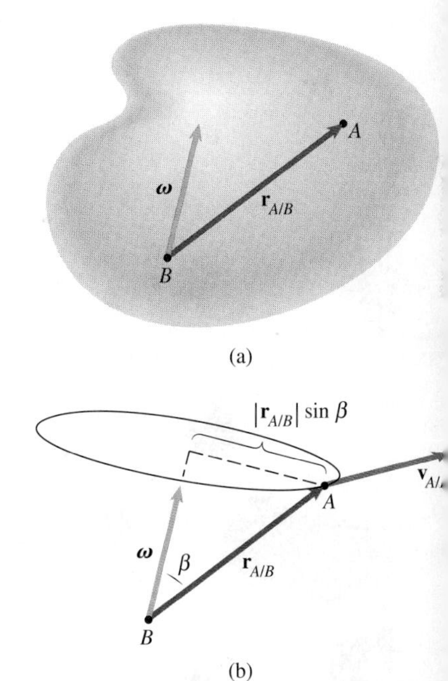

Figure 6.15
(a) Points A and B of a rotating rigid body.
(b) A is moving in a circular path relative to B.

Let's return to the example of a disc of radius R rolling with angular velocity ω (Figure 6.16), and use Equation (6.6) to determine the velocity of point A. The velocity of the centre of the disc is given in terms of its angular velocity by $\mathbf{v}_B = -R\omega\,\mathbf{i}$, the disc's angular velocity vector is $\boldsymbol{\omega} = \omega\,\mathbf{k}$, and the position vector of A relative to the centre is $\mathbf{r}_{A/B} = R\,\mathbf{i}$. The velocity of A is

$$\begin{aligned}\mathbf{v}_A = \mathbf{v}_B + \boldsymbol{\omega} \times \mathbf{r}_{A/B} &= -R\omega\,\mathbf{i} + (\omega\,\mathbf{k}) \times (R\,\mathbf{i}) \\ &= -R\omega\,\mathbf{i} + R\omega\,\mathbf{j}\end{aligned}$$

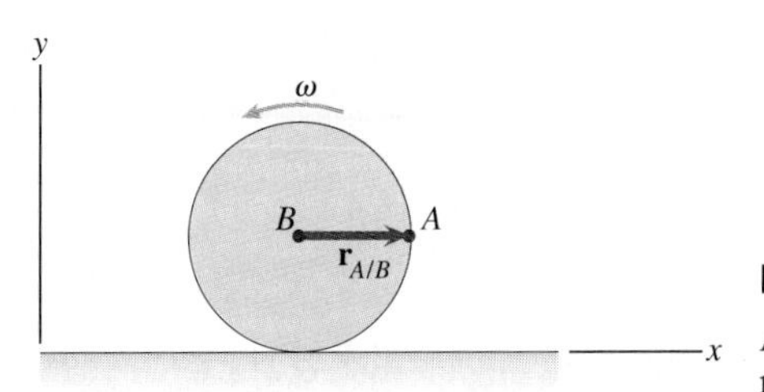

Figure 6.16
A rotating disc and the position vector of A relative to B.

The following examples show how you can apply Equation (6.6). Beginning with a point whose velocity is known, you can express the velocities of other points of a rigid body in terms of its angular velocity. By repeating this step for different points, you can analyse the motions of systems of connected rigid bodies.

Example 6.2

If the velocity v_R in Figure 6.17 is 900 mm/s, what is the velocity v_L?

Figure 6.17

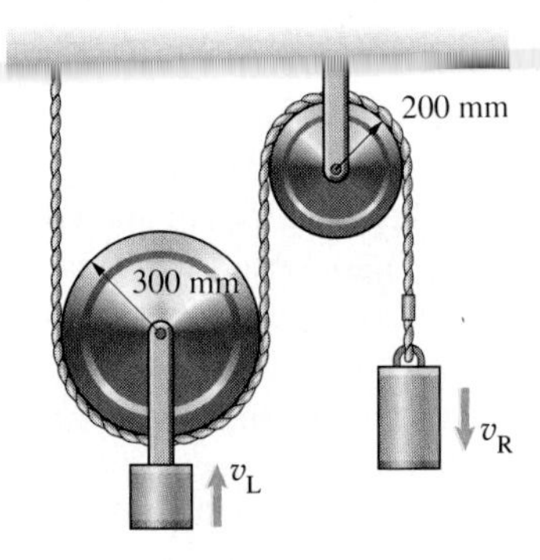

STRATEGY

The centre of the right pulley is fixed, so the vertical part of the cable between the two pulleys moves upwards with velocity v_R. The point of the left pulley in contact with that part of the cable moves upwards with the same velocity. The vertical part of the cable connected to the ceiling is stationary, so the point of the left pulley in contact with that part of the cable is also stationary. Thus we know the velocities of two points of the left pulley. Using this information, we can determine its angular velocity and then determine the velocity of its centre, which is equal to the velocity v_L.

SOLUTION

The velocity of point A of the left pulley in Figure (a) is $v_A = 900\,\text{mm/s}$ and the velocity of point B is zero. Relative to B, point A moves in a circular path with the angular velocity ω of the left pulley, so

$$v_A = 900\,\text{mm/s} = (600\,\text{mm})\omega$$

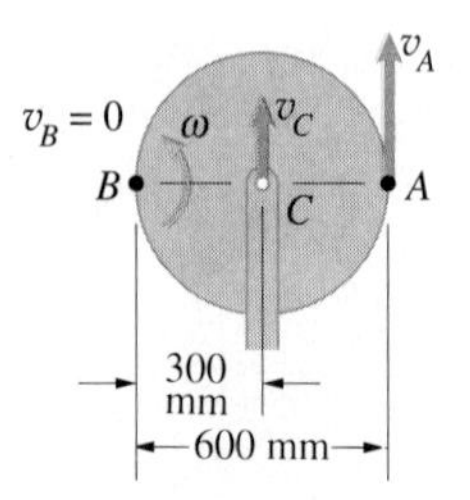

(a) Analysing the motion of the left pulley.

and the angular velocity of the left pulley is $\omega = 900/600 = 1.5\,\text{rad/s}$. Point C, the centre of the pulley, also moves relative to B in a circular path with angular velocity ω, so

$$v_C = v_L = (300\,\text{mm})\omega = (300)(1.5) = 450\,\text{mm/s}$$

We can also obtain this result with Equation (6.6). In terms of the coordinate system in Figure (b), $\mathbf{v}_A = 900\,\mathbf{j}\,\text{mm/s}$, $\mathbf{r}_{A/B} = 600\,\text{mm}$, and the angular velocity vector of the pulley is $\boldsymbol{\omega} = \omega\,\mathbf{k}$. Therefore

$$\mathbf{v}_A = \mathbf{v}_B + \boldsymbol{\omega} \times \mathbf{r}_{A/B}:$$

$$900\,\mathbf{j} = 0 + (\omega\,\mathbf{k}) \times (600\,\mathbf{i}) = 600\omega\,\mathbf{j}$$

From this equation we obtain $\omega = 900/600 = 1.5\,\text{rad/s}$. The velocity of the centre of the pulley is

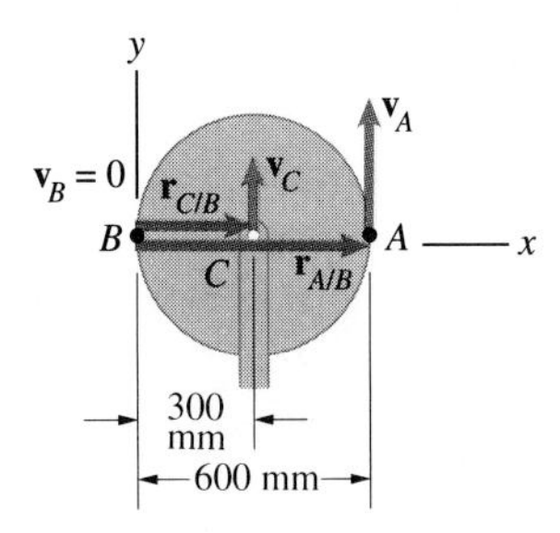

(b) Position vectors $\mathbf{r}_{A/B}$ and $\mathbf{r}_{C/B}$.

$$\begin{aligned}\mathbf{v}_C &= \mathbf{v}_B + \boldsymbol{\omega} \times \mathbf{r}_{C/B}\\ &= 0 + (1.5\,\mathbf{k}) \times (300\,\mathbf{i}) = 450\,\mathbf{j}\,(\text{mm/s})\end{aligned}$$

DISCUSSION

In this example, the geometry is sufficiently simple that we can easily relate the velocities of the cable to the angular velocities of the pulleys without using Equation (6.6). That is often not the case. The next example illustrates a situation that would be much more difficult to solve without using Equation (6.6).

Example 6.3

Bar AB in Figure 6.18 rotates with a clockwise angular velocity of 10 rad/s. Determine the angular velocity of bar BC and the velocity of point C.

Figure 6.18

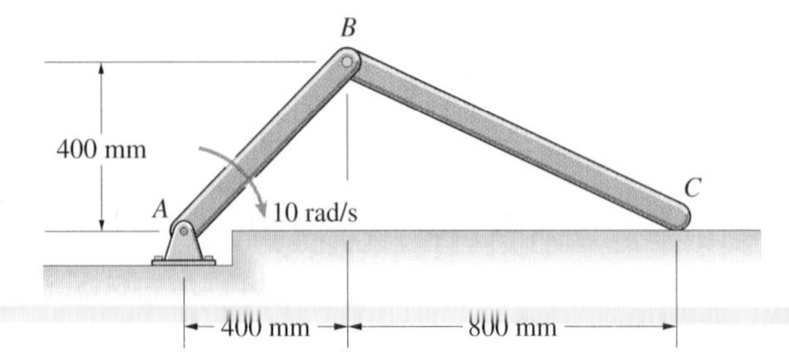

STRATEGY

Since we know the angular velocity of bar AB and point A is fixed, we can apply Equation (6.6) to points A and B to determine the velocity of B. Then by applying Equation (6.6) again to express the horizontal velocity of point C in terms of the velocity of B, we will obtain a vector equation in two unknowns: the velocity of C and the angular velocity of bar BC.

SOLUTION

In terms of the coordinate system in Figure (a), the position vector of B relative to A is $\mathbf{r}_{B/A} = (0.4\,\mathbf{i} + 0.4\,\mathbf{j})$ m. The angular velocity vector of bar AB is $\boldsymbol{\omega}_{AB} = -10\,\mathbf{k}$ rad/s, so the velocity of B is

$$\mathbf{v}_B = \mathbf{v}_A + \boldsymbol{\omega}_{AB} \times \mathbf{r}_{B/A} = 0 + \begin{vmatrix} \mathbf{i} & \mathbf{j} & \mathbf{k} \\ 0 & 0 & -10 \\ 0.4 & 0.4 & 0 \end{vmatrix}$$

$$= (4\,\mathbf{i} - 4\,\mathbf{j})\ \text{m/s}$$

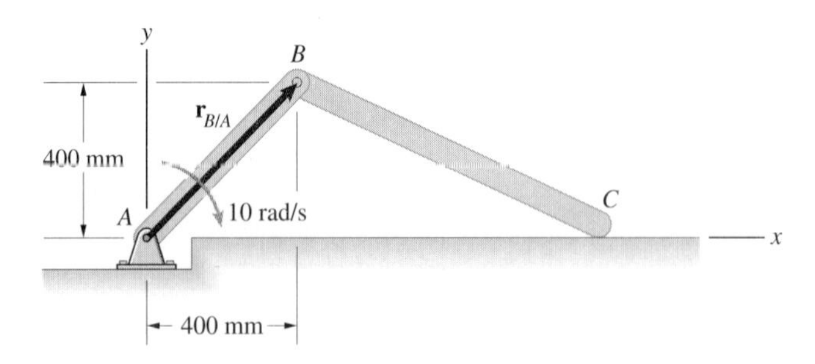

(a) Determining the velocity of B.

Let ω_{BC} be the unknown angular velocity of bar BC (Figure (b)), so that its angular velocity vector is $\boldsymbol{\omega}_{BC} = \omega_{BC}\,\mathbf{k}$. The position vector of C relative to B is $\mathbf{r}_{C/B} = (0.8\,\mathbf{i} - 0.4\,\mathbf{j})$ m. Although we don't know the velocity of C, we know it is in the horizontal direction, so we can write it in the form $\mathbf{v}_C = v_c\,\mathbf{i}$ (Figure (b)). We express the velocity of C in terms of the velocity of B:

$$\mathbf{v}_C = \mathbf{v}_B + \boldsymbol{\omega}_{BC} \times \mathbf{r}_{B/C}$$

Now we substitute the values of $\mathbf{v}_B$ and $\mathbf{r}_{B/C}$ and our expressions for $\mathbf{v}_C$ and $\boldsymbol{\omega}_{BC}$ into this equation, obtaining

$$v_C\,\mathbf{i} = 4\,\mathbf{i} - 4\,\mathbf{j} + \begin{vmatrix} \mathbf{i} & \mathbf{j} & \mathbf{k} \\ 0 & 0 & \omega_{BC} \\ 0.8 & -0.4 & 0 \end{vmatrix}$$

$$= 4\,\mathbf{i} - 4\,\mathbf{j} + 0.4\omega_{BC}\,\mathbf{i} + 0.8\omega_{BC}\,\mathbf{j}$$

Equating the $\mathbf{i}$ and $\mathbf{j}$ components in this equation yields two equations:

$$v_C = 4 + 0.4\omega_{BC}$$

$$0 = -4 + 0.8\omega_{BC}$$

Solving them, we obtain $\omega_{BC} = 5\,\text{rad/s}$ and $v_C = 6\,\text{m/s}$.

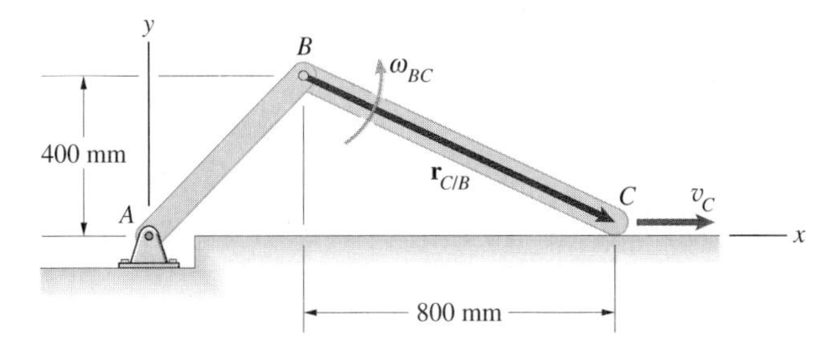

(b) Expressing $\mathbf{v}_C = v_C\,\mathbf{i}$ in terms of $\mathbf{v}_B$.

DISCUSSION

By expressing the velocity of C in terms of the velocity of B, we introduced into the solution the fact that point C is constrained to move horizontally. That is, we accounted for the presence of the floor. Our procedure in this example – applying Equation (6.6) systematically to relate the velocities of the joints to the angular velocities – applies to many problems in which you must determine velocities and angular velocities of connected rigid bodies. Some trial and error may be necessary to find the particular relationships you need.

Example 6.4

Bar AB in Figure 6.19 rotates with a clockwise angular velocity of 10 rad/s. What is the vertical velocity v_R of the rack of the rack and pinion gear?

Figure 6.19

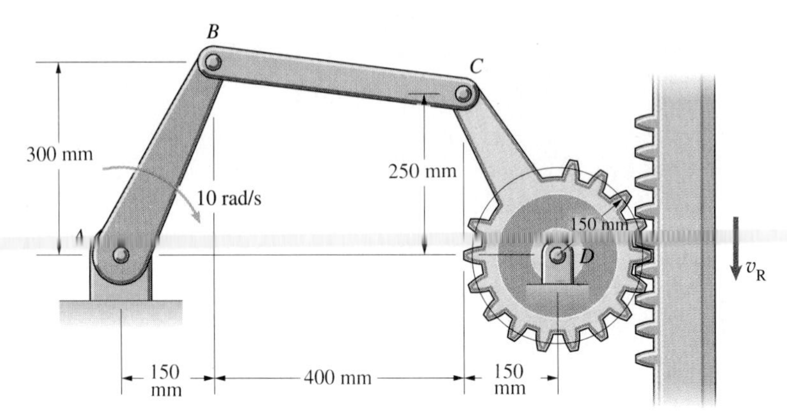

STRATEGY

To determine the velocity of the rack, we must determine the angular velocity of the member CD. Since we know the angular velocity of bar AB, we can apply Equation (6.6) to points A and B to determine the velocity of point B. Then we can apply Equation (6.6) to points C and D to obtain an equation for $\mathbf{v}_C$ in terms of the angular velocity of the member CD. We can also apply Equation (6.6) to points B and C to obtain an equation for $\mathbf{v}_C$ in terms of the angular velocity of bar BC. By equating the two expressions for $\mathbf{v}_C$, we will obtain a vector equation in two unknowns: the angular velocities of bars BC and CD.

SOLUTION

We first apply Equation (6.6) to points A and B (Figure (a)). In terms of the coordinate system shown, the position vector of B relative to A is $\mathbf{r}_{B/A} = (0.15\,\mathbf{i} + 0.3\,\mathbf{j})\,\text{m}$, and the angular velocity vector of bar AB is $\boldsymbol{\omega}_{AB} = -10\,\mathbf{k}\,\text{rad/s}$. The velocity of B is

$$\mathbf{v}_B = \mathbf{v}_A + \boldsymbol{\omega}_{AB} \times \mathbf{r}_{B/A} = 0 + \begin{vmatrix} \mathbf{i} & \mathbf{j} & \mathbf{k} \\ 0 & 0 & -10 \\ 0.15 & 0.3 & 0 \end{vmatrix}$$

$$= (3\,\mathbf{i} - 1.5\,\mathbf{j})\,\text{m/s}$$

We now apply Equation (6.6) to points C and D. Let ω_{CD} be the unknown angular velocity of member CD (Figure (a)). The position vector of C relative to D is $\mathbf{r}_{C/D} = (-0.15\,\mathbf{i} + 0.25\,\mathbf{j})\,\text{m}$, and the angular velocity vector of membrane CD is $\boldsymbol{\omega}_{CD} = -\omega_{CD}\,\mathbf{k}$. The velocity of C is

$$\mathbf{v}_C = \mathbf{v}_D + \boldsymbol{\omega}_{CD} \times \mathbf{r}_{C/D} = 0 + \begin{vmatrix} \mathbf{i} & \mathbf{j} & \mathbf{k} \\ 0 & 0 & -\omega_{CD} \\ -0.15 & 0.25 & 0 \end{vmatrix}$$

$$= 0.25\omega_{CD}\,\mathbf{i} + 0.15\omega_{CD}\,\mathbf{j}$$

Now we apply Equation (6.6) to points B and C (Figure (b)). We denote the unknown angular velocity of bar BC by ω_{BC}. The position vector of C relative to B is $\mathbf{r}_{C/B} = (0.4\,\mathbf{i} - 0.05\,\mathbf{j})\,\mathrm{m}$, and the angular velocity vector of bar BC is $\boldsymbol{\omega}_{BC} = \omega_{BC}\,\mathbf{k}$. Expressing the velocity of C in terms of the velocity of B, we obtain

$$\mathbf{v}_C = \mathbf{v}_B + \boldsymbol{\omega}_{BD} \times \mathbf{r}_{C/B} = \mathbf{v}_B + \begin{vmatrix} \mathbf{i} & \mathbf{j} & \mathbf{k} \\ 0 & 0 & \omega_{BC} \\ 0.4 & -0.05 & 0 \end{vmatrix}$$

$$= \mathbf{v}_B + 0.05\omega_{BC} 0.4\,\mathbf{i} + 0.4\omega_{BC}\,\mathbf{j}$$

Substituting our expressions for $\mathbf{v}_B$ and $\mathbf{v}_C$ into this equation, we obtain

$$0.25\omega_{CD}\,\mathbf{i} + 0.15\omega_{CD}\,\mathbf{j} = 3\,\mathbf{i} - 1.5\,\mathbf{j} + 0.05\omega_{BC}\,\mathbf{i} + 0.4\omega_{BC}\,\mathbf{j}$$

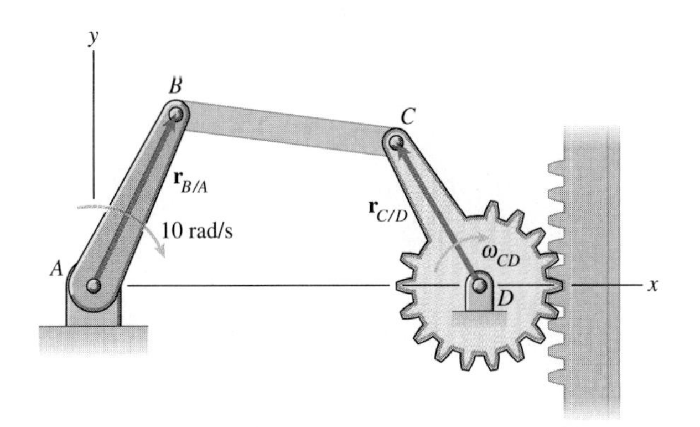

(a) Determining the velocities of points B and C.

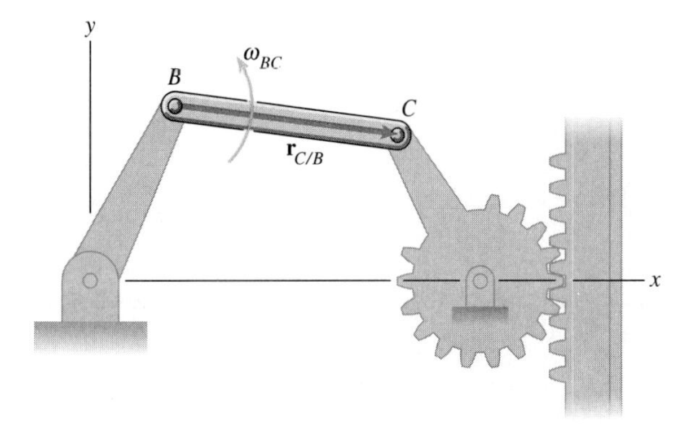

(b) Expressing the velocity of point C in terms of the velocity of point B.

Equating the $\mathbf{i}$ and $\mathbf{j}$ components yields two equations in terms of ω_{BC} and ω_{CD}:

$$0.25\omega_{CD} = 3 + 0.05\omega_{BC}$$

$$0.15\omega_{CD} = -1.5 + 0.4\omega_{BC}$$

Solving them, we obtain $\omega_{BC} = 8.92\,\mathrm{rad/s}$ and $\omega_{CD} = 13.78\,\mathrm{rad/s}$.

The vertical velocity of the rack is equal to the velocity of the gear where it contacts the rack:

$$v_{\mathrm{R}} = (0.15\,\mathrm{m})\omega_{CD} = (0.15)(13.78) = 2.07\,\mathrm{m/s}$$

Problems

6.12 A turbine rotates at 30 rad/s about a fixed axis coincident with the x axis. What is its angular velocity vector?

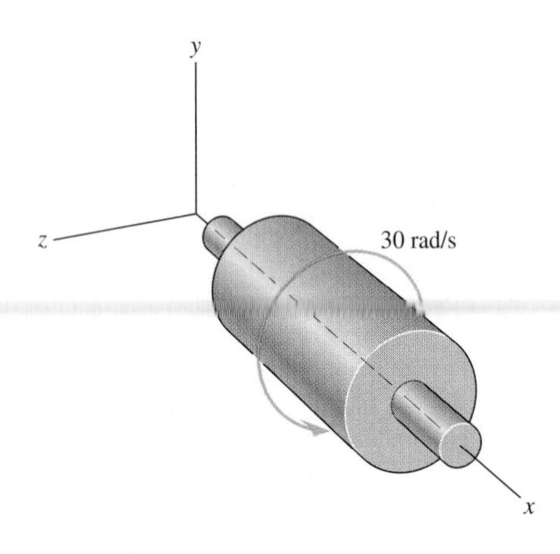

P6.12

6.13 The rectangular plate swings from arms of equal length. Determine the angular velocity vector of (a) the rectangular plate; (b) the bar AB.

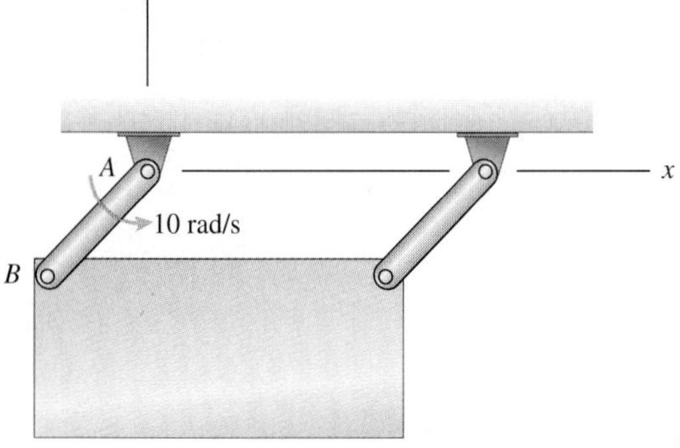

P6.13

6.14 What are the angular velocity vectors of each bar of the linkage?

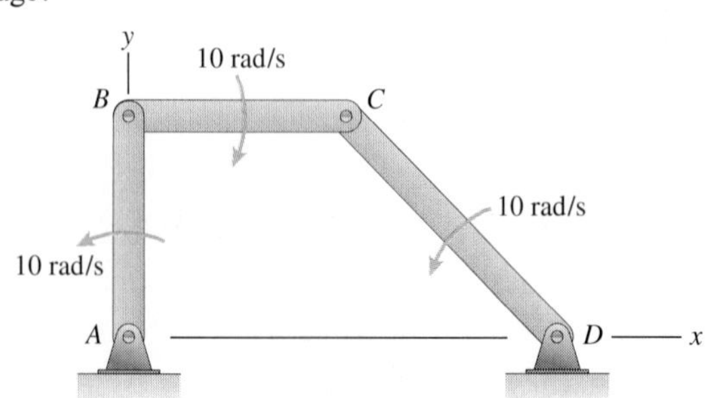

P6.14

6.15 If you model the earth as a rigid body, what is the magnitude of its angular velocity vector $\boldsymbol{\omega}_E$? Does $\boldsymbol{\omega}_E$ point north or south?

6.16 The rigid body rotates about the z axis with counterclockwise angular velocity ω.
(a) What is its angular velocity vector?
(b) Use Equation (6.6) to determine the velocity of point A relative to point B.

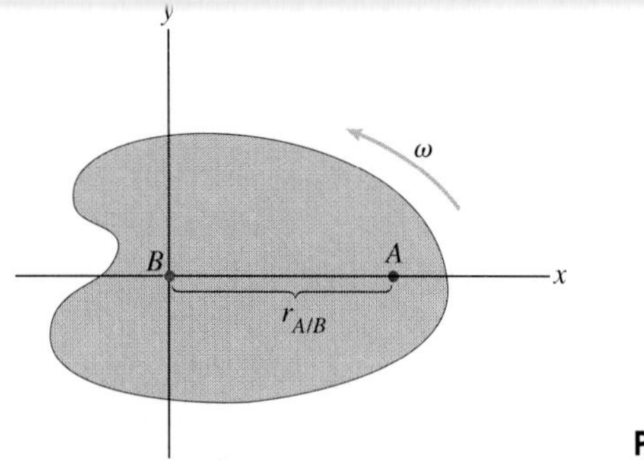

P6.16

6.17 (a) What is the annular velocity vector of the bar?
(b) Use Equation (6.6) to determine the velocity of point B relative to point O.
(c) Use Equation (6.6) to determine the velocity of point A relative to point B.

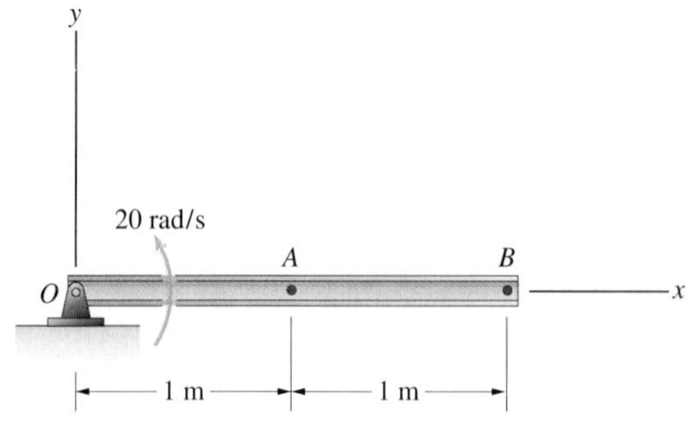

P6.17

6.18 (a) What is the angular velocity vector of the bar?
(b) Use Equation (6.6) to determine the velocity of point A.

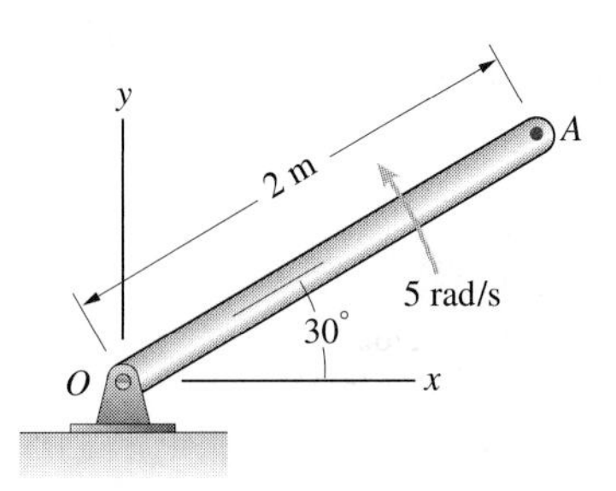

P6.18

6.19 The disc is rotating about the z axis at 50 rad/s in the clockwise direction. Use Equation (6.6) to determine the velocities of points A, B and C.

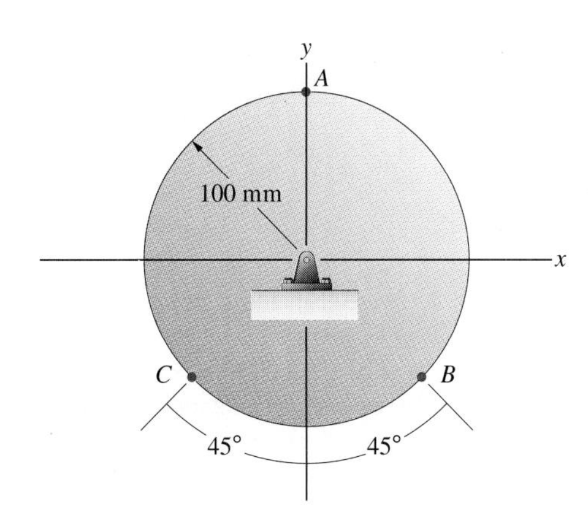

P6.19

6.20 The car is moving to the right at 100 km/hr, and its tyres are 600 mm in diameter.
(a) What is the angular velocity of its tyres?
(b) Which point on the tyre shown has the largest velocity relative to the road, and what is the magnitude of the velocity?

P6.20

6.21 The disc rolls on the plane surface. Point A is moving to the right at 6 m/s.
(a) What is the angular velocity vector of the disc?
(b) Use Equation (6.6) to determine the velocities of points B, C and D.

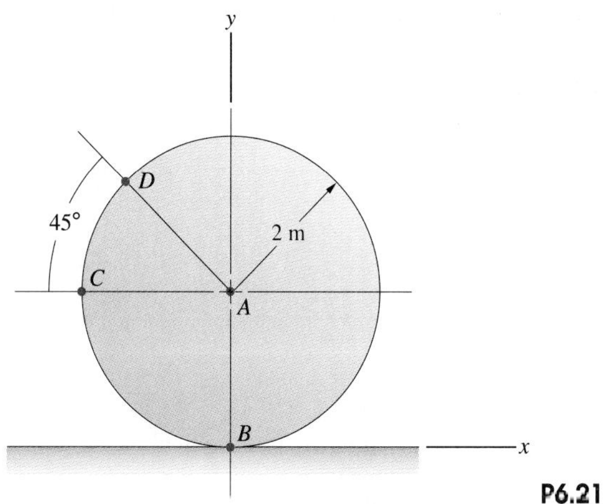

P6.21

6.22 The ring gear is stationary, and the sun gear rotates at 120 rpm (revolutions per minute) in the counterclockwise direction. Determine the angular velocity of the planet gears and the magnitude of the velocity of their centrepoints.

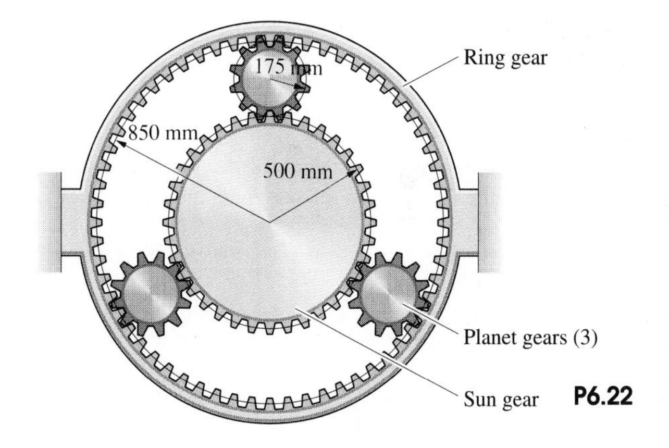

P6.22

6.23 The bar is in two-dimensional motion in the x-y plane. The velocity of point A is $8\mathbf{i}$ m/s. The x component of the velocity of point B is 6 m/s.
(a) What is the angular velocity vector of the bar?
(b) What is the velocity of point B?

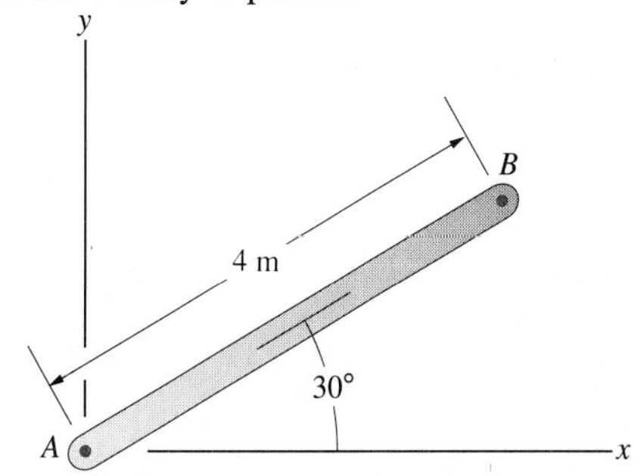

P6.23

6.24 Points A and B of the 1 m bar slide on the plane surfaces. The velocity of point B is $2\mathbf{i}$ m/s.
(a) What is the angular velocity vector of the bar?
(b) What is the velocity of point A?

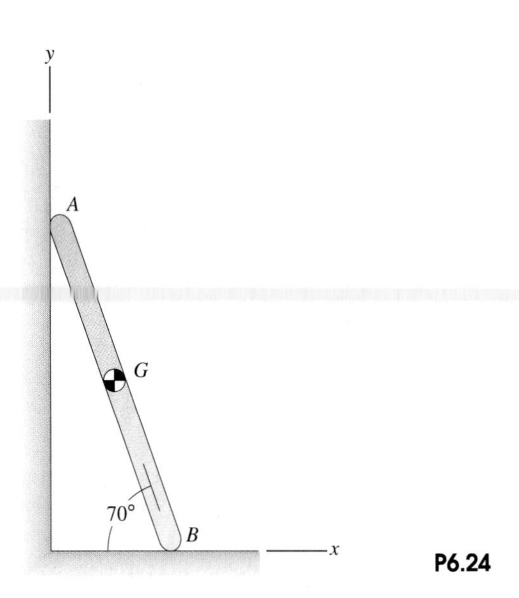

P6.24

6.25 In Problem 6.24, what is the velocity of the midpoint G of the bar?

6.26 Bar AB rotates in the counterclockwise direction at 6 rad/s. Determine the angular velocity of bar BCD and the velocity of point D.

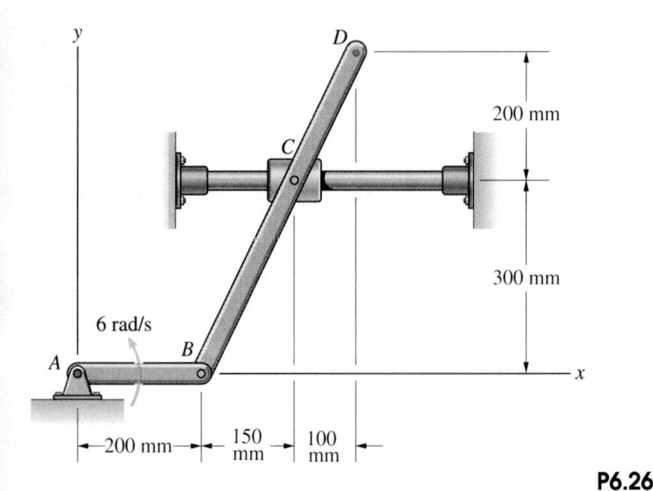

P6.26

6.27 If the crankshaft AB rotates at 6000 rpm (revolutions per minute) in the counterclockwise direction, what is the velocity of the piston at the instant shown?

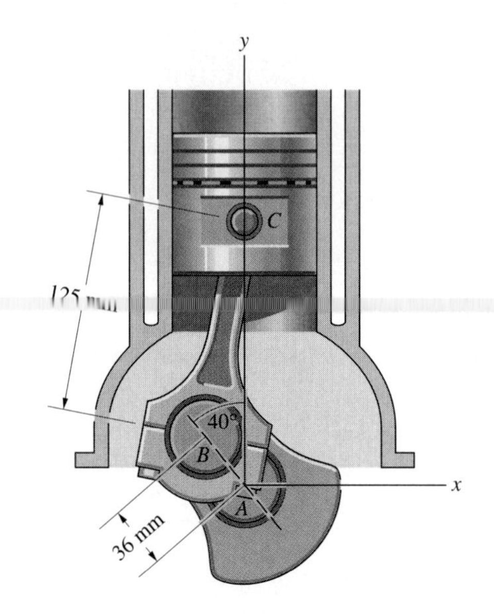

P6.27

6.28 Bar AB rotates at 10 rad/s in the counterclockwise direction. Determine the angular velocity of bar CD.

Strategy: Since you know the angular velocity of the bar AB, you can determine the velocity of B. Then apply Equation (6.6) to points B and C to obtain an equation for $\mathbf{v}_C$ in terms of the angular velocity of bar BC, and apply it to points C and D to obtain an equation for $\mathbf{v}_C$ in terms of the angular velocity of bar CD. By equating the two expressions, you will obtain a vector equation in two unknowns: the angular velocities of bars BC and CD.

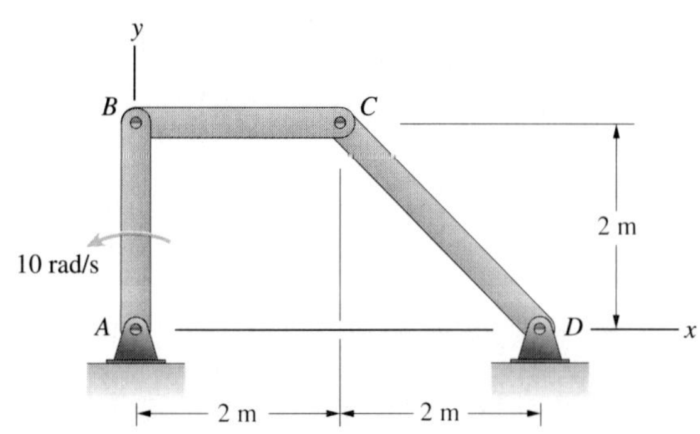

P6.28

6.29 Bar AB rotates at 12 rad/s in the clockwise direction. Determine the angular velocities of bars BC and CD.

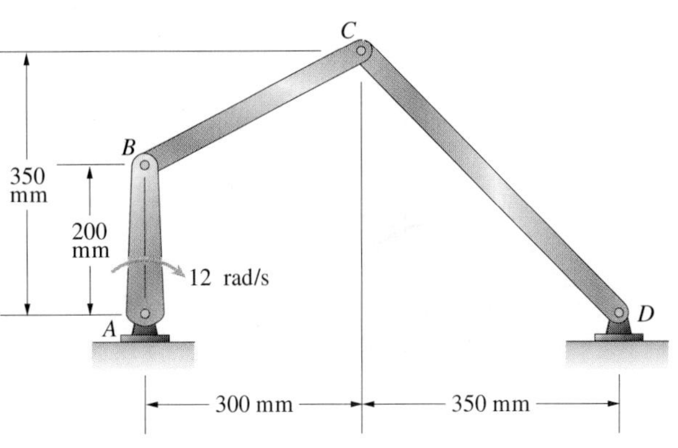

P6.29

6.30 Bar CD rotates at 2 rad/s in the clockwise direction. Determine the angular velocities of bars AB and BC.

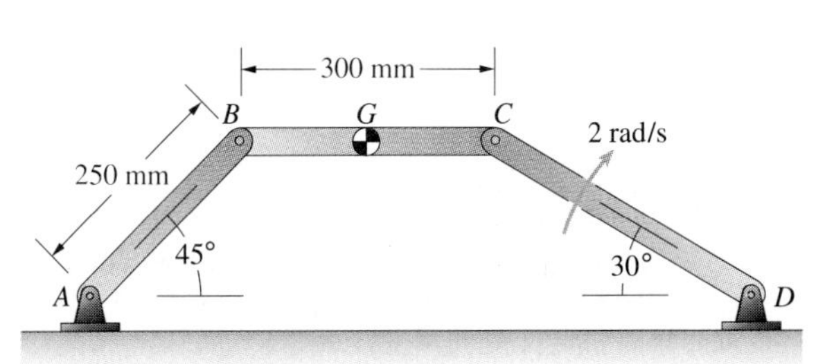

P6.30

6.31 In Problem 6.30, what is the magnitude of the velocity of the midpoint G of bar BC?

6.32 Bar AB rotates at 10 rad/s in the counterclockwise direction. Determine the velocity of point E.

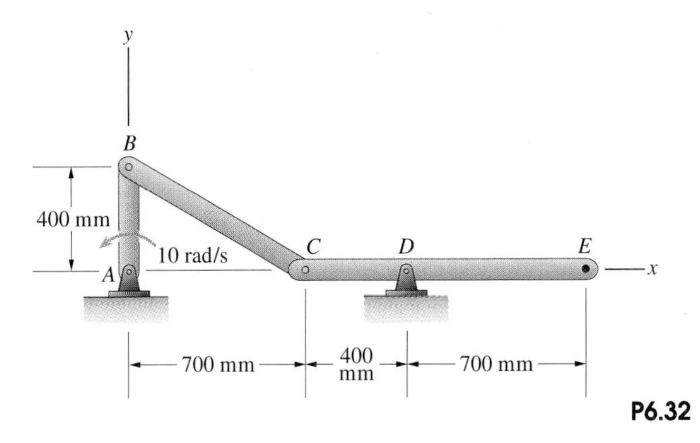

P6.32

6.33 Bar AB rotates at 4 rad/s in the counterclockwise direction. Determine the velocity of point C.

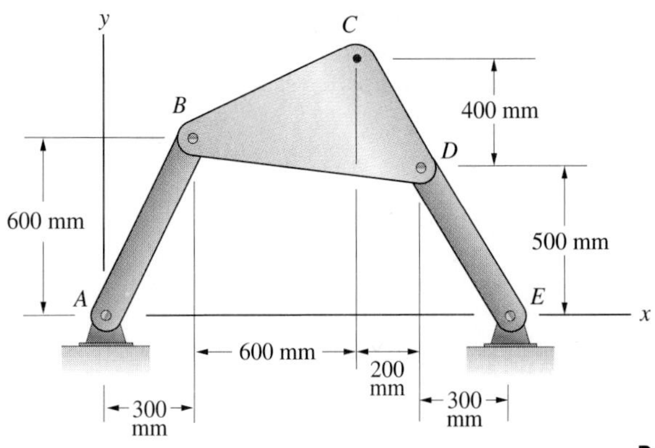

P6.33

6.34 In the system shown in Problem 6.33, if the magnitude of the velocity of point C is $|\mathbf{v}_C| = 2$ m/s, what are the magnitudes of the angular velocities of bars AB and DE?

6.35 Bars OA and AB are each 2 m long. Point B is sliding up the inclined surface at 10 m/s. Determine the angular velocities of the bars.

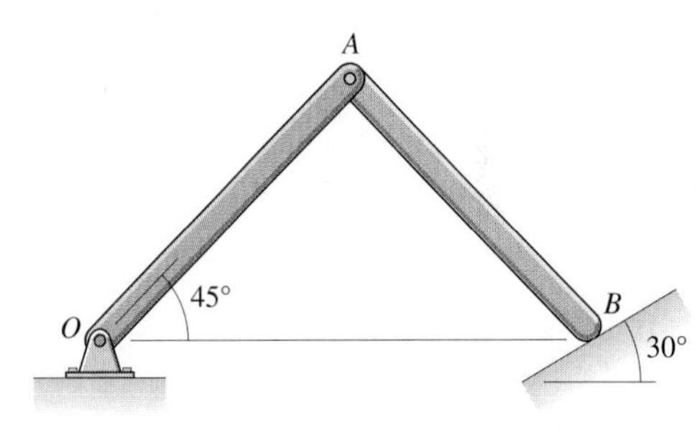

P6.35

6.36 The diameter of the disc is 1 m, and the length of bar AB is 1 m. The disc is rolling, and point B slides on the plane surface. Determine the angular velocity of bar AB and the velocity of point B.

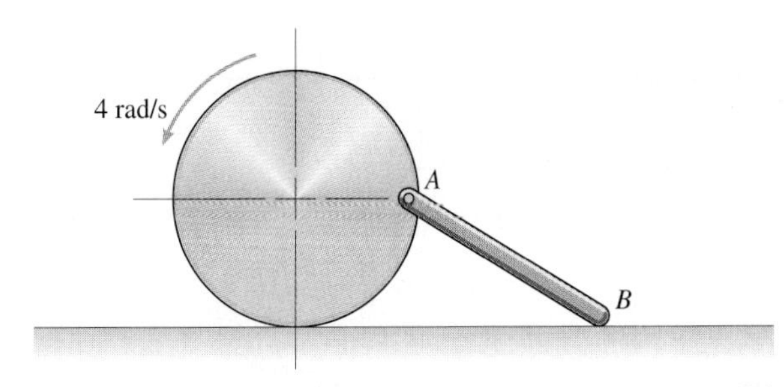

P6.36

6.37 A motor rotates the circular disc mounted at A, moving the saw back and forth. (The saw is supported by a horizontal slot so that point C moves horizontally.) The radius AB is 100 mm, and the link BC is 350 mm long. In the position shown, $\theta = 45^\circ$ and the link BC is horizontal. If the angular velocity of the disc is one revolution per second counterclockwise, what is the velocity of the saw?

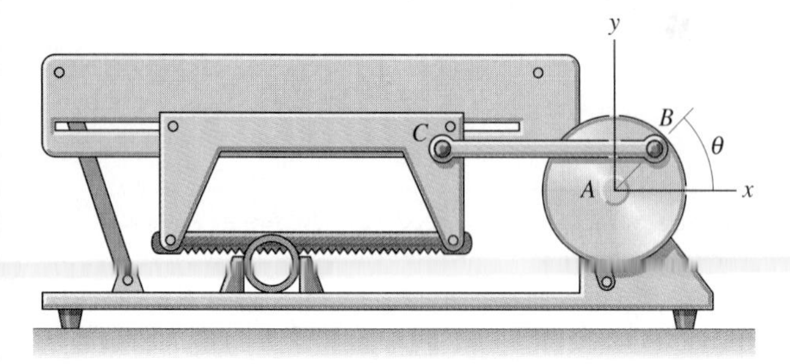

P6.37

6.38 In Problem 6.37, if the angular velocity of the disc is one revolution per second counterclockwise and $\theta = 270^\circ$, what is the velocity of the saw?

6.39 The discs roll on the plane surface. The angular velocity of the left disc is 2 rad/s in the clockwise direction. What is the angular velocity of the right disc?

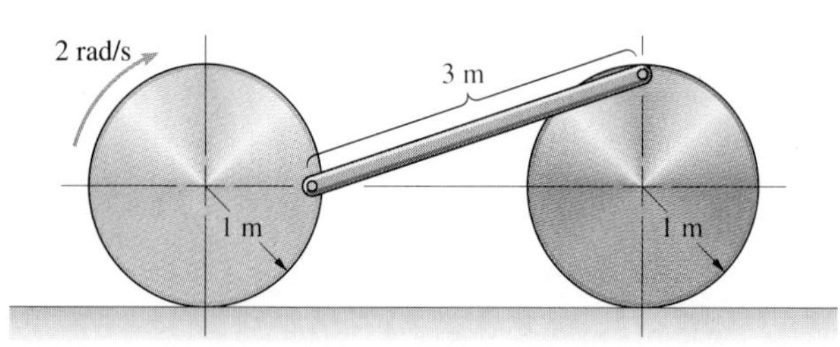

P6.39

6.40 The disc rolls on the curved surface. The bar rotates at 10 rad/s in the counterclockwise direction. Determine the velocity of point A.

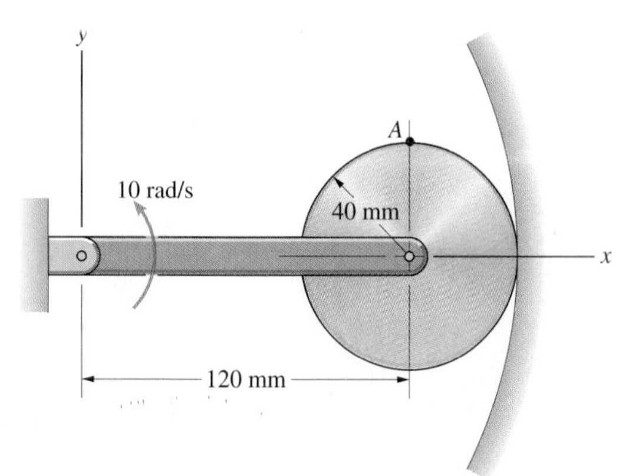

P6.40

6.41 If $\omega_{AB} = 2$ rad/s and $\omega_{BC} = 4$ rad/s, what is the velocity of point C, where the excavator's bucket is attached?

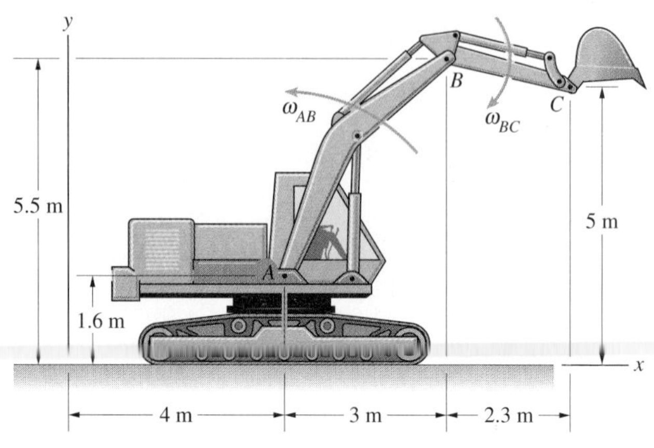

P6.41

6.42 In Problem 6.41, if $\omega_{AB} = 2$ rad/s, what clockwise angular velocity ω_{BC} will cause the vertical component of the velocity of point C to be zero? What is the resulting velocity of point C?

6.43 In Problem 6.41, if the velocity of point C is $\mathbf{v}_C = (-6\,\mathbf{i} - 4\,\mathbf{j})$ m/s, what are the angular velocities ω_{AB} and ω_{BC}?

6.44 An athlete exercises his arm by raising the mass m. The shoulder joint A is stationary. The distance AB is 300 mm, and the distance BC is 400 mm. At the instant shown, $\omega_{AB} = 1$ rad/s and $\omega_{BC} = 2$ rad/s. How fast is the mass m rising?

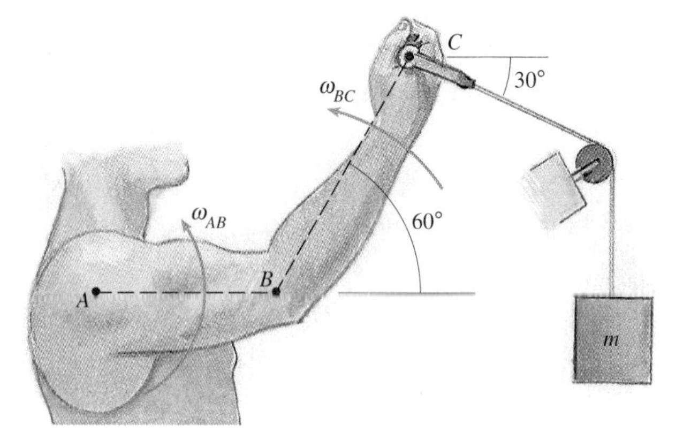

P6.44

6.45 In Problem 6.44, suppose that the distance AB is 300 mm, the distance BC is 400 mm, $\omega_{AB} = 0.6$ rad/s, and the mass m is rising at 800 mm/s. What is the angular velocity ω_{BC}?

6.46 Points B and C are in the x-y plane. The angular velocity vectors of the arms AB and BC are $\boldsymbol{\omega}_{AB} = -2\,\mathbf{k}$ rad/s and $\boldsymbol{\omega}_{BC} = 0.4\,\mathbf{k}$ rad/s. Determine the velocity of point C.

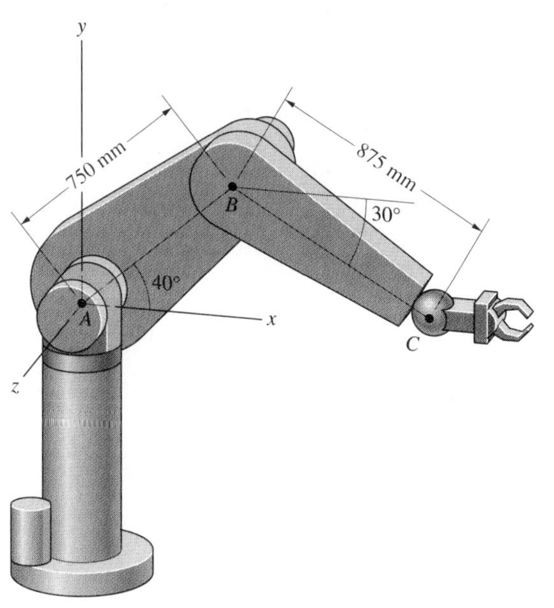

P6.46

6.47 In Problem 6.46, if the velocity of point C is $\mathbf{v}_C = 250\,\mathbf{j}$ mm/s, what are the angular velocity vectors of the arms AB and BC?

6.48 Determine the velocity v_W and the angular velocity of the small pulley.

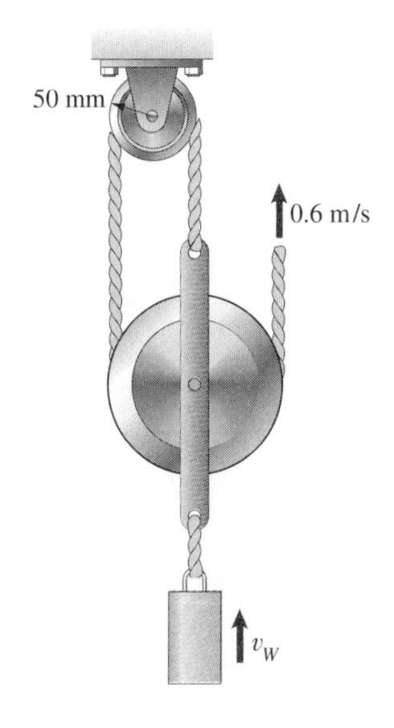

P6.48

6.49 Determine the velocity of the block and the angular velocity of the small pulley.

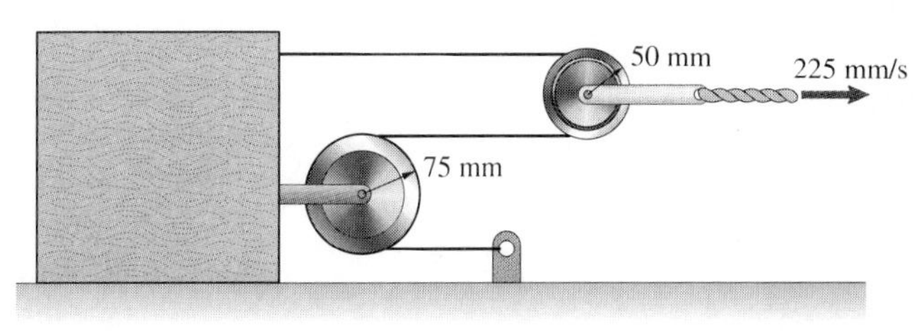

P6.49

6.50 The ring gear is fixed and the hub and planet gears are bonded together. The connecting rod rotates in the counterclockwise direction at 60 rpm (revolutions per minute). Determine the angular velocity of the sun gear and the magnitude of the velocity of point A.

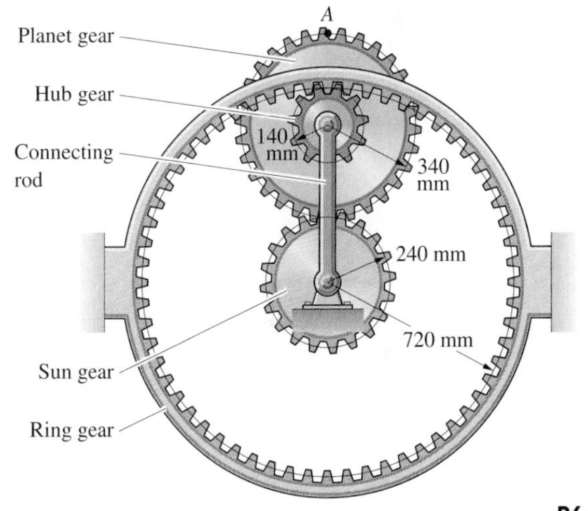

P6.50

6.51 The large gear is fixed. Bar AB has a counterclockwise angular velocity of 2 rad/s. What are the angular velocities of bars CE and DE?

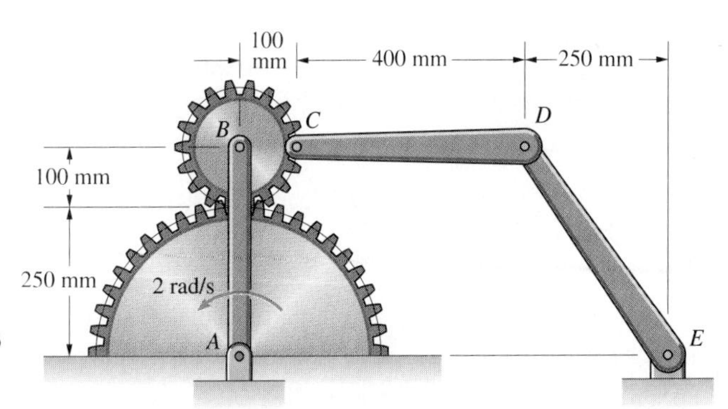

P6.51

Instantaneous Centres

By an **instantaneous centre**, we simply mean a point of a rigid body whose velocity is zero at a given instant. 'Instantaneous' means it may have zero velocity *only* at the instant under consideration, although we also refer to a fixed point, such as a point of a fixed axis about which a rigid body rotates, as an instantaneous centre.

When we know the location of an instantaneous centre of a rigid body in two-dimensional motion and we know its angular velocity, the velocities of other points are easy to determine. For example, suppose that point C in Figure 6.20(a) is the instantaneous centre of a rigid body in plane motion with angular velocity ω. Relative to C, a point A moves in a circular path. The velocity of A relative to C is tangent to the circular path and equal to the product of the distance from C to A and the angular velocity. But since C is stationary at this instant, the velocity of A relative to C is the velocity of A. At this instant, every point of the rigid body rotates about C (Figure 6.20(b)).

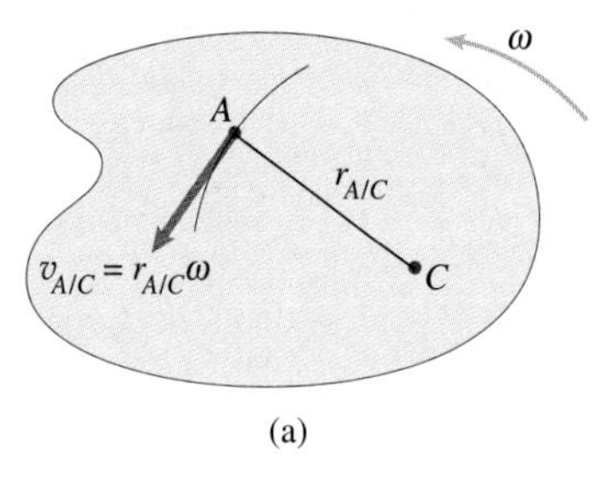

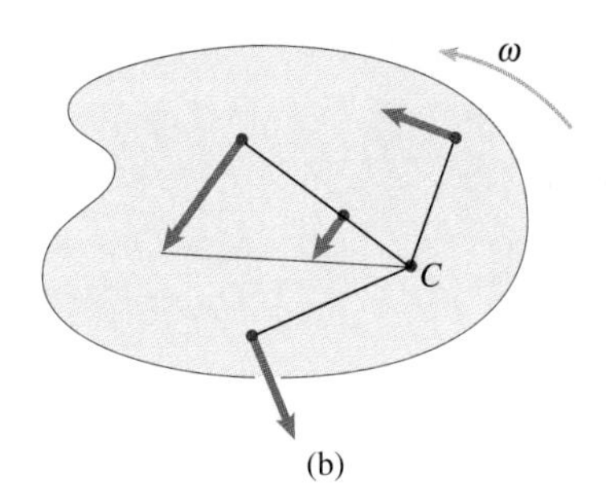

Figure 6.20
(a) An instantaneous centre C and a different point A.
(b) Every point rotates about the instantaneous centre.

You can often locate the instantaneous centre of a rigid body in two-dimensional motion in a simple way. Suppose that you know the directions of the motions of two points A and B (Figure 6.21(a)). If you draw lines through A and B perpendicular to their directions of motion, the point C where the lines intersect is the instantaneous centre.

To show that this is true, let us express the velocity of C in terms of the velocity of A (Figure 6.21(b)):

$$\mathbf{v}_C = \mathbf{v}_A + \boldsymbol{\omega} \times \mathbf{r}_{C/A}$$

Since the vector $\boldsymbol{\omega} \times \mathbf{r}_{C/A}$ is perpendicular to $\mathbf{r}_{C/A}$, this equation states that the direction of motion of C is parallel to the direction of motion of A. We can also express the velocity of C in terms of the velocity of B:

$$\mathbf{v}_C = \mathbf{v}_B + \boldsymbol{\omega} \times \mathbf{r}_{C/B}$$

The vector $\boldsymbol{\omega} \times \mathbf{r}_{C/B}$ is perpendicular to $\mathbf{r}_{C/B}$, so this equation states that the direction of motion of C is parallel to the direction of motion of B. But C cannot be moving parallel to A and parallel to B, so these equations are contradictory unless $\mathbf{v}_C = 0$.

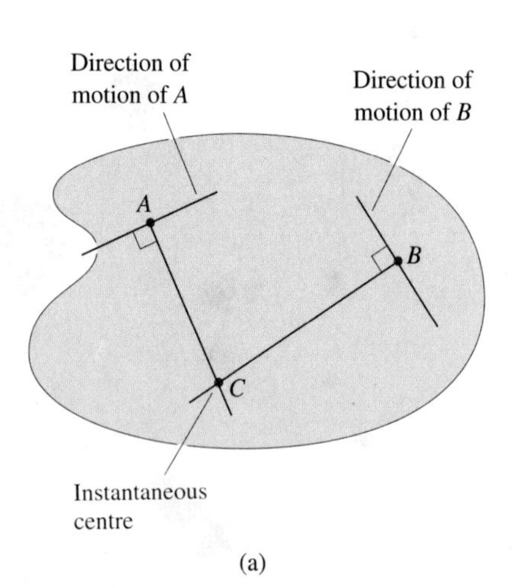

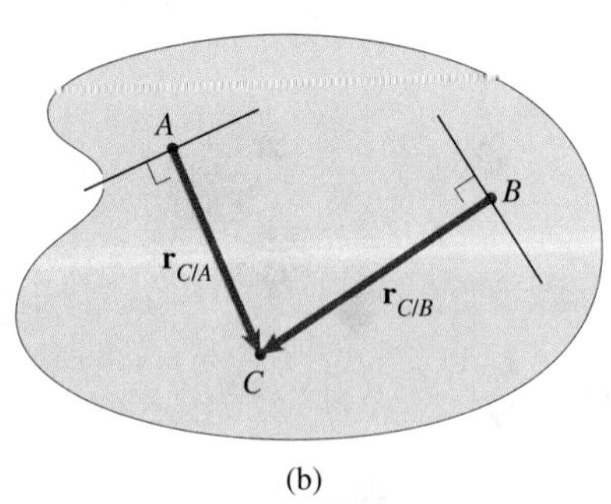

Figure 6.21
(a) Locating the instantaneous centre in planar motion.
(b) Proving that $\mathbf{v}_C = 0$.

The instantaneous centre may not be a point of the rigid body (Figure 6.22(a)). This simply means that at this instant, the rigid body is rotating about an external point. It's helpful to imagine extending the rigid body so that it includes the instantaneous centre (Figure 6.22(b)). The velocity of point C of the extended body would be zero at this instant.

Notice in Figure 6.22(a) that if you change the directions of motion of A and B so that the lines perpendicular to their directions of motion become parallel, C goes to infinity. In that case, the rigid body is in translation; its angular velocity is zero.

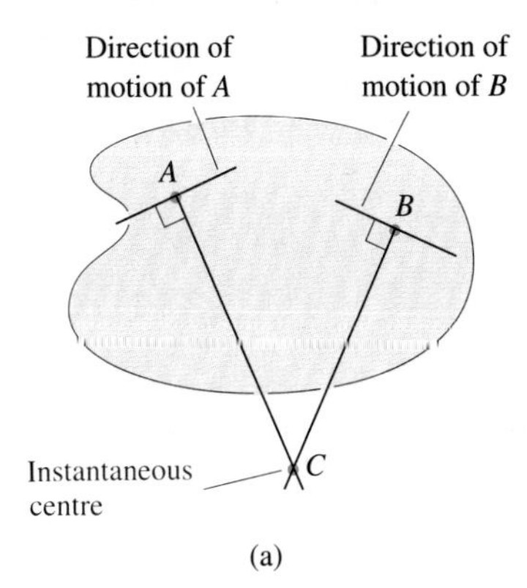

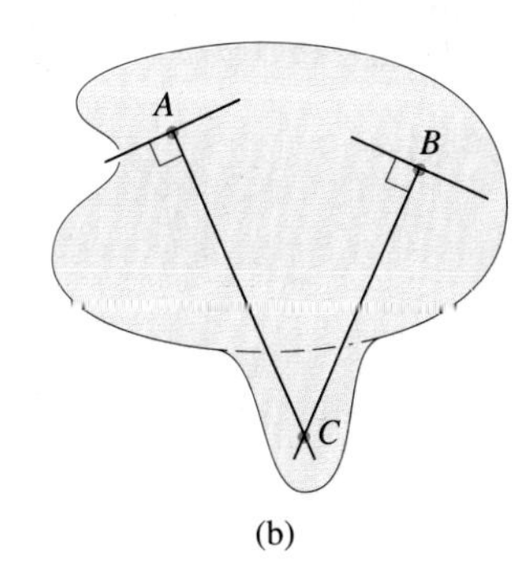

Figure 6.22

(a) An instantaneous centre external to the rigid body.
(b) A hypothetical extended body. Point C would be stationary.

Returning once again to our example of a disc of radius R rolling with angular velocity ω (Figure 6.23(a)), the point C in contact with the floor is stationary at that instant – it is the instantaneous centre of the disc. Therefore the velocity of any other point is perpendicular to the line from C to the point and its magnitude equals the product of ω and the distance from C to the point. In terms of the coordinate system shown in Figure 6.23(b)), the velocity of point A is

$$\begin{aligned}\mathbf{v}_A &= -\sqrt{2}R\omega\cos 45^\circ\,\mathbf{i} + \sqrt{2}R\omega\sin 45^\circ\,\mathbf{j}\\ &= -R\omega\,\mathbf{i} + R\omega\,\mathbf{j}\end{aligned}$$

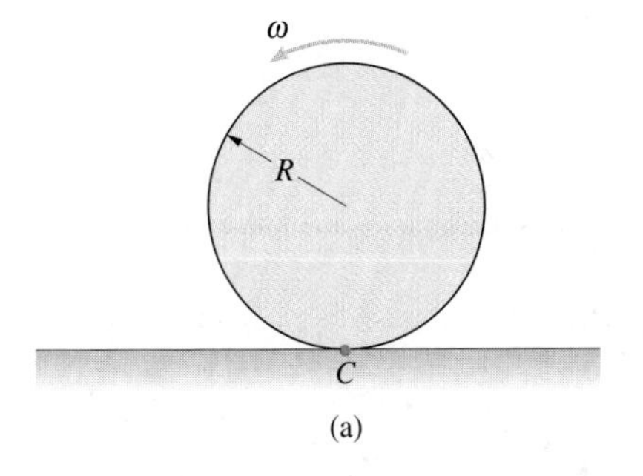

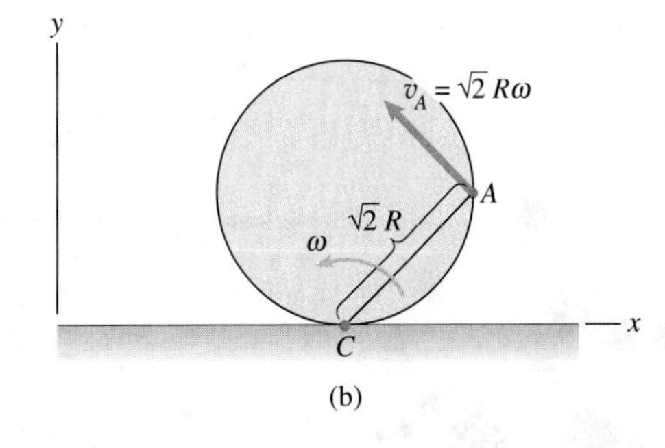

Figure 6.23

(a) Point C is the instantaneous centre of the rolling disc.
(b) Determining the velocity of point A.

In the following example we use instantaneous centres to analyse the motion of a linkage. By identifying the instantaneous centre of a rigid body in plane motion, you can express the velocities of its points as products of their distances from the instantaneous centre and the angular velocity of the rigid body.

Example 6.5

Bar AB in Figure 6.24 rotates with a counterclockwise angular velocity of 10 rad/s. What are the angular velocities of bars BC and CD?

Figure 6.24

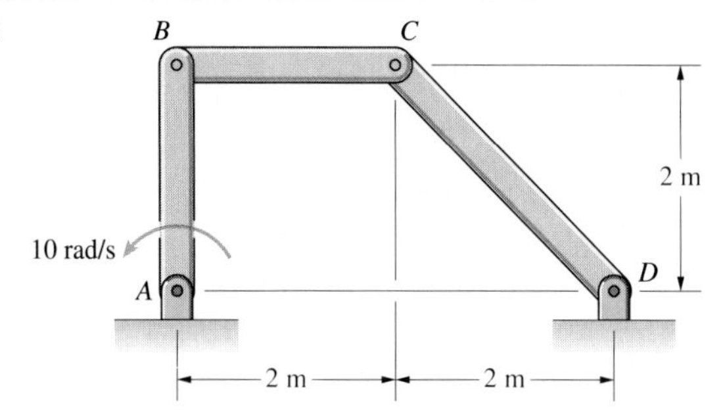

STRATEGY

Because bars AB and CD rotate about fixed axes, we know the directions of motion of points B and C and so can locate the instantaneous centre of bar BC. Beginning with bar AB (because we know its angular velocity), we can use the instantaneous centres of the bars to determine both the velocities of the points where they are connected and their angular velocities.

SOLUTION

The velocity of B due to the rotation of bar AB about A (Figure (a)) is

$$v_B = (2\,\text{m})(10\,\text{rad/s}) = 20\,\text{m/s}$$

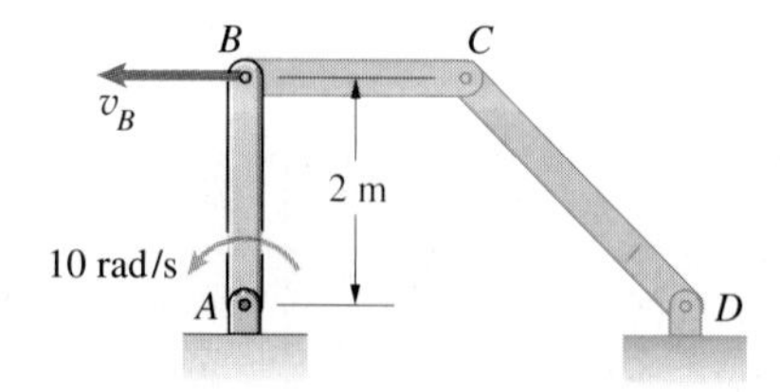

(a) Determining v_B.

Drawing lines perpendicular to the directions of motion of B and C, we locate the instantaneous centre of bar BC (Figure (b)). The velocity of B is equal to the product of its distance from the instantaneous centre of bar BC and the angular velocity ω_{BC},

$$v_B = 20\,\text{m/s} = (2\,\text{m})\omega_{BC}$$

so $\omega_{BC} = 10\,\text{rad/s}$. (Notice that bar BC rotates in the clockwise direction.) Using the instantaneous centre of bar BC and its angular velocity ω_{BC}, we can determine the velocity of point C:

$$v_C = (\sqrt{8}\,\text{m})\omega_{BC} = 10\sqrt{8}\,\text{m/s}$$

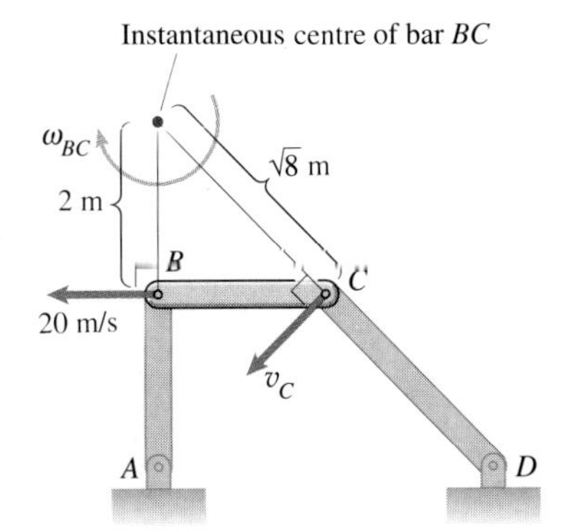

(b) Determining ω_{BC} and v_C.

Our last step is to use the velocity of point C to determine the angular velocity of bar CD about point D (Figure (c))

$$v_C = 10\sqrt{8}\,\text{m/s} = (\sqrt{8}\,\text{m})\omega_{CD}$$

obtaining $\omega_{CD} = 10\,\text{rad/s}$ counterclockwise.

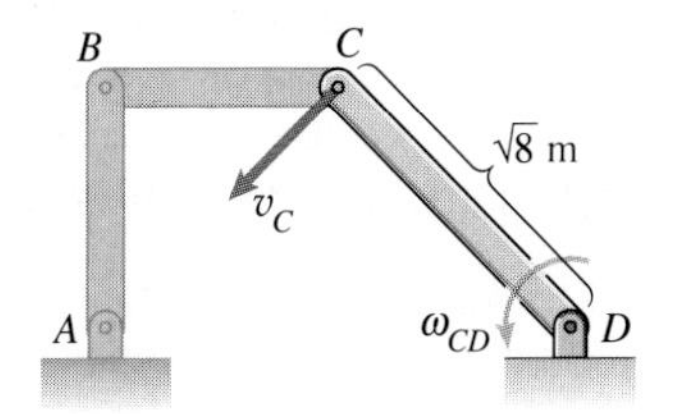

(c) Determining ω_{CD}.

DISCUSSION

In this example, the use of instantaneous centres greatly simplified determining the angular velocities of bars BC and CD in comparison with our previous approach. However, notice that the lengths and positions of the bars made it very easy for us to locate the instantaneous centre of bar BC. If the geometry is too complicated, the use of instantaneous centres can be impractical.

Problems

6.52 If the bar has a clockwise angular velocity of 10 rad/s and $v_A = 20$ m/s, what are the coordinates of its instantaneous centre and the value of v_B?

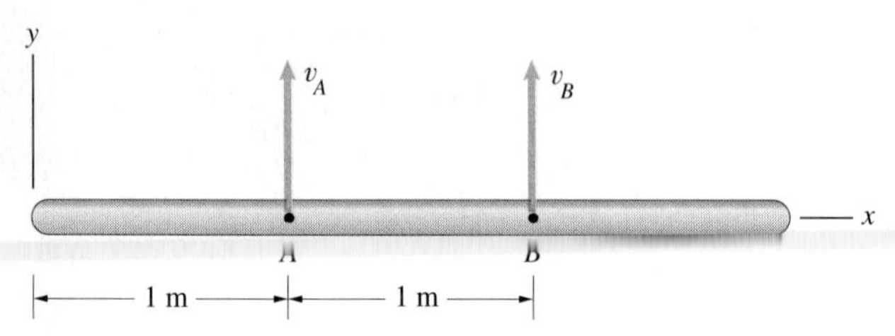

P6.52

6.53 In Problem 6.52, if $v_A = 24$ m/s and $v_B = 36$ m/s, what are the coordinates of the instantaneous centre of the bar and its angular velocity?

6.54 The velocity of point O of the bat is $\mathbf{v}_O = (-1.8\,\mathbf{i} - 0.42\,\mathbf{j})$ m/s, and the bat rotates about the z axis with a counterclockwise angular velocity of 4 rad/s. What are the x and y coordinates of its instantaneous centre?

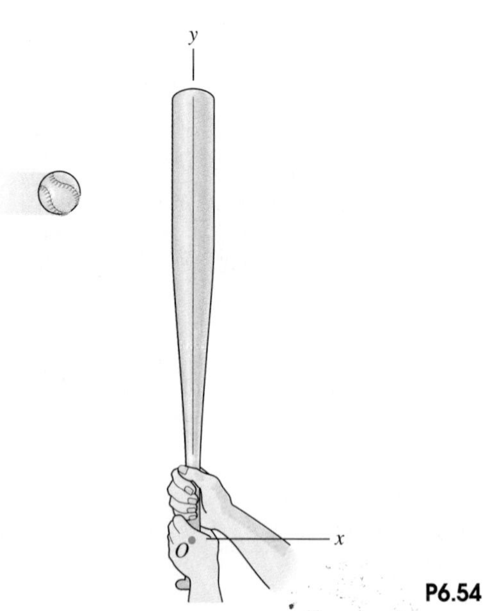

P6.54

6.55 Points A and B of the 1 m bar slide on the plane surfaces. The velocity of B is $\mathbf{v}_B = 2\,\mathbf{i}$ m/s.

(a) What are the coordinates of the instantaneous centre?

(b) Use the instantaneous centre to determine the velocity of A.

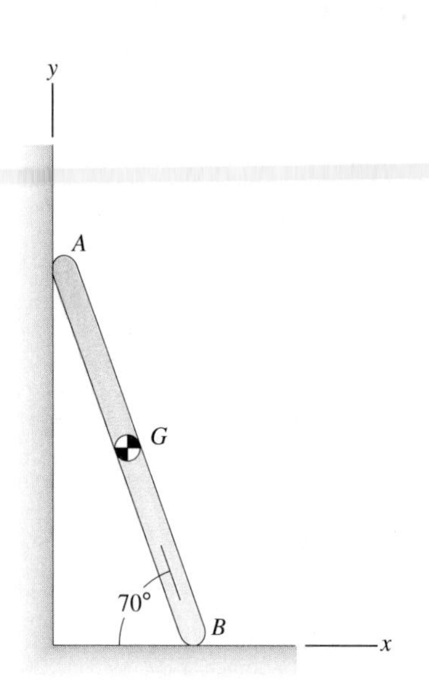

P6.55

6.56 In Problem 6.55, use the instantaneous centre to determine the velocity of the bar's midpoint G.

6.57 The bar is in two-dimensional motion in the x-y plane. The velocity of point A is $\mathbf{v}_A = 2.4\,\mathbf{i}$ m/s, and B is moving in the direction parallel to the bar. Determine the velocity of B (a) by using Equation (6.6); (b) by using the instantaneous centre.

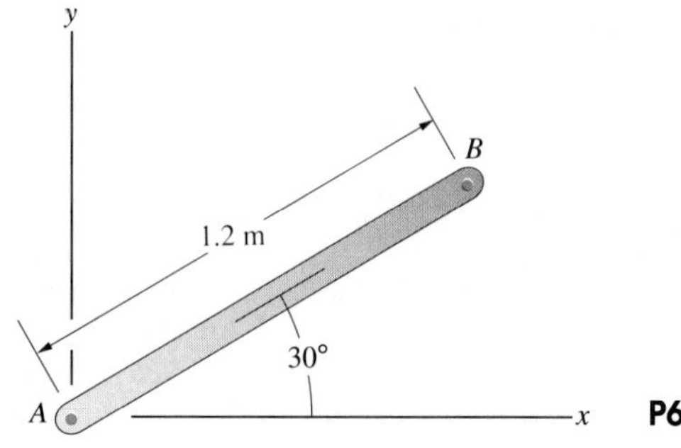

P6.57

6.58 Points A and B of the 1.2 m bar slide on the plane surfaces. Point B is sliding down the slanted surface at 0.6 m/s.
(a) What are the coordinates of the instantaneous centre?
(b) Use the instantaneous centre to determine the velocity of A.

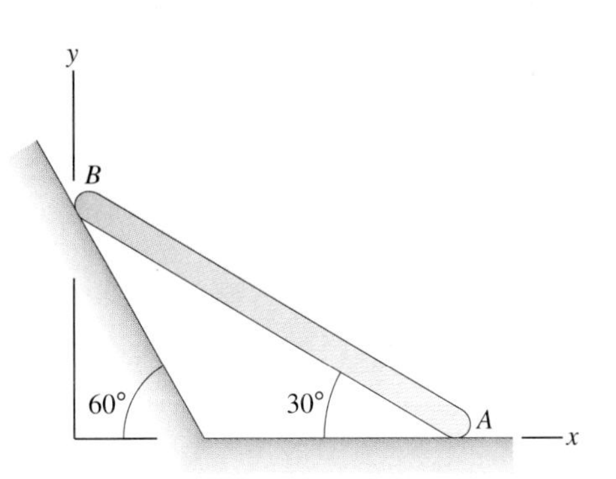

P6.58

6.59 Use instantaneous centres to determine the horizontal velocity of B.

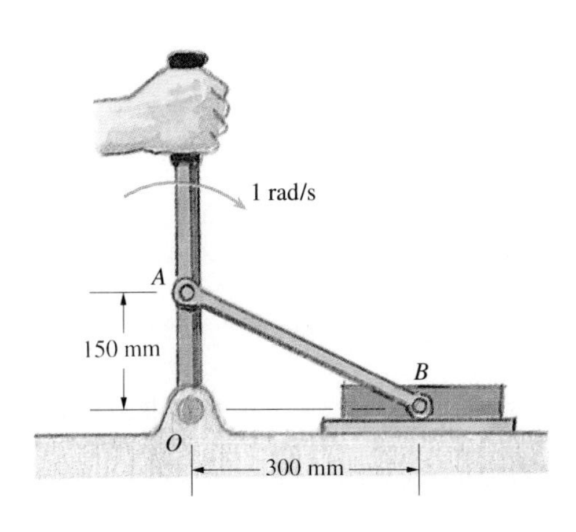

P6.59

6.60 When the mechanism in Problem 6.59 is in this position, use instantaneous centres to determine the horizontal velocity of B.

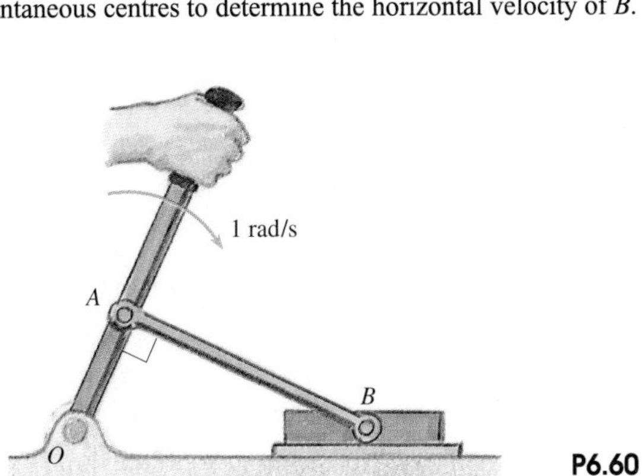

P6.60

6.61 Bar AB rotates at 6 rad/s in the clockwise direction. Use instantaneous centres to determine the angular velocity of bar BC.

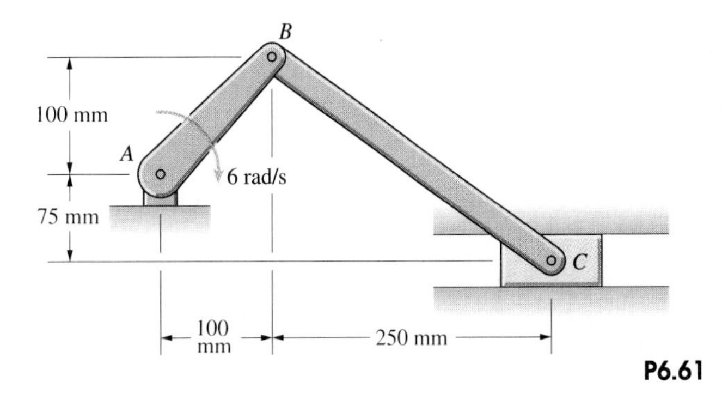

P6.61

6.62 Bar AB rotates at 10 rad/s in the counterclockwise direction. Use instantaneous centres to determine the velocity of point E.

P6.62

6.63 The discs roll on the plane surface. The left disc rotates at 2 rad/s in the clockwise direction. Use instantaneous centres to determine the angular velocities of the bar and the right disc.

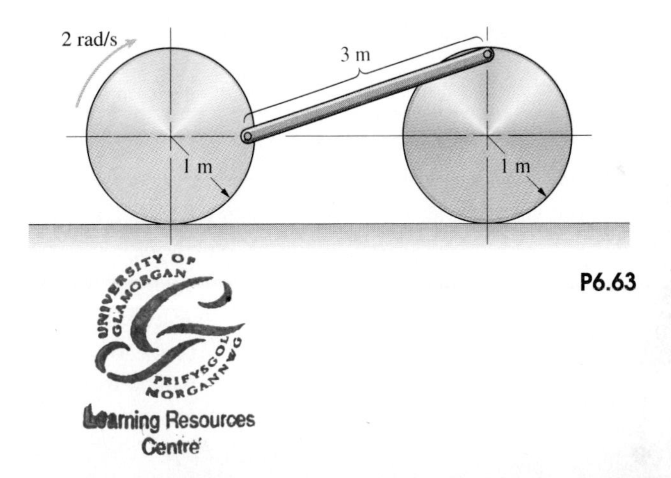

P6.63

6.64 Bar AB rotates at 12 rad/s in the clockwise direction. Use instantaneous centres to determine the angular velocities of bars BC and CD.

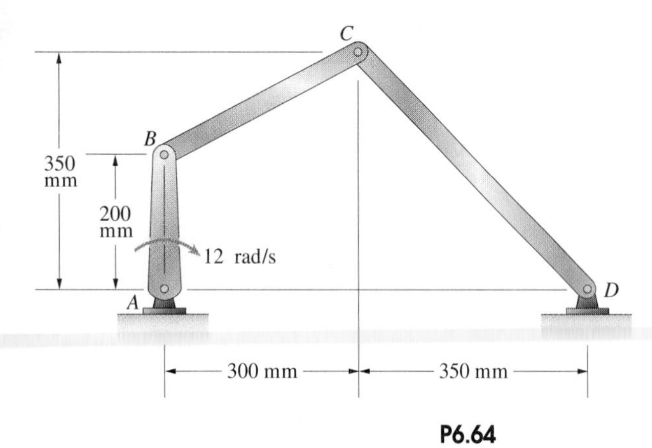

P6.64

6.65 A rigid body is in planar motion. A point A with coordinates $x = 200$ mm, $y = 600$ mm is moving parallel to the unit vector $-0.970\,\mathbf{i} + 0.243\,\mathbf{j}$, and a point B with coordinates $x = 800$ mm, $y = 400$ mm is moving parallel to the unit vector $-0.832\,\mathbf{i} + 0.555\,\mathbf{j}$.

(a) What are the coordinates of the instantaneous centre?

(b) Determine $|\mathbf{v}_A|/|\mathbf{v}_B|$.

6.66 Show that if a rigid body in planar motion has two instantaneous centres, it is stationary at that instant.

6.4 General Motions: Accelerations

In Chapter 7 you will be concerned with determining the motion of a rigid body when you know the external forces and couples acting on it. The governing equations are expressed in terms of the acceleration of the centre of mass of the rigid body and its angular acceleration. To solve such problems, you need to understand the relationship between the accelerations of points of a rigid body and its angular acceleration. In this section we extend the methods we have used to analyse velocities of rigid bodies to accelerations.

Consider points A and B in the plane of the motion of a rigid body in two-dimensional motion (Figure 6.25(a)). Their velocities are related by

$$\mathbf{v}_A = \mathbf{v}_B + \mathbf{v}_{A/B}$$

where $\mathbf{v}_A$ and $\mathbf{v}_B$ are velocities relative to a reference point O. Taking the time derivative of this equation, we obtain

$$\mathbf{a}_A = \mathbf{a}_B + \mathbf{a}_{A/B}$$

Because point A moves in a circular path relative to point B as the rigid body rotates, $\mathbf{a}_{A/B}$ has normal and tangential components (Figure 6.25(b)). The value of the tangential component is the product of $|\mathbf{r}_{A/B}|$ and the angular acceleration α of the rigid body. The normal component points towards the centre of the circular path, and its magnitude is $|\mathbf{v}_{A/B}|^2/|\mathbf{r}_{A/B}| = \omega^2|\mathbf{r}_{A/B}|$. Notice that because the normal component of acceleration points opposite to the direction of the vector $\mathbf{r}_{A/B}$, we can express it as a vector by writing it as $-\omega^2\mathbf{r}_{A/B}$.

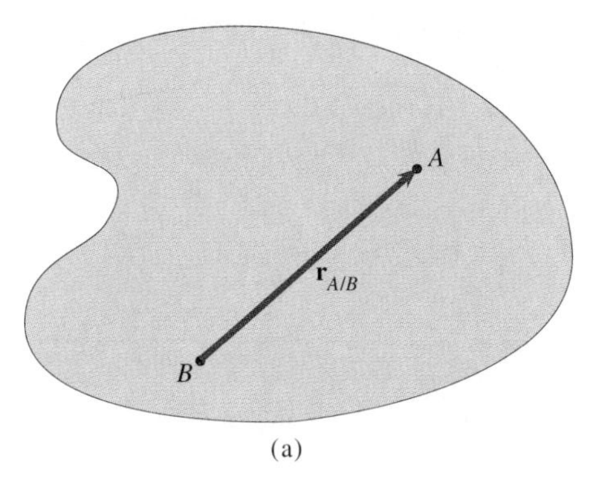

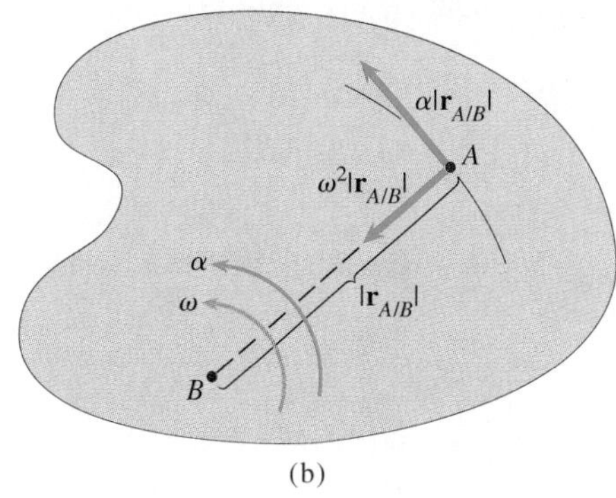

Figure 6.25

(a) Points of a rigid body in planar motion.
(b) Components of the acceleration of A relative to B.

Let's consider a circular disc of radius R rolling on a stationary plane surface with a counterclockwise angular velocity ω and counterclockwise angular acceleration α (Figure 6.26(a)). The disc's centre B moves in a straight line with velocity $R\omega$. Its velocity is to the left if ω is positive. The acceleration of the centre B is $d/dt(R\omega) = R\alpha$. Its acceleration is to the left if α is positive. *The magnitude of the acceleration of the centre of a round object rolling on a stationary surface is the product of the radius and the angular acceleration.*

Now that we know the acceleration of the disc's centre let us determine the acceleration of the point C in contact with the surface (Figure 6.26(b)). In terms of the coordinate system shown in Figure 6.26(c), the acceleration of the centre B is $-R\alpha\,\mathbf{i}$. Relative to B, point C moves in a circular path of radius R. The tangential component of the acceleration of C relative to B is $R\alpha\,\mathbf{i}$, and the normal component is $R\omega^2\,\mathbf{j}$. Therefore the acceleration of C is

$$\mathbf{a}_C = \mathbf{a}_B + \mathbf{a}_{C/B} = -R\alpha\,\mathbf{i} + R\alpha\,\mathbf{i} + R\omega^2\,\mathbf{j}$$
$$= R\omega^2\,\mathbf{j}$$

The acceleration of point C parallel to the surface is zero, but it does have an acceleration normal to the surface.

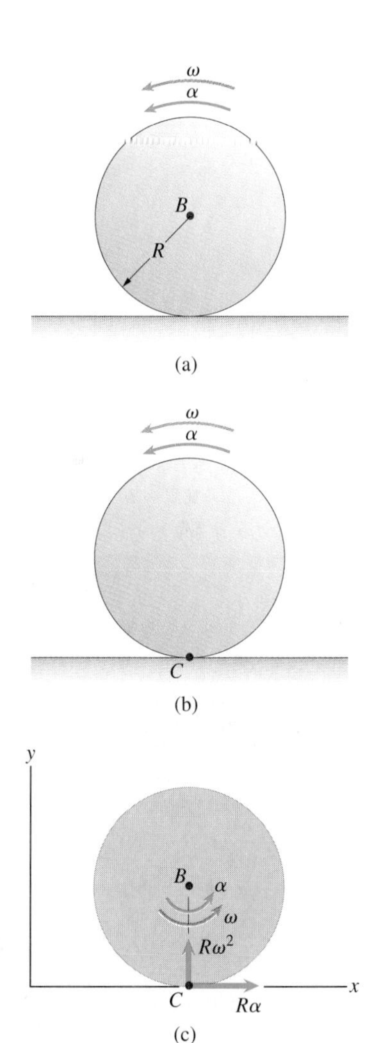

Figure 6.26

(a) A disk rolling with angular velocity ω and angular acceleration α.
(b) Point C is in contact with the surface.
(c) Determining the acceleration of C relative to B.

Expressing the acceleration of a point A relative to a point B in terms of A's circular path about B as we have done helps you visualize and understand it. However, just as we did in the case of the relative velocity, we can obtain $\mathbf{a}_{A/B}$ in a form more convenient for applications by using the angular velocity vector $\boldsymbol{\omega}$. The velocity of A relative to B is given in terms of $\boldsymbol{\omega}$ by Equation (6.5):

$$\mathbf{v}_{A/B} = \boldsymbol{\omega} \times \mathbf{r}_{A/B}$$

Taking the time derivative of this equation, we obtain

$$\mathbf{a}_{A/B} = \frac{d\boldsymbol{\omega}}{dt} \times \mathbf{r}_{A/B} + \boldsymbol{\omega} \times \mathbf{v}_{A/B}$$
$$= \frac{d\boldsymbol{\omega}}{dt} \times \mathbf{r}_{A/B} + \boldsymbol{\omega} \times (\boldsymbol{\omega} \times \mathbf{r}_{A/B})$$

Defining the **angular acceleration vector** α to be the rate of change of the angular velocity vector,

$$\alpha = \frac{d\boldsymbol{\omega}}{dt} \tag{6.7}$$

the acceleration of A relative to B is

$$\mathbf{a}_{A/B} = \alpha \times \mathbf{r}_{A/B} + \boldsymbol{\omega} \times (\boldsymbol{\omega} \times \mathbf{r}_{A/B})$$

Using this expression, we can write equations relating to the velocities and accelerations of two points of a rigid body in terms of its angular velocity and angular acceleration:

$$\mathbf{v}_A = \mathbf{v}_B + \boldsymbol{\omega} \times \mathbf{r}_{A/B} \tag{6.8}$$

$$\mathbf{a}_A = \mathbf{a}_B + \alpha \times \mathbf{r}_{A/B} + \boldsymbol{\omega} \times (\boldsymbol{\omega} \times \mathbf{r}_{A/B}) \tag{6.9}$$

In the case of two-dimensional motion, the term $\alpha \times \mathbf{r}_{A/B}$ in Equation (6.9) is the tangential component of the acceleration of A relative to B and $\boldsymbol{\omega} \times (\boldsymbol{\omega} \times \mathbf{r}_{A/B})$ is the normal component (Figure 6.27). Therefore, for two-dimensional motion, we can write Equation (6.9) in the simpler form

$$\mathbf{a}_A = \mathbf{a}_B + \alpha \times \mathbf{r}_{A/B} - \omega^2 \mathbf{r}_{A/B} \tag{6.10}$$

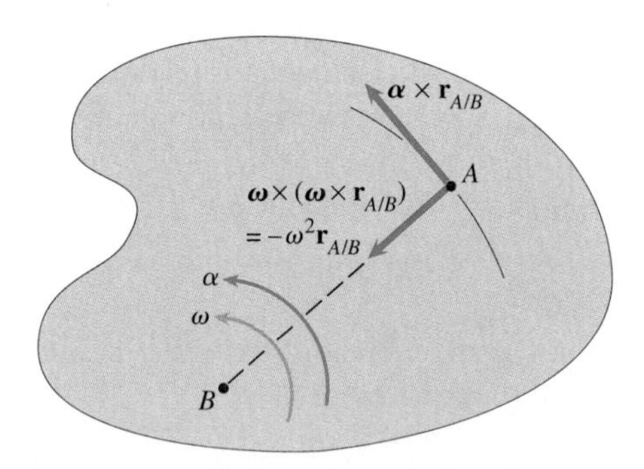

Figure 6.27
Vector components of the acceleration of A relative to B in planar motion.

In the following examples we use Equations (6.8)–(6.10) to analyse motions of rigid bodies. To determine accelerations of points and angular accelerations of rigid bodies, usually you must first determine the velocities of the points and the angular velocities of the rigid bodies, because Equations (6.9) and (6.10) contain the angular velocity. When you find a sequence of steps using Equation (6.8) that determines the velocities and angular velocities, the same sequence of steps using Equation (6.9) or (6.10) will determine the accelerations and angular accelerations.

Example 6.6

The rolling disc in Figure 6.28 has counterclockwise angular velocity ω and counterclockwise angular acceleration α. What is the acceleration of point A?

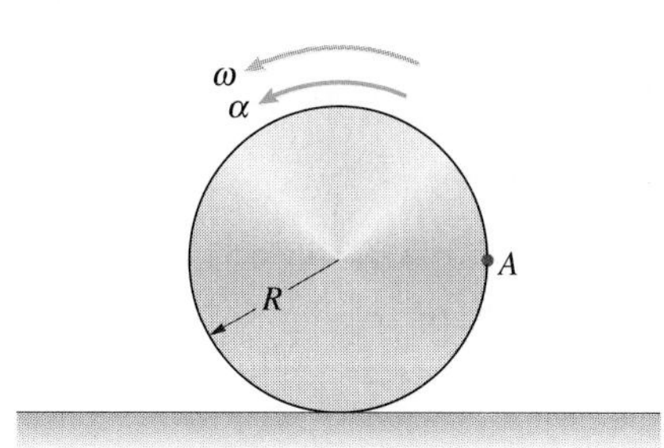

Figure 6.28

STRATEGY

We know that the magnitude of the acceleration of the centre of the disc is the product of the radius and the angular acceleration. Therefore we can express the acceleration of A as the sum of the acceleration of the centre and the acceleration of A relative to the centre. We will do so both by inspection and by using Equation (6.10).

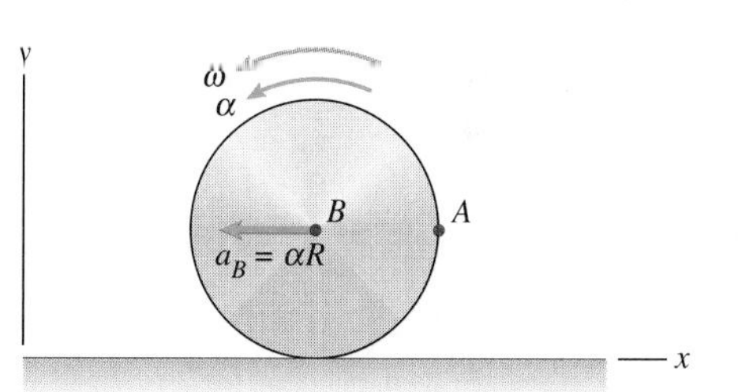

(a) Acceleration of the centre of the disc.

SOLUTION

In terms of the coordinate system in Figure (a), the acceleration of the centre B is $\mathbf{a}_B = -\alpha R\,\mathbf{i}$. A's motion in a circular path of radius R relative to B results in the tangential and normal components of relative acceleration shown in Figure (b):

$$\mathbf{a}_{A/B} = -\omega^2 R\,\mathbf{i} + \alpha R\,\mathbf{j}$$

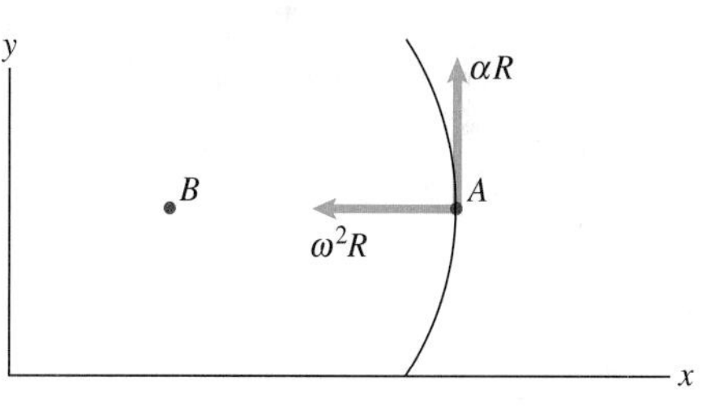

(b) Components of the acceleration of A relative to B.

Therefore the acceleration of A is

$$\begin{aligned}\mathbf{a}_A = \mathbf{a}_B + \mathbf{a}_{A/B} &= -\alpha R\,\mathbf{i} - \omega^2 R\,\mathbf{i} + \alpha R\,\mathbf{j} \\ &= (-\alpha R - \omega^2 R)\,\mathbf{i} + \alpha R\,\mathbf{j}\end{aligned}$$

Alternative Solution: The angular acceleration vector of the disc is $\boldsymbol{\alpha} = \alpha\,\mathbf{k}$, and the position of A relative to B is $\mathbf{r}_{A/B} = R\,\mathbf{i}$ (Figure (c)). From Equation (6.10), the acceleration of A is

$$\begin{aligned}\mathbf{a}_A &= \mathbf{a}_B + \boldsymbol{\alpha} \times \mathbf{r}_{A/B} - \omega^2 \mathbf{r}_{A/B} \\ &= -\alpha R\,\mathbf{i} + (\alpha\,\mathbf{k}) \times (R\,\mathbf{i}) - \omega^2 (R\,\mathbf{i}) \\ &= (-\alpha R - \omega^2 R)\,\mathbf{i} + \alpha R\,\mathbf{j}\end{aligned}$$

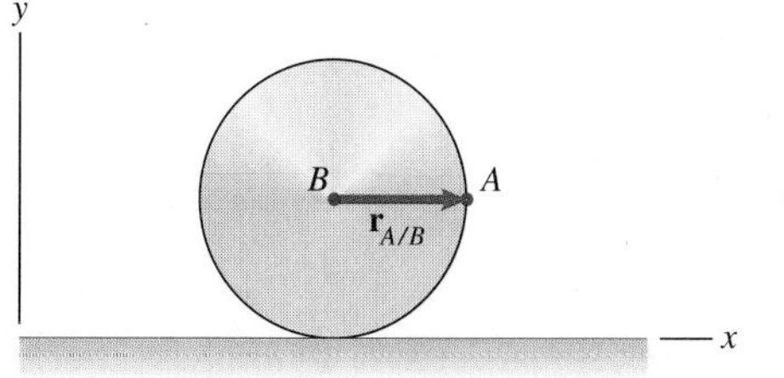

(c) Position of A relative to B.

Example 6.7

Bar AB in Figure 6.29 has a counterclockwise angular velocity of 10 rad/s and a clockwise angular acceleration of 300 rad/s^2. What are the angular accelerations of bars BC and CD?

Figure 6.29

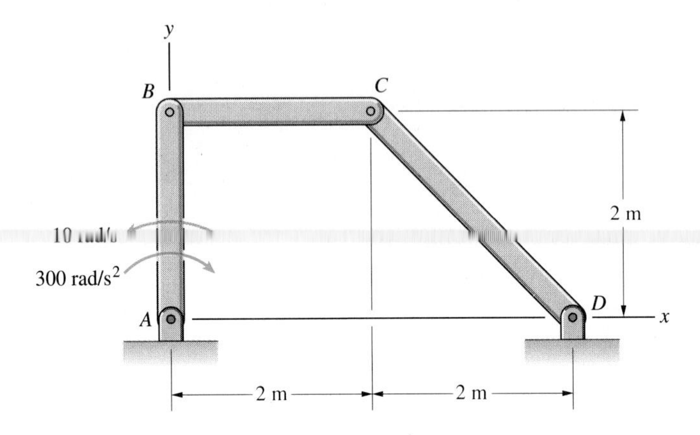

STRATEGY

Since we know the angular velocity of bar AB, we can determine the velocity of point B. Then we can apply Equation (6.8) to points C and D to obtain an equation for $\mathbf{v}_C$ in terms of the angular velocity of bar CD. We can also apply Equation (6.8) to points B and C to obtain an equation for $\mathbf{v}_C$ in terms of the angular velocity of bar BC. By equating the two expressions for $\mathbf{v}_C$, we will obtain a vector equation in two unknowns: the angular velocities of bars BC and CD. Then by following the same sequence of steps using Equation (6.10), we can obtain the angular accelerations of bars BC and CD.

SOLUTION

The velocity of B is (Figure (a))

$$\begin{aligned}\mathbf{v}_B &= \mathbf{v}_A + \boldsymbol{\omega}_{AB} \times \mathbf{r}_{B/A}\\ &= 0 + (10\,\mathbf{k}) \times (2\,\mathbf{j})\\ &= -20\,\mathbf{i}\ \text{m/s}\end{aligned}$$

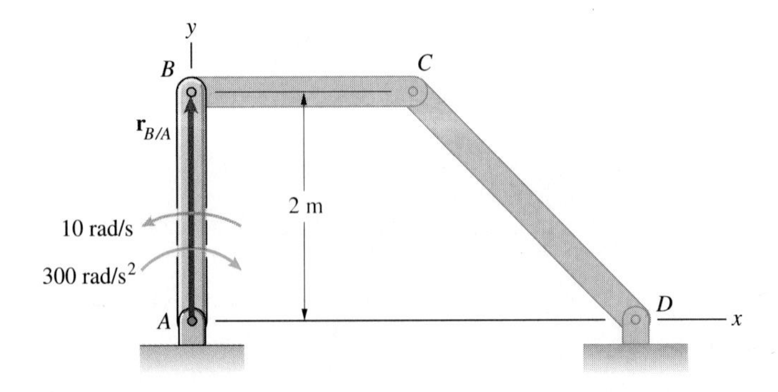

(a) Determining the motion of B.

Let ω_{CD} be the unknown angular velocity of bar CD (Figure (b)). The velocity of C in terms of the velocity of D is

$$\mathbf{v}_C = \mathbf{v}_D + \boldsymbol{\omega}_{CD} \times \mathbf{r}_{C/D}$$

$$= 0 + \begin{vmatrix} \mathbf{i} & \mathbf{j} & \mathbf{k} \\ 0 & 0 & \omega_{CD} \\ -2 & 2 & 0 \end{vmatrix}$$

$$= -2\omega_{CD}\,\mathbf{i} - 2\omega_{CD}\,\mathbf{j}$$

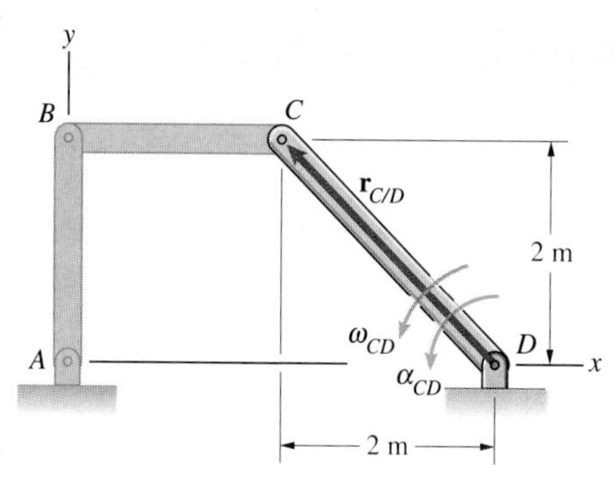

(b) Determining the motion of C in terms of the angular motion of bar CD.

Denoting the angular velocity of bar BC by ω_{BC} (Figure (c)), the velocity of C in terms of the velocity of B is

$$\mathbf{v}_C = \mathbf{v}_B + \boldsymbol{\omega}_{BC} \times \mathbf{r}_{C/B}$$

$$= -20\,\mathbf{i} + (\omega_{BC}\,\mathbf{k}) \times (2\,\mathbf{i})$$

$$= -20\,\mathbf{i} + 2\omega_{BC)}\,\mathbf{j}$$

Equating our two expressions for $\mathbf{v}_C$,

$$-2\omega_{CD}\,\mathbf{i} - 2\omega_{CD}\,\mathbf{j} = -20\,\mathbf{i} + 2\omega_{BC}\,\mathbf{j}$$

and equating the **i** and **j** components, we obtain $\omega_{CD} = 10$ rad/s and $\omega_{BC} = -10$ rad/s.

We can use the same sequence of steps to determine the angular accelerations. The acceleration of B is (Figure (a))

$$\mathbf{a}_B = \mathbf{a}_A + \alpha_{AB} \times \mathbf{r}_{B/A} - \omega_{AB}^2 \mathbf{r}_{B/A}$$

$$= 0 + (-300\,\mathbf{k}) \times (2\,\mathbf{j}) - (10)^2(2\,\mathbf{j})$$

$$= (600\,\mathbf{i} - 200\,\mathbf{j})\ \text{m/s}^2$$

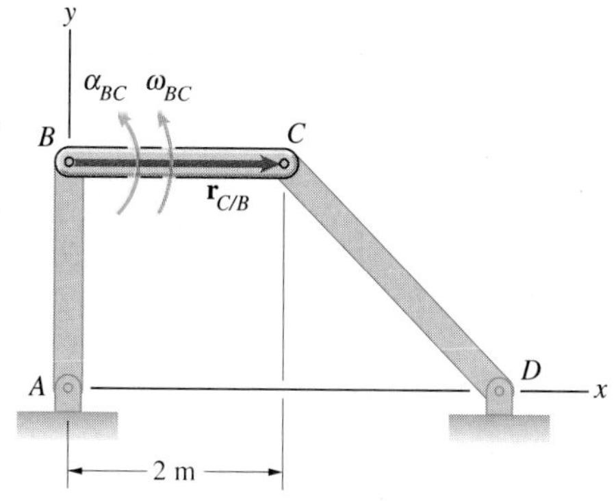

(c) Determining the motion of C in terms of the angular motion of bar BC.

The acceleration of C in terms of the acceleration of D is (Figure (b))

$$\mathbf{a}_C = \mathbf{a}_D + \alpha_{CD} \times \mathbf{r}_{C/D} - \omega_{CD}^2 \mathbf{r}_{C/D}$$

$$= 0 + \begin{vmatrix} \mathbf{i} & \mathbf{j} & \mathbf{k} \\ 0 & 0 & \alpha_{CD} \\ -2 & 2 & 0 \end{vmatrix} - (10)^2(-2\,\mathbf{i} + 2\,\mathbf{j})$$

$$= (200 - 2\alpha_{CD})\,\mathbf{i} - (200 + 2\alpha_{CD})\,\mathbf{j}$$

The acceleration of C in terms of the acceleration of B is (Figure (c))

$$\mathbf{a}_C = \mathbf{a}_B + \alpha_{BC} \times \mathbf{R}_{C/B} - \omega_{BC}^2 \mathbf{r}_{C/B}$$

$$= 600\,\mathbf{i} - 200\,\mathbf{j} + (\alpha_{BC}\,\mathbf{k}) \times (2\,\mathbf{i}) - (-10)^2(2\,\mathbf{i})$$

$$= 400\,\mathbf{i} - (200 - 2\alpha_{BC})\,\mathbf{j}$$

Equating the expressions for $\mathbf{a}_C$, we obtain

$$(200 - 2\alpha_{CD})\,\mathbf{i} - (200 + 2\alpha_{CD})\,\mathbf{j} = 400\,\mathbf{i} - (200 - 2\alpha_{BC})\,\mathbf{j}$$

and equating **i** and **j** components, we obtain the angular accelerations $\alpha_{BC} = 100\ \text{rad/s}^2$ and $\alpha_{CD} = -100\ \text{rad/s}^2$.

Problems

6.67 The rigid body rotates about the z axis with counterclockwise angular velocity ω and counterclockwise angular acceleration α. Determine the acceleration of point A relative to point B (a) by using Equation (6.9); (b) by using Equation (6.10).

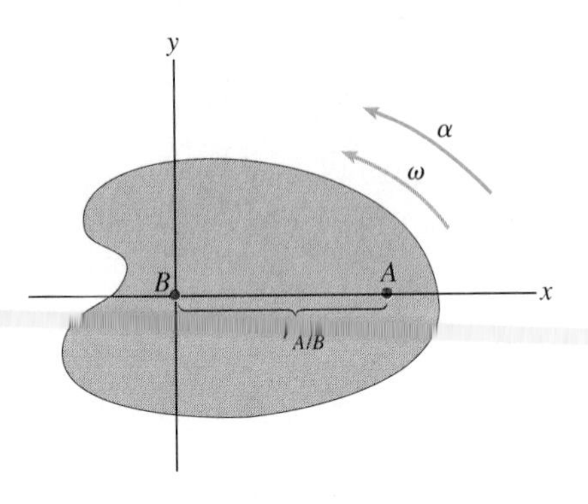

P6.67

6.68 The bar rotates with a counterclockwise angular velocity of 5 rad/s and a counterclockwise angular acceleration of 30 rad/s^2. Determine the acceleration of A (a) by expressing it in terms of polar coordinates; (b) by using Equation (6.9); (c) by using Equation (6.10).

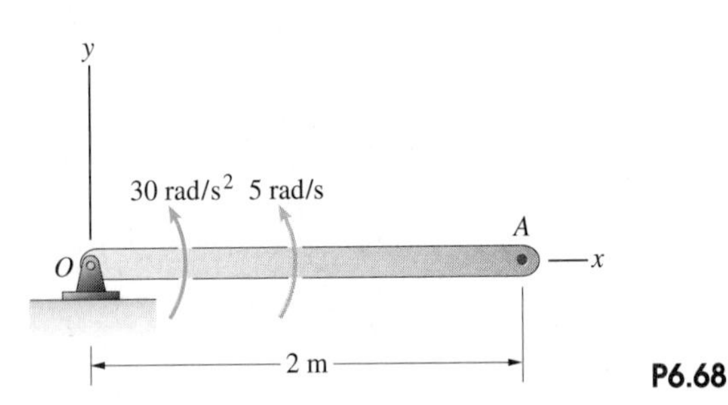

P6.68

6.69 The bar rotates with a counterclockwise angular velocity of 5 rad/s and a counterclockwise angular acceleration of 30 rad/s^2. Determine the acceleration of A (a) by using Equation (6.9); (b) by using Equation (6.10).

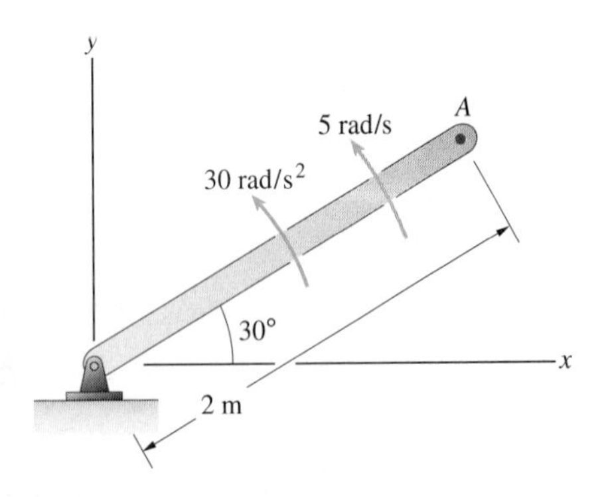

P6.69

6.70 The bar rotates with a constant angular velocity of 20 rad/s in the counterclockwise direction.
(a) Determine the acceleration of point B.
(b) Using your result from part (a) and Equation (6.10), determine the acceleration of point A.

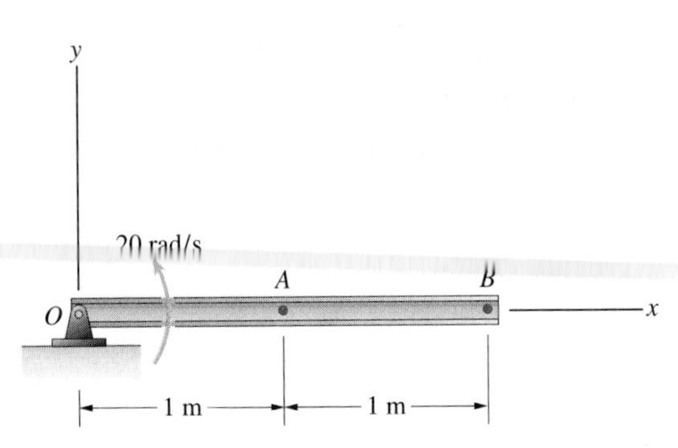

P6.70

6.71 The disc rolls on the plane surface. The velocity of point A is 6 m/s to the right, and its acceleration is 20 m/s^2 to the right.
(a) What is the angular acceleration vector of the disc?
(b) Determine the accelerations of points B, C and D.

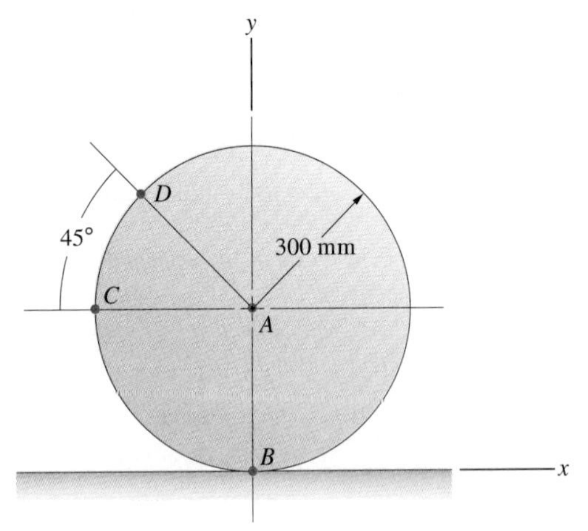

P6.71

6.72 The angular velocity and angular acceleration of bar AB are $\omega_{AB} = 2\,\text{rad/s}$, $\alpha_{AB} = 10\,\text{rad/s}^2$. The dimensions of the rectangular plate are 300 mm × 600 mm. What are the angular velocity and angular acceleration of the rectangular plate?

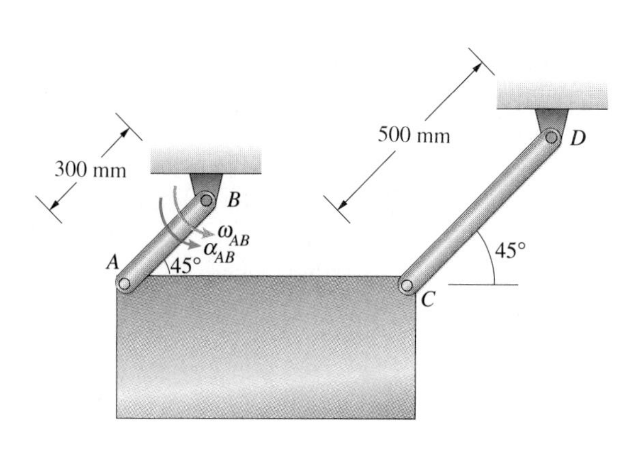

P6.72

6.73 The endpoints of the bar slide on the plane surfaces. Show that the acceleration of the midpoint G is related to the bar's angular velocity and angular acceleration by

$$\mathbf{a}_G = \frac{1}{2}L[(\alpha\cos\theta - \omega^2\sin\theta)\,\mathbf{i} - (\alpha\sin\theta + \omega^2\cos\theta)\,\mathbf{j}]$$

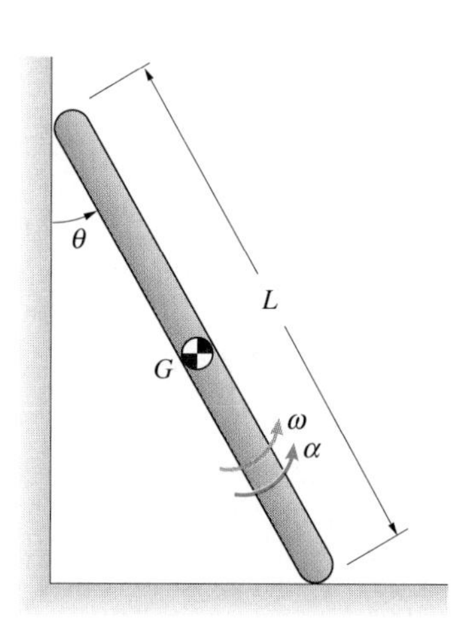

P6.73

6.74 The disc rolls on the circular surface with a constant clockwise angular velocity of 1 rad/s. What are the accelerations of points A and B?

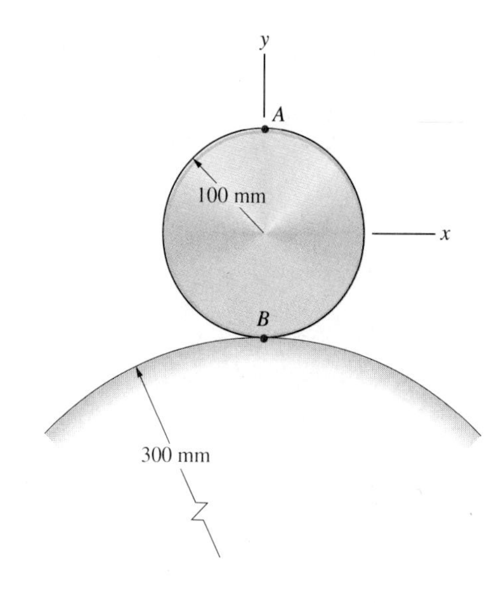

P6.74

6.75 The ring gear is stationary, and the sun gear has angular acceleration of $10\,\text{rad/s}^2$ in the counterclockwise direction. Determine the angular acceleration of the planet gears.

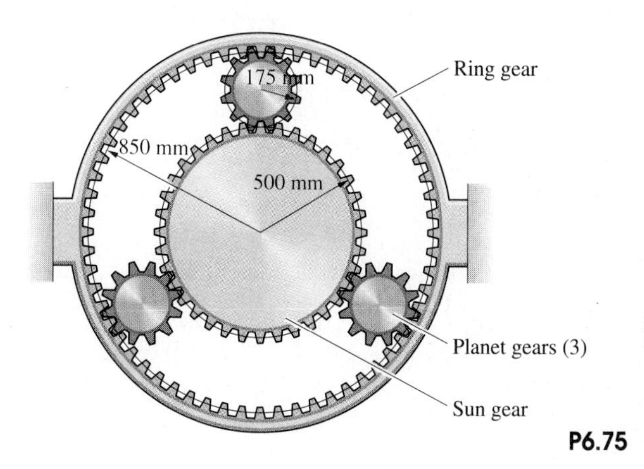

P6.75

6.76 The sun gear in Problem 6.75 has a counterclockwise angular velocity of 4 rad/s and a clockwise angular acceleration of $12\,\text{rad/s}^2$. What is the magnitude of the acceleration of the centrepoints of the planet gears?

6.77 The 1 m diameter disc rolls and point B of the 1 m long bar slides on the plane surface. Determine the angular acceleration of the bar and the acceleration of point B.

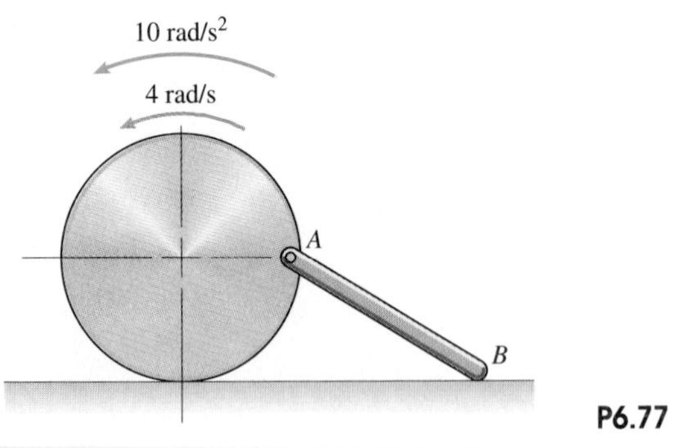

P6.77

6.78 The crank AB has a constant clockwise angular velocity of 200 rpm (revolutions per minute). What are the velocity and acceleration of the piston P?

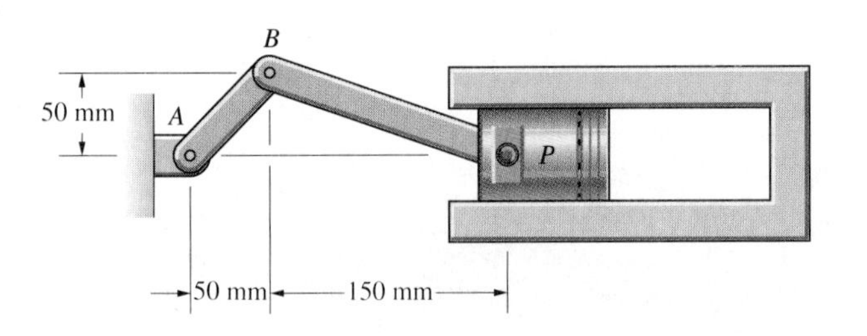

P6.78

6.79 Bar AB has a counterclockwise angular velocity of 10 rad/s and a clockwise angular acceleration of 20 rad/s^2. Determine the angular acceleration of bar BC and the acceleration of point C.

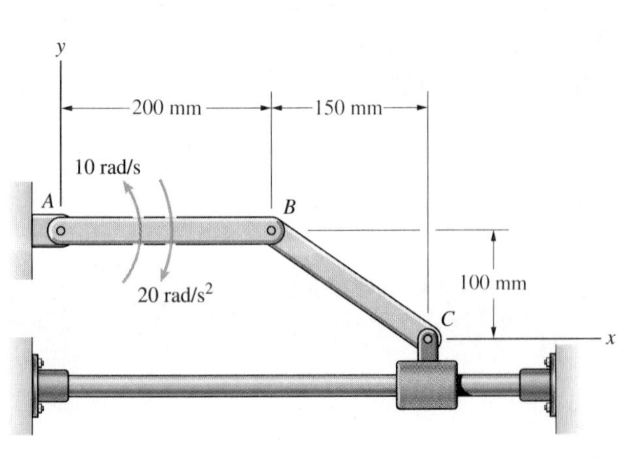

P6.79

6.80 The angular velocity and acceleration of bar AB are $\omega_{AB} = 2\,\text{rad/s}$, $\alpha_{AB} = 6\,\text{rad/s}^2$. What are the angular velocity and angular acceleration of bar BD?

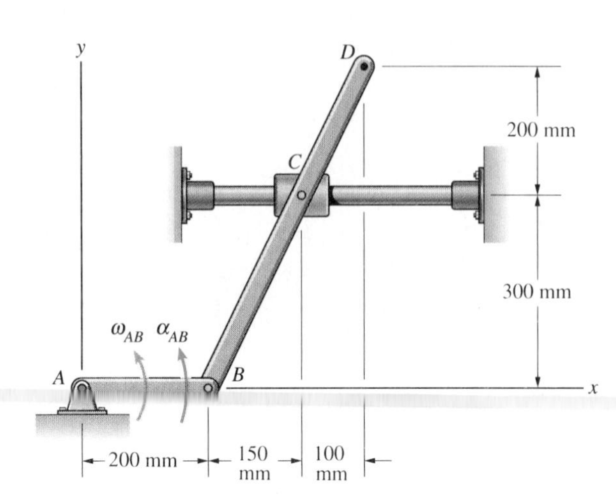

P6.80

6.81 In Problem 6.80, if the angular velocity and acceleration of bar AB are $\omega_{AB} = 2\,\text{rad/s}$, $\alpha_{AB} = -10\,\text{rad/s}^2$, what are the velocity and acceleration of point D?

6.82 If $\omega_{AB} = 6\,\text{rad/s}$ and $\alpha_{AB} = 20\,\text{rad/s}^2$, what are the velocity and acceleration of point C?

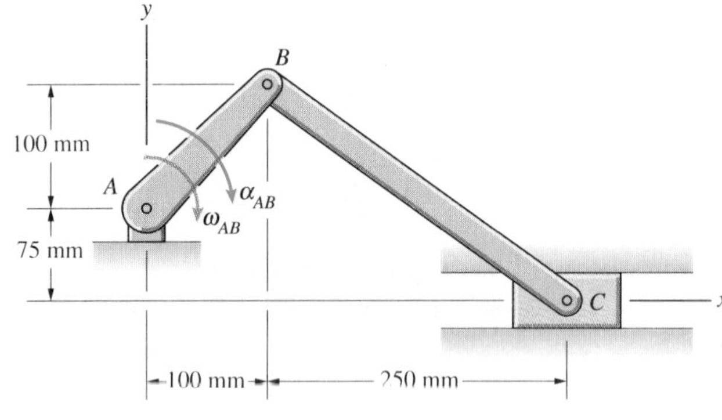

P6.82

6.83 A motor rotates the circular disc mounted at A, moving the saw back and forth. (The saw is supported by a horizontal slot so that point C moves horizontally.) The radius AB is 100 mm, and the link BC is 350 mm long. In the position shown, $\theta = 45^\circ$ and the link BC is horizontal. If the disc has a constant angular velocity of one revolution per second counterclockwise, what is the acceleration of the saw?

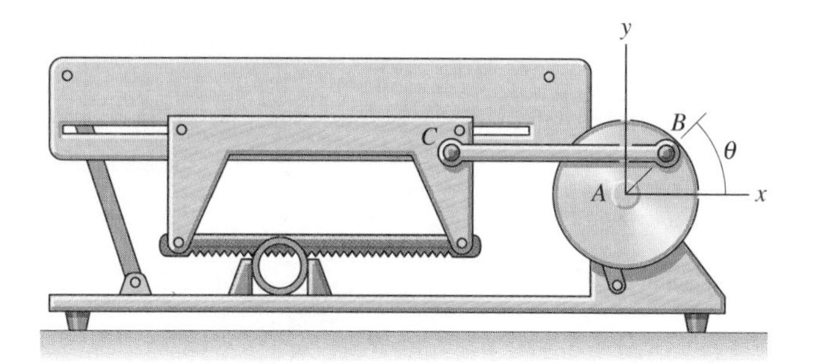

P6.83

6.84 In Problem 6.83, if the disc has a constant angular velocity of one revolution per second counterclockwise and $\theta = 180^\circ$, what is the acceleration of the saw?

6.85 If $\omega_{AB} = 2$ rad/s, $\alpha_{AB} = 2\text{ rad/s}^2$, $\omega_{BC} = 1$ rad/s and $\alpha_{BC} = 4\text{ rad/s}^2$, what is the acceleration of point C where the scoop of the excavator is attached?

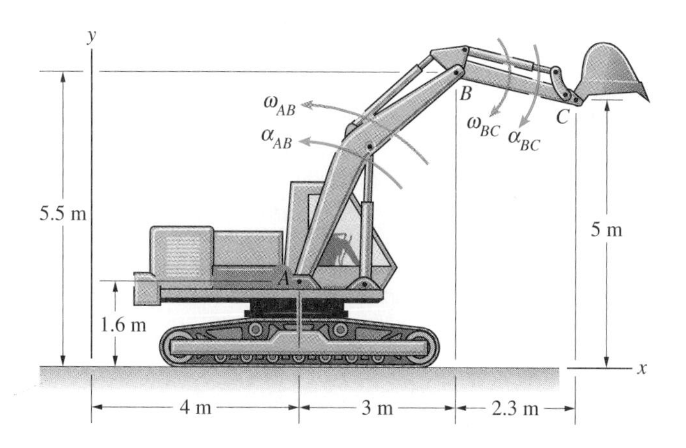

P6.85

6.86 If the velocity of point C of the excavator in Problem 6.85 is $\mathbf{v}_C = 4\,\mathbf{i}$ m/s and is constant at the instant shown, what are ω_{AB}, α_{AB}, ω_{BC} and α_{BC}?

6.87 Bar AB rotates in the counterclockwise direction with a constant angular velocity of 10 rad/s. What are the angular accelerations of bars BC and CD?

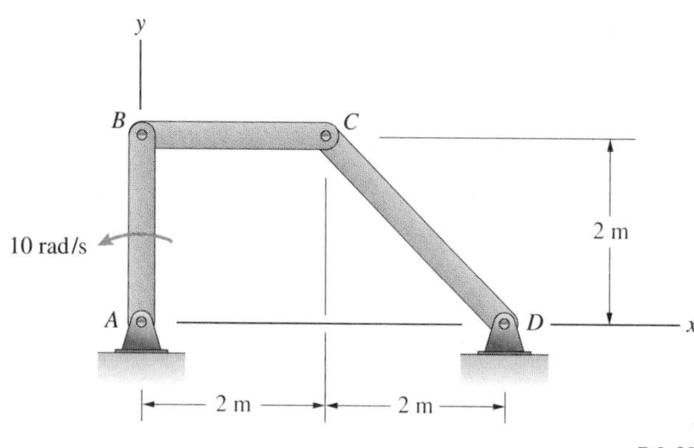

P6.87

6.88 At the instant shown, bar AB has no angular velocity but has a counterclockwise angular acceleration of 10 rad/s^2. Determine the acceleration of point E.

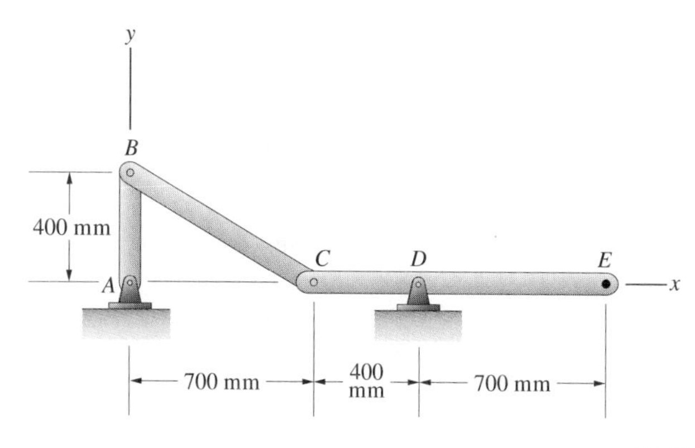

P6.88

6.89 If $\omega_{AB} = 12$ rad/s and $\alpha_{AB} = 100\text{ rad/s}^2$, what are the angular accelerations of bars BC and CD?

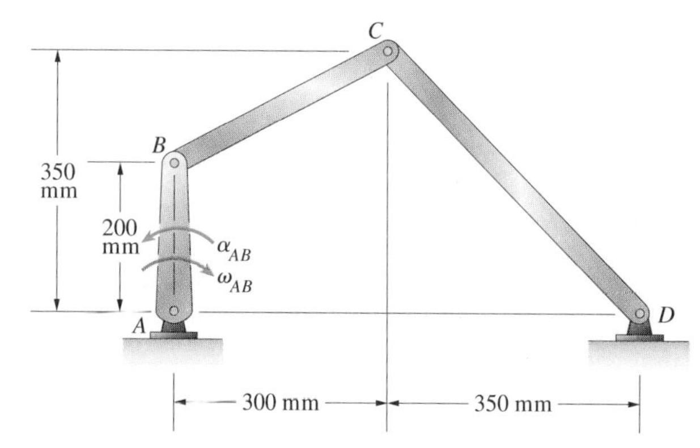

P6.89

6.90 If $\omega_{AB} = 4\,\text{rad/s}$ counterclockwise and $\alpha_{AB} = 12\,\text{rad/s}^2$ counterclockwise, what is the acceleration of point C?

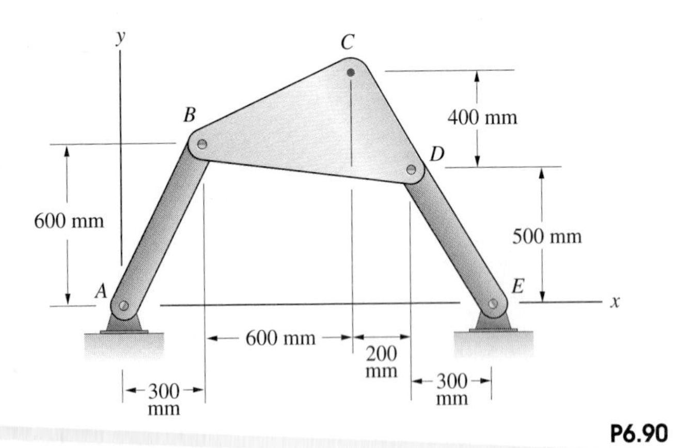

P6.90

6.91 In Problem 6.90, if $\omega_{AB} = 6\,\text{rad/s}$ clockwise and $\alpha_{DE} = 0$, what is the acceleration of point C?

6.92 If arm AB has a constant clockwise angular velocity of 0.8 rad/s, arm BC has a constant clockwise angular velocity of 0.2 rad/s, and arm CD remains vertical, what is the acceleration of the part D?

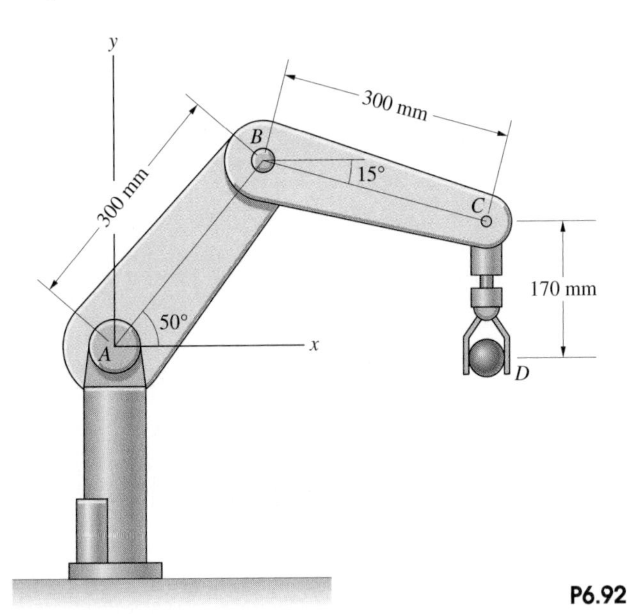

P6.92

6.93 In Problem 6.92, if arm AB has a constant clockwise angular velocity of 0.8 rad/s and you want part D to have zero velocity and acceleration, what are the necessary angular velocities and angular accelerations of arms BC and CD?

6.94 In Problem 6.92, if you want arm CD to remain vertical and you want part D to have velocity $\mathbf{v}_D = 1.0\,\mathbf{i}\,\text{m/s}$ and zero acceleration, what are the necessary angular velocities and angular accelerations of arms AB and BC?

6.95 If the velocity of point C of the excavator in Problem 6.85 is zero and its acceleration is $\mathbf{a}_C = 4\,\mathbf{i}\,\text{m/s}^2$ at the instant shown, what are ω_{AB}, ω_{BC} and α_{BC}?

6.96 The ring gear is fixed, and the hub and planet gears are bonded together. The connecting rod has a counterclockwise angular acceleration of $10\,\text{rad/s}^2$. Determine the angular accelerations of the planet and sun gears.

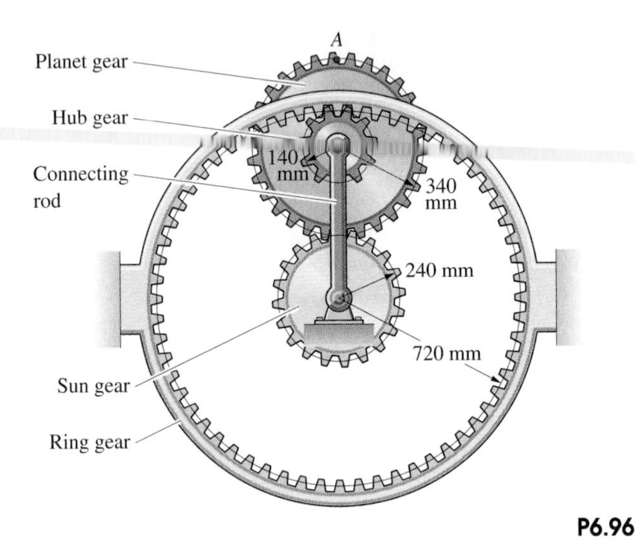

P6.96

6.97 The connecting rod in Problem 6.96 has a counterclockwise angular velocity of 4 rad/s and a clockwise angular acceleration of $12\,\text{rad/s}^2$. Determine the magnitude of the acceleration of point A.

6.98 The large gear is fixed. The angular velocity and angular acceleration of bar AB are $\omega_{AB} = 2\,\text{rad/s}$, $\alpha_{AB} = 4\,\text{rad/s}^2$. Determine the angular accelerations of bars CD and DE.

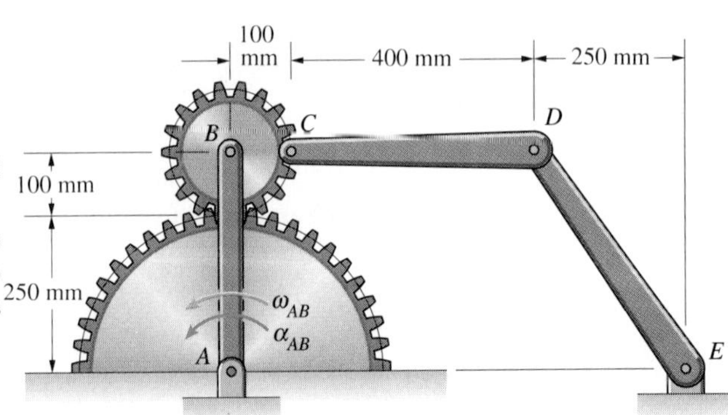

P6.98

6.5 Sliding Contacts

Here we consider a type of problem superficially similar to those we have discussed previously in this chapter, but which requires a different method of solution. For example, suppose that we know the angular velocity and angular acceleration of the bar AB in Figure 6.30, and we want to determine the angular velocity and angular acceleration of bar AC. We cannot use the equation $\mathbf{v}_A = \mathbf{v}_B + \boldsymbol{\omega} \times \mathbf{r}_{A/B}$ to express the velocity of point A in terms of the angular velocity of bar AB, because we derived it under the assumption that points A and B are points of the same rigid body. Point A is not a part of the bar AB, but moves relative to it as the pin slides along the slot. This is an example of a **sliding contact** between rigid bodies. To solve such problems, we must re-derive Equations (6.8)–(6.10) without making the assumption that A is a point of the rigid body.

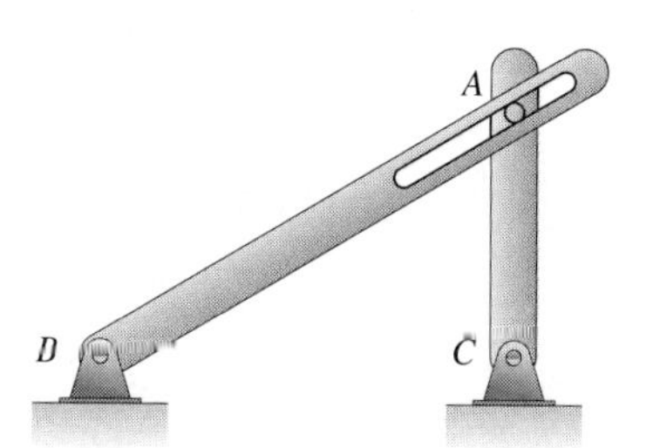

Figure 6.30
Linkage with a sliding contact.

In Figure 3.31, we assume the coordinate system is body-fixed and that B is a point of the rigid body, but we *do not* assume that A is a point of the rigid body. The position of A relative to O is

$$\mathbf{r}_A = \mathbf{r}_B + \underbrace{x\,\mathbf{i} + y\,\mathbf{j} + z\,\mathbf{k}}_{\mathbf{r}_{A/B}}$$

where x, y and z are the coordinates of A in terms of the body-fixed coordinate system. Our next step is to take the time derivative of this expression to obtain an equation for the velocity of A. In doing so, we recognize that the unit

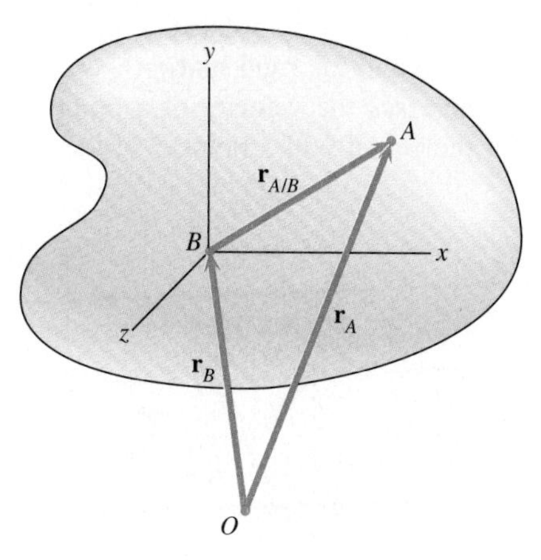

Figure 6.31
A point B of a rigid body, a body-fixed coordinate system, and an arbitrary point A.

vectors **i**, **j** and **k** are not constant, because they rotate with the body-fixed coordinate system:

$$\mathbf{v}_A = \mathbf{v}_B + \frac{dx}{dt}\mathbf{i} + x\frac{d\mathbf{i}}{dt} + \frac{dy}{dt}\mathbf{j} + y\frac{d\mathbf{j}}{dt} + \frac{dz}{dt}\mathbf{k} + z\frac{d\mathbf{k}}{dt}$$

What are the time derivatives of the unit vectors? in Section 6.3 we showed that if $\mathbf{r}_{P/B}$ is the position of a point P of a rigid body relative to another point B of the same rigid body, $d\mathbf{r}_{P/B}/dt = \mathbf{v}_{P/B} = \boldsymbol{\omega} \times \mathbf{r}_{P/B}$. Since we can regard the unit vector **i** as the position vector of a point P of the rigid body (Figure 6.32), its time derivative is $d\,\mathbf{i}/dt = \boldsymbol{\omega} \times \mathbf{i}$. Applying the same argument to the unit vectors **j** and **k**, we obtain

$$\frac{d\mathbf{i}}{dt} = \boldsymbol{\omega} \times \mathbf{i} \qquad \frac{d\mathbf{j}}{dt} = \boldsymbol{\omega} \times \mathbf{j} \qquad \frac{d\mathbf{k}}{dt} = \boldsymbol{\omega} \times \mathbf{k}$$

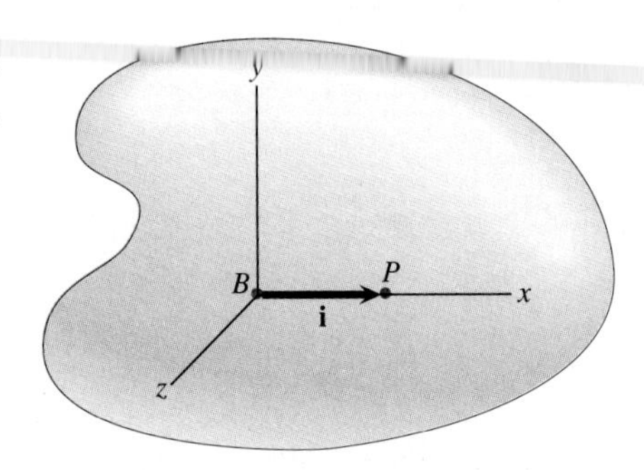

Figure 6.32
Interpreting **i** as the position vector of a point P relative to B.

Using these expressions, we can write the velocity of point A as

$$\boxed{\mathbf{v}_A = \mathbf{v}_B + \underbrace{\mathbf{v}_{A\,\text{rel}} + \boldsymbol{\omega} \times \mathbf{r}_{A/B}}_{\mathbf{v}_{A/B}}} \tag{6.11}$$

where

$$\mathbf{v}_{A\,\text{rel}} = \frac{dx}{dt}\mathbf{i} + \frac{dy}{dt}\mathbf{j} + \frac{dz}{dt}\mathbf{k} \tag{6.12}$$

is the velocity of A relative to the body-fixed coordinate system. That is, $\mathbf{v}_{A\,\text{rel}}$ is the velocity of A relative to the rigid body.

Equation (6.11) expresses the velocity of a point A as the sum of three terms (Figure 6.33): the velocity of a point B of the rigid body, the velocity

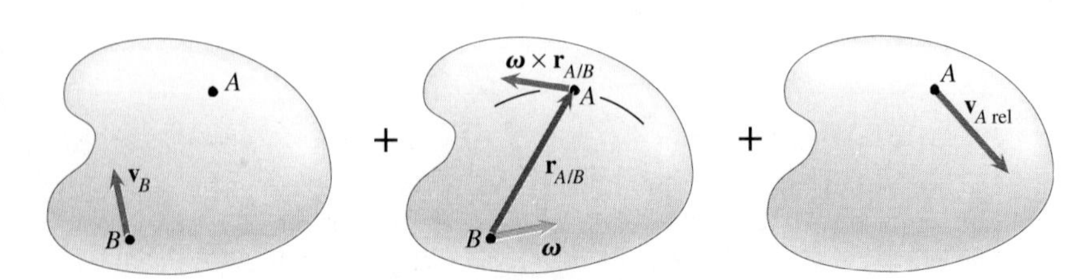

Figure 6.33
Expressing the velocity of A in terms of the velocity of a point B of the rigid body.

$\boldsymbol{\omega} \times \mathbf{r}_{A/B}$ of A relative to B due to the rotation of the rigid body, and the velocity $\mathbf{v}_{A\,\mathrm{rel}}$ of A relative to the rigid body.

To obtain an equation for the acceleration of point A, we take the time derivative of Equation (6.11) and use Equation (6.12). The result is

$$\mathbf{a}_A = \mathbf{a}_B + \underbrace{\mathbf{a}_{A\,\mathrm{rel}} + 2\boldsymbol{\omega} \times \mathbf{v}_{A\,\mathrm{rel}} + \boldsymbol{\alpha} \times \mathbf{r}_{A/B} + \boldsymbol{\omega} \times (\boldsymbol{\omega} \times \mathbf{r}_{A/B})}_{\mathbf{a}_{A/B}} \qquad (6.13)$$

where

$$\mathbf{a}_{A\,\mathrm{rel}} = \frac{d^2x}{dt^2}\,\mathbf{i} + \frac{d^2y}{dt^2}\,\mathbf{j} + \frac{d^2z}{dt^2}\,\mathbf{k} \qquad (6.14)$$

is the acceleration of A relative to the body-fixed coordinate system.

The terms $\mathbf{v}_A$ and $\mathbf{a}_A$ are the velocity and acceleration of point A relative to a non-rotating coordinate system that is stationary relative to point O. The terms $\mathbf{v}_{A\,\mathrm{rel}}$ and $\mathbf{a}_{A\,\mathrm{rel}}$ are the velocity and acceleration of point A measured by an observer moving with the rigid body (Figure 6.34). If A is a point of the rigid body, $\mathbf{v}_{A\,\mathrm{rel}}$ and $\mathbf{a}_{A\,\mathrm{rel}}$ are zero, and Equations (6.11) and (6.13) are identical to Equations (6.8) and (6.9).

In the case of two-dimensional motion, we can express Equation (6.13) in the simpler form

$$\mathbf{a}_A = \mathbf{a}_B + \underbrace{\mathbf{a}_{A\,\mathrm{rel}} + 2\boldsymbol{\omega} \times \mathbf{v}_{A\,\mathrm{rel}} + \boldsymbol{\alpha} \times \mathbf{r}_{A/B} - \omega^2\mathbf{r}_{A/B}}_{\mathbf{a}_{A/B}} \qquad (6.15)$$

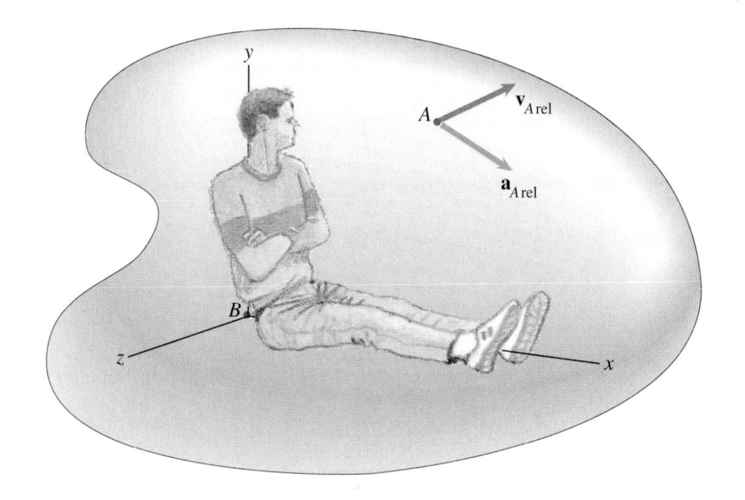

Figure 6.34

Imagine yourself to be stationary relative to the rigid body.

In the following examples we analyse the motions of linkages with sliding contacts. You can use the same approach that you applied to systems of pinned rigid bodies, beginning with points whose velocities and accelerations are known and applying Equations (6.11) and (6.15).

Example 6.8

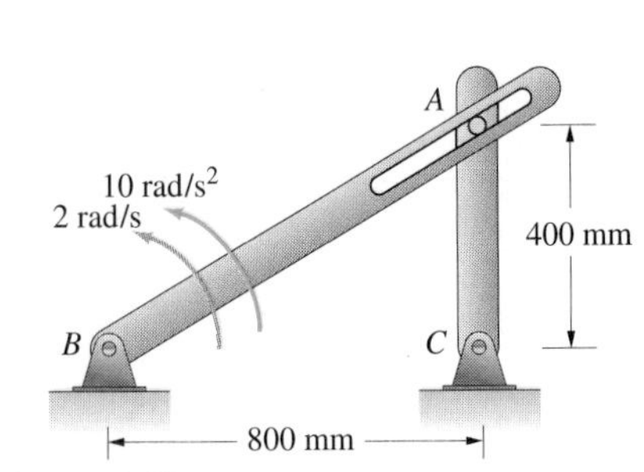

Figure 6.35

Bar AB in Figure 6.35 has a counterclockwise angular velocity of 2 rad/s and a counterclockwise angular acceleration of $10\,\text{rad/s}^2$.

(a) Determine the angular velocity of bar AC and the velocity of the pin A relative to the slot in bar AB.

(b) Determine the angular acceleration of bar AC and the acceleration of the pin A relative to the slot in bar AB.

STRATEGY

We can use Equation (6.11) to express v_A in terms of the velocity of A relative to the slot in the bar and the known angular velocity of bar AB. A and C are both points of the bar AC, so we can also express $\mathbf{v}_A$ in terms of the angular velocity of the bar AC in the usual way. By equating the resulting expressions for $\mathbf{v}_A$, we will obtain a vector equation in terms of the velocity of A relative to the slot and the angular velocity of bar AC. Then, by following the same sequence of steps but this time using Equation (6.15), we can obtain the acceleration of A relative to the slot and the angular acceleration of bar AC.

SOLUTION

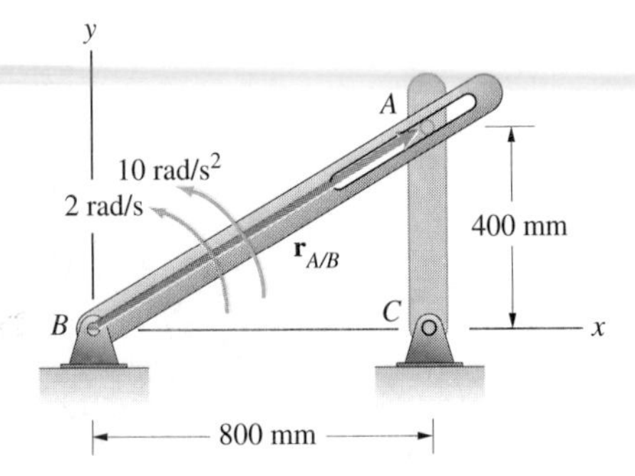

(a) Expressing the velocity and acceleration of A in terms of the angular velocity and acceleration of bar AB.

(a) Applying Equation (6.11) to bar AB (Figure (a)), the velocity of A is

$$\mathbf{v}_A = \mathbf{v}_B + \mathbf{v}_{A\,\text{rel}} + \boldsymbol{\omega}_{AB} \times \mathbf{r}_{A/B}$$

$$= 0 + \mathbf{v}_{A\,\text{rel}} + \begin{vmatrix} \mathbf{i} & \mathbf{j} & \mathbf{k} \\ 0 & 0 & 2 \\ 0.8 & 0.4 & 0 \end{vmatrix}$$

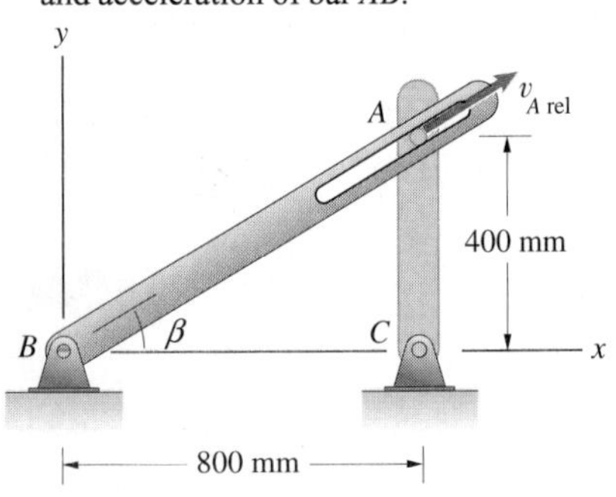

(b) Direction of the velocity of A relative to the body-fixed coordinate system.

Assuming the coordinate system in Figure (a) to be body-fixed with respect to bar AB, the velocity $\mathbf{v}_{A\,\text{rel}}$ is the velocity of A relative to this coordinate system. We don't know the magnitude of $\mathbf{v}_{A\,\text{rel}}$, but its direction is parallel to the slot (Figure (b)). Therefore we can express it as

$$\mathbf{v}_{A\,\text{rel}} = v_{A\,\text{rel}} \cos\beta\,\mathbf{i} + v_{A\,\text{rel}} \sin\beta\,\mathbf{j}$$

where $\beta = \arctan(0.4/0.8)$. Substituting this expression into our equation for $\mathbf{v}_A$, we obtain

$$\mathbf{v}_A = (v_{A\,\text{rel}} \cos\beta - 0.8)\,\mathbf{i} + (v_{A\,\text{rel}} \sin\beta + 1.6)\,\mathbf{j}$$

Let ω_{AC} be the angular velocity of bar AC (Figure (c)). Expressing the velocity of A in terms of the velocity of C, we obtain

$$\mathbf{v}_A = \mathbf{v}_C + \boldsymbol{\omega}_{AC} \times \mathbf{r}_{A/C}$$
$$= 0 + (\omega_{AC}\,\mathbf{k}) \times (0.4\,\mathbf{j})$$
$$= -0.4\omega_{AC}\,\mathbf{i}$$

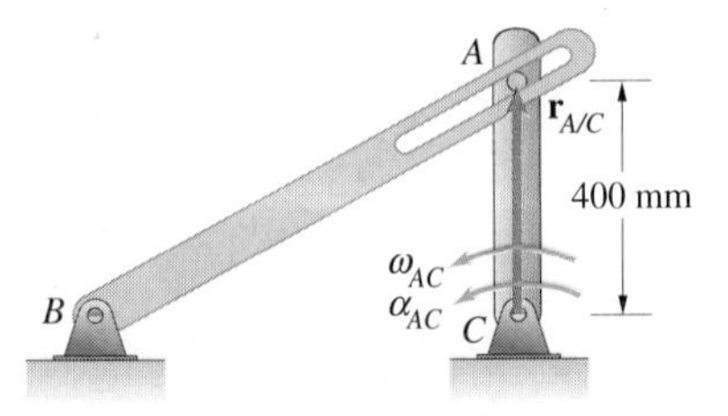

(c) Expressing the velocity and acceleration of A in terms of the angular velocity and acceleration of bar AC.

Notice that there is no relative velocity term in this equation, because A is a point of the bar AC. Equating our two expressions for $\mathbf{v}_A$, we obtain

$$(v_{A\,\text{rel}} \cos\beta - 0.8)\,\mathbf{i} + (v_{A\,\text{rel}} \sin\beta + 1.6)\,\mathbf{j} = -0.4\omega_{AC}\,\mathbf{i}$$

Equating **i** and **j** components yields the two equations

$$v_{A\,\mathrm{rel}} \cos\beta - 0.8 = -0.4\omega_{AC}$$
$$v_{A\,\mathrm{rel}} \sin\beta + 1.6 = 0$$

Solving them, we obtain $v_{A\,\mathrm{rel}} = -3.58\ \mathrm{m/s}$ and $\omega_{AC} = 10\ \mathrm{rad/s}$. At this instant, the pin A is moving relative to the slot at $3.58\ \mathrm{m/s}$ towards B. The vector $\mathbf{v}_{A\,\mathrm{rel}}$ is

$$\mathbf{v}_{A\,\mathrm{rel}} = -3.58(\cos\beta\,\mathbf{i} + \sin\beta\,\mathbf{j}) = (-3.21\,\mathbf{i} - 1.6\,\mathbf{j})\ \mathrm{m/s}$$

(b) Applying Equation (6.15) to bar AB (Figure (b)), the acceleration of A is

$$\mathbf{a}_A = \mathbf{a}_B + \mathbf{a}_{A\,\mathrm{rel}} + 2\boldsymbol{\omega}_{AB} \times \mathbf{v}_{A\,\mathrm{rel}} + \alpha_{AB} \times \mathbf{r}_{A/B} - \omega_{AB}^2 \mathbf{r}_{A/B}$$

$$= 0 + \mathbf{a}_{A\,\mathrm{rel}} + 2\begin{vmatrix} \mathbf{i} & \mathbf{j} & \mathbf{k} \\ 0 & 0 & 2 \\ -3.2 & -1.6 & 0 \end{vmatrix} + \begin{vmatrix} \mathbf{i} & \mathbf{j} & \mathbf{k} \\ 0 & 0 & 10 \\ 0.8 & 0.4 & 0 \end{vmatrix}$$

$$- (2)^2(0.8\,\mathbf{i} + 0.4\,\mathbf{j})$$

The acceleration of A relative to the body-fixed coordinate system is parallel to the slot (Figure (d)), so we can write it in the same way as we did $\mathbf{v}_{A\,\mathrm{rel}}$:

$$\mathbf{a}_{A\,\mathrm{rel}} = a_{A\,\mathrm{rel}} \cos\beta\,\mathbf{i} + a_{A\,\mathrm{rel}} \sin\beta\,\mathbf{j}$$

Substituting this expression into our equation for $\mathbf{a}_A$ gives

$$\mathbf{a}_A = (a_{A\,\mathrm{rel}} \cos\beta - 0.8)\,\mathbf{i} + (a_{A\,\mathrm{rel}} \sin\beta - 6.4)\,\mathbf{j}$$

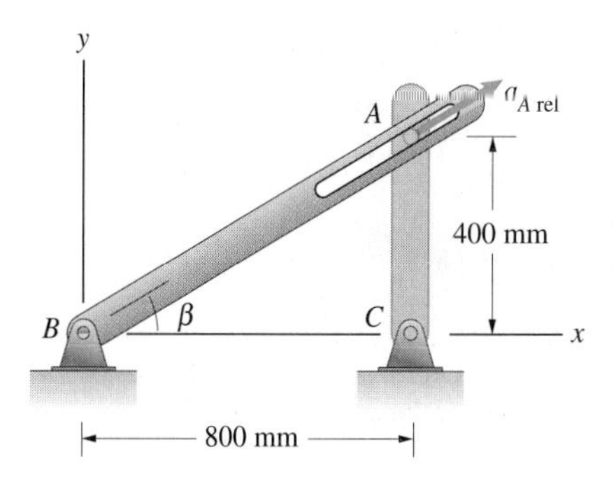

(d) Direction of the acceleration of A relative to the body-fixed coordinate system.

Expressing the acceleration of A in terms of the acceleration of C (Figure (c)), we obtain

$$\mathbf{a}_A = \mathbf{a}_C + \alpha_{AC} \times \mathbf{r}_{A/C} - \omega_{AC}^2 \mathbf{r}_{A/C}$$
$$= 0 + (\alpha_{AC}\,\mathbf{k}) \times (0.4\,\mathbf{j}) - (10)^2(0.4\,\mathbf{j})$$
$$= -0.4\alpha_{AC}\,\mathbf{i} - 40\,\mathbf{j}$$

Equating our expression for $\mathbf{a}_A$, we obtain

$$(a_{A\,\mathrm{rel}} \cos\beta - 0.8)\,\mathbf{i} + (a_{A\,\mathrm{rel}} \sin\beta - 6.4)\,\mathbf{j} = -0.4\alpha_{AC}\,\mathbf{i} - 40\,\mathbf{j}$$

Equating **i** and **j** components yields the two equations

$$a_{A\,\mathrm{rel}} \cos\beta - 0.8 = -0.4\alpha_{AC}$$
$$a_{A\mathrm{rel}} \sin\beta - 6.4 = -40$$

Solving them, we obtain $a_{A\,\mathrm{rel}} = -75.13\ \mathrm{m/s^2}$ and $\alpha_{AC} = 170\ \mathrm{rad/s^2}$. At this instant, the pin A is accelerating relative to the slot at $75.13\ \mathrm{m/s^2}$ towards B.

Example 6.9

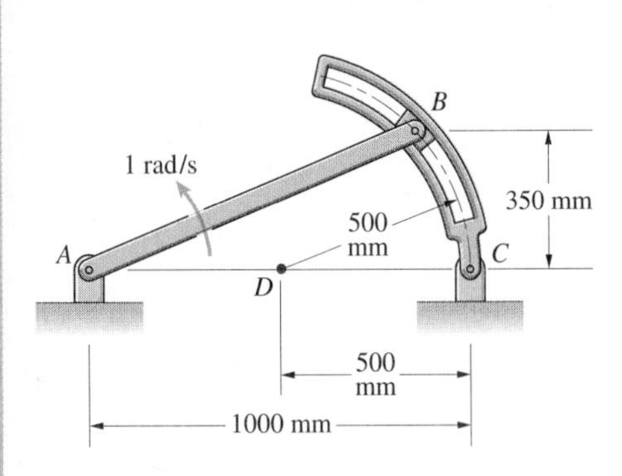

Figure 6.36

Bar AB in Figure 6.36 rotates with a constant counterclockwise angular velocity of 1 rad/s. The block B slides in a circular slot in the curved bar BC. At the instant shown, the centre of the circular slot is at D. Determine the angular velocity and angular acceleration of bar BC.

STRATEGY

Since we know the angular velocity of bar AB, we can determine the velocity of point B. Because B is not a point of bar BC, we must apply Equation (6.11) to points B and C. By equating our expressions for $\mathbf{v}_B$, we can solve for the angular velocity of bar BC. Then, by following the same sequence of steps but this time using Equation (6.15), we can determine the angular acceleration of bar BC.

SOLUTION

To determine the velocity of B, we express it in terms of the velocity of A and the angular velocity of bar AB: $\mathbf{v}_B = \mathbf{v}_A + \boldsymbol{\omega}_{AB} \times \mathbf{r}_{B/A}$. In terms of the coordinate system shown in Figure (a), the position vector of B relative to A is

$$\mathbf{r}_{B/A} = (0.500 + 0.500 \cos\beta)\,\mathbf{i} + 0.350\,\mathbf{j} = (0.857\,\mathbf{i} + 0.350\,\mathbf{j})\,\text{m}$$

where $\beta = \arcsin(350/500) = 44.4°$. Therefore the velocity of B is

$$\mathbf{v}_B = \mathbf{v}_A + \boldsymbol{\omega}_{AB} \times \mathbf{r}_{B/A} = 0 + \begin{vmatrix} \mathbf{i} & \mathbf{j} & \mathbf{k} \\ 0 & 0 & 1 \\ 0.857 & 0.350 & 0 \end{vmatrix} \tag{6.16}$$

$$= (-0.350\,\mathbf{i} + 0.857\,\mathbf{j})\,\text{m/s}$$

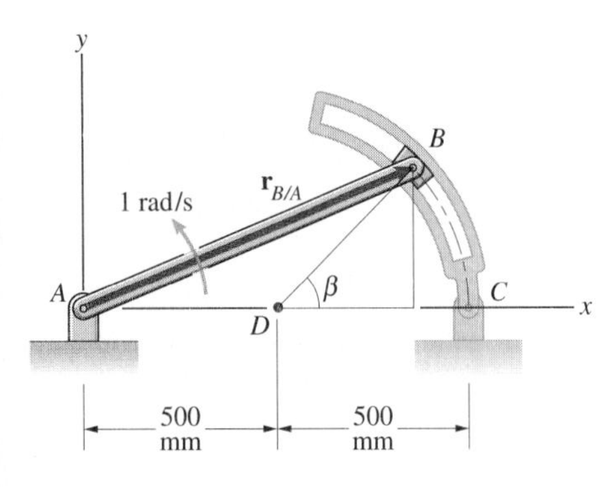

(a) Determining the velocity of point B.

To apply Equation (6.11) to points B and C, we introduce a coordinate system with its origin at C that rotates with the curved bar (Figure (b)). The velocity of B is

$$\mathbf{v}_B = \mathbf{v}_C + \mathbf{v}_{B\,\text{rel}} + \boldsymbol{\omega}_{BC} \times \mathbf{r}_{B/C} \tag{6.17}$$

The position vector of B relative to C is

$$\mathbf{r}_{B/C} = -(0.500 - 0.500 \cos\beta)\,\mathbf{i} + 0.350\,\mathbf{j} = (-0.143\,\mathbf{i} + 0.350\,\mathbf{j})\,\text{m}$$

Relative to the body-fixed coordinate system, point B moves in a circular path about point D (Figure (c)). In terms of the angle β, the vector $\mathbf{v}_{B\,\text{rel}}$ is

$$\mathbf{v}_{B\,\text{rel}} = -v_{B\,\text{rel}} \sin\beta\,\mathbf{i} + v_{B\,\text{rel}} \cos\beta\,\mathbf{j}$$

We substitute these expressions for $\mathbf{r}_{B/C}$ and $\mathbf{v}_{B\,\text{rel}}$ into Equation (6.17), obtaining

$$\mathbf{v}_B = -v_{B\,\text{rel}} \sin\beta\,\mathbf{i} + v_{B\,\text{rel}} \cos\beta\,\mathbf{j} + \begin{vmatrix} \mathbf{i} & \mathbf{j} & \mathbf{k} \\ 0 & 0 & \omega_{BC} \\ -0.143 & 0.350 & 0 \end{vmatrix}$$

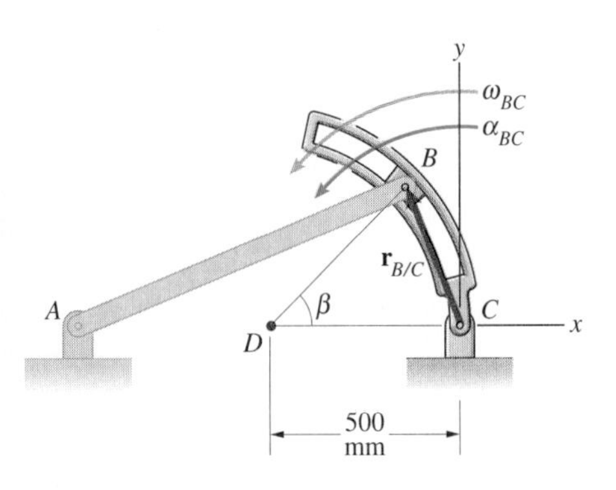

(b) A coordinate system fixed with respect to the curved bar.

Equating this expression for $\mathbf{v}_B$ to its value given in Equation (6.16) yields the two equations

$$-v_{B\,\mathrm{rel}} \sin\beta - 0.350\omega_{BC} = -0.350$$
$$v_{B\,\mathrm{rel}} \cos\beta - 0.143\omega_{BC} = 0.857$$

Solving them, we obtain $v_{B\,\mathrm{rel}} = 1.0\,\mathrm{m/s}$ and $\omega_{BC} = -1.0\,\mathrm{rad/s}$.

We follow the same sequence of steps to determine the angular acceleration of bar BC. The acceleration of point B is

$$\begin{aligned}\mathbf{a}_B &= \mathbf{a}_A + \alpha_{AB} \times \mathbf{r}_{B/A} - \omega_{AB}^2 \mathbf{r}_{B/A} \\ &= 0 + 0 - (1)^2(0.857\,\mathbf{i} + 0.350\,\mathbf{j}) \\ &= (-0.857\,\mathbf{i} - 0.350\,\mathbf{j})\,\mathrm{m/s^2}\end{aligned} \qquad (6.18)$$

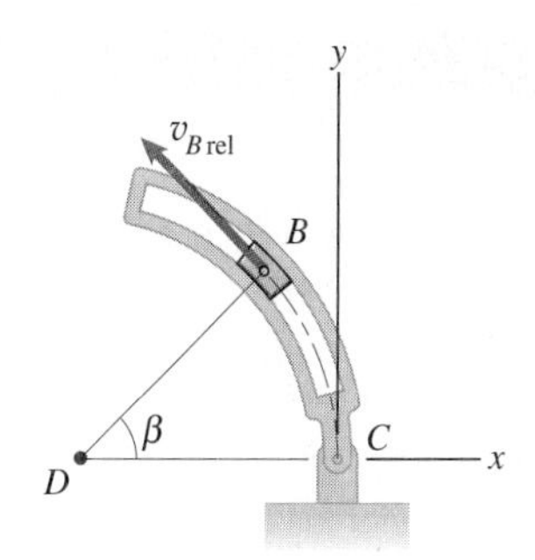

(c) Velocity of B relative to the body-fixed coordinate system.

Because the motion of point B relative to the body-fixed coordinate system is a circular path about point D, there is a tangential component of acceleration, which we denote a_{Bt}, and a normal component of acceleration $v_{B\,\mathrm{rel}}^2/(0.5\,\mathrm{m})$. These components are shown in Figure (d). In terms of the angle β, the vector $\mathbf{a}_{B\,\mathrm{rel}}$ is

$$\begin{aligned}\mathbf{a}_{B\,\mathrm{rel}} = &- a_{Bt} \sin\beta\,\mathbf{i} + a_{Bt} \cos\beta\,\mathbf{j} \\ &- (v_{B\,\mathrm{rel}}^2/0.5)\cos\beta\,\mathbf{i} - (v_{B\,\mathrm{rel}}^2/0.5)\sin\beta\,\mathbf{j}\end{aligned}$$

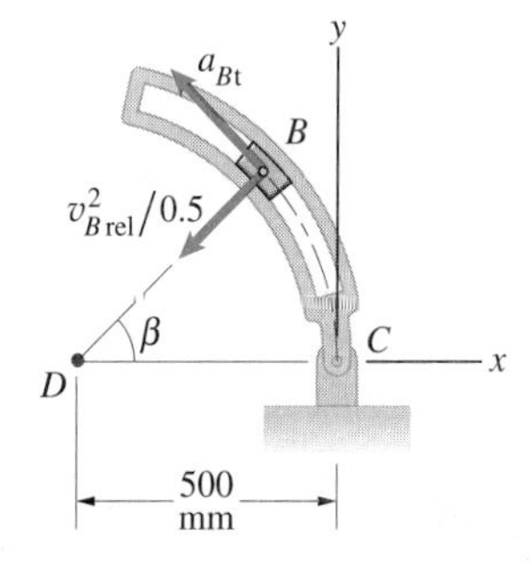

(d) Acceleration of B relative to the body-fixed coordinate system.

Applying Equation (6.15) to points B and C, the acceleration of B is

$$\begin{aligned}\mathbf{a}_B &= \mathbf{a}_C + \mathbf{a}_{B\,\mathrm{rel}} + 2\boldsymbol{\omega}_{BC} \times \mathbf{v}_{B\,\mathrm{rel}} \\ &\quad + \alpha_{BC} \times \mathbf{r}_{B/C} - \omega_{BC}^2 \mathbf{r}_{B/C} \\ &= 0 - a_{Bt} \sin\beta\,\mathbf{i} + a_{Bt} \cos\beta\,\mathbf{j} \\ &\quad - [(1)^2/0.5]\cos\beta\,\mathbf{i} - [(1)^2/0.5]\sin\beta\,\mathbf{j} \\ &\quad + 2\begin{vmatrix} \mathbf{i} & \mathbf{j} & \mathbf{k} \\ 0 & 0 & -1 \\ -(1)\sin\beta & (1)\cos\beta & 0 \end{vmatrix} \\ &\quad + \begin{vmatrix} \mathbf{i} & \mathbf{j} & \mathbf{k} \\ 0 & 0 & \alpha_{BC} \\ -0.143 & 0.350 & 0 \end{vmatrix} - (-1)^2(0.143\,\mathbf{i} + 0.350\,\mathbf{j})\end{aligned}$$

Equating this expression for $\mathbf{a}_B$ to its value given in Equation (6.18) yields the two equations

$$-a_{Bt} \sin\beta - 0.350\alpha_{BC} + 0.143 = -0.857$$
$$a_{Bt} \cos\beta - 0.143\alpha_{BC} - 0.350 = -0.350$$

Solving them, we obtain $a_{Bt} = 0.408\,\mathrm{m/s^2}$ and $\alpha_{BC} = 2.040\,\mathrm{rad/s^2}$.

Problems

6.99 The bar rotates with a constant counterclockwise angular velocity of 10 rad/s, and the sleeve *A* slides at 4 m/s relative to the bar. Use Equation (6.11) to determine the velocity of *A*.

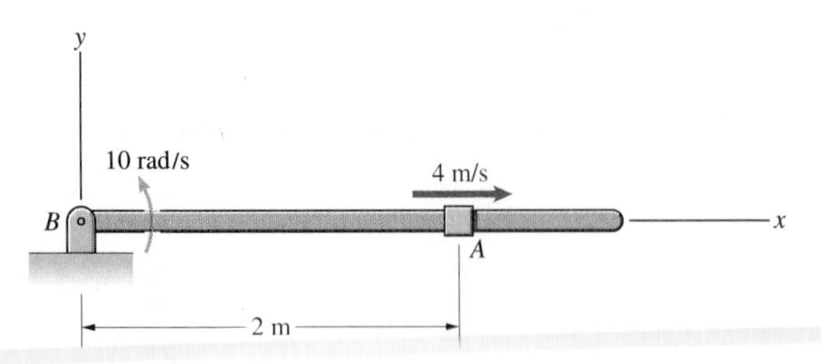

P6.99

6.100 The sleeve *A* in Problem 6.99 slides relative to the bar at a constant velocity of 4 m/s. Use Equation (6.15) to determine the acceleration of *A*.

6.101 The sleeve *C* slides at 1 m/s relative to the bar *BD*. What is its velocity?

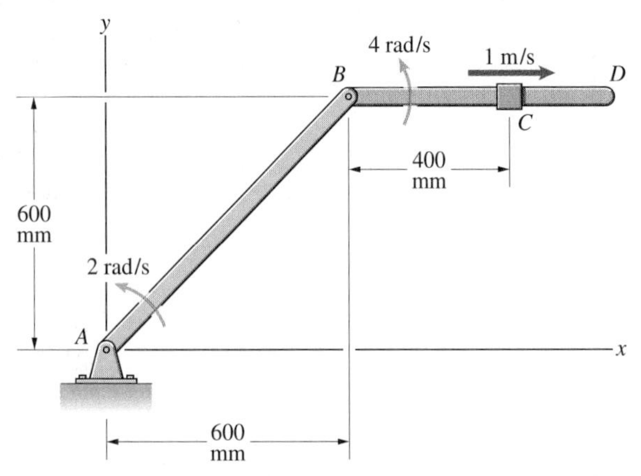

P6.101

6.102 In Problem 6.101, the angular accelerations of the two bars are zero and the sleeve *C* slides at a constant velocity of 1 m/s relative to bar *BD*. What is the acceleration of the sleeve *C*?

6.103 Bar *AC* has an angular velocity of 2 rad/s in the counterclockwise direction that is decreasing at 4 rad/s^2. The pin at *C* slides in the slot in bar *BD*.
(a) Determine the angular velocity of bar *BD* and the velocity of the pin relative to the slot.
(b) Determine the angular acceleration of bar *BD* and the acceleration of the pin relative to the slot.

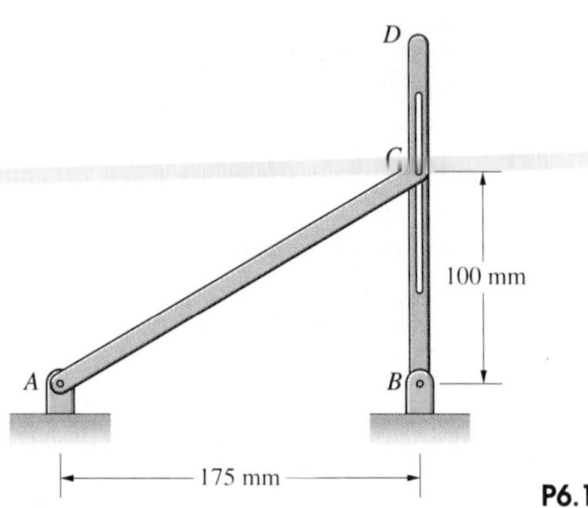

P6.103

6.104 In the system shown in Problem 6.103, the velocity of the pin *C* relative to the slot is 500 mm/s upwards and is decreasing at 1000 mm/s^2. What are the angular velocity and acceleration of bar *AC*?

6.105 In the system shown in Problem 6.103, what should the angular velocity and acceleration of bar *AC* be if you want the angular velocity and acceleration of bar *BD* to be 4 rad/s counterclockwise and 24 rad/s^2 counterclockwise, respectively?

6.106 Bar *AB* has an angular velocity of 4 rad/s in the clockwise direction. What is the velocity of the pin *B* relative to the slot?

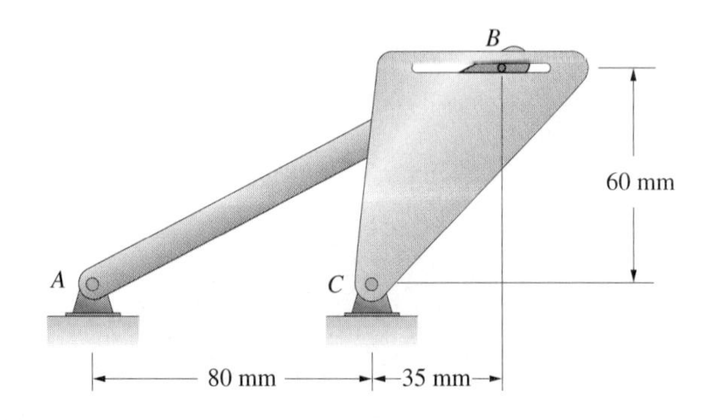

P6.106

6.107 In the system shown in Problem 6.106, bar AB has an angular velocity of 4 rad/s in the clockwise direction and an angular acceleration of 10 rad/s^2 in the counterclockwise direction. What is the acceleration of the pin B relative to the slot?

6.108 Arm AB is rotating at 4 rad/s in the clockwise direction. Determine the angular velocity of arm BC and the velocity of point B relative to the slot in arm BC.

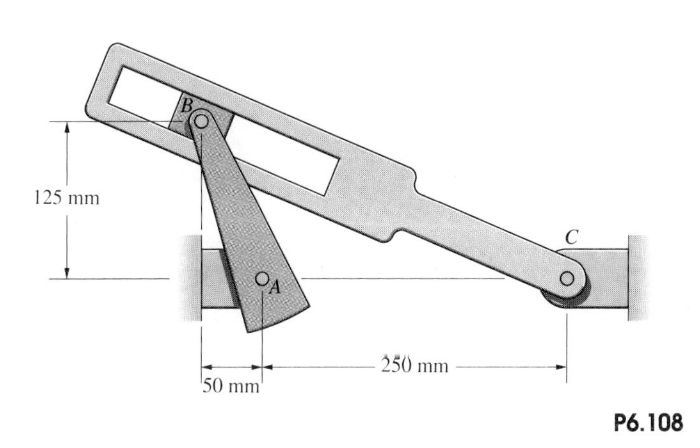

P6.108

6.109 Arm AB in Problem 6.108 is rotating with a constant angular velocity of 4 rad/s in the clockwise direction. Determine the angular acceleration of arm BC and the acceleration of point B relative to the slot in arm BC.

6.110 The angular velocity $\omega_{AC} = 5°$ per second. Determine the angular velocity of the hydraulic actuator BC and the rate at which it is extending.

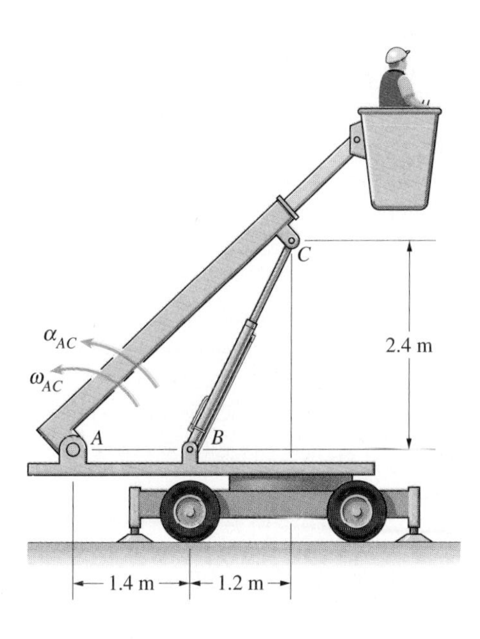

P6.110

6.111 In Problem 6.110, if the angular velocity $\omega_{AC} = 5°$ per second and the angular acceleration $\alpha_{AC} = -2°$ per second squared, determine the angular acceleration of the hydraulic actuator BC and the rate of change of its rate of extension.

6.112 The sleeve at A slides upwards at a constant velocity of 10 m/s. The bar AC slides through the sleeve at B. Determine the angular velocity of bar AC and the velocity at which it slides relative to the sleeve at B.

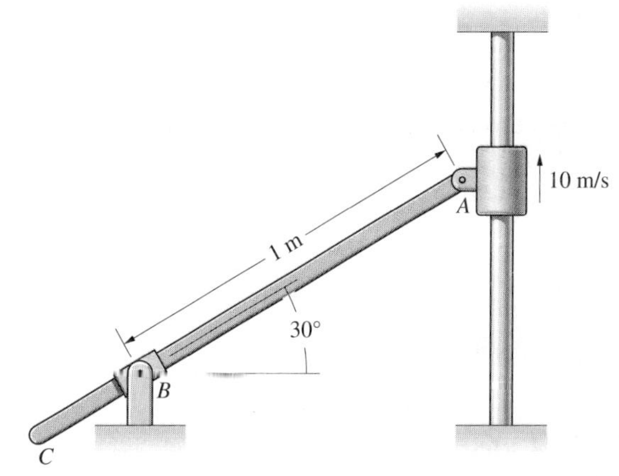

P6.112

6.113 In Problem 6.112, the sleeve at A slides upwards at a constant velocity of 10 m/s. Determine the angular acceleration of the bar AC and the rate of change of the velocity at which it slides relative to the sleeve at B.

6.114 The block A slides up the inclined surface at 0.6 m/s. Determine the angular velocity of bar AC and the velocity of point C.

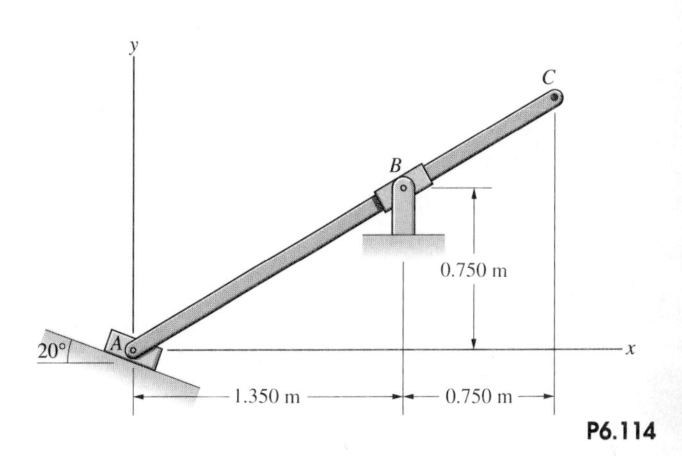

P6.114

6.115 In Problem 6.114, the block A slides up the inclined surface at a constant velocity of 0.6 m/s. Determine the angular acceleration of bar AC and the acceleration of point C.

6.116 The angular velocity of the scoop is 1.0 rad/s clockwise. Determine the rate at which the hydraulic actuator AB is extending.

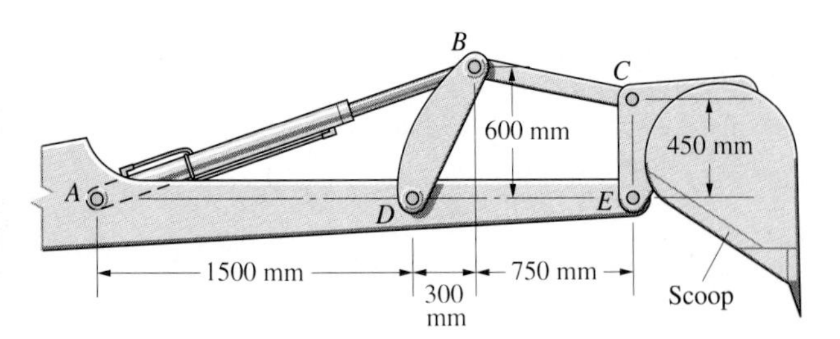

P6.116

6.117 The angular acceleration of the scoop in Problem 6.116 is zero. Determine the rate of change of the rate at which the hydraulic actuator AB is extending.

6.118 Suppose that the curved bar in Example 6.9 rotates with a counterclockwise angular velocity of 2 rad/s.
(a) What is the angular velocity of bar AB?
(b) What is the velocity of the block B relative to the slot?

6.119 Suppose that the curved bar in Example 6.9 has a clockwise angular velocity of 4 rad/s and a counterclockwise angular acceleration of 10 rad/s^2. What is the angular acceleration of bar AB?

6.120 The disc rolls on the plane surface with a counterclockwise angular velocity of 10 rad/s. Bar AB slides on the surface of the disc at A. Determine the angular velocity of bar AB.

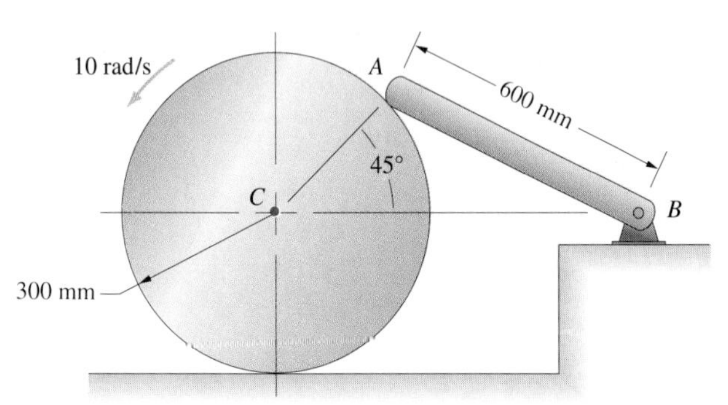

P6.120

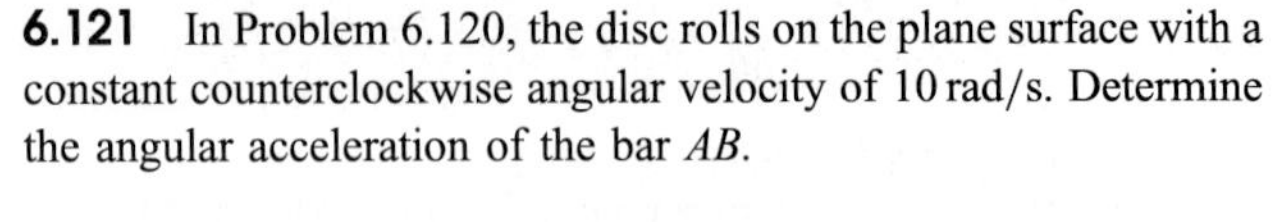

6.121 In Problem 6.120, the disc rolls on the plane surface with a constant counterclockwise angular velocity of 10 rad/s. Determine the angular acceleration of the bar AB.

6.122 Bar BC rotates with a counterclockwise angular velocity of 2 rad/s. A pin at B slides in a circular slot in the rectangular plate. Determine the angular velocity of the plate and the velocity at which the pin slides relative to the circular slot.

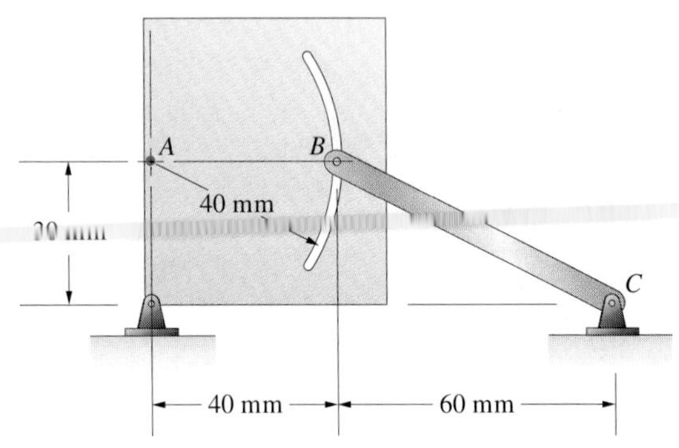

P6.122

6.123 The bar BC in Problem 6.122 rotates with a constant counterclockwise angular velocity of 2 rad/s. Determine the angular acceleration of the plate.

6.124 By taking the time derivative of Equation (6.11) and using Equation (6.12), derive Equation (6.13).

6.6 Rotating Coordinate Systems

In this section we revisit the subjects of Chapters 2 and 3 – the motion of a point and Newton's second law. In some situations it is convenient to describe the motion of a point by using a cooardinate system that rotates. For example, to measure the motion of a point relative to a moving vehicle, you might use a coordinate system that moves and rotates with the vehicle. Here we show how the velocity and acceleration of a point are related to their values relative to a rotating coordinate system. In Chapter 3 we mentioned the example of playing tennis on the deck of a cruise ship. If the ship translates with constant velocity, you can use the equation $\Sigma \mathbf{F} = m\mathbf{a}$ expressed in terms of a coordinate system fixed relative to the ship to analyse the ball's motion. You cannot do so if the ship is turning, or changing its speed. However, you *can* apply the second law using coordinate systems that accelerate and rotate by properly accounting for the acceleration and rotation. We explain how this is done.

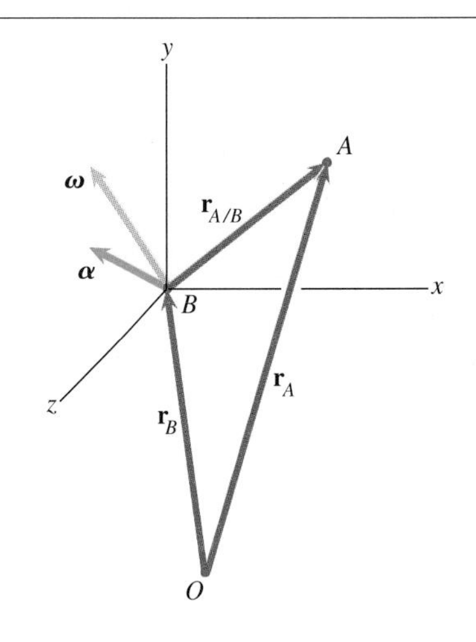

Figure 6.37
A rotating coordinate system with origin B and an arbitrary point A.

Motion of a Point Relative to a Rotating Coordinate System

Equations (6.11) and (6.13) give the velocity and acceleration of an arbitrary point A relative to a point B of a rigid body in terms of a body-fixed coordinate system:

$$\mathbf{v}_A = \mathbf{v}_B + \mathbf{v}_{A\,\mathrm{rel}} + \boldsymbol{\omega} \times \mathbf{r}_{A/B} \tag{6.19}$$

$$\mathbf{a}_A = \mathbf{a}_B + \mathbf{a}_{A\,\mathrm{rel}} + 2\boldsymbol{\omega} \times \mathbf{v}_{A\,\mathrm{rel}} + \boldsymbol{\alpha} \times \mathbf{r}_{A/B} + \boldsymbol{\omega} \times (\boldsymbol{\omega} \times \mathbf{r}_{A/B}) \tag{6.20}$$

But these results don't require us to assume that the coordinate system is connected to some rigid body. They apply to any coordinate system rotating with angular velocity $\boldsymbol{\omega}$ and angular acceleration $\boldsymbol{\alpha}$ (Figure 6.37). The terms $\mathbf{v}_A$ and $\mathbf{a}_A$ are the velocity and acceleration of A relative to a non-rotating coordinate system that is stationary relative to O. The terms $\mathbf{v}_{A\,\mathrm{rel}}$ and $\mathbf{a}_{A\,\mathrm{rel}}$ are the velocity and acceleration of A relative to the rotating coordinate system. That is, they are the velocity and acceleration measured by an 'observer' moving with the rotating coordinate system (Figure 6.38).

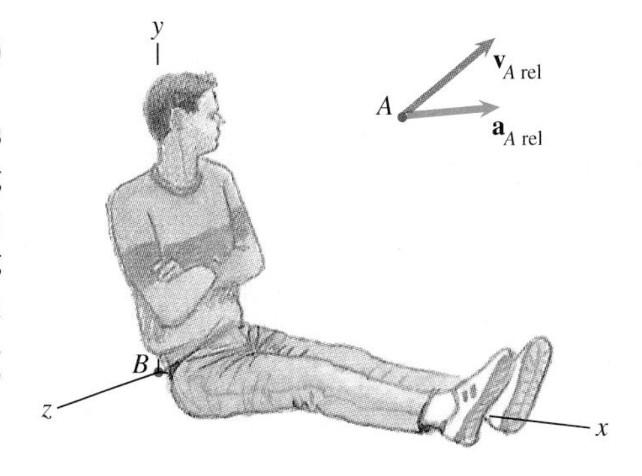

Figure 6.38
Imagine yourself to be stationary relative to the rotating coordinate system.

The following examples demonstrate applications of rotating coordinate systems. If you know the motion of a point A relative to a rotating coordinate system, you can use Equations (6.19) and (6.20) to determine $\mathbf{v}_A$ and $\mathbf{a}_A$. In other situations, you will know $\mathbf{v}_A$ and $\mathbf{a}_A$. and will want to use Equations (6.19) and (6.20) to determine the velocity and acceleration of A relative to a rotating coordinate system.

Example 6.10

The merry-go-round in Figure 6.39 rotates with constant angular velocity ω. Suppose that you are in the centre at B and observe the motion of a second person A, using a coordinate system that rotates with the merry-go-round. Consider two cases.

Case 1 The person A is not on the merry-go-round, but stands on the ground next to it. At the instant shown, what are his velocity and acceleration relative to your coordinate system?

Case 2 The person A is on the edge of the merry-go-round and moves with it. What are his velocity and acceleration relative to the earth?

Figure 6.39

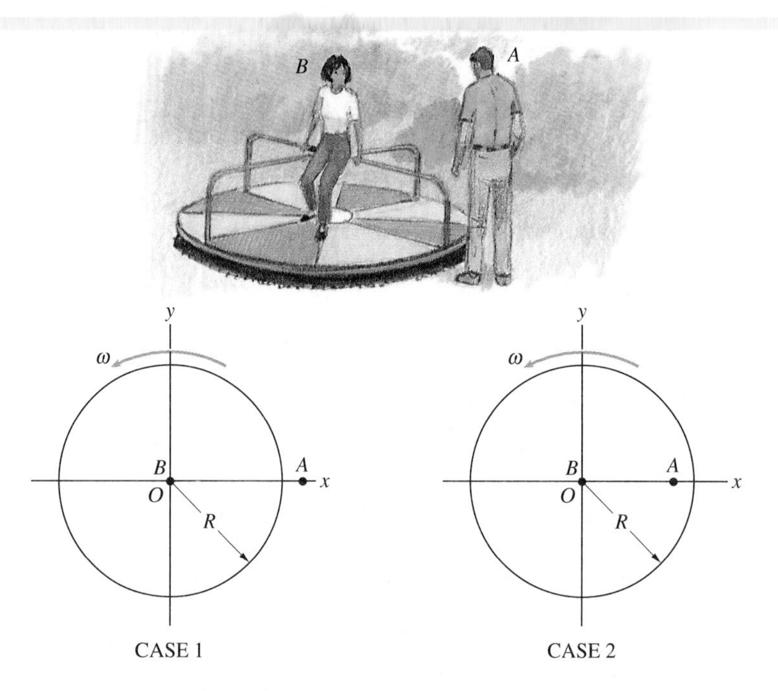

STRATEGY

This simple example clarifies the distinction between the terms $\mathbf{v}_A$, $\mathbf{a}_A$ and the terms $\mathbf{v}_{A\,\mathrm{rel}}$, $\mathbf{a}_{A\,\mathrm{rel}}$ in Equations (6.19) and (6.20). In case 1, A's velocity and acceleration relative to the earth, $\mathbf{v}_A$ and $\mathbf{a}_A$, are known: he is standing still. We can use Equation (6.19) and (6.20) to determine $\mathbf{v}_{A\,\mathrm{rel}}$ and $\mathbf{a}_{A\mathrm{rel}}$, which are his velocity and acceleration relative to your rotating coordinate system. In case 2, $\mathbf{v}_{A\,\mathrm{rel}}$ and $\mathbf{a}_{A\,\mathrm{rel}}$ are known: A is stationary relative to your coordinate system. We can use Equations (6.19) and (6.20) to determine $\mathbf{v}_A$ and $\mathbf{a}_A$.

SOLUTION

Case 1 A is standing on the ground, so his velocity relative to the earth is $\mathbf{v}_A = 0$. The angular velocity vector of your coordinate system is $\boldsymbol{\omega} = \omega\,\mathbf{k}$, and at the instant shown $\mathbf{r}_{A/B} = R\,\mathbf{i}$. From Equation (6.19),

$$\mathbf{v}_A = \mathbf{v}_B + \mathbf{v}_{A\,\text{rel}} + \boldsymbol{\omega} \times \mathbf{r}_{A/B}:$$

$$0 = 0 + \mathbf{v}_{A\,\text{rel}} + (\omega\,\mathbf{k}) \times (R\,\mathbf{i})$$

We find that $\mathbf{v}_{A\,\text{rel}} = -\omega R\,\mathbf{j}$. Although A is stationary relative to the earth, $\mathbf{v}_{A\,\text{rel}}$ is not zero. What does this term represent? As you sit at the centre of the merry-go-round, you see A moving around you in a circular path. *Relative to your rotating coordinate system*, A moves in a circular path of radius R in the clockwise direction with a velocity of constant magnitude ωR. At the instant shown, A's velocity relative to your coordinate system is $-\omega R\,\mathbf{j}$.

You know that a point moving in a circular path of radius R with velocity v has a normal component of acceleration equal to v^2/R. Relative to your coordinate system, person A moves in a circular path of radius R with velocity ωR. Therefore, *relative to your coordinate system*, A has a normal component of acceleration $(\omega R)^2/R = \omega^2 R$. At the instant shown, the normal acceleration points in the negative x direction. Therefore we conclude that A's acceleration relative to your coordinate system is $\mathbf{a}_{A\,\text{rel}} = -\omega^2 R\,\mathbf{i}$.

We can confirm this result with Equation (6.20). A's acceleration relative to the earth is $\mathbf{a}_A = 0$. The angular velocity vector of the coordinate system is constant, so $\alpha = 0$. From Equation (6.20),

$$\mathbf{a}_A = \mathbf{a}_B + \mathbf{a}_{A\,\text{rel}} + 2\boldsymbol{\omega} \times \mathbf{v}_{A\,\text{rel}} + \alpha \times \mathbf{r}_{A/B} + \boldsymbol{\omega} \times (\boldsymbol{\omega} \times \mathbf{r}_{A/B}):$$

$$0 = 0 + \mathbf{a}_{A\,\text{rel}} + 2(\omega\,\mathbf{k}) \times (-\omega R\,\mathbf{j}) + 0 + (\omega\,\mathbf{k}) \times [(\omega\,\mathbf{k}) \times (R\,\mathbf{i})]$$

Solving this equation for $\mathbf{a}_{A\,\text{rel}}$, we obtain $\mathbf{a}_{A\,\text{rel}} = -\omega^2 R\,\mathbf{i}$. A's velocity and acceleration relative to your coordinate system are shown in Figure (a).

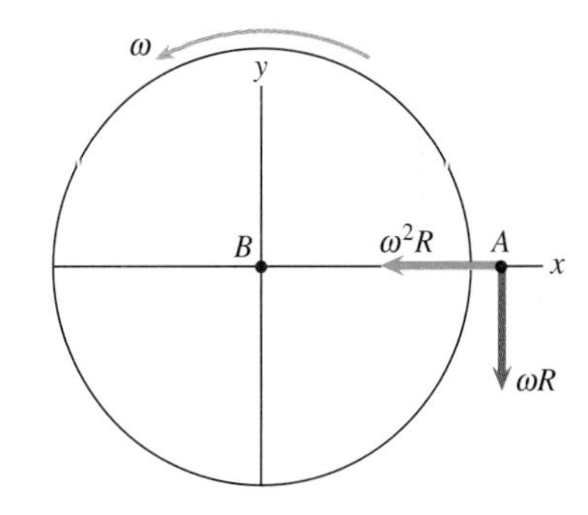

(a) The velocity and acceleration of A relative to the rotating coordinate system in case 1.

Case 2 *Relative to your coordinate system*, A is stationary, so $\mathbf{v}_{A\,\text{rel}} = 0$ and $\mathbf{a}_{A\,\text{rel}} = 0$. From Equation (6.19). A's velocity relative to the earth is

$$\mathbf{v}_A = \mathbf{v}_B + \mathbf{v}_{A\,\text{rel}} + \boldsymbol{\omega} \times \mathbf{r}_{A/B} = 0 + 0 + (\omega\,\mathbf{k}) \times (R\,\mathbf{i})$$
$$= \omega R\,\mathbf{j}$$

In this case, A is moving in a circular path of radius R with a velocity of constant magnitude ωR relative to the earth.

From Equation (6.20), A's acceleration relative to the earth is

$$\begin{aligned}\mathbf{a}_A &= \mathbf{a}_B + \mathbf{a}_{A\,\text{rel}} + 2\boldsymbol{\omega} \times \mathbf{v}_{A\,\text{rel}} + \alpha \times \mathbf{r}_{A/B} + \boldsymbol{\omega} \times (\boldsymbol{\omega} \times \mathbf{r}_{A/B})\\ &= 0 + 0 + 0 + 0 + (\omega\,\mathbf{k}) \times [(\omega\,\mathbf{k}) \times (R\,\mathbf{i})]\\ &= -\omega^2 R\,\mathbf{i}\end{aligned}$$

This is A's acceleration relative to the earth due to his circular motion. A's velocity and acceleration relative to the earth are shown in Figure (b).

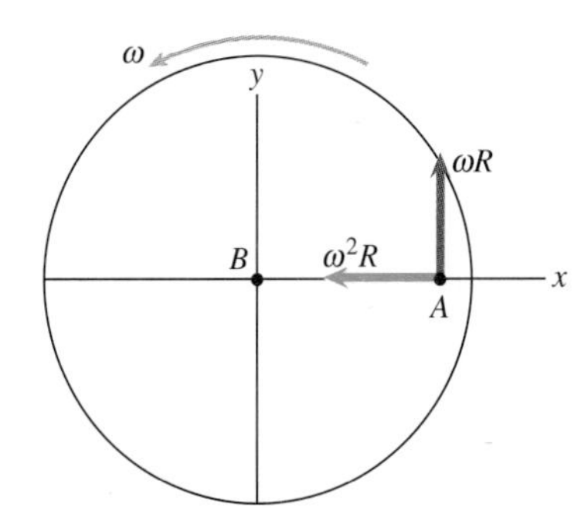

(b) The velocity and acceleration of A relative to the earth in case 2.

Example 6.11

At the instant shown, the ship in Figure 6.40 is moving north at a constant speed of 15.0 m/s relative to the earth and is turning towards the west at a constant rate of 5.0° per second. Relative to the ship's body-fixed coordinate system, its radar indicates that the position, velocity and acceleration of the helicopter are

$$\mathbf{r}_{A/B} = (420.0\,\mathbf{i} + 236.2\,\mathbf{j} + 212.0\,\mathbf{k})\,\text{m}$$

$$\mathbf{v}_{A\,\text{rel}} = (-53.5\,\mathbf{i} + 2.0\,\mathbf{j} + 6.6\,\mathbf{k})\,\text{m/s}$$

$$\mathbf{a}_{A\,\text{rel}} = (0.4\,\mathbf{i} - 0.2\mathbf{j} - \mathbf{13.0\,k})\,\text{m/s}^2$$

What are the helicopter's velocity and acceleration relative to the earth?

Figure 6.40

STRATEGY

We are given the ship's velocity relative to the earth and are given enough information to determine its acceleration, angular velocity and angular acceleration. Therefore we can use Equations (6.19) and (6.20) to determine the helicopter's velocity and acceleration relative to the earth.

SOLUTION

In terms of the body fixed coordinate system, the ship's velocity is $\mathbf{v}_B = 15.0\,\mathbf{i}$ m/s. The ship's angular velocity due to its rate of turning is $\omega = (5.0/180)\pi = 0.0873$ rad/s. The ship is rotating about the y axis. Pointing the arc of the fingers of the right hand around the y axis in the direction of the ship's rotation, the thumb points in the positive y direction, so the ship's angular velocity vector is $\boldsymbol{\omega} = 0.0873\,\mathbf{j}$ rad/s. The helicopter's velocity relative to the earth is

$$\mathbf{v}_A = \mathbf{v}_B + \mathbf{v}_{A\,\mathrm{rel}} + \boldsymbol{\omega} \times \mathbf{r}_{A/B}$$

$$= 15.0\,\mathbf{i} + (-53.5\,\mathbf{i} + 2.0\,\mathbf{j} + 6.6\,\mathbf{k}) + \begin{vmatrix} \mathbf{i} & \mathbf{j} & \mathbf{k} \\ 0 & 0.087 & 0 \\ 420.0 & 236.2 & 212.0 \end{vmatrix}$$

$$= -(20.0\,\mathbf{i} + 2.0\,\mathbf{j} - 30.1\,\mathbf{k})\ \mathrm{m/s}$$

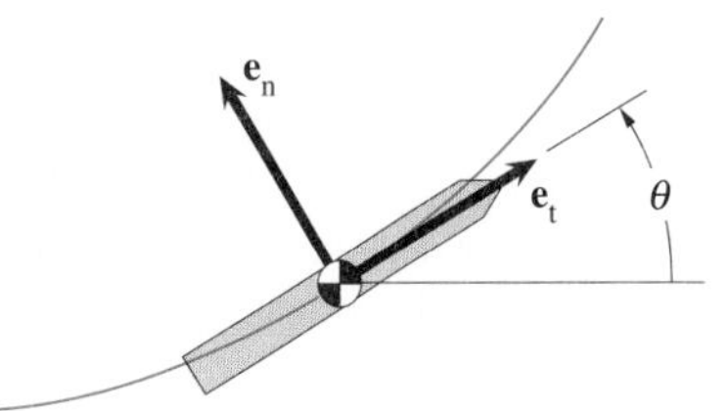

(a) Determining the ship's acceleration.

We can determine the ship's acceleration by expressing it in terms of normal and tangential components in the form given by Equation (2.37) (Figure (a)):

$$\mathbf{a}_B = \frac{dv}{dt}\mathbf{e}_t + v\frac{d\theta}{dt}\mathbf{e}_n = 0 + (15)(0.0873)\,\mathbf{e}_n$$

$$= 1.31\mathbf{e}_n\ \mathrm{m/s^2}$$

The z axis is perpendicular to the ship's path and points towards the convex side of the path (Figure (b)). Therefore, in terms of the body-fixed coordinate system, the ship's acceleration is $\mathbf{a}_B = -1.31\,\mathbf{k}\ \mathrm{m/s^2}$. The ship's angular velocity vector is constant, so $\alpha = 0$. The helicopter's acceleration relative to the earth is

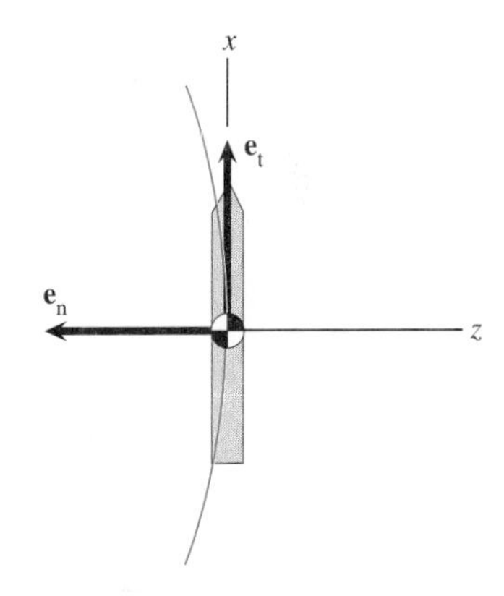

(b) Correspondence between the normal and tangential components and the body-fixed coordinate system.

$$\mathbf{a}_A = \mathbf{a}_B + \mathbf{a}_{A\,\mathrm{rel}} + 2\boldsymbol{\omega} \times \mathbf{v}_{A\,\mathrm{rel}} + \alpha \times \mathbf{r}_{A/B}$$

$$+ \boldsymbol{\omega} \times (\boldsymbol{\omega} \times \mathbf{r}_{A/B})$$

$$= -1.31\,\mathbf{k} + (0.4\,\mathbf{i} - 0.2\,\mathbf{j} - 13.0\,\mathbf{k})$$

$$+ 2\begin{vmatrix} \mathbf{i} & \mathbf{j} & \mathbf{k} \\ 0 & 0.0873 & 0 \\ -53.5 & 2.0 & 6.6 \end{vmatrix}$$

$$+ 0 + (0.0873\,\mathbf{j}) \times \begin{vmatrix} \mathbf{i} & \mathbf{j} & \mathbf{k} \\ 0 & 0.0873 & 0 \\ 420.0 & 236.2 & 212.0 \end{vmatrix}$$

$$= -1.65\,\mathbf{i} - 0.20\,\mathbf{j} - 6.59\,\mathbf{k}\ \mathrm{m/s^2}$$

DISCUSSION

Notice the substantial differences between the helicopter's velocity and acceleration relative to the earth and the values the ship measures using its body-fixed coordinate system.

Inertial Reference Frames

We say that a reference frame is inertial if you can use it to apply Newton's second law in the form $\Sigma \mathbf{F} = m\,\mathbf{a}$. Why can you usually assume that an earth-fixed reference frame is inertial, even though it both accelerates and rotates? How can you apply Newton's second law using a coordinate system that is fixed with respect to an accelerating, turning ship or aeroplane? We are now in a position to answer these questions.

Earth-Centred, Non-Rotating Coordinate System We begin by showing why a non-rotating reference frame fixed relative to the centre of the earth can be assumed to be inertial for the purpose of describing motions of objects near the earth. Figure 6.41(a) shows a hypothetical non-accelerating non-rotating coordinate system with origin O, and a second non-rotating, **earth-centred coordinate system**. The earth, and therefore the earth-centred coordinate system, accelerates due to the gravitational attractions of the sun, moon, and so on. We denote the earth's acceleration by the vector $\mathbf{g}_B$.

Suppose that we want to determine the motion of an object A of mass m (Figure 6.41(b)). A is also subject to the gravitational attractions of the sun, moon, and so on, and we denote the resulting gravitational acceleration by the vector $\mathbf{g}_A$. The vector $\Sigma \mathbf{F}$ is the sum of all *other* external forces acting on A, including the gravitational force exerted on it by the earth. The total external force acting on A is $\Sigma \mathbf{F} + m\mathbf{g}_A$. We can apply Newton's second law to A, using our hypothetical inertial coordinate system:

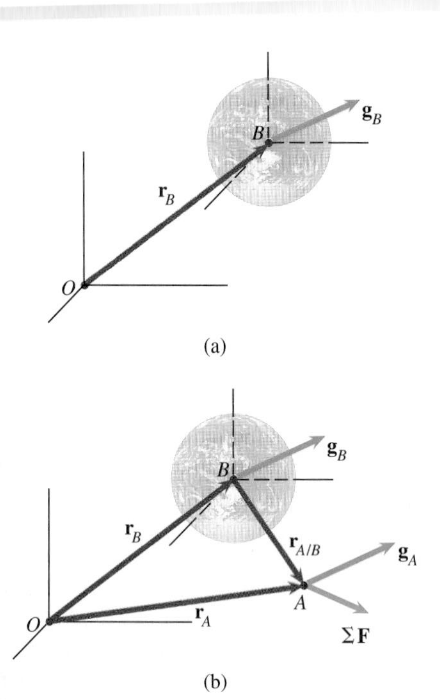

Figure 6.41
(a) An inertial reference frame and a non-rotating reference frame with its origin at the centre of the earth.
(b) Determining the motion of an object A.

$$\Sigma \mathbf{F} + m\,\mathbf{g}_A = m\,\mathbf{a}_A \tag{6.21}$$

where $\mathbf{a}_A$ is the acceleration of A relative to O. Since the earth-centred coordinate system does not rotate, we can use Equation (6.20) to write $\mathbf{a}_A$ as

$$\mathbf{a}_A = \mathbf{a}_B + \mathbf{a}_{A\,\mathrm{rel}}$$

where $\mathbf{a}_{A\,\mathrm{rel}}$ is the acceleration of A relative to the earth-centred coordinate system. Using this relation and our definition of the earth's acceleration $\mathbf{a}_B = \mathbf{g}_B$, Equation (6.21) becomes

$$\Sigma \mathbf{F} = m\,\mathbf{a}_{A\,\mathrm{rel}} + m(\mathbf{g}_B - \mathbf{g}_A) \tag{6.22}$$

If the object A is on or near the earth, its gravitational acceleration $\mathbf{g}_A$ due to the attraction of the sun and so on is very nearly equal to the earth's gravitational acceleration $\mathbf{g}_B$. If we neglect the difference, Equation (6.22) becomes

$$\Sigma \mathbf{F} = m\mathbf{a}_{A\mathrm{rel}} \tag{6.23}$$

Thus you can apply Newton's second law using a non-rotating, earth-centred reference frame. Even though this reference frame accelerates, *virtually the same gravitational acceleration acts on the object*. Notice that this argument does not hold if the object is not near the earth. If you wanted to analyse the motion of a spacecraft travelling to another planet, for example, you would need to use a non-rotating, sun-centred reference frame.

Earth-Fixed Coordinate System For 'down to earth' applications, the most convenient reference frame is a local, **earth-fixed coordinate system**. Why can we usually assume that an earth-fixed coordinate system is inertial? Figure 6.42 shows a non-rotating coordinate system with its origin at the centre of the earth O and an earth-fixed coordinate system with its origin at a point B. Since we can assume that the earth-centred, non-rotating coordinate system is inertial, we can write Newton's second law for an object A of mass m as

$$\Sigma \mathbf{F} = m\,\mathbf{a}_A \tag{6.24}$$

where $\mathbf{a}_A$ is A's acceleration relative to O. The earth-fixed reference frame rotates with the angular velocity of the earth, which we denote by $\boldsymbol{\omega}_{\mathrm{E}}$. We can use Equation (6.20) to write Equation (6.24) in the form

$$\Sigma \mathbf{F} = m\,\mathbf{a}_{A\,\mathrm{rel}} + m[\mathbf{a}_B + 2\boldsymbol{\omega}_{\mathrm{E}} \times \mathbf{v}_{A\,\mathrm{rel}} + \boldsymbol{\omega}_{\mathrm{E}} \times (\boldsymbol{\omega}_{\mathrm{E}} \times \mathbf{r}_{A/B})] \tag{6.25}$$

where $\mathbf{a}_{A\,\mathrm{rel}}$ is A's acceleration relative to the earth-fixed coordinate system. If we can neglect the terms in brackets on the right side of Equation (6.25), the earth-fixed coordinate system is inertial. Let's consider each term. (Recall from the definition of the cross product that $|\mathbf{U} \times \mathbf{V}| = |\mathbf{U}\,||\,\mathbf{V}|\sin\theta$, where θ is the angle between the two vectors. Therefore the magnitude of the cross product is bounded by the product of the magnitudes of the vectors.)

- *The term* $\boldsymbol{\omega}_{\mathrm{E}} \times (\boldsymbol{\omega}_{\mathrm{E}} \times \mathbf{r}_{A/B})$: The earth's angular velocity ω_{E} is approximately one revolution per day $= 7.27 \times 10^{-5}$ rad/s. Therefore the magnitude of this term is bounded by $\omega_{\mathrm{E}}^2|\mathbf{r}_{A/B}| = (5.29 \times 10^{-9})|\mathbf{r}_{A/B}|$. For example, if the distance $|\mathbf{r}_{A/B}|$ from the origin of the earth-fixed coordinate system to the object A is 10 000 m, this term is no larger than $5.3 \times 10^{-5}\,\mathrm{m/s^2}$.
- *The term* $\mathbf{a}_B$: This term is the acceleration of the origin B of the earth-fixed coordinate system relative to the centre of the earth. B moves in a circular path due to the earth's rotation. If B lies on the earth's surface, this term is bounded by $\omega_{\mathrm{E}}^2 R_{\mathrm{E}}$, where R_{E} is the radius of the earth. Using the value $R_{\mathrm{E}} = 6370$ km, we find that $\omega_{\mathrm{E}}^2 R_{\mathrm{E}} = 0.0337\,\mathrm{m/s^2}$. This value is too large to neglect for many purposes. However, under normal circumstances this term is accounted for as a part of the local value of the acceleration due to gravity.
- *The term* $2\omega_{\mathrm{E}} \times \mathbf{v}_{A\,\mathrm{rel}}$: This term is called the **Coriolis acceleration**. Its magnitude is bounded by $2\omega_{\mathrm{E}}|\mathbf{v}_{A\,\mathrm{rel}}| = (1.45 \times 10^{-4})|\mathbf{v}_{A\,\mathrm{rel}}|$. For example, if the magnitude of the velocity of A relative to the earth-fixed coordinate system is 10 m/s, this term is no larger than $1.45 \times 10^{-3}\,\mathrm{m/s^2}$.

We see that in most applications, the terms in brackets in Equation (6.25) can be neglected. However, in some cases this is not possible. The Coriolis acceleration becomes significant if an object's velocity relative to the earth is large, and even very small accelerations becomes significant if an object's motion must be predicted over a large period of time. In such cases, you can still use Equation (6.25) to determine the motion, but you must retain the significant terms. When this is done, the terms in brackets are usually moved to the left side:

$$\begin{aligned}\Sigma \mathbf{F} - m\,\mathbf{a}_B - 2m\boldsymbol{\omega}_{\mathrm{E}} \times \mathbf{v}_{A\,\mathrm{rel}} - m\boldsymbol{\omega}_{\mathrm{E}} \times (\boldsymbol{\omega}_{\mathrm{E}} \times \mathbf{r}_{A/B})\\ = m\,\mathbf{a}_{A\,\mathrm{rel}}\end{aligned} \tag{6.26}$$

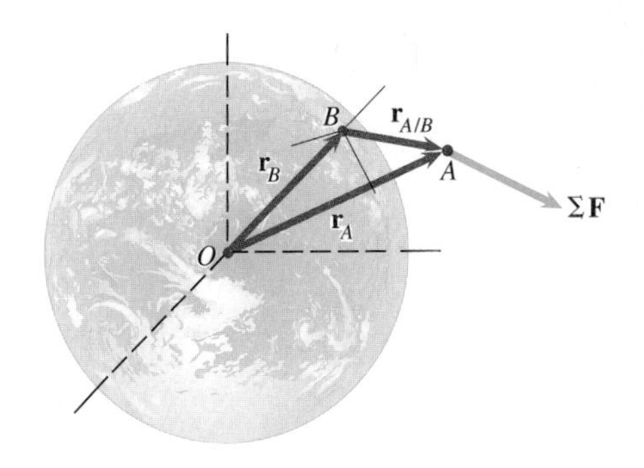

Figure 6.42

An earth-centred, nonrotating reference frame (origin O), an earth-fixed reference frame (origin B), and an object A.

Written in this way, the equation has the usual form of Newton's second law except that the left side contains additional 'forces'. We use quotation marks because these terms are not forces, but are artefacts arising from the motion of the earth-fixed reference frame.

The term $-2m\boldsymbol{\omega}_{\mathrm{E}} \times \mathbf{v}_{A\,\mathrm{rel}}$ in Equation (6.26) is called the **Coriolis force**. It explains a number of physical phenomena that exhibit different behaviours in the northern and southern hemispheres, such as the direction a liquid tends to rotate when going down a drain, the direction a vine tends to grow around a vertical shaft, and the direction of rotation of a storm. The earth's angular velocity vector $\boldsymbol{\omega}_{\mathrm{E}}$ points north. When an object in the northern hemisphere that is moving at a tangent to the earth's surface travels north (Figure 6.43(a)), the cross product $\boldsymbol{\omega}_{\mathrm{E}} \times \mathbf{v}_{A\,\mathrm{rel}}$ points west (Figure 6.43(b)). Therefore the Coriolis force points east – it causes an object moving north to turn to the right (Figure 6.43(c)). If the object is moving south, the direction of $\mathbf{v}_{A\,\mathrm{rel}}$ is reversed and the Coriolis force points west; its effect is to cause the object moving south to turn to the right (Figure 6.43(c)). For example, in the northern hemisphere winds converging on a centre of low pressure tend to rotate about it in the counterclockwise direction (Figure 6.44(a)).

When an object in the southern hemisphere travels north (Figure 6.43(d)), the cross product $\boldsymbol{\omega}_{\mathrm{E}} \times \mathbf{v}_{A\,\mathrm{rel}}$ points east (Figure 6.43(e)). The Coriolis force points west and tends to cause the object to turn to the left (Figure 6.43(f)). If the object is moving south, the Coriolis force points east and tends to cause the object to turn to the left (Figure 6.43(f)). In the southern hemisphere, winds converging on a centre of low pressure tend to rotate about it in the clockwise direction (Figure 6.44(b)).

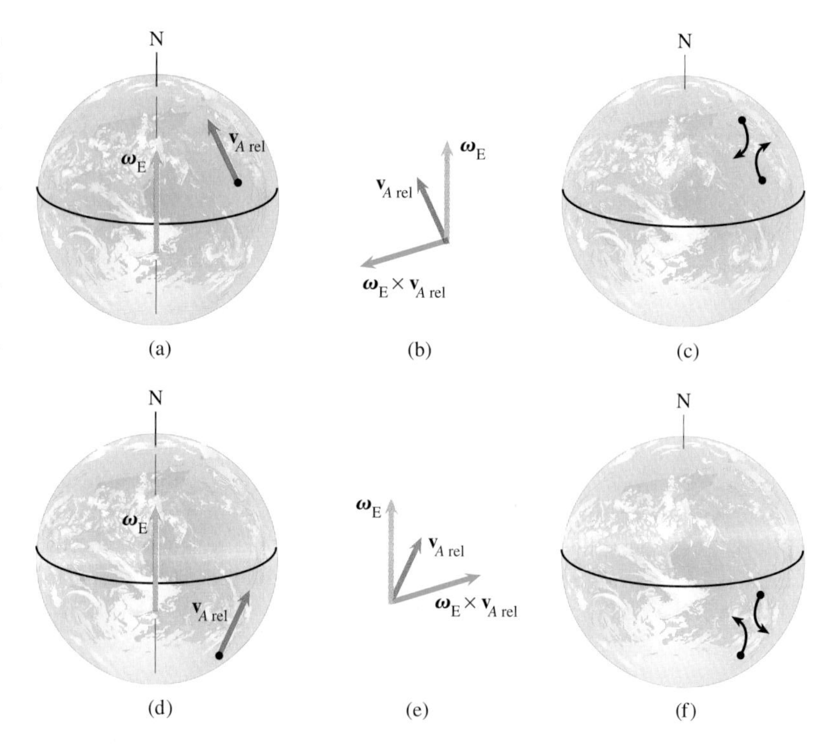

Figure 6.43
(a) An object in the northern hemisphere moving north.
(b) Cross product of the earth's angular velocity with the object's velocity.
(c) Effects of the Coriolis force in the northern hemisphere.
(d) An object in the southern hemisphere moving north.
(e) Cross product of the earth's angular velocity with the object's velocity.
(f) Effects of the Coriolis force in the southern hemisphere.

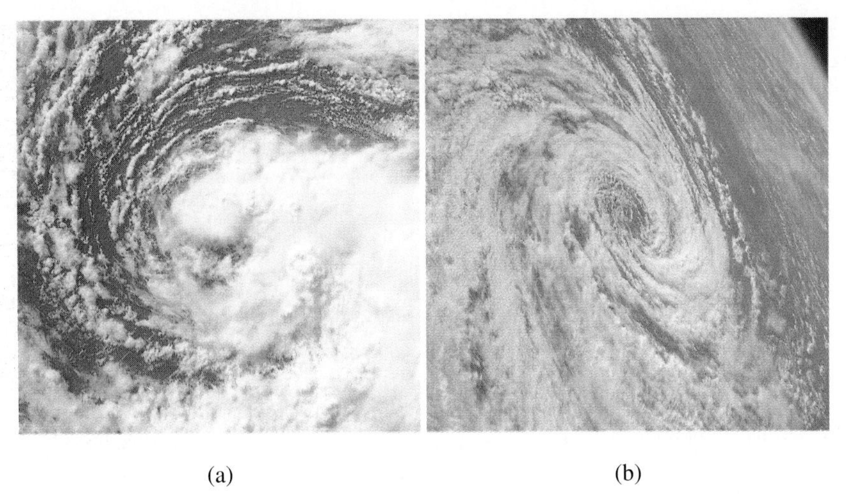

Figure 6.44

Storms in the (a) northern and (b) southern hemispheres.

Arbitrary Coordinate System How can you analyse an object's motion relative to a coordinate system that undergoes an arbitrary motion, such as a coordinate system attached to a moving vehicle? Suppose that the coordinate system with its origin at O in Figure 6.45 is inertial, and the coordinate system with its origin at B undergoes an arbitrary motion with angular velocity $\boldsymbol{\omega}$ and angular acceleration α. We can write Newton's second law for an object A of mass m as

$$\Sigma \mathbf{F} = m\,\mathbf{a}_A \tag{6.27}$$

where $\mathbf{a}_A$ is A's acceleration relative to O. We use Equation (6.20) to write Equation (6.27) in the form

$$\Sigma\mathbf{F} - m[\mathbf{a}_B = 2\boldsymbol{\omega} \times \mathbf{v}_{A\,\mathrm{rel}} + \alpha \times \mathbf{r}_{A/B} + \boldsymbol{\omega} \times (\boldsymbol{\omega} \times \mathbf{r}_{A/B})] = m\mathbf{a}_{A\,\mathrm{rel}} \tag{6.28}$$

where $\mathbf{a}_{A\,\mathrm{rel}}$ is A's acceleration relative to the coordinate system undergoing an arbitrary motion. This is Newton's second law expressed in terms of a reference frame undergoing an arbitrary motion relative to an inertial reference frame: if you know the forces acting on A and the coordinate system's motion, you can use this equation to determine $\mathbf{a}_{A\,\mathrm{rel}}$.

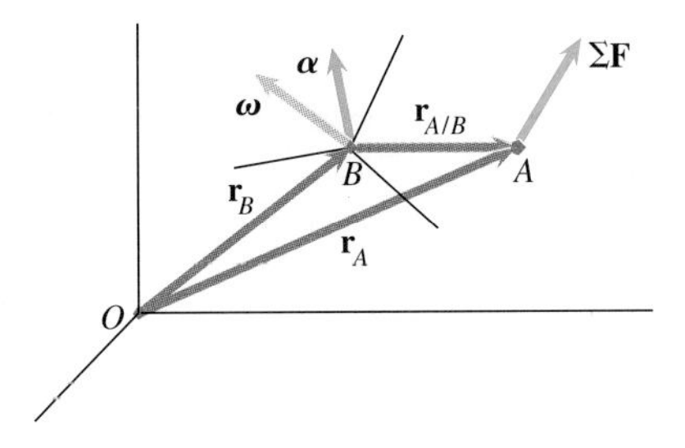

Figure 6.45

An inertial reference frame (origin O) and a reference frame undergoing an arbitrary motion (origin B).

Example 6.12

Suppose that you and a friend play tennis on the deck of a cruise ship (Figure 6.46), and use the ship-fixed coordinate system with origin B to analyse the motion of the ball A. At the instant shown, the ball's position and velocity relative to the ship-fixed coordinate system are $\mathbf{r}_{A/B} = (4.5\,\mathbf{i} + 2.4\,\mathbf{j} + 10.8\,\mathbf{k})\,\mathrm{m}$ and $\mathbf{v}_{A\,\mathrm{rel}} = (0.6\,\mathbf{i} - 2.4\,\mathbf{i} + 6.6\,\mathbf{k})\,\mathrm{m/s}$. The ball weighs 0.5 N, and the aerodynamic force acting on it at the instant shown is $\mathbf{F} = (0.1\,\mathbf{i} + 0.004\,\mathbf{j} + 0.01\,\mathbf{k})\,\mathrm{N}$. The ship is turning at a constant rate, and as a result the acceleration of point B relative to the earth is $\mathbf{a}_B = (-0.9\,\mathbf{i} + 0.06\,\mathbf{k})\,\mathrm{m/s^2}$ and the ship's angular velocity is $\boldsymbol{\omega} = 0.1\,\mathbf{j}\,\mathrm{rad/s}$. Determine the ball's acceleration relative to the ship-fixed coordinate system: (a) assuming that the ship-fixed coordinate system is inertial; (b) not assuming that the ship-fixed coordinate system is inertial, but assuming that a local earth-fixed coordinate system is inertial.

Figure 6.46

STRATEGY

In part (a), we know the ball's mass and the external forces acting on it, so we can simply apply Newton's second law to determine the acceleration. In part (b), we can express Newton's second law in the form given by Equation (6.28), which applies to a coordinate system undergoing an arbitrary motion relative to an inertial coordinate system.

SOLUTION

(a) Assuming that the ship-fixed coordinate system is inertial, Newton's second law is

$$\Sigma \mathbf{F} = m\,\mathbf{a}_{A\,\text{rel}}:$$

$$-0.5\,\mathbf{j} + (0.1\,\mathbf{i} + 0.04\,\mathbf{j} + 0.01\,\mathbf{k}) = \left(\frac{0.5}{9.81}\right)\mathbf{a}_{A\,\text{rel}}$$

Solving this equation, we obtain the ball's acceleration under the assumption that the ship-fixed coordinate system is inertial:

$$\mathbf{a}_{A\,\text{rel}} = (1.96\,\mathbf{i} - 9.73\,\mathbf{j} + 0.20\mathbf{k})\,\text{m/s}^2$$

(b) Dividing Equation (6.28) by m gives

$$\left(\frac{1}{m}\right)\Sigma\,\mathbf{F} - \mathbf{a}_B - 2\boldsymbol{\omega} \times \mathbf{v}_{A\,\text{rel}} - \alpha \times \mathbf{r}_{A/B} - \boldsymbol{\omega} \times (\boldsymbol{\omega} \times \mathbf{r}_{A/B}) = \mathbf{a}_{A\,\text{rel}}:$$

$$\left[\frac{1}{0.5/9.81}\right][-0.5\,\mathbf{j} + (0.1\,\mathbf{i} + 0.04\,\mathbf{j} + 0.01\,\mathbf{k})]$$

$$-(-0.9\,\mathbf{i} + 0.06\,\mathbf{k}) - 2\begin{vmatrix} \mathbf{i} & \mathbf{j} & \mathbf{k} \\ 0 & 0.1 & 0 \\ 0.6 & -2.4 & 6.6 \end{vmatrix} - 0$$

$$-(0.1\,\mathbf{j}) \times \begin{vmatrix} \mathbf{i} & \mathbf{j} & \mathbf{k} \\ 0 & 0.1 & 0 \\ 4.5 & 2.4 & 10.8 \end{vmatrix} = \mathbf{a}_{A\,\text{rel}}$$

The ball's acceleration under the assumption that an earth-fixed coördinate system is inertial is

$$\mathbf{a}_{A\,\text{rel}} = (1.59\,\mathbf{i} - 9.73\,\mathbf{j} + 0.36\,\mathbf{k})\,\text{m/s}^2$$

DISCUSSION

This example illustrates the care that you must exercise in applying Newton's second law. The acceleration we predicted by assuming that the ship-fixed coordinate system is inertial does not even approximate the correct value.

Problems

6.125 A merry-go-round rotates at a constant angular velocity of 0.5 rad/s. The person A walks at a constant speed of 1 m/s along a radial line. Determine A's velocity and acceleration *relative to the earth* when she is 2 m from the centre of the merry-go-round, using two methods:

(a) Express the velocity and acceleration in terms of polar coordinates.

(b) Use Equations (6.19) and (6.20) to express the velocity and acceleration in terms of a body-fixed coordinate system with its x axis aligned with the line along which A walks and its z axis perpendicular to the merry-go-round.

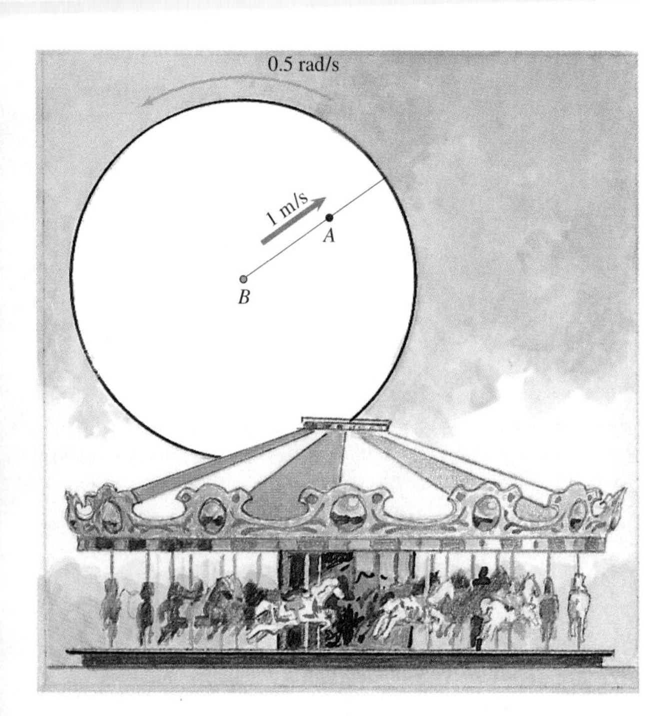

P6.125

6.126 A disc-shaped space station of radius R rotates with constant angular velocity ω about the axis perpendicular to the page. Two persons are stationary relative to the station at A and B, and O is the centre of the station. Using Equations (6.19) and (6.20) and the body-fixed coordinate system shown, (a) determine A's velocity and acceleration relative to a non-rotating reference frame with its origin at O; (b) determine A's velocity and acceleration relative to a non-rotating reference frame whose origin moves with point B.

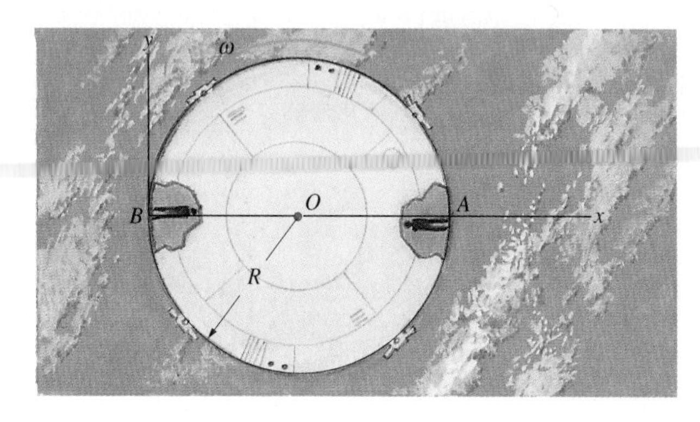

P6.126

6.127 The metal plate is attached to a fixed ball and socket support at O. The pin A slides in a slot in the plate. At the instant shown, $x_A = 1$ m, $dx_A/dt = 2$ m/s, and $d^2x_A/dt^2 = 0$, and the plate's angular velocity and angular acceleration are $\boldsymbol{\omega} = 2\,\mathbf{k}$ rad/s and $\alpha = 0$. What are the x, y, z components of the velocity and acceleration of A relative to a non-rotating reference frame that is stationary with respect to O?

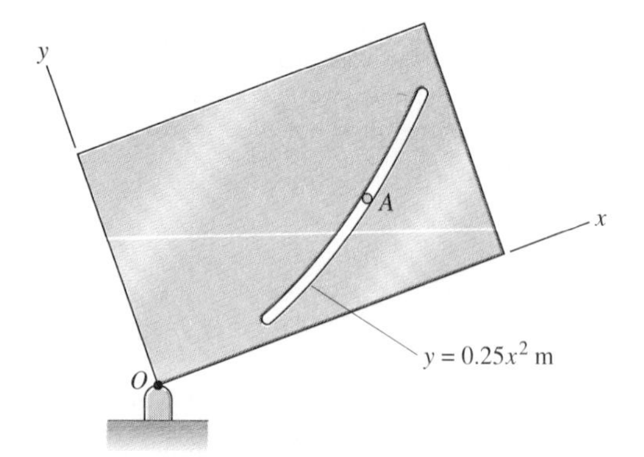

P6.127

6.128 Suppose that at the instant shown in Problem 6.127, $x_A = 1$ m, $dx_A/dt = -3$ m/s, and $d^2x_A/dt^2 = 4$ m/s^2, and the plate's angular velocity and angular acceleration are $\boldsymbol{\omega} = (-4\mathbf{j} + 2\mathbf{k})$ rad/s and $\boldsymbol{\alpha} = (3\mathbf{i} - 6\mathbf{j})$ rad/s^2. What are the x, y, z components of the velocity and acceleration of A relative to a non-rotating reference frame that is stationary with respect to O?

6.129 The coordinate system shown is fixed relative to the ship B. At the instant shown, the ship is sailing north at 3 m/s relative to the earth and its angular velocity is 0.02 rad/s clockwise. The aeroplane is flying east at 120 m/s relative to the earth, and its position relative to the ship is $\mathbf{r}_{A/B} = (600\mathbf{i} + 600\mathbf{j} + 300\mathbf{k})$ m. If the ship uses its radar to measure the plane's velocity relative to its body-fixed coordinate system, what is the result?

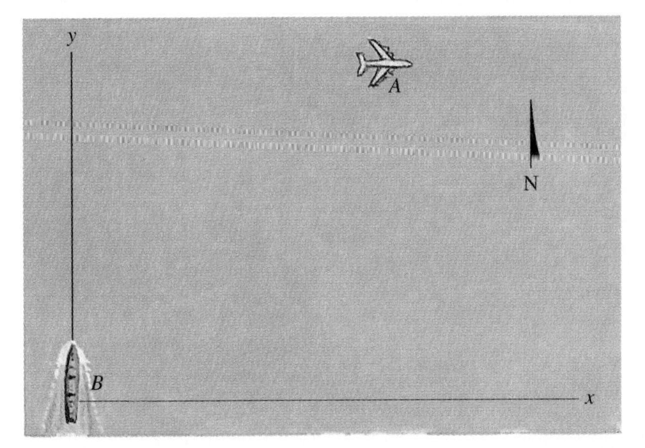

P6.129

6.130 The space shuttle is attempting to recover a satellite for repair. At the current time, the satellite's position relative to a coordinate system fixed to the shuttle is $50\mathbf{i}$ m. The rate-gyros on the shuttle indicate that its current angular velocity is $(0.05\mathbf{j} + 0.03\mathbf{k})$ rad/s. The shuttle pilot measures the velocity of the satellite relative to the body-fixed coordinate system and determines it to be $(-2\mathbf{i} - 1.5\mathbf{j} + 2.5\mathbf{k})$ m/s. What are the x, y, z components of the satellite's velocity relative to a non-rotating coordinate system with its origin at the shuttle?

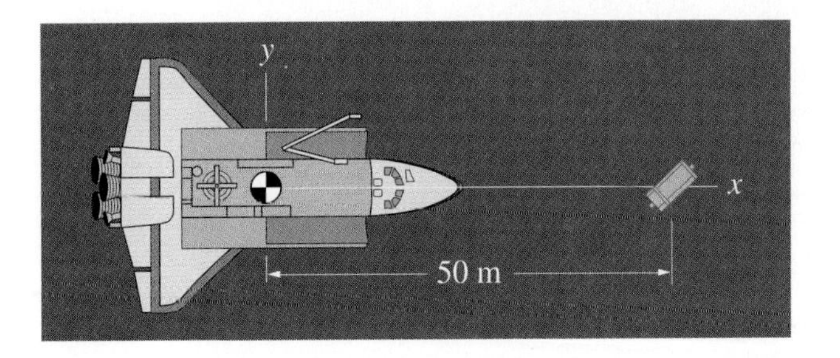

P6.130

6.131 The train on the circular track is travelling at a constant speed of 15 m/s in the direction shown. The train on the straight track is travelling at 6 m/s in the direction shown and is increasing its speed at 0.6 m/s^2. Determine the velocity of passenger A that passenger B observes relative to the coordinate system shown, which is fixed to the car in which B is riding.

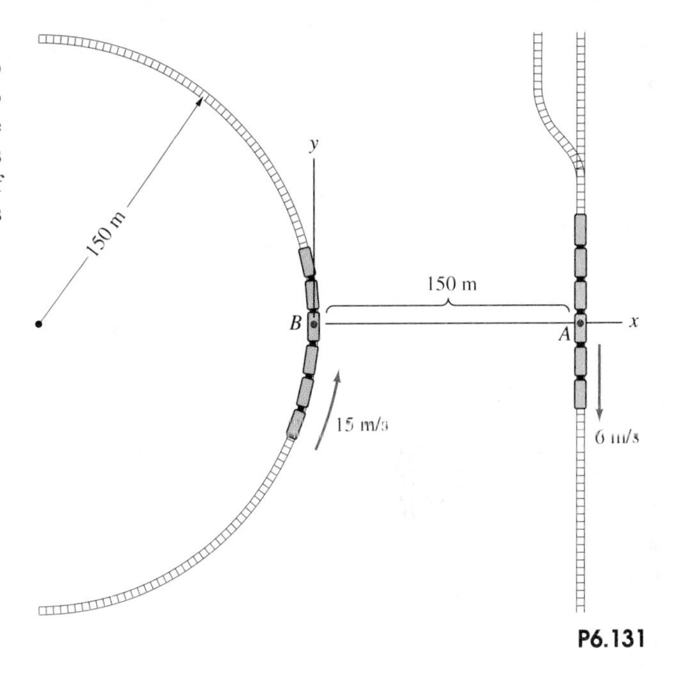

P6.131

6.132 In Problem 6.131, determine the acceleration of passenger A that passenger B observes relative to the coordinate system fixed to the car in which B is riding.

6.133 The satellite A is in a circular polar orbit (a circular orbit that intersects the poles). The radius of the orbit is R, and the magnitude of the satellite's velocity relative to a non-rotating reference frame with its origin at the centre of the earth is v_A. At the instant shown, the satellite is above the equator. An observer B on the earth directly below the satellite measures its motion using the earth-fixed coordinate system shown. What are the velocity and acceleration of the satellite relative to B's earth-fixed coordinate system? The radius of the earth is R_E and its angular velocity is ω_E.

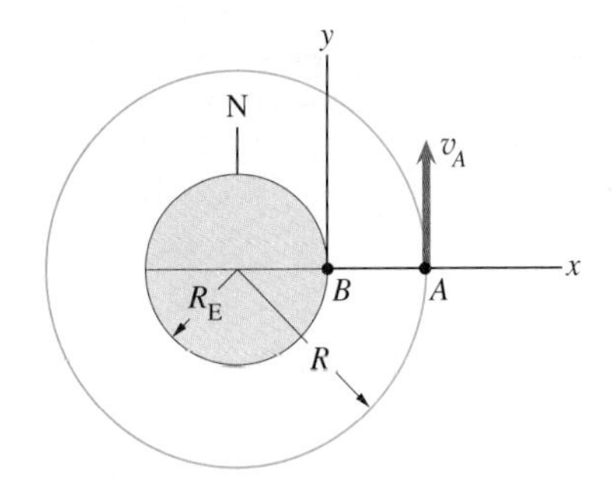

P6.133

6.134 A car A at north latitude L drives north on a north–south highway with constant velocity v. The earth's radius is R_E and its angular velocity is ω_E. Determine the x, y, z components of the car's velocity and acceleration (a) relative to the earth-fixed coordinate system shown; (b) relative to a non-rotating coordinate system with its origin at the centre of the earth.

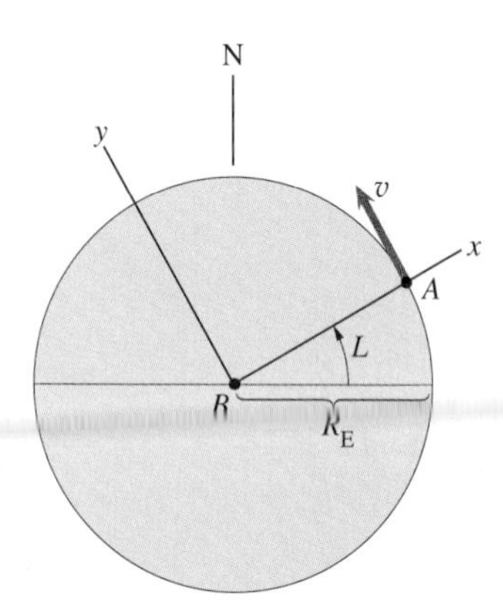

P6.134

6.135 The aeroplane B conducts flight tests of a missile. At the instant shown, the aeroplane is travelling at 200 m/s relative to the earth in a circular path of 200 m radius *in the horizontal plane*. The coordinate system is fixed relative to the aeroplane. The x axis is tangent to the plane's path and points forward. The y axis points out of the plane's right side, and the z axis points out of the bottom of the plane. The plane's bank angle (the inclination of the z axis from the vertical) is constant and equal to 20°. *Relative to the aeroplane's coordinate system*, the pilot measures the missile's position and velocity and determines them to be $\mathbf{r}_{A/B} = 1000\,\mathbf{i}$ m and $\mathbf{v}_{A/B} = (100.0\,\mathbf{i} + 94.0\,\mathbf{j} + 34.2\,\mathbf{k})$ m/s.
(a) What are the x, y, z components of the aeroplane's angular velocity vector?
(b) What are the x, y, z components of the missile's velocity relative to the earth?

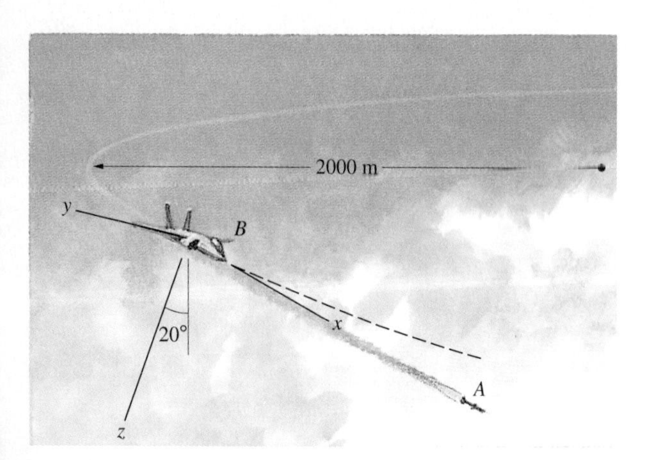

P6.135

6.136 To conduct experiments related to long-term space flight, engineers construct a laboratory on earth that rotates about the vertical axis at B with a constant angular velocity ω of one revolution every 6 seconds. They establish a laboratory-fixed coordinate system with its origin at B and the z axis upwards. An engineer holds an object at point A, 3 m from the axis of rotation, and releases it. At the instant he drops the object, determine its acceleration relative to the laboratory-fixed coordinate system (a) assuming that the laboratory-fixed coordinate system is inertial; (b) not assuming that the laboratory-fixed coordinate system is inertial, but assuming that an earth-fixed coordinate system with its origin at B is inertial.

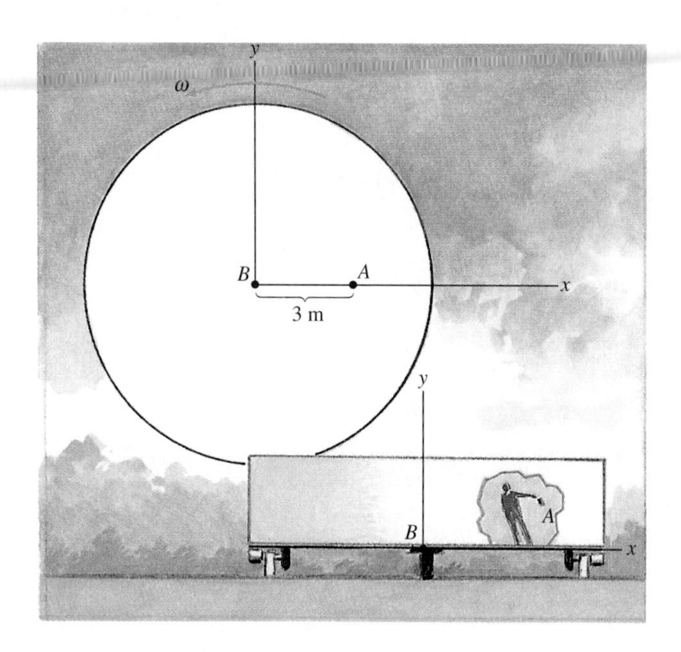

P6.136

6.137 A disc *lying in the horizontal plane* rotates about a fixed shaft at the origin with constant angular velocity ω. The slider A of mass m moves in a smooth slot in the disc. The spring is unstretched when $x = 0$.

(a) By expressing Newton's second law in terms of the body-fixed coordinate system, show that the slider's motion is governed by the equation

$$\frac{d^2x}{dt^2} + \left(\frac{k}{m} - \omega^2\right)x = 0$$

(b) The slider is given an initial velocity $dx/dt = v_0$ at $x = 0$. Determine its velocity as a function of x.

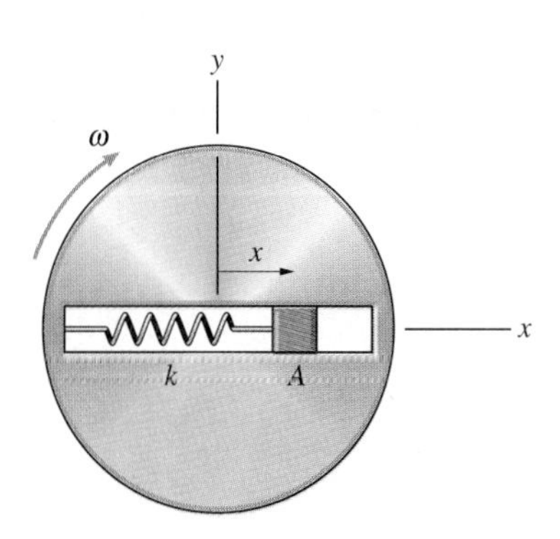

P6.137

6.138 Engineers conduct flight tests of a rocket at 30° north latitude. They measure the rocket's motion using an earth-fixed coordinate system with the x axis upwards and the y axis northwards. At a particular instant, the mass of the rocket is 4000 kg, its velocity relative to their coordinate system is $(2000\mathbf{i} + 2000\mathbf{j})$ m/s, and the sum of the forces exerted on the rocket by its thrust, weight and aerodynamic forces is $(400\mathbf{i} + 400\mathbf{j})$ N. Determine the rocket's acceleration relative to their coordinate system (a) assuming that their earth-fixed coordinate system is inertial; (b) not assuming that their earth-fixed coordinate system is inertial.

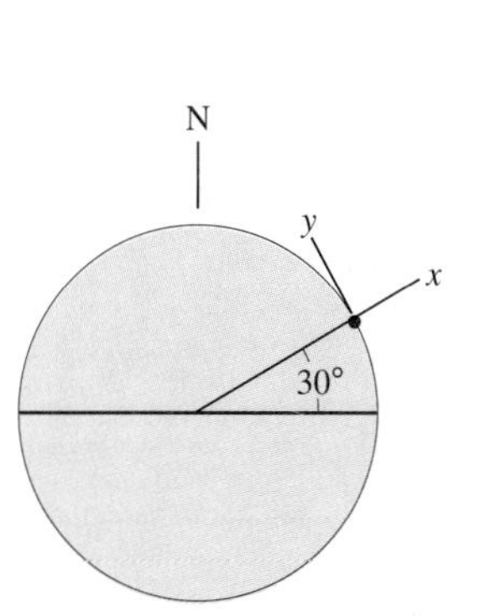

P6.138

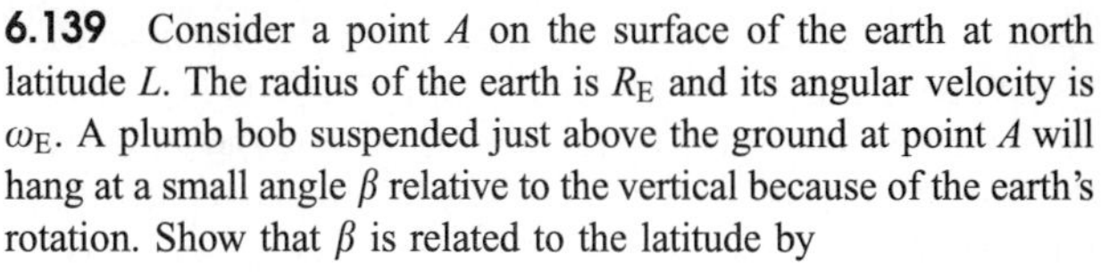

6.139 Consider a point A on the surface of the earth at north latitude L. The radius of the earth is R_E and its angular velocity is ω_E. A plumb bob suspended just above the ground at point A will hang at a small angle β relative to the vertical because of the earth's rotation. Show that β is related to the latitude by

$$\tan\beta = \frac{\omega_E^2 R_E \sin L \cos L}{g - \omega_E^2 R_E \cos^2 L}$$

Strategy: Using the earth-fixed coordinate system shown, express Newton's second law in the form given by Equation (6.25).

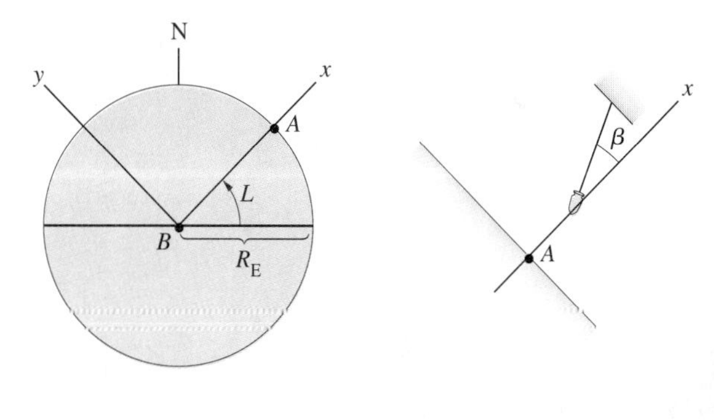

P6.139

6.140 Suppose that a space station is in orbit around the earth and two astronauts on the station toss a ball back and forth. They observe that the ball appears to travel between them in a straight line at constant velocity.

(a) Write Newton's second law for the ball as it travels between them in terms of a non-rotating coordinate system that is stationary relative to the station. What is the term $\Sigma\mathbf{F}$? Use the equation to explain the behaviour of the ball observed by the astronauts.

(b) Write Newton's second law for the ball as it travels between them in terms of a non-rotating coordinate system that is stationary relative to the centre of the earth. What is the term $\Sigma\mathbf{F}$? Explain the difference between this equation and the one you obtained in part (a).

Chapter Summary

A **rigid body** is an idealized model of an object in which the distance between every pair of points of the object remains constant. If a rigid body in motion does not rotate, it is said to be in **translation**. If the centre of mass moves in a fixed plane and an axis of a body-fixed coordinate system remains perpendicular to the plane, it is said to undergo **two-dimensional**, or **planar**, motion.

Relative Velocities and Accelerations

The **angular velocity vector** $\boldsymbol{\omega}$ of a rigid body is parallel to the axis of rotation and its magnitude $|\boldsymbol{\omega}|$ is the rate of rotation. If the thumb of the right hand points in the direction of $\boldsymbol{\omega}$, the fingers curl around $\boldsymbol{\omega}$ in the direction of the rotation. The **angular acceleration vector** $\alpha = d\boldsymbol{\omega}/dt$ is the rate of change of the angular velocity vector.

Consider a point B of a rigid body, a body-fixed coordinate system, and an arbitrary point A (Figure (a)). The velocities $\mathbf{v}_A$ and $\mathbf{v}_B$ of the points relative to O are related by

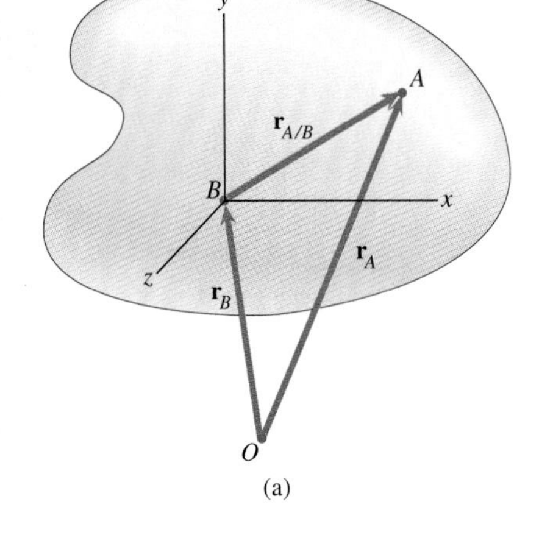

(a)

$$\mathbf{v}_A = \mathbf{v}_B + \mathbf{v}_{A\,\mathrm{rel}} + \boldsymbol{\omega} \times \mathbf{r}_{A/B} \qquad \textbf{Equation (6.11)}$$

where $\mathbf{v}_{A\,\mathrm{rel}}$ is the velocity of A relative to the body-fixed coordinate system. If A is a point of the rigid body, $\mathbf{v}_{A\,\mathrm{rel}}$ is zero.

The accelerations $\mathbf{a}_A$ and $\mathbf{a}_B$ of the points relative to O are related by

$$\mathbf{a}_A = \mathbf{a}_B + \mathbf{a}_{A\,\mathrm{rel}} + 2\boldsymbol{\omega} \times \mathbf{v}_{A\,\mathrm{rel}} + \alpha \times \mathbf{r}_{A/B} + \boldsymbol{\omega} \times (\boldsymbol{\omega} \times \mathbf{r}_{A/B}) \qquad \textbf{Equation (6.13)}$$

where $\mathbf{a}_{A\,\mathrm{rel}}$ is the acceleration of A relative to the body-fixed coordinate system. In plane motion, the term $\boldsymbol{\omega} \times (\boldsymbol{\omega} \times \mathbf{r}_{A/B})$ can be written in the simpler form $-\omega^2\mathbf{r}_{A/B}$.

If A is a point of the rigid body, $\mathbf{v}_{A\,\mathrm{rel}}$ and $\mathbf{a}_{A\,\mathrm{rel}}$ are zero.

Instantaneous Centres

An **instantaneous centre** is a point of a rigid body whose velocity is zero at a given instant. Consider a rigid body in plane motion, and suppose that C is an instantaneous centre. The velocity of a point A is perpendicular to the line from C to A and its magnitude is the product of the distance from C to A and the angular velocity (Figure (b)).

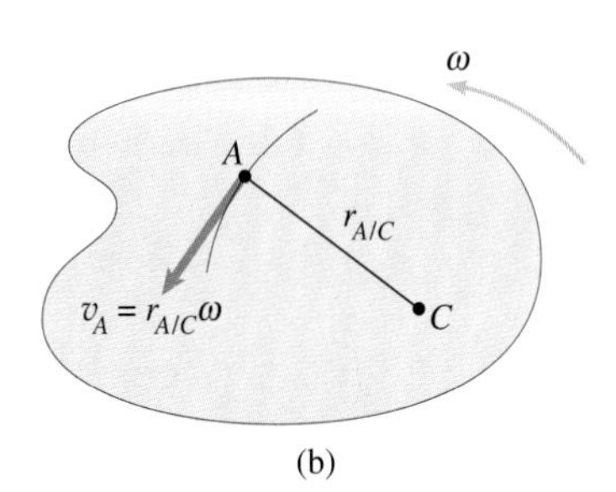

(b)

If you know the directions of the motions of two points A and B of a rigid body in planar motion, lines drawn through A and B perpendicular to their directions of motion intersect at the instantaneous centre (Figure (c)).

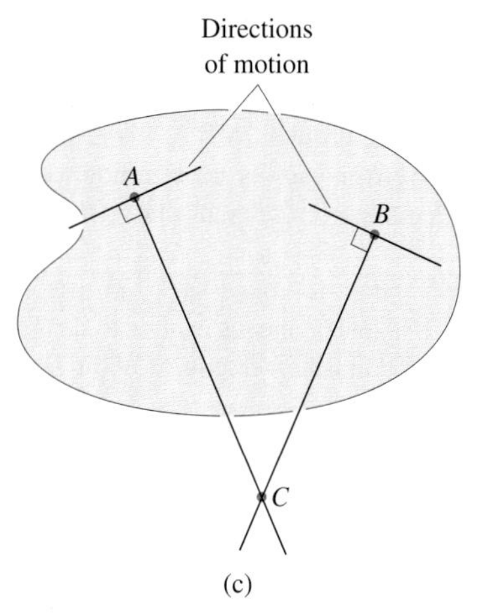

(c)

Rotating Coordinate Systems

Consider a point A and a coordinate system with origin B that rotates with angular velocity $\boldsymbol{\omega}$ and angular acceleration α (Figure (d)). The velocities of A and B relative to a non-rotating coordinate system that is stationary with respect to the reference point O are related by

$$\mathbf{v}_A = \mathbf{v}_B + \mathbf{v}_{A\,\mathrm{rel}} + \boldsymbol{\omega} \times \mathbf{r}_{A/B} \qquad \textbf{Equation (6.19)}$$

where $\mathbf{v}_{A\,\mathrm{rel}}$ is the velocity of A relative to the rotating coordinate system. The accelerations of A and B relative to a non-rotating coordinate system that is stationary with respect to the reference point O are related by

$$\begin{aligned}\mathbf{a}_A &= \mathbf{a}_B + \mathbf{a}_{A\,\mathrm{rel}} + 2\boldsymbol{\omega} \times \mathbf{v}_{A\,\mathrm{rel}} \\ &\quad + \alpha \times \mathbf{r}_{A/B} + \boldsymbol{\omega} \times (\boldsymbol{\omega} \times \mathbf{r}_{A/B})\end{aligned} \qquad \textbf{Equation (6.20)}$$

where $\mathbf{a}_{A\,\mathrm{rel}}$ is the acceleration of A relative to the rotating coordinate system.

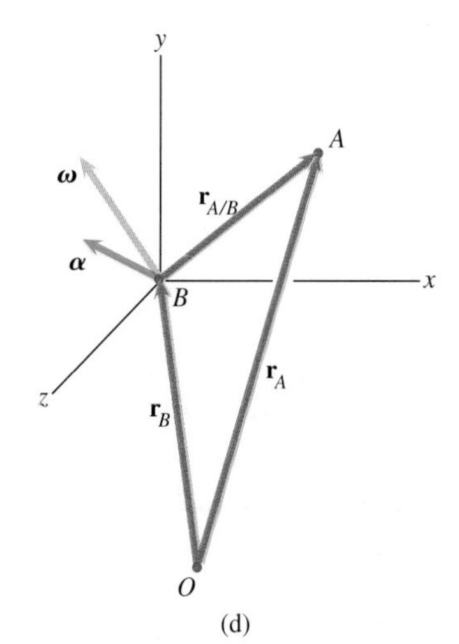

(d)

Review Problems

6.141 Determine the vertical velocity v_H of the hook and the angular velocity of the small pulley.

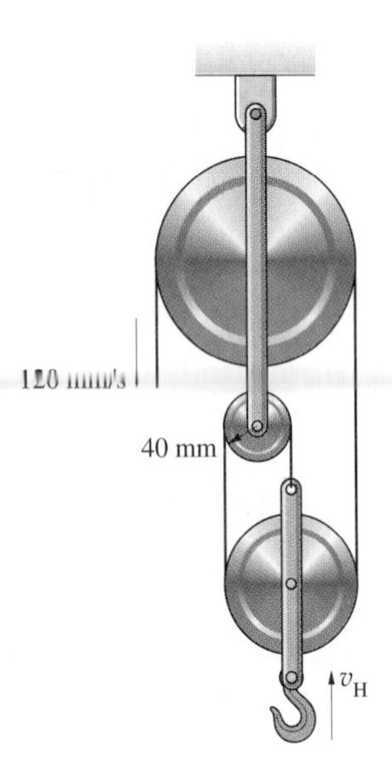

P6.141

6.142 If the crankshaft AB is turning in the counterclockwise direction at 2000 rpm (revolutions per minute), what is the velocity of the piston?

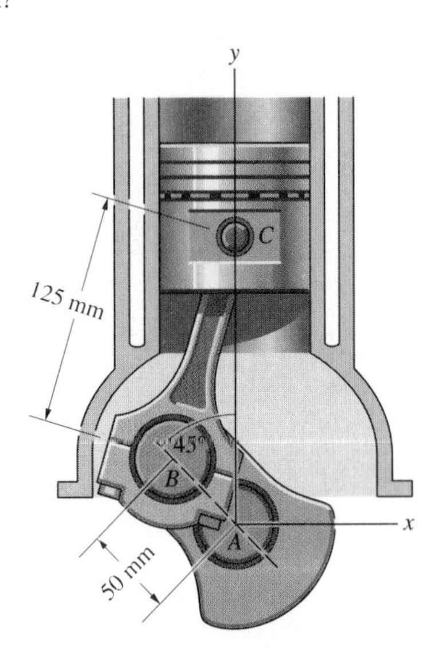

P6.142

6.143 In Problem 6.142, if the piston is moving with velocity $\mathbf{v}_C = 6\mathbf{j}$ m/s, what are the angular velocities of the crankshaft AB and the connecting rod BC?

6.144 In Problem 6.142, if the piston is moving with velocity $\mathbf{v}_C = 6\mathbf{j}$ m/s and its acceleration is zero, what are the angular accelerations of the crankshaft AB and the connecting rod BC?

6.145 Bar AB rotates at 6 rad/s in the counterclockwise direction. Use instantaneous centres to determine the angular velocity of bar BCD and the velocity of point D.

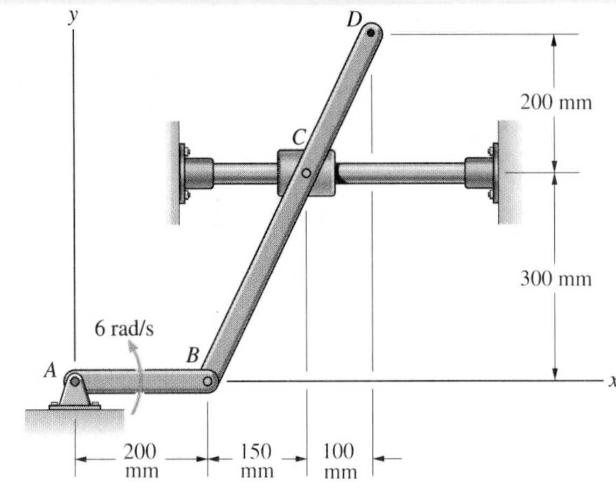

P6.145

6.146 In problem 6.145, bar AB rotates with a constant angular velocity of 6 rad/s in the counterclockwise direction. Determine the acceleration of point D.

6.147 Point C is moving to the right at 500 mm/s. What is the velocity of the midpoint G of bar BC?

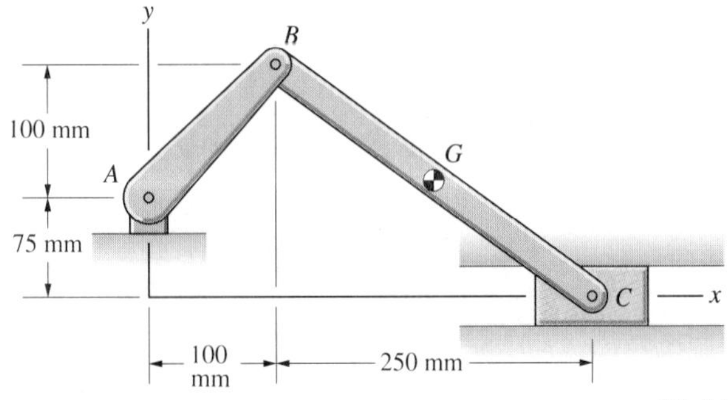

P6.147

6.148 In Problem 6.147, point C is moving to the right with a constant velocity of 500 mm/s. What is the acceleration of the midpoint G of bar BC?

6.149 In Problem 6.147, if the velocity of point C is $\mathbf{v}_C = 25\,\mathbf{i}\,(\text{mm/s})$, what are the angular velocity vectors of arms AB and BC?

6.150 Points B and C are the x-y plane. The angular velocity vectors of arms AB and BC are $\boldsymbol{\omega}_{AB} = -0.5\,\mathbf{k}\,(\text{rad/s})$, $\boldsymbol{\omega}_{BC} = 2.0\,\mathbf{k}\,(\text{rad/s})$. Determine the velocity of point C.

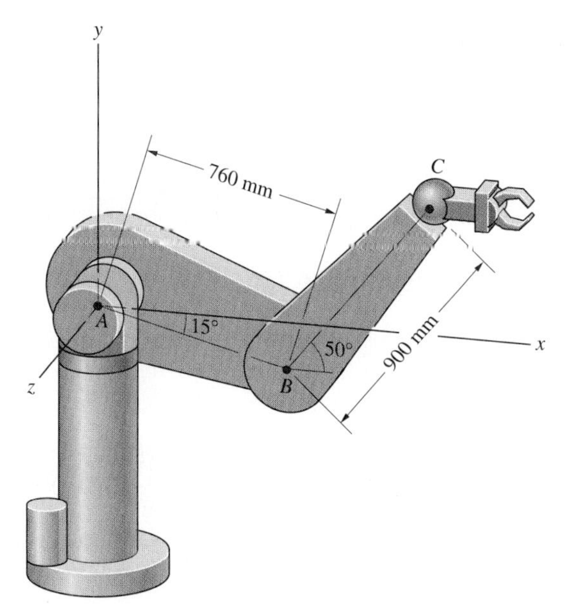

P6.150

6.151 In Problem 6.150, if the velocity of point C is $\mathbf{v}_C = 1.0\,\mathbf{i}\,(\text{m/s})$, what are the angular velocity vectors of arms AB and BC?

6.152 In Problem 6.150, if the angular velocity vectors of arms AB and BC are $\boldsymbol{\omega}_{AB} = -0.5\,\mathbf{k}\,(\text{rad/s})$, $\boldsymbol{\omega}_{BC} = 2.0\,\mathbf{k}\,(\text{rad/s})$, and their angular acceleration vectors are $\boldsymbol{\alpha}_{AB} = 1.0\,\mathbf{k}\,(\text{rad/s}^2)$, $\boldsymbol{\alpha}_{BC} = 1.0\,\mathbf{k}\,(\text{rad/s}^2)$, what is the acceleration of point C?

6.153 In Problem 6.150, if the velocity of point C is $\mathbf{v}_C = 1.0\,\mathbf{i}\,(\text{m/s})$ and $\mathbf{a}_C = 0$, what are the angular velocity and angular acceleration vectors of arm BC?

6.154 The angular velocity of arm AC is 1 rad/s counterclockwise. What is the angular velocity of the scoop?

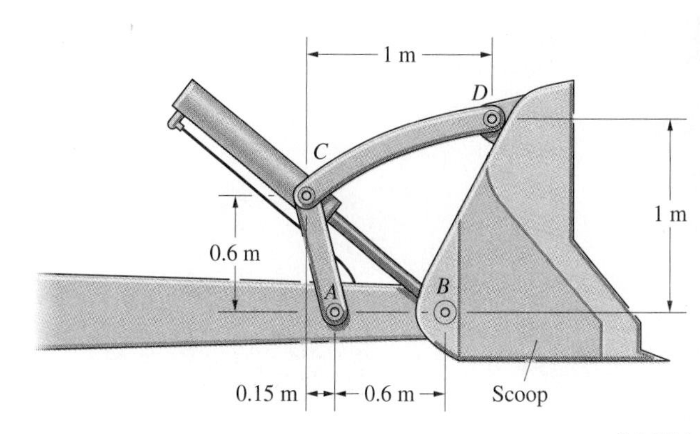

P6.154

6.155 The angular velocity of arm AC in Problem 6.154 is 2 rad/s counterclockwise and its angular acceleration is 4 rad/s^2 clockwise. What is the angular acceleration of the scoop?

6.156 If you want to program the robot so that, at the instant shown, the velocity of point D is $\mathbf{v}_D = (0.2\,\mathbf{i} + 0.8\,\mathbf{j})\,\text{m/s}$ and the angular velocity of arm CD is 0.3 rad/s counterclockwise, what are the necessary angular velocities of arms AB and BC?

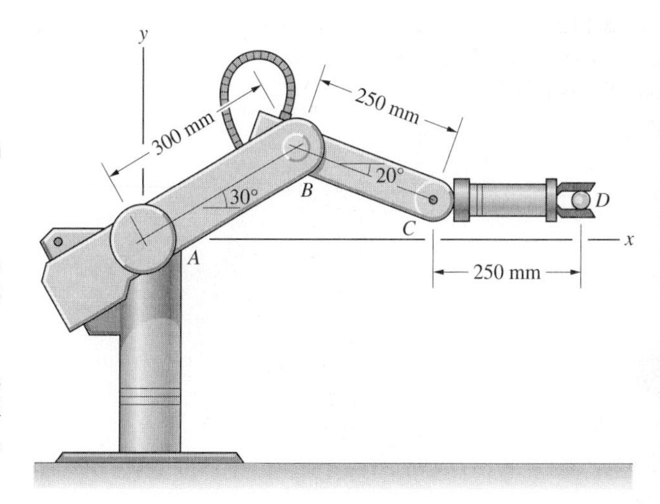

P6.156

6.157 In Problem 6.156, if the acceleration of point D and the angular acceleration of arm CD are zero at the instant shown, what are the angular accelerations of arms AB and BC?

6.158 Arm AB is rotating at 10 rad/s in the clockwise direction. Determine the angular velocity of arm BC and the velocity at which it slides relative to the sleeve at C.

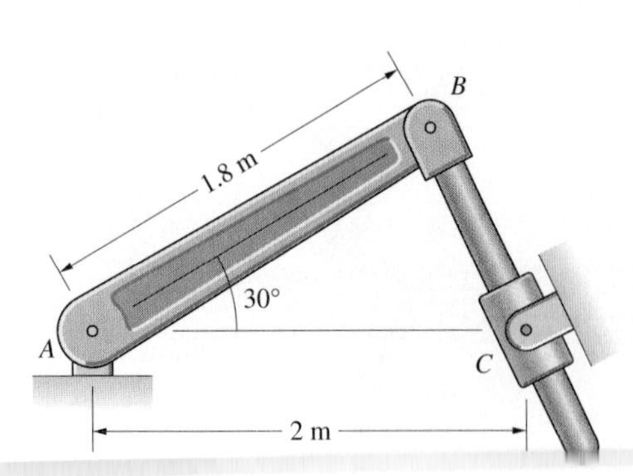

P6.158

6.159 In Problem 6.158, arm AB is rotating with an angular velocity of 10 rad/s and an angular acceleration of 20 rad/s^2, both in the clockwise direction. Determine the angular acceleration of arm BC.

6.160 Arm AB is rotating with a constant counterclockwise angular velocity of 10 rad/s. Determine the vertical velocity and acceleration of the rack R of the rack and pinion gear.

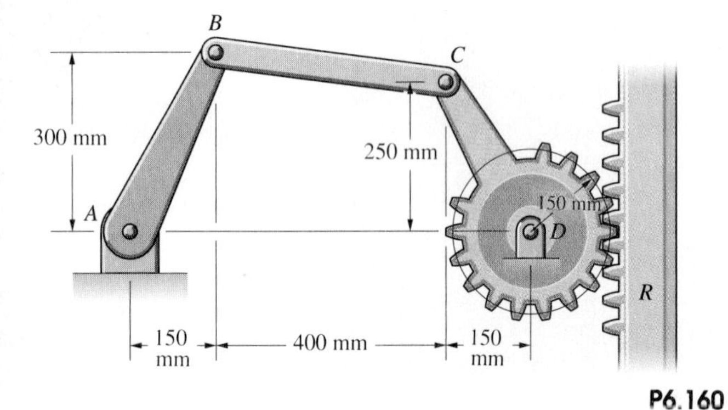

P6.160

6.161 In Problem 6.160, if the rack R of the rack and pinion gear is moving upwards with a constant velocity of 3 m/s, what are the angular velocity and angular acceleration of bar BC?

6.162 The bar AB has a constant counterclockwise angular velocity of 2 rad/s. The 1 kg collar C slides on the smooth horizontal bar. At the instant shown, what is the tension in the cable BC?

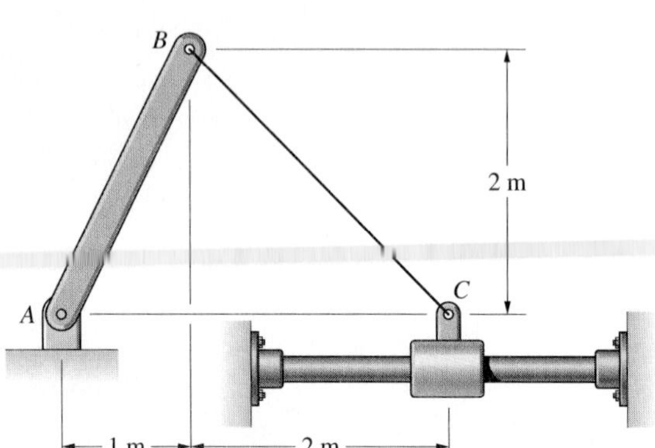

P6.162

6.163 An athlete exercises his arm by raising the 8 kg mass m. The shoulder joint A is stationary. The distance AB is 300 mm, the distance BC is 400 mm, and the distance from C to the pulley is 340 mm. The angular velocities $\omega_{AB} = 1.5$ rad/s and $\omega_{BC} = 2$ rad/s are constant. What is the tension in the cable?

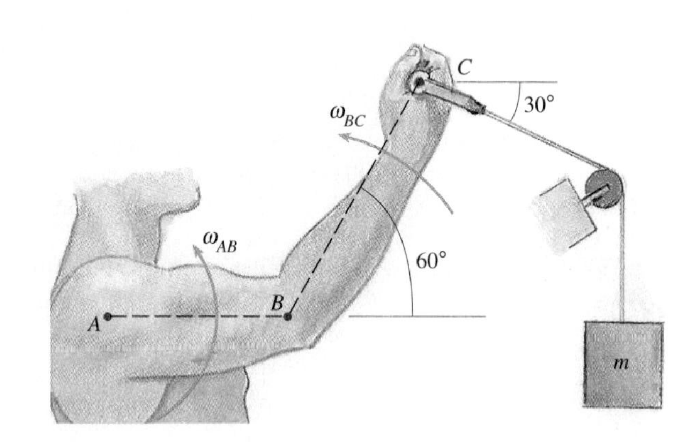

P6.163

6.164 The coordinate system rotates with a constant angular velocity $\boldsymbol{\omega} = 2\,\mathbf{k}$ rad/s. The point A moves outwards along the x axis at a constant rate of 5 m/s.
(a) What are the velocity and acceleration of A relative to the coordinate system?
(b) What are the velocity and acceleration of A relative to a non-rotating coordinate system with its origin at B, when A is at the position $x = 1$ m?

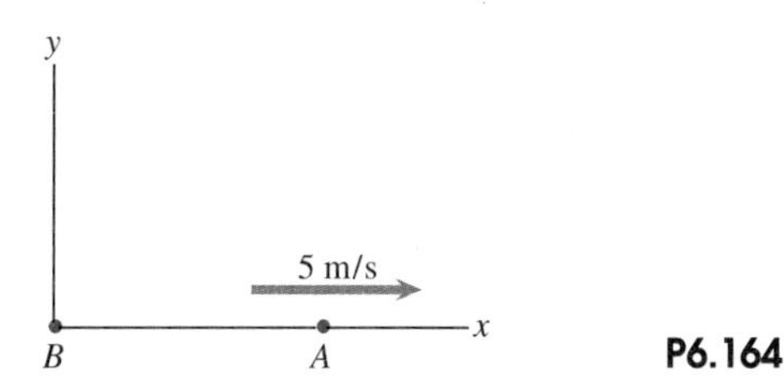

P6.164

6.165 The coordinate system shown is fixed relative to the ship B. The ship uses its radar to measure the position of a stationary buoy A and determines it to be $(400\,\mathbf{i} + 200\,\mathbf{j})$ m. The ship also measures the velocity of the buoy relative to its body-fixed coordinate system and determines it to be $(2\,\mathbf{i} - 8\,\mathbf{j})$ m/s. What are the ship's velocity and angular velocity relative to the earth? (Assume that the ship's velocity is in the direction of the y axis.)

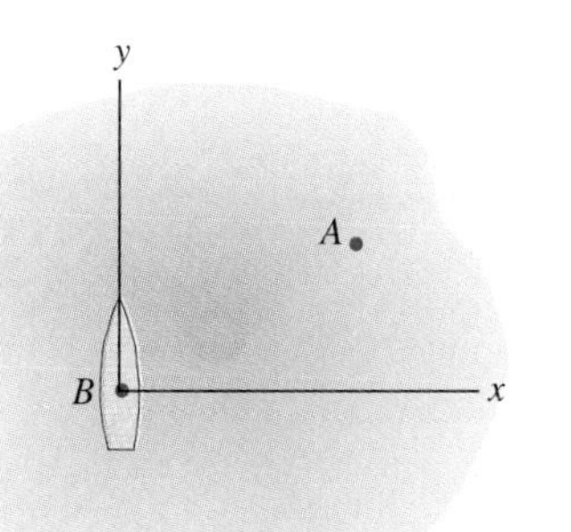

P6.165

The front-end loader shovel undergoes two-dimensional motion as the hydraulic cylinder and suppoting members raise it and rotate it in the vertical plane. Newton's second law relates the sum of the forces on the shovel to the acceleration of its centre of mass, and an equation of angular motion relates the sum of the moments about the shovel's centre of mass to its angular acceleration. In this chapter we use free-body diagrams and the equations of motion for rigid bodies to determine the motions of objects resulting from the forces and couples acting on them.

Chapter 7

Two-Dimensional Dynamics of Rigid Bodies

IN Chapter 6 we analysed two-dimensional motions of rigid bodies without considering the forces and couples causing them. You have used Newton's second law to determine the motions of the centres of mass of objects, but how can you determine their *rotational* motions? In this chapter we derive two-dimensional equations of angular motion for a rigid body. By drawing the free-body diagram of an object such as an excavator's shovel, we can determine both the acceleration of its centre of mass and its angular acceleration in terms of the forces and couples to which it is subjected.

7.1 Preview of the Equations of Motion

The two-dimensional equations of angular motion for a rigid body are quite simple, but you can easily lose sight of the forest among the trees as we derive them. To help you follow the derivations, we summarize the equations in this section.

The equations of motion of a rigid body include Newton's second law,

$$\Sigma \mathbf{F} = m\mathbf{a}$$

which states that the sum of the external forces acting on the body equals the product of its mass and the acceleration of its centre of mass. The equations of motion are completed by an equation of angular motion. If the rigid body rotates about a fixed axis O (Figure 7.1(a)), the sum of the moments about the axis due to external forces and couples acting on the body is related to its angular acceleration by

$$\Sigma M_0 = I_0\alpha$$

where I_0 is the mass moment of inertia of the rigid body about O. Just as an object's mass determines the acceleration resulting from the forces acting on it, its mass moment of inertia I_0 about a fixed axis determines the angular acceleration resulting from the sum of the moments about the axis.

In the case of general planar motion (Figure 7.1(b)), the sum of the moments about the centre of mass of a rigid body is related to its angular acceleration by

$$\Sigma M = I\alpha$$

where I is the mass moment of inertia of the rigid body about its centre of mass. If we know the external forces and couples acting on a rigid body in planar motion, we can use these equations to determine the acceleration of its centre of mass and its angular acceleration.

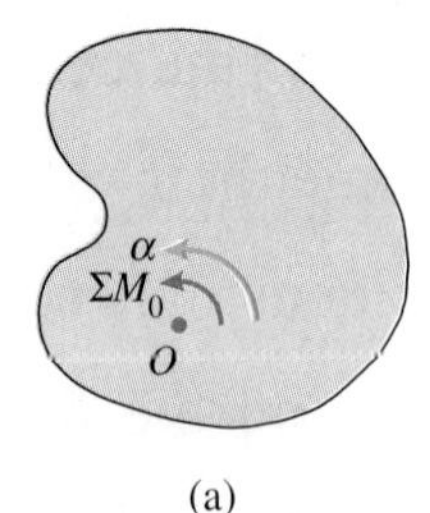

(a)

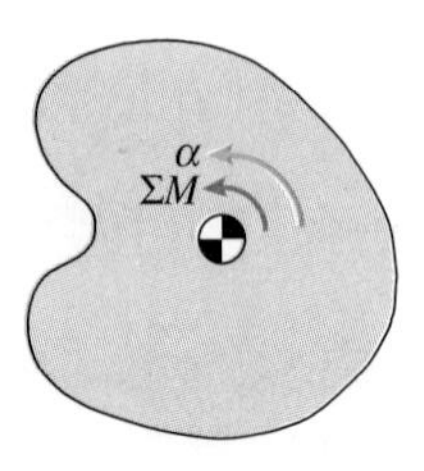

(b)

Figure 7.1

(a) A rigid body rotating about a fixed axis O.

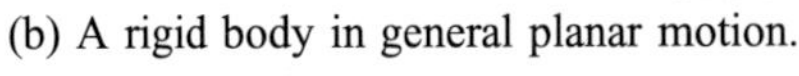

(b) A rigid body in general planar motion.

7.2 Momentum Principles for a System of Particles

In this chapter and in our discussion of three-dimensional dynamics of rigid bodies in Chapter 9, our derivations of the equations of motion begin with principles governing the motion of a system of particles. We summarize these general and important principles in this section.

Force-Linear Momentum Principle

We begin by showing that the sum of the external forces on a system of particles equals the rate of change of its total linear momentum. Let us consider a system of N particles. We denote the mass of the ith particle by m_i and denote its position vector relative to a fixed point O by $\mathbf{r}_i$ (Figure 7.2). Let $\mathbf{f}_{ij}$ be the force exerted on the ith particle by the jth particle, and let the external force on the ith particle (that is, the total force exerted by objects other than the system of particles we are considering) be $\mathbf{f}_i^{\mathrm{E}}$. Newton's second law states that the total force on the ith particle equals the product of its mass and the rate of change of its linear momentum,

$$\sum_j \mathbf{f}_{ij} + \mathbf{f}_i^{\mathrm{E}} = \frac{d}{dt}(m_i \mathbf{v}_i) \tag{7.1}$$

where $\mathbf{v}_i = d\,\mathbf{r}_i/dt$ is the velocity of the ith particle. Writing this equation for each particle of the system and summing from $i = 1$ to N, we obtain

$$\sum_i \sum_j \mathbf{f}_{ij} + \sum_i \mathbf{f}_i^{\mathrm{E}} = \frac{d}{dt} \sum_i m_i \mathbf{v}_i \tag{7.2}$$

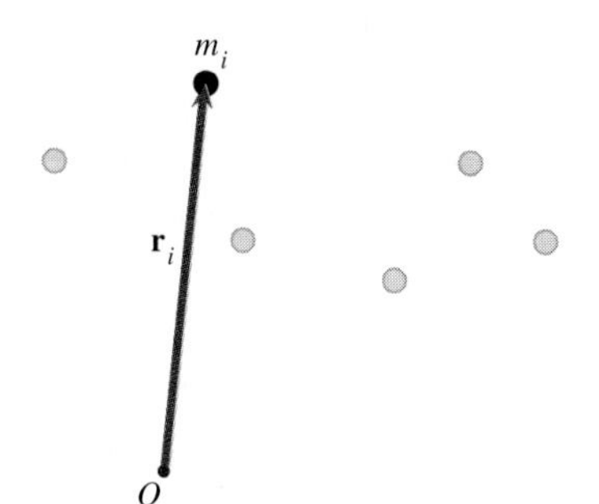

Figure 7.2
A system of particles. The vector $\mathbf{r}_i$ is the position vector of the ith particle.

The first term on the left side of this equation is the sum of the internal forces on the system of particles. As a consequence of Newton's third law ($\mathbf{f}_{ji} + \mathbf{f}_{ij} = 0$), this term equals zero:

$$\sum_i \sum_j \mathbf{f}_{ij} = \mathbf{f}_{12} + \mathbf{f}_{21} + \mathbf{f}_{13} + \mathbf{f}_{31} + \cdots = \mathbf{0}$$

The second term on the left side of Equation (7.2) is the sum of the external forces on the system. Denoting it by $\Sigma \mathbf{F}$, we conclude that the sum of the external forces on the system equals the rate of change of its total linear momentum:

$$\Sigma \mathbf{F} = \frac{d}{dt} \sum_i m_i \mathbf{v}_i \tag{7.3}$$

Let m be the sum of the masses of the particles:

$$m = \sum_i m_i$$

The position of the centre of mass of the system is

$$\mathbf{r} = \frac{\sum_i m_i \mathbf{r}_i}{m} \tag{7.4}$$

so the velocity of the centre of mass is

$$\mathbf{v} = \frac{d\mathbf{r}}{dt} = \frac{\sum_i m_i \mathbf{v}_i}{m}$$

By using this expression, we can write Equation (7.3) as

$$\Sigma \mathbf{F} = \frac{d}{dt}(m\mathbf{v})$$

The total external force on a system of particles equals the rate of change of the product of its total mass and the velocity of its centre of mass. Since any object or collection of objects, including a rigid body, can be regarded as a system of particles, this result is one of the most general and elegant in mechanics. Furthermore, if the total mass m is constant, we obtain

$$\Sigma \mathbf{F} = m\mathbf{a}$$

where $\mathbf{a} = d\mathbf{v}/dt$ is the acceleration of the centre of mass. The total external force equals the product of the total mass and the acceleration of the centre of mass.

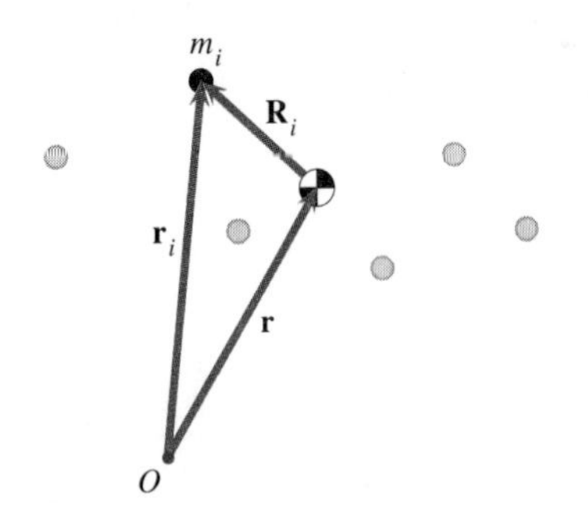

Figure 7.3
The vector $\mathbf{R}_i$ is the position vector of the ith particle relative to the centre of mass.

Moment-Angular Momentum Principles

We now obtain relations between the sum of the moments due to the external forces on a system of particles and the rate of change of its total angular momentum. We follow the same procedure used in Section 5.4 to relate the angular impulse to the change in the angular momentum.

The position of the ith particle of the system relative to O is related to its position relative to the centre of mass (Figure 7.3) by

$$\mathbf{r}_i = r + \mathbf{R}_i \tag{7.5}$$

Multiplying this equation by m_i, summing from 1 to N, and using Equation

(7.4), we find that the positions of the particles relative to the centre of mass are related by

$$\sum_i m_i \mathbf{R}_i = 0 \tag{7.6}$$

The total angular momentum of the system about O is the sum of the angular momenta of the particles

$$\mathbf{H}_0 \sum_i \mathbf{r}_i \times m_i \mathbf{v}_i \tag{7.7}$$

where $\mathbf{v}_i = d\,\mathbf{r}_i/dt$. The angular momentum of the system about its centre of mass (that is, the angular momentum about the fixed point coincident with the centre of mass at the present instant) is

$$\mathbf{H} = \sum_i \mathbf{R}_i \times m_i \mathbf{v}_i \tag{7.8}$$

By using Equations (7.5) and (7.6), it can be shown that

$$\mathbf{H}_0 = \mathbf{r} \times m\mathbf{v} + \mathbf{H} \tag{7.9}$$

This equation expresses the total angular momentum about O as the sum of the angular momentum about O due to the velocity $\mathbf{v}$ of the system's centre of mass and the total angular momentum about the centre of mass (Figure 7.4).

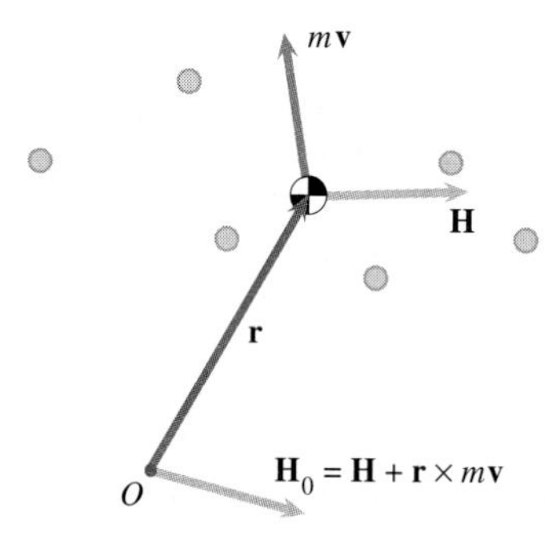

Figure 7.4

The angular momentum about O equals the sum of angular momentum about the centre of mass and the angular momentum about O due to the velocity of the centre of mass.

To obtain relations between the total moment exerted on the system and its total angular momentum, we begin with Newton's second law. We take the cross product of Equation (7.1) with the position vector $\mathbf{r}_i$ and sum from $i = 1$ to N:

$$\sum_i \sum_j \mathbf{r}_i \times \mathbf{f}_{ij} + \sum_i \mathbf{f}_i^E = \sum_i \mathbf{r}_i \times \frac{d}{dt}(m_i \mathbf{v}_i) \tag{7.10}$$

The term on the right side of this equation is the rate of change of the system's total angular momentum about O:

$$\sum_i \mathbf{r}_i \times \frac{d}{dt}(m_i \mathbf{v}_i) = \sum_i \left[\frac{d}{dt}(\mathbf{r}_i \times m_i \mathbf{v}_i) - \underbrace{\mathbf{v}_i \times m_i \mathbf{v}_i}_{=0} \right] = \frac{d\,\mathbf{H}_0}{dt}$$

(The second term in brackets vanishes because the cross product of two parallel vectors equals zero.)

The first term on the left side of Equation (7.10) is the sum of the moments about O due to internal forces. This term vanishes if we assume that the internal forces between each pair of particles are not only equal and opposite, but *are directed along the straight line between the two particles.* (This assumption holds except in the case of systems involving electromagnetic forces between charged particles.) For example, consider particles 1 and 2 in Figure 7.5. If the internal forces are directed along the straight line between the particles, we can write the moment about O due to $\mathbf{f}_{21}$ as $\mathbf{r}_1 \times \mathbf{f}_{21}$, and the total moment about O due to the forces the two particles exert on each other is

$$\mathbf{r}_1 \times \mathbf{f}_{12} + \mathbf{r}_1 \times \mathbf{f}_{21} = \mathbf{r}_1 \times (\mathbf{f}_{12} + \mathbf{f}_{21}) = \mathbf{0}$$

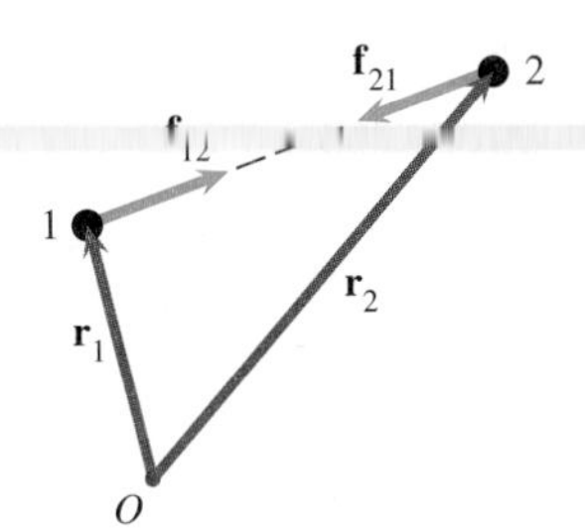

Figure 7.5

Particles 1 and 2 and the forces they exert on each other. If the forces act along the line between the particles, their total moment about O is zero.

The second term on the left side of Equation (7.10) is the sum of the moments about O due to external forces and couples, which we denote by $\Sigma\,(\mathbf{M}_0)$. Therefore Equation (7.10) states that the sum of the moments about O due to external forces and couples equals the rate of change of the system's angular momentum about O:

$$\Sigma\,\mathbf{M}_0 = \frac{d\,\mathbf{H}_0}{dt} \tag{7.11}$$

By using Equation (7.9), we can also write this result in terms of the total angular momentum relative to the centre of mass,

$$\Sigma\,\mathbf{M}_0 = \frac{d}{dt}(\mathbf{r} \times m\,\mathbf{v} + \mathbf{H}) = \mathbf{r} \times m\,\mathbf{a} + \frac{d\,\mathbf{H}}{dt} \tag{7.12}$$

where $\mathbf{a}$ is the acceleration of the centre of mass.

We also need to determine the relation between the sum of the moments about the system's centre of mass, which we denote by $\Sigma\,\mathbf{M}$, and the angular momentum about its centre of mass. We can obtain this result from Equation (7.12) by letting the fixed point O be coincident with the centre of mass at the present instant. In that case $\Sigma\,\mathbf{M}_0 = \Sigma\,\mathbf{M}$ and $\mathbf{r} = \mathbf{0}$, and we see that the sum of the moments about the centre of mass equals the rate of change of the angular momentum about the centre of mass:

$$\Sigma\,\mathbf{M} = \frac{d\,\mathbf{H}}{di} \tag{7.13}$$

7.3 Derivation of the Equations of Motion

We now derive the equations of motion for a rigid body in two-dimensional motion. We have already shown that the total exernal force on any object equals the product of its mass and the acceleration of its centre of mass:

$$\Sigma \mathbf{F} = m\,\mathbf{a}$$

Therefore this equation, which we refer to as Newton's second law, describes the motion of the centre of mass of a rigid body. To derive the equations of angular motion, we first consider rotation about a fixed axis, then general planar motion.

Rotation About a Fixed Axis

Suppose that a rigid body rotates about a fixed axis L_0 through a fixed point O. In terms of a coordinate system with the z axis aligned with L_0 (Figure 7.6(a)), we can express the angular velocity vector as $\boldsymbol{\omega} = \omega\,\mathbf{k}$, and the velocity of the ith particle is $d\,\mathbf{r}_i/dt = \boldsymbol{\omega} \times \mathbf{r}_i = \omega\,\mathbf{k} \times \mathbf{r}_i$. Let $\Sigma M_0 = \Sigma \mathbf{M}_0 \cdot \mathbf{k}$ be the sum of the moments about L_0. From Equations (7.7) and (7.11),

$$\Sigma M_0 = \frac{dH_0}{dt} \tag{7.14}$$

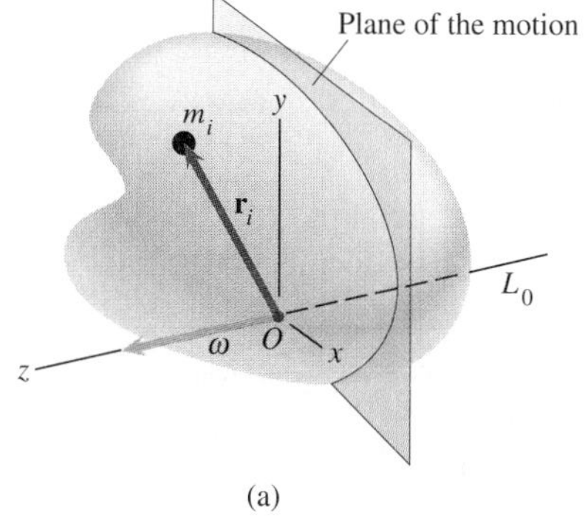

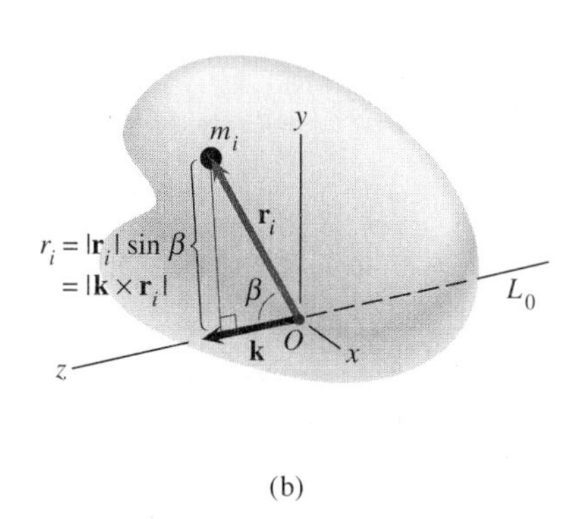

Figure 7.6

(a) A coordinate system with the z axis aligned with the axis of rotation L_0.
(b) The magnitude of $\mathbf{k} \times \mathbf{r}_i$ is the perpendicular distance from the axis of rotation to m_i.

where

$$H_0 = \mathbf{H}_0 \cdot \mathbf{k} = \sum_i \left[\mathbf{r}_i \times m_i(\omega\,\mathbf{k} \times \mathbf{r}_i)\right] \cdot \mathbf{k} \tag{7.15}$$

is the angular momentum about L_0. Using the identity $\mathbf{U} \cdot (\mathbf{V} \times \mathbf{W}) = (\mathbf{U} \times \mathbf{V}) \cdot \mathbf{W}$, we can write Equation (7.15) as

$$H_0 = \sum_i m_i(\mathbf{k} \times \mathbf{r}_i) \cdot (\mathbf{k} \times \mathbf{r}_i)\omega = \sum_i m_i |\,\mathbf{k} \times \mathbf{r}_i|^2 \omega \tag{7.16}$$

In Figure 7.6(b), we show that $|\,\mathbf{k} \times \mathbf{r}_i|$ is the perpendicular distance from L_0 to the ith particle, which we denote by r_i. Using the definition of the mass moment of inertia of the rigid body about L_0,

$$I_0 = \sum_i m_i r_i^2$$

we can write Equation (7.16) as

$$H_0 = I_0\omega$$

Substituting this expression into Equation (7.14), we obtain the equation of angular motion for a rigid body rotating about a fixed axis O:

$$\Sigma M_0 = I_0\alpha \tag{7.17}$$

General Planar Motion

Let L_0 be the axis through a fixed point O that is perpendicular to the plane of the motion of a rigid body, and let L be the parallel axis through the centre of mass (Figure 7.7(a)). We do *not* assume that the rigid body rotates about L_0. In terms of the coordinate system shown, we can express the angular velocity vector as $\boldsymbol{\omega} = \omega\mathbf{k}$, and the velocity of the ith particle relative to the centre of mass is $d\mathbf{R}_i/dt = \omega\mathbf{k}\times\mathbf{R}_i$. From Equations (7.8) and (7.12),

$$\Sigma M_0 = \frac{d}{dt}[(\mathbf{r}\times m\mathbf{v})\cdot\mathbf{k} + H] \tag{7.18}$$

where

$$H = \mathbf{H}\cdot\mathbf{k} = \sum_i [\mathbf{R}_i \times m_i(\omega\mathbf{k}\times\mathbf{R}_i)]\cdot\mathbf{k}$$

is the angular momentum about L. Using the same identity we applied to Equation (7.15), we can write this equation for H as

$$H = \sum_i m_i(\mathbf{k}\times\mathbf{R}_i)\cdot(\mathbf{k}\times\mathbf{R}_i)\omega = \sum_i m_i|\mathbf{k}\times\mathbf{R}_i|^2\omega \tag{7.19}$$

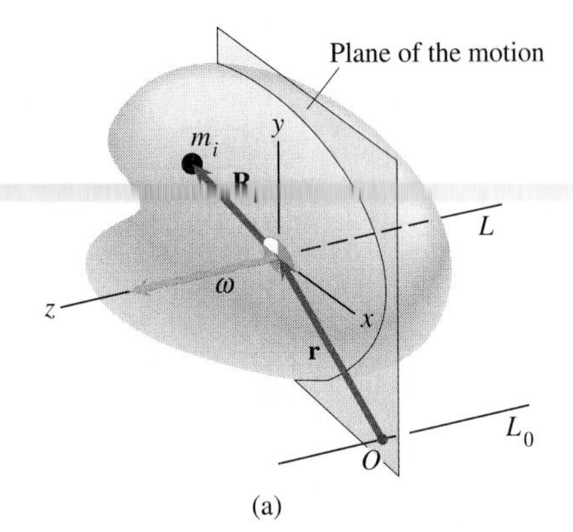

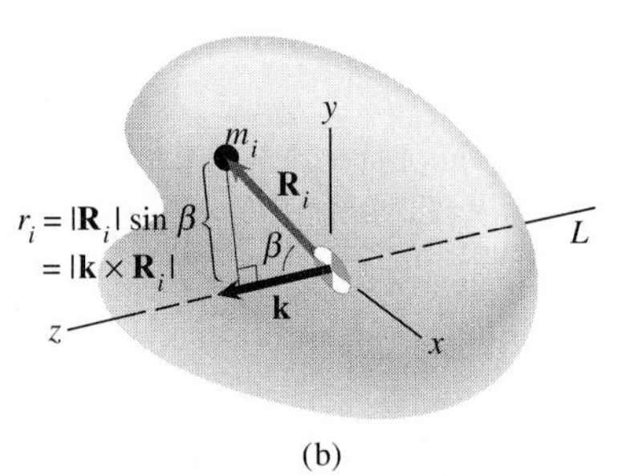

Figure 7.7
(a) A coordinate system with the z axis aligned with L.
(b) The magnitude of $\mathbf{k}\times\mathbf{R}_i$ is the perpendicular distance from L to m_i.

The term $|\mathbf{k}\times\mathbf{R}_i| = r_i$ is the perpendicular distance from L to the ith particle (Figure 7.7(b)). In terms of the mass moment of inertia of the rigid body about L,

$$I = \sum_i m_i r_i^2$$

Equation (7.19) states that the rigid body's angular momentum about L is

$$H = I\omega$$

Substituting this expression into Equation (7.18), we obtain

$$\Sigma M_0 = \frac{d}{dt}[(\mathbf{r}\times m\mathbf{v})\cdot\mathbf{k} + I\omega] = (\mathbf{r}\times m\mathbf{a})\cdot\mathbf{k} + I\alpha \tag{7.20}$$

With this equation we can obtain the relation between the sum of the moments about L, which we denote by ΣM, and the angular acceleration. If we let the fixed axis L_0 be coincident with L at the present instant, $\Sigma M_0 = \Sigma M$ and $\mathbf{r} = \mathbf{0}$, and from Equation (7.20) we obtain

$$\Sigma M = I\alpha$$

The sum of the moments about L equals the product of the moment of inertia about L and the angular acceleration.

7.4 Applications

We have seen that the equations of motion for a rigid body in planar motion include Newton's second law,

$$\boxed{\Sigma \mathbf{F} = m\,\mathbf{a}} \tag{7.21}$$

where **a** is the acceleration of the centre of mass, and an equation relating the moments due to forces and couples to the angular acceleration. If the rigid body rotates about a fixed axis O, the total moment about O equals the product of the moment of inertia about O and the angular acceleration:

$$\boxed{\Sigma M_0 = I_0 \alpha} \tag{7.22}$$

In *any* planar motion, the total moment about the centre of mass equals the product of the mass moment of inertia about the centre of mass and the angular acceleration:

$$\boxed{\Sigma M = I \alpha} \tag{7.23}$$

Of course, this equation applies to the case of rotation about a fixed axis, but for that type of motion you will usually find it more convenient to use Equation (7.22).

When you apply these equations, your objective may be to obtain information about an object's motion, or to determine the values of unknown forces or couples acting on it, or both. This typically involves three steps:

(1) **Draw the free-body diagram** – Isolate the object and identify the external forces and couples acting on it.

(2) **Apply the equations of motion** – Write equations of motion suitable for the type of motion. You should choose an appropriate coordinate system for applying Newton's second law. For example, if the centre of mass moves in a circular path, you may find it advantageous to use normal and tangential components.

(3) **Determine kinematic relationships** – If necessary, supplement the equations of motion with relationships between the acceleration of the centre of mass and the angular acceleration.

As we show in the following sections, your approach will depend in part on the type of motion involved.

Translation

If a rigid body is in translation (Figure 7.8), you need only Newton's second law to determine its motion. There is no rotational motion to determine. Nevertheless, you may need to apply the angular equation of motion to determine unknown forces or couples. Since $\alpha = 0$, Equation (7.23) states that the total moment **about the centre of mass** equals zero:

$$\Sigma M = 0$$

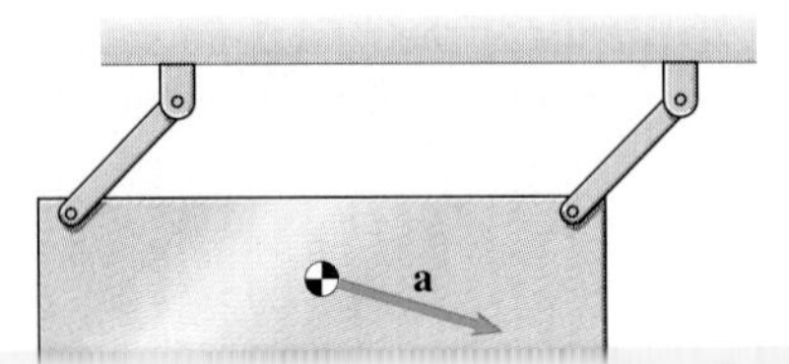

Figure 7.8
A rigid body in translation. There is no rotational motion to determine.

Example 7.1

The mass of the aeroplane in Figure 7.9 is $m = 250$ Mg (megagrams), and the thrust of its engines during its takeoff roll is $T = 700$ kN. Determine the aeroplane's acceleration and the normal forces exerted on its wheels at A and B. Neglect the horizontal forces exerted on its wheels.

Figure 7.9

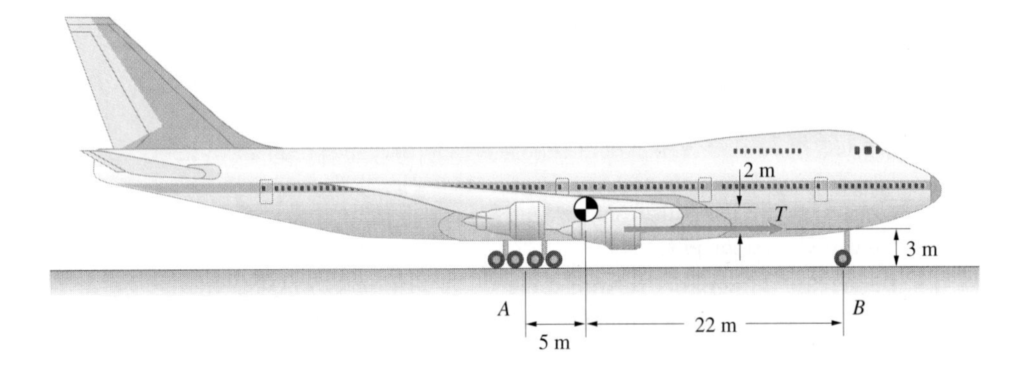

STRATEGY

The aeroplane is in translation during its takeoff roll, so the sum of the moments about its centre of mass equals zero. Using this condition and Newton's second law, we can determine the aeroplane's acceleration and the normal forces exerted on its wheels.

SOLUTION

Draw the Free-Body Diagram We draw the free-body diagram in Figure (a), showing the aeroplane's weight and the normal forces A and B exerted on the wheels.

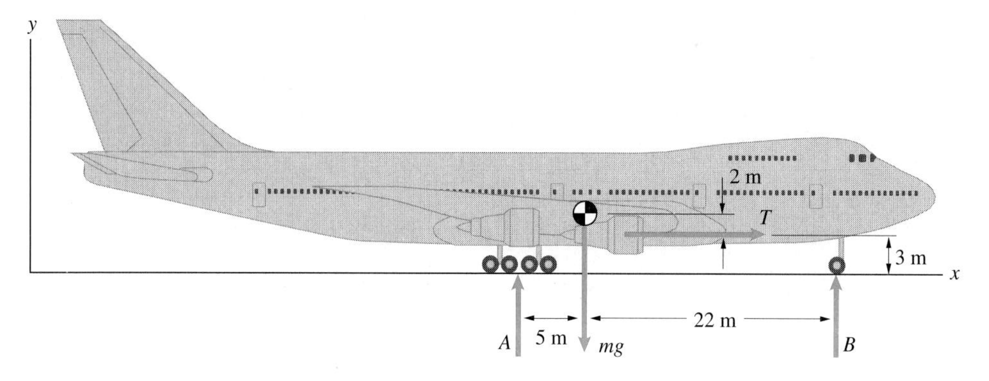

(a) Free-body diagram of the aeroplane.

Apply the Equations of Motion In terms of the coordinate system in Figure (a), Newton's second law is

$$\Sigma F_x = T = ma_x$$

$$\Sigma F_y = A + B - mg = 0$$

From the first equation, the aeroplane's acceleration is

$$a_x = \frac{T}{m} = \frac{700\,000\text{ N}}{250\,000\text{ kg}} = 2.8\text{ m/s}^2$$

The angular equation of motion is

$$\Sigma M = (2)T + (22)B - (5)A = 0$$

Solving this equation together with the second equation we obtained from Newton's second law for A and B, we obtain $A = 2050$ kN, $B = 402$ kN.

DISCUSSION

When an object is in equilibrium, the sum of the moments about any point due to the external forces and couples acting on it is zero. But you must remember that when a translating rigid body is not in equilibrium, you know only that the sum of the moments *about the centre of mass* is zero. It would be instructive for you to try reworking this example by assuming that the sum of the moments about A or B is zero. You will not obtain the correct values for the normal forces exerted on the wheels.

Rotation About a Fixed Axis

In the case of rotation about a fixed axis (Figure 7.10), you need only Equation (7.22) to determine the rotational motion, although you may also need Newton's second law to determine unknown forces or couples.

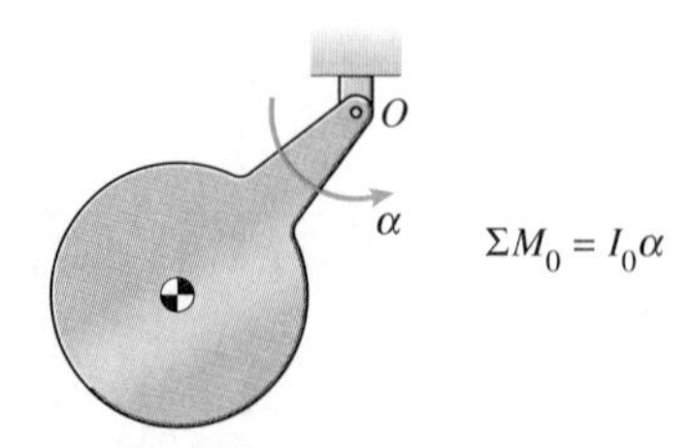

Figure 7.10
A rigid body rotating about O. You need only the equation of angular motion about O to determine its angular acceleration.

Example 7.2

The 50 kg crate in Figure 7.11 is pulled up the inclined surface by the winch. The coefficient of kinetic friction between the crate and the surface is $\mu_k = 0.4$. The mass moment of inertia of the drum on which the cable is wound, including the cable wound on the drum, is $I_A = 4\,\text{kg.m}^2$. If the motor exerts a couple $M = 60$ N.m on the drum, what is the crate's acceleration?

Figure 7.11

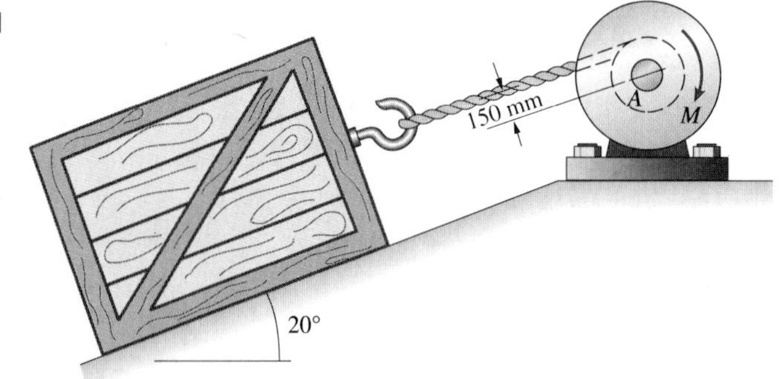

STRATEGY

We will draw separate free-body diagrams of the crate and drum and apply the equations of motion to them individually. The drum rotates about a fixed axis, so we can use the equation of angular motion about the axis to determine its angular acceleration. To complete the solution, we must determine the relationship between the crate's acceleration and the drum's angular acceleration.

SOLUTION

Draw the Free-Body Diagrams We draw the free-body diagrams in Figure (a), showing the equal forces exerted on the crate and the drum by the cable.

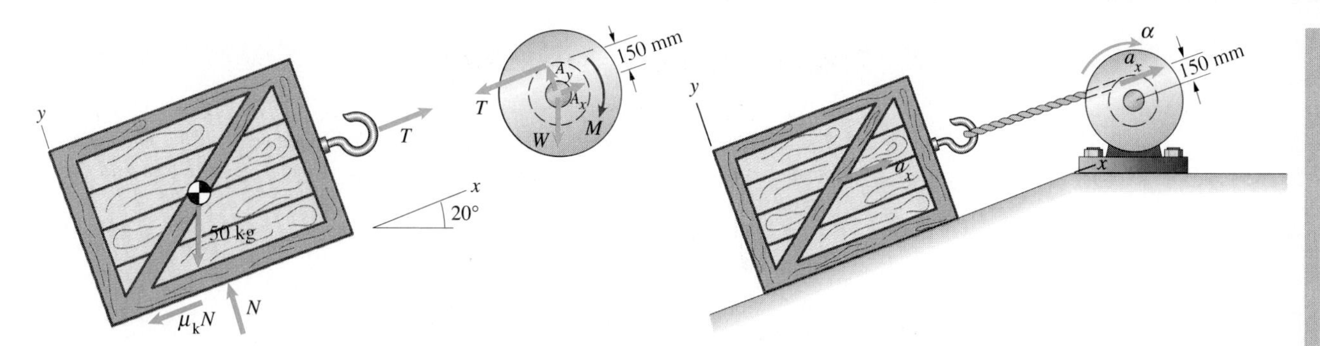

(a) Free-body diagrams of the crate and the drum.

(b) Relation between the crate's acceleration and the angular acceleration of the drum.

Apply the Equations of Motion We denote the crate's acceleration up the inclined surface by $a_x\mathbf{i}$ and the *clockwise* angular acceleration of the drum by α (Figure (b)). Newton's second law for the crate is

$$\Sigma F_x = T - 490.5 \sin 20^\circ - \mu_k N = (50)a_x$$

$$\Sigma F_y = N - 490.5 \cos 20^\circ = 0$$

Solving the second equation for N and substituting it into the first one, we obtain

$$T - 490.5 \sin 20^\circ - (0.4)(490.9 \cos 20^\circ) = (50)a_x$$

The equation of angular motion for the drum is

$$\Sigma M_A = M - (0.15\,\text{m})T = I_A \alpha$$

We eliminate T between these two equations, obtaining

$$6.67M - 490.5 \sin 20^\circ - (0.4)(490.5 \cos 20^\circ) = (50)a_x + 6.67 I_A \alpha \qquad (7.24)$$

Our last step is to determine the relation between a_x and α.

Determine Kinematic Relationships The tangential component of acceleration of the drum at the point where the cable begins winding onto it is equal to the crate's acceleration (Figure (b)):

$$a_x = (0.15\,\text{m})\alpha$$

Using this relation, the solution of Equation (7.24) for a_x is

$$a_x = \frac{6.67M - 490.5 \sin 20^\circ - (0.4)(490.5 \cos 20^\circ)}{(50) + 44.4 I_A} = 0.21\,\text{m/s}^2$$

DISCUSSION

Notice that, for convenience, we defined the angular acceleration α to be positive in the clockwise direction so that a positive α would correspond to a positive a_x.

Example 7.3

The slender bar of mass m in Figure 7.12 is released from rest in the horizontal position shown. At that instant, determine the bar's angular acceleration and the force exerted on the bar by the support A.

Figure 7.12

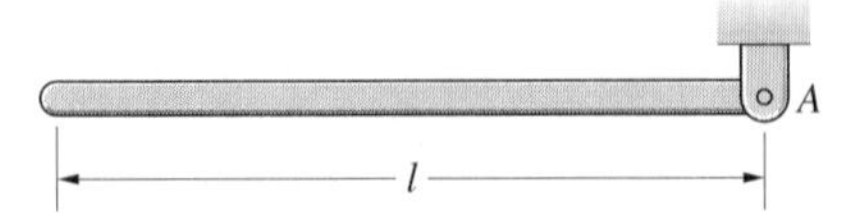

STRATEGY

Since the bar rotates about a fixed point, we can use Equation (7.22) to determine its angular acceleration. The advantage of using this equation instead of Equation (7.23) is that the unknown reactions at A will not appear in the equation of angular motion. Once we know the angular acceleration, we can determine the acceleration of the centre of mass and use Newton's second law to obtain the reactions at A.

SOLUTION

Draw the Free-Body Diagram In Figure (a) we draw the free-body diagram of the bar, showing the reactions at the pin support.

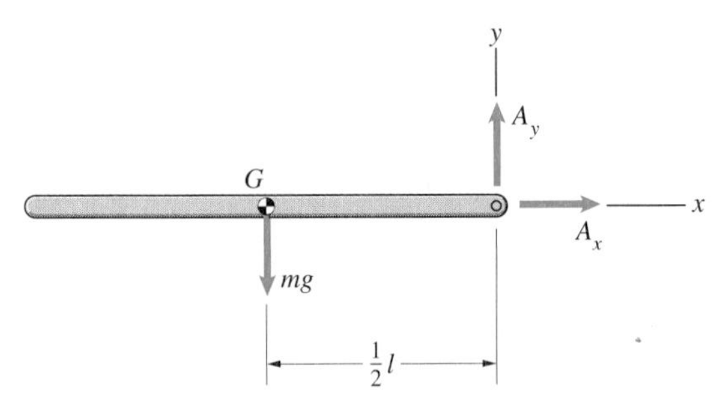

(a) Free-body diagram of the bar.

Apply the Equations of Motion Let the acceleration of the centre of mass G of the bar be $\mathbf{a}_G = a_x\,\mathbf{i} + a_y\,\mathbf{j}$, and let its counterclockwise angular acceleration be α (Figure (b)). Newton's second law for the bar is

$$\Sigma F_x = A_x = ma_x$$

$$\Sigma F_y = A_y - mg = ma_y$$

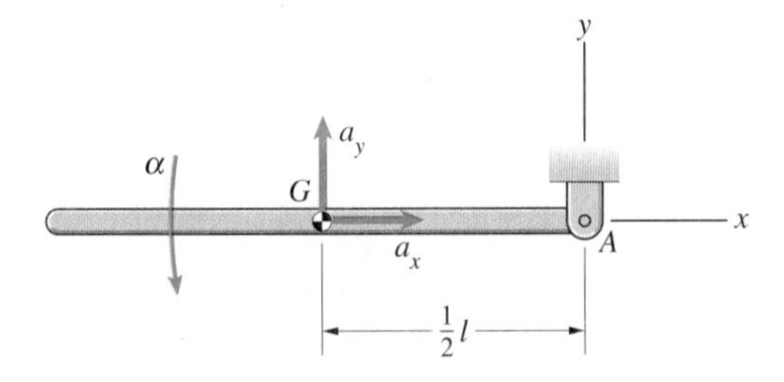

(b) The angular acceleration and components of the acceleration of the centre of mass.

The equation of angular motion about the fixed point A is

$$\Sigma M_A = \left(\frac{1}{2}l\right)mg = I_A\alpha \tag{7.25}$$

The mass moment of inertia of a slender bar about its centre of mass is $I = \frac{1}{12}ml^2$. (See Appendix C.) Using the parallel-axis theorem, the mass moment of inertia of the bar about A is

$$I_A = 1 + d^2m = \frac{1}{12}ml^2 + \left(\frac{1}{2}l\right)^2 m = \frac{1}{3}ml^2$$

Substituting this expression into Equation (7.25), we obtain the angular acceleration:

$$\alpha = \frac{(1/2)mgl}{(1/3)ml^2} = \frac{3g}{2l}$$

Determine Kinematic Relationships To determine the reactions A_x and A_y, we need to determine the acceleration components a_x and a_y. We can do so by expressing the acceleration of G in terms of the acceleration of A:

$$\mathbf{a}_G = \mathbf{a}_A + \alpha \times \mathbf{r}_{G/A} - \omega^2\mathbf{r}_{G/A}$$

At the instant the bar is released, its angular velocity $\omega = 0$. Also, $\mathbf{a}_A = \mathbf{0}$, so we obtain

$$\mathbf{a}_G = a_x\mathbf{i} + a_y\mathbf{j} = (\alpha\,\mathbf{k}) \times \left(-\frac{1}{2}l\,\mathbf{i}\right) = -\frac{1}{2}l\alpha\,\mathbf{j}$$

Equating $\mathbf{i}$ and $\mathbf{j}$ components, we obtain

$$a_x = 0$$

$$a_y = -\frac{1}{2}l\alpha = -\frac{3}{4}g$$

Substituting these acceleration components into Newton's second law, the reactions at A at the instant the bar is released are

$$A_x = 0$$

$$A_y = mg + m\left(-\frac{3}{4}g\right) = \frac{1}{4}mg$$

DISCUSSION

We could have determined the acceleration of G in a less formal way. Since G describes a circular path about A, we know the magnitude of the tangential component of acceleration equals the product of the radial distance from A to G and the angular acceleration. Because of the directions in which we define α and a_x to be positive, $a_y = -(\frac{1}{2}l)\alpha$. Also, the normal component of the acceleration of G equals the square of its velocity divided by the radius of its circular path. Since its velocity equals zero at the instant the bar is released, $a_x = 0$.

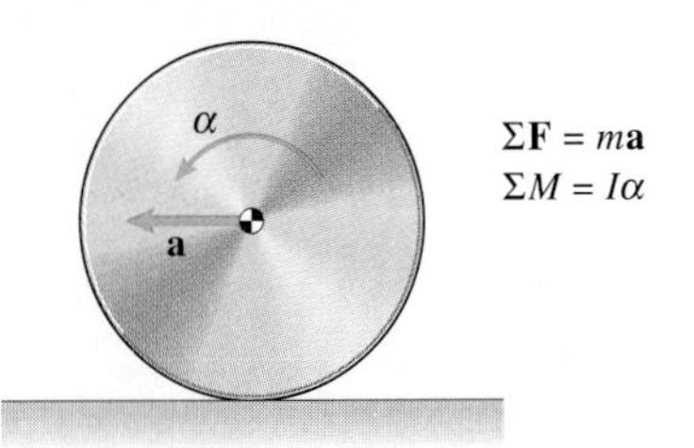

Figure 7.13
A rigid body in planar motion. You must apply both Newton's second law and the equation of angular motion about the centre of mass.

General Planar Motion

If a rigid body undergoes both translation and rotation (Figure 7.13), you need to use both Newton's second law and the equation of angular motion. If the motion of the centre of mass and the rotational motion are not independent – for example, when an object rolls – you will find that there are more unknown quantities than equations of motion. In such cases, you can obtain additional equations by relating the acceleration of the centre of mass to the angular acceleration.

Example 7.4

The slender bar of mass m in Figure 7.14 slides on the smooth floor and wall and has counterclockwise angular velocity ω at the instant shown. What is the bar's angular acceleration?

Figure 7.14

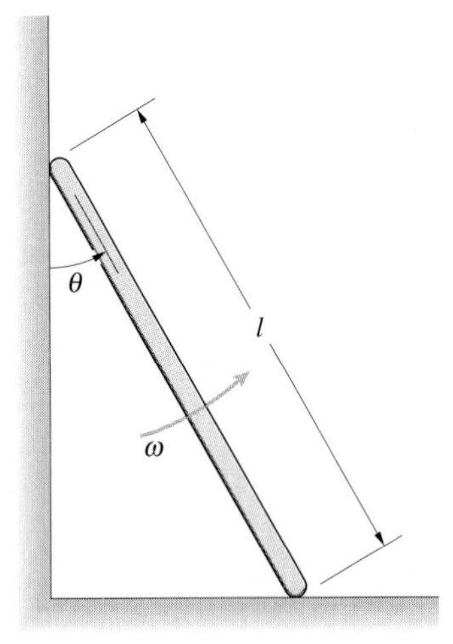

SOLUTION

Draw the Free-Body Diagram We draw the free-body diagram in Figure (a), showing the bar's weight and the normal forces exerted by the floor and wall.

Apply the Equations of Motion Writing the acceleration of the centre of mass G as $\mathbf{a}_G = a_x\,\mathbf{i} + a_y\,\mathbf{j}$, Newton's second law is

$$\Sigma F_x = P = ma_x$$

$$\Sigma F_y = N - mg = ma_y$$

Let α be the bar's counterclockwise angular acceleration. The equation of angular motion is

$$\Sigma M = N\left(\frac{1}{2}l \sin\theta\right) - P\left(\frac{1}{2}l \cos\theta\right) = I\alpha$$

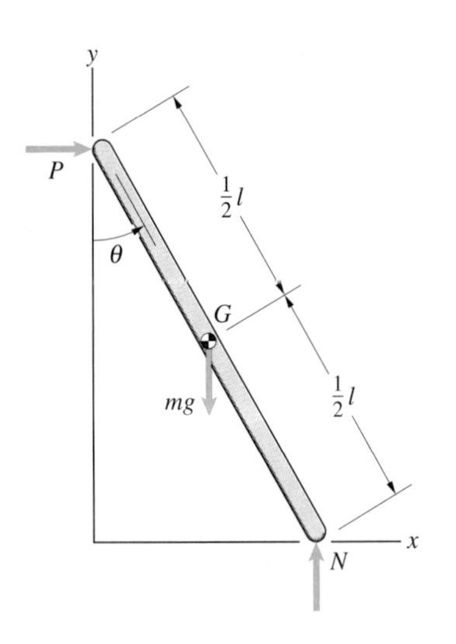

(a) Free-body diagram of the bar.

where I is the mass moment of inertia of the bar about its centre of mass. We have three equations of motion in terms of the five unknowns P, N, a_x, a_y and α. To complete the solution, we must relate the acceleration of the centre of mass of the bar to its angular acceleration.

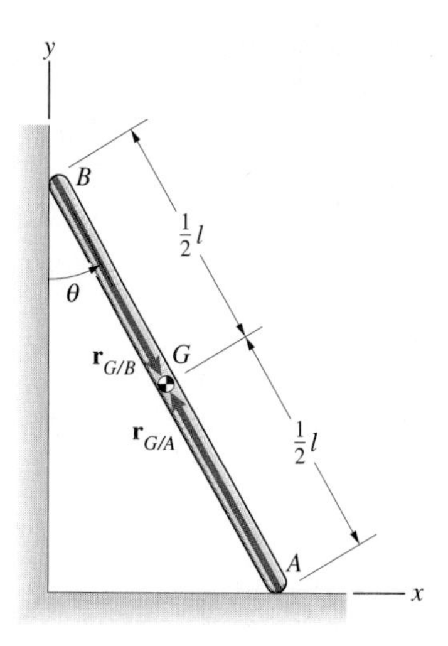

(b) Expressing the acceleration of G in terms of the accelerations of the endpoints A and B.

Determine Kinematic Relationships Although we don't know the accelerations of the endpoints A and B (Figure (b)), we know that A moves horizontally and B moves vertically. We can use this information to obtain the needed relations between the acceleration of the centre of mass and the angular acceleration. Expressing the acceleration of A as $\mathbf{a}_A = a_A\mathbf{i}$, we can write the acceleration of the centre of mass as

$$\mathbf{a}_G = \mathbf{a}_A + \boldsymbol{\alpha} \times \mathbf{r}_{G/A} - \omega^2 \mathbf{r}_{G/A}:$$

$$a_x\mathbf{i} + a_y\mathbf{j} = a_A\mathbf{i} + \begin{vmatrix} \mathbf{i} & \mathbf{j} & \mathbf{k} \\ 0 & 0 & \alpha \\ -\frac{1}{2}l\sin\theta & \frac{1}{2}l\cos\theta & 0 \end{vmatrix} - \omega^2\left(-\frac{1}{2}l\sin\theta\,\mathbf{i} + \frac{1}{2}l\cos\theta\,\mathbf{j}\right)$$

Taking advantage of the fact that $\mathbf{a}_A$ has no $\mathbf{j}$ component, we equate the $\mathbf{j}$ components in this equation, obtaining

$$a_y = -\frac{1}{2}l(\alpha\sin\theta + \omega^2\cos\theta)$$

Now we express the acceleration of B as $\mathbf{a}_B = a_B\mathbf{j}$ and write the acceleration of the centre of mass as

$$\mathbf{a}_G = \mathbf{a}_B + \boldsymbol{\alpha} \times \mathbf{r}_{G/B} - \omega^2 \mathbf{r}_{G/B}:$$

$$a_x\mathbf{i} + a_y\mathbf{j} = a_B\mathbf{j} + \begin{vmatrix} \mathbf{i} & \mathbf{j} & \mathbf{k} \\ 0 & 0 & \alpha \\ \frac{1}{2}l\sin\theta & -\frac{1}{2}l\cos\theta & 0 \end{vmatrix} - \omega^2\left(\frac{1}{2}l\sin\theta\,\mathbf{i} - \frac{1}{2}l\cos\theta\,\mathbf{j}\right)$$

We equate the $\mathbf{i}$ components in this equation, obtaining

$$a_x = \frac{1}{2}l(\alpha\cos\theta - \omega^2\sin\theta)$$

With these two kinematic relationships, we have five equations in five unknowns. Solving them for the angular acceleration and using the relation $I = \frac{1}{12}ml^2$ for the bar's mass moment of inertia (Appendix C), we obtain

$$\alpha = \frac{3}{2}\frac{g}{l}\sin\theta$$

DISCUSSION

Notice that by expressing the acceleration of G in terms of the accelerations of the endpoints, we introduced into the solution the constraints imposed on the bar by the floor and wall: we required that point A move horizontally and that point B move vertically.

Example 7.5

The slender bar in Figure 7.15 has mass m and is pinned at A to a metal block of mass m_B that rests on a smooth level surface. The system is released from rest in the position shown. What is the bar's angular acceleration at the instant of release?

Figure 7.15

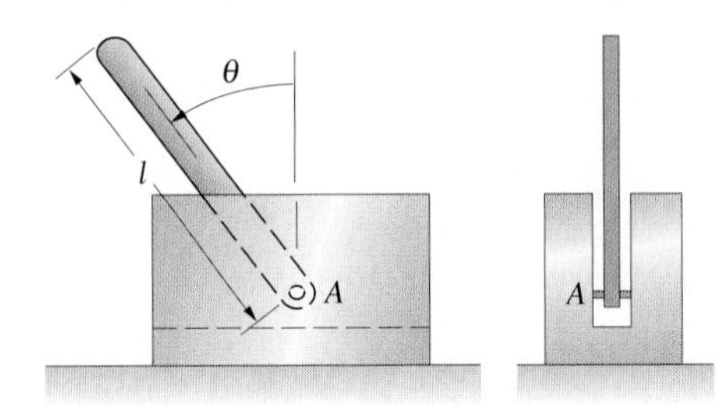

STRATEGY

We must draw free-body diagrams of the bar and the block and apply the equations of motion to them individually. To complete the solution, we must also relate the acceleration of the bar's centre of mass and its angular acceleration to the acceleration of the block.

SOLUTION

Draw the Free-Body Diagrams We draw the free-body diagrams of the bar and block in Figure (a). Notice the opposite forces that they exert on each other where they are pinned together.

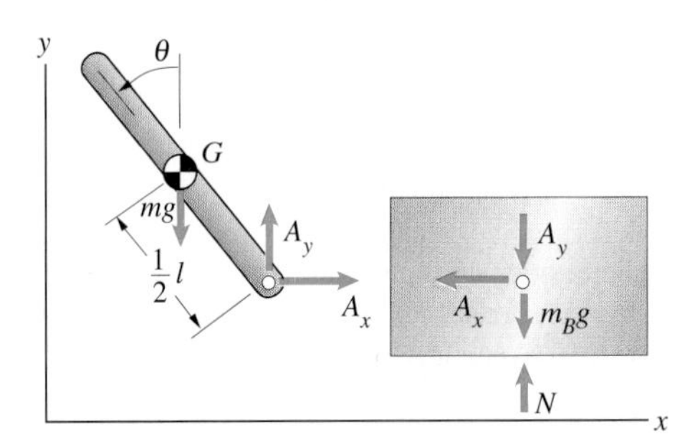

(a) Free-body diagrams of the bar and block.

Apply the Equations of Motion Writing the acceleration of the centre of mass of the bar as $\mathbf{a}_G = a_x\,\mathbf{i} + a_y\,\mathbf{j}$, Newton's second law for the bar is

$$\Sigma F_x = A_x = ma_x$$

$$\Sigma F_y = A_y - mg = ma_y$$

Letting α be the bar's counterclockwise angular acceleration, its equation of angular motion is

$$\Sigma M = A_x\left(\frac{1}{2}l\cos\theta\right) + A_y\left(\frac{1}{2}l\sin\theta\right) = I\alpha$$

We express the block's acceleration as $a_B\mathbf{i}$ and write Newton's second law for the block:

$$\Sigma F_x = -A_x = m_B a_B$$

$$\Sigma F_y = N - A_y - m_B g = 0$$

Determine Kinematic Relationships To relate the bar's motion to that of the block, we express the acceleration of the bar's centre of mass in terms of the acceleration of point A (Figure (b)):

$$\mathbf{a}_G = \mathbf{a}_A + \alpha \times \mathbf{r}_{G/A} - \omega^2\mathbf{r}_{G/A}$$

$$a_x\mathbf{i} + a_y\mathbf{j} = a_B\mathbf{i} + \begin{vmatrix} \mathbf{i} & \mathbf{j} & \mathbf{k} \\ 0 & 0 & \alpha \\ -\frac{1}{2}l\sin\theta & \frac{1}{2}l\cos\theta & 0 \end{vmatrix} - \mathbf{0}$$

Equating $\mathbf{i}$ and $\mathbf{j}$ components, we obtain

$$a_x = a_B - \frac{1}{2}l\alpha\cos\theta$$

$$a_y = -\frac{1}{2}l\alpha\sin\theta$$

We have five equations of motion and two kinematic relations in terms of seven unknowns: A_x, A_y, N, a_x, a_y, α and a_B. Solving them for the angular acceleration and using the relation $I = \frac{1}{12}ml^2$ for the bar's mass moment of inertia, we obtain

$$\alpha = \frac{(3/2)(g/l)\sin\theta}{1 - (3/4)\left(\dfrac{m}{m+m_B}\right)\cos^2\theta}$$

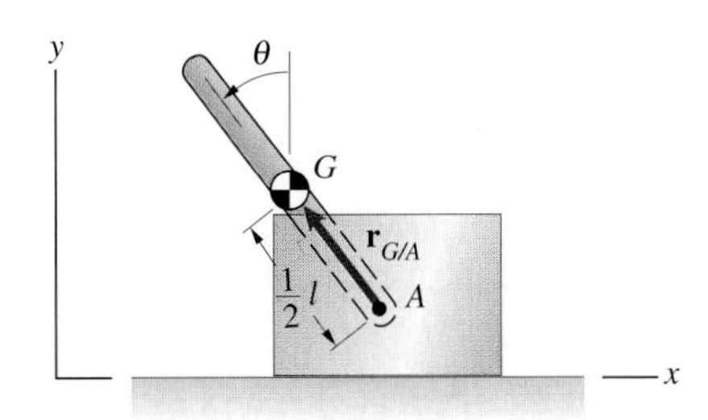

(b) Expressing the acceleration of G in terms of the acceleration of A.

Example 7.6

The drive wheel in Figure 7.16 rolls on the horizontal track. The wheel is subjected to a downward force F_A by its axle A and a horizontal force F_C by the connecting rod. The mass of the wheel is m and the mass moment of inertia about its centre of mass is I. The centre of mass G is offset a distance b from the wheel's centre. At the instant shown, the wheel has a counterclockwise angular velocity ω. What is the wheel's angular acceleration?

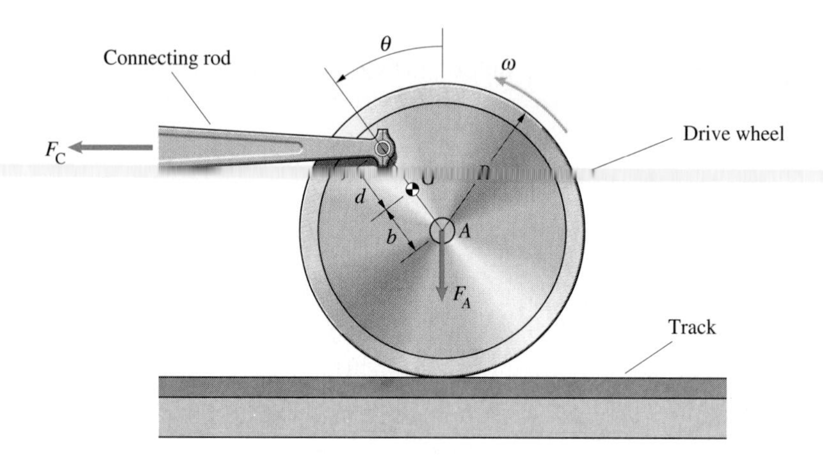

Figure 7.16

SOLUTION

Draw the Free-Body Diagram We draw the free-body diagram of the drive wheel in Figure (a), showing its weight and the normal and friction forces exerted by the track.

Apply the Equations of Motion Writing the acceleration of the centre of mass G as $\mathbf{a}_G = a_x\,\mathbf{i} + a_y\,\mathbf{j}$, Newton's second law is

$$\Sigma F_x = f - F_C = ma_x$$

$$\Sigma F_y = N - F_A - mg = ma_y$$

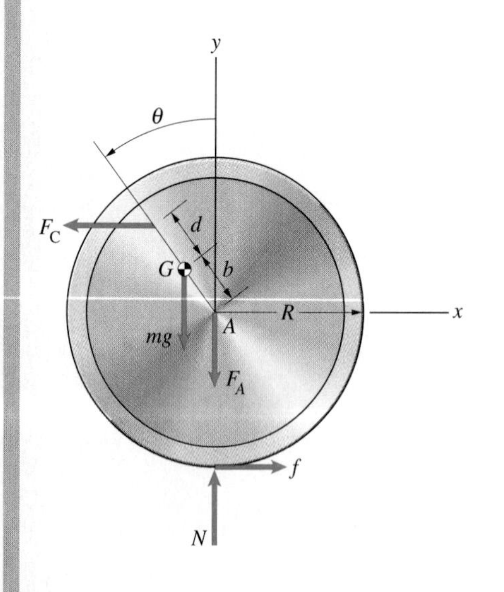

(a) Free-body diagram of the wheel.

Remember that we must express the equation of angular motion in terms of the sum of the moments about the centre of mass G, *not the centre of the wheel.* The equation of angular motion is

$$\Sigma M = F_C(d\cos\theta) - F_A(b\sin\theta) + N(b\sin\theta) + f(b\cos\theta + R) = I\alpha$$

We have three equations of motion in terms of the five unknowns N, f, a_x, a_y and α. To complete the solution, we must relate the acceleration of the wheel's centre of mass to its angular acceleration.

Determine Kinematic Relationships The acceleration of the centre A of the rolling wheel is $\mathbf{a}_A = -R\alpha\,\mathbf{i}$. By expressing the acceleration of the centre of mass, $\mathbf{a}_G$, in terms of $\mathbf{a}_A$ (Figure (b)), we can obtain relations between the components of $\mathbf{a}_G$ and α:

$$\mathbf{a}_G = \mathbf{a}_A + \alpha \times \mathbf{r}_{G/A} - \omega^2\,\mathbf{r}_{G/A}$$

$$a_x\,\mathbf{i} + a_y\,\mathbf{j} = -R\alpha\,\mathbf{i} + \begin{vmatrix} \mathbf{i} & \mathbf{j} & \mathbf{k} \\ 0 & 0 & \alpha \\ -b\sin\theta & b\cos\theta & 0 \end{vmatrix}$$

$$-\omega^2(-b\sin\theta\,\mathbf{i} + b\cos\theta\,\mathbf{j})$$

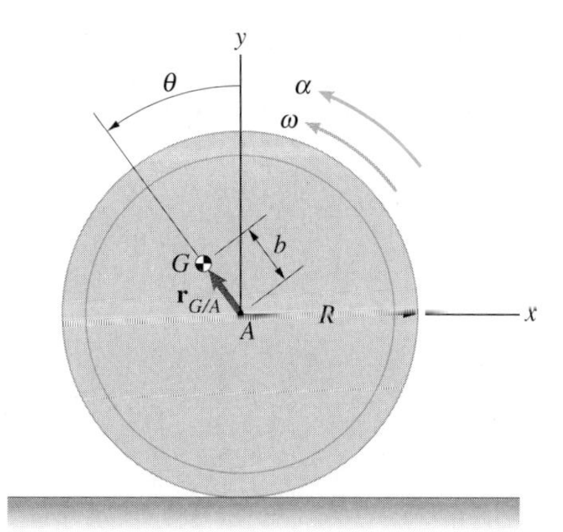

(b) Expressing the acceleration of the centre of mass G in terms of the acceleration of the centre A.

We equate the **i** and **j** components in this equation, obtaining

$$a_x = -R\alpha - b\alpha\cos\theta + b\omega^2\sin\theta$$

$$a_y = -b\alpha\sin\theta - b\omega^2\cos\theta$$

With these two kinematic relationships, we have five equations in five unknowns. Solving them for the angular acceleration, we obtain

$$\alpha = \frac{F_C(R + b\cos\theta + d\cos\theta) + mgb\sin\theta + mbR\omega^2\sin\theta}{m(b^2 + 2bR\cos\theta + R^2) + I}$$

Example 7.7

Application to Engineering

Internal Forces and Moments in Beams

The slender bar of mass m in Figure 7.17 starts from rest in the position shown and falls. When it has rotated through an angle θ, what is the maximum bending moment in the bar and where does it occur?

Figure 7.17

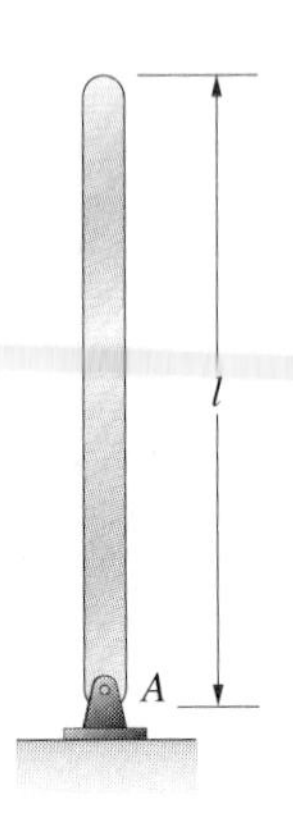

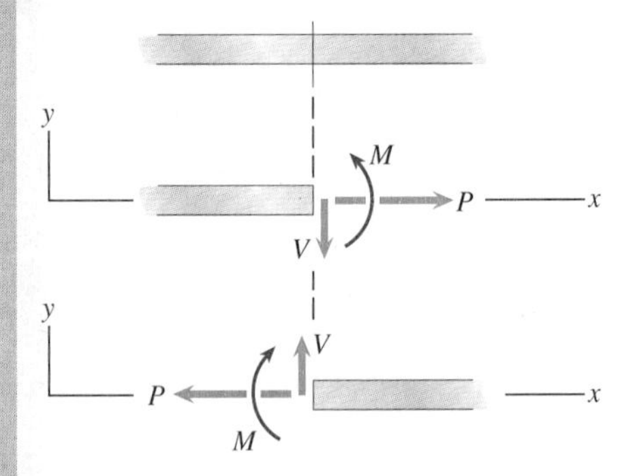

(a) The axial force, shear force and bending moment in a beam.

STRATEGY

The internal forces and moments in a beam subjected to two-dimensional loading are the axial force P, shear force V, and bending moment M (Figure (a)). We must first use the equation of angular motion to determine the bar's angular acceleration. Then we can cut the bar at an arbitrary distance x from one end and apply *the equations of motion* to determine the bending moment as a function of x.

SOLUTION

The mass moment of inertia of the bar about A is

$$I_A = I + d^2 m = \frac{1}{12} ml^2 + \left(\frac{1}{2} l\right)^2 m = \frac{1}{3} ml^2$$

When the bar has rotated through an angle θ (Figure (b)), the total moment about A is $\Sigma M_A = mg(\frac{1}{2} l \sin\theta)$. Point A is fixed, so we can write the equation of angular motion as

$$\Sigma M_A = I_A \alpha :$$

$$\frac{1}{2} mgl \sin\theta = \frac{1}{3} ml^2 \alpha$$

Solving for the angular acceleration, we obtain

$$\alpha = \frac{3g}{2l} \sin\theta$$

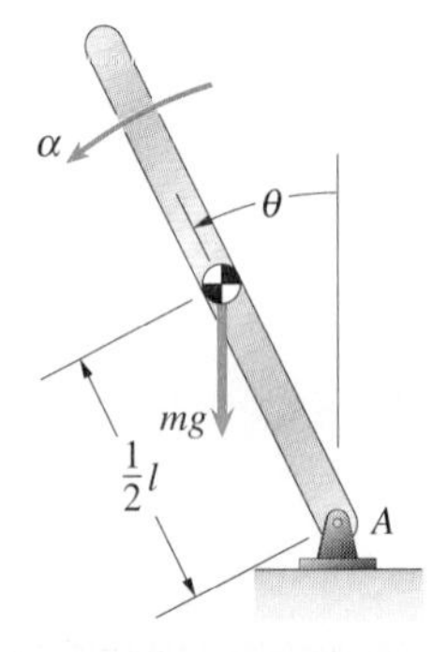

(b) Determining the moment about A.

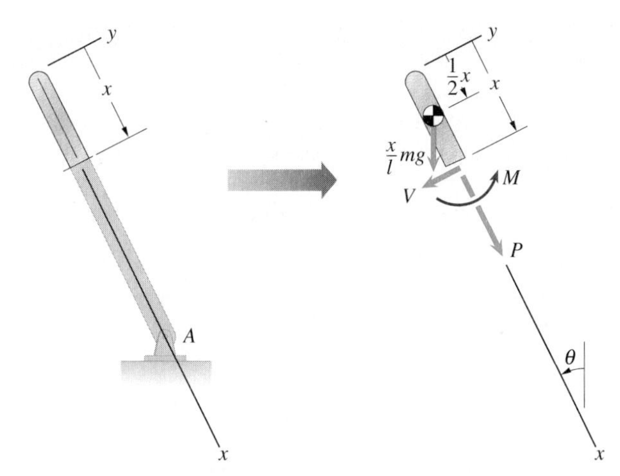

(c) Cutting the bar at an arbitrary distance x.

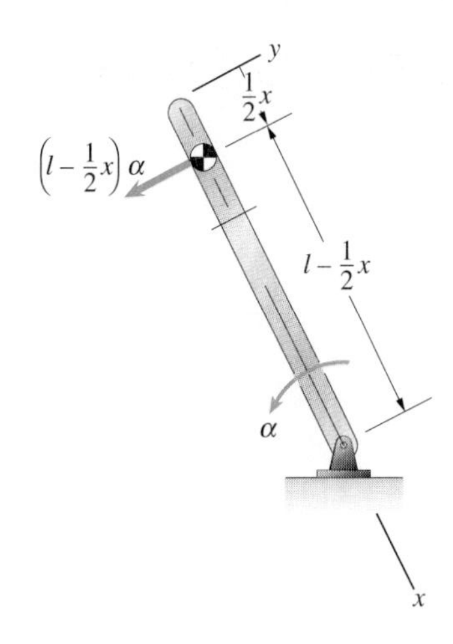

(d) Determining the acceleration of the centre of mass of the free body.

In Figure (c) we introduce a coordinate system, cut the bar at a distance x from the top, and draw the free-body diagram of the top part. The centre of mass is at the midpoint, and we determine the mass by multiplying the bar's mass by the ratio of the length of the free body to that of the bar. Applying Newton's second law in the y direction, we obtain

$$\Sigma F_y = -V - \frac{x}{l}mg \sin\theta = \frac{x}{l}ma_y$$

The mass moment of inertia of the free body about its centre of mass is $\frac{1}{12}[(x/l)m]x^2$, so the equation of angular motion is

$$\Sigma M = I\alpha :$$

$$M - \left(\frac{1}{2}x\right)V = \frac{1}{12}\left(\frac{x}{l}m\right)x^2 \frac{3}{2}\frac{g}{l} \sin\theta$$

The y component of the acceleration of the centre of mass is equal to the product of its radial distance from A and the angular acceleration (Figure (d)):

$$a_y = -\left(l - \frac{1}{2}x\right)\alpha = -\left(l - \frac{1}{2}x\right)\frac{3}{2}\frac{g}{l} \sin\theta$$

Using this expression, we can solve the two equations of motion for V and M in terms of θ. The solution for M is

$$M = \frac{1}{4}mgl \sin\theta\left(\frac{x}{l}\right)^2\left(1 - \frac{x}{l}\right) \tag{7.26}$$

The bending moment equals zero at both ends of the bar. Taking the derivative of this expression with respect to x and equating it to zero to determine where M is a maximum, we obtain $x = \frac{2}{3}l$. Substituting this value of x into Equation (7.26), we obtain the maximum bending moment:

$$M_{\max} = \frac{1}{27}mgl \sin\theta$$

The distribution of M is shown in Figure 7.18.

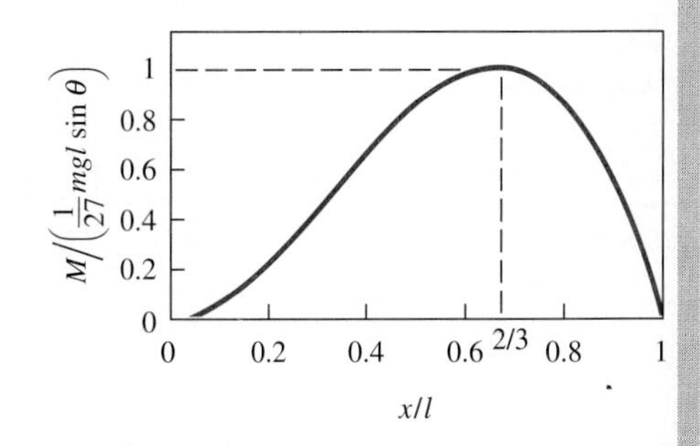

Figure 7.18
Distribution of the bending moment in a falling bar.

DESIGN ISSUES

To design a member of a structure, engineers must consider both the external and internal forces and moments it will be subjected to. In the case of a beam, they must determine the distributions of the axial force P, shear force V, and bending moment M as the first step in determining whether the beam will support its design loads without failing. If they know the external loads and reactions and the beam is in equilibrium, they can apply the equilibrium equations to determine the internal forces and moment at a given cross-section. But in many situations, a beam will not be in equilibrium. It could be a member of a structure, such as the internal frame of an aeroplane, that is accelerating, or it could be a connecting rod in an internal combustion engine. In such cases, the maximum internal forces and moments can far exceed the values predicted by a static analysis, and the procedure we describe in this example must be used.

The dynamic bending moment distribution we obtained in Example 7.7 (Figure 7.18) explains a phenomenon that has been observed during the demolition of masonry chimneys. An explosive charge at the base of the chimney causes it to fall, initially rotating as a rigid body about its base. As the chimney falls, it is observed to fracture near the location of the maximum bending moment (Figure 7.19).

Figure 7.19

A falling chimney fractures as it falls due to the bending moment it is subjected to.

7.5 D'Alembert's Principle

In this section we describe an alternative approach to rigid-body dynamics known as D'Alembert's principle. By writing Newton's second law as

$$\Sigma \mathbf{F} + (-m\,\mathbf{a}) = \mathbf{0} \tag{7.27}$$

we can regard it as an 'equilibrium' equation stating that the sum of the external forces, including an **inertial force** $-m\mathbf{a}$, equals zero (Figure 7.20(a)). To state the equation of angular motion in an equivalent way, we use Equation (7.20), which relates the total moment about a fixed point O to the angular acceleration in general plane motion:

$$\Sigma M_0 = (\mathbf{r} \times m\,\mathbf{a}) \cdot \mathbf{k} + I\alpha$$

Writing this equation as

$$\Sigma M_0 + [\mathbf{r} \times (-m\,\mathbf{a})] \cdot \mathbf{k} + (-I\alpha) = 0 \tag{7.28}$$

we can regard it as an equilibrium equation stating that the sum of the moments *about any point* due to external forces and couples, including the moment due to the inertial force $-m\mathbf{a}$ acting at the centre of mass and an **inertial couple** $-I\alpha$, equals zero.

Instead of using the expression in Equation (7.28) to determine the moment due to the internal force, you can often determine it more easily by multiplying the magnitude of the inertial force and the perpendicular distance from the line of action of the force to point O (Figure 7.20(b)). Also, remember that the sense of the inertial couple is opposite to that of the angular acceleration (Figure 7.20(c)).

Figure 7.20

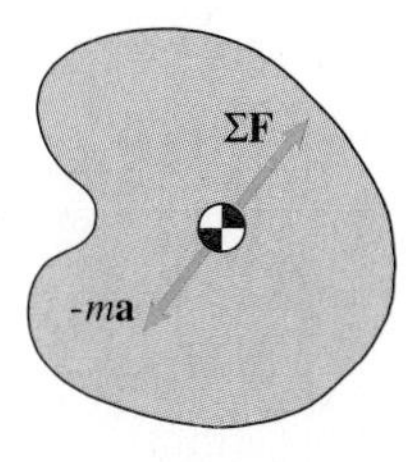

(a) The sum of the external forces and the inertial force is zero.

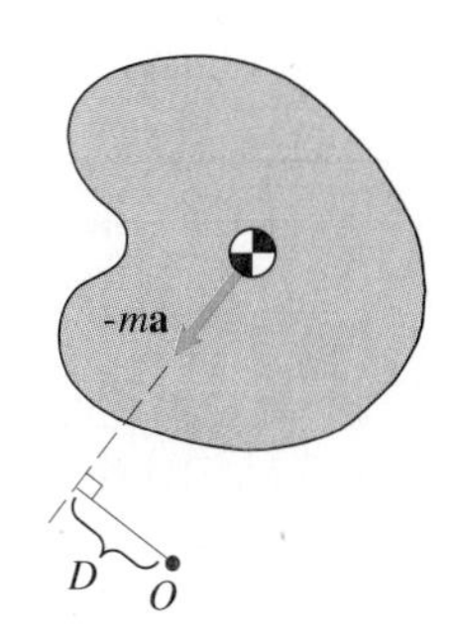

(b) The magnitude of the moment due to the inertial force is $|-m\,\mathbf{a}|D$.

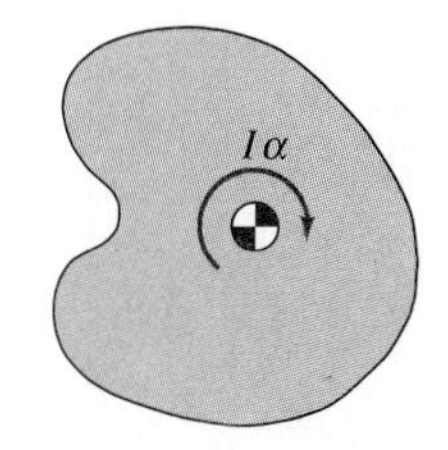

(c) A clockwise inertial couple results from a counterclockwise angular acceleration.

In the following examples we apply D'Alembert's principle to plane motions of rigid bodies. The sequence of steps – draw the free-body diagram, apply the 'equilibrium' equations, and determine kinematic relationships if necessary – is the same as in applying the equations of motion. However, in using D'Alembert's principle, you must be particularly careful to assign the correct signs to the terms in your equations. For example, if you define the angular acceleration to be positive in the counterclockwise direction, that is also the positive direction for the moment exerted by the inertial force, and the inertial couple is clockwise.

Example 7.8

The mass of the aeroplane in Figure 7.21 is $m = 250$ Mg (megagrams), and the thrust of its engines during its takeoff roll is $T = 700$ kN. Use D'Alembert's principle to determine the aeroplane's acceleration and the normal forces exerted on its wheels at A and B. Neglect the horizontal forces exerted on its wheels.

Figure 7.21

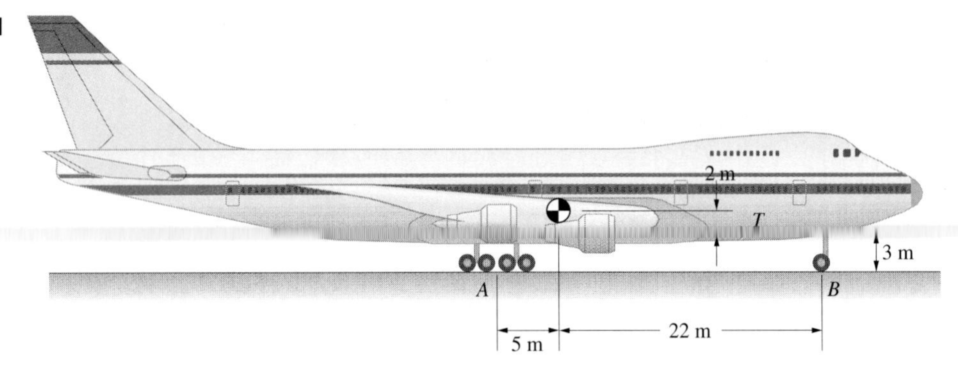

SOLUTION

Draw the Free-Body Diagram In terms of the coordinate system in Figure (a), we can write the aeroplane's acceleration as $\mathbf{a} = a_x\mathbf{i}$. On the free-body diagram we show the aeroplane's weight, the normal forces A and B exerted on the wheels, and the inertial force $-m\,\mathbf{a} = -ma_x\,\mathbf{i}$.

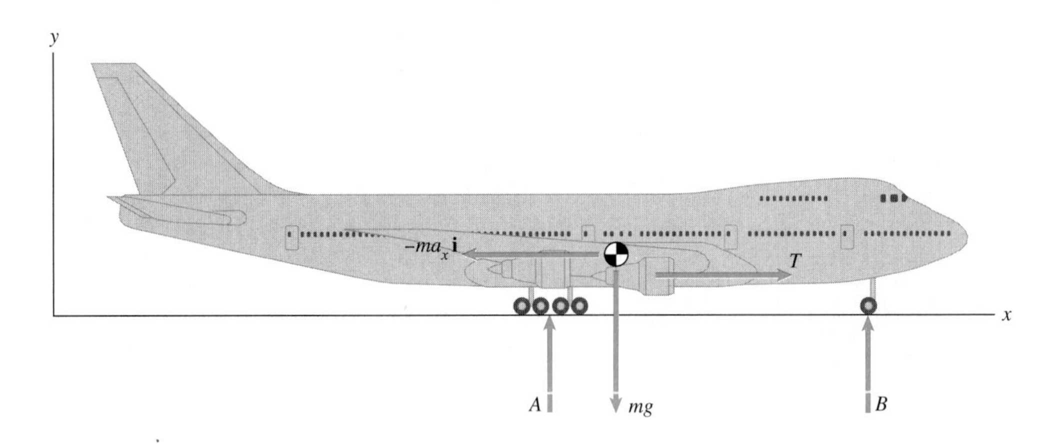

(a) Free-body diagram of the aeroplane.

Apply the 'Equilibrium' Equations Equation (7.27) is

$$F + (-m\,\mathbf{a}) = \mathbf{0}:$$

$$T\,\mathbf{i} + (A + B - mg)\,\mathbf{j} + (-ma_x\,\mathbf{i}) = \mathbf{0}$$

Equating **i** and **j** components, we obtain

$$T = ma_x$$
$$A + B = mg$$

From the first equation, the aeroplane's acceleration is

$$a_x = \frac{T}{m} = \frac{700\,000 \text{ N}}{250\,000 \text{ kg}} = 2.8 \text{ m/s}^2$$

and the inertial force is $-ma_x\mathbf{i} = -700\mathbf{i}$ kN. (See Figure (b).)

In applying Equation (7.28), we can select any point we wish as the point O. By placing it at A (Figure (b)), we will obtain an equation in which the only unknown is the force B. The aeroplane is translating, so $\alpha = 0$ and there is no inertial couple. Defining counterclockwise moments to be positive, the sum of the moments about O is

$$(5)(700\,000) - (3)T - (5)mg + (27)B = 0$$

From this equation we obtain $B = 402$ kN, and then $A = mg - B = 2050$ kN.

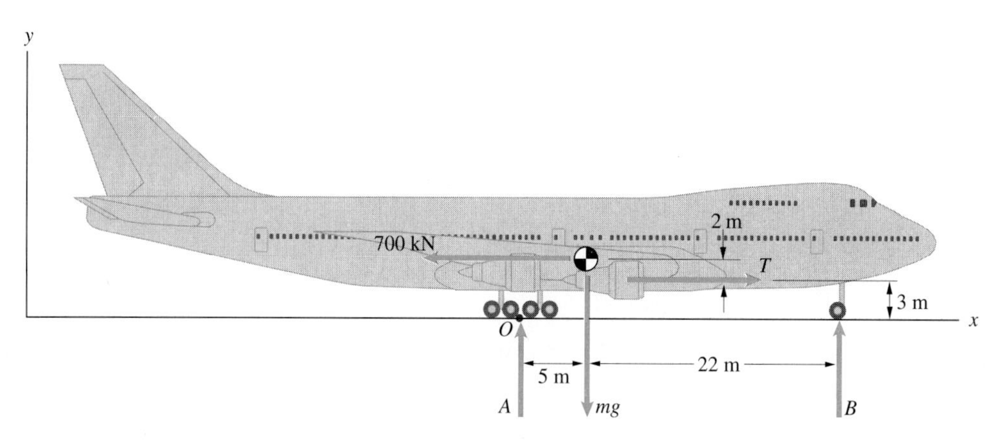

(b) Placing point O at the rear wheels.

DISCUSSION

Notice that we calculated the moment due to the inertial force by multiplying the magnitude of the inertial force by the perpendicular distance to its line of action, $(5)(700\,000) = 3\,500\,000$ N.m counterclockwise. In this particular example that method is simpler than using the cross product,

$$[\mathbf{r} \times (-m\,\mathbf{a})] \cdot \mathbf{k} = [(5\,\mathbf{i} + 5\,\mathbf{j}) \times (-700\,000\,\mathbf{i})] \cdot \mathbf{k}$$
$$= 3\,500\,000\ N - m \quad \text{counterclockwise}$$

but in some situations you may find that using the cross product is simpler.

You should compare this application of D'Alembert's principle with our determination of the aeroplane's acceleration and the normal forces exerted on its wheels in Example 7.1.

Example 7.9

A disc of mass m and moment of inertia I is released from rest on an inclined surface (Figure 7.22). Assuming that the disc rolls, use D'Alembert's principle to determine its angular acceleration.

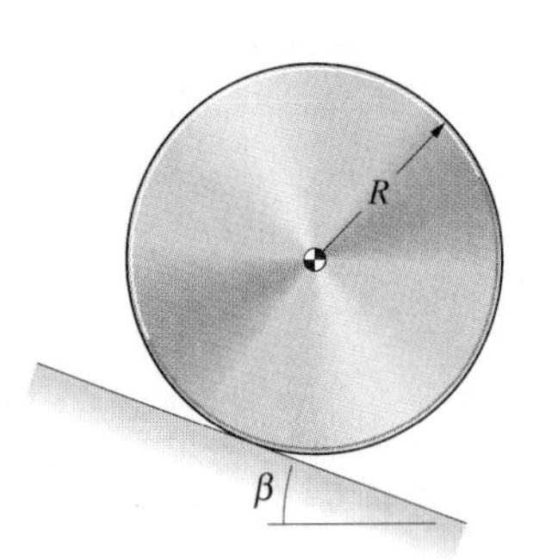

Figure 7.22

SOLUTION

Draw the Free-Body Diagram In terms of the coordinate system in Figure (a), the acceleration of the centre of the disc is $\mathbf{a} = a_x\mathbf{i}$. We define the angular acceleration α to be positive in the clockwise direction. In Figure (b) we draw the free-body diagram of the disc, showing its weight, the normal and friction forces exerted by the surface, and the inertial force and couple.

Apply the 'Equilibrium' Equations We apply Equation (7.28), evaluating moments about the point where the disc is in contact with the surface to eliminate f and N from the resulting equation:

$$-R(mg \sin \beta) + R(ma_x) + I\alpha = 0 \tag{7.29}$$

Determine Kinematic Relationships The acceleration of the centre of the rolling disc is related to the angular acceleration by $a_x = R\alpha$. Substituting this relation into Equation (7.29) and solving for α, we obtain

$$\alpha = \frac{mgR \sin \beta}{mR^2 + I}$$

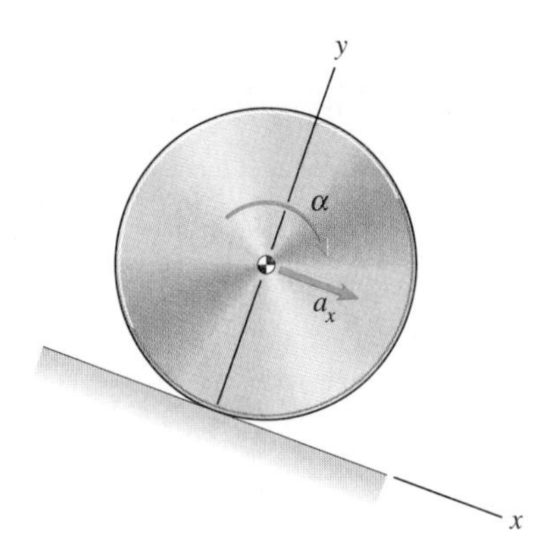

(a) Acceleration of the centre of the disc and its angular acceleration.

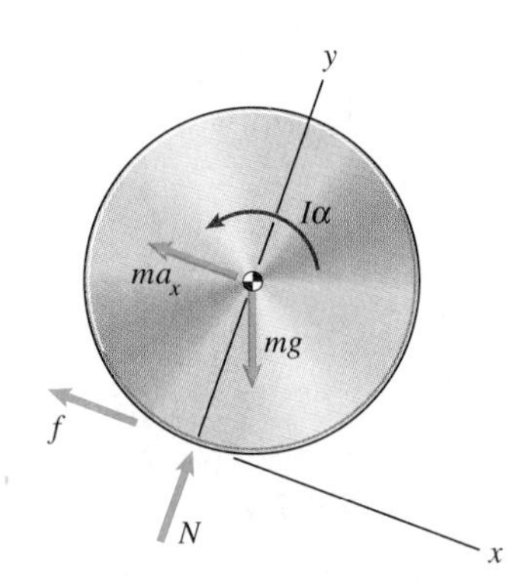

(b) Free-body diagram including the inertial force and couple.

DISCUSSION

As a consequence of summing moments about the disc's point of contact, we did not need to use the equation $\Sigma\mathbf{F} + (-m\mathbf{a}) = 0$ in determining the angular acceleration.

Problems

7.1 A refrigerator of mass m rests on castors at A and B. Suppose that you push on it with a horizontal force F as shown and that the castors remain on the smooth floor.
(a) What is the refrigerator's acceleration?
(b) What normal forces are exerted on the castors at A and B?

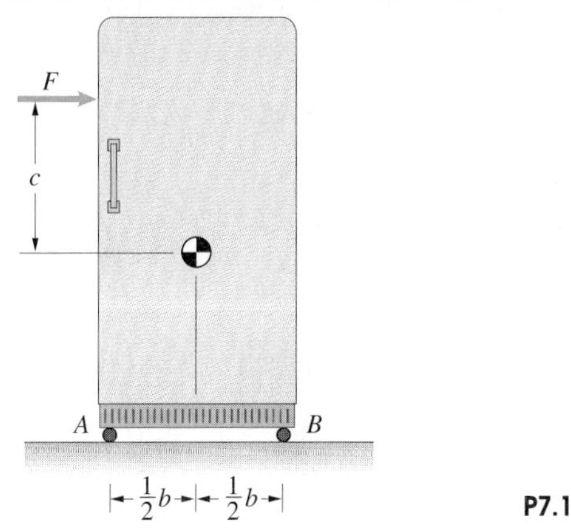

P7.1

7.2 In Problem 7.1, what is the largest force F you can apply if you want the refrigerator to remain on the floor at A and B? (Assume that c is positive.)

7.3 The combined mass of the person and bicycle is m. The location of their combined centre of mass is shown.
(a) If they have acceleration a, what are the normal forces exerted on the wheels by the ground? (Neglect the horizontal force exerted on the ground by the front wheel.)
(b) Based on the results of part (a), what is the largest acceleration that can be achieved without causing the front wheel to leave the ground?

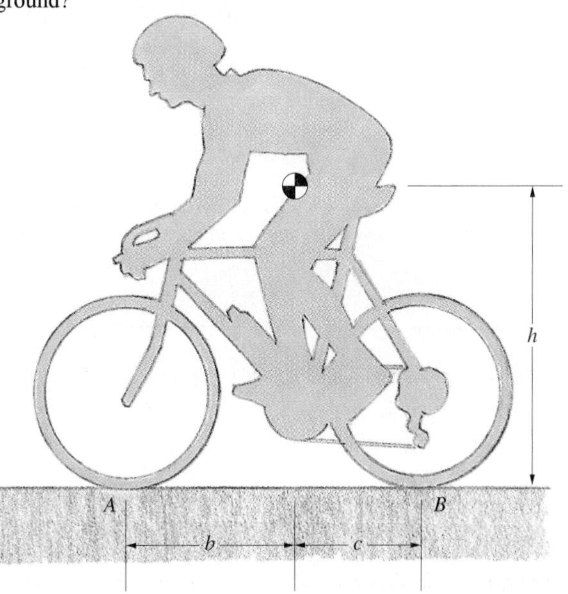

P7.3

7.4 In Problem 7.3, $b = 615$ mm, $c = 445$ mm, $h = 985$ mm and $m = 77$ kg. If the bicycle is travelling at 6 m/s and the person engages the brakes, achieving the largest deceleration for which the rear wheel will not leave the ground, how long does it take the bicycle to stop, and what distance does it travel during that time?

7.5 The 6350 kg aeroplane's arresting hook exerts the force F and causes the plane to decelerate at 6 g's. The horizontal forces exerted by the landing gear are negligible. Determine F and the normal forces exerted on the landing gear.

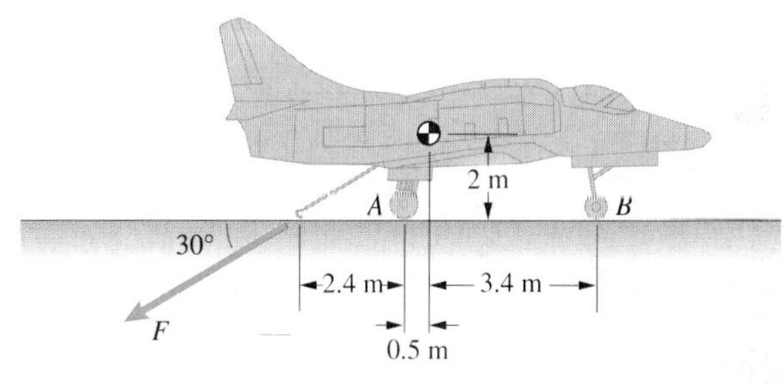

P7.5

7.6 A student catching a ride to his summer job unwisely supports himself in the back of an accelerating truck by exerting a horizontal force F on the truck's cab at A. Determine the horizontal force he must exert in terms of his weight W, the truck's acceleration a, and the dimensions shown.

P7.6

7.7 The crane moves to the right with constant acceleration, and the 800 kg load moves without swinging.
(a) What is the acceleration of the crane and load?
(b) What are the tensions in the cables attached at A and B?

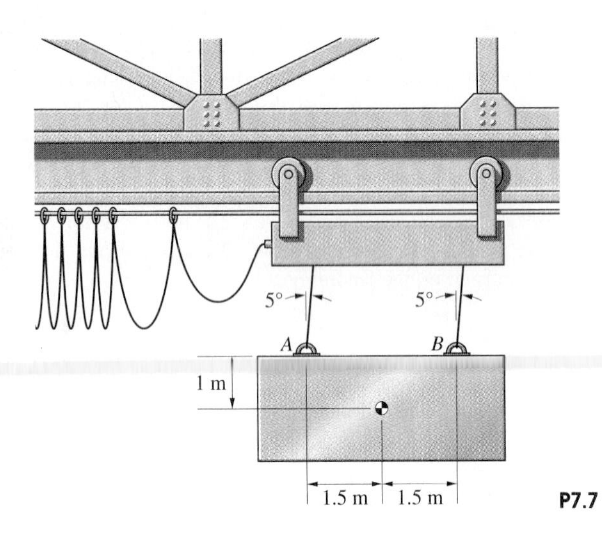

P7.7

7.8 If the acceleration of the crane in Problem 7.7 suddenly decreases to zero, what are the tensions in the cables attached at A and B immediately afterwards?

7.9 The combined mass of the motorcycle and rider is 160 kg. The rear wheel exerts a 400 N horizontal force on the road, and you can neglect the horizontal force exerted on the road by the front wheel. Modelling the motorcycle and its wheels as a rigid body, determine (a) the motorcycle's acceleration; (b) the normal forces exerted on the road by the rear and front wheels.

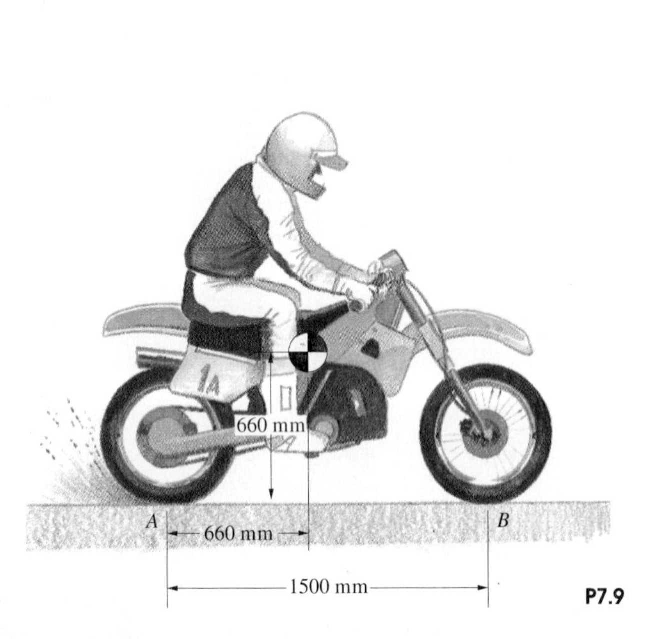

P7.9

7.10 In Problem 7.9, the coefficient of kinetic friction between the motorcycle's rear wheel and the road is $\mu_k = 0.8$. If the rider spins the rear wheel, what is the motorcycle's acceleration and what are the normal forces exerted on the road by the rear and front wheels?

7.11 During extravehicular activity, an astronaut fires a thruster of his manoeuvring unit, exerting a force $T = 14.2$ N for 1 s. It requires 60 s from the time the thruster is fired for him to rotate through one revolution. If you model the astronaut and manoeuvring unit as a rigid body, what is the moment of inertia about their centre of mass?

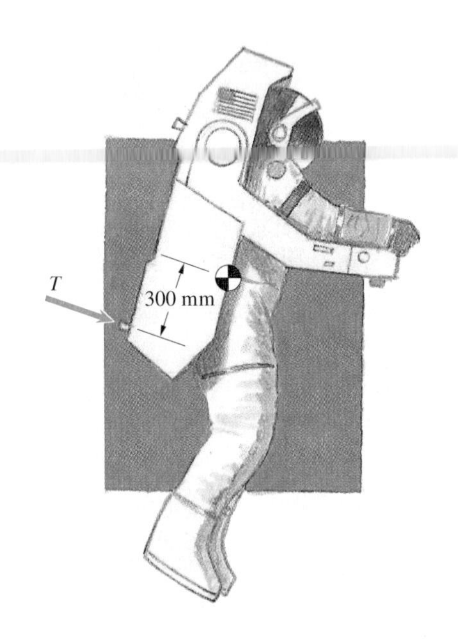

P7.11

7.12 The mass moment of inertia of the helicopter's rotor is 500 kg.m^2. If the rotor starts from rest at $t = 0$, the engine exerts a constant torque of 625 N.m on the rotor, and aerodynamic drag is neglected, what is the rotor's angular velocity ω at $t = 6$ s?

P7.12

7.13 In Problem 7.12, if aerodynamic drag exerts a torque on the helicopter's rotor of magnitude $20\omega^2$ N.m, what is the rotor's angular velocity at $t = 6$ s?

7.14 The mass moment of inertia of the robotic manipulator arm about the vertical y axis is $10\,\text{kg.m}^2$. The mass moment of inertia of the 14 kg casting held by the arm about the y' axis is $0.8\,\text{kg.m}^2$. What couple about the y axis is necessary to give the manipulator arm an angular acceleration of $2\,\text{rad/s}^2$?

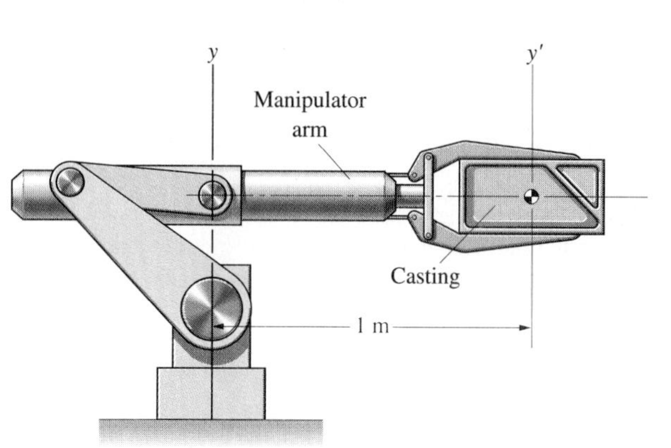

P7.14

7.15 The gears A and B can turn freely on their pin supports. Their mass moments of inertia are $I_A = 0.002\,\text{kg.m}^2$ and $I_B = 0.006\,\text{kg.m}^2$. They are initially stationary, and at $t = 0$ a constant couple $M = 2\,\text{N.m}$ is applied to gear B. How many revolutions has gear A turned at $t = 4\,\text{s}$?

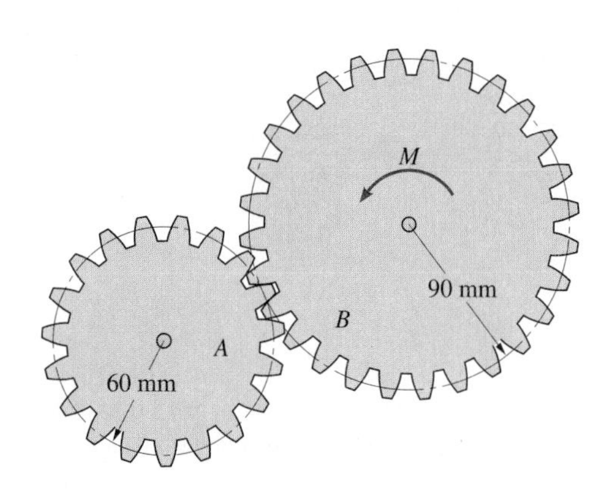

P7.15

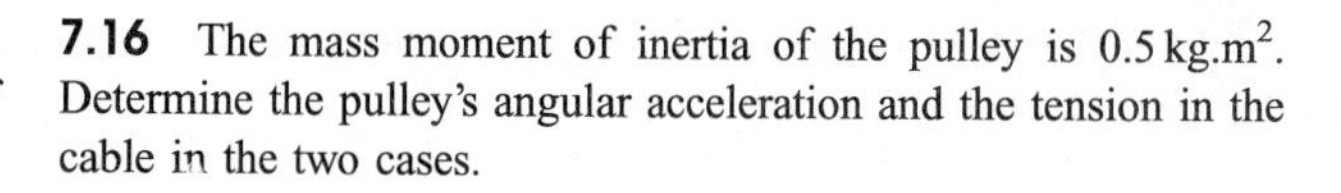

7.16 The mass moment of inertia of the pulley is $0.5\,\text{kg.m}^2$. Determine the pulley's angular acceleration and the tension in the cable in the two cases.

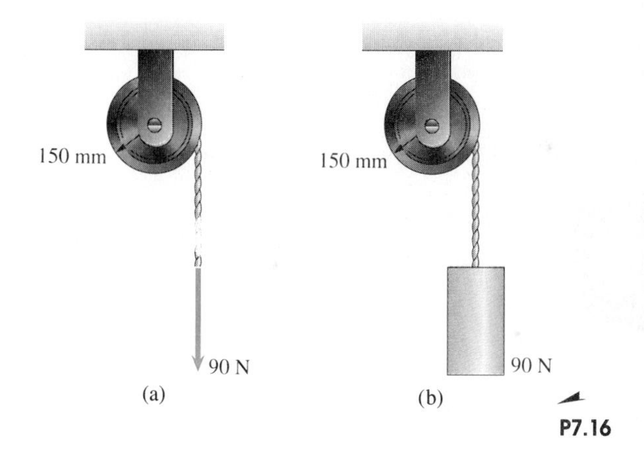

P7.16

7.17 Each box weighs 250 N, the mass moment of inertia of the pulley is $0.8\,\text{kg.m}^2$, and friction can be neglected. If the boxes start from rest at $t = 0$, determine the magnitude of their velocity and the distance they have moved from their initial position at $t = 1\,\text{s}$.

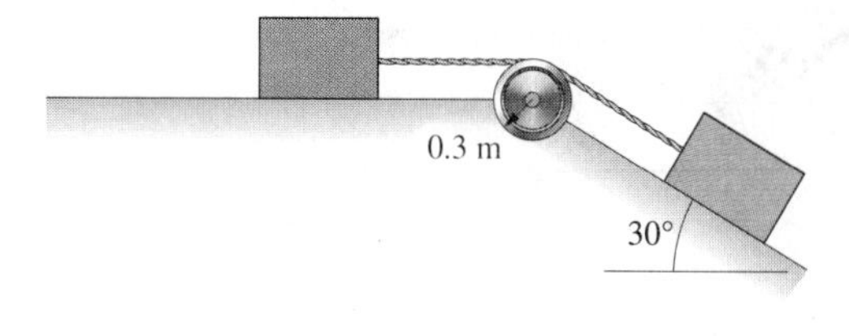

P7.17

7.18 The slender bar weighs 50 N and the disc weighs 100 N. The coefficient of kinetic friction between the disc and the horizontal surface is $\mu_k = 0.1$. If the disc has an initial counterclockwise angular velocity of 10 rad/s, how long does it take to stop spinning?

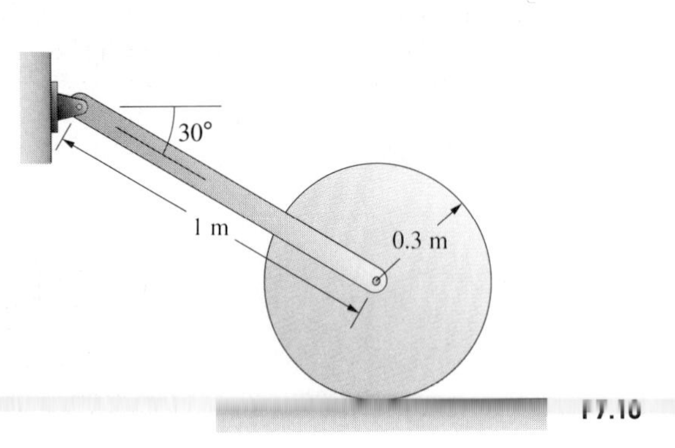

P7.18

7.19 In Problem 7.18, how long does it take the disc to stop spinning if it has an initial clockwise angular velocity of 10 rad/s?

7.20 The objects consist of identical 1 m, 5 kg bars welded together. If they are released from rest in the positions shown, what are their angular accelerations and what are the components of the reactions at A at that instant? (The y axes are vertical.)

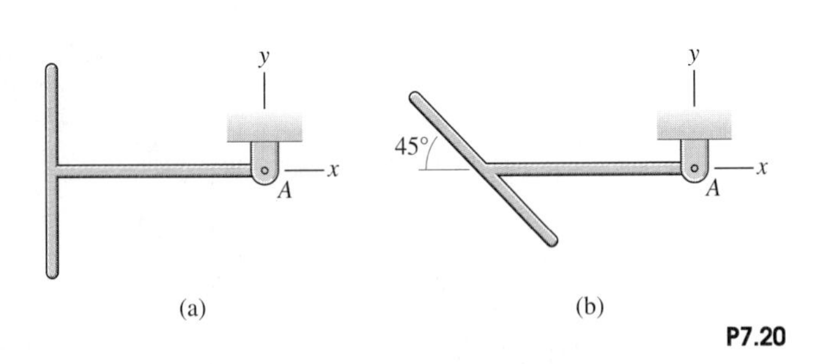

P7.20

7.21 The object consists of identical 1 m, 5 kg bars welded together. If it is released from rest in the position shown, what is its angular acceleration and what are the components of the reaction at A at that instant? (The y axis is vertical.)

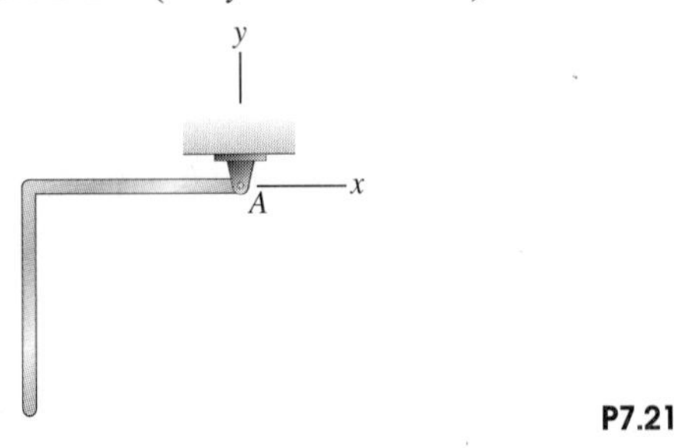

P7.21

7.22 For what value of x is the horizontal bar's angular acceleration a maximum, and what is the maximum angular acceleration?

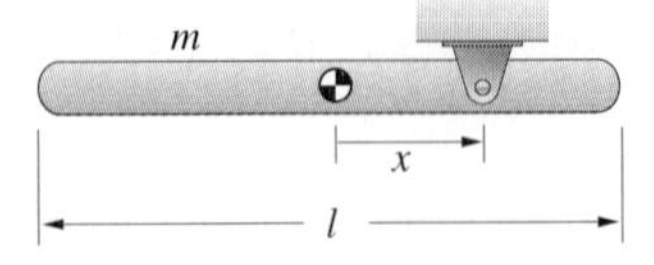

P7.22

7.23 Model the arm ABC as a single rigid body. Its mass is 300 kg, and the mass moment of inertia about its centre of mass is $I = 360\ \text{kg.m}^2$. If point A is stationary and the angular acceleration of the arm is $0.6\ \text{rad/s}^2$ counterclockwise, what force does the hydraulic cylinder exert on the arm at B? (The arm is actuated by two hydraulic cylinders, one on each side of the vehicle. You are to determine the total force exerted by the two cylinders.)

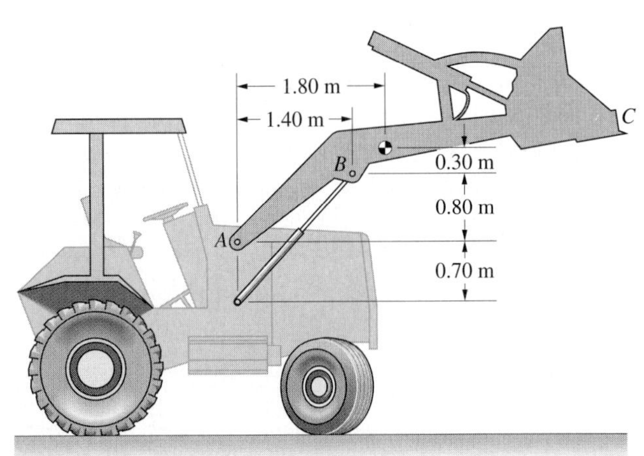

P7.23

7.24 In Problem 7.23, if the angular acceleration of arm ABC is $0.6\ \text{rad/s}^2$ counterclockwise and its angular velocity is 1.4 rad/s clockwise, what are the components of the force exerted on the arm at A? (There are two pin supports, one on each side of the vehicle. You are to determine the components of the total force exerted by the two supports.)

7.25 To lower the drawbridge, the gears that raised it are disengaged and a fraction of a second later a second set of gears that lower it are engaged. At the instant the gears that raised it are disengaged, what are the components of force exerted by the bridge on its support at O? The drawbridge weighs 1.6 MN, its mass moment of inertia about O is $I_O = 1.0 \times 10^7$ kg.m^2, and the coordinates of its centre of mass at the instant the gears are disengaged are $\bar{x} = 2.5$ m, $\bar{y} = 5$ m.

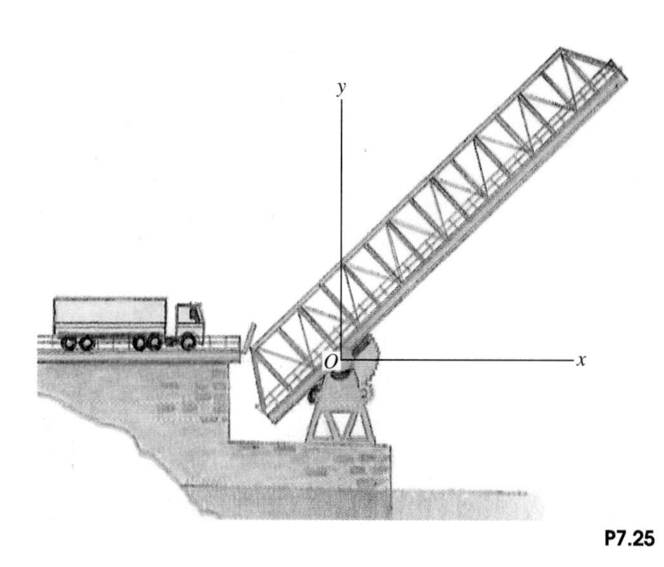

P7.25

7.26 Arm BC has a mass of 12 kg and the mass moment of inertia about its centre of mass is 3 kg.m^2. If B is stationary and arm BC has a constant counterclockwise angular velocity of 2 rad/s at the instant shown, determine the couple and the components of force exerted on arm BC at B.

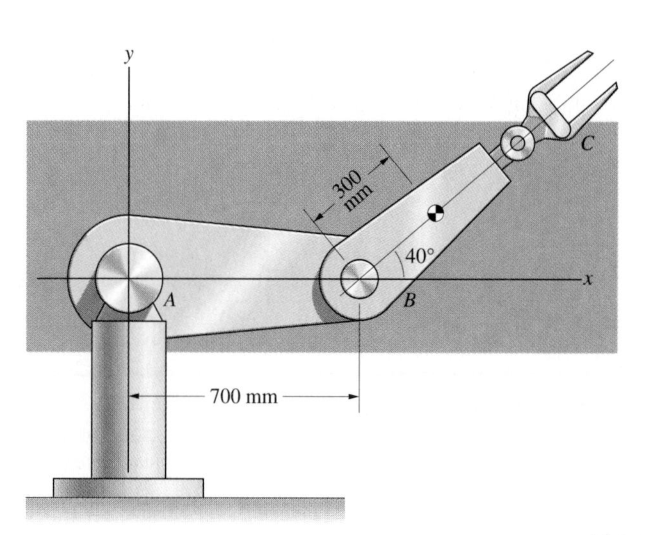

P7.26

7.27 In Problem 7.26, what are the couple and the components of force exerted on arm BC at B if arm AB has a constant clockwise angular velocity of 2 rad/s and arm BC has a counterclockwise angular velocity of 2 rad/s and a clockwise angular acceleration of 4 rad/s^2 at the instant shown?

7.28 A thin ring and a circular disc, each of mass m and radius R, are released from rest on an inclined surface and allowed to roll a distance D. Determine the ratio of the times required.

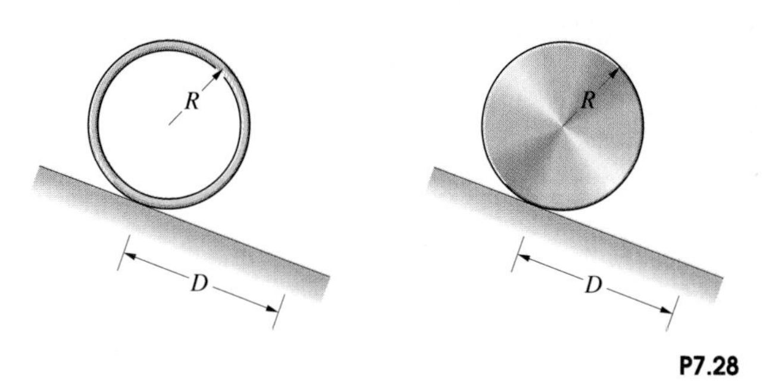

P7.28

7.29 The stepped disc weighs 180 N and its mass moment of inertia is $I = 0.2$ kg.m^2. If it is released from rest, how long does it take the centre of the disc to fall 1 m? (Assume that the string remains vertical.)

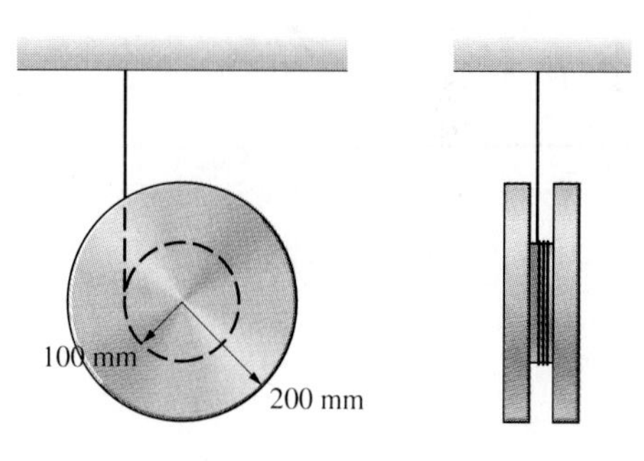

P7.29

7.30 At $t = 0$, a sphere of mass m and radius R ($I = \frac{2}{3}mR^2$) on a flat surface has angular velocity ω_0 and the velocity of its centre is zero. The coefficient of kinetic friction between the sphere and the surface is μ_k. What is the maximum velocity the centre of the sphere will attain, and how long does it take to reach it?

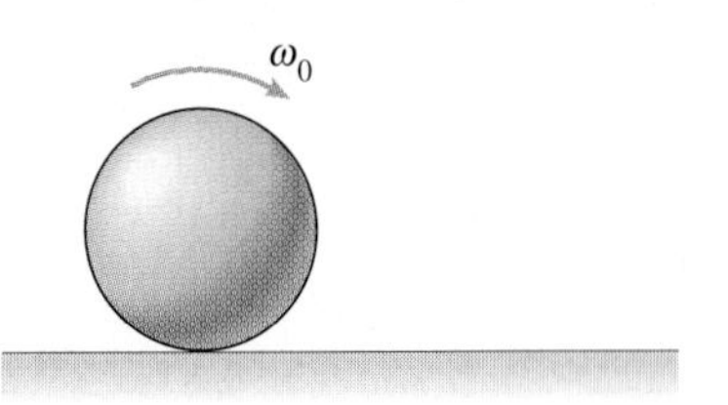

P7.30

7.31 A soccer player kicks the ball to a teammate 6 m away. The ball leaves his foot moving parallel to the ground at 6 m/s with no initial angular velocity. The coefficient of kinetic friction between the ball and the grass is $\mu_k = 0.4$. How long does it take the ball to reach his teammate? (The ball is 0.7 m in circumference and weighs 4 N. Estimate its mass moment of inertia by using the equation for a thin spherical shell: $I = \frac{2}{3}mR^2$.)

P7.31

7.32 The 100 kg cylindrical disc is at rest when the force F is applied to a cord wrapped around it. The static and kinetic coefficients of friction between the disc and the surface equal 0.2. Determine the angular acceleration of the disc if (a) $F = 500$ N; (b) $F = 1000$ N.

Strategy: First solve the problem by assuming that the disc does not slip, but rolls on the surface. Determine the friction force and find out whether it exceeds the product of the friction coefficient and the normal force. If it does, you must rework the problem assuming that the disc slips.

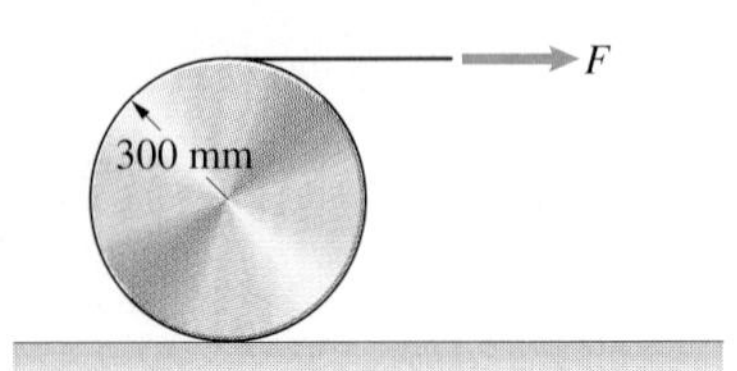

P7.32

7.33 The 18 kg ladder is released from rest in the position shown. Model it as a slender bar and neglect friction. At the instant of release, determine (a) the angular acceleration; (b) the normal force exerted on the ladder by the floor.

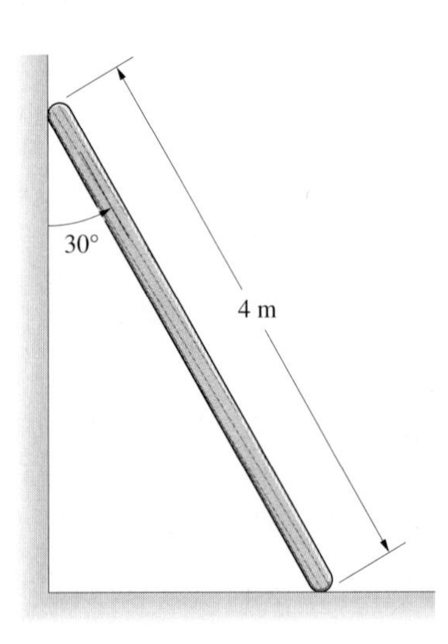

P7.33

7.34 Suppose that the ladder in Problem 7.33 has a counterclockwise angular velocity of 1.0 rad/s in the position shown. Determine (a) the angular acceleration; (b) the normal force exerted on the ladder by the floor.

7.35 Suppose that the ladder in Problem 7.33 has a counterclockwise angular velocity of 1.0 rad/s in the position shown and that the coefficient of kinetic friction at the floor and the wall is $\mu_k = 0.2$. Determine (a) the angular acceleration; (b) the normal force exerted on the ladder by the floor.

7.36 The slender bar weighs 150 N and the cylindrical disc weighs 100 N. The system is released from rest with the bar horizontal. Determine the bar's angular acceleration at the instant of release if the bar and disc are welded together at A.

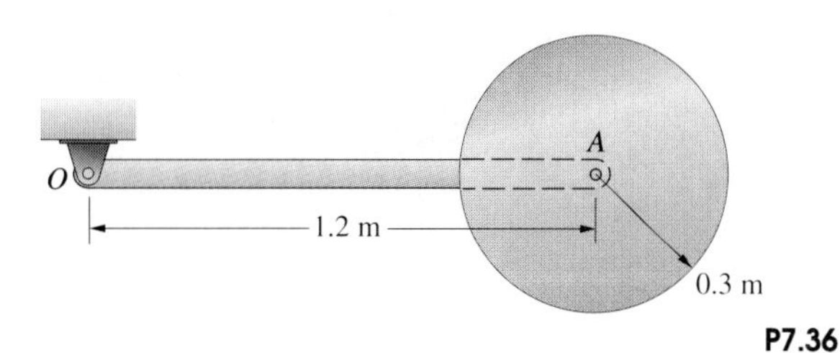

P7.36

7.37 In Problem 7.36, determine the bar's angular acceleration if the bar and disc are pinned together at A.

7.38 The 0.1 kg slender bar and 0.2 kg cylindrical disc are released from rest with the bar horizontal. The disc rolls on the curved surface. What is the bar's angular acceleration at that instant?

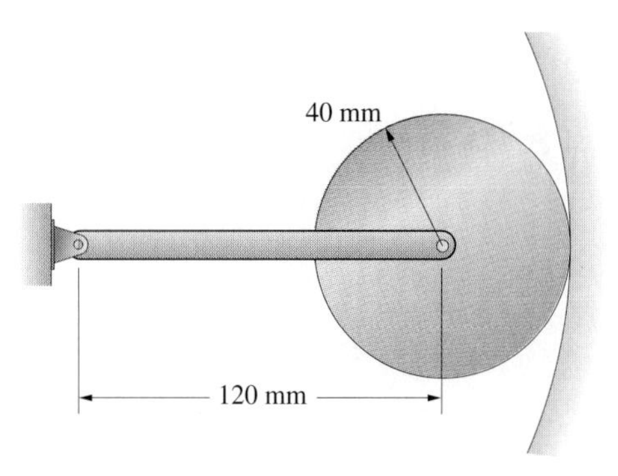

P7.38

7.39 The 2 kg slender bar and 5 kg block are released from rest in the position shown. If friction is negligible, what is the block's acceleration at that instant?

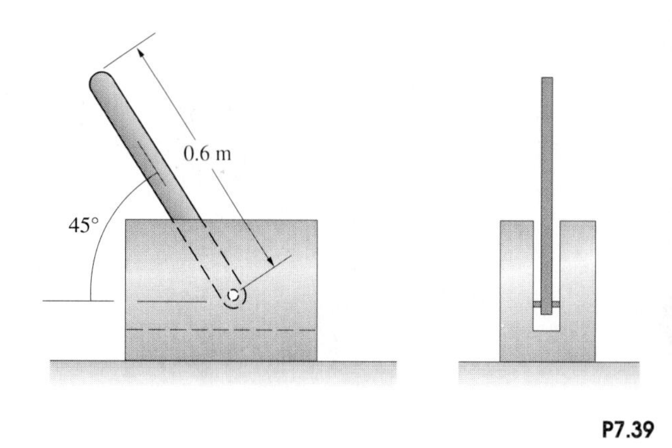

P7.39

7.40 In Problem 7.39, suppose that the velocity of the block is zero and the bar has an angular velocity of 4 rad/s at the instant shown. What is the block's acceleration?

7.41 The 0.4 kg slender bar and 1 kg disc are released from rest in the position shown. If the disc rolls, what is the bar's angular acceleration at that instant?

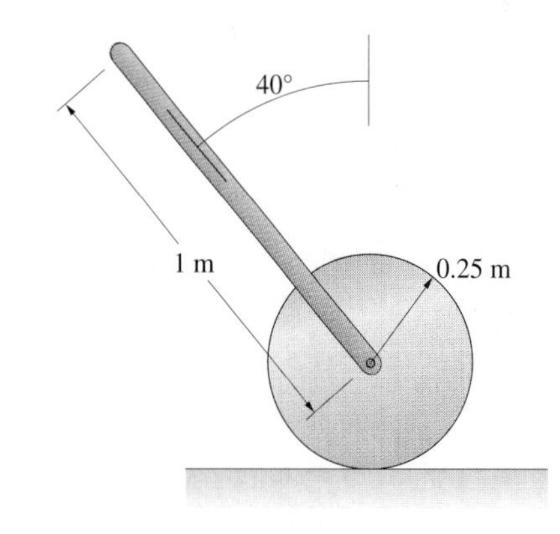

P7.41

7.42 In Problem 7.41, what is the smallest value of the coefficient of static friction for which the disc will roll when the system is released instead of slipping?

7.43 Pulley A weighs 20 N, $I_A = 0.08$ kg.m^2, and $I_B = 0.02$ kg.m^2. If the system is released from rest, what distance does the 80 N weight fall in one-half second?

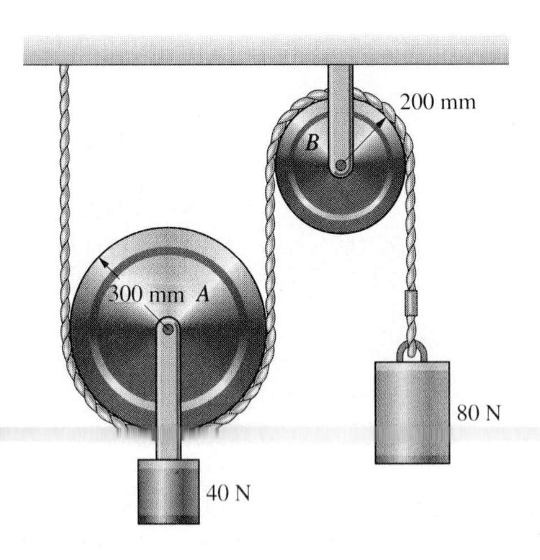

P7.43

7.44 The slender bar weighs 100 N and the crate weighs 400 N. The surface the crate rests on is smooth. If the system is stationary at the instant shown, what couple M will cause the crate to accelerate to the left at 1 m/s^2 at that instant?

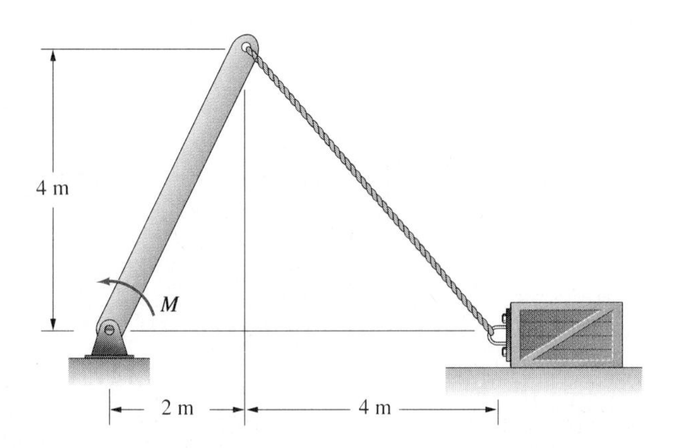

P7.44

7.45 Suppose that the slender bar in Problem 7.44 is rotating in the counterclockwise direction at 2 rad/s at the instant shown and that the coefficient of kinetic friction between the crate and the horizontal surface is $\mu_k = 0.2$. What couple M will cause the crate to accelerate to the left at 1 m/s^2 at that instant?

7.46 Bar AB rotates with a constant angular velocity of 6 rad/s in the counterclockwise direction. The slender bar BCD weighs 50 N and the collar that bar BCD is attached to at C weighs 10 N. The y axis points upwards. Neglecting friction, determine the components of the forces exerted on bar BCD by the pins at B and C at the instant shown.

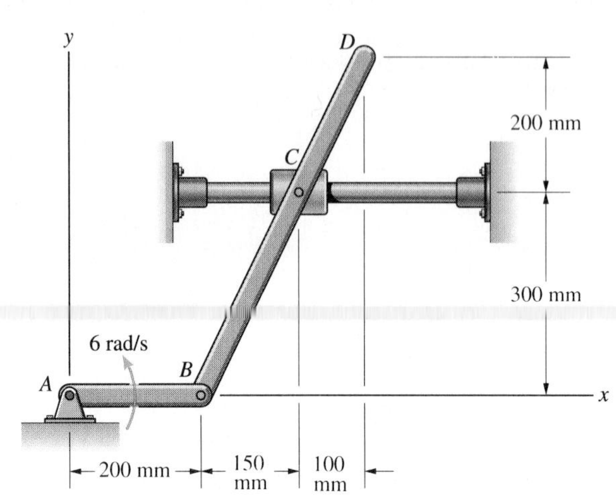

P7.46

7.47 Bar AB weighs 50 N and bar BC weighs 30 N. If the system is released from rest in the position shown, what are the angular acceleration of bar AB and the normal force exerted by the floor at C at that instant? Neglect friction.

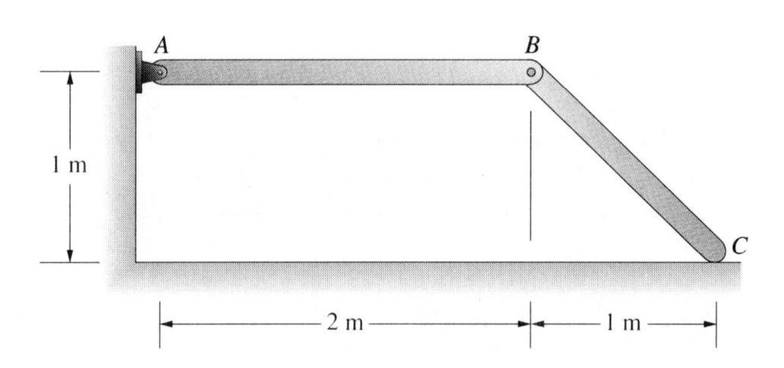

P7.47

7.48 In Problem 7.47, if the angular velocity of bar AB is 1.0 rad/s clockwise at the instant shown, what are the angular acceleration of bar BC and the normal force exerted by the floor at C at that instant?

7.49 The combined mass of the motorcycle and rider is 160 kg. Each 9 kg wheel has a 330 mm radius and mass moment of inertia $I = 0.8$ kg.m^2. The engine drives the rear wheel. If the rear wheel exerts a 400 N horizontal force on the road and you do *not* neglect the horizontal force exerted on the road by the front wheel, determine (a) the motorcycle's acceleration; (b) the normal forces exerted on the road by the rear and front wheels. (The location of the centre of mass of the motorcycle *not including* its wheels is shown.)

Strategy: Isolate the wheels and draw three free-body diagrams. The motorcycle's engine drives the rear wheel by exerting a couple on it.

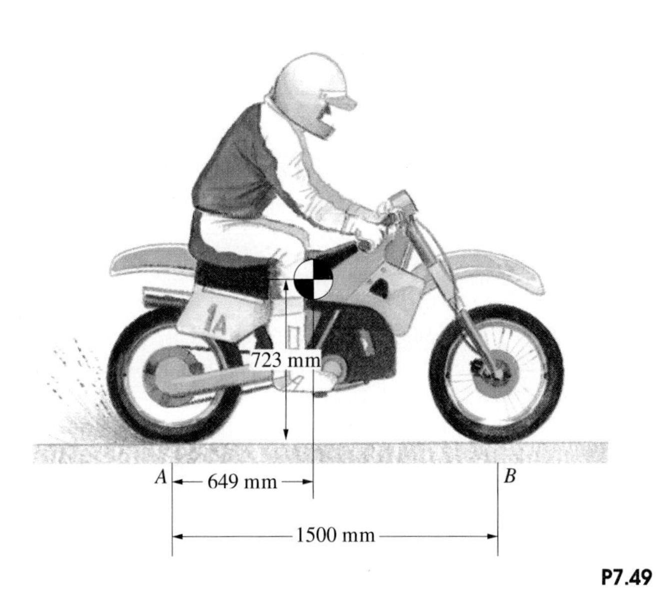

P7.49

7.50 In Problem 7.49, if the front wheel lifts slightly off the road when the rider accelerates, determine (a) the motorcycle's acceleration; (b) the torque exerted by the engine on the rear wheel.

7.51 By using Equations (7.5)–(7.8), show that the angular momentum of a rigid body about a fixed point O is the sum of the angular momentum about O due to the motion of its centre of mass and the angular momentum about its centre of mass: $\mathbf{H}_0 = \mathbf{r} \times m\mathbf{v} + \mathbf{H}$.

7.52 The mass of the slender bar is m and the mass of the homogeneous disc is $4m$. The system is released from rest in the position shown. If the disc rolls and friction between the bar and the horizontal surface is negligible, show that the disc's angular acceleration is $\alpha = 6g/95R$ counterclockwise.

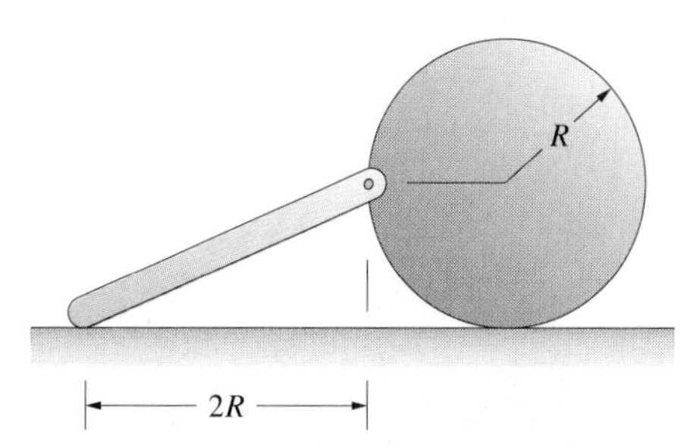

P7.52

7.53 If the disk in Problem 7.52 rolls and the coefficient of kinetic friction between the bar and the horizontal surface is μ_k, what is the disc's angular acceleration at the instant the system is released?

7.54 The ring gear is fixed. The mass and mass moment of inertia of the sun gear are $m_S = 320$ kg, $I_S = 6000$ kg.m^2. The mass and mass moment of inertia of each planet gear are $m_P = 40$ kg, $I_P = 88$ kg.m^2. If a couple $M = 800$ N.m is applied to the sun gear, what is the resulting angular acceleration of the planet gears, and what tangential force is exerted on the sun gear by each planet gear?

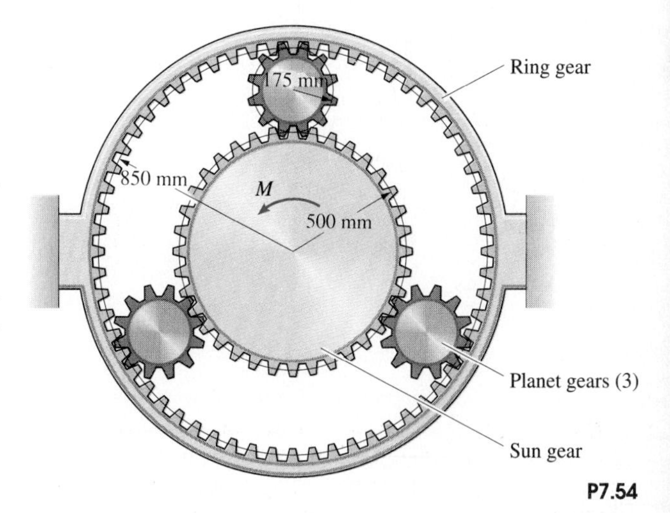

P7.54

7.55 If the system in Problem 7.54 starts from rest, what constant couple M exerted on the sun gear will cause it to accelerate to 120 rpm (revolutions per minute) in 1 min?

Problems 7.56–7.62 are related to Example 7.7.

7.56 The 3 Mg rocket is accelerating upwards at 2 g's. If you model it as a homogeneous bar, what is the magnitude of the axial force at the midpoint?

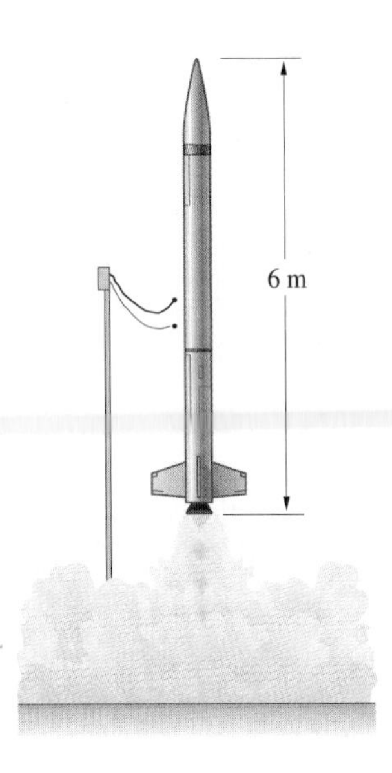

P7.56

7.57 The 20 kg slender bar is attached to a vertical shaft at A and rotates *in the horizontal plane* with a constant angular velocity of 10 rad/s. What is the axial force at the bar's midpoint?

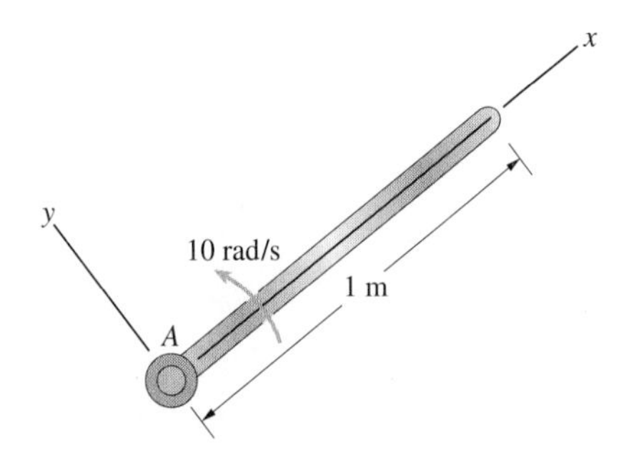

P7.57

7.58 For the rotating bar in Problem 7.57, draw the graph of the axial force as a function of x.

7.59 The 50 kg slender bar AB has a built-in support at A. The y axis points upwards. Determine the magnitudes of the shear force and bending moment at the bar's midpoint if (a) the support is stationary; (b) the support is accelerating upwards at 3 m/s^2.

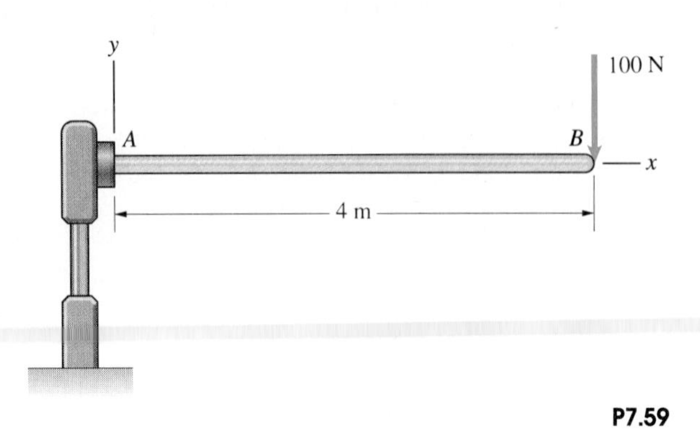

P7.59

7.60 For the bar in Problem 7.59, draw the shear force and bending moment diagrams for the two cases.

7.61 The 18 kg ladder is held in equilibrium in the position shown by the force F. Model the ladder as a slender bar and neglect friction.

(a) What are the axial force, shear force and bending moment at the ladder's midpoint?

(b) If the force F is suddenly removed, what are the axial force, shear force and bending moment at the ladder's midpoint at that instant?

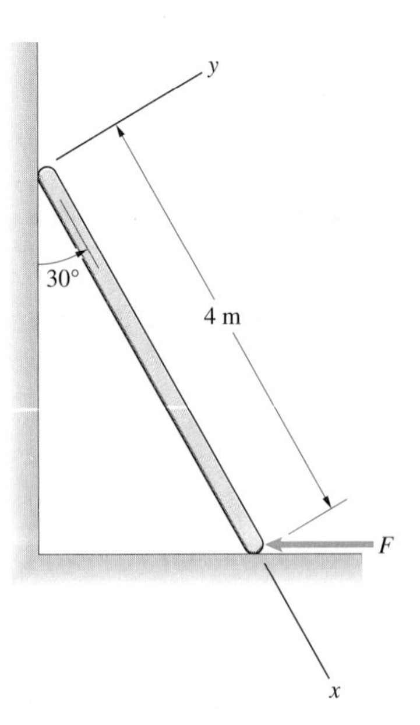

P7.61

7.62 For the ladder in Problem 7.61, draw the shear force and bending moment diagrams for the two cases.

Computational Mechanics

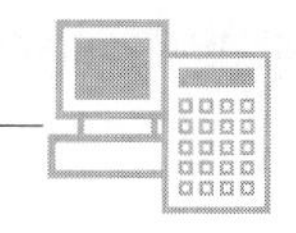

The material in this section is designed for the use of a programmable calculator or computer.

When you know the forces and couples acting on a rigid body, you can use the equations of motion to determine the acceleration of its centre of mass and its angular acceleration. In some situations, you can then integrate to obtain closed-form expressions for the velocity and position of its centre of mass and for its angular velocity and angular position as functions of time. But if the functions describing the accelerations are too complicated, or the forces and couples are known in terms of continuous or analogue data instead of equations, you must use a numerical method to determine the velocities and positions as functions of time.

In Chapter 3, we described a simple finite-difference method for determining the position and velocity of the centre of mass as functions of time. You can determine the angular position and angular velocity in the same way. Let's suppose that the angular acceleration of a rigid body depends on time, its angular position and its angular velocity:

$$\alpha = \alpha(t, \theta, \omega) \tag{7.30}$$

Suppose that at a particular time t_0, we know the angle $\theta(t_0)$ and angular velocity $\omega(t_0)$. The angular acceleration at t_0 is

$$\frac{d\omega}{dt}(t_0) = \alpha[t_0, \theta(t_0), \omega(t_0)] \tag{7.31}$$

where

$$\frac{d\omega}{dt}(t_0) = \lim_{\Delta t \to 0} \frac{\omega(t_0 + \Delta t) - \omega(t_0)}{\Delta t}$$

Choosing a sufficiently small value of Δt, we can approximate this derivative by

$$\frac{d\omega}{dt}(t_0) = \frac{\omega(t_0 + \Delta t) - \omega(t_0)}{\Delta t}$$

and substitute it into Equation (7.31) to obtain an approximate expression for the angular velocity at time $t_0 + \Delta t$:

$$\omega(t_0 + \Delta t) = \omega(t_0) + \alpha[t_0, \theta(t_0), \omega(t_0)]\Delta t \tag{7.32}$$

The relation between the angular velocity and angular position at t_0 is

$$\frac{d\theta}{dt}(t_0) = \omega(t_0)$$

Approximating this derivative by

$$\frac{d\theta}{dt}(t_0) = \frac{\theta(t_0 + \Delta t) - \theta(t_0)}{\Delta t}$$

we obtain an approximate expression for the angular position at time $t_0 + \Delta t$:

$$\theta(t_0 + \Delta t) = \theta(t_0) + \omega(t_0)\Delta t \tag{7.33}$$

With Equations (7.32) and (7.33), we can determine the approximate values of the angular velocity and position at $t_0 + \Delta t$. Using these values as initial conditions, we can repeat the procedure to determine the angular velocity and position at time $t_0 + 2\Delta t$, and so forth.

Example 7.10

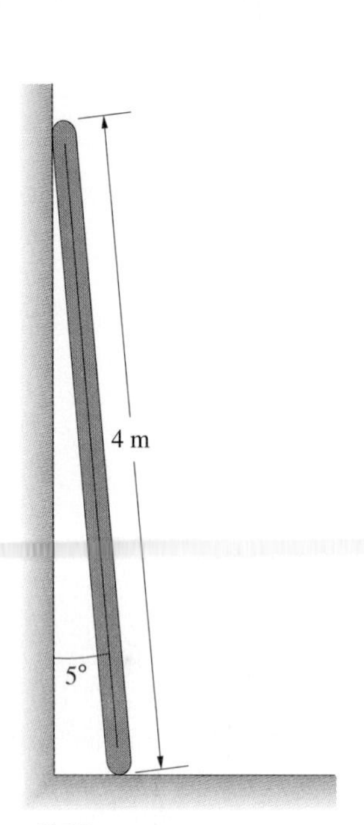

Figure 7.23

The 18 kg ladder in Figure 7.23 is released from rest in the position shown at $t = 0$. Neglecting friction, determine its angular position and angular velocity as functions of time. Use time increments Δt of 0.1 s, 0.01 s and 0.001 s.

STRATEGY

The initial steps – drawing the free-body diagram of the ladder, applying the equations of motion, and determining the angular acceleration – are presented in Example 7.4. The ladder's angular acceleration is

$$\alpha = \frac{3g}{2l} \sin\theta$$

where θ is the angle between the ladder and the wall and l is its length. With this expression, we can use Equations (7.32) and (7.33) to approximate the ladder's angular position and angular velocity as functions of time.

SOLUTION

The angular acceleration is

$$\alpha = \frac{(3)(9.81)}{(2)(4)} \sin\theta = 3.68 \sin\theta \text{ rad/s}^2$$

Let $\Delta t = 0.1$ s. At the initial time $t_0 = 0$, $\theta(t_0) = 5° = 0.0873$ rad and $\omega(t_0) = 0$. We can use Equations (7.32) and (7.33) to determine the angular velocity and position at time $t_0 + \Delta t = 0.1$ s. The angular position is

$$\begin{aligned} \theta(t_0 + \Delta t) &= \theta(t_0) + \omega(t_0)\Delta t: \\ \theta(0.1) &= \theta(0) + \omega(0)\Delta t \\ &= 0.0873 + (0)(0.1) = 0.0873 \text{ rad} \end{aligned}$$

The angular velocity is

$$\begin{aligned} \omega(t_0 + \Delta t) &= \omega(t_0) + \alpha(t_0)\Delta t: \\ \omega(0.1) &= 0 + [3.68 \sin(0.0873)](0.1) = 0.0321 \text{ rad/s} \end{aligned}$$

Using these values as the initial conditions for the next time step, the angular position at $t = 0.2$ s is

$$\begin{aligned} \theta(0.2) &= \theta(0.1) + \omega(0.1)\Delta t \\ &= 0.0873 + (0.0321)(0.1) = 0.0905 \text{ rad} \end{aligned}$$

and the angular velocity is

$$\begin{aligned} \omega(0.2) &= \omega(0.1) + \alpha(0.1)\Delta t \\ &= 0.0321 + [3.68 \sin(0.0873)](0.1) = 0.0641 \text{ rad/s} \end{aligned}$$

Continuing in this way, we obtain the following values for the first five time steps:

Time, s	**θ, rad**	ω, rad/s
0.0	0.0873	0.0000
0.1	0.0873	0.0321
0.2	0.0905	0.0641
0.3	0.0969	0.0974
0.4	0.1066	0.1329
0.5	0.1199	0.1721

Figures 7.24 and 7.25 show the numerical solutions for the angular position and angular velocity obtained using $\Delta t = 0.1$ s, $\Delta t = 0.01$ s and $\Delta t = 0.001$ s. Trials with smaller time intervals indicate that $\Delta t = 0.001$ s closely approximates the exact solution. We show the positions of the falling ladder at 0.2 s intervals in Figure 7.26.

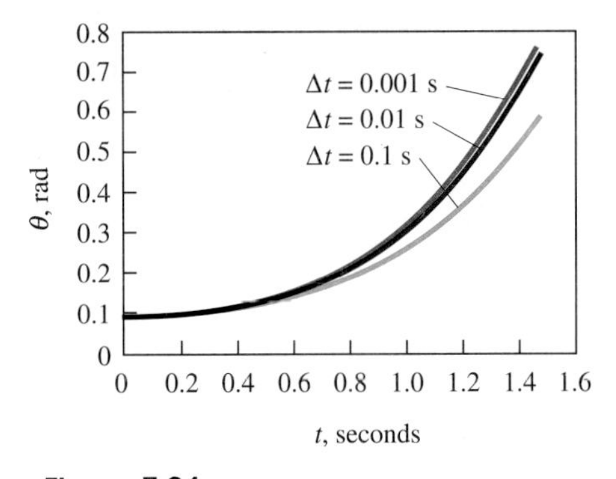

Figure 7.24

Numerical solutions for the ladder's angular position.

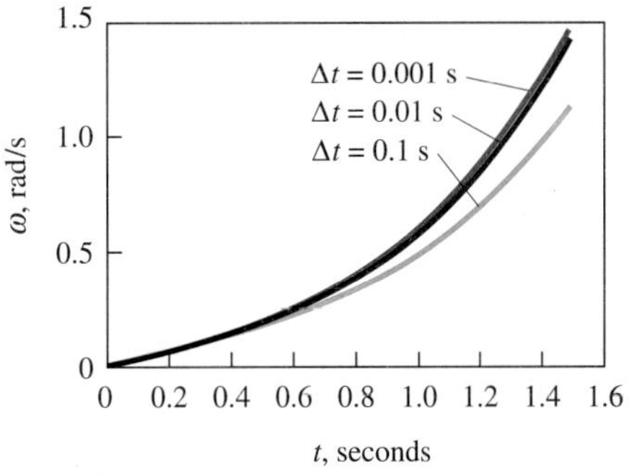

Figure 7.25

Numerical solutions for the ladder's angular velocity.

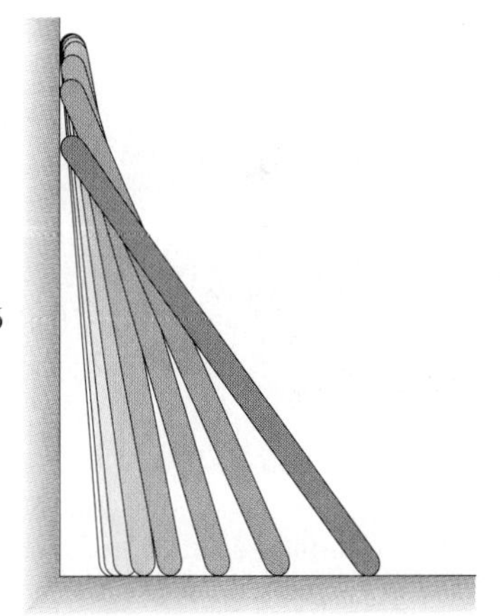

Figure 7.26

Position of the falling ladder at 0.2 s intervals from $t = 0$ to $t = 1.4$ s.

DISCUSSION

By using the chain rule, we can write the ladder's angular acceleration as

$$\alpha = \frac{d\omega}{dt} = \frac{d\omega}{d\theta}\omega = \frac{3g}{2l}\sin\theta$$

Separating variables, we can integrate to determine the angular velocity as a function of the angular position:

$$\int_0^{\omega} \omega\, d\omega = \int_{5^\circ}^{\theta} \frac{3g}{2l}\sin\theta\, d\theta$$

We obtain

$$\omega = \sqrt{(3g/l)(\cos 5^\circ - \cos\theta)}$$

This closed-form result is compared with the graph of our numerical solution (using $\Delta t = 0.001$ s) in Figure 7.27. The curves are indistinguishable.

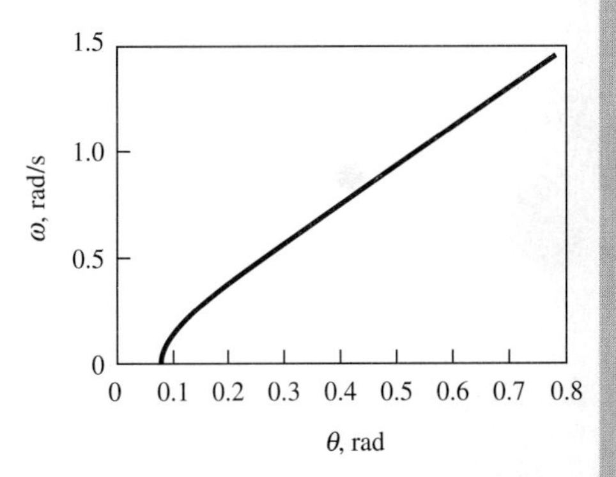

Figure 7.27

Analytical and numerical solutions for the ladder's angular velocity as a function of its angular position.

Problems

7.63 Continue the calculations presented in Example 7.10, using $\Delta t = 0.1$ s, and determine the ladder's angular position and angular velocity at $t = 0.6$ s and $t = 0.7$ s.

7.64 The mass moment of inertia of the helicopter's rotor is 500 kg.m^2. It starts from rest at $t = 0$, the engine exerts a constant torque of 625 N.m and aerodynamic drag exerts a torque of magnitude $25\omega^2$ N.m where ω is the rotor's angular velocity in radians per second. Using $\delta t = 0.2$ s, determine the rotor's angular position and angular velocity for the first five time steps. Compare your results for the angular velocity with the closed-form solution.

P7.64

7.65 In Problem 7.64, draw a graph of the rotor's angular velocity as a function of time from $t = 0$ to $t = 10$ s, comparing the closed-form solution, the numerical solution using $\delta t = 1.0$ s, and the numerical solution using $\delta t = 0.2$ s.

7.66 The slender 10 kg bar is released from rest in the horizontal position shown. Using $\delta t = 0.1$ s, determine the bar's angular position and angular velocity for the first five time steps.

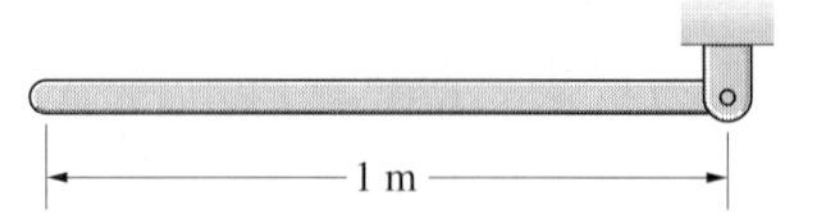

P7.66

7.67 In Problem 7.66, determine the bar's angular position and angular velocity as functions of time from $t = 0$ to $t =$ 0.8 s using $\Delta t = 0.1$ s, $\Delta t = 0.01$ s and $\Delta t = 0.001$ s. Draw the graphs of the angular velocity as a function of the angular position for these three cases and compare them with the graph of the closed-form solution for the angular velocity as a function of the angular position.

7.68 In Problem 7.66, suppose that the bar's pin support contains a damping device that exerts a resisting couple on the bar of magnitude $c\omega$ (N.m), where ω is the angular velocity in radians per second. Using $\Delta t = 0.001$ s, draw graphs of the bar's angular velocity as a function of time from $t = 0$ to $t = 0.8$ s for the cases $c = 0$, $c = 2$, $c = 4$ and $c = 8$.

7.69 The falling ladder in Example 7.10 will lose contact with the wall before it hits the floor. Using $\delta t = 0.001$ s, estimate the time and the value of the angle between the wall and the ladder when this occurs.

Appendix: Moments of Inertia

When a rigid body is subjected to forces and couples, the rotational motion that results depends not only on its mass, but also on how its mass is *distributed*. Although the two objects in Figure 7.28 have the same mass, the angular accelerations caused by the couple M are different. This difference is reflected in the equation of angular motion $M = I\alpha$ through the mass moment of inertia I. The object in Figure 7.28(a) has a smaller mass moment of inertia about the axis L, so its angular acceleration is greater.

In deriving the equations of motion of a rigid body in Sections 7.2 and 7.3, we regarded it as a finite number of particles and expressed its mass moment of inertia about an axis L_0 as

$$I_0 = \sum_i m_i r_i^2$$

where m_i is the mass of the ith particle and r_i is the perpendicular distance from L_0 to the ith particle (Figure 7.29(a)). To calculate the moments of inertia of objects, it is often more convenient to model them as continuous distribu-

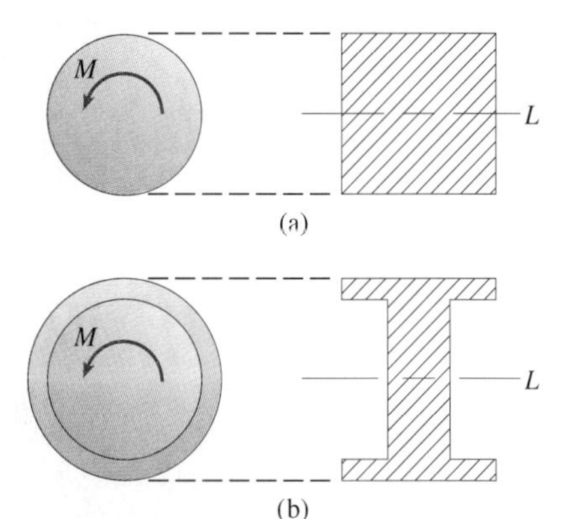

Figure 7.28
Objects of equal mass that have different mass moments of inertia about L.

tions of mass and express the mass moment of inertia about L_0 as

$$I_0 = \int_m r^2 dm \qquad (7.34)$$

where r is the perpendicular distance from L_0 to the differential element of mass dm (Figure 7.29(b)). When the axis passes through the centre of mass of the object, we denote the axis by L and the mass moment of inertia about L by I.

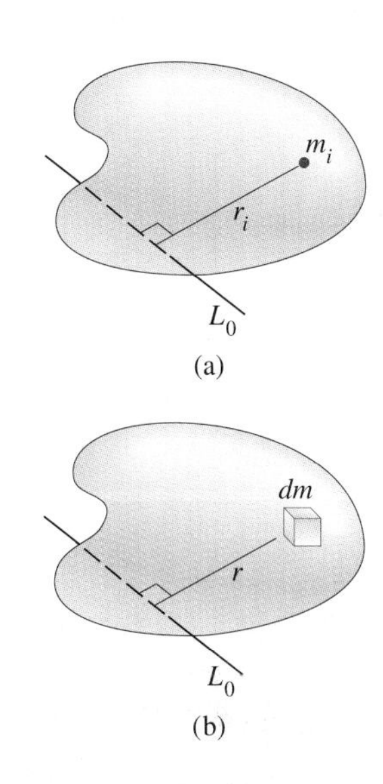

Figure 7.29
Determining the mass moment of inertia by modelling an object as (a) a finite number of particles and (b) a continuous distribution of mass.

Simple Objects

You can determine the mass moments of inertia of complicated objects by summing the moments of inertia of their individual parts. We therefore begin by determining mass moments of inertia of some simple objects. Then in the next section we describe the parallel-axis theorem, which makes it possible for you to determine mass moments of inertia of objects composed of combinations of simple parts.

Slender Bars We will determine the mass moment of inertia of a straight slender bar about a perpendicular axis L through the centre of mass of the bar (Figure 7.30(a)). 'Slender' means we assume that the bar's length is much greater than its width. Let the bar have length l, cross-sectional area A, and mass m. We assume that A is uniform along the length of the bar and that the material is homogeneous.

Consider a differential element of the bar of length dr at a distance r from the centre of mass (Figure 7.30(b)). The element's mass is equal to the product of its volume and the mass density: $dm = \rho A \, dr$. Substituting this expression into Equation (7.34), we obtain the mass moment of inertia of the bar about a perpendicular axis through its centre of mass:

$$I = \int_m r^2 dm = \int_{-l/2}^{l/2} \rho A r^2 \, dr = \frac{1}{12} \rho A l^3$$

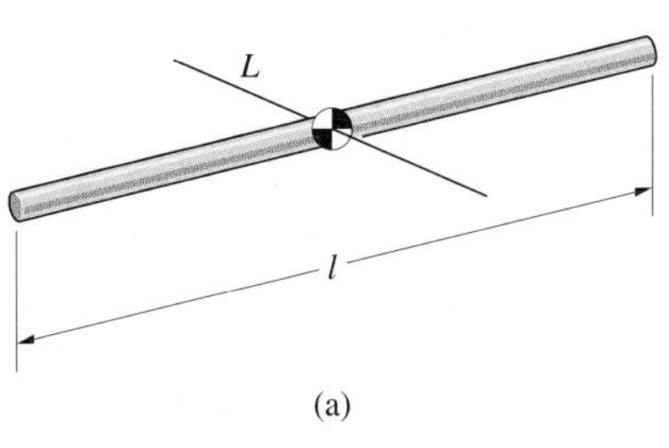

Figure 7.30
(a) A slender bar.
(b) A differential element of length dr.

The mass of the bar equals the product of the mass density and the volume of the bar, $m = \rho A l$, so we can express the mass moment of inertia as

$$I = \frac{1}{12} m l^2 \qquad (7.35)$$

We have neglected the lateral dimensions of the bar in obtaining this result. That is, we treated the differential element of mass dm as if it were concentrated on the axis of the bar. As a consequence, Equation (7.35) is an approximation for the mass moment of inertia of a bar. Later in this section, we will determine the moments of inertia for a bar of finite lateral dimension and show that Equation (7.35) is a good approximation when the width of the bar is small in comparison to its length.

Thin Plates Consider a homogeneous flat plate that has mass m and uniform thickness T. We will leave the shape of the cross-sectional area of the plate unspecified. Let a cartesian coordinate system be oriented so that the plate lies in the x-y plane (Figure 7.31(a)). Our objective is to determine the mass moments of inertia of the plate about the x, y and z axes.

We can obtain a differential element of volume of the plate by projecting an element of area dA through the thickness T of the plate (Figure 7.31(b)). The resulting volume is $T\,dA$. The mass of this element of volume is equal to the product of the mass density and the volume: $dm = \rho T\,dA$. Substituting this expression into Equation (7.34), we obtain the mass moment of inertia of the plate about the z axis in the form

$$I_{(z\text{ axis})} = \int_m r^2\,dm = \rho T \int_A r^2\,dA$$

where r is the distance from the z axis to dA. Since the mass of the plate is $m = \rho T A$, where A is the cross-sectional area of the plate, the product $\rho T = m/A$. The integral on the right is the polar moment of inertia J_0 of the cross-sectional area of the plate. Therefore we can write the mass moment of inertia of the plate about the z axis as

$$I_{(z\text{ axis})} = \frac{m}{A} J_0 \qquad (7.36)$$

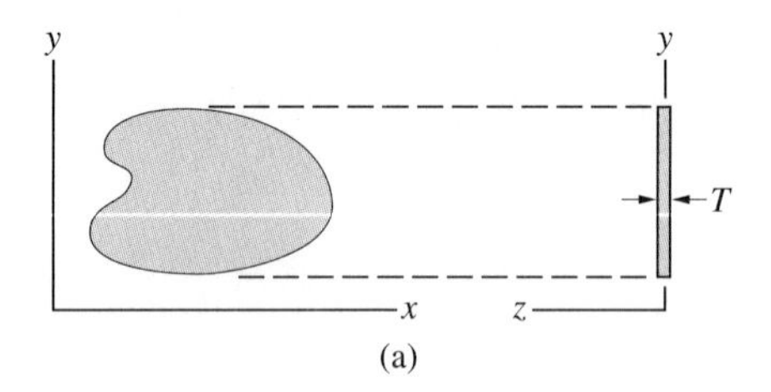

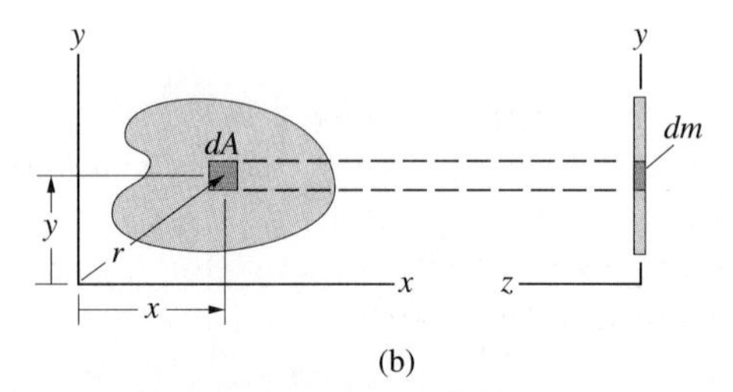

Figure 7.31
(a) A plate of arbitrary shape and uniform thickness T.
(b) An element of volume obtained by projecting an element of area dA through the plate.

From Figure 7.31(b), we see that the perpendicular distance from the x axis to the element of area dA is the y coordinate of dA. Therefore the mass moment of inertia of the plate about the x axis is

$$I_{(x \text{ axis})} = \int_m y^2 dm = \rho T \int_A y^2 dA = \frac{m}{A} I_x \tag{7.37}$$

where I_x is the moment of inertia of the cross-sectional area of the plate about the x axis. The mass moment of inertia of the plate about the y axis is

$$I_{(y \text{ axis})} = \int_m x^2 dm = \rho T \int_A x^2 dA = \frac{m}{A} I_y \tag{7.38}$$

where I_y is the moment of inertia of the cross-sectional area of the plate about the y axis.

Thus we have expressed the mass moments of inertia of a thin homogeneous plate of uniform thickness in terms of the moments of inertia of the cross-sectional area of the plate. In fact, these results explain why the area integrals I_x, I_y and J_O are called moments of inertia.

Since the sum of the area moments of inertia I_x and I_y is equal to the polar moment of inertia J_O, the mass moment of inertia of the thin plate about the z axis is equal to the sum of its moments of inertia about the x and y axes:

$$I_{(z \text{ axis})} = I_{(x \text{ axis})} + I_{(y \text{ axis})} \tag{7.39}$$

In the following example we use integration to determine the mass moment of inertia of an object consisting of two slender bars welded together. We then present an example that demonstrates the use of Equations (7.36)–(7.38) to determine the mass moments of inertia of a thin, homogeneous plate with a specific cross-sectional area.

Example 7.11

Two homogeneous, slender bars, each of length l, mass m, and cross-sectional area A, are welded together to form an L-shaped object (Figure 7.32). Determine the mass moment of inertia of the object about the axis L_O through point O. (The axis L_O is perpendicular to the two bars.)

Figure 7.32

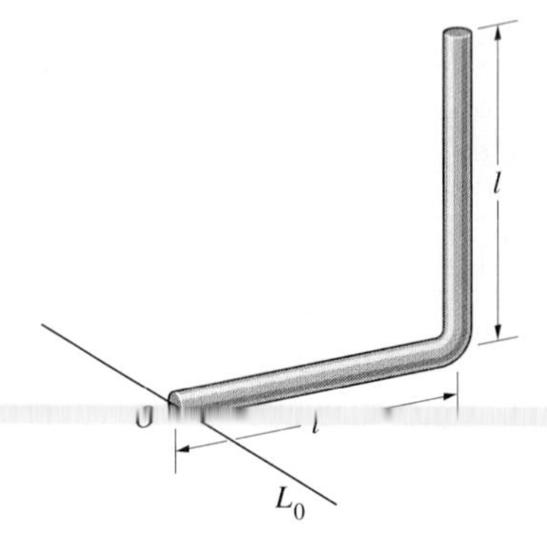

STRATEGY

Using the same integration procedure we used for a single bar, we can determine the mass moment of inertia of each bar about L_O and sum the results.

SOLUTION

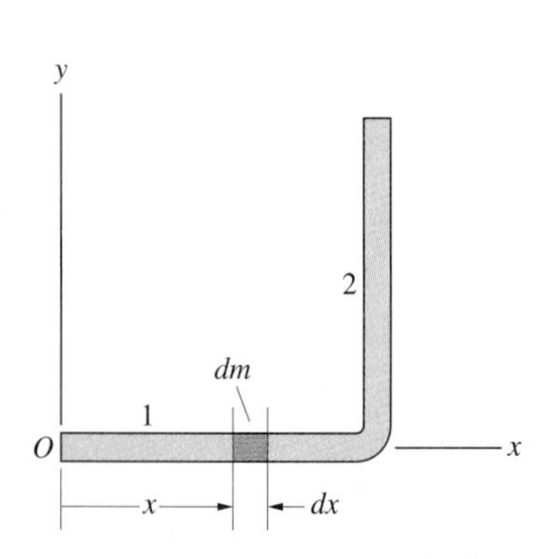

(a) Differential element of bar 1.

We orient a coordinate system with the z axis along L_O and the x axis colinear with bar 1 (Figure (a)). The mass of the differential element of bar 1 of length dx is $dm = \rho A dx$. The mass moment of inertia of bar 1 about L_O is

$$(I_O)_1 = \int_m r^2 dm = \int_0^l \rho A x^2 dx = \frac{1}{3}\rho A l^3$$

In terms of the mass of the bar, $m = \rho A l$, we can write this result as

$$(I_O)_1 = \frac{1}{3}ml^2$$

The mass of the element of bar 2 of length dy shown in Figure (b) is $dm = \rho A dy$. From the figure we see that the perpendicular distance from L_O to the element is $r = \sqrt{l^2 + y^2}$. Therefore the mass moment of inertia of bar 2 about L_O is

$$(I_O)_2 = \int_m r^2 dm = \int_0^l \rho A(l^2 + y^2) dy = \frac{4}{3}\rho A l^3$$

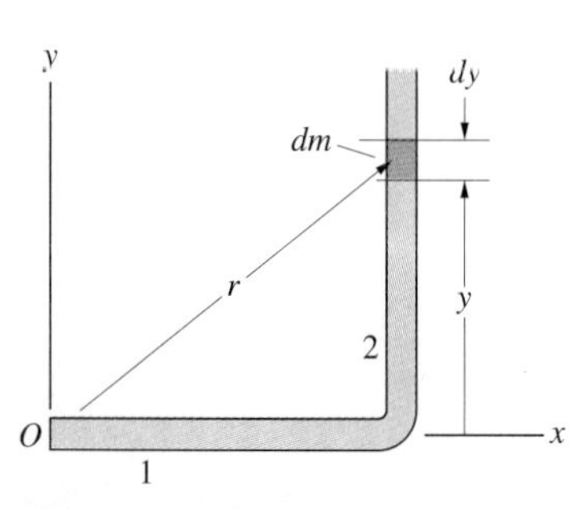

(b) Differential element of bar 2.

In terms of the mass of the bar, we obtain

$$(I_O)_2 = \frac{4}{3}ml^2$$

The mass moment of inertia of the L-shaped object about L_O is

$$I_O = (I_O)_1 + (I_O)_2 = \frac{1}{3}ml^2 + \frac{4}{3}ml^2 = \frac{5}{3}ml^2$$

Example 7.12

The thin, homogeneous plate in Figure 7.33 is of uniform thickness and mass m. Determine its mass moments of inertia about the x, y and z axes.

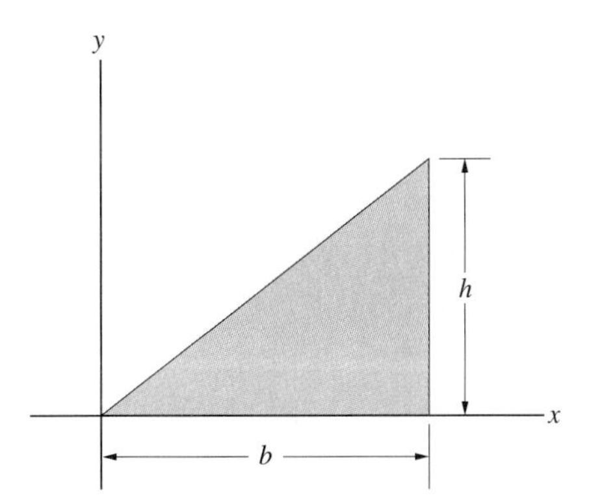

Figure 7.33

STRATEGY

The mass moments of inertia about the x and y axes are given by Equations (7.37) and (7.38) in terms of the moments of inertia of the cross-sectional area of the plate. We can determine the mass moment of inertia of the plate about the z axis from Equation (7.39).

SOLUTION

From Appendix B, the moments of inertia of the triangular area about the x and y axes are $I_x = \frac{1}{12}bh^3$ and $I_y = \frac{1}{4}hb^3$. Therefore the mass moments of inertia about the x and y axes are

$$I_{(x\text{ axis})} = \frac{m}{A}I_x = \frac{m}{(1/2)bh}\left(\frac{1}{12}bh^3\right) = \frac{1}{6}mh^2$$

$$I_{(y\text{ axis})} = \frac{m}{A}I_y = \frac{m}{(1/2)bh}\left(\frac{1}{4}hb^3\right) = \frac{1}{2}mb^2$$

The moment of inertia about the z axis is

$$I_{(z\text{ axis})} = I_{(x\text{ axis})} + I_{(y\text{ axis})} = m\left(\frac{1}{6}h^2 \frac{1}{2}b^2\right)$$

Parallel-Axis Theorem

This theorem allows us to determine mass moments of inertia of composite objects when we know the mass moments of inertia of its parts. Suppose that we know the mass moment of inertia I about an axis L through the centre of mass of an object, and we wish to determine its mass moment of inertia I_0 about a parallel axis L_0 (Figure 7.34(a)). To determine I_0, we introduce parallel coordinate systems xyz and $x'y'z'$ with the z axis along L_0 and the z' axis along L, as shown in Figure 7.34(b). (In this figure the axes L_0 and L are perpendicular to the page.) The origin O of the xyz coordinate system is contained in the x'-y' plane. The terms d_x and d_y are the coordinates of the centre of mass relative to the xyz coordinate system.

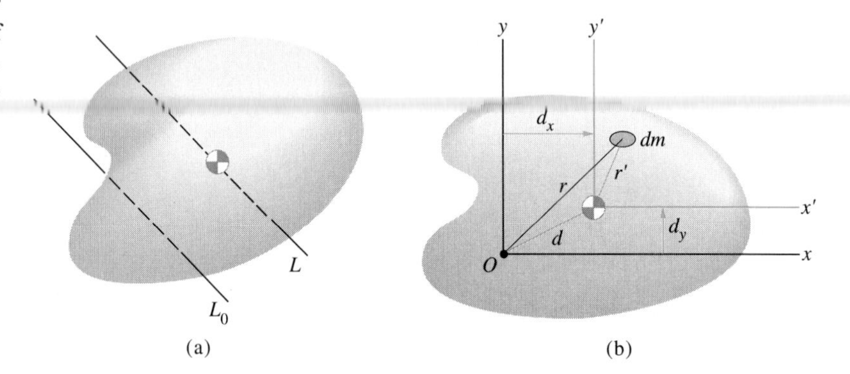

Figure 7.34
(a) An axis L through the centre of mass of an object and a parallel axis L_0.
(b) The xyz and $x'y'z'$ coordinate systems.

The mass moment of inertia of the object about L_0 is

$$I_0 = \int_m r^2 dm = \int_m (x^2 + y^2)\, dm \tag{7.40}$$

where r is the perpendicular distance from L_0 to the differential element of mass dm, and x, y are the coordinates of dm in the x-y plane. The x-y coordinates of dm are related to its x'-y' coordinates by

$$x = x' + d_x \qquad y = y' + d_y$$

By substituting these expressions into Equation (7.40), we can write it as

$$I_0 = \int_m [(x')^2 + (y')^2]\, dm + 2d_x \int_m x'\, dm + 2d_y \int_m y' dm + \int_m (d_x^2 + d_y^2)\, dm$$

Since $(x')^2 + (y')^2 = (r')^2$, where r' is the perpendicular distance from L to dm, the first integral on the right side of this equation is the mass moment of inertia I of the object about L. Recall that the x' and y' coordinates of the centre of mass of the object relative to the $x'y'z'$ coordinate system are defined by

$$\bar{x}' = \frac{\int_m x'\, dm}{\int_m dm} \qquad \bar{y}' = \frac{\int_m y'\, dm}{\int_m dm}$$

Because the centre of mass of the object is at the origin of the $x'y'z'$ system, $\bar{x}' = 0$ and $\bar{y}' = 0$. Therefore the integrals in the second and third terms on the right side of Equation (7.41) are equal to zero. From Figure 7.34(b), we see that $d_x^2 + d_y^2 = d^2$, where d is the perpendicular distance between the axes L and L_0. Therefore we obtain the theorem

$$I_0 = I + d^2 m \tag{7.42}$$

where m is the mass of the object. This is the **parallel-axis theorem**. If you know the mass moment of inertia of an object about a given axis, you can use this theorem to determine its mass moment of inertia about any parallel axis.

In the next two examples we use the parallel-axis theorem to determine mass moments of inertia of composite objects. Determining the mass moment of inertia about a given axis L_0 typically requires three steps:

(1) Choose the parts – *Try to divide the object into parts whose mass moments of inertia you know or can easily determine.*

(2) Determine the mass moments of inertia of the parts – *You must first determine the mass moment of inertia of each part about the axis through its centre of mass parallel to L_0. Then you can use the parallel-axis theorem to determine its mass moment of inertia about L_0.*

(3) Sum the results – *Sum the mass moments of inertia of the parts (or subtract in the case of a hole or cutout) to obtain the mass moment of inertia of the composite object.*

Example 7.13

Two homogeneous, slender bars, each of length l and mass m, are welded together to form an L-shaped object (Figure 7.35). Determine the mass moment of inertia of the object about the axis L_0 through point O. (The axis L_0 is perpendicular to the two bars.)

Figure 7.35

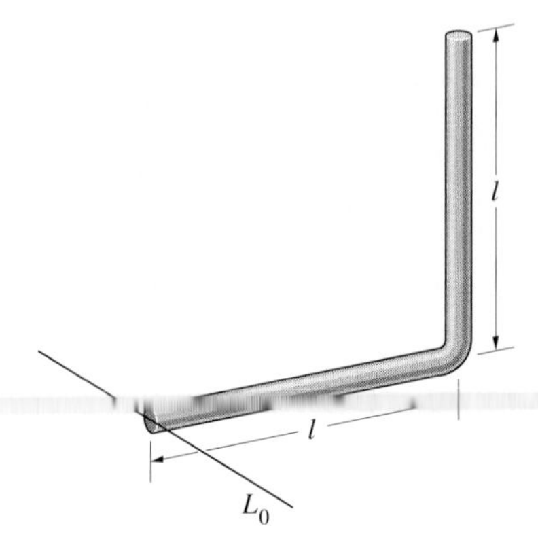

SOLUTION

Choose the Parts The parts are the two bars, which we call bar 1 and bar 2 (Figure (a)).

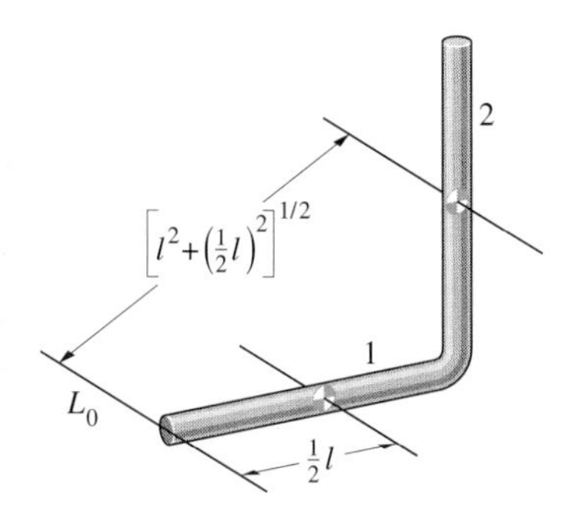

(a) The distances from L_0 to parallel axes through the centres of mass of bars 1 and 2.

Determine the Mass Moments of Inertia of the Parts From Equation (7.35), the mass moment of inertia of each bar about a perpendicular axis through its centre of mass is $I = \frac{1}{12}ml^2$. The distance from L_0 to the parallel axis through the centre of mass of bar 1 is $\frac{1}{2}l$ (Figure (a)). Therefore the mass moment of inertia of bar 1 about L_0 is

$$(I_0)_1 = I + d^2m = \frac{1}{12}ml^2 + \left(\frac{1}{2}l\right)^2 m = \frac{1}{3}ml^2$$

The distance from L_0 to the parallel axis through the centre of mass of bar 2 is $[l^2 + (\frac{1}{2}l)^2]^{1/2}$. The mass moment of inertia of bar 2 about L_0 is

$$(I_0)_2 = I + d^2m = \frac{1}{12}ml^2 + \left[l^2 + \left(\frac{1}{2}l\right)^2\right]m = \frac{4}{3}ml^2$$

Sum the Results The mass moment of inertia of the L-shaped object about L_0 is

$$I_0 = (I_0)_1 + (I_0)_2 = \frac{1}{3}ml^2 + \frac{4}{3}ml^2 = \frac{5}{3}ml^2$$

DISCUSSION

Compare this solution with Example 7.11, in which we used integration to determine the mass moment of inertia of this object about L_0. We obtained the result much more easily with the parallel-axis theorem, but of course we needed to know the mass moments of inertia of the bars about the axes through their centres of mass.

Example 7.14

The object in Figure 7.36 consists of a slender, 3 kg bar welded to a thin, circular, 2 kg disc. Determine its mass moment of inertia about the axis L through its centre of mass. (The axis L is perpendicular to the bar and disc.)

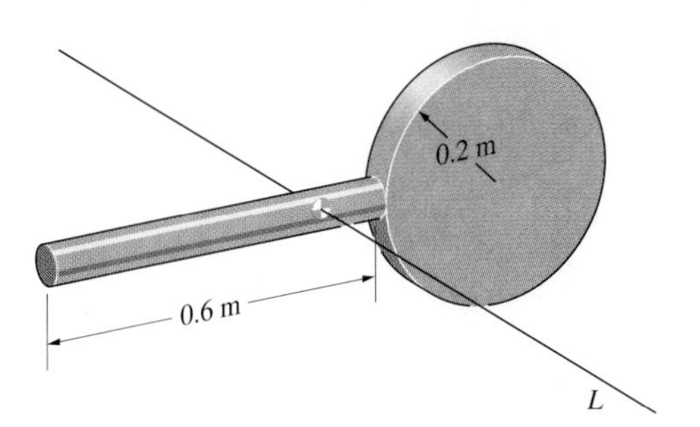

Figure 7.36

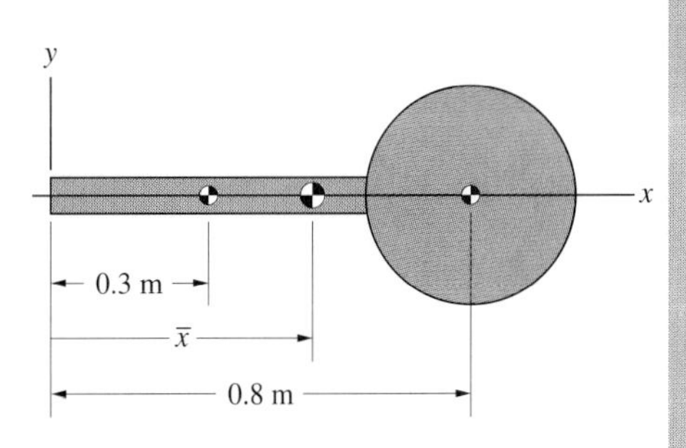

(a) The coordinate $\bar{x}$ of the centre of mass of the object.

STRATEGY

We must locate the centre of mass of the composite object, then apply the parallel-axis theorem. We can obtain the mass moments of inertia of the bar and disc from Appendix C.

SOLUTION

Choose the Parts The parts are the bar and the disc. Introducing the coordinate system in Figure (a), the x coordinate of the centre of mass of the composite object is

$$\bar{x} = \frac{\bar{x}_{(\text{bar})}m_{(\text{bar})} + \bar{x}_{(\text{disc})}m_{(\text{disc})}}{m_{(\text{bar})} + m_{(\text{disc})}} = \frac{(0.3)(3) + (0.6 + 0.2)(2)}{3 + 2} = 0.5\,\text{m}$$

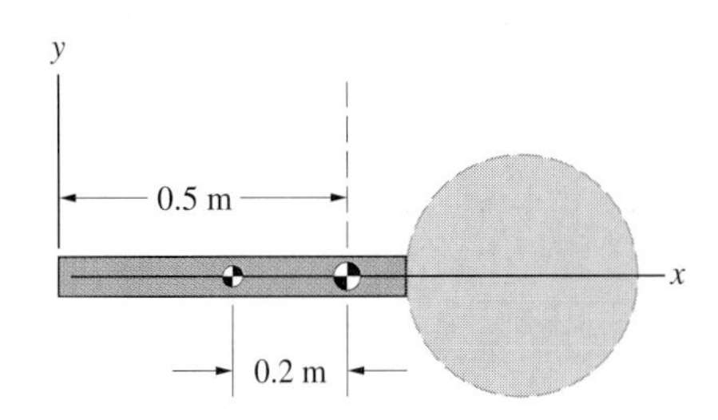

(b) Distance from L to the centre of mass of the bar.

Determine the Mass Moments of Inertia of the Parts The distance from the centre of mass of the bar to the centre of mass of the composite object is 0.2 m (Figure (b)). Therefore the mass moment of inertia of the bar about L is

$$I_{(\text{bar})} = \frac{1}{12}(3)(0.6)^2 + (0.2)^2(3) = 0.210\,\text{kg.m}^2$$

The distance from the centre of mass of the disc to the centre of mass of the composite object is 0.3 m (Figure (c)). The mass moment of inertia of the disc about L is

$$I_{(\text{disc})} = \frac{1}{2}(2)(0.2)^2 + (0.3)^2(2) = 0.220\ \text{kg.m}^2$$

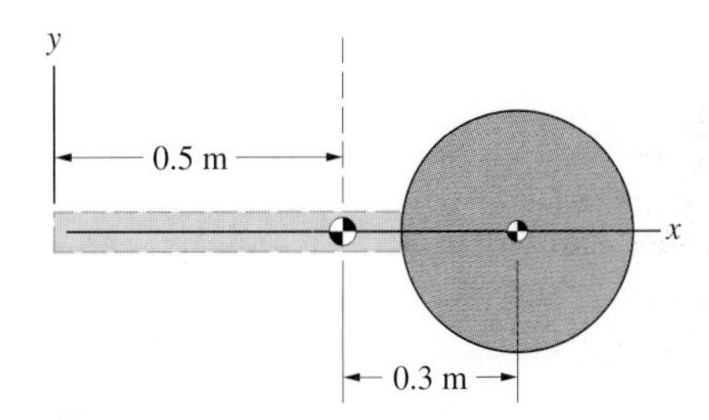

(c) Distance from L to the centre of mass of the disc.

Sum the Results The mass moment of inertia of the composite object about L is

$$I = I_{(\text{bar})} + I_{(\text{disc})} = 0.430\ \text{kg.m}^2$$

Example 7.15

The homogeneous cylinder in Figure 7.37 has mass m, length l, and radius R. Determine its mass moments of inertia about the x, y and z axes.

Figure 7.37

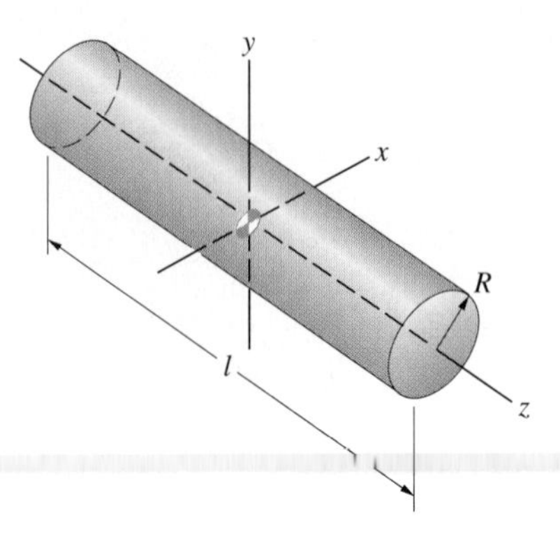

STRATEGY

We can determine the mass moments of inertia of the cylinder by an interesting application of the parallel-axis theorem. We first use it to determine the mass moments of inertia about the x, y and z axes of an infinitesimal element of the cylinder consisting of a disc of thickness dz. Then we integrate the results with respect to z to obtain the moments of inertia of the cylinder.

SOLUTION

Consider an element of the cylinder of thickness dz at a distance z from the centre of the cylinder (Figure (a)). (You can imagine obtaining this element by 'slicing' the cylinder perpendicular to its axis.) The mass of the element is equal to the product of the mass density and the volume of the element, $dm = \rho(\pi R^2 dz)$. We obtain the mass moments of inertia of the element by using the values for a thin circular plate given in Appendix C. The mass moment of inertia about the z axis is

$$dI_{(z\text{ axis})} = \frac{1}{2} dm\, R^2 = \frac{1}{2}(\rho\pi R^2\, dz)R^2$$

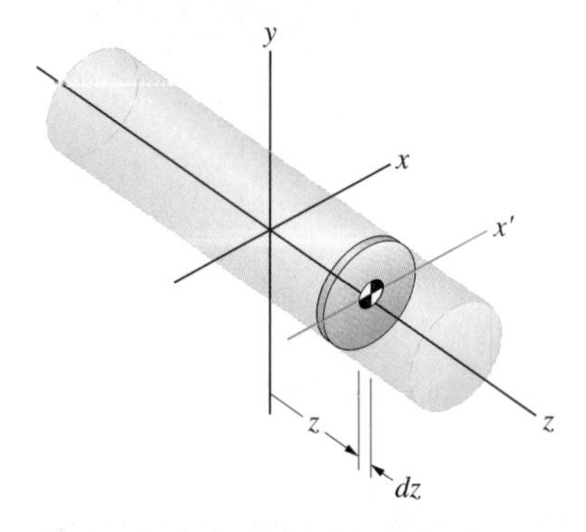

(a) A differential element of the cylinder in the form of a disc.

We integrate this result with respect to z from $-l/2$ to $l/2$, thereby summing the mass moments of inertia of the infinitesimal disc elements that make up the cylinder. The result is the moment of inertia of the cylinder about the z axis:

$$I_{(z\text{ axis})} = \int_{-l/2}^{l/2} \frac{1}{2}\rho\pi R^4 dz = \frac{1}{2}\rho\pi R^4 l$$

We can write this result in terms of the mass of the cylinder, $m = \rho(\pi R^2 l)$, as

$$I_{(z\text{ axis})} = \frac{1}{2}mR^2$$

The mass moment of inertia of the disc element about the x' axis is

$$dI_{(x'\text{ axis})} = \frac{1}{4}dmR^2 = \frac{1}{4}(\rho\pi R^2 dz)R^2$$

We use this result and the parallel-axis theorem to determine the mass moment of inertia of the element about the x axis:

$$dI_{(x\text{ axis})} = dI_{(x'\text{ axis})} + z^2 dm = \frac{1}{4}(\rho\pi R^2 dz)R^2 + z^2(\rho\pi R^2 dz)$$

Integrating this expression with respect to z from $-l/2$ to $l/2$, we obtain the mass moment of inertia of the cylinder about the x axis:

$$I_{(x\text{ axis})} = \int_{-l/2}^{l/2}\left(\frac{1}{4}\rho\pi R^4 + \rho\pi R^2 z^2\right)dz = \frac{1}{4}\rho\pi R^4 l + \frac{1}{12}\rho\pi R^2 l^3$$

In terms of the mass of the cylinder,

$$I_{(x\text{ axis})} = \frac{1}{4}mR^2 + \frac{1}{12}ml^2$$

Due to the symmetry of the cylinder,

$$I_{(y\text{ axis})} = I_{(x\text{ axis})}$$

DISCUSSION

When the cylinder is very long in comparison to its width, $l \gg R$, the first term in the equation for $I_{(x\text{ axis})}$ can be neglected and we obtain the mass moment of inertia of a slender bar about a perpendicular axis, Equation (7.35). On the other hand, when the radius of the cylinder is much greater than its length, $R \gg l$, the second term in the equation for $I_{(x\text{ axis})}$ can be neglected and we obtain the moment of inertia for a thin circular disc about an axis parallel to the disc. This indicates the sizes of the terms you neglect when you use the approximate expressions for the moments of inertia of a 'slender' bar and a 'thin' disc.

Problems

7.70 The homogeneous, slender bar has mass m and length l. Use integration to determine its mass moment of inertia about the perpendicular axis L_0.

Strategy: Use the same approach we used to obtain Equation (7.35). You need only to change the limits of integration.

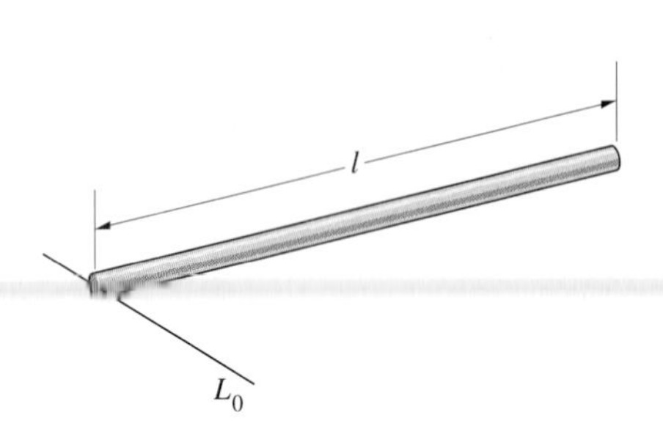

P7.70

7.71 Two homogeneous, slender bars, each of mass m and length l, are welded together to form the T-shaped object. Use integration to determine the mass moment of inertia of the object about the axis through point O that is perpendicular to the bars.

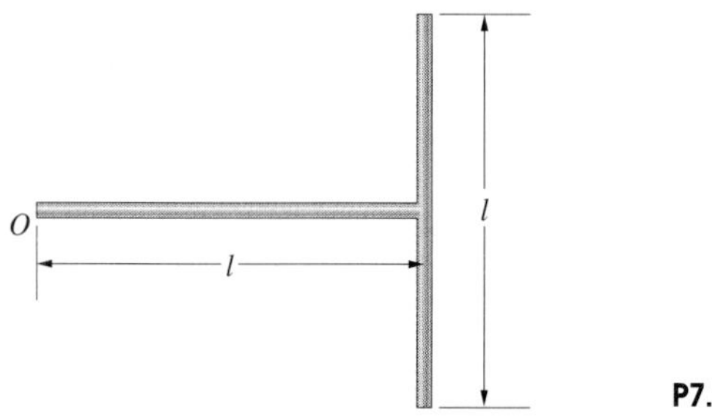

P7.71

7.72 The homogeneous, slender bar has mass m and length l. Use integration to determine the mass moment of inertia of the bar about the axis L.

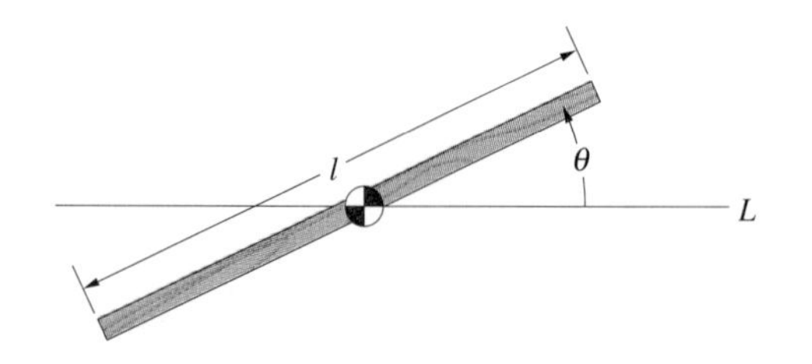

P7.72

7.73 A homogeneous, slender bar is bent into a circular ring of mass m and radius R. Determine the mass moment of inertia of the ring (a) about the axis through its centre of mass that is perpendicular to the ring; (b) about the axis L.

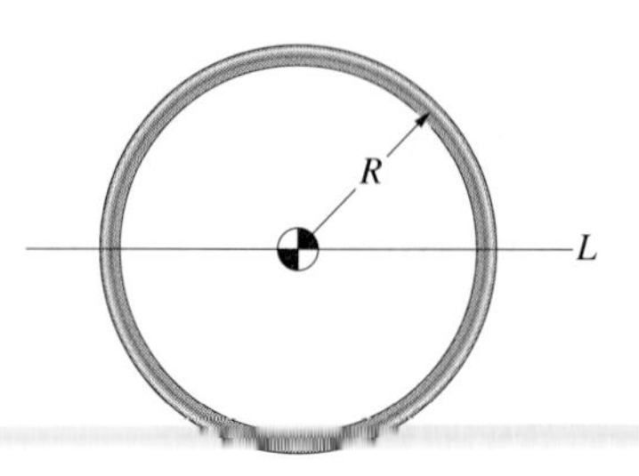

P7.73

7.74 The homogeneous, thin plate is of uniform thickness and mass m. Determine its mass moment of inertia about the x, y and z axes.

Strategy: The mass moments of inertia of a thin plate of arbitrary shape are given by Equations (7.37)–(7.39) in terms of the moments of inertia of the cross-sectional area of the plate. You can obtain the moments of inertia of the rectangular area from Appendix B.

y

h

x

b

P7.74

7.75 The brass washer is of uniform thickness and mass m.
(a) Determine its mass moments of inertia about the x and z axes.
(b) Let $R_i = 0$ and compare your results with the values given in Appendix C for a thin circular plate.
(c) Let $R_i \to R_o$, and compare your results with the solutions of Problem 7.73.

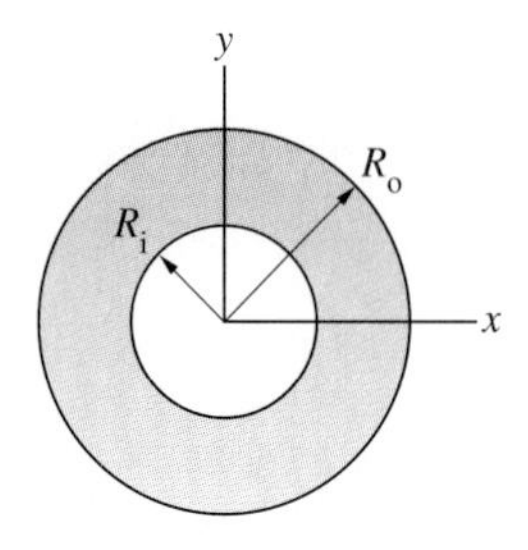

P7.75

7.76 The homogeneous, thin plate is of uniform thickness and weighs 100 N. Determine its mass moment of inertia about the y axis.

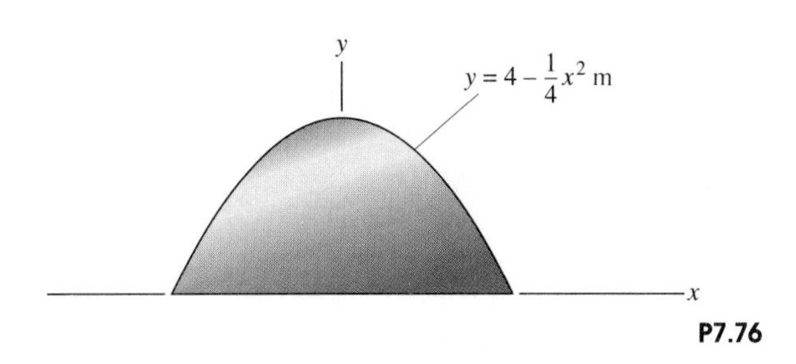

P7.76

7.77 Determine the mass moment of inertia of the plate in Problem 7.76 about the x axis.

7.78 The mass of the object is 10 kg. Its mass moment of inertia about L_1 is 10 kg.m^2. What is its mass moment of inertia about L_2? (The three axes lie in the same plane.)

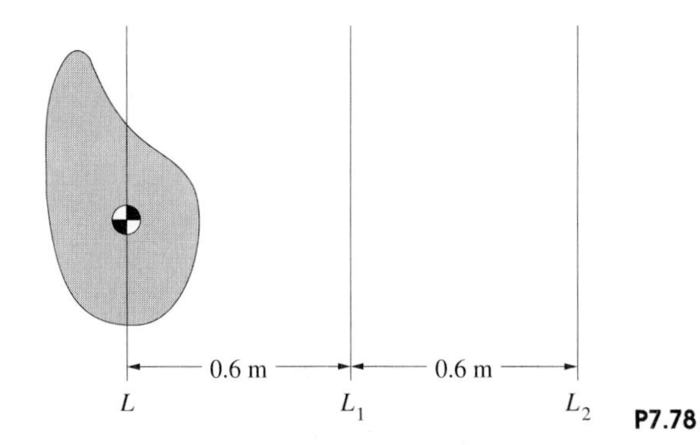

P7.78

7.79 An engineer gathering data for the design of a manoeuvring unit determines that the astronaut's centre of mass is at $x = 1.01$ m, $y = 0.16$ m and that his mass moment of inertia about the z axis is 105.6 kg.m^2. His mass is 81.6 kg. What is his mass moment of inertia about the z' axis through his centre of mass?

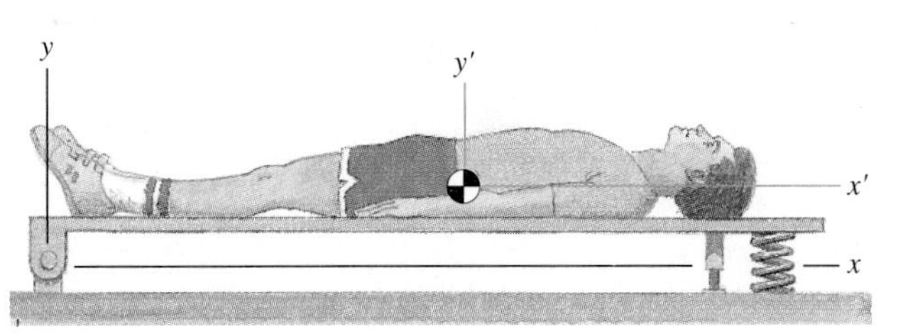

P7.79

7.80 Two homogeneous, slender bars, each of mass m and length l, are welded together to form the T-shaped object. Use the parallel-axis theorem to determine the mass moment of inertia of the object about the axis through point O that is perpendicular to the bars.

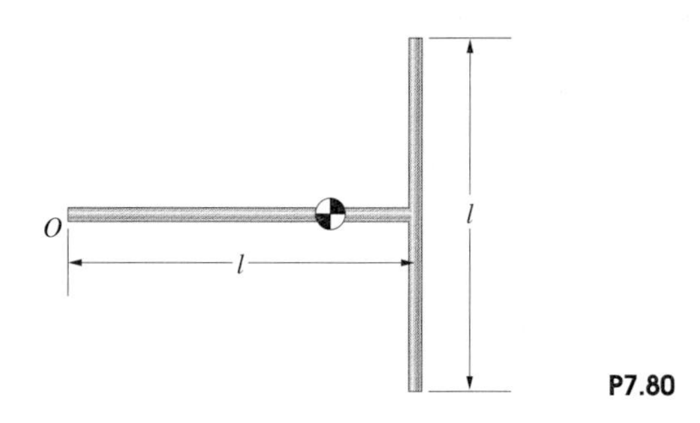

P7.80

7.81 Use the parallel-axis theorem to determine the mass moment of inertia of the T-shaped object in Problem 7.80 about the axis through the centre of mass of the object that is perpendicular to the two bars.

7.82 The mass of the homogeneous, slender bar is 20 kg. Determine its mass moment of inertia about the z axis.

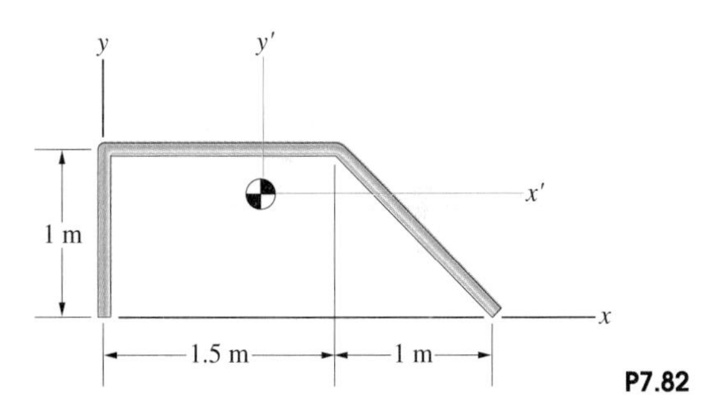

P7.82

7.83 Determine the mass moment of inertia of the bar in Problem 7.82 about the z' axis through its centre of mass.

7.84 The rocket is used for atmospheric research. Its weight and its mass moment of inertia about the z axis through its centre of mass (including its fuel) are 45 kN and 14 000 kg.m^2, respectively. The rocket's fuel weighs 27 kN, its centre of mass is located at $x = -1$ m, $y = 0$, $z = 0$, and the mass moment of inertia of the fuel about the axis through the fuel's centre of mass parallel to the z axis is 3000 kg.m^2. When the fuel is exhausted, what is the rocket's mass moment of inertia about the axis through its new centre of mass parallel to the z axis?

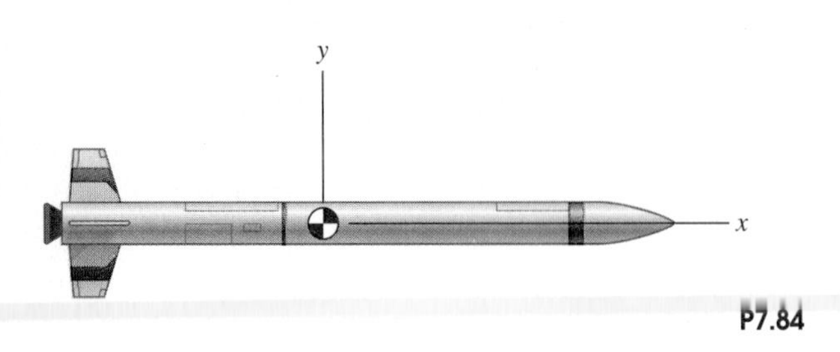

P7.84

7.85 The mass of the homogeneous, thin plate is 36 kg. Determine its mass moment of inertia about the x axis.

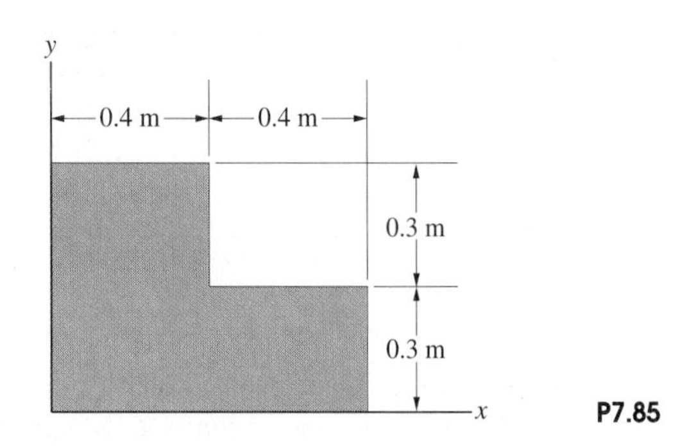

P7.85

7.86 Determine the mass moment of inertia of the plate in Problem 7.85 about the z axis.

7.87 The homogeneous, thin plate weighs 50 N. Determine its mass moment of inertia about the x axis.

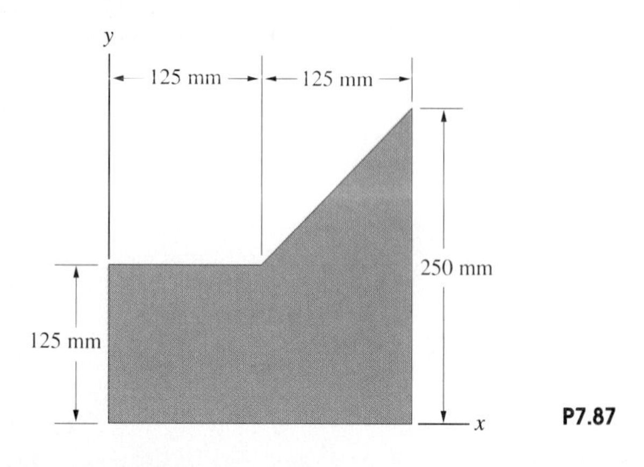

P7.87

7.88 Determine the mass moment of inertia of the plate in Problem 7.87 about the y axis.

7.89 The thermal radiator (used to eliminate excess heat from a satellite) can be modelled as a homogeneous, thin, rectangular plate. Its mass is 75 kg. Determine its mass moments of inertia about the x, y and z axes.

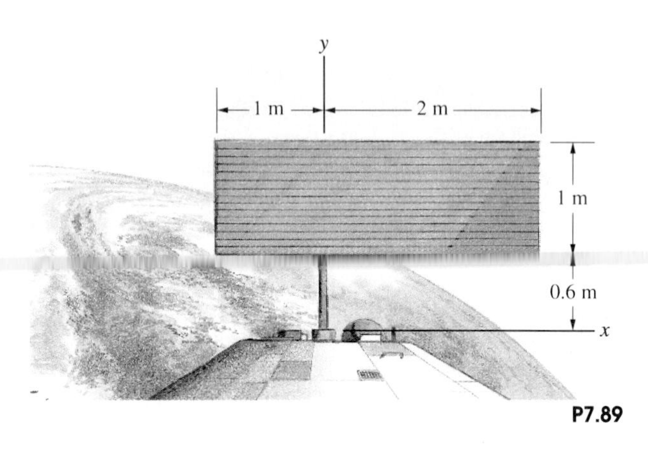

P7.89

7.90 The mass of the homogeneous, thin plate is 2 kg. Determine its mass moment of inertia about the axis through point O that is perpendicular to the plate.

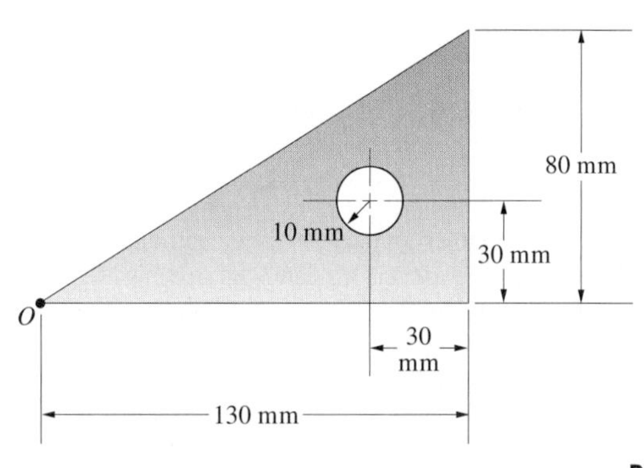

P7.90

7.91 The homogeneous cone is of mass m. Determine its mass moment of inertia about the z axis and compare your result with the value given in Appendix C.

Strategy: Use the same approach we used in Example 7.15 to obtain the moments of inertia of a homogeneous cylinder.

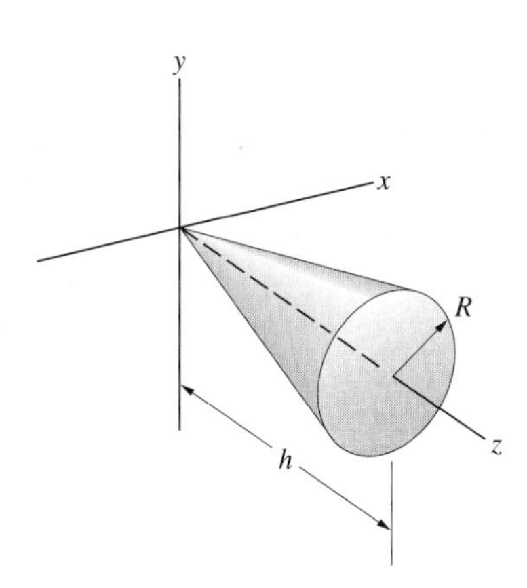

P7.91

7.92 Determine the mass moments of inertia of the homogeneous cone in Problem 7.91 about the x and y axes and compare your results with the values given in Appendix C.

7.93 The homogeneous pyramid is of mass m. Determine its mass moment of inertia about the z axis.

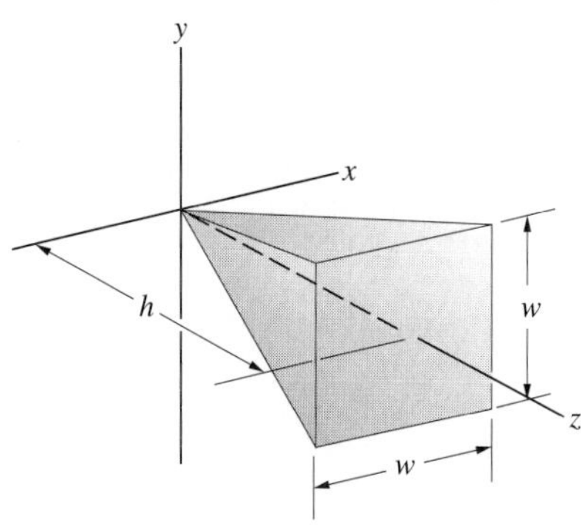

P7.93

7.94 Determine the mass moments of inertia of the homogeneous pyramid in Problem 7.93 about the x and y axes.

7.95 The homogeneous, rectangular parallelepiped is of mass m. Determine its mass moments of inertia about the x, y and z axes and compare your results with the values given in Appendix C.

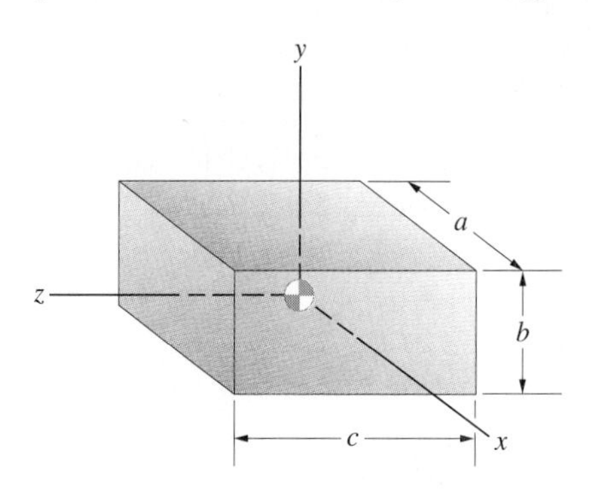

P7.95

7.96 The homogeneous ring consists of steel of density $\rho = 7800\,\text{kg/m}^3$. Determine its mass moment of inertia about the axis L through its centre of mass.

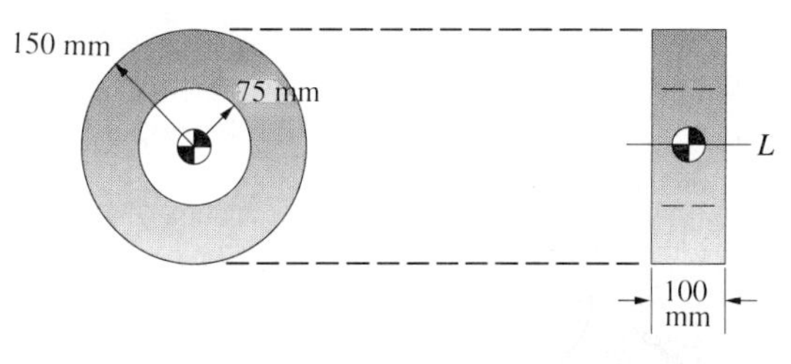

P7.96

7.97 The homogeneous half-cylinder is of mass m. Determine its mass moment of inertia about the axis L through its centre of mass.

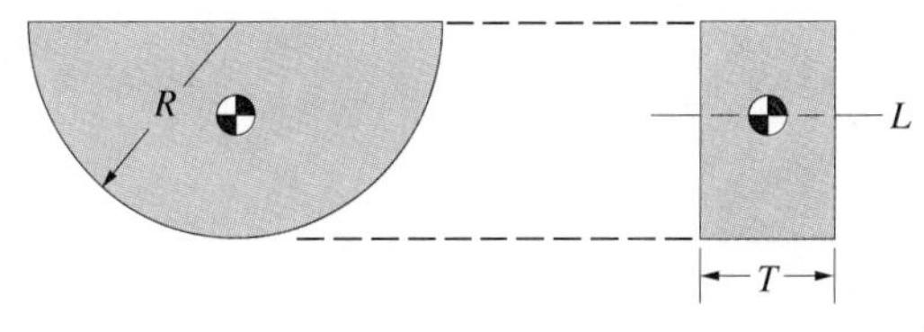

P7.97

7.98 The object shown consists of steel of density $\rho = 7800$ kg/m^3. Determine its mass moment of inertia about the axis L_O through point O.

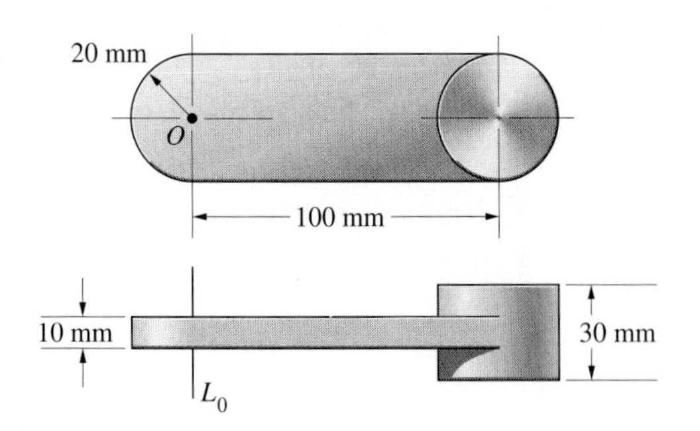

P7.98

7.99 Determine the mass moment of inertia of the object in Problem 7.98 about the axis through the centre of mass of the object parallel to L_O.

7.100 The thick plate consists of steel of density $\rho = 7800$ kg/m^3. Determine its mass moment of inertia about the z axis.

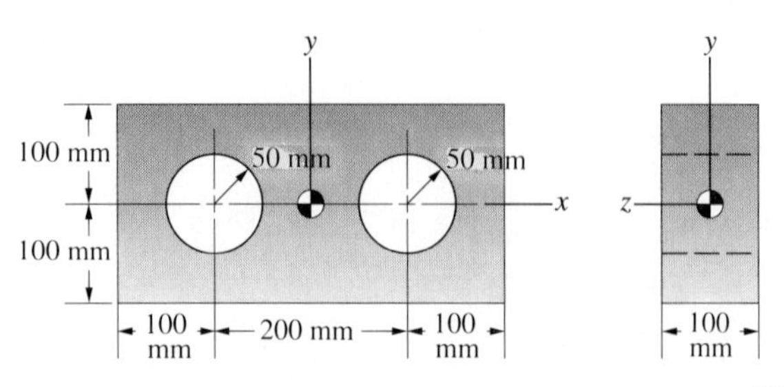

P7.100

7.101 Determine the mass moment of inertia of the object in Problem 7.100 about the x axis.

Chapter Summary

Moment-Angular Momentum Relations

Let $\mathbf{r}_i$ be the position of the ith particle of a system of particles, and let $\mathbf{R}_i$ be its position relative to the centre of mass. The angular momentum of the system about a point O is the sum of the angular momenta of the particles,

$$\mathbf{H}_O = \sum_i \mathbf{r}_i \times m_i \mathbf{v}_i \qquad \textbf{Equation (7.7)}$$

where $\mathbf{v}_i = d\mathbf{r}_i/dt$, and the angular momentum about the centre of mass is

$$\mathbf{H} = \sum_i \mathbf{R}_i \times m_i \mathbf{v}_i \qquad \textbf{Equation (7.8)}$$

These angular momenta are related by

$$\mathbf{H}_O = \mathbf{r} \times m\mathbf{v} + \mathbf{H} \qquad \textbf{Equation (7.9)}$$

where $\mathbf{v} = d\mathbf{r}/dt$ is the velocity of the centre of mass.

The total moment about a fixed point O equals the rate of change of the angular momentum about O:

$$\Sigma \mathbf{M}_O = \frac{d\mathbf{H}_O}{dt} \qquad \textbf{Equation (7.11)}$$

This result can also be expressed in terms of the angular momentum about the centre of mass:

$$\Sigma \mathbf{M}_O = \frac{d}{dt}(\mathbf{r} \times m\mathbf{v} + \mathbf{H}) = \mathbf{r} \times m\mathbf{a} + \frac{d\mathbf{H}}{dt} \qquad \textbf{Equation (7.12)}$$

where **a** is the acceleration of the centre of mass.

The total moment about the centre of mass equals the rate of change of the angular momentum about the centre of mass:

$$\Sigma \mathbf{M} = \frac{d\mathbf{H}}{dt} \qquad \textbf{Equation (7.13)}$$

Equations of Planar Motion

The equations of motion for a rigid body in planar motion include Newton's second law,

$$\Sigma \mathbf{F} = m\,\mathbf{a} \qquad \textbf{Equation (7.21)}$$

where **a** is the acceleration of the centre of mass. If the rigid body rotates about a fixed axis O, the total moment about O equals the product of the moment of inertia about O and the angular acceleration:

$$\Sigma M_0 = I_0\alpha \qquad \textbf{Equation (7.22)}$$

In any planar motion, the total moment about the centre of mass equals the product of the moment of inertia about the centre of mass and the angular acceleration:

$$\Sigma M = I\alpha \qquad \textbf{Equation (7.23)}$$

If a rigid body is in translation, Newton's second law is sufficient to determine its motion. Nevertheless, the angular equation of motion may be needed to determine unknown forces or couples. Since $\alpha = 0$, the total moment *about the centre of mass* equals zero. In the case of rotation about a fixed axis, Equation (7.22) is sufficient to determine the rotational motion, although Newton's second law may be needed to determine unknown forces or couples. If a rigid body undergoes general planar motion, both Newton's second law and the equation of angular motion are needed.

D'Alembert's Principle

By writing Newton's second law as

$$\Sigma \mathbf{F} + (-m\,\mathbf{a}) = 0 \qquad \textbf{Equation (7.27)}$$

it can be regarded as an 'equilibrium' equation stating that the sum of the external forces, including an **inertial force** $-m\mathbf{a}$, equals zero. The equation of angular motion can be written as

$$\Sigma M_0 + [\mathbf{r} \times (-m\,\mathbf{a})] \cdot \mathbf{k} + (-I\alpha) = 0 \qquad \textbf{Equation (7.28)}$$

stating that the sum of the moments about *any point* due to external forces and couples, including the moment due to the inertial force $-m\mathbf{a}$ acting at the centre of mass and an **inertial couple** $-I\alpha$, equals zero. Stated in this way, the equations of motion of a rigid body are analogous to the equations for static equilibrium.

Moments of Inertia

The mass moment of inertia of an object about an axis L_0 is

$$I_0 = \int_m r^2 dm \qquad \textbf{Equation (7.34)}$$

where r is the perpendicular distance from L_0 to the differential element of mass dm.

Let L be an axis through the centre of mass of an object, and let L_0 be a parallel axis. The mass moment of inertia I_0 about L_0 is given in terms of the mass moment of inertia I about L by the **parallel-axis theorem**

$$I_0 = I + d^2 m \qquad \textbf{Equation (7.42)}$$

where m is the mass of the object and d is the distance between L and L_0.

Review Problems

7.102 The aeroplane is at the beginning of its takeoff run. Its weight is 4.5 kN, and the initial thrust T exerted by its engine is 1.39 kN. Assume that the thrust is horizontal, and neglect the tangential forces exerted on its wheels.
(a) If the acceleration of the aeroplane remains constant, how long will it take to reach its takeoff speed of 130 km/hr?
(b) Determine the normal force exerted on the forward landing gear at the beginning of the takeoff run.

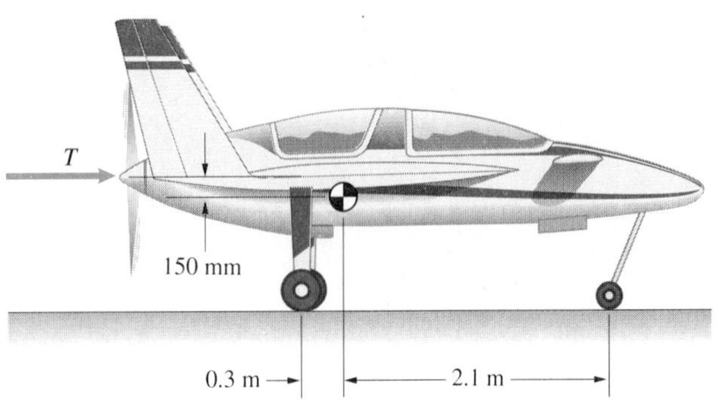

P7.102

7.103 The pulleys can turn freely on their pin supports. Their mass moments of inertia are $I_A = 0.002$ kg.m^2, $I_B = 0.036$ kg.m^2 and $I_C = 0.032$ kg.m^2. They are initially stationary, and at $t = 0$ a constant couple $M = 2$ N.m is applied to pulley A. What is the angular velocity of pulley C and how many revolutions has it turned at $t = 2$ s?

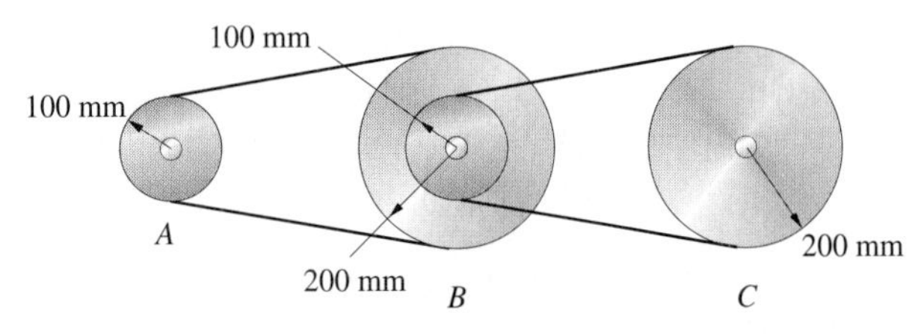

P7.103

7.104 A 2 kg box is subjected to a 40 N horizontal force. Neglect friction.
(a) If the box remains on the floor, what is its acceleration?
(b) Determine the range of values of c for which the box will remain on the floor when the force is applied.

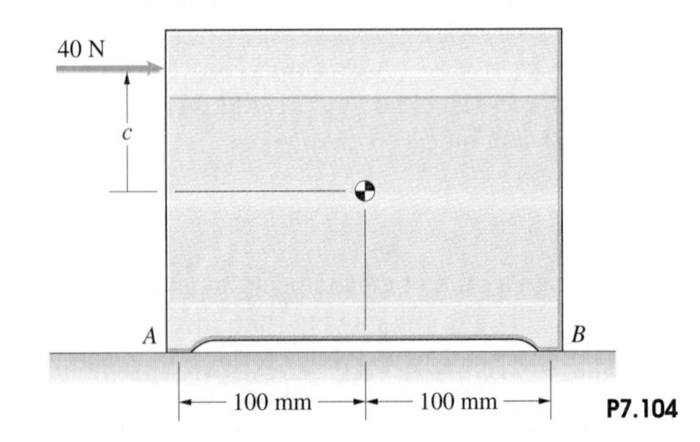

P7.104

7.105 The slender, 30 kg bar AB is 1 m long. It is pinned to the cart at A and leans against it at B.
(a) If the acceleration of the cart is $a = 6\,\text{m/s}^2$, what normal force is exerted on the bar by the cart at B?
(b) What is the largest acceleration a for which the bar will remain in contact with the surface at B?

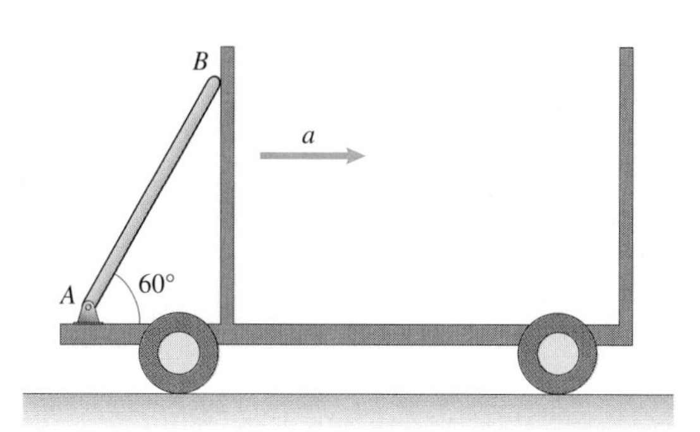

P7.105

7.106 To determine a 4.5 kg tyre's mass moment of inertia, an engineer lets it roll down an inclined surface. If it takes 3.5 s to start from rest and roll 3 m down the surface, what is the tyre's mass moment of inertia about its centre of mass?

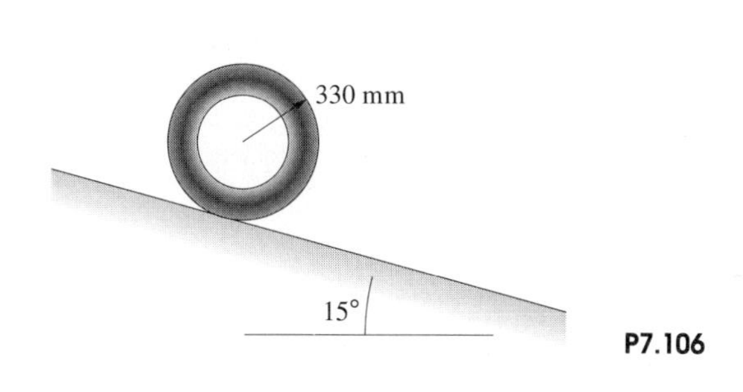

P7.106

7.107 The mass moment of inertia of the disc is $0.2\,\text{kg.m}^2$. What is the smallest coefficient of static friction between the rope and the disc for which the rope will not slip when the system is released from rest?

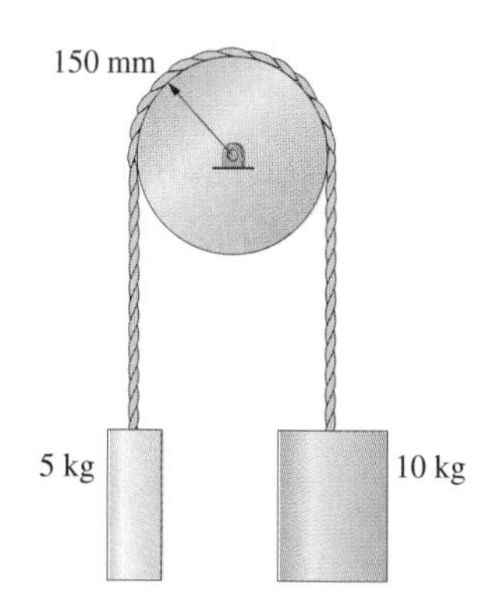

P7.107

7.108 Model the excavator's arm ABC as a single rigid body. Its mass is 1200 kg, and the mass moment of inertia about its centre of mass is $I = 3600\,\text{kg.m}^2$. If point A is stationary and the angular acceleration of the arm is $1.0\,\text{rad/s}^2$ counterclockwise, what force does the vertical hydraulic cylinder exert on the arm at B?

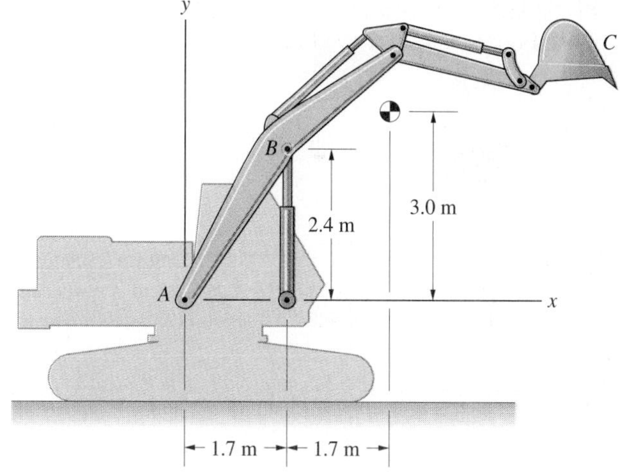

P7.108

7.109 In Problem 7.108, if the angular acceleration of arm ABC is $1.0\,\text{rad/s}^2$ counterclockwise and its angular velocity is $2.0\,\text{rad/s}$ counterclockwise, what are the components of the force exerted on the arm at A?

7.110 To decrease the angle of elevation of the stationary 200 kg ladder, the gears that raised it are disengaged and a fraction of a second later a second set of gears that lower it are engaged. At the instant the gears that raised it are disengaged, what is the ladder's angular acceleration and what are the components of force exerted on the ladder by its support at *O*? The mass moment of inertia of the ladder about *O* is $I_O = 14\,000\ \text{kg.m}^2$, and the coordinates of its centre of mass at the instant the gears are disengaged are $\bar{x} = 3$ m, $\bar{y} = 4$ m.

P7.110

7.111 The slender bars each weigh 20 N and are 250 mm long. The homogeneous plate weighs 50 N. If the system is released from rest in the position shown, what is the angular acceleration of the bars at that instant?

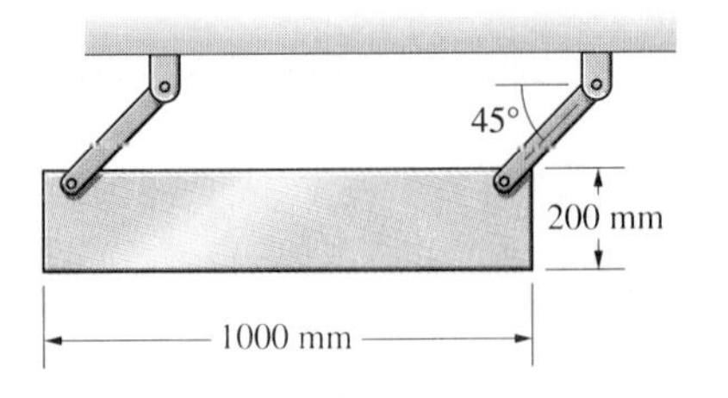

P7.111

7.112 A slender bar of mass m is released from rest in the position shown. The static and kinetic coefficients of friction at the floor and wall have the same value μ. If the bar slips, what is its angular acceleration at the instant of release?

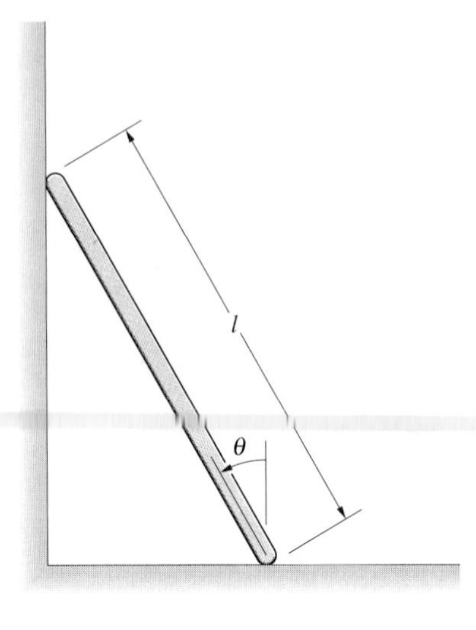

P7.112

7.113 Each of the go-cart's front wheels weighs 25 N and has a mass moment of inertia of $0.014\ \text{kg.m}^2$. The two rear wheels and rear axle form a single rigid body weighing 50 N and having a mass moment of inertia of $0.1\ \text{kg.m}^2$. The total weight of the go-cart and rider is 1200 N. (The location of the centre of mass of the go-cart and driver *not including* the front wheels or the rear wheels and rear axle is shown.) If the engine exerts a torque of 16 N.m on the rear axle, what is the go-cart's acceleration?

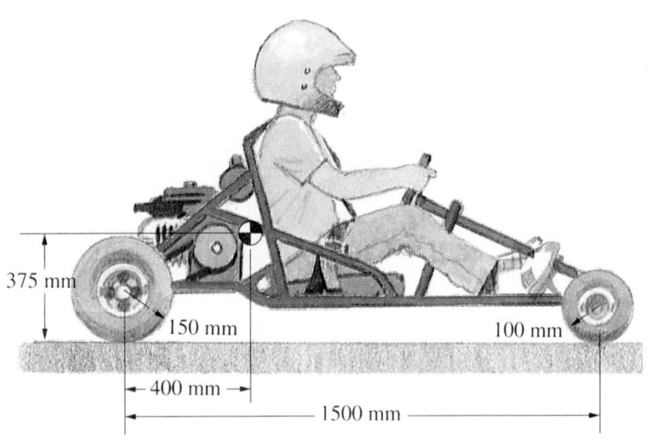

P7.113

7.114 Bar AB rotates with a constant angular velocity of 10 rad/s in the counterclockwise direction. The masses of the slender bars BC and CDE are 2 kg and 3.6 kg, respectively. The y axis points upwards. Determine the components of the forces exerted on bar BC by the pins at B and C at the instant shown.

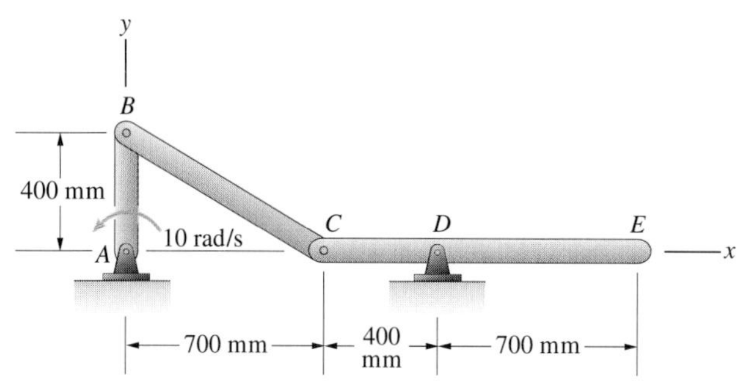

P7.114

7.115 At the instant shown, the arms of the robotic manipulator have constant counterclockwise angular velocities $\omega_{AB} = -0.5$ rad/s, $\omega_{BC} = 2$ rad/s and $\omega_{CD} = 4$ rad/s. The mass of arm CD is 10 kg, and its centre of mass is at its midpoint. At this instant, what force and couple are exerted on arm CD at C?

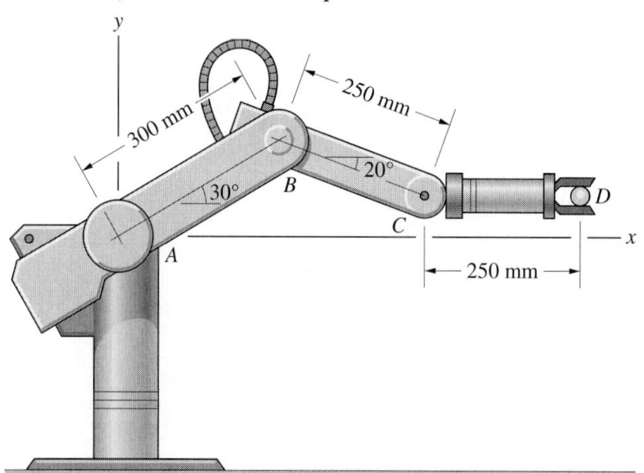

P7.115

7.116 Each bar is 1 m in length and has a mass of 4 kg. The inclined surface is smooth. If the system is released from rest in the position shown, what are the angular accelerations of the bars at that instant?

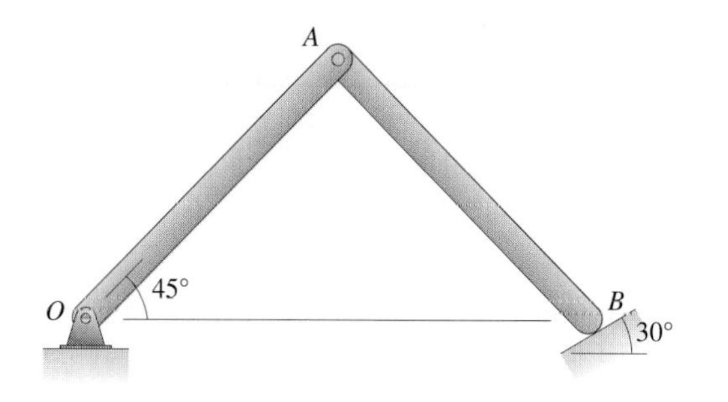

P7.116

7.117 At the instant the system in Problem 7.116 is released, what is the magnitude of the force exerted on bar OA by the support at O?

7.118 The fixed ring gear lies *in the horizontal plane*. The hub and planet gears are bonded together. The mass and mass moment of inertia of the combined hub and planet gears are $m_{HP} = 130$ kg and $I_{HP} = 130$ kg.m^2. The mass moment of inertia of the sun gear is $I_S = 60$ kg.m^2. The mass of the connecting rod is 5 kg, and it can be modelled as a slender bar. If a 1 kN.m counterclockwise couple is applied to the sun gear, what is the resulting angular acceleration of the bonded hub and planet gears?

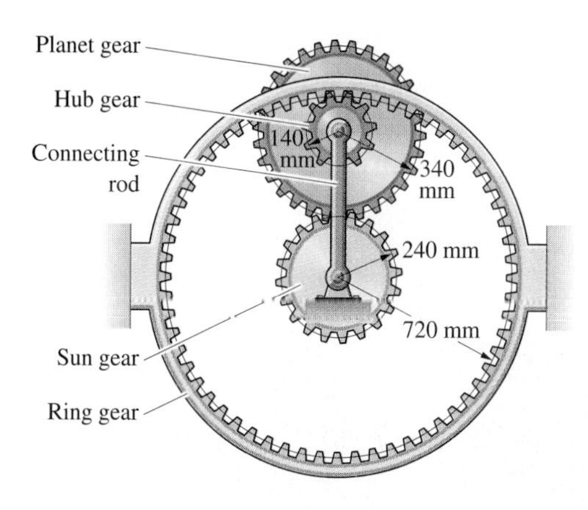

P7.118

7.119 The system is stationary at the instant shown. The net force exerted on the piston by the exploding fuel–air mixture and friction is 5 kN to the left. A clockwise couple $M = 200$ N.m acts on the crank AB. The mass moment of inertia of the crank about A is 0.0003 kg.m^2. The mass of the connecting rod BC is 0.36 kg, and its centre of mass is 40 mm from B on the line from B to C. The connecting rod's mass moment of inertia about its centre of mass is 0.0004 kg.m^2. The mass of the piston is 4.6 kg. What is the piston's acceleration at this instant? (Neglect the gravitational forces on the crank and connecting rod.)

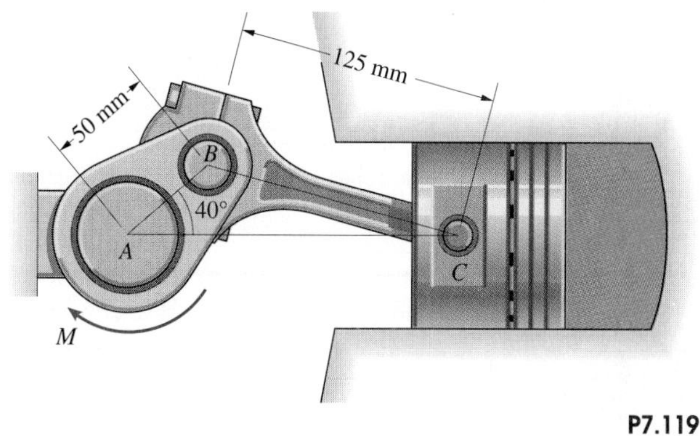

P7.119

7.120 If the crank AB in Problem 7.119 has a counterclockwise angular velocity of 2000 rpm (revolutions per minute) at the instant shown, what is the piston's acceleration?

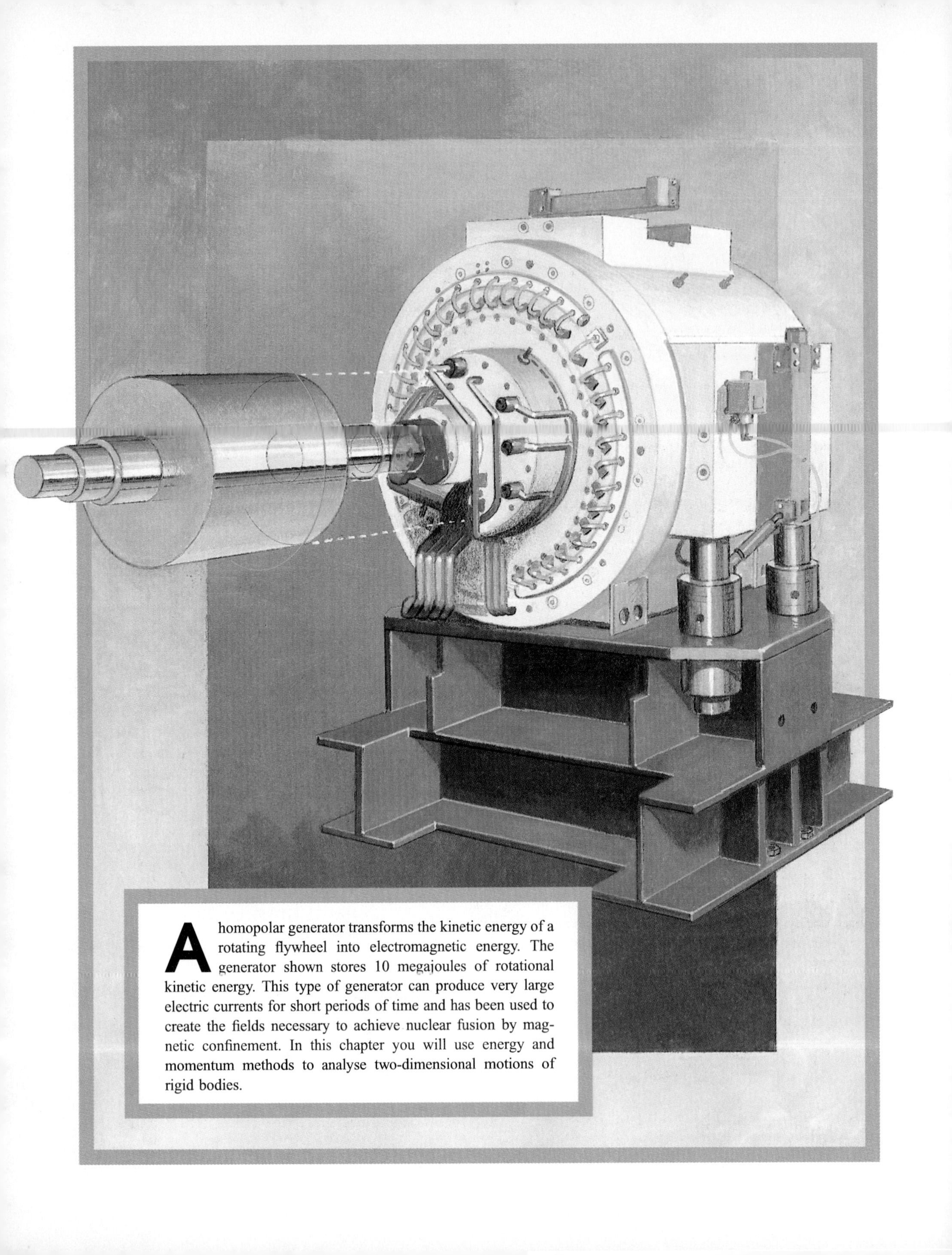

A homopolar generator transforms the kinetic energy of a rotating flywheel into electromagnetic energy. The generator shown stores 10 megajoules of rotational kinetic energy. This type of generator can produce very large electric currents for short periods of time and has been used to create the fields necessary to achieve nuclear fusion by magnetic confinement. In this chapter you will use energy and momentum methods to analyse two-dimensional motions of rigid bodies.

Chapter 8

Energy and Momentum in Planar Rigid-Body Dynamics

YOU have seen in Chapters 4 and 5 that energy and momentum methods are very useful for particular types of problems in dynamics. If the forces on an object are known functions of position, you can use the principle of work and energy to relate the change in the magnitude of the object's velocity to the change in its position. If the forces are known functions of time, you can use the principle of impulse and momentum to determine the change in the object's velocity during an interval of time. In this chapter we extend these methods to situations in which you must consider both the translational and rotational motions of objects.

8.1 Principle of Work and Energy

The principle of work and energy for a rigid body in planar motion is a simple statement and involves simple equations, although its derivation is rather involved. To help you follow our derivation, we begin by summarizing the principle. Let T be the kinetic energy of a rigid body. The principle of work and energy states that the work U done by external forces and couples as the rigid body moves between two positions 1 and 2 equals the change in its kinetic energy:

$$U = T_2 - T_1$$

In general planar motion, the kinetic energy is

$$T = \frac{1}{2}mv^2 + \frac{1}{2}I\omega^2$$

where v is the magnitude of the velocity of the centre of mass and I is the mass moment of inertia about the centre of mass. In the case of rotation about a fixed axis O, the kinetic energy can also be expressed as

$$T = \frac{1}{2}I_O\omega^2$$

To derive these results, we adopt the same approach used in Chapter 7 to derive the equations of motion for a rigid body. We obtain the principle of work and energy for a system of particles and use it to obtain the principle for a rigid body.

System of Particles

Let m_i be the mass of the ith particle of a system of N particles, and let $\mathbf{r}_i$ be its position relative to a fixed reference point O (Figure 8.1). We denote the sum of the kinetic energies of the particles by T,

$$T = \sum_i \frac{1}{2}m_i\mathbf{v}_i \cdot \mathbf{v}_i \tag{8.1}$$

where $\mathbf{v}_i = d\mathbf{r}_i/dt$ is the velocity of the ith particle. Our objective is to relate the

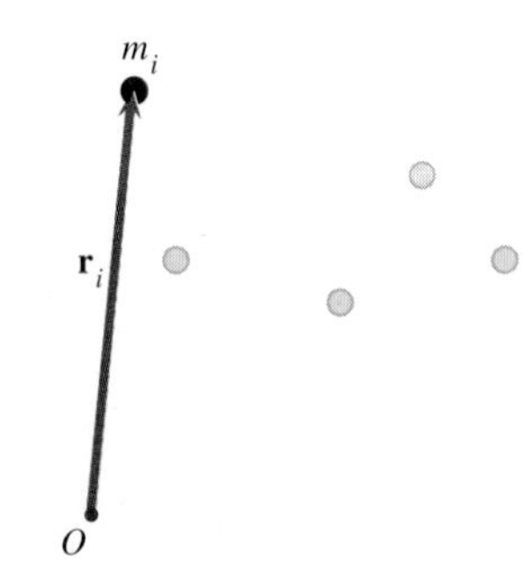

Figure 8.1
A system of particles. The vector $\mathbf{r}_i$ is the position vector of the ith particle.

work done on the system of particles to the change in T. We begin with Newton's second law for the ith particle,

$$\sum_j \mathbf{f}_{ij} + \mathbf{f}_i^{\mathrm{E}} = \frac{d}{dt}(m_i \mathbf{v}_i) \tag{8.2}$$

where $\mathbf{f}_{ij}$ is the force exerted on the ith particle by the jth particle and $\mathbf{f}_i^{\mathrm{E}}$ is the external force on the ith particle. We take the dot product of this equation with $\mathbf{v}_i$ and sum from $i = 1$ to N:

$$\sum_i \sum_j \mathbf{f}_{ij} \cdot \mathbf{v}_i + \sum_i \mathbf{f}_i^{\mathrm{E}} \cdot \mathbf{v}_i = \sum_i \mathbf{v}_i \cdot \frac{d}{dt}(m_i \mathbf{v}_i) \tag{8.3}$$

We can express the term on the right side of this equation as the rate of change of the total kinetic energy:

$$\sum_i \mathbf{v}_i \cdot \frac{d}{dt}(m_i \mathbf{v}_i) = \frac{d}{dt} \sum_i \frac{1}{2} m_i \mathbf{v}_i \cdot \mathbf{v}_i = \frac{dT}{dt}$$

Therefore multiplying Equation (8.3) by dt yields

$$\sum_i \sum_j \mathbf{f}_{ij} \cdot d\mathbf{r}_i + \sum_i \mathbf{f}_i^{\mathrm{E}} \cdot d\mathbf{r}_i = dT$$

We integrate this equation, obtaining

$$\sum_i \sum_j \int_{(\mathbf{r}_i)_1}^{(\mathbf{r}_i)_2} \mathbf{f}_{ij} \cdot d\mathbf{r}_i + \sum_i \int_{(\mathbf{r}_i)_1}^{(\mathbf{r}_i)_2} \mathbf{f}_i^{\mathrm{E}} \cdot d\mathbf{r}_i = T_2 - T_1$$

The terms on the left side are the work done on the system by internal and external forces as the particles move from positions $(\mathbf{r}_i)_1$ to positions $(\mathbf{r}_i)_2$. Denoting the work by U, we obtain the **principle of work and energy for a system of particles**: the work done by internal and external forces equals the change in the total kinetic energy:

$$\boxed{U = T_2 - T_1} \tag{8.4}$$

This result applies to any object or collection of objects, including a rigid body.

Rigid Body in Planar Motion

We have shown that the work done on a rigid body by internal and external forces as it moves between two positions equals the change in its kinetic energy. If we assume that the internal forces between each pair of particles are directed along the straight line between the two particles, *the work done on a rigid body by internal forces is zero*. To show that this is true, we consider two particles of a rigid body designated 1 and 2 (Figure 8.2). The sum of the forces that the two particles exert on each other is zero, $\mathbf{f}_{12} + \mathbf{f}_{21} = \mathbf{0}$, so the rate at which the forces do work (thc power) is

$$\mathbf{f}_{12} \cdot \mathbf{v}_1 + \mathbf{f}_{21} \cdot \mathbf{v}_2 = \mathbf{f}_{21} \cdot (\mathbf{v}_2 - \mathbf{v}_1)$$

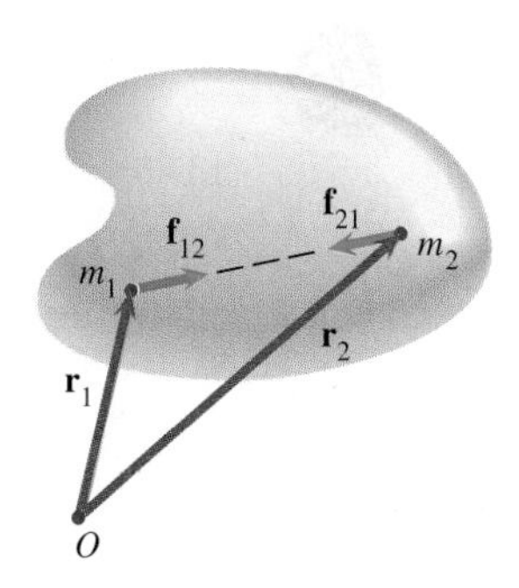

Figure 8.2
Particles 1 and 2 and the forces they exert on each other.

We can show that $\mathbf{f}_{21}$ is perpendicular to $\mathbf{v}_2 - \mathbf{v}_1$, and therefore the rate at which work is done by the internal forces between these two particles is zero. Because the particles are points of a rigid body, we can express their relative velocity in terms of the rigid body's angular velocity ω as

$$\mathbf{v}_2 - \mathbf{v}_1 = \omega \times (\mathbf{r}_2 - \mathbf{r}_1) \tag{8.5}$$

This equation shows that the relative velocity $\mathbf{v}_2 - \mathbf{v}_1$ is perpendicular to $\mathbf{r}_2 - \mathbf{r}_1$, which is the position vector from particle 1 to particle 2. Since the force $\mathbf{f}_{21}$ is parallel to $\mathbf{r}_2 - \mathbf{r}_1$, it is perpendicular to $\mathbf{v}_2 - \mathbf{v}_1$. We can repeat this argument for each pair of particles of the rigid body, so the total rate at which work is done by internal forces is zero. This implies that the work done by internal forces as the rigid body moves between two positions is zero.

The system of external forces on a rigid body may be represented as forces and couples, so we obtain the **principle of work and energy for a rigid body**: the work done by external forces and couples as a rigid body moves between two positions equals the change in its kinetic energy. We can also state this principle for a system of rigid bodies: the work done by external *and internal* forces and couples as a system of rigid bodies moves between two positions equals the change in their total kinetic energy.

To complete our derivation of the principle of work and energy for a rigid body in planar motion, we must express the kinetic energy in terms of the velocity of the centre of mass of the rigid body and its angular velocity. We first consider general planar motion, then rotation about a fixed axis.

Kinetic Energy in General Planar Motion Let us represent a rigid body as a system of particles, and let $\mathbf{R}_i$ be the position vector of the ith particle relative to the centre of mass (Figure 8.3). The position of the centre of mass is

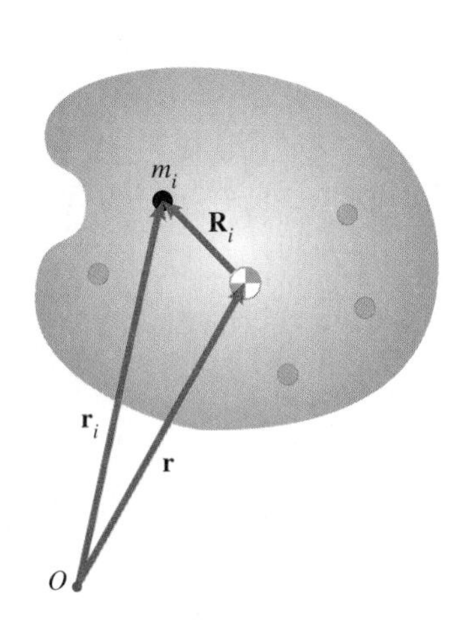

Figure 8.3
Representing a rigid body as a system of particles.

$$\mathbf{r} = \frac{\sum_i m_i \mathbf{r}_i}{m} \tag{8.6}$$

where m is the mass of the rigid body. The position of the ith particle relative to O is related to its position relative to the centre of mass by

$$\mathbf{r}_i = \mathbf{r} + \mathbf{R}_i \tag{8.7}$$

and the vectors $\mathbf{R}_i$ satisfy the relation

$$\sum_i m_i \mathbf{R}_i = \mathbf{0} \tag{8.8}$$

The kinetic energy of the rigid body is the sum of the kinetic energies of its particles, given by Equation (8.1):

$$T = \sum_i \frac{1}{2} m_i \mathbf{v}_i \cdot \mathbf{v}_i \tag{8.9}$$

By taking the time derivative of Equation (8.7), we obtain

$$\mathbf{v}_i = \mathbf{v} + \frac{d\mathbf{R}_i}{dt}$$

where $\mathbf{v}$ is the velocity of the centre of mass. Substituting this expression into Equation (8.9) and using Equation (8.8), we obtain the kinetic energy in the form

$$T = \frac{1}{2}mv^2 + \sum_i \frac{1}{2}m_i \frac{d\mathbf{R}_i}{dt} \cdot \frac{d\mathbf{R}_i}{dt} \tag{8.10}$$

where v is the magnitude of the velocity of the centre of mass.

Let L_0 be the axis through a fixed point O that is perpendicular to the plane of the motion, and let L be the parallel axis through the centre of mass (Figure 8.4(a)). In terms of the coordinate system shown, we can express the angular velocity vector as $\boldsymbol{\omega} = \omega\,\mathbf{k}$. The velocity of the ith particle relative to the centre of mass is $d\mathbf{R}_i/dt = \omega\,\mathbf{k} \times \mathbf{R}_i$, so we can write Equation (8.10) as

$$T = \frac{1}{2}mv^2 + \frac{1}{2}\left[\sum_i m_i(\mathbf{k} \times \mathbf{R}_i)\cdot(\mathbf{k} \times \mathbf{R}_i)\right]\omega^2 \tag{8.11}$$

The magnitude of the vector $\mathbf{k} \times \mathbf{R}_i$ is the perpendicular distance r_i from L to the ith particle (Figure 8.4(b)), so the term in brackets in Equation (8.11) is the mass moment of inertia about L:

$$\sum_i m_i(\mathbf{k} \times \mathbf{R}_i)\cdot(\mathbf{k} \times \mathbf{R}_i) = \sum_i m_i|\mathbf{k} \times \mathbf{R}_i|^2 = \sum_i m_i r_i^2 = I$$

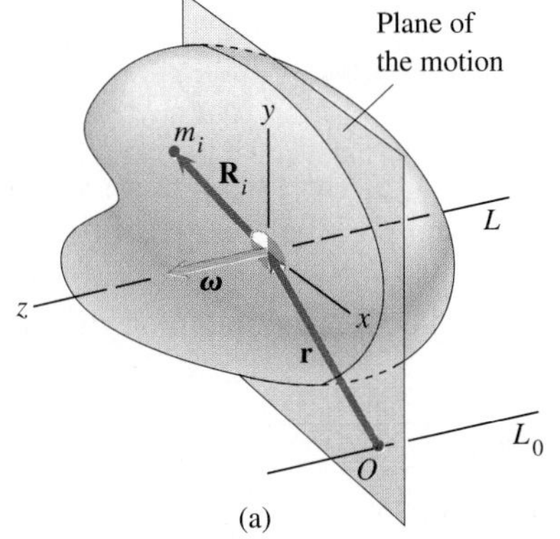

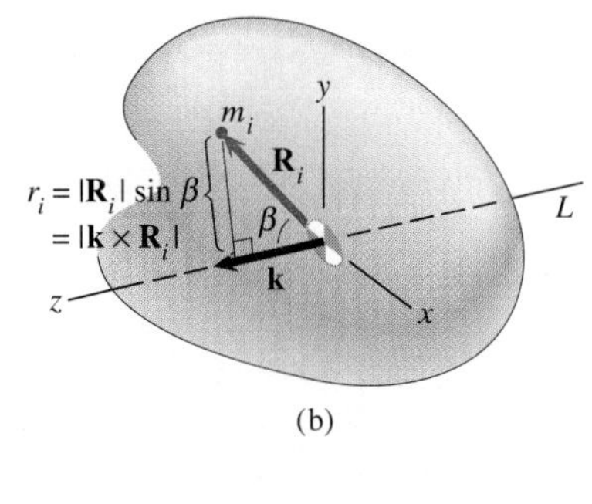

Figure 8.4

(a) A coordinate system with the z axis aligned with L.
(b) The magnitude of $\mathbf{k} \times \mathbf{R}_i$ is the perpendicular distance from L to m_i.

Thus we obtain the kinetic energy of a rigid body in general planar motion in the form

$$\boxed{T = \frac{1}{2}mv^2 + \frac{1}{2}I\omega^2} \tag{8.12}$$

The kinetic energy consists of two terms: the **translational kinetic energy**, expressed in terms of the velocity of the centre of mass, and the **rotational kinetic energy** (Figure 8.5).

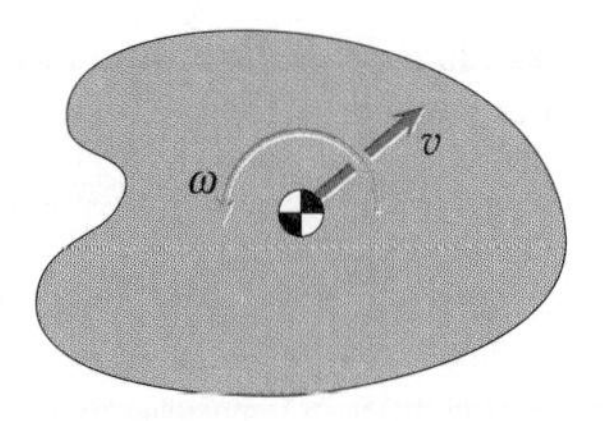

Figure 8.5

Kinetic energy in general planar motion.

Kinetic Energy in Fixed Axis Rotation An object rotating about a fixed axis is in general planar motion, and its kinetic energy is given by Equation (8.12). But there is another expression for the kinetic energy that you will often find convenient. Suppose that a rigid body rotates with angular velocity ω about a fixed axis O. In terms of the distance d from O to the centre of mass, the velocity of the centre of mass is $v = \omega d$ (Figure 8.6(a)). From Equation (8.12), the kinetic energy is

$$T = \frac{1}{2}m(\omega d)^2 + \frac{1}{2}I\omega^2 = \frac{1}{2}(I + d^2 m)\omega^2$$

According to the parallel-axis theorem, the mass moment of inertia about O is $I_0 = I + d^2 m$, so we obtain the kinetic energy of a rigid body rotating about a fixed axis O in the form (Figure 8.6(b))

$$\boxed{T = \frac{1}{2}I_0\omega^2} \tag{8.13}$$

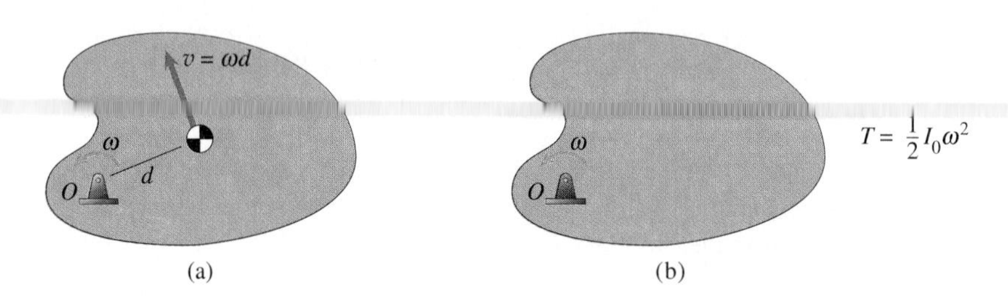

Figure 8.6
(a) Velocity of the centre of mass.
(b) Kinetic energy of a rigid body rotating about a fixed axis.

8.2 Work and Potential Energy

The procedures for determining the work done by different types of forces and the expressions you learned in Chapter 4 for the potential energies of forces provide you with the essential tools for applying the principle of work and energy to a rigid body. The work done on a rigid body by a force $\mathbf{F}$ is given by

$$\boxed{U = \int_{(\mathbf{r}_p)_1}^{(\mathbf{r}_p)_2} \mathbf{F} \cdot d\mathbf{r}_p} \tag{8.14}$$

where $\mathbf{r}_p$ is the position of the *point of application* of $\mathbf{F}$ (Figure 8.7). If the point of application is stationary, or if its direction of motion is perpendicular to $\mathbf{F}$, no work is done.

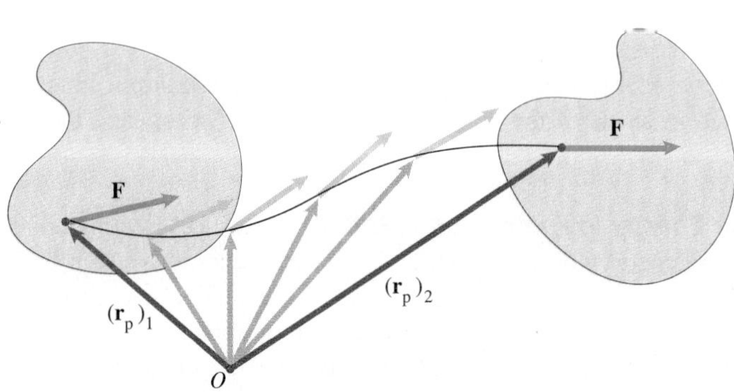

Figure 8.7
The work done by a force on a rigid body is determined by the path of the point of application.

A force $\mathbf{F}$ is conservative if a potential energy V exists such that

$$\mathbf{F} \cdot d\mathbf{r}_{\mathrm{p}} = -dV \tag{8.15}$$

In terms of its potential energy, the work done by a conservative force $\mathbf{F}$ is

$$U = \int_{(\mathbf{r}_{\mathrm{p}} 1}^{(\mathbf{r}_{\mathrm{p}})_2} \mathbf{F} \cdot d\mathbf{r}_{\mathrm{p}} = \int_{V_1}^{V_2} -dV = V_1 - V_2$$

where V_1 and V_2 are the values of V at $(\mathbf{r}_{\mathrm{p}})_1$ and $(\mathbf{r}_{\mathrm{p}})_2$.

If a rigid body is subjected to a couple M (Figure 8.8(a)), what work is done as it moves between two positions? We can evaluate the work by representing the couple by forces (Figure 8.8(b)) and determining the work done by the forces. If the rigid body rotates through an angle $d\theta$ in the direction of the couple (Figure 8.8(c)), the work done by each force is $(\frac{1}{2}D\,d\theta)F$, so the total work is $DF\,d\theta = M\,d\theta$. Integrating this expression, we obtain the work done by a couple M as the rigid body rotates from θ_1 to θ_2 in the direction of M:

$$\boxed{U = \int_{\theta_1}^{\theta_2} M d\theta} \tag{8.16}$$

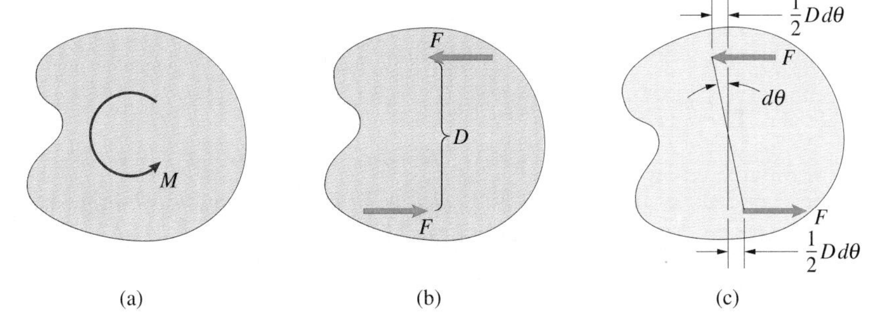

Figure 8.8
(a) A rigid body subjected to a couple.
(b) An equivalent couple consisting of two forces: $DF = M$.
(c) Determining the work done by the forces.

If M is constant between θ_1 and θ_2, the work is simply the product of the couple and the angular displacement:

$$U = M(\theta_2 - \theta_1) \qquad \textbf{Constant couple}$$

A couple M is conservative if a potential energy V exists such that

$$M\,d\theta = -dV \tag{8.17}$$

We can express the work done by a conservative couple in terms of its potential energy:

$$U = \int_{\theta_1}^{\theta_2} M d\theta = \int_{V_1}^{V_2} -dV = V_1 - V_2$$

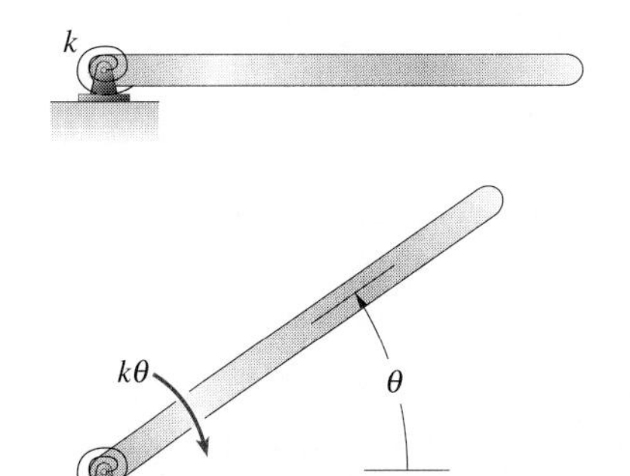

Figure 8.9
(a) A linear torsional spring connected to a bar.
(b) The spring exerts a couple of magnitude $k\theta$ in the direction opposite to the bar's rotation.

For example, in Figure 8.9 a torsional spring exerts a couple on a bar that is proportional to the bar's angle of rotation: $M = -k\theta$. From the relation

$$M\,d\theta = -k\theta\,d\theta = -dV$$

we see that the potential energy must satisfy the relation

$$\frac{dV}{d\theta} = k\theta$$

Integrating this equation, we find that the potential energy of the torsional spring is

$$V = \frac{1}{2}k\theta^2 \tag{8.18}$$

If *all* the forces and couples that do work on a rigid body are conservative, we can express the total work done as it moves between two positions 1 and 2 in terms of the total potential energy of the forces and couples:

$$U = V_1 - V_2$$

Combining this relation with the principle of work and energy, we conclude that the sum of the kinetic energy and the total potential energy is constant – energy is conserved:

$$\boxed{T + V = \text{constant}} \tag{8.19}$$

8.3 Power

The work done on a rigid body by a force $\mathbf{F}$ during an infinitesimal displacement $d\mathbf{r}_\text{p}$ of its point of application is

$$\mathbf{F} \cdot d\mathbf{r}_\text{p}$$

We obtain the power P transmitted to the rigid body, the rate at which work is done on it, by dividing this expression by the interval of time dt during which the displacement takes place:

$$P = \mathbf{F} \cdot \mathbf{v}_\text{p} \tag{8.20}$$

where $\mathbf{v}_\text{p}$ is the velocity of the point of application of $\mathbf{F}$.

Similarly, the work done on a rigid body in planar motion by a couple M during an infinitesimal rotation $d\theta$ in the direction of M is

$$M\, d\theta$$

Dividing this expression by dt, the power transmitted to the rigid body is the product of the couple and the angular velocity:

$$P = M\omega \tag{8.21}$$

The total work done on a rigid body during an interval of time equals the change in its kinetic energy, so the total power transmitted equals the rate of change of its kinetic energy:

$$P = \frac{dT}{dt}$$

The average with respect to time of the power during an interval of time from t_1 to t_2 is

$$P_{av} = \frac{1}{t_2 - t_1}\int_{t_1}^{t_2} P dt = \frac{1}{t_2 - t_1}\int_{T_1}^{T_2} dT = \frac{T_2 - T_1}{t_2 - t_1}$$

This expression shows that you can determine the average power transferred to or from a rigid body during an interval of time by dividing the change in its kinetic energy, or the total work done, by the interval of time:

$$P_{av} = \frac{T_2 - T_1}{t_2 - t_1} = \frac{U}{t_2 - t_1} \tag{8.22}$$

In the following examples we use energy methods to analyse motions of rigid bodies and systems of rigid bodies. You should consider using energy methods when you want to relate changes in the translational and angular velocities of an object to a change in its position. This typically involves three steps:

(1) Identify the forces and couples that do work – *You must use free-body diagrams to determine which external forces and couples do work.*

(2) Apply work and energy or conservation of energy – *Either equate the total work done during a change in position to the change in the kinetic energy or equate the sum of the kinetic and potential energies at two positions.*

(3) Determine kinematic relationships – *To complete your solution, you will often need to relate the velocity of the centre of mass of a rigid body to its angular velocity.*

Example 8.1

A disc of mass m and moment of inertia I is released from rest on an inclined surface (Figure 8.10). Assuming that it rolls, what is the velocity of the disc's centre when it has moved a distance b?

Figure 8.10

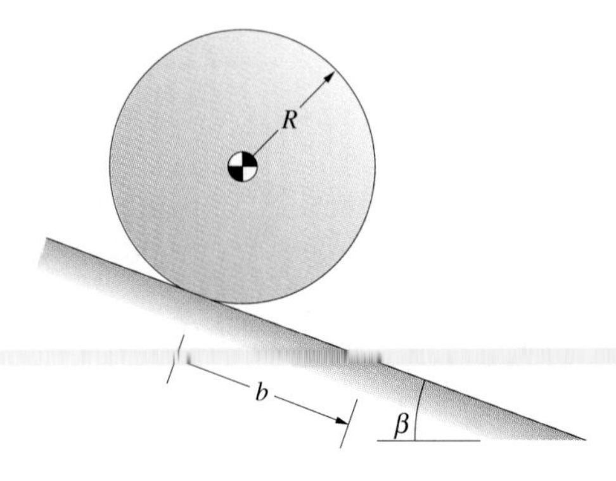

STRATEGY

We can determine the velocity by equating the total work done as the disc rolls a distance b to the change in its kinetic energy.

SOLUTION

Identify the Forces and Couples That Do Work We draw the free-body diagram of the disc in Figure (a). The disc's weight does work as it rolls, but the normal force N and the friction force f do not. To help you understand why the friction force does no work, we can write the work done by a force $\mathbf{F}$ as

$$\int_{(\mathbf{r}_\mathrm{P})_1}^{(\mathbf{r}_\mathrm{P})_2} \mathbf{F} \cdot d\mathbf{r}_\mathrm{P} = \int_{t_1}^{t_2} \mathbf{F} \cdot \frac{d\mathbf{r}_\mathrm{P}}{dt} dt = \int_{t_1}^{t_2} \mathbf{F} \cdot \mathbf{v}_\mathrm{P} dt$$

where $\mathbf{v}_\mathrm{P}$ is the velocity of the point of application of $\mathbf{F}$. Since the velocity of the point where f acts is zero as the disc rolls, the work done by f is zero.

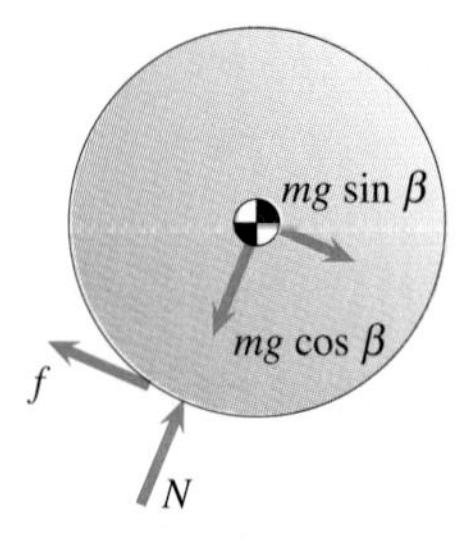

(a) Free-body diagram of the disc.

Apply Work and Energy We can determine the work done by the weight by multiplying the component in the direction of the motion of the centre of the disc by the distance b:

$$U = (mg \sin \beta)b$$

Letting v and ω be the velocity and angular velocity of the centre of the disc when it has moved a distance b (Figure (b)), we equate the work to the change in the disc's kinetic energy:

$$mgb \sin \beta = \frac{1}{2}mv^2 + \frac{1}{2}I\omega^2 - 0 \tag{8.23}$$

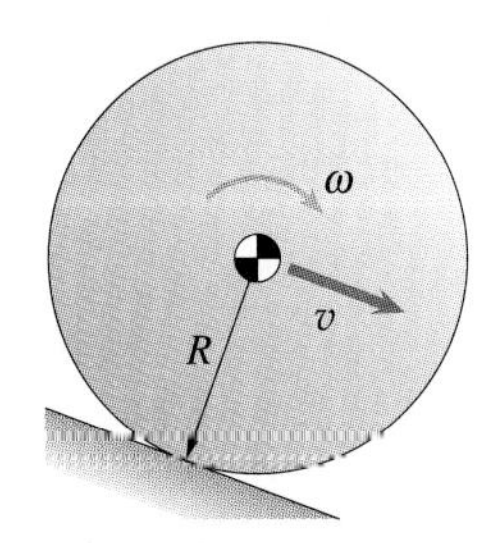

(b) Velocity of the centre and angular velocity when the disc has moved a distance b.

Determine Kinematic Relationships The angular velocity ω of the rolling disc is related to the velocity v by $\omega = v/R$. Substituting this relation into Equation (8.23) and solving for v, we obtain

$$v = \sqrt{\frac{2gb \sin \beta}{1 + I/mR^2}}$$

DISCUSSION

Suppose that the surface is smooth, so that the disc slides instead of rolling. In this case, the disc has no angular velocity, so Equation (8.23) becomes

$$mgb \sin \beta = \frac{1}{2}mv^2 - 0$$

and the velocity of the centre of the disc is

$$v = \sqrt{2gb \sin \beta}$$

The velocity is greater when the disc slides. You can see why by comparing the two expressions for the principle of work and energy. The work done by the disc's weight is the same in each case. When the disc rolls, part of the work increases the disc's translational kinetic energy and part increases its rotational kinetic energy. When the disc slides, all of the work increases its translational kinetic energy.

Example 8.2

Each wheel of the motorcycle in Figure 8.11 has mass $m_W = 9$ kg, radius $R = 330$ mm, and moment of inertia $I = 0.8$ kg.m^2. The combined mass of the rider and the motorcycle, not including the wheels, is $m_C = 142$ kg. The motorcycle starts from rest, and its engine exerts a constant couple $M = 140$ N.m on the rear wheel. Assume that the wheels do not slip.

(a) What horizontal distance b must the motorcycle travel to reach a velocity of 25 m/s?

(b) What is the maximum power transmitted to the motorcycle by its engine during the motion described in part (a)?

Figure 8.11

STRATEGY

(a) We can apply the principle of work and energy to the system consisting of the rider and the motorcycle, including its wheels, to determine the distance b.

(b) The power transmitted by the couple exerted on the rear wheel is given by Equation (8.21). To determine the maximum power, we must determine the wheel's maximum angular velocity.

SOLUTION

(a) Determining the distance b by energy methods requires three steps.

Identify the Forces and Couples That Do Work We draw the free-body diagram of the system in Figure (a). The weights do no work because the motion is horizontal, and the forces exerted on the wheels by the road do no work because the velocity of their point of application is zero. (See Example 8.1.) No work is done by external forces and couples! However, work is done by the couple M exerted on the rear wheel by the engine (Figure (b)). Although this is an internal couple for the system we are considering – the wheel exerts an opposite couple on the body of the motorcycle – net work is done because the wheel rotates whereas the body does not.

Apply Work and Energy If the motorcycle moves a horizontal distance b, the wheels turn through an angle b/R rad and the work done by the constant couple M is

$$U = M(\theta_2 - \theta_1) = M\left(\frac{b}{R}\right)$$

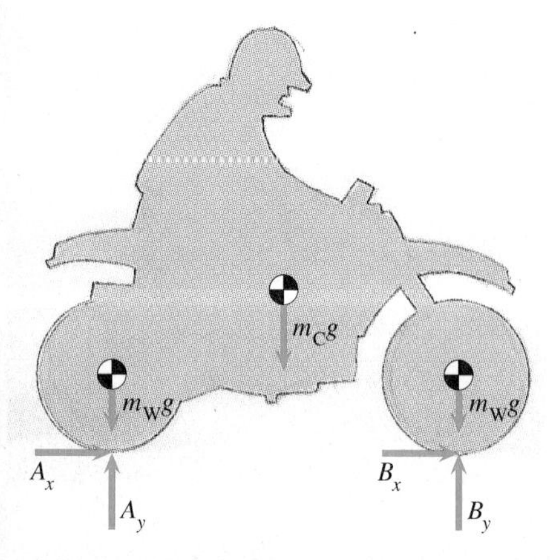

(a) Free-body diagram of the system.

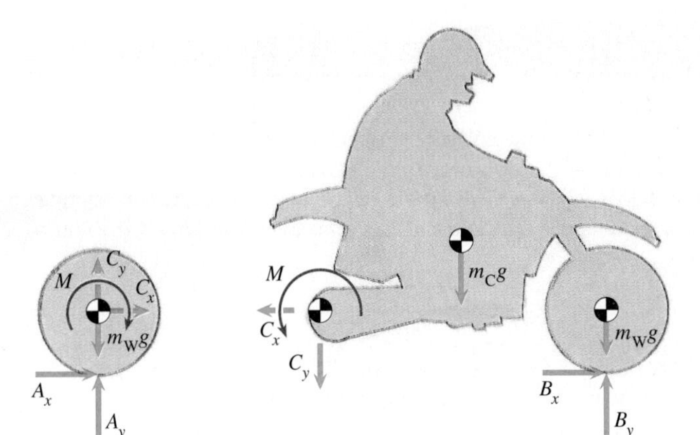

(b) Isolating the rear wheel.

Let v be the motorcycle's velocity and ω the angular velocity of the wheels when the motorcycle has moved a distance b. The work equals the change in the total kinetic energy:

$$M\left(\frac{b}{R}\right) = \frac{1}{2}m_C v^2 + 2\left[\frac{1}{2}m_W v^2 + \frac{1}{2}I\omega^2\right] \quad (8.24)$$

Determine Kinematic Relationships The angular velocity of the rolling wheels is related to the velocity v by $\omega = v/R$. Substituting this relation into Equation (8.24) and solving for b, we obtain

$$\begin{aligned} b &= \left(\frac{1}{2}m_C + m_W + \frac{I}{R^2}\right)\frac{Rv^2}{M} \\ &= \left[\frac{1}{2}(142) + (9) + \frac{(0.8)}{(0.33)^2}\right]\frac{(0.33)(25)^2}{(140)} \\ &= 128.7 \text{ m} \end{aligned}$$

(b) The angular velocity of the wheels when the motorcycle reaches its maximum velocity is

$$\omega = \frac{v}{R} = \frac{25}{0.33} = 75.8 \text{ rad/s}$$

From Equation (8.21), the maximum power is

$$P = M\omega = (140)(75.8) = 10\,600 \text{ W}$$

DISCUSSION

Although we drew separate free-body diagrams of the motorcycle and its rear wheel to clarify the work done by the couple exerted by the engine, notice that we treated the motorcycle, including its wheels, as a single system in applying the principle of work and energy. By doing so, we did not need to consider the work done by the internal forces between the motorcycle's body and its wheels. When applying the principle of work and energy to a system of rigid bodies, you will usually find it simplest to express the principle for the system as a whole. This is in contrast to determining the motion of a system of rigid bodies by using the equations of motion, which usually requires that you draw free-body diagrams of each rigid body.

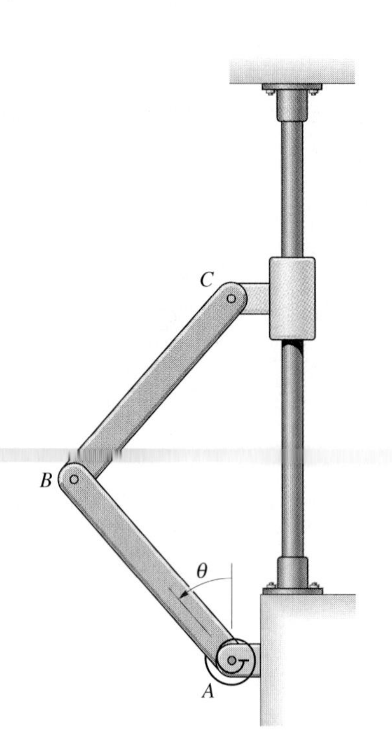

Figure 8.12

Example 8.3

The slender bars AB and BC of the linkage in Figure 8.12 have mass m and length l, and the collar C has mass m_C. A torsional spring at A exerts a clockwise couple $k\theta$ on bar AB. The system is released from rest in the position $\theta = 0$ and allowed to fall. Neglecting friction, determine the angular velocity $\omega = d\theta/dt$ of bar AB as a function of θ.

SOLUTION

Identify the Forces and Couples That Do Work We draw the free-body diagram of the system in Figure (a). The forces and couples that do work – the weights of the bars and collar and the couple exerted by the torsional spring – are conservative. We can use conservation of energy and the kinematic relationships between the angular velocities of the bars and the velocity of the collar to determine ω as a function of θ.

Apply Conservation of Energy We denote the centre of mass of bar BC by G and the angular velocity of bar BC by ω_{BC} (Figure (b)). The moment of inertia of each bar about its centre of mass is $I = \frac{1}{12}ml^2$. Since bar AB rotates about the fixed point A, we can write its kinetic energy as

$$T_{\text{bar }AB} = \frac{1}{2}I_A\omega^2 = \frac{1}{2}\left[I + \left(\frac{1}{2}l\right)^2 m\right]\omega^2 = \frac{1}{6}ml^2\omega^2$$

The kinetic energy of bar BC is

$$T_{\text{bar }BC} = \frac{1}{2}mv_G^2 + \frac{1}{2}I\omega_{BC}^2 = \frac{1}{2}mv_G^2 + \frac{1}{24}ml^2\omega_{BC}^2$$

The kinetic energy of the collar C is

$$T_{\text{collar}} = \frac{1}{2}m_C v_C^2$$

Using the datum in Figure (a), we obtain the potential energies of the weights:

$$V_{\text{bar }AB} + V_{\text{bar }BC} + V_{\text{collar}} = mg\left(\frac{1}{2}l\cos\theta\right) + mg\left(\frac{3}{2}l\cos\theta\right) + m_C g(2l\cos\theta)$$

The potential energy of the torsional spring is given by Equation (8.18):

$$V_{\text{spring}} = \frac{1}{2}k\theta^2$$

We now have all the ingredients to apply conservation of energy. We equate the sum of the kinetic and potential energies at the position $\theta = 0$ to the sum of the kinetic and potential energies at an arbitrary value of θ:

$$T_1 + V_1 = T_2 + V_2:$$

$$0 + 2mgl + 2m_C gl = \frac{1}{6}ml^2\omega^2 + \frac{1}{2}mv_G^2 + \frac{1}{24}ml^2\omega_{BC}^2 + \frac{1}{2}m_C v_C^2$$
$$+ 2mgl\cos\theta + 2m_C gl\cos\theta + \frac{1}{2}k\theta^2$$

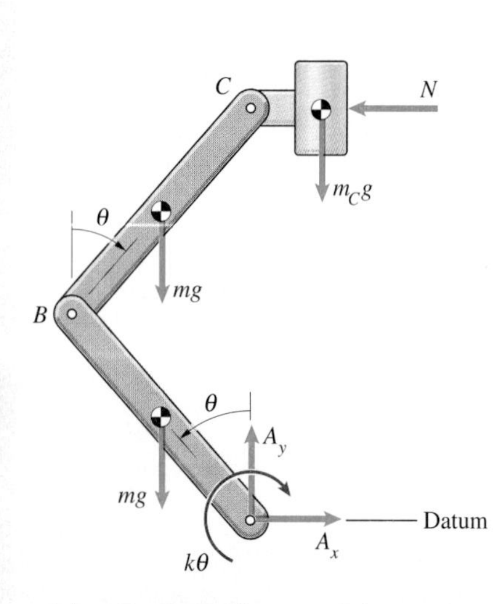

(a) Free-body diagram of the system.

To determine ω from this equation, we must express the velocities v_G, v_C and ω_{BC} in terms of ω.

Determine Kinematic Relationships We can determine the velocity of point B in terms of ω and then express the velocity of point C in terms of the velocity of point B and the angular velocity ω_{BC}.

The velocity of B is

$$\mathbf{v}_B = \mathbf{v}_A + \boldsymbol{\omega}_{AB} \times \mathbf{r}_{B/A}$$

$$= 0 + \begin{vmatrix} \mathbf{i} & \mathbf{j} & \mathbf{k} \\ 0 & 0 & \omega \\ -l \sin\theta & l \cos\theta & 0 \end{vmatrix}$$

$$= -l\omega \cos\theta\,\mathbf{i} - l\omega \sin\theta\,\mathbf{j}$$

The velocity of C expressed in terms of the velocity of B is

$$v_C\,\mathbf{j} = \mathbf{v}_B + \boldsymbol{\omega}_{BC} \times \mathbf{r}_{C/B}$$

$$= -l\omega \cos\theta\,\mathbf{i} - l\omega \sin\theta\,\mathbf{j} + \begin{vmatrix} \mathbf{i} & \mathbf{j} & \mathbf{k} \\ 0 & 0 & \omega_{BC} \\ l \sin\theta & l \cos\theta & 0 \end{vmatrix}$$

Equating $\mathbf{i}$ and $\mathbf{j}$ components, we obtain

$$\omega_{BC} = -\omega \qquad v_C = -2l\omega \sin\theta$$

(The minus signs indicate that the directions of the velocities are opposite to the directions we assumed in Figure (b).) Now that we know the angular velocity of bar BC in terms of ω, we can determine the velocity of its centre of mass in terms of ω by expressing it in terms of $\mathbf{v}_B$:

$$\mathbf{v}_G = \mathbf{v}_B + \boldsymbol{\omega}_{BC} \times \mathbf{r}_{G/B}$$

$$= -l\omega \cos\theta\,\mathbf{i} - l\omega \sin\theta\,\mathbf{j} + \begin{vmatrix} \mathbf{i} & \mathbf{j} & \mathbf{k} \\ 0 & 0 & -\omega \\ \frac{1}{2}l \sin\theta & \frac{1}{2}l \cos\theta & 0 \end{vmatrix}$$

$$= -\frac{1}{2}l\omega \cos\theta\,\mathbf{i} - \frac{3}{2}l\omega \sin\theta\,\mathbf{j}$$

Substituting these expressions for ω_{BC}, v_C and $\mathbf{v}_G$ into our equation of conservation of energy and solving for ω, we obtain

$$\omega = \left[\frac{2gl(m + m_C)(1 - \cos\theta) - \frac{1}{2}k\theta^2}{\frac{1}{3}ml^2 + (m + 2m_C)l^2 \sin^2\theta}\right]^{1/2}$$

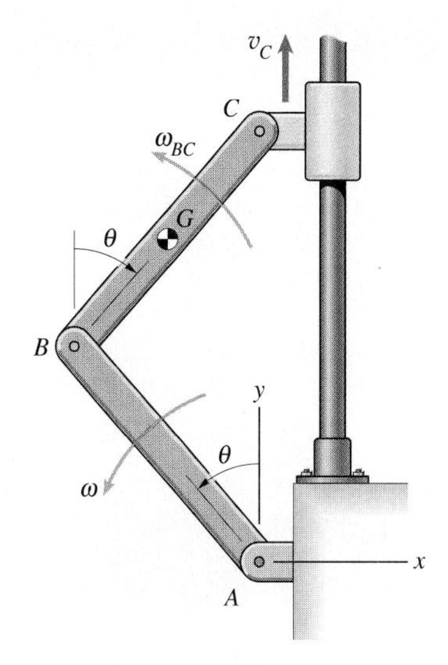

(b) Angular velocities of the bars and the velocity of the collar.

Problems

8.1 A main landing gear wheel of a Boeing 747 weighs 1068 N, has a mass moment of inertia of 23 kg.m^2, and has a radius of 0.6 m. If the aeroplane is moving at 75 m/s and the wheel rolls, what is the wheel's kinetic energy?

P8.1

8.2 The flywheel of the homopolar generator shown on page 366 can be modelled as a 1200 kg homogeneous cylinder with a 280 mm radius. Suppose that the flywheel is turning at 6200 rpm. By converting part of the flywheel's kinetic energy into an electric current, the generator is used to power an electromagnetic rail gun that accelerates a 0.28 kg projectile to a velocity of 5.3 km/s. If you assume that all of the energy obtained from the flywheel is converted into the projectile's kinetic energy, what is the flywheel's final angular velocity?

8.3 The angular velocity of the space station is 1 rpm (revolution per minute). Use work and energy to determine the constant couple the station's reaction control system would have to exert to reduce its angular velocity to zero in 100 revolutions. The mass moment of inertia of the station is $I = 1.5 \times 10^{10}$ kg.m^2.

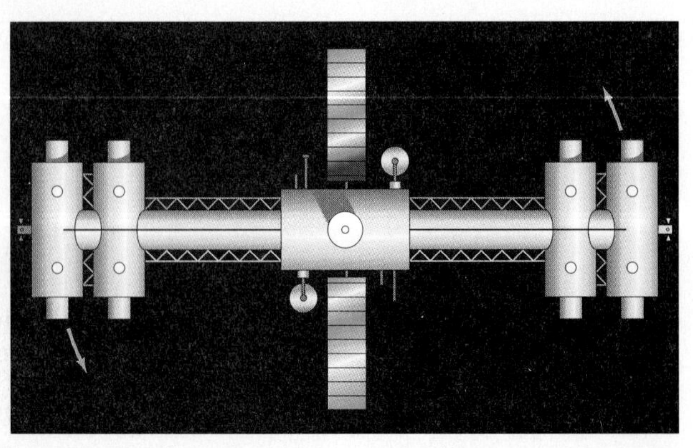

P8.3

8.4 The mass moment of inertia of the helicopter's rotor is 400 kg.m^2. If the rotor starts from rest, the engine exerts a constant torque of 500 N.m on it, and aerodynamic drag is neglected, use the principle of work and energy to determine how many revolutions the rotor must turn to reach an angular velocity of 2 revolutions per second.

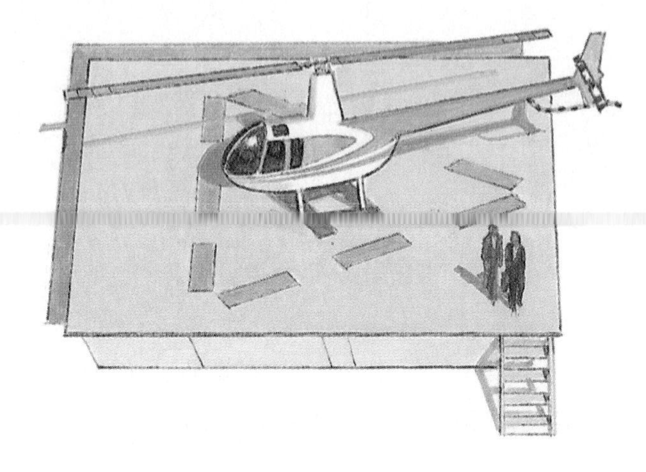

P8.4

8.5 What average power is transmitted to the rotor in Problem 8.4 in accelerating it from rest to 2 revolutions per second?

8.6 During extravehicular activity, an astronaut fires a thruster of her manoeuvring unit, exerting a constant force $T = 20$ N. The mass moment of inertia of the astronaut and her equipment about their centre of mass is 45 kg.m^2. Using the principle of work and energy, determine her rate of rotation in revolutions per second when she has rotated one quarter of a revolution from her initial orientation.

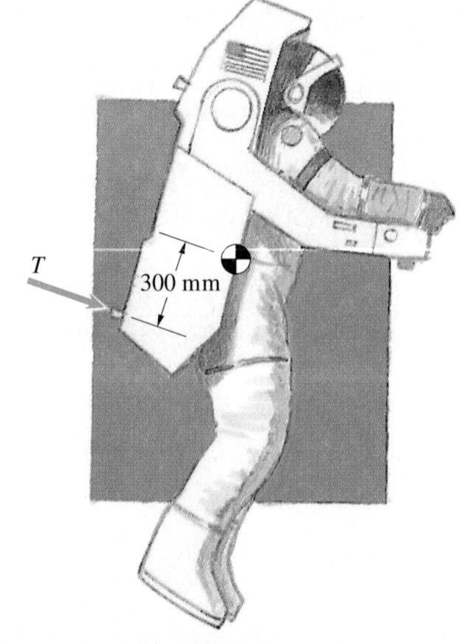

P8.6

8.7 A slender bar of mass m is released from rest in the horizontal position shown. Determine its angular velocity when it is vertical (a) by using the principle of work and energy; (b) by using conservation of energy.

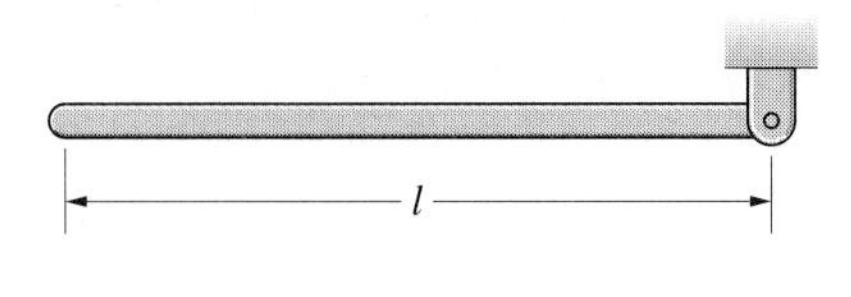

P8.7

8.8 The mass moment of inertia of the pulley is 0.4 kg.m^2. The pulley starts from rest. For both cases, use the principle of work and energy to determine the pulley's angular velocity when it has turned one revolution.

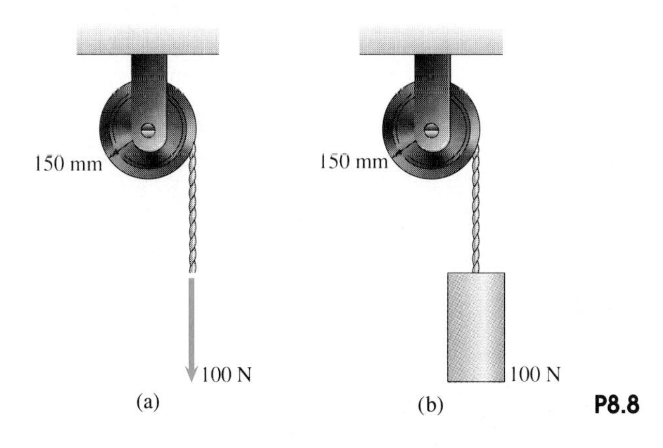

P8.8

8.9 The object consists of identical 1 m, 5 kg bars welded together. If it is released from rest in the position shown, what is its angular velocity when the bar attached at A is vertical?

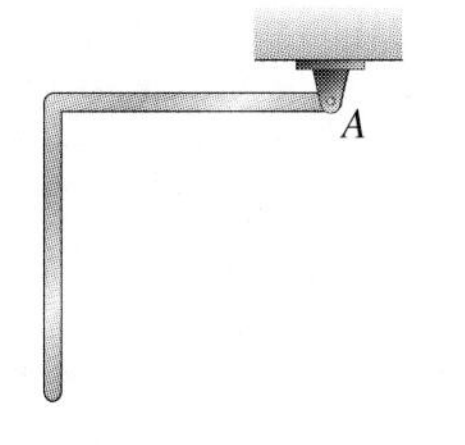

P8.9

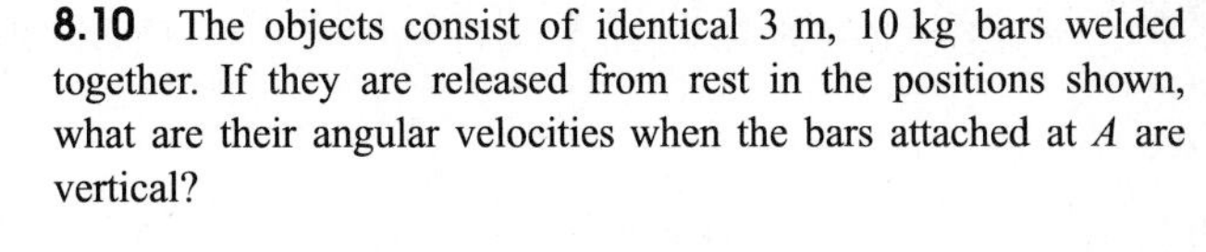
8.10 The objects consist of identical 3 m, 10 kg bars welded together. If they are released from rest in the positions shown, what are their angular velocities when the bars attached at A are vertical?

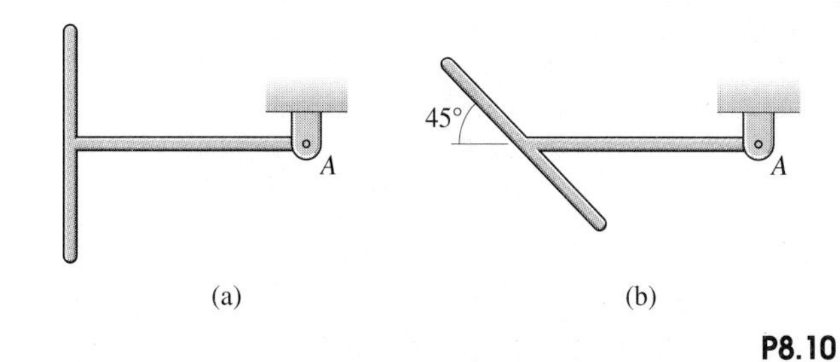

P8.10

8.11 The 8 kg slender bar is released from rest in the horizontal position. When it has fallen to the position shown, what are the x and y components of force exerted on the bar by the pin support A?

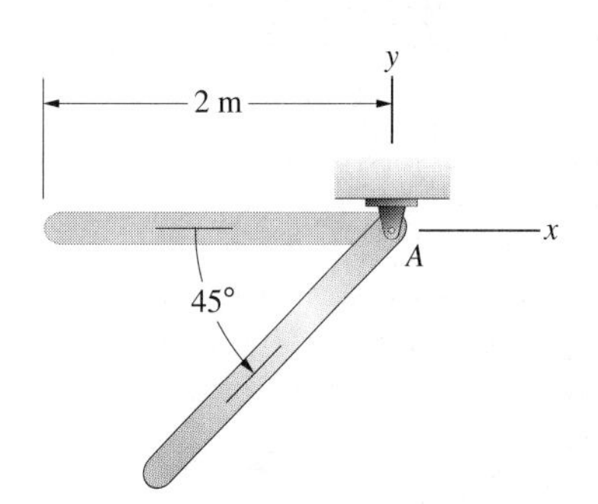

P8.11

8.12 The slender bar is released from rest in the position shown. (a) Use conservation of energy to determine the angular velocity when the bar is vertical.

(b) For what value of x is the angular velocity determined in part (a) a maximum?

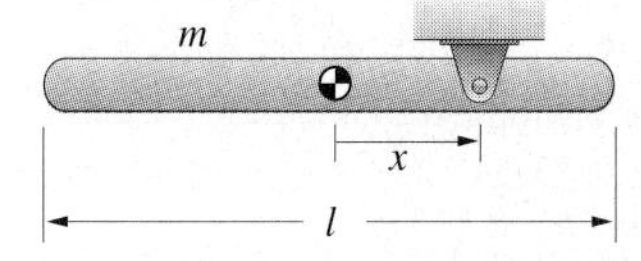

P8.12

8.13 The gears can turn freely on their pin supports. Their mass moments of inertia are $I_A = 0.002$ kg.m^2 and $I_B = 0.006$ kg.m^2. They are at rest when a constant couple $M = 2$ N.m is applied to gear B. Neglecting friction, use the principle of work and energy to determine the angular velocities of the gears when gear A has turned 100 revolutions.

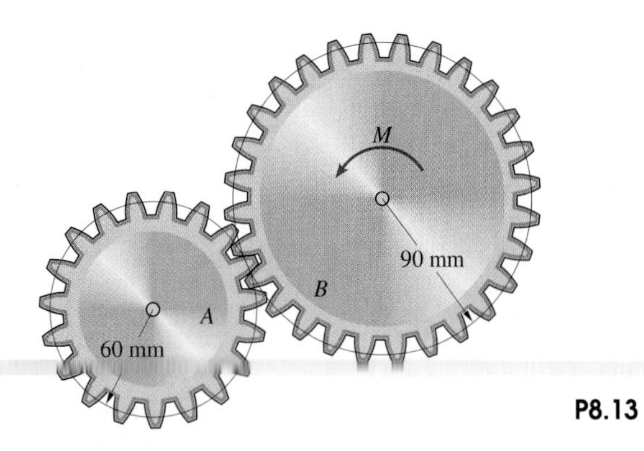

P8.13

8.14 The pulleys can turn freely on their pin supports. Their mass moments of inertia are $I_A = 0.002$ kg.m^2, $I_B = 0.036$ kg.m^2 and $I_C = 0.032$ kg.m^2. They are stationary when a constant couple $M = 2$ N.m is applied to pulley A. What is the angular velocity of pulley A when it has turned 10 revolutions?

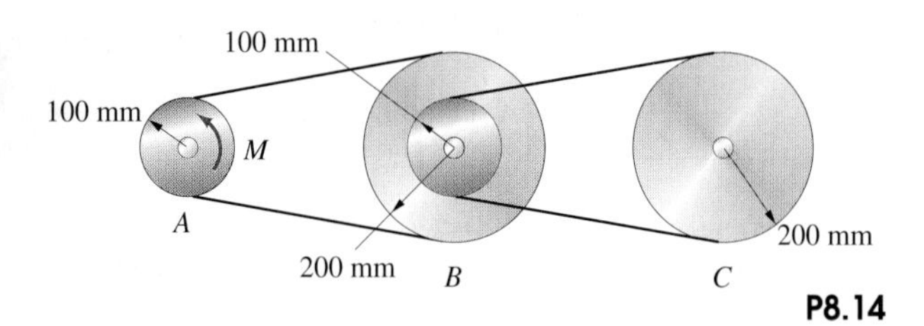

P8.14

8.15 The mass moments of inertia of gears A and B are $I_A = 0.014$ kg.m^2 and $I_B = 0.100$ kg.m^2. Gear A is connected to a torsional spring with constant $k = 0.2$ N.m/rad. If the spring is unstretched, and the surface supporting the 2 kg mass is removed, what is the mass's velocity when it has fallen 75 mm?

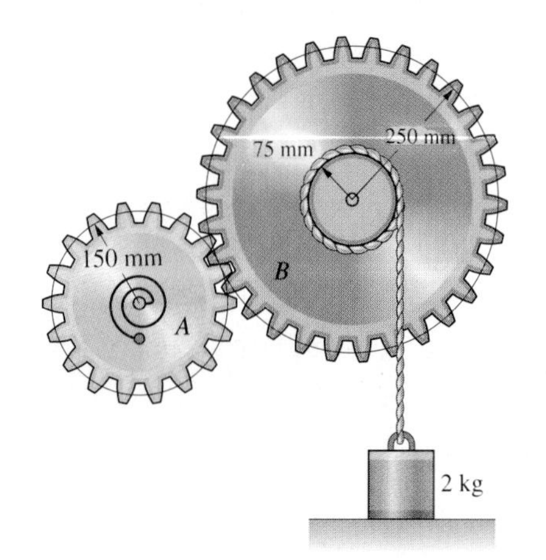

P8.15

8.16 Consider the system in Problem 8.15.
(a) What maximum distance does the 2 kg mass fall when the supporting surface is removed?
(b) What maximum velocity does the mass achieve?

8.17 Consider the system in Problem 8.15. Suppose that the torsional spring is not unstretched in the position shown, and the resulting tension in the string is 8 N. If the surface supporting the 2 kg weight is removed, what is the weight's velocity when it has fallen 75 mm?

8.18 Model the arm ABC as a single rigid body. Its mass is 300 kg, and the mass moment of inertia about its centre of mass is $I = 360$ kg.m^2. Starting from rest with its centre of mass 2 m above the ground (position 1), the hydraulic cylinders push arm ABC upwards. When it is in the position shown (position 2), its counterclockwise angular velocity is 1.4 rad/s. How much work do the hydraulic cylinders do on the arm in moving it from position 1 to position 2?

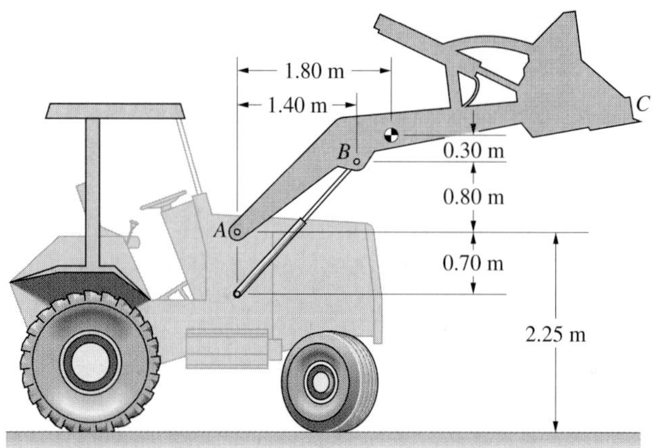

P8.18

8.19 The mass of the homogeneous cylindrical disc is m, and its radius is R. The disc is stationary when a constant clockwise couple M is applied to it. Use work and energy to determine the disc's angular velocity when it has rolled a distance b.

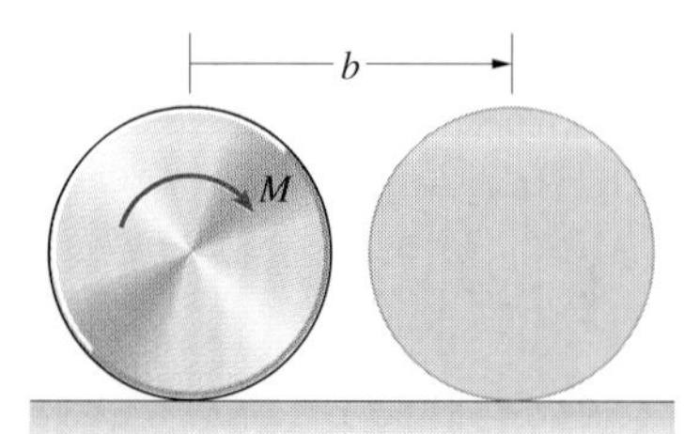

P8.19

8.20 A disc of mass m and moment of inertia I starts from rest on an inclined surface and is subjected to a constant clockwise couple M. Assuming that it rolls, what is the angular velocity of the disc when it has moved a distance b?

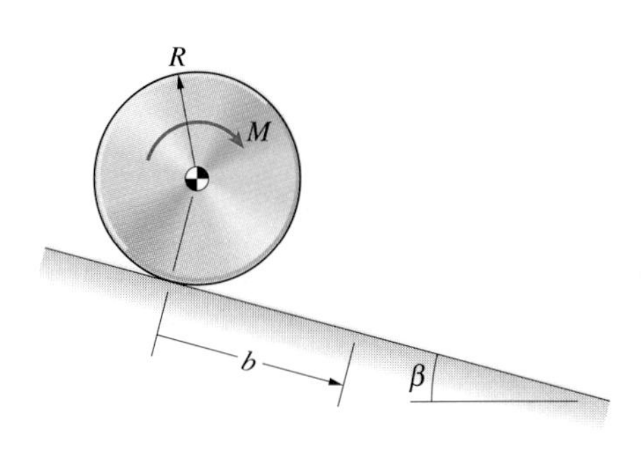

P8.20

8.21 The stepped disc weighs 130 N, and its mass moment of inertia is $I = 0.2$ kg.m^2. If it is released from rest, what is its angular velocity when the centre of the disc has fallen 1 m?

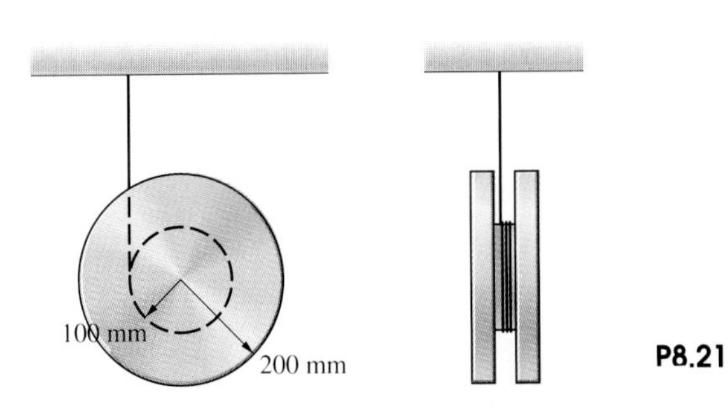

P8.21

8.22 The 100 kg homogeneous cylindrical disc is at rest when the force $F = 500$ N is applied to a cord wrapped around it, causing the disc to roll. Use the principle of work and energy to determine the disc's angular velocity when it has turned one revolution.

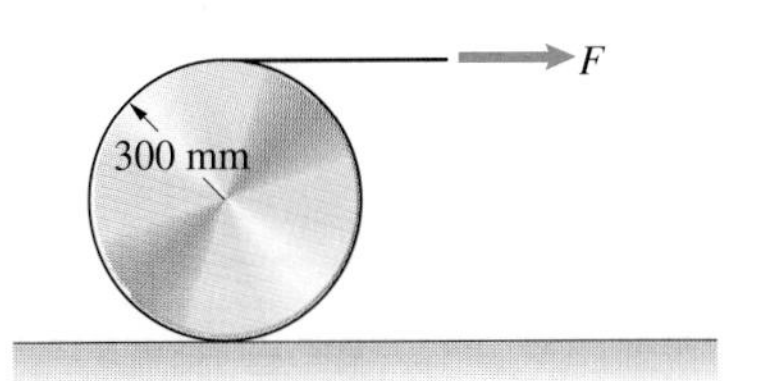

P8.22

8.23 The 15 kg homogeneous cylindrical disc is given a clockwise angular velocity of 2 rad/s with the spring unstretched. The spring constant is $k = 45$ N/m. If the disc rolls, how far will its centre move to the right?

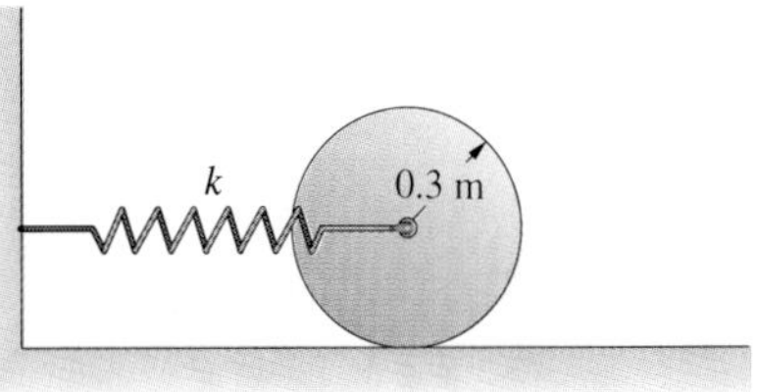

P8.23

8.24 The 22 kg platen P rests on four roller bearings. The roller bearings can be modelled as 1 kg homogeneous cylinders with 30 mm radii. The platen is stationary and the spring ($k = 900$ N/m) is unstretched when a constant horizontal force $F = 100$ N is applied as shown. What is the platen's velocity when it has moved 200 mm to the right?

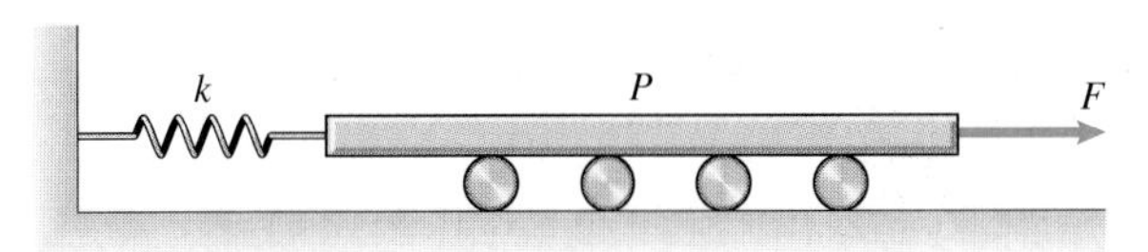

P8.24

8.25 Consider the system described in Problem 8.24.
(a) What maximum distance does the platen move to the right when the force F is applied?
(b) What maximum velocity does the platen achieve, and how far has the platen moved to the right when it occurs?

8.26 The rules of a soapbox derby specify the required combined weight of the car and driver and the radius of the wheels. A young contestant designing her car ponders two possibilities: (a) use heavy wheels; (b) use light wheels, making up the weight by adding ballast. Analyse this problem using the principle of work and energy, and explain the advice you would give her.

P8.26

8.27 Each of the go-cart's front wheels weighs 20 N and has a mass moment of inertia of 0.01 kg.m^2. The two rear wheels and rear axle form a single rigid body weighing 160 N and having a mass moment of inertia of 0.1 kg.m^2. The total weight of the rider and go-cart, including its wheels, is 1000 N. The go-cart starts from rest, its engine exerts a constant torque of 20 N.m on the rear axle, and its wheels do not slip. If you neglect friction and aerodynamic drag, how fast is it moving when it has travelled 15 m?

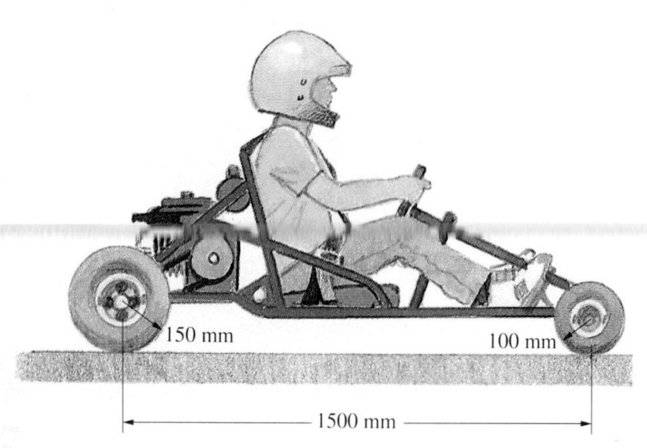

P8.27

8.28 Determine the maximum power and the average power transmitted to the go-cart in Problem 8.27 by its engine.

8.29 Each box weighs 200 N, the mass moment of inertia of the pulley is 0.6 kg.m^2, and friction can be neglected. If the boxes start from rest, determine the magnitude of their velocity when they have moved 1.5 m from their initial positions.

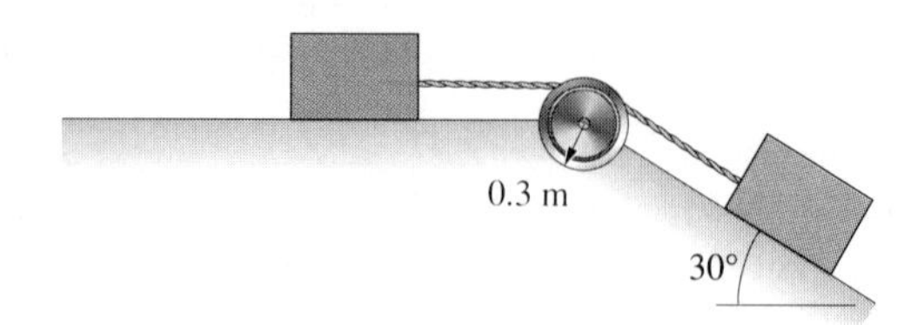

P8.29

8.30 The slender bar weighs 120 N and the cylindrical disc weighs 80 N. The system is released from rest with the bar horizontal. Determine the magnitude of the bar's angular velocity when it is vertical if the bar and disc are welded together at A.

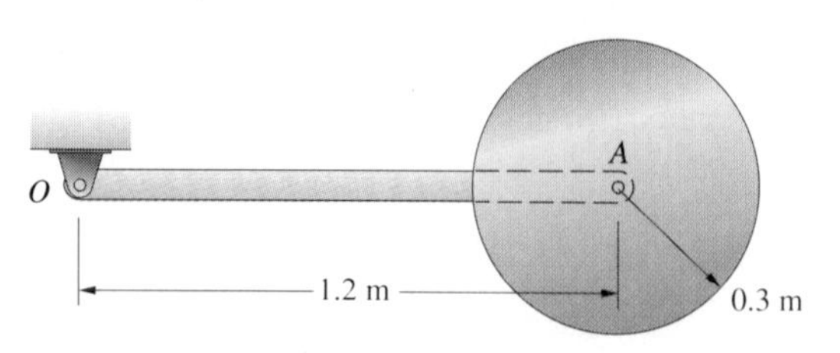

P8.30

8.31 In Problem 8.30, determine the magnitude of the bar's angular velocity when it has reached the vertical position if the bar and disc are connected by a smooth pin at A.

8.32 The 45 kg crate is pulled up the inclined surface by the winch. The coefficient of kinetic friction between the crate and the surface is $\mu_k = 0.4$. The mass moment of inertia of the drum on which the cable is wound, including the cable wound on the drum, is $I_A = 4$ kg.m^2. The motor exerts a constant couple $M = 55$ N.m on the drum. If the crate starts from rest, use the principle of work and energy to determine its velocity when it has moved 0.6 m.

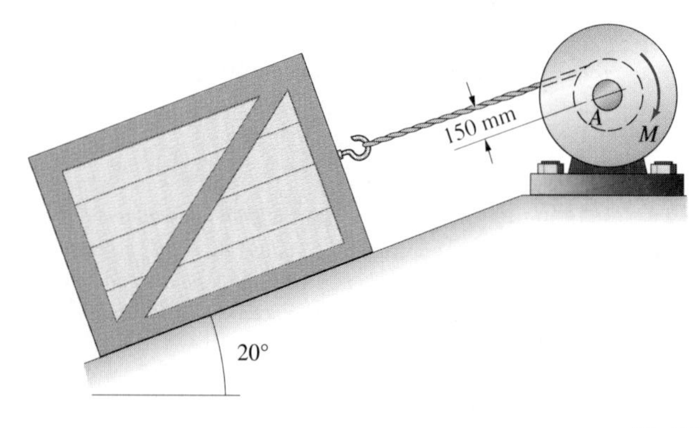

P8.32

8.33 The 2 m slender bars each weigh 40 N, and the rectangular plate weighs 200 N. If the system is released from rest in the position shown, what is the velocity of the plate when the bars are vertical?

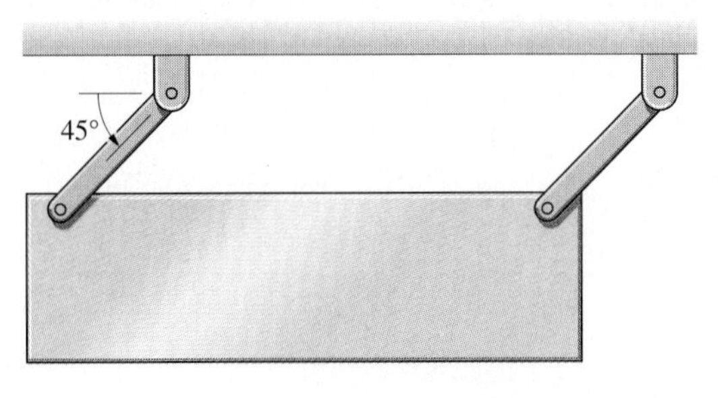

P8.33

8.34 The system starts from rest with the 4 kg slender bar horizontal. The mass of the suspended cylinder is 10 kg. What is the bar's angular velocity when it is in the position shown?

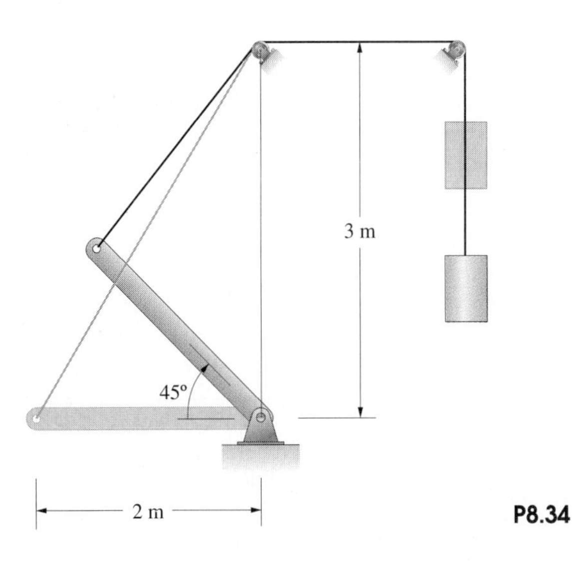

P8.34

8.35 The unstretched length of the spring is 1.5 m, and its constant is $k = 50$ N/m. When the 15 kg slender bar is horizontal, its angular velocity is 0.1 rad/s. What is its angular velocity when it is in the position shown?

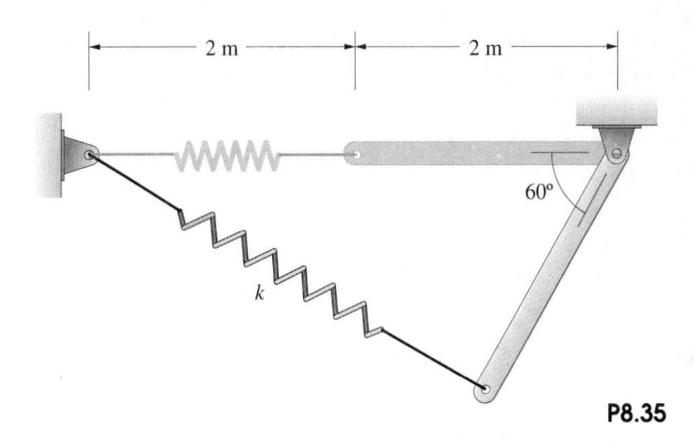

P8.35

8.36 Pulley A weighs 16 N, $I_A = 0.060$ kg.m^2, and $I_B = 0.014$ kg.m^2. If the system is released from rest, what is the velocity of the 8 kg mass when it has fallen 0.6 m?

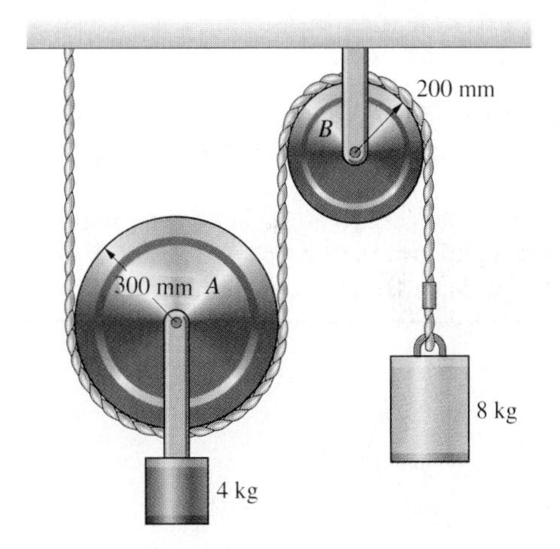

P8.36

8.37 The 18 kg ladder is released from rest with $\theta = 10°$. The wall and floor are smooth. Modelling the ladder as a slender bar, use conservation of energy to determine its angular velocity when $\theta = 40°$.

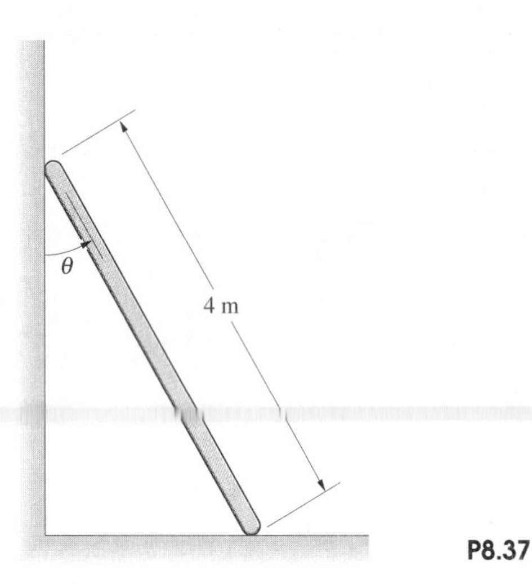

P8.37

8.38 The 4 kg slender bar is pinned to a 2 kg slider at A and to a 4 kg homogeneous cylindrical disc at B. Neglect the friction force on the slider and assume that the disc rolls. If the system is released from rest with $\theta = 60°$, what is the bar's angular velocity when $\theta = 0$?

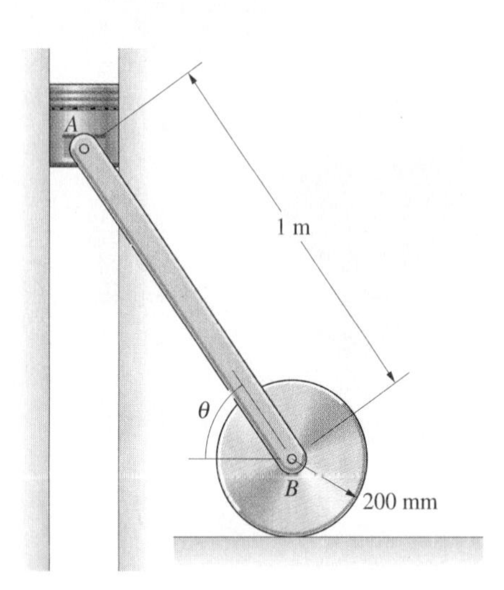

P8.38

8.39 If the system in Problem 8.38 is released from rest with $\theta = 80°$, what is the bar's angular velocity when $\theta = 20°$?

8.40 The system is in equilibrium in the position shown. The mass of the slender bar ABC is 6 kg, the mass of the slender bar BD is 3 kg, and the mass of the slider at C is 1 kg. The spring constant is $k = 200$ N/m. If a constant 100 N downward force is applied at A, what is the angular velocity of bar ABC when it has rotated 20° from its initial position?

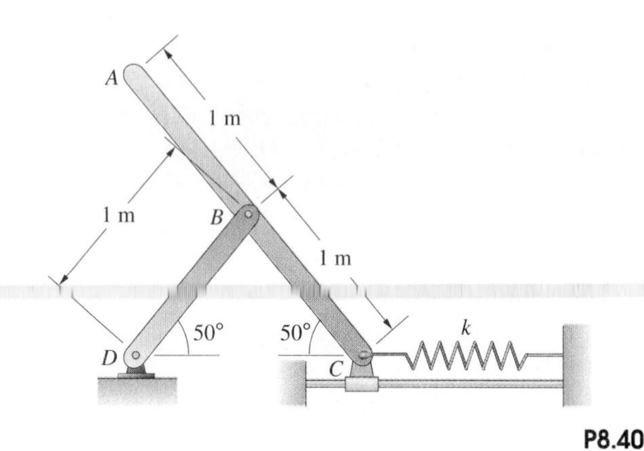

P8.40

8.41 Bar AB weighs 45 N and bar BC weighs 25 N. If the system is released from rest in the position shown, what are the angular velocities of the bars at the instant just before the joint B hits the smooth floor?

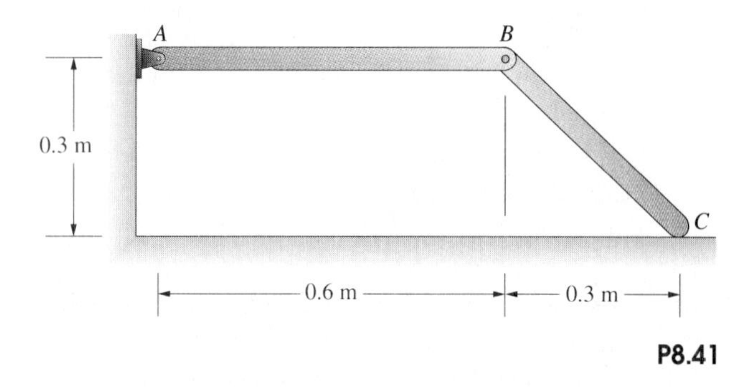

P8.41

8.42 If bar AB in Problem 8.41 is rotating at 1 rad/s in the clockwise direction at the instant shown, what are the angular velocities of the bars at the instant just before the joint B hits the smooth floor?

8.4 Principles of Impulse and Momentum

In this section we review our discussion of the principle of linear impulse and momentum from Chapter 5 and then derive the principle of angular impulse and momentum for a rigid body. These principles relate time integrals of the forces and couples acting on a rigid body to changes in the velocity of its centre of mass and its angular velocity. You can use these principles to determine the effects of impulsive forces and couples on the motion of a rigid body. They also allow you to determine both the velocities of the centres of mass and the angular velocities of objects after they undergo collisions.

Linear Momentum

Integrating Newton's second law with respect to time yields the principle of linear impulse and momentum for a rigid body:

$$\boxed{\int_{t_1}^{t_2} \Sigma \mathbf{F}\, dt = m\mathbf{v}_2 - m\mathbf{v}_1} \tag{8.25}$$

where $\mathbf{v}_1$ and $\mathbf{v}_2$ are the velocities of the centre of mass at the times t_1 and t_2 (Figure 8.13). If you know the external forces acting on a rigid body as functions of time, this principle allows you to determine the change in the velocity of its centre of mass during an interval of time. In terms of the average with respect to time of the total force from t_1 to t_2,

$$\Sigma \mathbf{F}_{\text{av}} = \frac{1}{t_2 - t_1} \int_{t_1}^{t_2} \Sigma \mathbf{F}\, dt$$

we can write Equation (8.25) as

$$(t_2 - t_1)\Sigma \mathbf{F}_{\text{av}} = m\mathbf{v}_2 - m\mathbf{v}_1 \tag{8.26}$$

This form of the principle of linear impulse and momentum is often useful when an object is subjected to impulsive forces.

If the only forces acting on two rigid bodies A and B are the forces they exert on each other, or if other forces are negligible, their total linear momentum is conserved:

$$m_A\mathbf{v}_A + m_B\mathbf{v}_B = \text{constant} \tag{8.27}$$

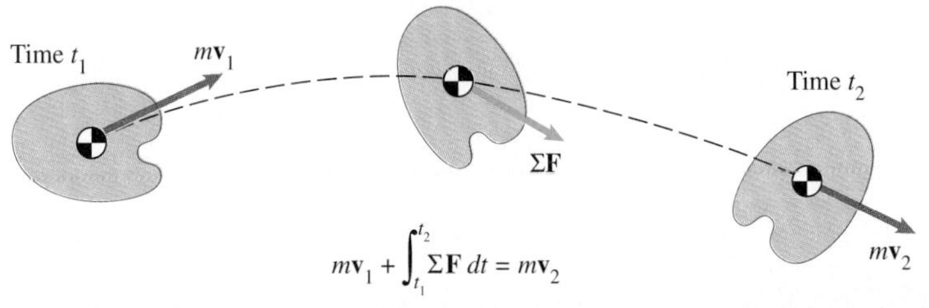

Figure 8.13
Principle of linear impulse and momentum.

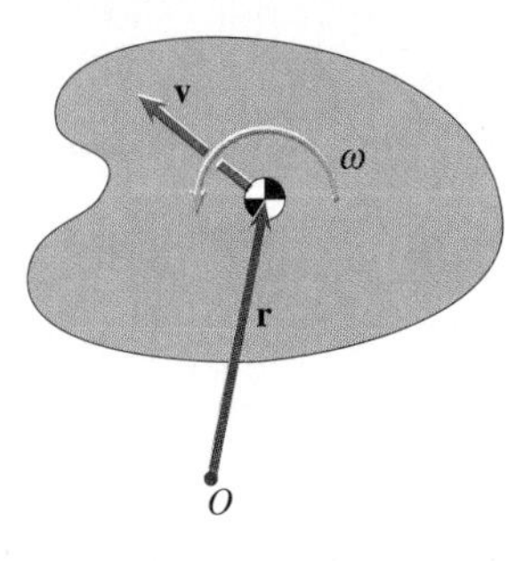

Figure 8.14

A rigid body in planar motion with velocity **v** and angular velocity ω.

Angular Momentum

When you apply momentum principles to rigid bodies, you will often be interested in determining both the velocities of their centres of mass and their angular velocities. For this task, the principle of linear impulse and momentum alone is not sufficient. In this section we derive the principle of angular impulse and momentum for a rigid body in planar motion.

Let's consider a rigid body in general planar motion relative to a fixed reference point O (Figure 8.14). In Chapter 7, we expressed the relation between the total moment about O due to external forces and couples and the rate of change of the rigid body's angular momentum about O in the form (see Equations 7.11 and 7.20)

$$\Sigma M_0 = \frac{d}{dt}[(\mathbf{r} \times m\mathbf{v}) \cdot \mathbf{k} + I\omega] \tag{8.28}$$

where I is the rigid body's mass moment of inertia about its centre of mass. This equation expresses the angular momentum as the sum of the angular momentum about O due to the velocity of the centre of mass and the angular momentum about the centre of mass. The unit vector **k** is perpendicular to the plane of the motion, and its direction is defined by your choice of the positive direction for the moment and angular velocity. If you define counterclockwise moments and angular velocities to be positive, **k** points out of the page (Figure 8.15).

Instead of using the cross product to evaluate the angular momentum of a rigid body in planar motion, you can evaluate it in the same way that moments

Figure 8.15

Determining the direction of **k**.

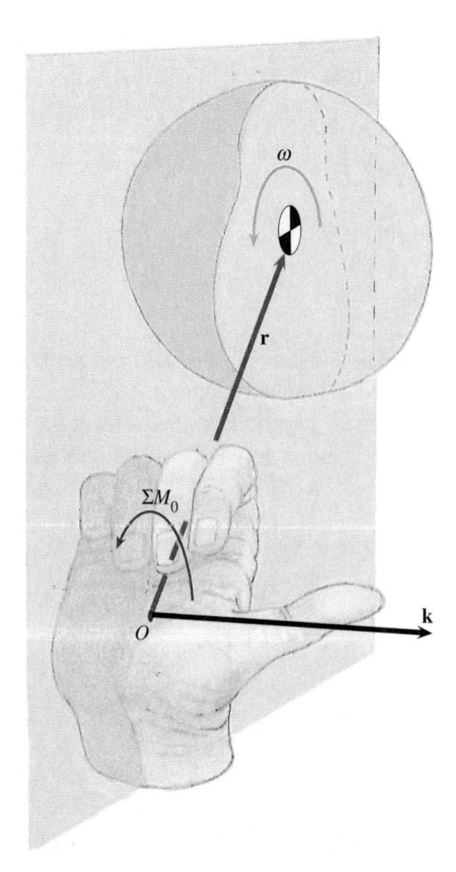

of forces are determined in two-dimensional problems. The magnitude of the 'moment' of the linear momentum $(\mathbf{r} \times m\mathbf{v}) \cdot \mathbf{k}$ equals the product of the magnitude of the linear momentum and the perpendicular distance from O to the line of action of the linear momentum (Figure 8.16). It is positive if the 'moment' is in the direction of positive ω and negative if it is in the opposite direction.

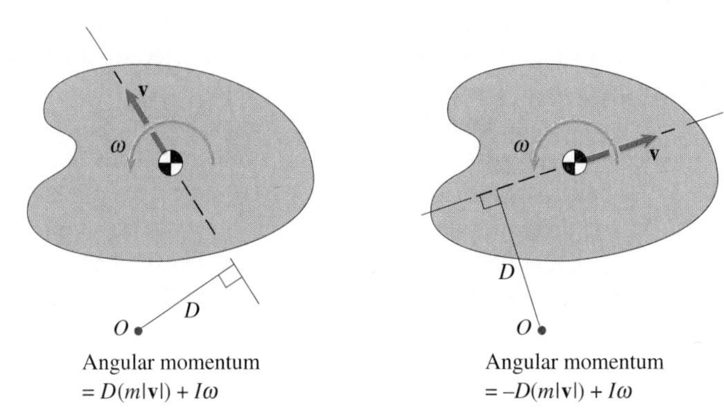

Figure 8.16

Determining the angular momentum about O by calculating the 'moment' of the linear momentum.

Integrating Equation (8.28) with respect to time, we obtain the **principle of angular impulse and momentum**:

$$\int_{t_1}^{t_2} \Sigma M_0 \, dt = [(\mathbf{r} \times m\mathbf{v}) \cdot \mathbf{k} + I\omega]_2 - [(\mathbf{r} \times m\mathbf{v}) \cdot \mathbf{k} + I\omega]_1 \quad (8.29)$$

The angular impulse about O during the interval of time from t_1 to t_2 is equal to the change in the rigid body's angular momentum about O (Figure 8.17). We can also express the principle of angular impulse and momentum in terms of the total moment about the centre of mass. By integrating Equation (7.23) with respect to time, we obtain

$$\int_{t_1}^{t_2} \Sigma M \, dt = I\omega_2 - I\omega_1 \quad (8.30)$$

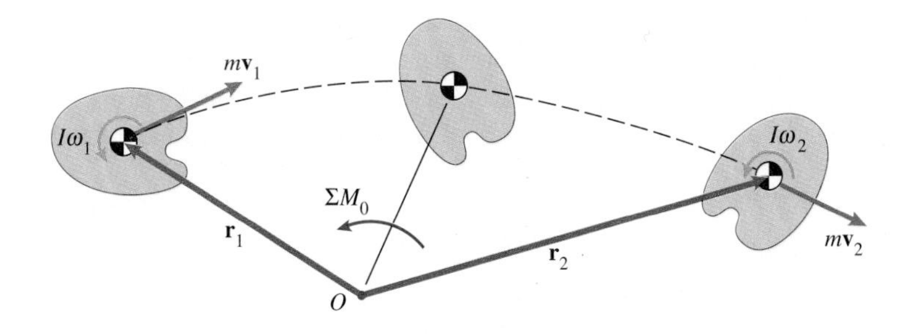

Figure 8.17

Principle of angular impulse and momentum.

The average with respect to time of the moment about O from t_1 to t_2 is

$$(\Sigma M_0)_{\text{av}} = \frac{1}{t_2 - t_1} \int_{t_1}^{t_2} \Sigma M_0 \, dt$$

so we can write Equation (8.29) as

$$(t_2 - t_1)(\Sigma M_0)_{av} = [(\mathbf{r} \times m\mathbf{v}) \cdot \mathbf{k} + I\omega]_2 - [(\mathbf{r} \times m\mathbf{v}) \cdot \mathbf{k} + I\omega]_1 \quad (8.31)$$

You can use this equation to determine the change in the angular momentum of a rigid body subjected to impulsive forces and couples.

We can also use Equation (8.29) to obtain an equation of conservation of total angular momentum for two rigid bodies. Let A and B be rigid bodies in two-dimensional motion in the same plane, and suppose that they are subjected only to the forces and couples they exert on each other, or that other forces and couples are negligible. Let M_{OA} be the moment about a fixed point O due to the forces and couples acting on A, and let M_{OB} be the moment about O due to the forces and couples acting on B. Under the same assumption we made in deriving the equations of motion – the forces between each pair of particles are directed along the line between the particles – the moment $M_{OA} = -M_{OB}$. For example, in Figure 8.18, A and B exert forces on each other by contact. The resulting moments about O are $\mathbf{M}_{OA} = \mathbf{r}_p \times \mathbf{R}$ and $\mathbf{M}_{OB} = \mathbf{r}_p \times (-\mathbf{R}) = -\mathbf{r}_p \times \mathbf{R}$. We apply the principle of angular impulse and momentum to A and B for arbitrary times t_1 and t_2, obtaining

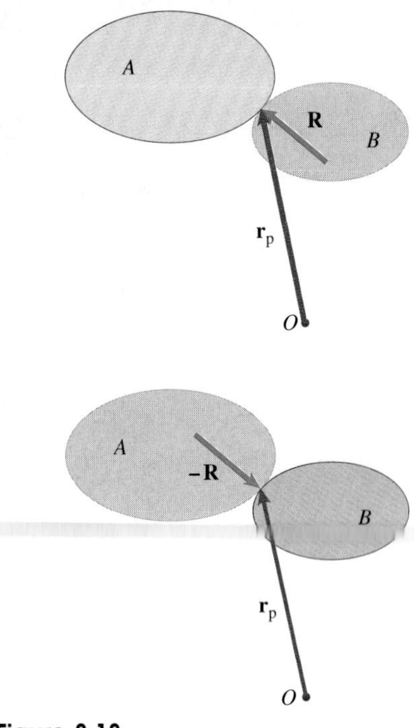

Figure 8.18

Rigid bodies A and B exerting forces on each other by contact.

$$\int_{t_1}^{t_2} M_{OA} dt = [(\mathbf{r}_A \times m_A\mathbf{v}_A) \cdot \mathbf{k} + I_A\omega_A]_2 - [(\mathbf{r}_A \times m_A\mathbf{v}_A) \cdot \mathbf{k} + I_A\omega_A]_1$$

$$\int_{t_1}^{t_2} M_{OB} dt = [(\mathbf{r}_B \times m_B\mathbf{v}_B) \cdot \mathbf{k} + I_B\omega_B]_2 - [(\mathbf{r}_B \times m_B\mathbf{v}_B) \cdot \mathbf{k} + I_B\omega_B]_1$$

Summing these equations, the terms on the left cancel and we obtain

$$\begin{aligned}[(\mathbf{r}_A \times m_A\mathbf{v}_A) \cdot \mathbf{k} + I_A\omega_A]_1 + [(\mathbf{r}_B \times m_B\mathbf{v}_B) \cdot \mathbf{k} + I_B\omega_B]_1 \\ = [(\mathbf{r}_A \times m_A\mathbf{v}_A) \cdot \mathbf{k} + I_A\omega_A]_2 + [(\mathbf{r}_B \times m_B\mathbf{v}_B) \cdot \mathbf{k} + I_B\omega_B]_2\end{aligned}$$

The total angular momentum of A and B about O is conserved:

$$(\mathbf{r}_A \times m_A\mathbf{v}_A) \cdot \mathbf{k} + I_A\omega_A + (\mathbf{r}_B \times m_B\mathbf{v}_B) \cdot \mathbf{k} + I_B\omega_B = \text{constant} \quad (8.32)$$

Notice that this result holds even when A and B are subjected to significant external forces and couples *if the total moment about O due to the external forces and couples is zero*. You can sometimes choose the point O so that this condition is satisfied. This result also applies to an arbitrary number of rigid bodies: their total angular momentum about O is conserved if the total moment about O due to external forces and couples is zero.

In the following examples we demonstrate the use of the principles of linear and angular impulse and momentum to analyse motions of rigid bodies. You should consider using momentum methods when you know the forces and couples acting on an object as functions of time and want to relate them to changes in the velocity of its centre of mass and its angular velocity.

Example 8.4

Disc A in Figure 8.19 initially has a counterclockwise angular velocity ω_0, and disc B is stationary. At $t = 0$, the discs are moved into contact. As a result of friction at the point of contact, the angular velocity of A decreases and the angular velocity of B increases until there is no slip between them. What are their final angular velocities ω_A and ω_B? The discs are supported at their centres of mass and their mass moments of inertia are I_A, I_B.

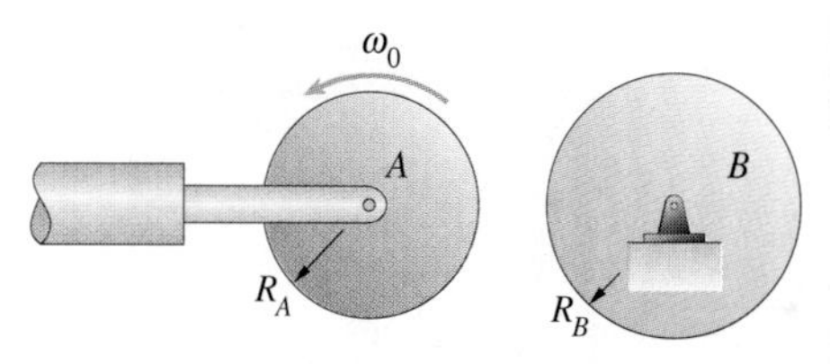

Initial position

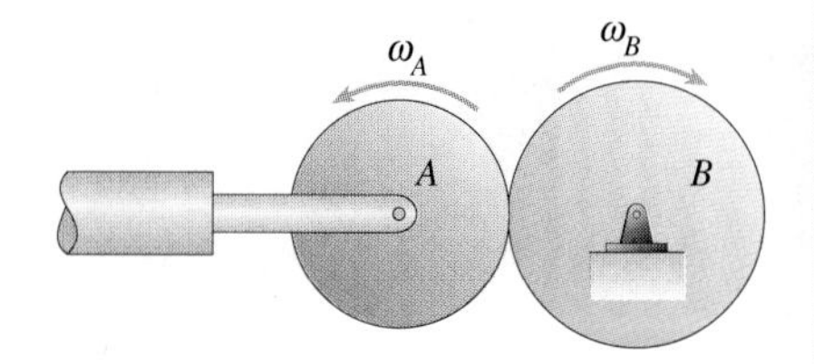

Time $t = 0$

Figure 8.19

STRATEGY

Since the discs rotate about fixed axes through their centres of mass while they are in contact, we can apply the principle of angular impulse and momentum in the form given by Equation (8.30) to each disc. When there is no longer any slip between the discs, their velocities are equal at their point of contact. With this kinematic relationship and the expressions we obtain with the principle of angular impulse and momentum, we can determine the final angular velocities.

SOLUTION

We draw the free-body diagrams of the discs while slip occurs in Figure (a), showing the normal and friction forces they exert on each other. To apply Equation (8.30) to disc A, we let the fixed point O be coincident with its centre. Letting t_f be the time at which slip ceases, we obtain

$$\int_{t_1}^{t_2} \Sigma M_0 \, dt = I\omega_2 - I\omega_1 :$$

$$\int_0^{t_f} -R_A f \, dt = I_A\omega_A - I_A\omega_0$$

We also apply Equation (8.30) to disc B:

$$\int_{t_1}^{t_2} \Sigma M_0 dt = I\omega_2 - I\omega_1 :$$

$$\int_0^{t_f} -R_B f dt = -I_B\omega_B - 0$$

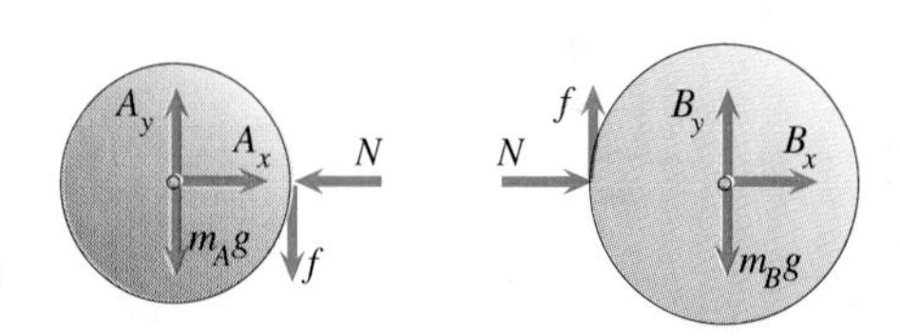

(a) Free-body diagrams of the discs.

(Notice that we have defined ω_B to be positive in the clockwise direction.) We divide the first equation by the second one and write the resulting equation as

$$\omega_A + \frac{R_A I_B}{R_B I_A}\omega_B = \omega_0$$

When there is no slip, the velocities of the discs are equal at their point of contact:

$$R_A\omega_A = R_B\omega_B$$

Solving these two equations, we obtain

$$\omega_A = \omega_0 \left[\frac{1}{1 + \dfrac{R_A^2 I_B}{R_B^2 I_A}} \right] \qquad \omega_B = \omega_0 \left[\frac{R_A/R_B}{1 + \dfrac{R_A^2 I_B}{R_B^2 I_A}} \right]$$

Notice that if the discs have the same radius and mass moment of inertia, $\omega_A = \frac{1}{2}\omega_0$ and $\omega_B = \frac{1}{2}\omega_0$.

Example 8.5

Engineers design a street light to shear off at ground level when struck by a vehicle, to help prevent injuries to passengers (Figure 8.20). From videotape of a test impact, they estimate the angular velocity of the pole to be 0.74 rad/s and the horizontal velocity of the its centrepoint to be 6.7 m/s after the impact, and they estimate the duration of the impact to be $\Delta t = 0.01$ s. If the pole can be modelled as a 6 m, 64 kg slender bar, the car strikes it 0.6 m above the ground, and the couple exerted on the pole by its support can be neglected, what average force was required to shear off the bolts supporting the pole?

Figure 8.20

STRATEGY

We can use the principles of linear and angular impulse and momentum, expressed in terms of the average force and moment exerted on the pole, to determine the average shear force.

SOLUTION

We draw the free-body diagram of the pole in Figure (a), where F is the average force exerted by the car and S is the average shearing force exerted on the pole by the bolts. Let m be the mass of the pole, and let v and ω be the velocity of its centre of mass and its angular velocity at the end of the impact (Figure (b)). The principle of linear impulse and momentum expressed in terms of the average horizontal force is

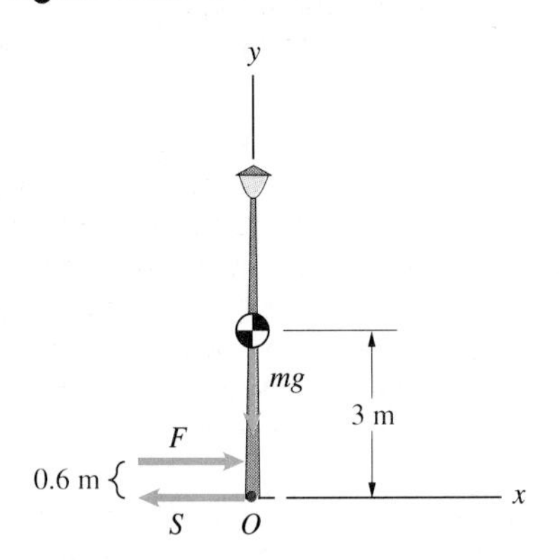

(a) Free-body diagram of the pole.

$$(t_2 - t_1)(\Sigma F_x)_{\text{av}} = (mv_x)_2 - (mv_x)_1 :$$
$$\Delta t(F - S) = mv - 0 \qquad (8.33)$$

To apply the principle of angular impulse and momentum, we use Equation (8.31), placing the fixed point O at the bottom of the pole (Figure (a)). The pole's angular momentum about O at the end of the impact is

$$[(\mathbf{r} \times m\mathbf{v}) \cdot \mathbf{k} + I\omega]_2 = [(3\,\mathbf{j}) \times m(v\,\mathbf{i})] \cdot k + I\omega = -3mv + I\omega$$

(We can also obtain this result by calculating the 'moment' of the linear momentum about O and adding the term $I\omega$. The magnitude of the 'moment' is the product of magnitude of the linear momentum (mv) and the perpendicular distance from O to the line of action of the linear momentum (3 m), and it is negative because the 'moment' is in the direction opposite to that in which we define ω to be positive. See Figure 8.16.) We express the principle of angular impulse and momentum in terms of the average moment about O:

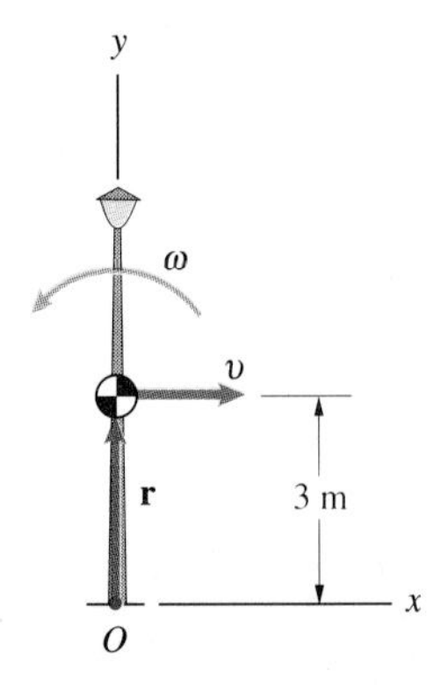

(b) Velocity and angular velocity at the end of the impact.

$$(t_2 - t_1)(\Sigma M_O)_{\text{av}} = [(\mathbf{r} \times m\mathbf{v}) \cdot \mathbf{k} + I\omega]_2 - [(\mathbf{r} \times m\mathbf{v}) \cdot \mathbf{k} + I\omega]_1 :$$
$$\Delta t(-0.6F) = -3mv + I\omega - 0$$

Solving this equation together with Equation (8.33) for the average shear force S, we obtain

$$S = \frac{2.4mv - I\omega}{0.6\Delta t} = \frac{2.4(64)(6.7) - \frac{1}{12}(64)(6)^2(0.74)}{0.6(0.01)}$$
$$= 147.8 \text{ kN}$$

Problems

8.43 The mass moment of inertia of the pulley is 0.5 kg.m^2. The pulley starts from rest at $t = 0$. For both cases, use momentum principles to determine the pulley's angular velocity at $t = 1$ s.

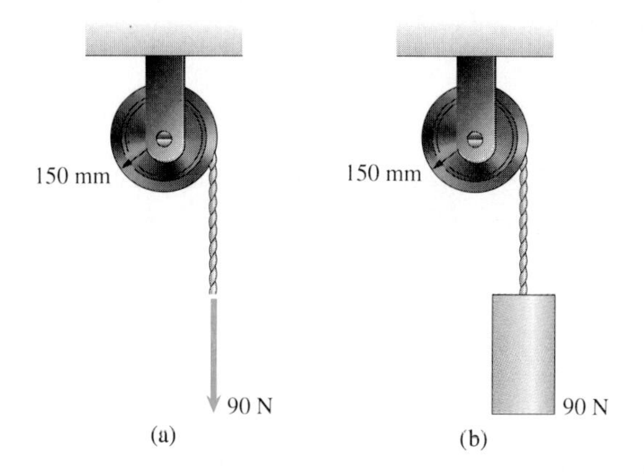

P8.43

8.44 An astronaut fires a thruster of his manoeuvring unit, exerting a force $T = 2(1 + t)$ N, where t is in seconds. The combined mass of the astronaut and his equipment is 122 kg, and the mass moment of inertia about their centre of mass is 45 kg.m^2. Modelling the astronaut and his equipment as a rigid body, use the principle of angular impulse and momentum to determine how long it takes for his angular velocity to reach 0.1 rad/s.

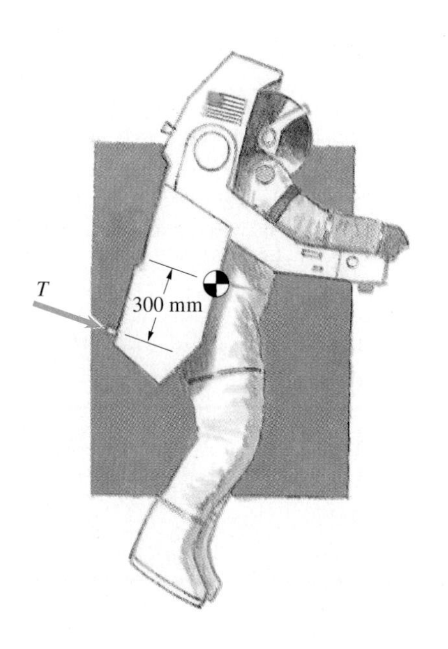

P8.44

8.45 The manoeuvring unit in Problem 8.44 exerts an impulsive force T of 0.2 s duration, giving the astronaut a counterclockwise angular velocity of one revolution per minute.
(a) What is the average value of the impulsive force?
(b) What is the magnitude of the change in the velocity of his centre of mass?

8.46 A flywheel attached to an electric motor is initially at rest. At $t = 0$, the motor exerts a couple $M = 200e^{-0.1t}$ N.m on the flywheel. The mass moment of inertia of the flywheel is 10 kg.m^2.
(a) What is the flywheel's angular velocity at $t = 10$ s?
(b) What maximum angular velocity will the flywheel attain?

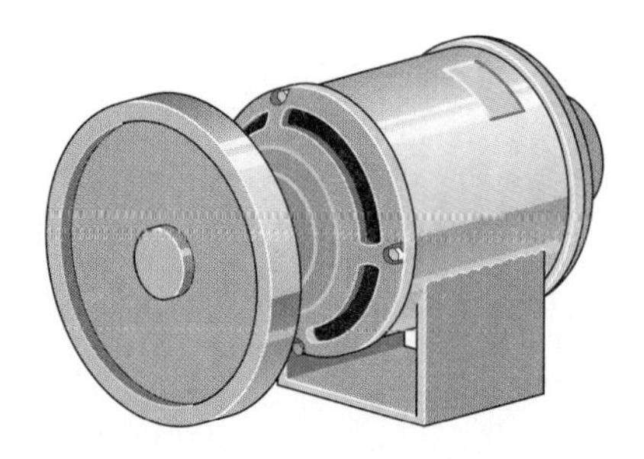

P8.46

8.47 A main landing gear wheel of a Boeing 747 has a mass moment of inertia of 23 kg.m^2 and a 0.6 m radius. The aeroplane is moving at 75 m/s when it touches down. Suppose that you measure the skid marks where the plane touches down and find that they are 10 m long. Assuming that the aeroplane's velocity and the normal force on the wheel are constant while the wheel skids, use the principle of angular impulse and momentum to estimate the friction force exerted on the wheel while it skids.

8.48 The force a club exerts on a 0.046 kg golf ball is shown. The ball is 42.7 mm in diameter and can be modelled as a homogeneous sphere. The club is in contact with the ball for 0.0006 s, and the magnitude of the velocity of the ball's centre of mass after it is struck is 50 m/s. What is the ball's angular velocity after it is struck?

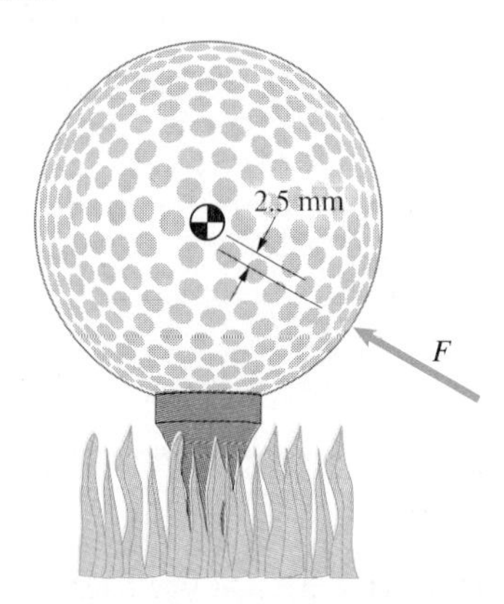

P8.48

8.49 The suspended 8 kg slender bar is subjected to a horizontal impulsive force at B. The average value of the force is 1000 N, and its duration is 0.03 s. If the force causes the bar to swing to the horizontal position before coming to a stop, what is the distance h?

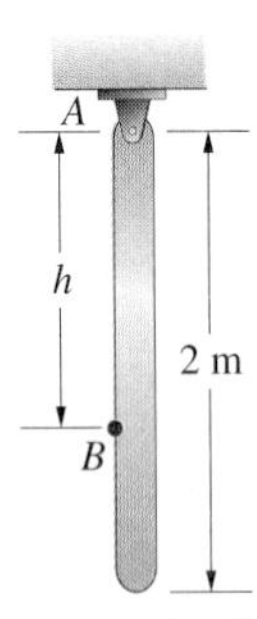

P8.49

8.50 For what value of the distance h in Problem 8.49 will no average horizontal force be exerted on the bar by the support A when the horizontal impulsive force is applied at B? What is the angular velocity of the bar just after the impulsive force is applied?

8.51 The force exerted on the cue ball by the cue is horizontal. Determine the value of h for which the ball rolls without slipping. (Assume that the average friction force exerted on the ball by the table is negligible.)

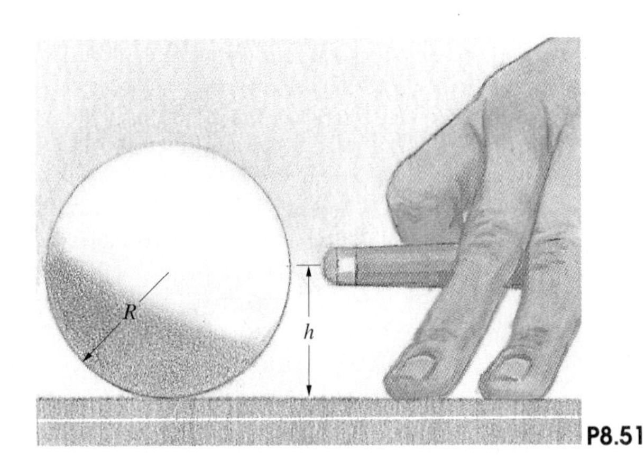

P8.51

8.52 In a well-known demonstration of conservation of angular momentum, a person stands on a rotating platform holding a weight in each hand. Suppose that the mass moment of inertia of the person and platform is 0.4 kg.m^2, and the mass moment of inertia of each 4 kg mass about its centre of mass is 0.001 kg.m^2. If her angular velocity with her arms extended is $\omega_1 = 1$ revolution per second, what is her angular velocity ω_2 when she pulls the weights inwards? (You have observed figure skaters using this phenomenon to control their angular velocity in a spin by altering the positions of their arms.)

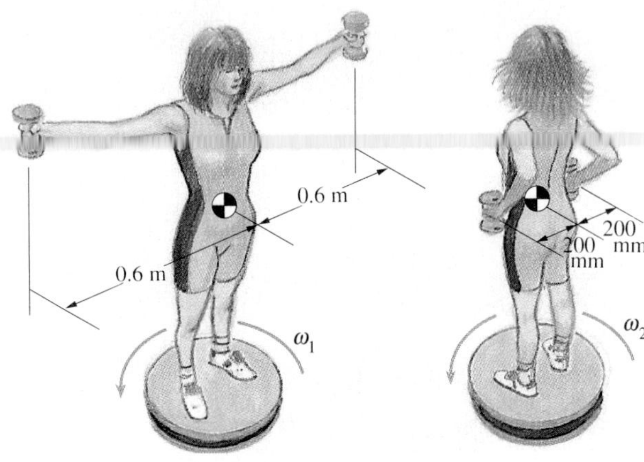

P8.52

8.53 The space shuttle simultaneously deploys two satellites by sending them into space connected together and then allowing them to separate. The satellites can be modelled as identical homogeneous cylinders of mass m, radius R, and length l. Before separation, they are spinning about an axis perpendicular to the axis of the cylinders with angular velocity ω (Figure (a)). The attachments are then released and they drift apart (Figure (b)).

(a) Use conservation of angular momentum to determine the angular velocity ω' of the individual satellites.

(b) What is the magnitude of the velocity **v** of their centres of mass relative to the velocity of their centre of mass before separation?

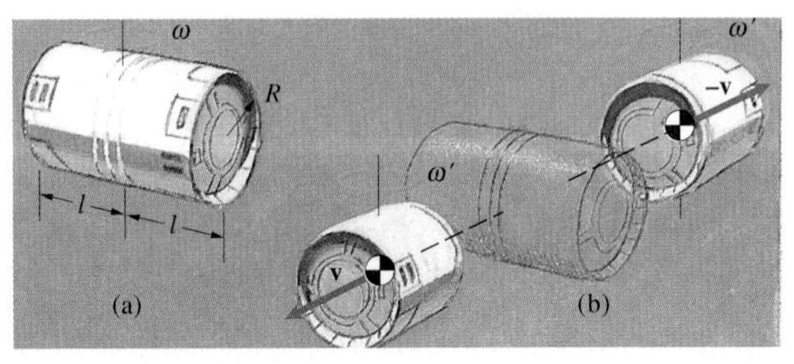

P8.53

8.54 A satellite is deployed with angular velocity $\omega = 1$ rad/s (Figure (a)). Two internally stored antennas that span the diameter of the satellite are then extended, and the satellite's angular velocity decreases to ω' (Figure (b)). By modelling the satellite as a 500 kg sphere of 1.2 m radius and each antenna as a 10 kg slender bar, determine ω'.

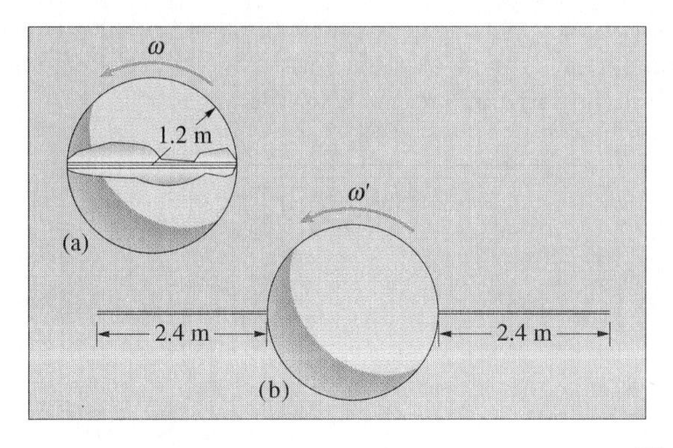

P8.54

8.55 The slender bar rotates freely *in the horizontal plane* about a vertical shaft at O. The bar weighs 90 N and its length is 2 m. The slider A weighs 9 N. If the bar's angular velocity is $\omega = 10$ rad/s and the radial component of the velocity of A is zero when $r = 0.3$ m, what is the angular velocity of the bar when $r = 1.2$ m? (The mass moment of inertia of A about its centre of mass is negligible; that is, treat A as a particle.)

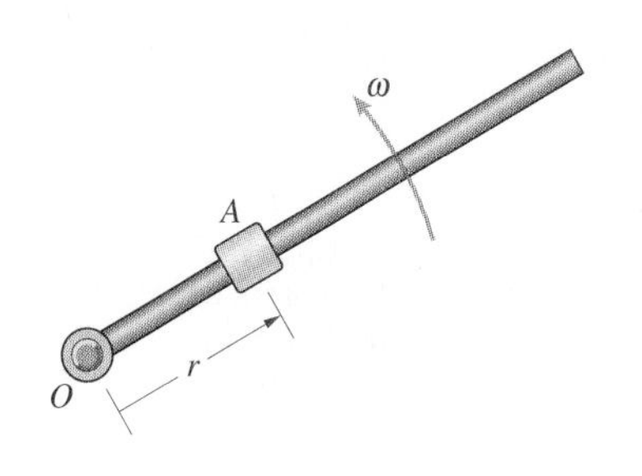

P8.55

8.5 Impacts

In Chapter 5, we analysed impacts between objects with the objective of determining their velocities – the velocities of their centres of mass – after the collision. We now discuss how you can determine the velocities of the centres of mass *and the angular velocities* of rigid bodies after they collide.

Conservation of Momentum

Suppose that two rigid bodies A and B, in two-dimensional motion in the same plane, collide. What do the principles of linear and angular momentum tell us about their motions after the collision?

Linear Momentum If other forces are negligible in comparison with the impact forces that A and B exert on each other, their total linear momentum is the same before and after the impact. But you must use care in applying this result. For example, if one of the rigid bodies has a pin support (Figure 8.21), the reactions exerted by the support cannot be neglected and linear momentum is not conserved.

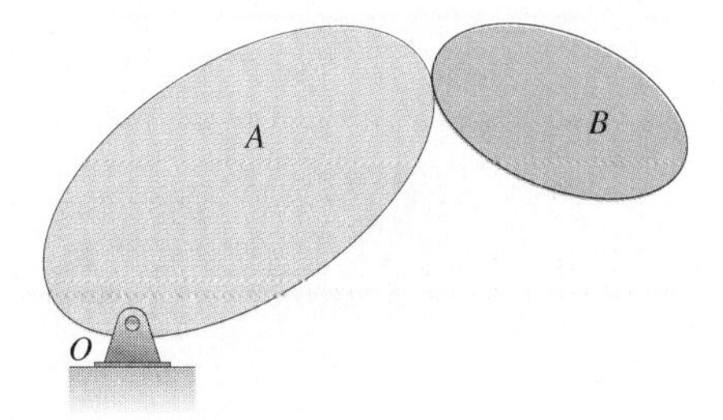

Figure 8.21

Rigid bodies A and B colliding. Because of the pin support, their total linear momentum is *not* conserved, but their total angular momentum about O is conserved.

Angular Momentum If other forces and couples are negligible in comparison with the impact forces and couples that A and B exert on each other, their total angular momentum about *any* fixed point O is the same before and after the impact. (See Equation 8.32.) If, in addition, A and B exert only forces on each other at their point of impact P, and no couples, the angular momentum about P of *each* rigid body is the same before and after the impact (Figure 8.22). This result follows from the principle of angular impulse and momentum, Equation (8.29), because the impact forces on A and B exert no moment about P. If one of the rigid bodies has a pin support at a point O, as in Figure 8.21, their total angular momentum about O is the same before and after the impact.

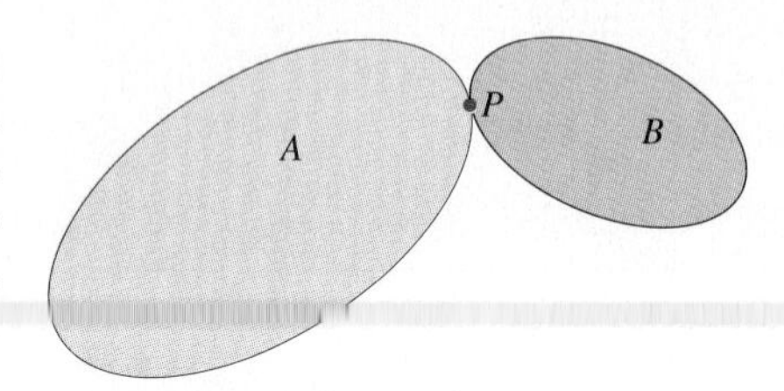

Figure 8.22

Rigid bodies A and B colliding at P. If only forces are exerted at P, the angular momentum of A about P and the angular momentum of B about P are each conserved.

Coefficient of Restitution

If two rigid bodies adhere and move as a single rigid body after colliding, you can determine their velocities and angular velocities using momentum conservation and kinematic relationships alone. These relationships are not sufficient if the objects do not adhere. But you can analyse some impacts of this type by also using the concept of the coefficient of restitution.

Let P be the point of contact of rigid bodies A and B during an impact (Figure 8.23), and let their velocities at P be $\mathbf{v}_{AP}$ and $\mathbf{v}_{BP}$ just before the impact and $\mathbf{v}'_{AP}$ and $\mathbf{v}'_{BP}$ just afterwards. The x axis is perpendicular to the contacting surfaces at P. If the friction forces resulting from the impact are negligible, we can show that the components of the velocities normal to the surfaces at P are related to the coefficient of restitution e by

$$e = \frac{(\mathbf{v}'_{BP})_x - (\mathbf{v}'_{AP})_x}{(\mathbf{v}_{AP})_x - (\mathbf{v}_{BP})_x}$$

To derive this result, we must consider the effects of the impact on the individual objects. Let t_1 be the time at which they first come into contact. The objects are not actually rigid, but will deform as a result of the collision. At a time t_C, the maximum deformation will occur and the objects will begin a 'recovery' phase in which they tend to resume their original shapes. Let t_2 be the time at which they separate.

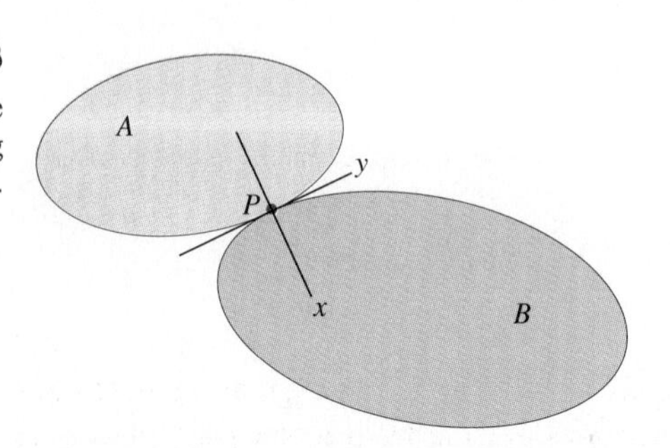

Figure 8.23

Rigid bodies A and B colliding at P. The x axis is perpendicular to the contacting surfaces.

Our first step is to apply the principle of linear impulse and momentum to A and B for the intervals of time from t_1 to t_C and from t_C to t_2. Let R be the magnitude of the normal force exerted during the impact (Figure 8.24). We denote the velocity of the centre of mass of A at the times t_1, t_C and t_2 by $\mathbf{v}_A$, $\mathbf{v}_{AC}$ and $\mathbf{v}'_A$, and denote the corresponding velocities of the centre of mass of B by $\mathbf{v}_B$, $\mathbf{v}_{BC}$ and $\mathbf{v}'_B$. For A, we have

$$\int_{t_1}^{t_C} -R\,dt = m_A(\mathbf{v}_{AC})_x - m_A(\mathbf{v}_A)_x \tag{8.34}$$

$$\int_{t_C}^{t_2} -R\,dt = m_A(\mathbf{v}'_A)_x - m_A(\mathbf{v}_{AC})_x \tag{8.35}$$

and for B,

$$\int_{t_1}^{t_C} R\,dt = m_B(\mathbf{v}_{BC})_x - m_B(\mathbf{v}_B)_x \tag{8.36}$$

$$\int_{t_C}^{t_2} R\,dt = m_B(\mathbf{v}'_B)_x - m_B(\mathbf{v}_{BC})_x \tag{8.37}$$

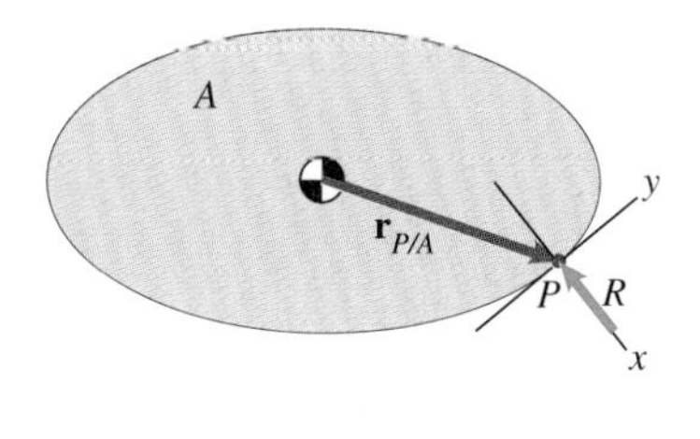

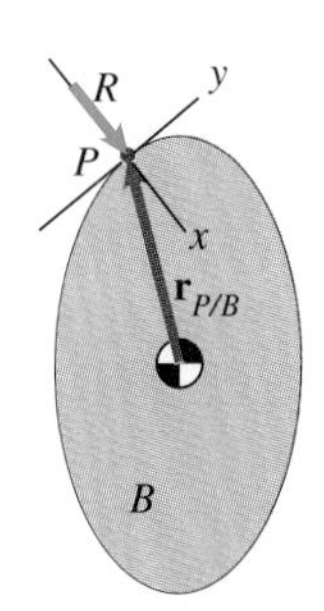

Figure 8.24
The normal force R resulting from the impact.

The coefficient of restitution is the ratio of the linear impulse during the recovery phase to the linear impulse during the deformation phase:

$$e = \frac{\displaystyle\int_{t_C}^{t_2} R\,dt}{\displaystyle\int_{t_1}^{t_C} R\,dt}$$

If we divide Equation (8.35) by Equation (8.34) and divide Equation (8.37) by Equation (8.36), we can write the resulting equations as

$$\begin{aligned} (\mathbf{v}'_A)_x &= -(\mathbf{v}_A)_x e + (\mathbf{v}_{AC})_x(1+e) \\ (\mathbf{v}'_B)_x &= -(\mathbf{v}_B)_x e + (\mathbf{v}_{BC})_x(1+e) \end{aligned} \tag{8.38}$$

We now apply the principle of angular impulse and momentum to A and B for the intervals of time from t_1 to t_C and from t_C to t_2. We denote the counterclockwise angular velocity of A at the times t_1, t_C and t_2 by ω_A, ω_{AC} and ω'_A, and denote the corresponding angular velocities of B by ω_B, ω_{BC} and ω'_B. We write the position vectors of P relative to the centres of mass of A and B as (Figure 8.24)

$$\mathbf{r}_{P/A} = x_A\,\mathbf{i} + y_A\,\mathbf{j}$$
$$\mathbf{r}_{P/B} = x_B\,\mathbf{i} + y_B\,\mathbf{j}$$

To apply the principle of angular impulse and momentum to A, we place the fixed point O at the centre of mass of A. Then the moment about O due to the force exerted on A by the impact is $\mathbf{r}_{P/A} \times (-R\,\mathbf{i}) = y_A R\,\mathbf{k}$, and we obtain

$$\int_{t_1}^{t_C} y_A R\,dt = I_A\omega_{AC} - I_A\omega_A \tag{8.39}$$

$$\int_{t_C}^{t_2} y_A R\,dt = I_A\omega'_A - I_A\omega_{AC} \tag{8.40}$$

The corresponding equations for B, obtained by placing the fixed point O at the centre of mass of B, are

$$\int_{t_1}^{t_C} -y_B R\,dt = I_B\omega_{BC} - I_B\omega_B \tag{8.41}$$

$$\int_{t_C}^{t_2} y_B R\,dt = I_B\omega'_B - I_B\omega_{BC} \tag{8.42}$$

Dividing Equation (8.40) by Equation (8.39) and dividing Equation (8.42) by Equation (8.41), we can write the resulting equations as

$$\begin{aligned} \omega'_A &= -\omega_A e + \omega_{AC}(1+e) \\ \omega'_B &= -\omega_B e + \omega_{BC}(1+e) \end{aligned} \tag{8.43}$$

By expressing the velocity of the point of A at P in terms of the velocity of the centre of mass of A and the angular velocity of A, and expressing the velocity of the point of B at P in terms of the velocity of the centre of mass of B and the angular velocity of B, we obtain

$$\begin{aligned} (\mathbf{v}_{AP})_x &= (\mathbf{v}_A)_x - \omega_A y_A \\ (\mathbf{v}'_{AP})_x &= (\mathbf{v}'_A)_x - \omega'_A y_A \\ (\mathbf{v}_{BP})_x &= (\mathbf{v}_B)_x - \omega_B y_B \\ (\mathbf{v}'_{BP})_x &= (\mathbf{v}'_B)_x - \omega'_B y_B \end{aligned} \tag{8.44}$$

At time t_C, the x components of the velocities of the two objects are equal at P, which yields the relation

$$(\mathbf{v}_{AC})_x - \omega_{AC} y_A = (\mathbf{v}_{BC})_x - \omega_{BC} y_B \tag{8.45}$$

From Equations (8.44),

$$\frac{(\mathbf{v}'_{BP})_x - (\mathbf{v}'_{AP})_x}{(\mathbf{v}_{AP})_x - (\mathbf{v}_{BP})_x} = \frac{(\mathbf{v}'_B)_x - \omega'_B y_B - (\mathbf{v}'_A)_x + \omega'_A y_a}{(\mathbf{v}_A)_x - \omega_A y_A - (\mathbf{v}_B)_x + \omega_B y_B}$$

Substituting Equations (8.38) and (8.43) into this equation and collecting terms, we obtain

$$\frac{(\mathbf{v}'_{BP})_x - (\mathbf{v}'_{AP})_x}{(\mathbf{v}_{AP})_x - (\mathbf{v}_{BP})_x} = e - \left[\frac{(\mathbf{v}_{AC})_x - \omega_{AC} y_A - (\mathbf{v}_{BC})_x + \omega_{BC} y_B}{(\mathbf{v}_A)_x - \omega_A y_A - (\mathbf{v}_B)_x + \omega_B y_B}\right](e + 1)$$

The term in brackets vanishes due to Equation (8.45), and we obtain the equation relating the normal components of the velocities at the point of contact to the coefficient of restitution:

$$e = \frac{(\mathbf{v}'_{BP})_x - (\mathbf{v}'_{AP})_x}{(\mathbf{v}_{AP})_x - (\mathbf{v}_{BP})_x} \tag{8.46}$$

In obtaining this equation, we assumed that the contacting surfaces are smooth, so **the collision exerts no force on A or B in the direction tangential to their contacting surfaces.**

Although we derived Equation (8.46) under the assumption that the motions of A and B are unconstrained, it also holds if they are not, for example if one of them is connected to a pin support.

In the following examples we analyse collisions of rigid bodies in planar motion. Your approach will depend on the type of collision. If other forces are negligible in comparison with the impact forces, total linear momentum is conserved. If other forces and couples are negligible in comparison with the impact forces and couples, total angular momentum about any fixed point is conserved. If, in addition, only forces are exerted at the point of impact P, the angular momentum about P of each rigid body is conserved. If one of the rigid bodies has a pin support at a point O, the total angular momentum about O is conserved. If impact is assumed to exert no forces on the colliding objects in the direction tangential to their surface of contact, the coefficient of restitution e relates the normal components of the velocities at the point of contact through Equation (8.46).

Example 8.6

The homogeneous sphere in Figure 8.25 is moving horizontally with velocity v_A and no angular velocity when it strikes the stationary slender bar. The sphere has mass m_A, and the bar has mass m_B and length l. The coefficient of restitution of the impact is e.

(a) What is the angular velocity of the bar after the impact?

(b) If the duration of the impact is Δt, what average horizontal force is exerted on the bar by the pin support C as a result of the impact?

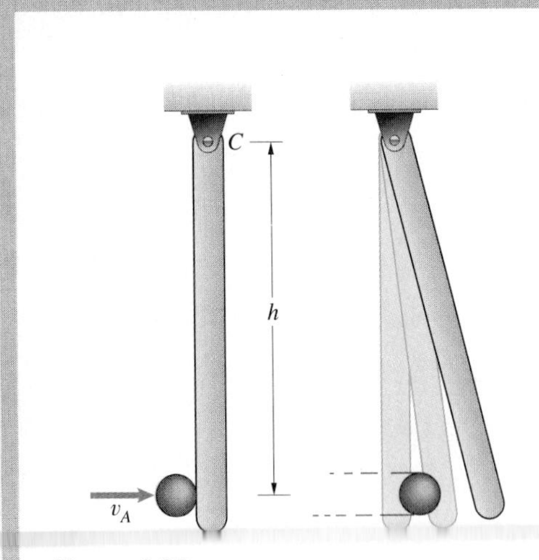

Figure 8.25

STRATEGY

(a) From the definition of the coefficient of restitution, we can obtain an equation relating the horizontal velocity of the sphere and the velocity of the bar at the point of impact after the collision occurs. In addition, the total angular momentum of the sphere and bar about the pin C is conserved. With these two equations and kinematic relationships, we can determine the velocity of the sphere and the angular velocity of the bar. (b) We can determine the average force exerted on the bar by the support by applying the principle of angular impulse and momentum to the bar.

SOLUTION

(a) In Figure (a) we show the velocities just after the impact, where v'_{BP} is the bar's velocity at the point of impact. From the definition of the coefficient of restitution, we obtain

$$e = \frac{v'_{BP} - v'_A}{v_A - 0}$$

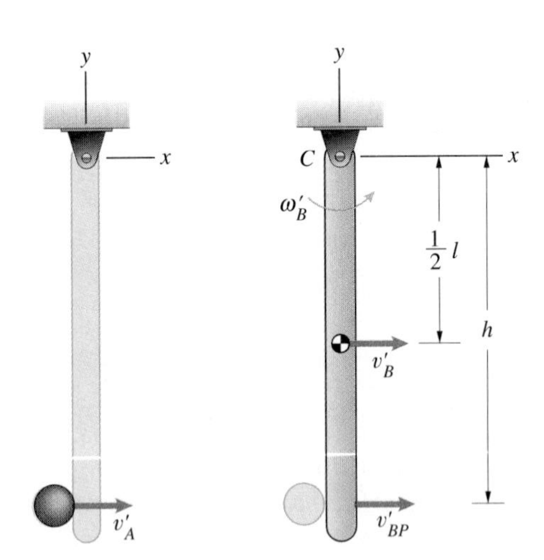

(a) Velocities of the sphere and bar after the impact.

The equation of conservation of total angular momentum about C is

$$(\mathbf{r}_A \times m_A \mathbf{v}_A) \cdot \mathbf{k} + I_A \omega_A + (\mathbf{r}_B \times m_B \mathbf{v}_B) \cdot \mathbf{k} + I_B \omega_B$$

$$= (\mathbf{r}_A \times m_A \mathbf{v}'_A) \cdot \mathbf{k} + I_A \omega'_A + (\mathbf{r}_B \times m_B \mathbf{v}'_B) \cdot \mathbf{k}_I + I_B B \omega'_B :$$

$$(-h\mathbf{j} \times m_A v_A \mathbf{i}) \cdot \mathbf{k} = (-h\mathbf{j} \times m_A v'_A \mathbf{i}) \cdot \mathbf{k} + \left(-\frac{1}{2} l \mathbf{j} \times m_B v'_B \mathbf{i}\right) \cdot \mathbf{k} + I_B \omega'_B$$

Carrying out the vector operations, we obtain

$$h m_A v_A = h m_A v'_A + \frac{1}{2} l m_B v'_B + I_B \omega'_B$$

Notice in Figure (a) that the velocities v'_B and v'_{BP} are related to the angular velocity of the bar ω'_B by

$$v'_B = \frac{1}{2} l \omega'_B \qquad v'_{BP} = h \omega'_B$$

We now have four equations in the four unknowns v'_A, v'_B, v'_{BP} and ω'_B. Solving them for the angular velocity of the bar and using the expression $I_B = \frac{1}{12} m_B l^2$, we obtain

$$\omega'_B = \frac{(1+e) h m_A v_A}{h^2 m_A + \frac{1}{3} m_B l^2}$$

(b) Let the forces on the free-body diagram of the bar in Figure (b) represent the average forces exerted during the impact. We apply the principle of angular impulse and momentum, in the form given by Equation (8.31), about the point of impact:

$$(t_2 - t_1)(\Sigma M_P)_{\text{av}} = [(\mathbf{r}_B \times m_B \mathbf{v}'_B) \cdot \mathbf{k} + I_B \omega'_B] - [(\mathbf{r}_B \times m_B \mathbf{v}_B) \cdot \mathbf{k} + I_B \omega_B]:$$

$$\Delta t(-hC_x) = \left[\left(h - \frac{1}{2} l\right) \mathbf{j} \times m_B v'_B \mathbf{i}\right] \cdot \mathbf{k} + I_B \omega'_B - 0$$

Solving for C_x, we obtain

$$C_x = \frac{(h - \frac{1}{2} l) m_B v'_B - I_B \omega'_B}{h \Delta t}$$

Using our solution for ω'_B from part (a) and the relation $v'_B = \frac{1}{2} \omega'_B$, we obtain the average horizontal force exerted by the support:

$$C_x = \frac{(1+e)(\frac{1}{2} h - \frac{1}{3} l) l m_A m_B v_A}{(h^2 m_A + \frac{1}{3} m_B l^2) \Delta t}$$

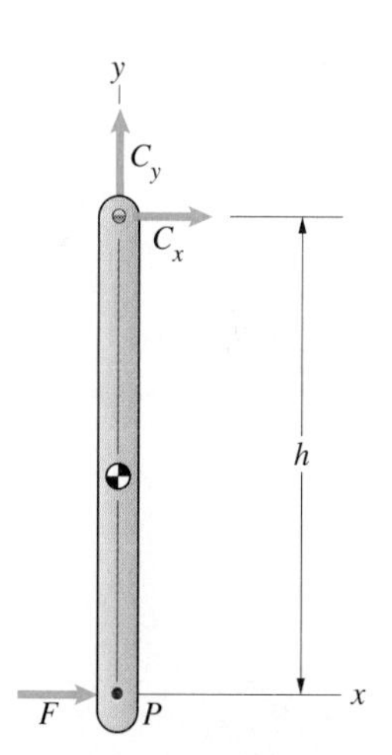

(b) Average forces exerted on the bar during the impact.

DISCUSSION

Notice that the average horizontal force exerted on the bar by the support can be in either direction or can be zero, depending on where the impact occurs. The force is zero if $h = \frac{2}{3} l$.

Example 8.7

The combined mass of the motorcycle and rider in Figure 8.26 is $m = 170$ kg, and their combined moment of inertia about their centre of mass is 22 kg.m^2. Following a jump over an obstacle, the motorcycle and rider are in the position shown just before the rear wheel contacts the ground. The velocity of their centre of mass is of magnitude $|\mathbf{v}_G| = 8.8$ m/s and their angular velocity is $\omega = 0.2$ rad/s. If the motorcycle and rider are modelled as a single rigid body and the coefficient of restitution of the impact is $e = 0.8$, what are the angular velocity ω' and velocity $\mathbf{v}'_G$ after the impact? Neglect the tangential component of force exerted on the motorcycle's wheel during the impact.

Figure 8.26

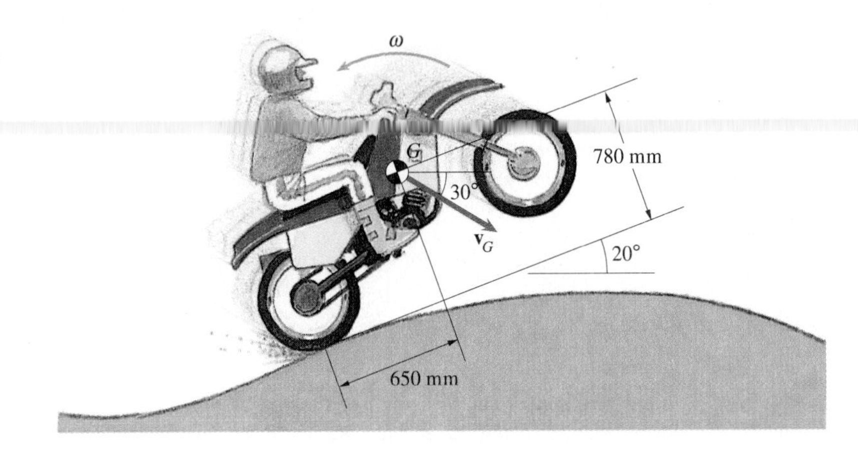

STRATEGY

Since the tangential component of force on the motorcycle's wheel during the impact is neglected, the component of the velocity of the centre of mass parallel to the ground is unchanged by the impact. The coefficient of restitution relates the motorcycle's velocity normal to the ground *at the point of impact* before the impact to its value after the impact. Also, the force of the impact exerts no moment about the point of impact, so the motorcycle's angular momentum about that point is conserved. (We assume the impact to be so brief that the angular impulse due to the weight is negligible.) With these three relations we can determine the two components of the velocity of the centre of mass and the angular velocity after the impact.

SOLUTION

In Figure (a) we align a coordinate system parallel and perpendicular to the ground at the point P where the impact occurs. Let the components of the velocity of the centre of mass before and after the impact be $\mathbf{v}_G = v_x\,\mathbf{i} + v_y\,\mathbf{j}$ and $\mathbf{v}'_G = v'_x\,\mathbf{i} + v'_y\,\mathbf{j}$, respectively. The components v_x and v_y are

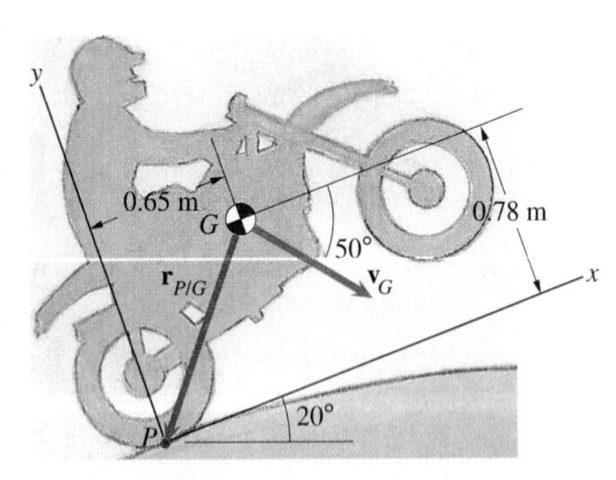

(a) Aligning the x axis of the coordinate system tangent to the ground at P.

$$v_x = 8.8 \cos 50^\circ = 5.66 \text{ m/s}$$

$$v_y = -8.8 \sin 50^\circ = -6.74 \text{ m/s}$$

Because the component of the impact force tangential to the ground is neglected, the x component of the velocity of the centre of mass is unchanged:

$$v'_x = v_x = 5.66 \text{ m/s}$$

We can express the y component of the wheel's velocity at P before the impact in terms of the velocity of the centre of mass and the angular velocity (Figure (a)):

$$\begin{aligned}\mathbf{j}\cdot\mathbf{v}_P &= \mathbf{j}\cdot(\mathbf{v}_G + \boldsymbol{\omega}\times\mathbf{r}_{P/G}) \\ &= \mathbf{j}\cdot\left\{v_x\,\mathbf{i} + v_y\,\mathbf{j} + \begin{vmatrix}\mathbf{i} & \mathbf{j} & \mathbf{k} \\ 0 & 0 & \omega \\ -0.65 & -0.78 & 0\end{vmatrix}\right\} \\ &= v_y - 0.65\omega\end{aligned}$$

(Notice that this expression gives the y component of the velocity at P even though the wheel is spinning.) The y component of the wheel's velocity at P after the impact is

$$\begin{aligned}\mathbf{j}\cdot\mathbf{v}'_P &= \mathbf{j}\cdot(\mathbf{v}'_G + \boldsymbol{\omega}'\times\mathbf{r}_{P/G}) \\ &= v'_y - 0.65\omega'\end{aligned}$$

The coefficient of restitution relates the y components of the wheel's velocity at P before and after the impact:

$$e = \frac{-(\mathbf{j}\cdot\mathbf{v}'_P)}{(\mathbf{j}\cdot\mathbf{v}_P)} = \frac{-(v'_y - 0.65\omega')}{(v_y - 0.65\omega)} \tag{8.47}$$

The force of the impact exerts no moment about P, so angular momentum about P is conserved:

$$[(\mathbf{r}_{G/P}\times m\mathbf{v}_G)\cdot\mathbf{k} + I\omega] = [(\mathbf{r}_{G/P}\times m\mathbf{v}'_G)\cdot\mathbf{k} + I\omega']$$

$$\begin{vmatrix}\mathbf{i} & \mathbf{j} & \mathbf{k} \\ 0.65 & 0.78 & 0 \\ mv_x & mv_y & 0\end{vmatrix}\cdot\mathbf{k} + I\omega = \begin{vmatrix}\mathbf{i} & \mathbf{j} & \mathbf{k} \\ 0.65 & 0.78 & 0 \\ mv'_x & mv'_y & 0\end{vmatrix}\cdot\mathbf{k} + I\omega'$$

Expanding the determinants and evaluating the dot products, we obtain

$$0.65mv_y - 0.78mv_x + I\omega = 0.65mv'_y - 0.78mv'_x + I\omega' \tag{8.48}$$

Since we have already determined v'_x, we can solve Equations (8.47) and (8.48) for v'_y and ω'. The results are

$$v'_y = -3.84 \text{ m/s} \qquad \omega' = -14.37 \text{ rad/s}$$

The velocity of the centre of mass after the impact is $\mathbf{v}'_G = 5.66\,\mathbf{i} - 3.84\,\mathbf{j}$ m/s, and the angular velocity is 14.37 rad/s in the clockwise direction.

Example 8.8

An engineer simulates a collision between two 1600 kg cars by modelling them as rigid bodies (Figure 8.27). The mass moment of inertia of each car about its centre of mass is 960 kg.m^2. He assumes the contacting surfaces at P to be smooth and parallel to the x axis and assumes the coefficient of restitution to be $e = 0.2$. What are the angular velocities of the cars and the velocities of their centres of mass after the collision?

Figure 8.27

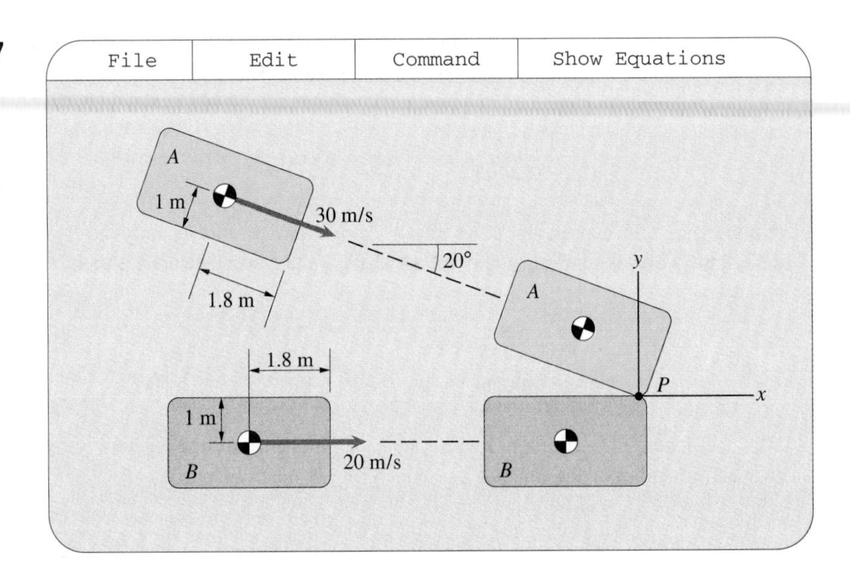

STRATEGY

Since the contacting surfaces are smooth, the x components of the velocities of the centres of mass are unchanged by the collision. The y components of the velocities must satisfy conservation of linear momentum, and the y components of the velocities *at the point of impact* before and after the impact are related by the coefficient of restitution. The force of the impact exerts no moment about P on either car, so the angular momentum of each car about P is conserved. From these conditions and kinematic relations between the velocities of the centres of mass and the velocities at P, we can determine the angular velocities and velocities of the centres of mass after the impact.

SOLUTION

The components of the velocities of the centres of mass before the impact are

$$\mathbf{v}_A = 30 \cos 20^\circ \, \mathbf{i} - 30 \sin 20^\circ \, \mathbf{j}$$
$$= (28.2\,\mathbf{i} - 10.3\,\mathbf{j}) \text{ m/s}$$

and

$$\mathbf{v}_B = 20\,\mathbf{i} \text{ m/s}$$

The x components of the velocities are unchanged by the impact:

$$v'_{Ax} = v_{Ax} = 28.2 \text{ m/s} \qquad v'_{Bx} = v_{Bx} = 20 \text{ m/s}$$

The y components of the velocities must satisfy conservation of linear momentum:

$$m_A v_{Ay} + m_B v_{By} = m_A v'_{Ay} + m_B v'_{By} \tag{8.49}$$

Let the velocities of the two cars at P before the collision be $\mathbf{v}_{AP}$ and $\mathbf{v}_{BP}$. The coefficient of restitution $e = 0.2$ relates the y components of the velocities at P:

$$0.2 = \frac{v'_{BPy} - v'_{APy}}{v_{APy} - v_{BPy}} \tag{8.50}$$

We can express the velocities at P after the impact in terms of the velocities of the centres of mass and the angular velocities after the impact (Figure (a)). The position of P relative to the centre of mass of car A is

$$\begin{aligned}\mathbf{r}_{P/A} &= [(1.8)\cos 20^\circ - (1)\sin 20^\circ]\mathbf{i} - (1.8)\sin 20^\circ + (1)\cos 20^\circ]\mathbf{j}\\ &= 1.35\,\mathbf{i} - 1.56\,\mathbf{j}(m)\end{aligned}$$

Therefore we can express the velocity of point P of car A after the impact as

$$\mathbf{v}'_{AP} = \mathbf{v}'_A + \boldsymbol{\omega}'_A \times \mathbf{r}_{P/A}:$$

$$v'_{APx}\,\mathbf{i} + v'_{APy}\,\mathbf{j} = v'_{Ax}\,\mathbf{i} + v'_{Ay}\,\mathbf{j} + \begin{vmatrix} \mathbf{i} & \mathbf{j} & \mathbf{k} \\ 0 & 0 & \omega'_A \\ 1.35 & -1.56 & 0 \end{vmatrix}$$

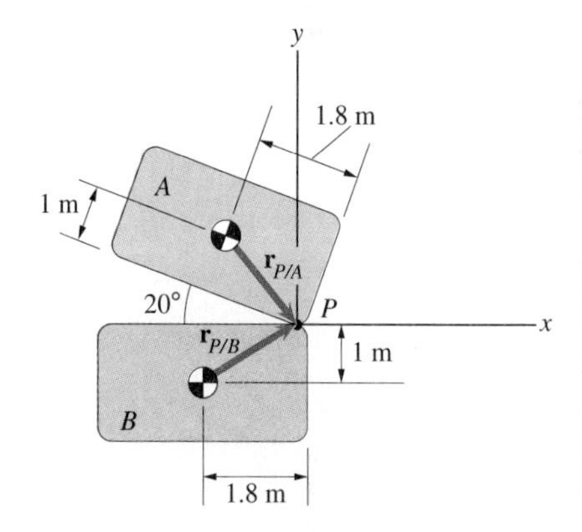

(a) Position vectors of P relative to the centres of mass.

Equating $\mathbf{i}$ and $\mathbf{j}$ components in this equation, we obtain

$$v'_{APx} = v'_{Ax} + 1.56\omega'_A \qquad v'_{APy} = v_{Ay} + 1.35\omega'_A \tag{8.51}$$

The position of P relative to the centre of mass of car B is

$$\mathbf{r}_{P/B} = (1.8\,\mathbf{i} + \mathbf{j})\text{ m}$$

We can express the velocity of point P of car B after the impact as

$$\mathbf{v}'_{BP} = \mathbf{v}'_B + \boldsymbol{\omega}'_B \times \mathbf{r}_{P/B}:$$

$$v'_{BPx}\,\mathbf{i} + v'_{BPy}\,\mathbf{j} = v'_{Bx}\,\mathbf{i} + v'_{By}\,\mathbf{j} + \begin{vmatrix} \mathbf{i} & \mathbf{j} & \mathbf{k} \\ 0 & 0 & \omega'_B \\ 1.8 & 1 & 0 \end{vmatrix}$$

Equating $\mathbf{i}$ and $\mathbf{j}$ components, we obtain

$$v'_{BPx} = v'_{Bx} - \omega'_B \qquad v'_{BPy} = v'_{By} + 1.8\omega'_B \tag{8.52}$$

The angular momentum of car A about P is conserved:

$$[(\mathbf{r}_{A/P} \times m_A\mathbf{v}_A)\cdot\mathbf{k} + I_A\omega_A] = [(\mathbf{r}_{A/P} \times m_A\mathbf{v}'_A)\cdot\mathbf{k} + I_A\omega'_A]$$

$$\begin{vmatrix} \mathbf{i} & \mathbf{j} & \mathbf{k} \\ -1.35 & 1.56 & 0 \\ m_A v_{Ax} & m_A v_{Ay} & 0 \end{vmatrix}\cdot\mathbf{k} + 0 = \begin{vmatrix} \mathbf{i} & \mathbf{j} & \mathbf{k} \\ -1.35 & 1.56 & 0 \\ m_A v'_{Ax} & m_A v'_{Ay} & 0 \end{vmatrix}\cdot\mathbf{k} + I_A\omega'_A$$

Expanding the determinants and evaluating the dot products, we obtain

$$-1.35m_A v_{Ay} - 1.56m_A v_{Ax}$$

$$= -1.35m_A v'_{Ay} - 1.56m_A v'_{Ax} + I_A\omega'_A \qquad (8.53)$$

The angular momentum of car B about P is also conserved,

$$[(\mathbf{r}_{B/P} \times m_B\mathbf{v}_B)\cdot\mathbf{k} + I_B\omega_B] = [(\mathbf{r}_{B/P} \times m_B\mathbf{v}'_B)\cdot\mathbf{k} + I_B\omega'_B]:$$

$$\begin{vmatrix} \mathbf{i} & \mathbf{j} & \mathbf{k} \\ -1.8 & -1 & 0 \\ m_B v_{Bx} & 0 & 0 \end{vmatrix} \cdot \mathbf{k} + 0 = \begin{vmatrix} \mathbf{i} & \mathbf{j} & \mathbf{k} \\ -1.8 & -1 & 0 \\ m_B v'_{Bx} & m_B v'_{By} & 0 \end{vmatrix} \cdot \mathbf{k} + I_B\omega'_B$$

From this equation we obtain

$$m_B v_{Bx} = -1.8m_B v'_{By} + m_B v'_{Bx} + I_B\omega'_B \qquad (8.54)$$

We can solve Equations (8.49)–(8.54) for $\mathbf{v}'_A$, $\mathbf{v}'_{AP}$, ω'_A, $\mathbf{v}'_B$, $\mathbf{v}'_{BP}$ and ω'_B. The results for the velocities of the centres of mass of the cars and their angular velocities are

$$\mathbf{v}'_A = (28.2\,\mathbf{i} - 9.08\,\mathbf{j})\ \text{m/s} \qquad \omega'_A = 2.65\ \text{rad/s}$$

$$\mathbf{v}'_B = (20.0\,\mathbf{i} - 1.18\,\mathbf{j})\ \text{m/s} \qquad \omega'_B = -3.54\ \text{rad/s}$$

Problems

8.56 The 2 kg slender bar starts from rest in the vertical position and falls, striking the smooth surface at P. The coefficient of restitution of the impact is $e = 0.5$. When the bar rebounds, through what angle relative to the horizontal will it rotate?

Strategy: Use the coefficient of restitution to relate the bar's velocity at P just after the impact to its value just before the impact.

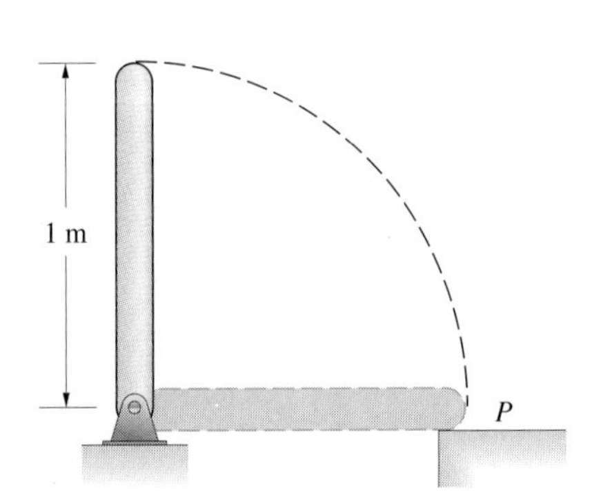

P8.56

8.57 The slender bar of mass m falls from rest in the position shown and hits the smooth projection at A. The coefficient of restitution is e. Show that the velocity of the centre of mass of the bar is zero immediately after the impact if $b^2 = el^2/12$.

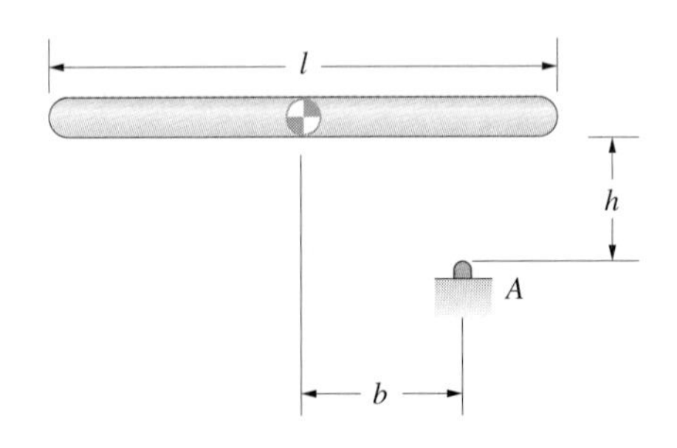

P8.57

8.58 In Problem 8.57, if $m = 2$ kg, $l = 1$ m, $b = 350$ mm, $h = 200$ mm, and the coefficient of restitution of the impact is $e = 0.4$, determine the bar's angular velocity after the impact.

8.59 If the duration of the impact described in Problem 8.58 is 0.02 s, what average force is exerted on the bar by the projection at A during the impact?

8.60 Wind causes the 600 tonne ship to drift sideways at 0.3 m/s and strike the stationary quay at P. The ship's mass moment of inertia about its centre of mass is 4×10^8 kg.m^2, and the coefficient of restitution of the impact is $e = 0.2$. What is the ship's angular velocity after the impact?

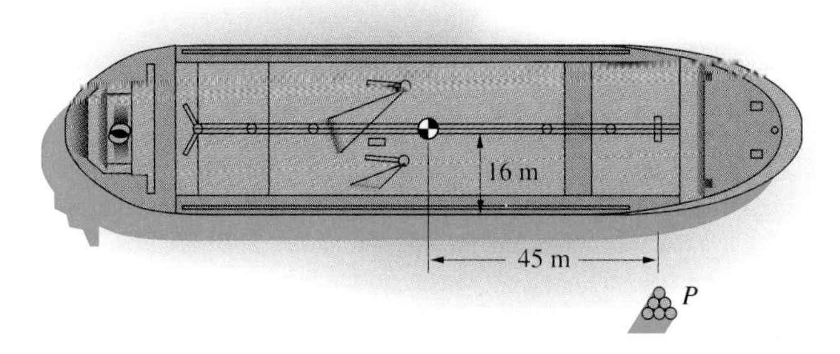

P8.60

8.61 In Problem 8.60, if the duration of the ship's impact with the quay is 10 s, what is the average value of the force exerted on the ship by the impact?

8.62 A 1 kg sphere A translating at 6 m/s strikes the end of a stationary 10 kg slender bar B. The bar is pinned to a fixed support at O. What is the angular velocity of the bar after the impact if the sphere adheres to the bar?

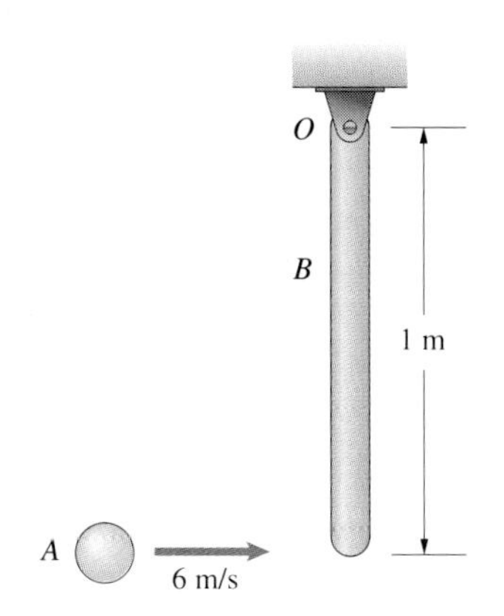

P8.62

8.63 In Problem 8.62, determine the velocity of the smooth sphere and the angular velocity of the bar after the impact if the coefficient of restitution is $e = 0.8$.

8.64 The 1 kg sphere A is moving at 10 m/s when it strikes the end of the 4 kg stationary slender bar B. If the sphere adheres to the bar, what is the bar's angular velocity after the impact?

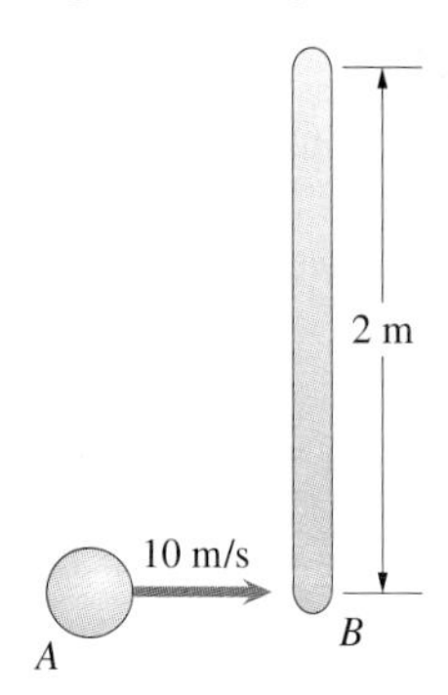

P8.64

8.65 In Problem 8.64, what is the bar's angular velocity after the impact if the coefficient of restitution is $e = 0.5$?

8.66 In Problem 8.64, determine the total kinetic energy of the sphere and bar before and after the impact if (a) $e = 0.5$; (b) $e = 1$.

8.67 The 0.14 kg ball is translating with velocity $v_A = 25$ m/s perpendicular to the bat just before impact. The player is swinging the 0.88 kg bat with angular velocity $\omega = 6\pi$ rad/s before the impact. Point C is the bat's instantaneous centre both before and after the impact. The distances $b = 350$ mm and $\bar{y} = 650$ mm. The bat's mass moment of inertia about its centre of mass is $I_B = 0.045$ kg.m^2. The coefficient of restitution is $e = 0.6$, and the duration of the impact is 0.008 s. Determine the magnitude of velocity of the ball after the impact and the average force A_x exerted on the bat by the player during the impact if (a) $d = 0$; (b) $d = 75$ mm; (c) $d = 200$ mm.

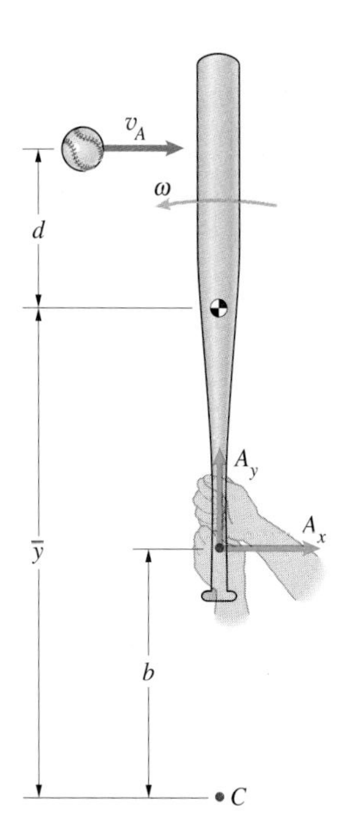

P8.67

8.68 In Problem 8.67, show that the force A_x is zero if $d = I_B/(m_B\bar{y})$, where m_B is the mass of the bat.

8.69 A slender bar of mass m is released from rest in the horizontal position at a height h above a peg (Figure (a)). A small hook at the end of the bar engages the peg, and the bar swings from the peg (Figure (b)). What minimum height h is necessary for the bar to swing 270° from its position when it engages the peg?

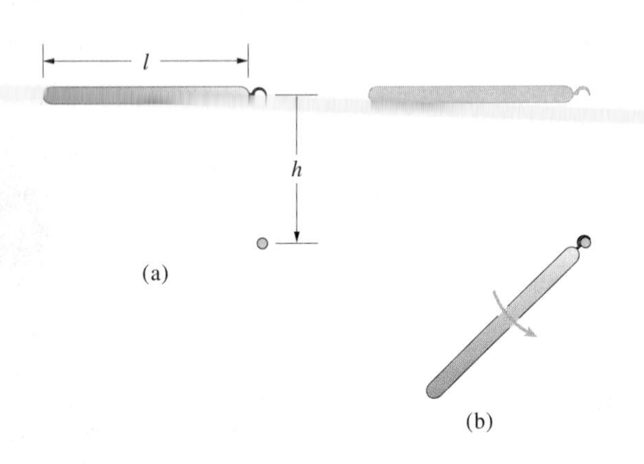

P8.69

8.70 Is energy conserved in Problem 8.69? If not, how much energy is lost?

8.71 A wheel that can be modelled as a 15 kg homogeneous cylindrical disc rolls at 3 m/s on a horizontal surface towards a 150 mm step. If the wheel remains in contact with the step and does not slip while rolling up onto it, what is the wheel's velocity once it is on the step?

P8.71

8.72 In Problem 8.71, what is the minimum velocity the wheel must have rolling towards the step in order to climb up onto it?

8.73 The slender bar is released from rest in the position shown in Figure (a) and falls a distance $h = 1$ m. When the bar hits the floor, its tip is supported by a depression and remains on the floor (Figure (b)). The length of the bar is 1 m and its weight is 1 N. What is its angular velocity ω just after it hits the floor?

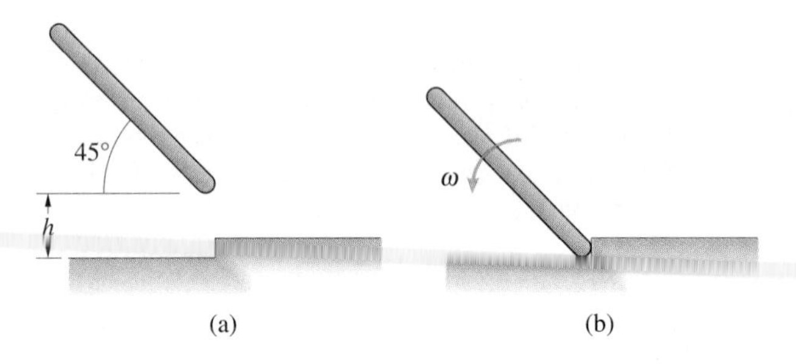

P8.73

8.74 During her parallel-bars routine, the velocity of the 40 kg gymnast's centre of mass is $(1.2\,\mathbf{i} - 3\,\mathbf{j})$ m/s and her angular velocity is zero just before she grasps the bar at A. In the position shown, her mass moment of inertia about her centre of mass is 2.4 kg.m^2. If she stiffens her shoulders and hips so that she can be modelled as a rigid body, what is the velocity of her centre of mass and her angular velocity just after she grasps the bar?

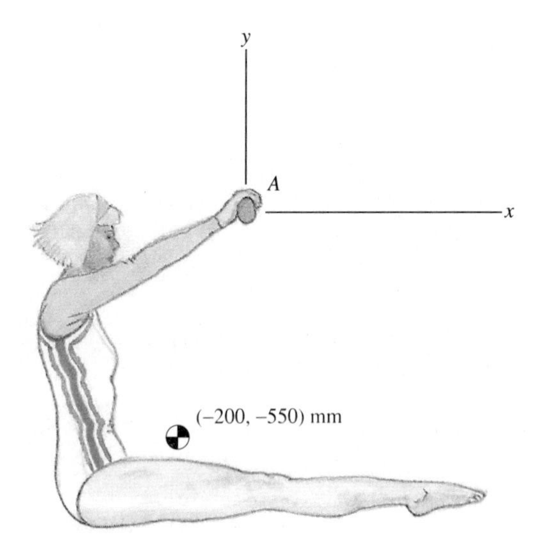

P8.74

8.75 The 20 kg homogeneous rectangular plate is released from rest (Figure (a)) and falls 200 mm before coming to the end of the string attached at the corner A (Figure (b)). Assuming that the vertical component of the velocity of A is zero just after the plate reaches the end of the string, determine the angular velocity of the plate and the magnitude of the velocity of the corner B at that instant.

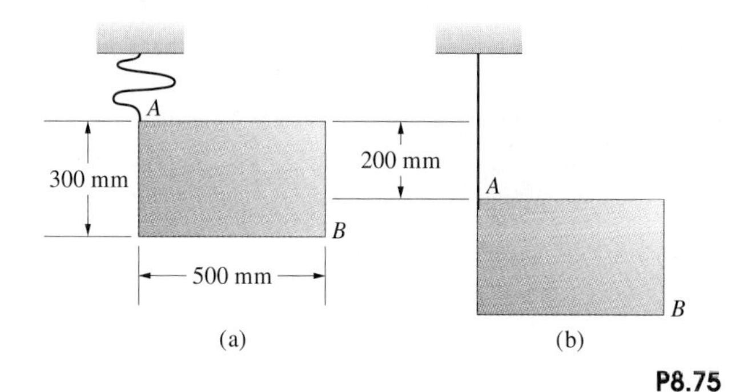

P8.75

8.76 A non-rotating slender bar A moving with velocity v_0 strikes a stationary slender bar B. Each bar has mass m and length l. If the bars adhere when they collide, what is their angular velocity after the impact?

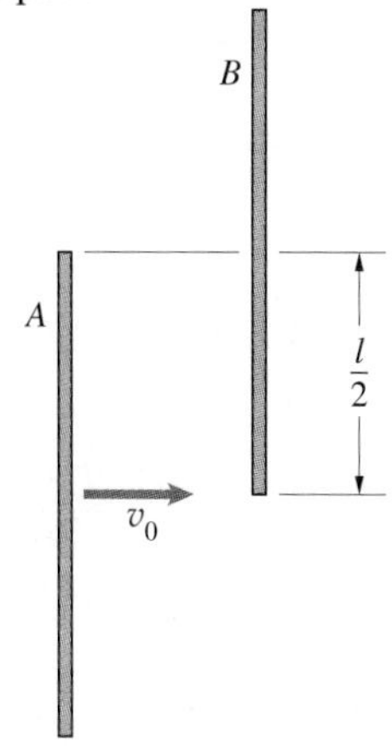

P8.76

8.77 The horizontal velocity of the landing aeroplane is 50 m/s, its vertical velocity (rate of descent) is 2 m/s, and its angular velocity is zero. The mass of the aeroplane is 12 Mg and the mass moment of inertia about its centre of mass is 1×10^5 kg.m^2. When the rear wheels touch the runway, they remain in contact with it. Neglecting the horizontal force exerted on the wheels by the runway, determine the aeroplane's angular velocity just after it touches down.

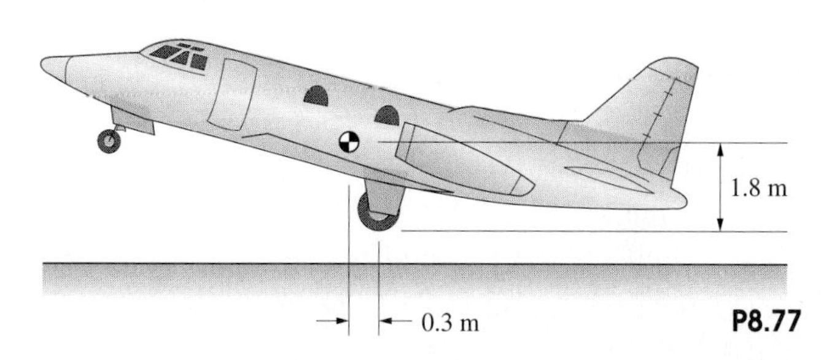

P8.77

8.78 Determine the angular velocity of the aeroplane in Problem 8.77 just after it touches down if its wheels don't stay in contact with the runway and the coefficient of restitution of the impact is $e = 0.4$.

8.79 A 1270 kg car skidding on ice strikes a concrete post at 5 km/hr. The car's moment of inertia about its centre of mass is 2440 kg.m^2. Assume that the impacting surfaces are smooth and parallel to the y axis and that the coefficient of restitution of the impact is $e = 0.8$. What is the car's angular velocity and the velocity of its centre of mass after the impact?

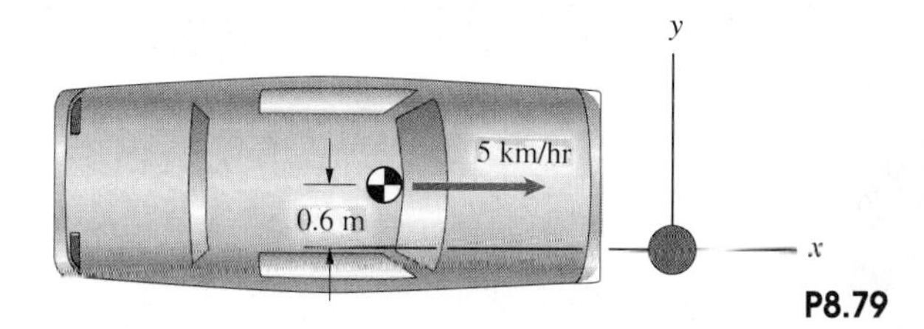

P8.79

8.80 While attempting to drive on an icy street for the first time, a student skids his 1270 kg car (A) into the university chancellor's unoccupied 2720 kg Rolls-Royce Corniche (B). The point of impact is P. Assume that the impacting surfaces are smooth and parallel to the y axis and that the coefficient of restitution of the impact is $e = 0.5$. The moments of inertia of the cars about their centres of mass are $I_A = 2440$ kg.m^2 and $I_B = 7600$ kg.m^2. Determine the angular velocities of the cars and the velocities of their centres of mass after the collision.

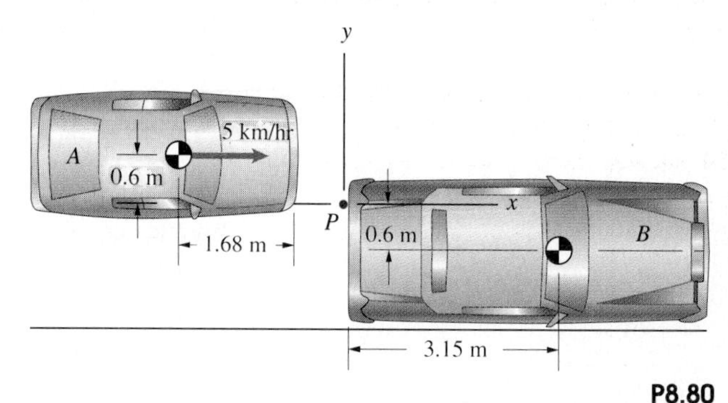

P8.80

8.81 Each slender bar is 1.22 m long and weighs 90 N. Bar A is released in the horizontal position shown. The bars are smooth and the coefficient of restitution of their impact is $e = 0.8$. Determine the angle through which B swings afterwards.

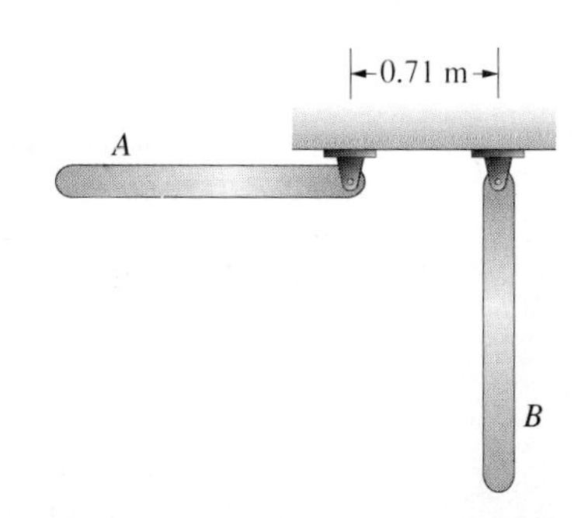

P8.81

8.82 The *Apollo* CSM (A) approaches the *Soyuz* space station (B). The mass of the *Apollo* is $m_A = 18$ Mg, and the mass moment of inertia about the axis through its centre of mass parallel to the z axis is $I_A = 114$ Mg.m^2. The mass of the *Soyuz* is $m_B = 6.6$ Mg, and the mass moment of inertia about the axis through its centre of mass parallel to the z axis is $I_B = 70$ Mg.m^2. The *Soyuz* is stationary relative to the reference frame shown and the CSM approaches with velocity $\mathbf{v}_A = (0.2\,\mathbf{i} + 0.05\,\mathbf{j})$ m/s and no angular velocity. What is their angular velocity after docking?

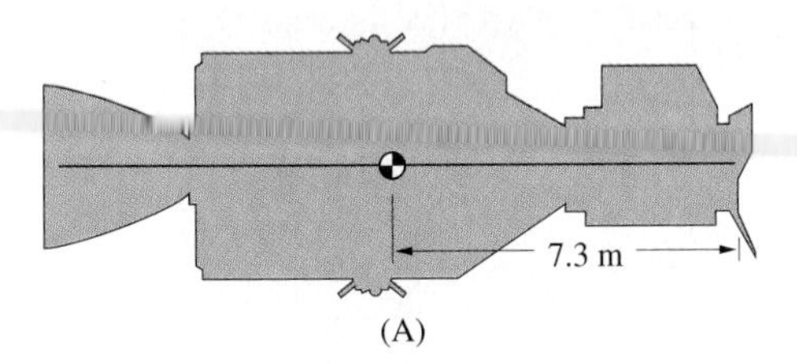

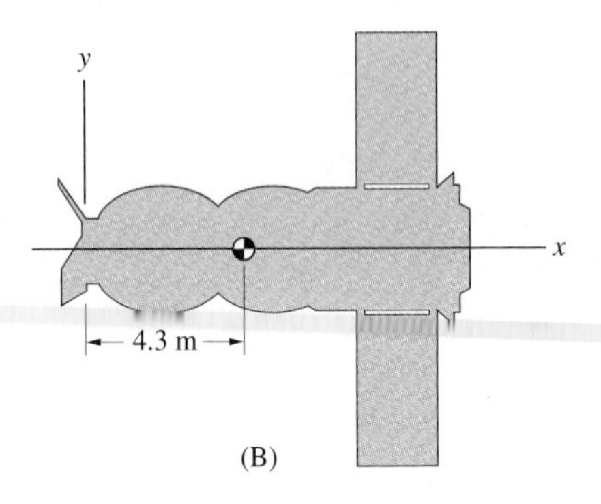

P8.82

Chapter Summary

Work and Energy

The work done by external forces and couples as a rigid body moves between two positions is equal to the change in its kinetic energy;

$$U = T_2 - T_1 \qquad \textbf{Equation (8.4)}$$

The work done on a system of rigid bodies by external and internal forces and couples equals the change in the total kinetic energy.

The kinetic energy of a rigid body in general planar motion is

$$T = \frac{1}{2}mv^2 + \frac{1}{2}I\omega^2 \qquad \textbf{Equation (8.12)}$$

where v is the magnitude of the velocity of the centre of mass and I is the mass moment of inertia about the centre of mass. If a rigid body rotates about a fixed axis O, its kinetic energy can also be expressed as

$$T = \frac{1}{2}I_0\omega^2 \qquad \textbf{Equation (8.13)}$$

The work done on a rigid body by a force $\mathbf{F}$ is

$$U = \int_{(\mathbf{r}_\mathrm{p})_1}^{(\mathbf{r}_\mathrm{p})_2} \mathbf{F} \cdot d\mathbf{r}_\mathrm{p} \qquad \textbf{Equation (8.14)}$$

where $\mathbf{r}_\mathrm{p}$ is the position of the *point of application* of $\mathbf{F}$. If the point of application is stationary, or if its direction of motion is perpendicular to $\mathbf{F}$, no work is done.

The work done by a couple M on a rigid body in planar motion as it rotates from θ_1 to θ_2 in the direction of M is

$$U = \int_{\theta_1}^{\theta_2} M\,d\theta \qquad \textbf{Equation (8.16)}$$

A couple M is conservative if a potential energy V exists such that

$$M\,d\theta = -dV \qquad \textbf{Equation (8.17)}$$

The potential energy of a linear torsional spring that exerts a couple $k\theta$ in the direction opposite to its angular displacement θ (Figure (a)) is $\frac{1}{2}k\theta^2$.

If all the forces and couples that do work on a rigid body are conservative, the sum of the kinetic energy and the total potential energy is constant:

$$T + V = \text{constant} \qquad \textbf{Equation (8.19)}$$

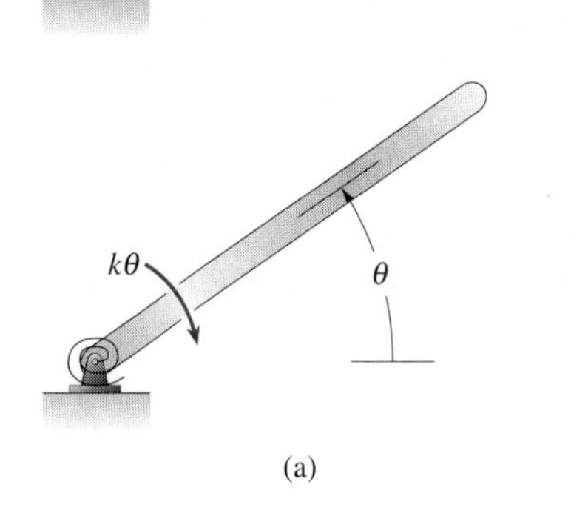

(a)

Power

The power transmitted to a rigid body by a force $\mathbf{F}$ is

$$P = \mathbf{F} \cdot \mathbf{v}_{\mathrm{p}} \qquad \textbf{Equation (8.20)}$$

where $\mathbf{v}_{\mathrm{p}}$ is the velocity of the point of application of $\mathbf{F}$. The power transmitted to a rigid body in planar motion by a couple M is

$$P = M\omega \qquad \textbf{Equation (8.21)}$$

The average power transferred to a rigid body during an interval of time is equal to the change in its kinetic energy, or the total work done, divided by the interval of time:

$$P_{\mathrm{av}} = \frac{T_2 - T_1}{t_2 - t_1} = \frac{U}{t_2 - t_1} \qquad \textbf{Equation (8.22)}$$

Impulse and Momentum

The principle of linear impulse and momentum states that the linear impulse applied to a rigid body during an interval of time is equal to the change in its linear momentum:

$$\int_{t_1}^{t_2} \Sigma\mathbf{F}\,dt = m\mathbf{v}_2 - m\mathbf{v}_1 \qquad \textbf{Equation (8.25)}$$

This result can also be expressed in terms of the average with respect to time of the total force:

$$(t_2 - t_1)\,\Sigma\mathbf{F}_{\mathrm{av}} = m\mathbf{v}_2 - m\mathbf{v}_1 \qquad \textbf{Equation (8.26)}$$

If the only forces acting on two rigid bodies A and B are the forces they exert on each other, or if other forces are negligible, their total linear momentum is conserved:

$$m_A\mathbf{v}_A + m_B\mathbf{v}_B = \text{constant} \qquad \textbf{Equation (8.27)}$$

The principle of angular impulse and momentum states that the angular impulse about a fixed point O applied to a rigid body during an interval of time is equal to the change in its angular momentum about O:

$$\int_{t_1}^{t_2} \Sigma M_O \, dt = [(\mathbf{r} \times m\mathbf{v}) \cdot \mathbf{k} + I\omega]_2 - [(\mathbf{r} \times m\mathbf{v}) \cdot \mathbf{k} + I\omega]_1 \qquad \textbf{Equation (8.29)}$$

The unit vector $\mathbf{k}$ is perpendicular to the plane of the motion. Pointing the thumb of the right hand in the direction of $\mathbf{k}$, the fingers curl in the positive direction for ω. This principle can also be expressed in terms of the average moment about O:

$$(t_2 - t_1)(\Sigma M_O)_{\text{av}} = [(\mathbf{r} \times m\mathbf{v}) \cdot \mathbf{k} + I\omega]_2 - [(\mathbf{r} \times m\mathbf{v}) \cdot \mathbf{k} + I\omega]_1 \qquad \textbf{Equation (8.31)}$$

If the only forces and couples acting on two rigid bodies A and B in planar motion are the forces and couples that they exert on each other, or if other forces and couples are negligible, their total angular momentum about any fixed point O is conserved:

$$(\mathbf{r}_A \times m_A\mathbf{v}_A) \cdot \mathbf{k} + I_A\omega_A + (\mathbf{r}_B \times m_B\mathbf{v}_B) \cdot \mathbf{k} + I_B\omega_B = \text{constant} \qquad \textbf{Equation (8.32)}$$

This result holds even when A and B are subject to significant external forces and couples if the total moment about O due to the external forces and couples is zero.

Impacts

Suppose that two rigid bodies A and B, in two-dimensional motion in the same plane, collide. If other forces and couples are negligible in comparison with the impact forces and couples that A and B exert on each other, their total linear momentum and their total angular momentum about any fixed point O are conserved. If, in addition, A and B exert only forces on each other at their point of impact P, the angular momentum about P of *each* rigid body is conserved. If one of the rigid bodies has a pin support at a point O, their total angular momentum about that point is conserved.

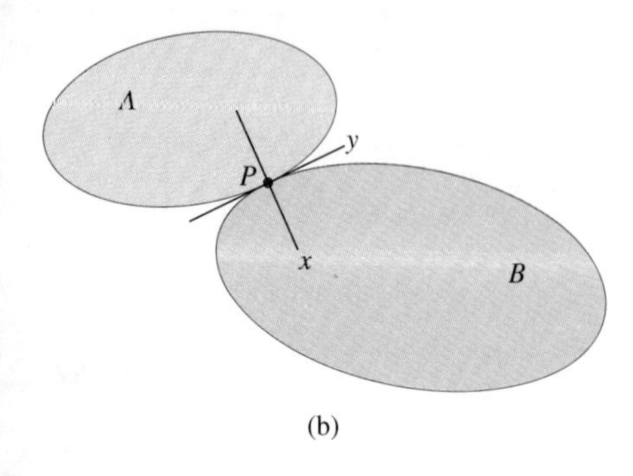

(b)

Let P be the point of impact (Figure (b)). The normal components of the velocities at P are related to the coefficient of restitution e by

$$e = \frac{(\mathbf{v}'_{BP})_x - (\mathbf{v}'_{AP})_x}{(\mathbf{v}_{AP})_x - (\mathbf{v}_{BP})_x} \qquad \textbf{Equation (8.46)}$$

Review Problems

8.83 The mass moment of inertia of the pulley is 0.2 kg.m^2. The system is released from rest. Use the principle of work and energy to determine the velocity of the 10 kg cylinder when it has fallen 1 m.

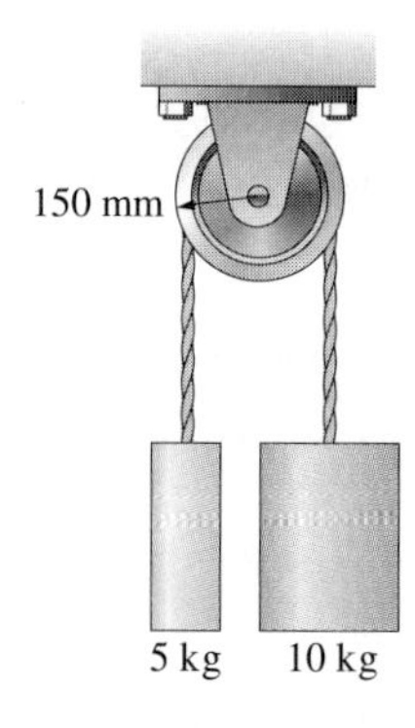

P8.83

8.84 Use momentum principles to determine the velocity of the 10 kg cylinder in Problem 8.83 one second after the system is released from rest.

8.85 Arm BC has a mass of 12 kg and the mass moment of inertia about its centre of mass is 3 kg.m^2. Point B is stationary. Arm BC is initially aligned with the (horizontal) x axis with zero angular velocity and a constant couple M applied at B causes it to rotate upwards. When it is in the position shown, its counter-clockwise angular velocity is 2 rad/s. Determine M.

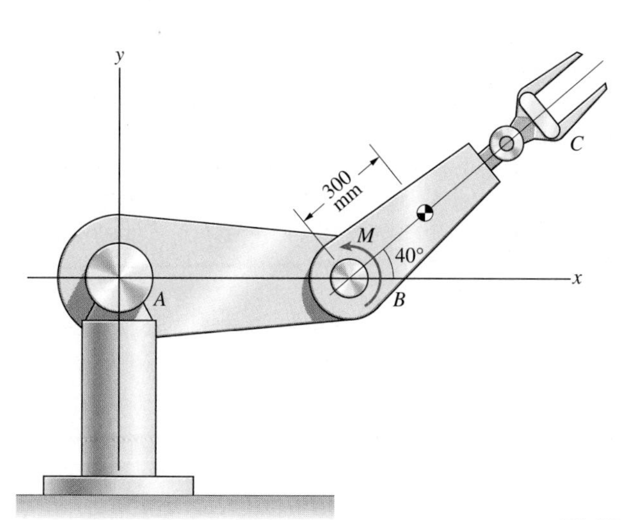

P8.85

8.86 The cart is stationary when a constant force F is applied to it. What will its velocity be when it has rolled a distance b? The mass of the body of the cart is m_c and each of the four wheels has mass m, radius R, and mass moment of inertia I.

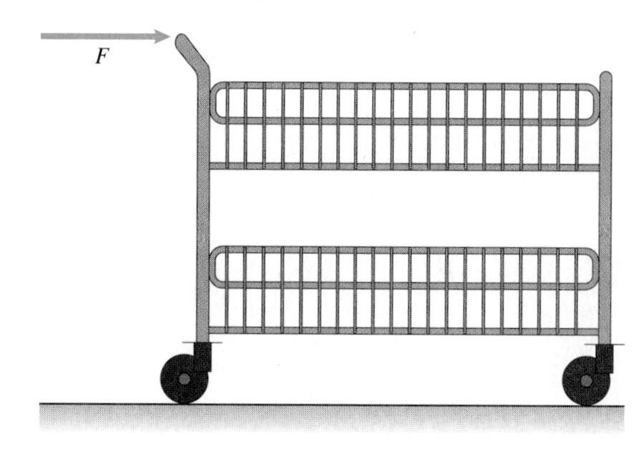

P8.86

8.87 Each pulley has mass moment of inertia $I = 0.003$ kg.m^2, and the mass of the belt is 0.2 kg. If a constant couple $M = 4$ N.m is applied to the bottom pulley, what will its angular velocity be when it has turned 10 revolutions?

100 mm
M

P8.87

8.88 The ring gear is fixed. The mass and mass moment of inertia of the sun gear are $m_S = 320$ kg, $I_S = 6000$ kg.m^2. The mass and mass moment of inertia of each planet gear are $m_P = 40$ kg, $I_P = 90$ kg.m^2. A couple $M = 200$ N.m is applied to the sun gear. Use work and energy to determine the sun gear's angular velocity after it has turned 100 revolutions.

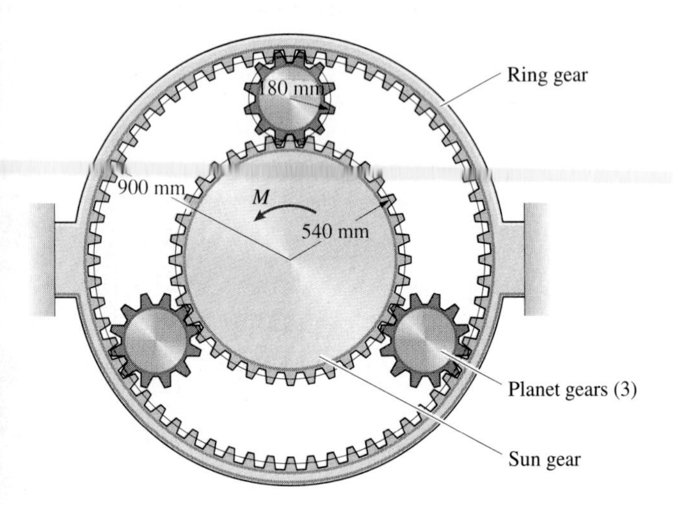

P8.88

8.89 The mass moment of inertia of the crank AB about A is 0.0003 kg.m^2. The mass of the connecting rod BC is 0.36 kg, its centre of mass is at its midpoint, and its mass moment of inertia about the centre of mass is 0.0004 kg.m^2. The radius of the piston at C is 55 mm, and its mass is 4.6 kg. In the position shown, the air in the cylinder is at atmospheric pressure, $p_{atm} = 1 \times 10^5$ Pa. Assume that as the piston moves in the cylinder, the pressure p within the cylinder is inversely proportional to the volume. The net force towards the left exerted on the piston by pressure is $(p - p_{atm})A$, where A is the piston's cross-sectional area. If a constant couple $M = 100$ N.m is applied to the crank AB, what is its angular velocity when it has rotated 45° in the clockwise direction? (Neglect friction and the work done by the gravitational forces on the crank and connecting rod.)

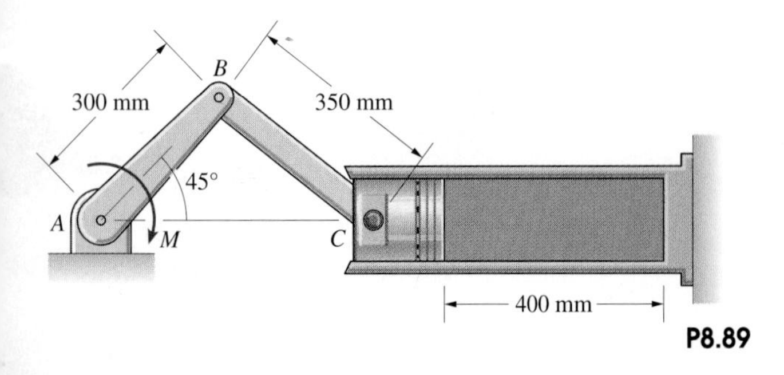

P8.89

8.90 In Problem 8.89, determine the angular velocity of the crank AB and the piston's velocity when the crank AB has rotated 20° in the clockwise direction.

8.91 In Problem 8.89, what is the minimum constant couple M necessary to rotate the crank AB 45° in the clockwise direction?

8.92 The 0.1 kg slender bar and 0.2 kg cylindrical disc are released from rest with the bar horizontal. The disc rolls on the curved surface. What is the bar's angular velocity when it is vertical?

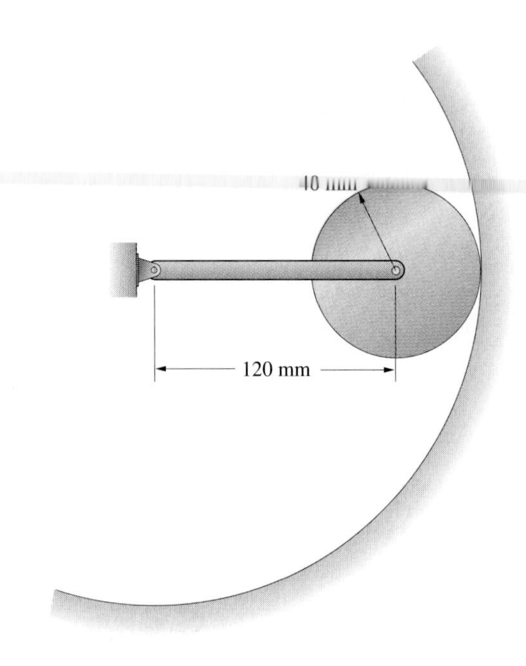

P8.92

8.93 A slender bar of mass m is released from rest in the vertical position and allowed to fall. Neglecting friction and assuming that it remains in contact with the floor and wall, determine its angular velocity as a function of θ.

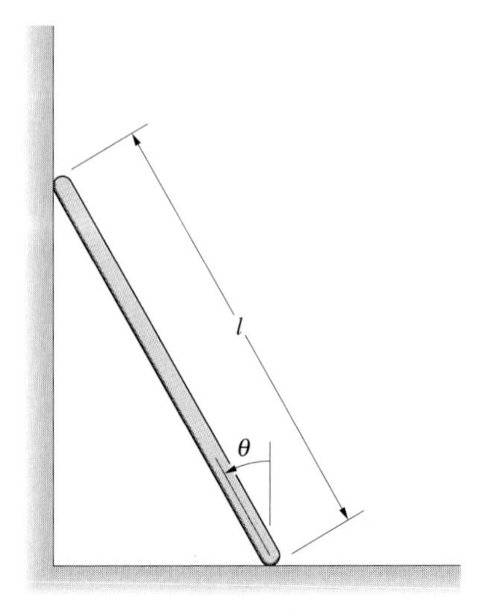

P8.93

8.94 The 4 kg slender bar is pinned to 2 kg sliders at A and B. If friction is negligible and the system starts from rest in the position shown, what is the bar's angular velocity when the slider at A has fallen 0.5 m?

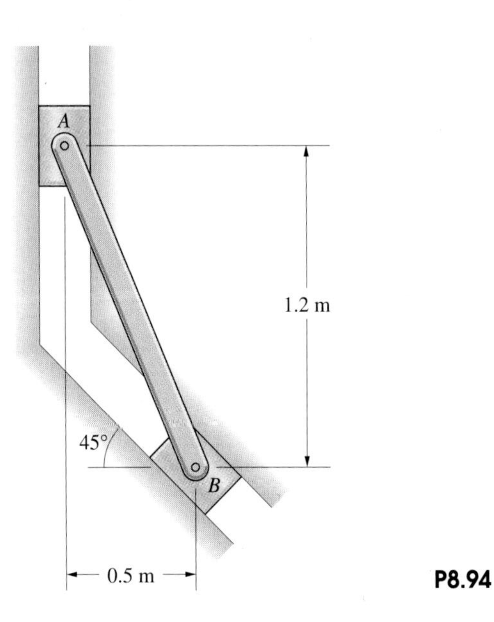

P8.94

8.95 A homogeneous hemisphere of mass m is released from rest in the position shown. If it rolls on the horizontal surface, what is its angular velocity when its flat surface is horizontal?

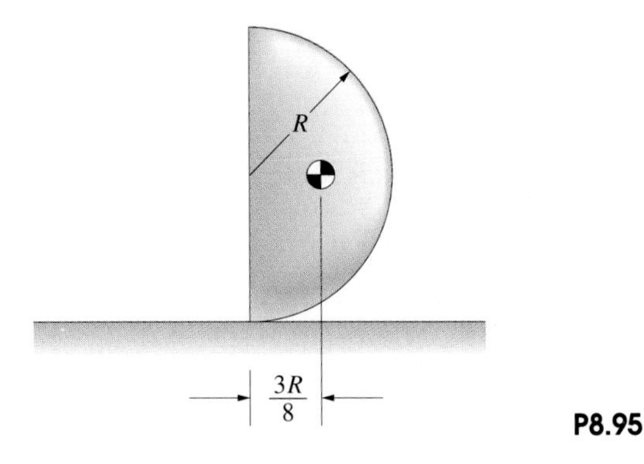

P8.95

8.96 What normal force is exerted on the hemisphere in Problem 8.95 by the horizontal surface at the instant its flat surface is horizontal?

8.97 An engineer decides to control the angular velocity of a satellite by deploying small masses attached to cables. If the angular velocity of the satellite in configuration (a) is 4 rpm, determine the distance d in configuration (b) that will cause the angular velocity to be 1 rpm. The moment of inertia of the satellite is $I = 500$ kg.m^2 and each mass is 2 kg. (Assume that the cables and masses rotate with the same angular velocity as the satellite. Neglect the masses of the cables and the mass moments of inertia of the masses about their centres of mass.)

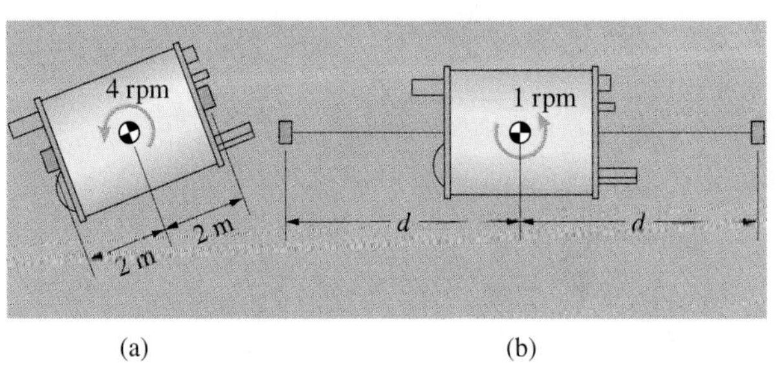

P8.97

8.98 A homogeneous cylindrical disc of mass m rolls on the horizontal surface with angular velocity ω. If it does not slip or leave the slanted surface when it comes into contact with it, what is the angular velocity ω' of the disc immediately afterwards?

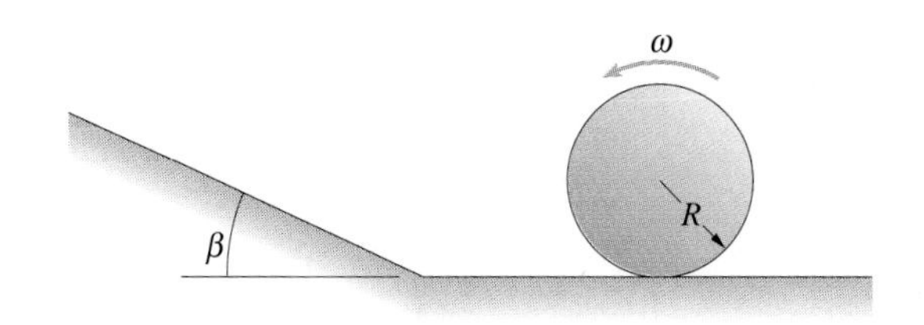

P8.98

8.99 The 10 kg slender bar falls from rest in the vertical position and hits the smooth projection at B. The coefficient of restitution of the impact is $e = 0.6$, the duration of the impact is 0.1 s, and $b = 1$ m. Determine the average force exerted on the bar at B as a result of the impact.

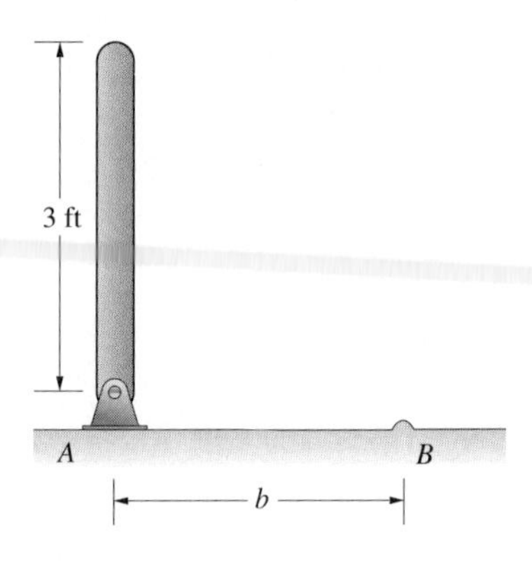

P8.99

8.100 In Problem 8.99, determine the value of b for which the average force exerted on the bar at A as a result of the impact is zero.

8.101 The 1 kg sphere A is moving at 2 m/s when it strikes the end of the 2 kg stationary slender bar B. If the velocity of the sphere after the impact is 0.8 m/s to the right, what is the coefficient of restitution?

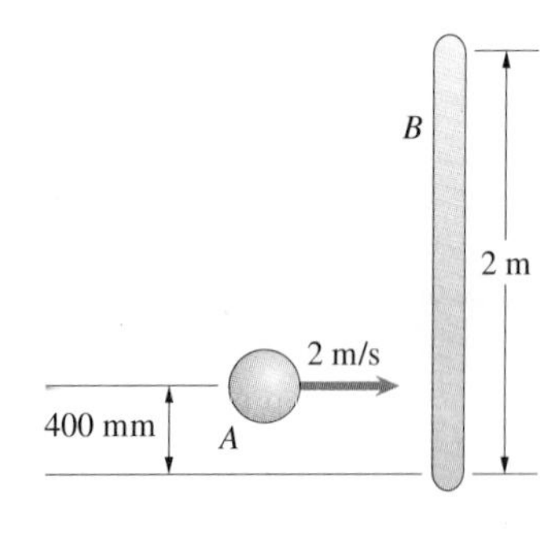

P8.101

8.102 The slender bar is released from rest with $\theta = 45°$ and falls a distance $h = 1$ m onto the smooth floor. The length of the bar is 1 m and its mass is 2 kg. If the coefficient of restitution of the impact is $e = 0.4$, what is the bar's angular velocity just after it hits the floor?

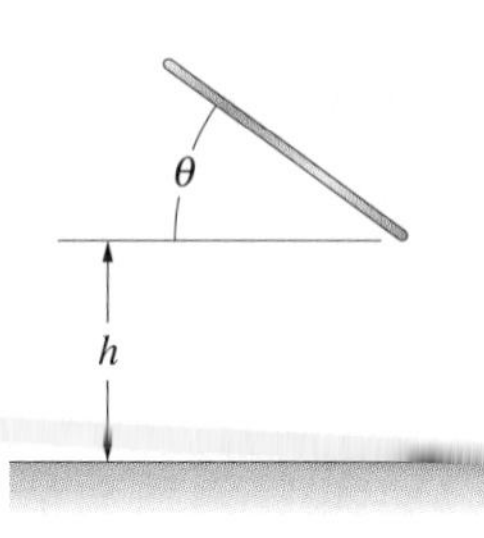

P8.102

8.103 In Problem 8.102, determine the angle θ for which the angular velocity of the bar just after it hits the floor is a maximum. What is the maximum angular velocity?

8.104 An astronaut translates towards a non-rotating satellite at $1.0\,\mathbf{i}$ m/s relative to the satellite. Her mass is 136 kg, and the mass moment of inertia about the axis through her centre of mass parallel to the z axis is 45 kg.m^2. The mass of the satellite is 450 kg and its mass moment of inertia about the z axis is 675 kg.m^2. At the instant she attaches to the satellite and begins moving with it, the position of her centre of mass is $(-1.8, -0.9, 0)$ m. The axis of rotation of the satellite after she attaches is parallel to the z axis. What is their angular velocity?

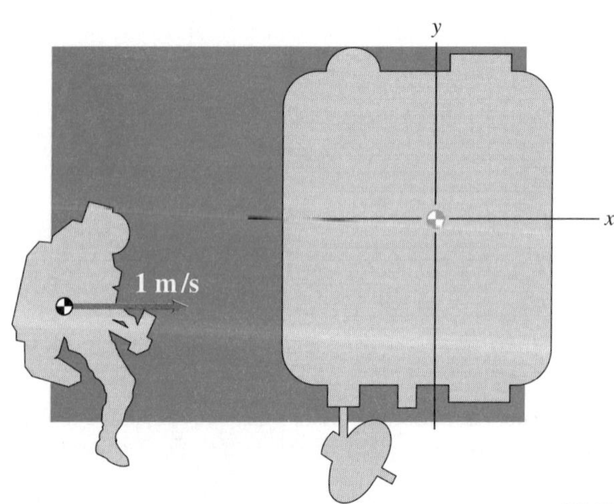

P8.104

8.105 In Problem 8.104, suppose that the design parameters of the satellite's control system require that its angular velocity not exceed 0.02 rad/s. If the astronaut is moving parallel to the x axis and the position of her centre of mass when she attaches is $(-1.8, -0.9, 0)$ m, what is the maximum relative velocity at which she should approach the satellite?

8.106 A 77 kg wide receiver jumps vertically to receive a pass and is stationary at the instant he catches the ball. At the same instant, he is hit at P by an 82 kg linebacker moving horizontally at 4.6 m/s. The wide receiver's mass moment of inertia about his centre of mass is 9.5 kg.m^2. If you model the players as rigid bodies and assume that the coefficient of restitution is $e = 0$, what is the wide receiver's angular velocity immediately after the impact?

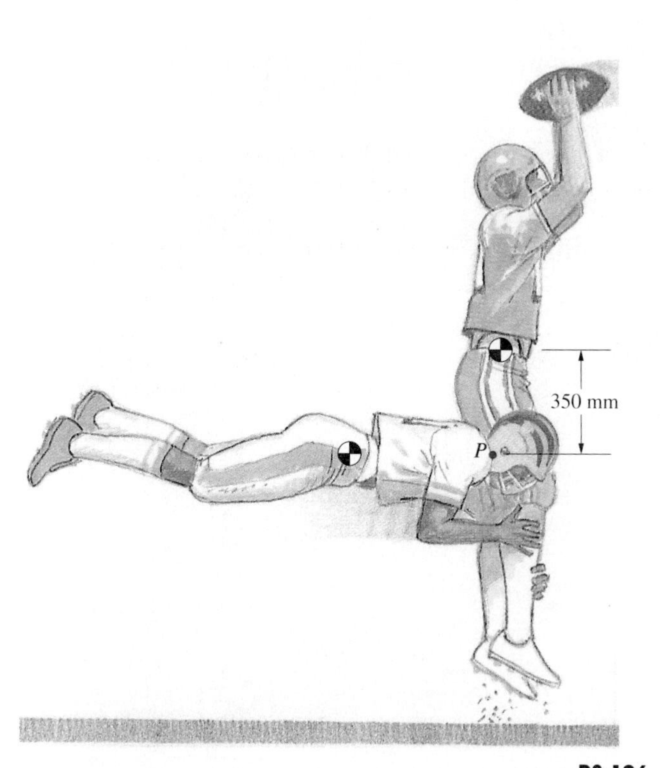

P8.106

8.107 If the football players in Problem 8.106 do not remain in contact after the collision, and the linebacker's horizontal velocity immediately after the impact is 2.75 m/s, what is the wide receiver's angular velocity?

8.108 The 2 kg slender bar is pinned at A to a 10 kg metal block that rests on a smooth level surface. The system is released from rest with the bar vertical. Friction is negligible. When the system is in the position shown, determine (a) the magnitude of the bar's angular velocity; (b) the magnitude of the velocity of the block.

Strategy: Use conservation of energy and conservation of linear momentum.

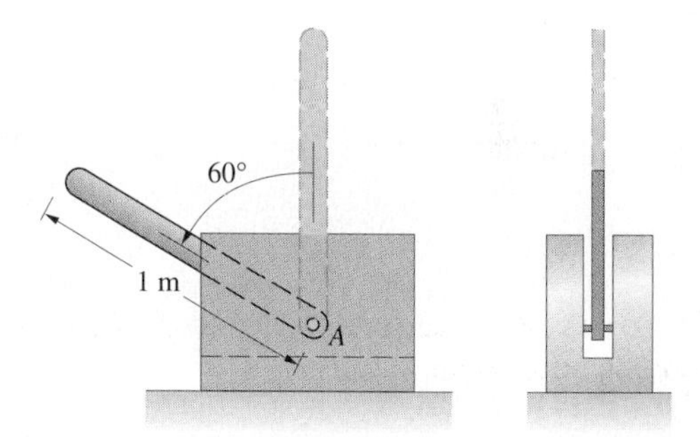

P8.108

The orientation of the biplane, or any rigid body, can be described by three angles specifying rotations of a body-fixed coordinate system relative to a fixed reference frame. The change in the biplane's orientation as a function of time is governed by equations of three-dimensional angular motion that relate the forces and couples acting on the biplane to its angular acceleration. In this chapter you will learn to analyse three-dimensional motions of rigid bodies.

Chapter 9

Three-Dimensional Kinematics and Dynamics of Rigid Bodies

UNTIL now our discussion of the dynamics of rigid bodies has dealt only with two-dimensional motion. But for many engineering applications of dynamics, such as the design of aeroplanes and other vehicles, we must consider three-dimensional motion. Our first step is to explain how three-dimensional motion of a rigid body is described. We then derive the equations of motion and use them to analyse simple three-dimensional motions. Finally, we introduce the Eulerian angles used to specify the orientation of a rigid body in three dimensions and express the equations of angular motion in terms of them.

9.1 Kinematics

If you ride a bicycle in a straight path, the wheels undergo planar motions; but if you are turning, their motions are three-dimensional (Figure 9.1(a)). An aeroplane can remain in planar motion while in level flight, descending, climbing, or performing loops. But if it banks and turns, it is in three-dimensional motion (Figure 9.1(b)). If you spin a top, it may remain in planar motion for a brief period, rotating about a fixed vertical axis; but eventually the top's axis begins to tilt and rotate. The top is then in three-dimensional motion and exhibits interesting, apparently gravity-defying behaviour (Figure 9.1(c)). In this section we begin the analysis of such motions by discussing kinematics of rigid bodies in three-dimensional motion.

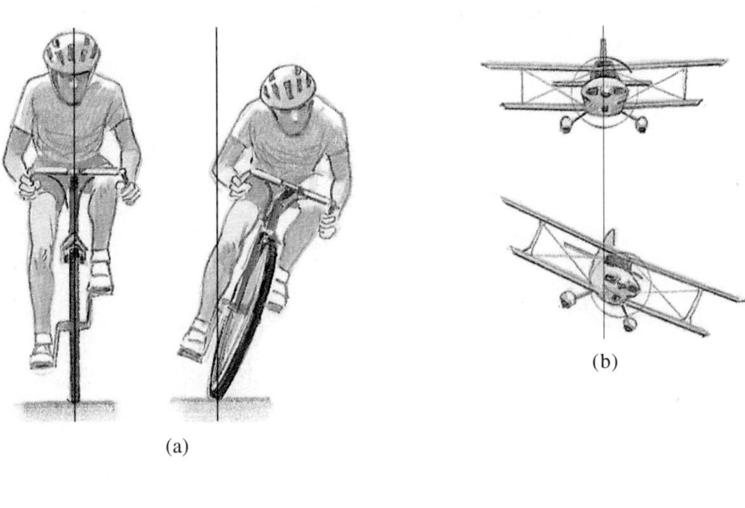

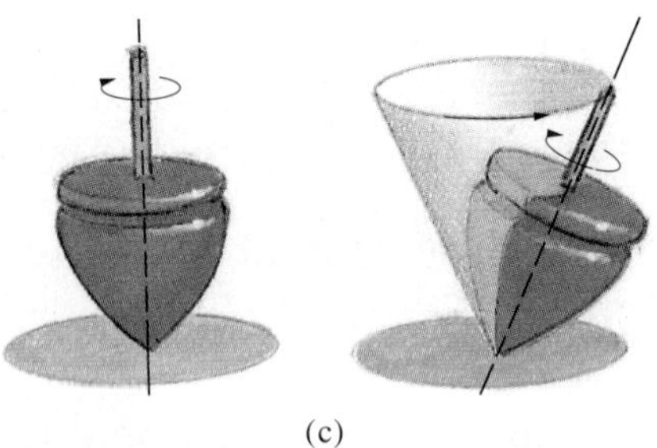

Figure 9.1
Example of planar and three-dimensional motions.

You are already familiar with some of the concepts involved in describing three-dimensional motion of a rigid body. In Chapter 6 we showed that Euler's theorem implies that a rigid body undergoing any motion other than translation has an instantaneous axis of rotation. The direction of this axis at a particular instant, and the rate at which the rigid body rotates about the axis, can be specified by the angular velocity vector $\boldsymbol{\omega}$.

Furthermore, a rigid body's velocity is completely specified by its angular velocity vector and the velocity vector of a single point of the rigid body. For the rigid body in Figure 9.2, suppose that we know $\boldsymbol{\omega}$ and the velocity $\mathbf{v}_B$ of point B. (The vector $\mathbf{v}_B = d\,\mathbf{r}_B/dt$ is the velocity of B relative to the reference point O.) Then the velocity of any point A is given by Equation (6.8):

$$\boxed{\mathbf{v}_A = \mathbf{v}_B + \boldsymbol{\omega} \times \mathbf{r}_{A/B}} \tag{9.1}$$

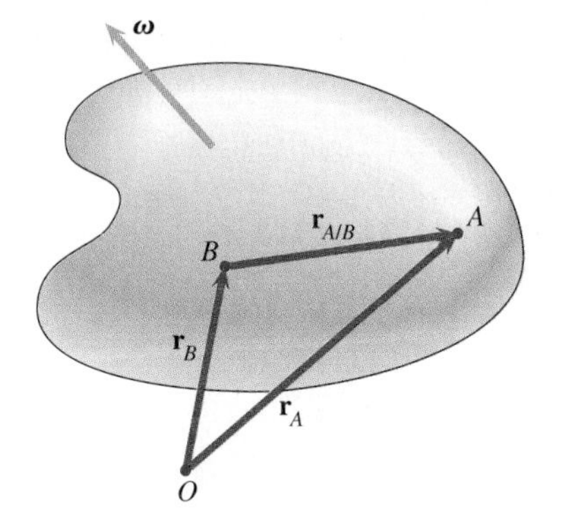

Figure 9.2

Points A and B of a rigid body. The velocity of A can be determined if the velocity of B and the rigid body's angular velocity are known. The acceleration of A can be determined if the acceleration of B, the angular velocity and the angular acceleration are known.

A rigid body's acceleration is completely specified by its angular acceleration vector $\boldsymbol{\alpha} = d\boldsymbol{\omega}/dt$, its angular velocity vector, and the acceleration vector of a single point. If we know $\boldsymbol{\alpha}$, $\boldsymbol{\omega}$ and the acceleration $\mathbf{a}_B$ of point B in Figure 9.2, the acceleration of any point A is given by Equation (6.9):

$$\boxed{\mathbf{a}_A = \mathbf{a}_B + \boldsymbol{\alpha} \times \mathbf{r}_{A/B} + \boldsymbol{\omega} \times \mathbf{r}_{A/B})} \tag{9.2}$$

We have assumed that the velocities and accelerations of points, the angular velocity $\boldsymbol{\omega}$, and the angular acceleration $\boldsymbol{\alpha}$ in Equations (9.1) and (9.2) are relative to a non-rotating reference frame fixed with respect to point O. In this chapter we will also use body-fixed coordinate systems, which are stationary relative to moving rigid bodies. In addition, in some situations we will use coordinate systems that rotate but are not body-fixed. We emphasize that *even though a vector specifies a velocity or acceleration relative to a fixed reference frame, at any given instant we can express it in terms of its components in a rotating coordinate system.*

Suppose that the angular velocity of a rotating coordinate system xyz relative to a fixed reference frame is described by an angular velocity vector $\boldsymbol{\Omega}$, and the angular velocity of a rigid body *relative to* the xyz system is $\boldsymbol{\omega}_{\text{rel}}$. Then the rigid body's angular velocity and angular acceleration relative to the fixed reference frame are

$$\begin{aligned} \boldsymbol{\omega} &= \boldsymbol{\Omega} + \boldsymbol{\omega}_{\text{rel}} \\ \boldsymbol{\alpha} &= \frac{d\boldsymbol{\Omega}}{dt} + \frac{d\omega_{\text{rel}\,x}}{dt} + \frac{d\omega_{\text{rel}\,y}}{dt} + \frac{d\omega_{\text{rel}\,z}}{dt} + \boldsymbol{\Omega} \times \boldsymbol{\omega}_{\text{rel}} \end{aligned} \tag{9.3}$$

The following examples demonstrate the use of Equations (9.1)–(9.3) to analyse rigid bodies in three-dimensional motion. You will often find that the simplest way to determine the angular velocity and angular acceleration vectors of a rotating rigid body is first to determine its angular velocity $\boldsymbol{\omega}_{\text{rel}}$ relative to a rotating coordinate system and then use Equations (9.3).

Example 9.1

The tyre in Figure 9.3 is rolling on the level surface. Its midpoint is moving at 5 m/s. The car is turning, and the perpendicular line through the tyre's midpoint rotates about the fixed point P shown in the top view.

(a) What is the tyre's angular velocity vector $\boldsymbol{\omega}$?

(b) Determine the velocity of point A, the rearmost point of the tyre at the instant shown.

Figure 9.3

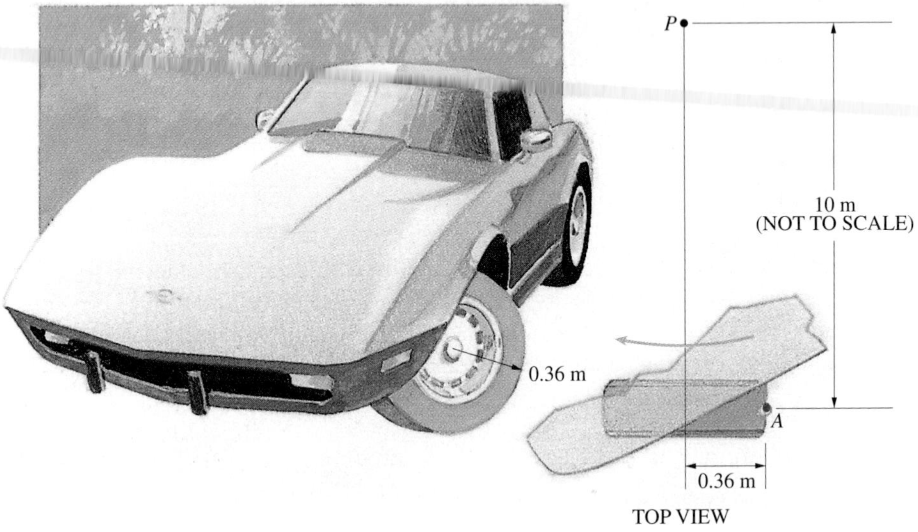

STRATEGY

(a) Let B be the tyre's midpoint. We introduce a rotating system with its origin at B and its y axis along the line from B to P (Figure (a)). We assume that the x axis *remains horizontal*. Knowing the velocity of point B, we can determine the angular velocity vector of the coordinate system and also the tyre's angular velocity relative to the coordinate system. Then we can use Equation (9.3) to determine the tyre's angular velocity.

(b) Knowing the velocity of point B and the tyre's angular velocity, we can use Equation (9.1) to determine the velocity of point A.

SOLUTION

(a) The angular velocity of the line PB is $(5\,\text{m/s})/(10\,\text{m}) = 0.5\,\text{rad/s}$. The xyz coordinate system rotates about its z axis in the clockwise direction as viewed in Figure (a), so the angular velocity vector of the coordinate system is

$$\Omega = -0.5\,\mathbf{k}\,\text{rad/s}$$

Relative to the xyz coordinate system, the tyre rotates about the y axis with angular velocity $(5\,\text{m/s})/(0.36\,\text{m}) = 13.9\,\text{rad/s}$. The tyre's angular velocity vector relative to the coordinate system is

$$\omega_{\text{rel}} = -13.9\,\mathbf{j}\,\text{rad/s}$$

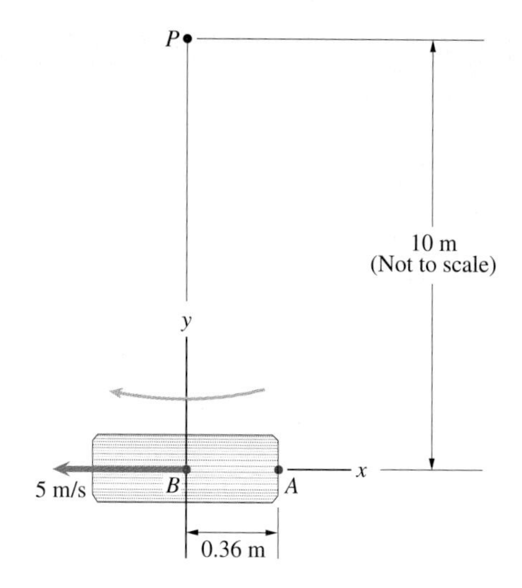

(a) A rotating coordinate system. The y axis remains aligned with BP, and the x axis remains horizontal.

The tyre's angular velocity vector is

$$\boldsymbol{\omega} = \boldsymbol{\Omega} + \boldsymbol{\omega}_{\text{rel}} = (-13.9\,\mathbf{j} - 0.5\,\mathbf{k})\ \text{rad/s}$$

(b) The position vector of point A relative to point B is $\mathbf{r}_{A/B} = 0.36\,\mathbf{i}$ m. Therefore the velocity of point A is

$$\begin{aligned}
\mathbf{v}_A &= \mathbf{v}_B + \boldsymbol{\omega} \times \mathbf{r}_{A/B} \\
&= -5\,\mathbf{i} + \begin{vmatrix} \mathbf{i} & \mathbf{j} & \mathbf{k} \\ 0 & -13.9 & -0.5 \\ 0.36 & 0 & 0 \end{vmatrix} \\
&= (-5\,\mathbf{i} - 0.18\,\mathbf{j} + 5\,\mathbf{k})\ \text{m/s}
\end{aligned}$$

DISCUSSION

The tyre's angular velocity vector $\boldsymbol{\omega}$ rotates in space as the car turns. But by expressing it in terms of a coordinate system that rotates with the car, we obtained a very simple expression for $\boldsymbol{\omega}$. (Notice that point A has a component of velocity in the negative y direction due to the rotation of the tyre's axis.)

Example 9.2

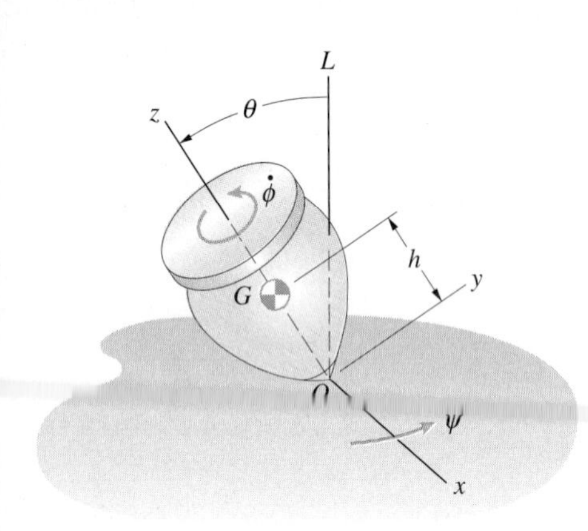

Figure 9.4

The point of the spinning top in Figure 9.4 remains at the fixed point O on the floor. The angle θ between the vertical axis L and the top's axis (the z axis) is constant. The x axis of the coordinate system *remains parallel to the floor* and rotates about L with constant angular velocity $\dot{\psi}$. Relative to the rotating coordinate system, the top spins about the z axis with constant angular velocity $\dot{\phi}$.

(a) What are the top's angular velocity and angular acceleration vectors?

(b) Determine the velocity of the centre of mass G.

STRATEGY

(a) We know the top's angular velocity relative to the rotating coordinate system, and we are given sufficient information to determine the angular velocity vector of the coordinate system. We can therefore use Equation (9.3) to determine the top's angular velocity and angular acceleration vectors.

(b) Once we know the top's angular velocity vector, we can determine the velocity of its centre of mass by applying Equation (9.1) to points O and G.

SOLUTION

(a) The coordinate system rotates about the vertical axis L with angular velocity $\dot{\psi}$. Therefore the coordinate system's angular velocity vector is parallel to L, and the right-hand rule indicates that it points upwards (Figure a). Resolving $\dot{\psi}$ into its y and z components, we obtain the angular velocity vector of the coordinate system:

$$\boldsymbol{\Omega} = \dot{\psi}\sin\theta\,\mathbf{j} = \dot{\psi}\,\mathbf{k}.$$

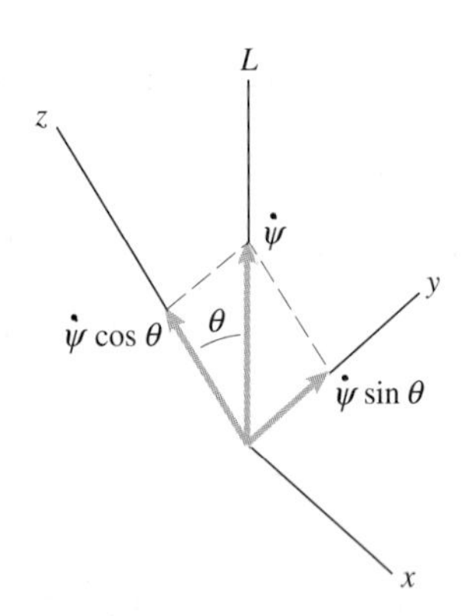

(a) Resolving the angular velocity vector of the coordinate system into components.

The top's angular velocity relative to the rotating coordinate system is $\boldsymbol{\omega}_{\rm rel} = \dot{\phi}\,\mathbf{k}$. Therefore the top's angular velocity vector is

$$\boldsymbol{\omega} = \boldsymbol{\Omega} + \boldsymbol{\omega}_{\rm rel} = \dot{\psi}\sin\theta\,\mathbf{j} + (\dot{\psi}\cos\theta + \dot{\phi})\,\mathbf{k}.$$

The x, y, and z components of the angular velocities $\boldsymbol{\Omega}$ and $\boldsymbol{\omega}_{\rm rel}$ are constant, so the top's angular acceleration vector is

$$\boldsymbol{\alpha} = \boldsymbol{\Omega} \times \boldsymbol{\omega}_{\rm rel} = \begin{vmatrix} \mathbf{i} & \mathbf{j} & \mathbf{k} \\ 0 & \dot{\psi}\sin\theta & \dot{\psi}\cos\theta \\ 0 & 0 & \dot{\phi} \end{vmatrix}$$

$$= \dot{\phi}\dot{\psi}\sin\theta\,\mathbf{i}.$$

(a) The position vector of the centre of mass relative to O is $\mathbf{r}_{G/O} = h\,\mathbf{k}$. The velocity of the centre of mass is

$$\mathbf{v}_G = \mathbf{v}_O + \boldsymbol{\omega} \times \mathbf{r}_{G/O}$$

$$= 0 + \begin{vmatrix} \mathbf{i} & \mathbf{j} & \mathbf{k} \\ 0 & \dot{\psi}\sin\theta & \dot{\psi}\cos\theta + \dot{\phi} \\ 0 & 0 & h \end{vmatrix}$$

$$= h\dot{\psi}\sin\theta\,\mathbf{i}$$

Problems

9.1 A rigid body's angular velocity vector is $\boldsymbol{\omega} = (200\,\mathbf{i} + 900\,\mathbf{j} - 600\mathbf{k})$ rad/s. Relative to a reference point O, the position and velocity of its centre of mass are $\mathbf{r}_G = (6\,\mathbf{i} + 6\,\mathbf{j} + 2\,\mathbf{k})$ m and $\mathbf{v}_G = 100\,\mathbf{i} + 80\,\mathbf{j} = 60\,\mathbf{k}$ m/s. What is the velocity relative to O of a point A of the rigid body with position $\mathbf{r}_A = (5.8\,\mathbf{i} + 6.4\,\mathbf{j} + 1.6\,\mathbf{k})$ m?

9.2 The angular acceleration vector of the rigid body in Problem 9.1 is $\boldsymbol{\alpha} = (8000\,\mathbf{i} - 8000\,\mathbf{j} - 4000\,\mathbf{k})$ rad/s^2 and the acceleration of its centre of mass relative to O is zero. What is the acceleration of point A relative to O?

9.3 The aeroplane's rate gyros indicate that its angular velocity is $\boldsymbol{\omega} = (4.0\,\mathbf{i} + 6.4\,\mathbf{j} + 0.2\,\mathbf{k})$ rad/s. What is the velocity relative to the centre of mass of the point A with coordinates $(8, 2, 2)$ m?

P9.3

9.4 The rate gyros of the aeroplane in Problem 9.3 indicate that its angular acceleration is $\boldsymbol{\alpha} = (-4\,\mathbf{i} + 12\,\mathbf{j} + 2\,\mathbf{k})$ rad/s^2. What is the acceleration of point A relative to the centre of mass?

9.5 The rectangular parallelepiped is rotating about a fixed axis through points A and B. Its direction of rotation is clockwise when the axis of rotation is viewed from point A towards point B.
(a) What is its angular velocity vector $\boldsymbol{\omega}$?
(b) What are the velocities of points C and D?

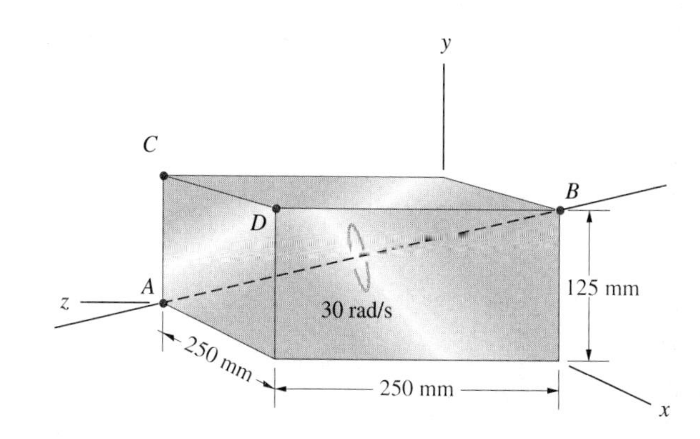

P9.5

9.6 If the angular velocity of the rectangular parallelepiped in Problem 9.5 is constant, what is the acceleration of point C?

9.7 The turbine is rotating about a fixed axis coincident with the line OA.
(a) What is its angular velocity vector?
(b) What is the velocity of the point of the turbine with coordinates $(3, 2, 2)$ m?

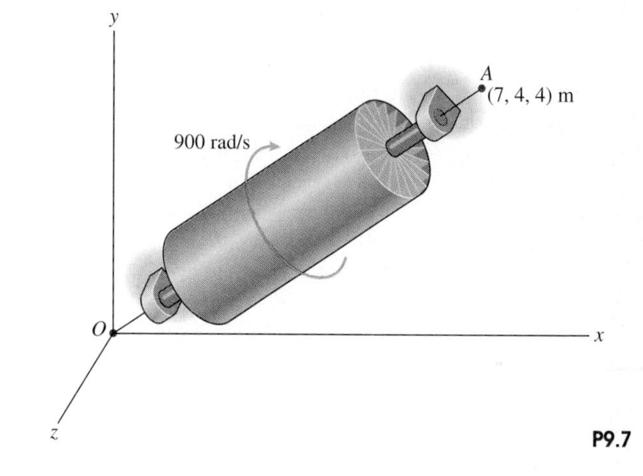

P9.7

9.8 The 900 rad/s angular velocity of the turbine in Problem 9.7 is decreasing at $100\ \text{rad/s}^2$.
(a) What is the turbine's angular acceleration vector?
(b) What is the acceleration of the point of the turbine with coordinate $(3, 2, 2)$ m?

9.9 The base of the dish antenna is rotating at 1 rad/s. The angle $\theta = 30°$ and is increasing at 20°/s.
(a) What are the components of the antenna's angular velocity vector $\boldsymbol{\omega}$ in terms of the body-fixed coordinate system shown?
(b) What is the velocity of the point of the antenna with coordinates $(2, 2, -2)$ m?

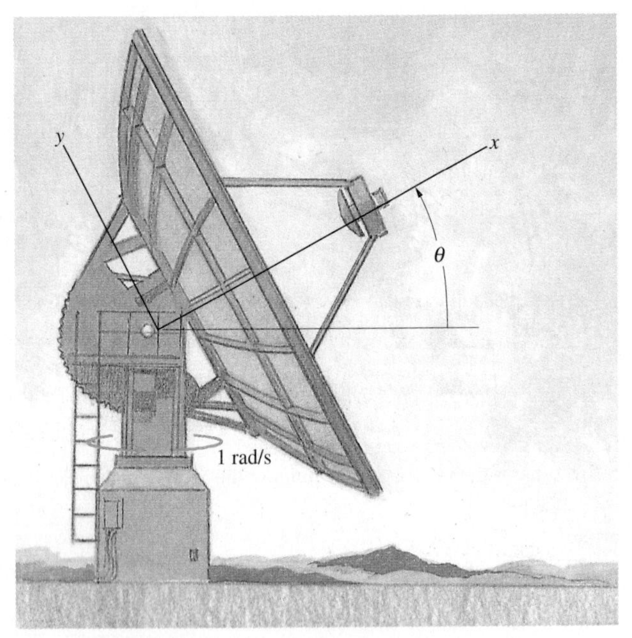

P9.9

9.10 The circular disc remains perpendicular to the horizontal shaft and rotates relative to it with angular velocity ω_d. The horizontal shaft is rigidly attached to a vertical shaft rotating with angular velocity ω_0.
(a) What is the disc's angular velocity vector $\boldsymbol{\omega}$?
(b) What is the velocity of point A of the disc?

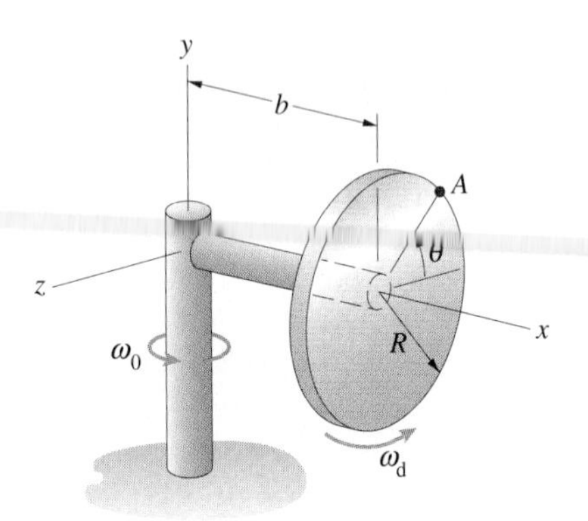

P9.10

9.11 If the angular velocities ω_d and ω_0 in Problem 9.10 are constant, what is the acceleration of point A of the disc?

9.12 The gyroscope's circular frame rotates about the vertical axis at 2 rad/s in the counterclockwise direction when viewed from above. The 60 mm diameter wheel rotates relative to the frame at 10 rad/s. Determine the velocities of points A and B relative to the origin.

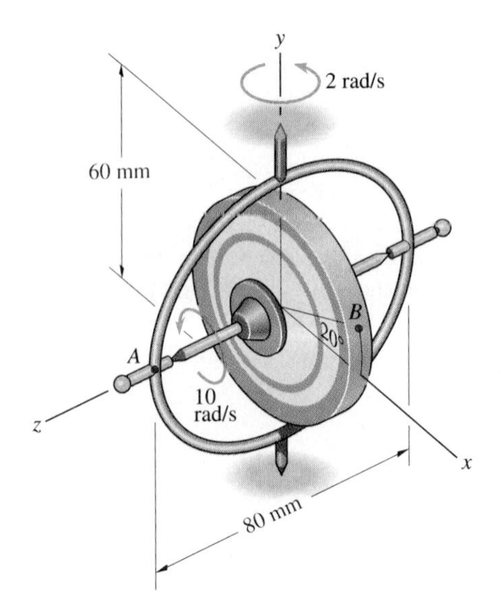

P9.12

9.13 If the angular velocities of the frame and wheel of the gyroscope in Problem 9.12 are constant, what are the accelerations of points A and B relative to the origin?

9.14 The manipulator rotates about the vertical axis with angular velocity $\Omega_y = 0.1\,\text{rad/s}$. The y axis of the coordinate system remains vertical and the coordinate system rotates with the manipulator so that points A, B and C remain in the x-y plane. The angular velocity vectors of the arms AB and BC *relative to the rotating coordinate system* are $-0.2\mathbf{k}\,\text{rad/s}$ and $0.4\mathbf{k}\,\text{rad/s}$, respectively.
(a) What is the angular velocity vector $\boldsymbol{\omega}_{BC}$ of arm BC?
(b) What is the velocity of point C?

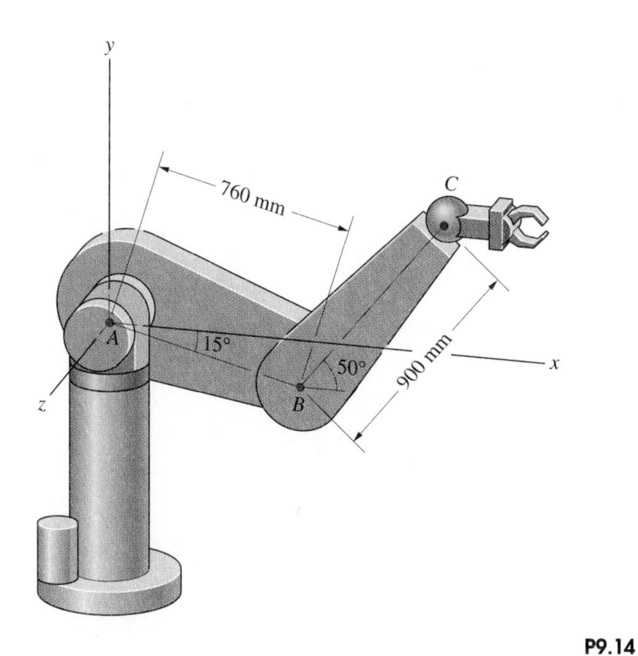

P9.14

9.15 The angular velocity of the manipulator in Problem 9.14 about the vertical axis is constant. The angular accelerations of the arms AB and BC relative to the rotating coordinate system are zero. What is the acceleration of point C?

9.16 The cone is connected by a ball and socket joint at its vertex to a 100 mm post. The radius of its base is 100 mm, and the base rolls on the floor. The velocity of the centre of the base is $\mathbf{v}_C = 2\,\mathbf{k}\,\text{m/s}$.
(a) What is the cone's angular velocity vector $\boldsymbol{\omega}$?
(b) What is the velocity of point A?

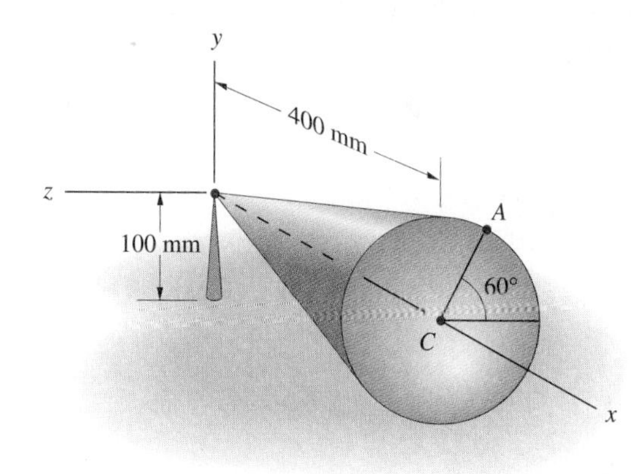

P9.16

9.17 The mechanism shown is a type of universal joint called a yoke and spider. The axis L lies in the x-z plane. Determine the angular velocity ω_L and the angular velocity vector $\boldsymbol{\omega}_S$ of the cross-shaped 'spider' in terms of the angular velocity ω_R at the instant shown.

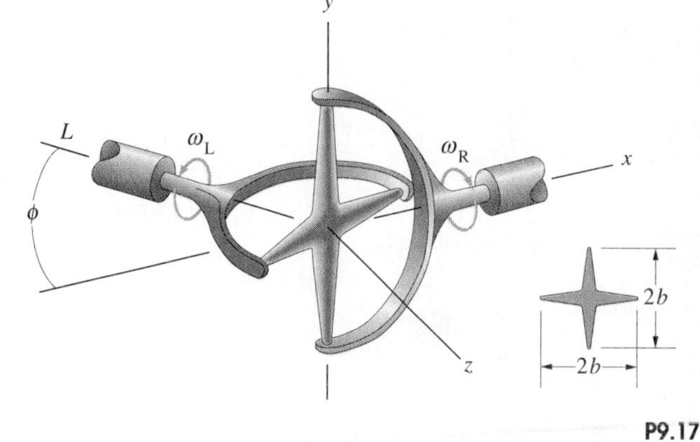

P9.17

9.2 Angular Momentum

Just as in the case of planar motion, the equations governing three-dimensional motion of a rigid body consists of Newton's second law and equations of angular motion. In comparison with the simple equation governing angular motion in two dimensions, the equations of angular motion in three dimensions are more complicated – three equations relate the components of the total moment about each coordinate axis to the components of the rigid body's angular acceleration and angular velocity. In this section we begin deriving the equations of angular motion by obtaining expressions for the angular momentum of a rigid body in three-dimensional motion. We first consider a rigid body rotating about a fixed point, and then a rigid body in general three-dimensional motion.

Rotation About a Fixed Point

Let m_i be the mass of the ith particle of a rigid body, and let $\mathbf{r}_i$ be its position relative to a fixed reference point O (Figure 9.5). The angular momentum of the rigid body about O is the sum of the angular momenta of its particles,

$$\mathbf{H}_O = \sum_i \mathbf{r}_i \times m_i \mathbf{v}_i$$

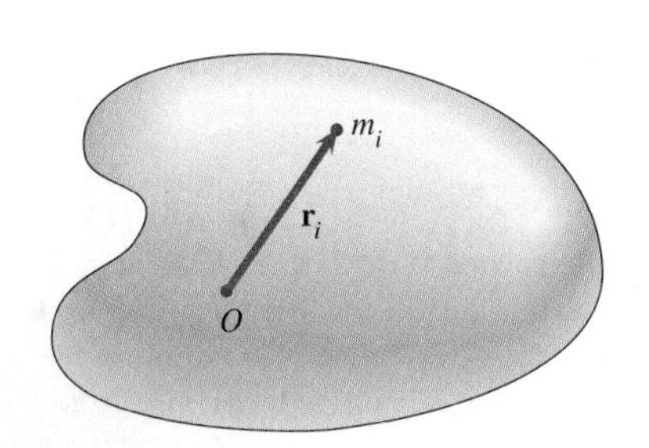

Figure 9.5
Mass and position of the ith particle of a rigid body.

where $\mathbf{v}_i = d\mathbf{r}_i/dt$. Let's assume that the rigid body rotates about the fixed point O with angular velocity $\boldsymbol{\omega}$. Then we can express the velocity of the ith particle as $\mathbf{v}_i = \boldsymbol{\omega} \times \mathbf{r}_i$, and the angular momentum is

$$\mathbf{H}_O = \sum_i \mathbf{r}_i \times m_i(\boldsymbol{\omega} \times \mathbf{r}_i) \tag{9.4}$$

In terms of a coordinate system with its origin at O (Figure 9.6), we can express the vectors $\boldsymbol{\omega}$ and $\mathbf{r}_i$ in terms of their components as

$$\boldsymbol{\omega} = \omega_x \mathbf{i} + \omega_y \mathbf{j} + \omega_z \mathbf{k}$$

$$\mathbf{r}_i = x_i \mathbf{i} + y_i \mathbf{j} + z_i \mathbf{k}$$

where (x_i, y_i, z_i) are the coordinates of the ith particle. Substituting these expressions into Equation (9.4) and evaluating the cross products, we can write the resulting components of the angular momentum vector in the forms

$$\begin{aligned} H_{Ox} &= I_{xx}\omega_x - I_{xy}\omega_y - I_{xz}\omega_z \\ H_{Oy} &= -I_{yx}\omega_x + I_{yy}\omega_y - I_{yz}\omega_z \\ H_{Oz} &= -I_{zx}\omega_x - I_{zy}\omega_y + I_{zz}\omega_z \end{aligned} \tag{9.5}$$

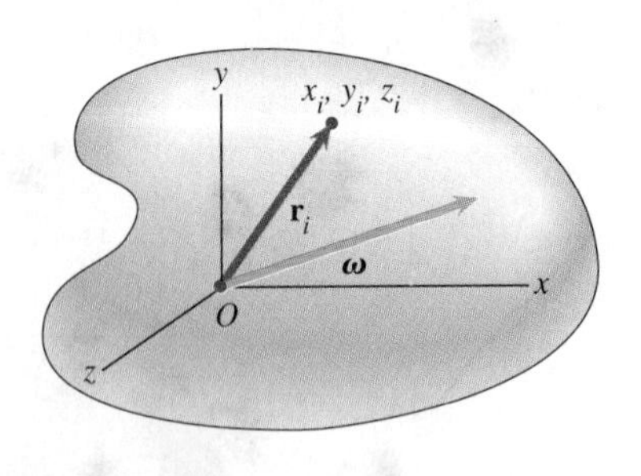

Figure 9.6
Introducing a coordinate system with its origin at O.

The coefficients

$$I_{xx} = \sum_i m_i(y_i^2 + z_i^2) \quad I_{yy} = \sum_i m_i(x_i^2 + z_i^2)$$
$$I_{zz} = \sum_i m_i(x_i^2 + y_i^2) \tag{9.6}$$

are called the **moments of inertia** about the x, y and z axes. The coefficients

$$I_{xy} = I_{yx} = \sum_i m_i x_i y_i \quad I_{yz} = I_{zy} = \sum_i m_i y_i z_i$$
$$I_{xz} = I_{zx} = \sum_i m_i x_i z_i \tag{9.7}$$

are called the **products of inertia**. We can write Equations (9.5) as the matrix equation

$$\begin{bmatrix} H_{Ox} \\ H_{Oy} \\ H_{Oz} \end{bmatrix} = \begin{bmatrix} I_{xx} & -I_{xy} & -I_{xz} \\ -I_{yx} & I_{yy} & I_{yz} \\ -I_{zx} & -I_{zy} & I_{zz} \end{bmatrix} \begin{bmatrix} \omega_x \\ \omega_y \\ \omega_z \end{bmatrix} \tag{9.8}$$

where

$$\begin{bmatrix} I_{xx} & -I_{xy} & -I_{xz} \\ -I_{yx} & I_{yy} & -I_{yz} \\ -I_{zx} & -I_{zy} & I_{zz} \end{bmatrix} = [I]$$

is called the **inertia matrix** of the rigid body.

Although Equations (9.5) appear to be complicated in comparison with the simple equation $H_O = I_O\omega$ for the angular momentum of a rigid body in planar motion about a fixed axis, we can point out simple correspondences between them. Suppose that the rigid body rotates about a fixed axis L_O coinciding with the x axis (Figure 9.7). From Equations (9.5), the angular momentum about the x axis is

$$H_{Ox} = I_{xx}\omega_x$$

The term $y_i^2 + z_i^2$ appearing in the definition of I_{xx} is the square of the perpendicular distance from the x axis to the ith particle, so $I_{xx} = I_O$. Therefore this equation relating H_{Ox} to ω_x is equivalent to the planar equation. Notice from Equations (9.5), however, that if the axis of rotation does not coincide with one of the coordinate axes, the component of the angular momentum about a coordinate axis depends in general not only on the component of the angular velocity about that axis, but also on the components of the angular velocity about the other axes through the products of inertia I_{xy}, I_{yz} and I_{zx}.

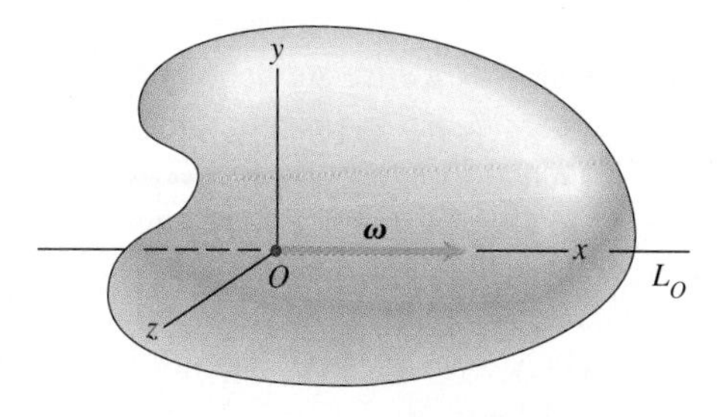

Figure 9.7
The axis of rotation L_O coincident with the x axis.

General Motion

Here we obtain the angular momentum for a rigid body undergoing general three-dimensional motion. The derivation and the resulting equations are very similar to those for rotation about a fixed point.

Let $\mathbf{R}_i$ be the position of the ith particle of a rigid body relative to the centre of mass (Figure 9.8). The rigid body's angular momentum about its centre of mass is

$$\mathbf{H} = \sum_i \mathbf{R}_i \times m_i \frac{d\,\mathbf{R}_i}{dt} = \sum_i \mathbf{R}_i \times m_i(\boldsymbol{\omega} \times \mathbf{R}_i)$$

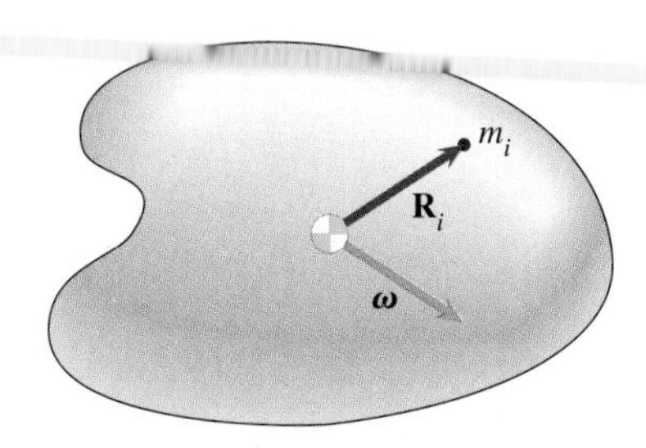

Figure 9.8
Position of the ith particle of a rigid body relative to the centre of mass.

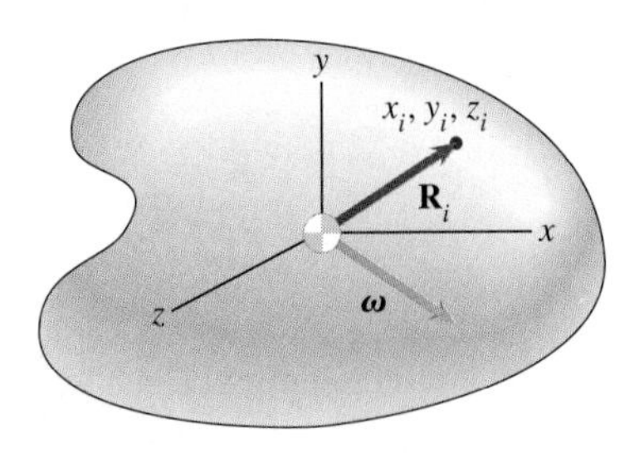

Figure 9.9
Introducing a coordinate system with its origin at the centre of mass.

Introducing a coordinate system with its origin at the centre of mass (Figure 9.9), we express $\boldsymbol{\omega}$ and $\mathbf{R}_i$ in terms of their components as

$$\boldsymbol{\omega} = \omega_x\,\mathbf{i} + \omega_y\,\mathbf{j} + \omega_z\,\mathbf{k}$$

$$\mathbf{R}_i = x_i\,\mathbf{i} + y_i\,\mathbf{j} + z_i\,\mathbf{k}$$

where (x_i, y_i, z_i) are the coordinates of the ith particle relative to the centre of mass. The resulting components of the angular momentum vector are

$$\begin{aligned} H_x &= I_{xx}\omega_x - I_{xy}\omega_y - I_{xz}\omega_z \\ H_y &= -I_{yx}\omega_x + I_{yy}\omega_y - I_{yz}\omega_z \\ H_z &= -I_{zx}\omega_x - I_{zy}\omega_y + I_{zz}\omega_z \end{aligned} \qquad (9.9)$$

or

$$\begin{bmatrix} H_x \\ H_y \\ H_z \end{bmatrix} = \begin{bmatrix} I_{xx} & -I_{xy} & -I_{xz} \\ -I_{yx} & I_{yy} & -I_{yz} \\ -I_{zx} & -I_{zy} & I_{zz} \end{bmatrix} \begin{bmatrix} \omega_x \\ \omega_y \\ \omega_z \end{bmatrix} \qquad (9.10)$$

The moments and products of inertia are defined by Equation (9.6).

These equations for the angular momentum in general motion are identical in form to those we obtained for rotation about a fixed point. The expressions for the moments and products of inertia are the same. However, in the case of rotation about a fixed point, the moments and products of inertia are expressed in terms of a coordinate system with its origin at the fixed point, whereas in the case of general motion they are expressed in terms of a coordinate system with its origin at the centre of mass.

9.3 Moments and Products of Inertia

To determine the angular momentum of a given rigid body in three-dimensional motion, you must know its moments and products of inertia. From your experience with planar rigid-body dynamics, you are familiar with evaluating an object's moment of inertia about a given axis. The same techniques apply in three-dimensional problems. We demonstrate this by evaluating the moments and products of inertia for a slender bar and a thin plate. We then extend the parallel-axis theorem to three dimensions to permit the evaluation of the moments and products of inertia of composite objects.

Simple Objects

If we model an object as a continuous distribution of mass, we can express the inertia matrix as

$$\lfloor I \rfloor = \begin{bmatrix} I_{xx} & -I_{xy} & -I_{xz} \\ -I_{yx} & I_{yy} & -I_{yz} \\ -I_{zx} & -I_{zy} & I_{zz} \end{bmatrix} = \begin{bmatrix} \int_m (y^2+z^2)dm & -\int_m xy\,dm & -\int_m xz\,dm \\ -\int_m yx\,dm & \int_m (x^2+z^2)dm & -\int_m yz\,dm \\ -\int_m zx\,dm & -\int_m zy\,dm & \int_m (x^2+y^2)dm \end{bmatrix} \quad (9.11)$$

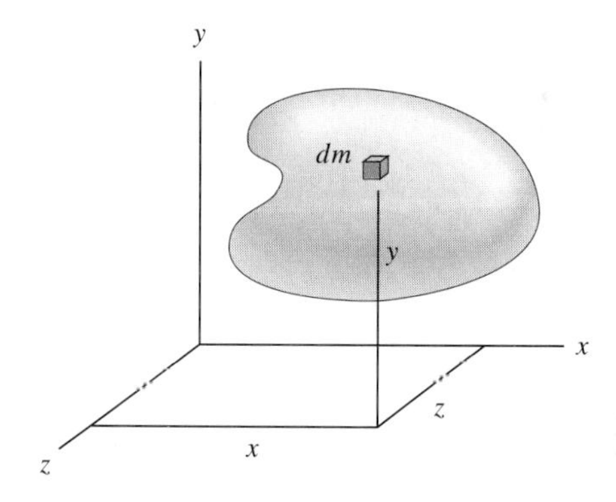

Figure 9.10
Determining the moments and products of inertia by modelling an object as a continuous distribution of mass.

where x, y and z are the coordinates of the differential elements of mass dm (Figure 9.10).

Slender Bars Let the origin of the coordinate system be at a slender bar's centre of mass with the x axis along the bar (Figure 9.11(a)). The bar has length l, cross-sectional area A, and mass m. We assume that A is uniform along the length of the bar and that the material is homogeneous.

Consider a differential element of the bar of length dx at a distance x from the centre of mass (Figure 9.11(b)). The mass of the element is $dm = \rho A dx$, where ρ is the mass density. We neglect the lateral dimensions of the bar, assuming the coordinates of the differential element dm to be $(x, 0, 0)$. As a consequence of this approximation, the moment of inertia of the bar about the x axis is zero:

$$I_{xx} = \int_m (y^2 + z^2)\,dm = 0$$

The moment of inertia about the y axis is

$$I_{yy} = \int_m (x^2 + z^2)\,dm = \int_{-l/2}^{l/2} \rho A x^2\,dx = \frac{1}{12}\rho A l^3$$

Expressing this result in terms of the mass of the bar $m = \rho A l$, we obtain

$$I_{yy} = \frac{1}{12} m l^2$$

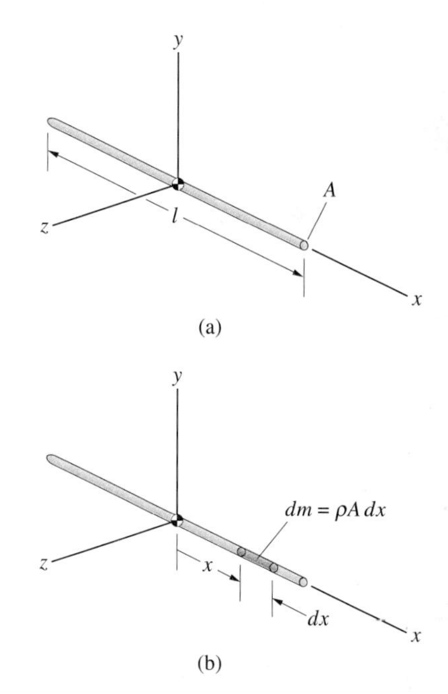

Figure 9.11
(a) A slender bar and a coordinate system with the x axis aligned with the bar.
(b) A differential element of mass of length dx.

The moment of inertia about the z axis is equal to the moment of inertia about the y axis:

$$I_{zz} = \int_m (x^2 + y^2)\, dm = \frac{1}{12} ml^2$$

Because the y and z coordinates of dm are zero, the products of inertia are zero, so the inertia matrix for the slender bar is

$$[I] = \begin{bmatrix} 0 & 0 & 0 \\ 0 & \frac{1}{12} ml^2 & 0 \\ 0 & 0 & \frac{1}{12} ml^2 \end{bmatrix} \tag{9.12}$$

You must remember that the moments and products of inertia depend on the orientation of the coordinate system relative to the object. In terms of the alternative coordinate system shown in Figure 9.12, the bar's inertia matrix is

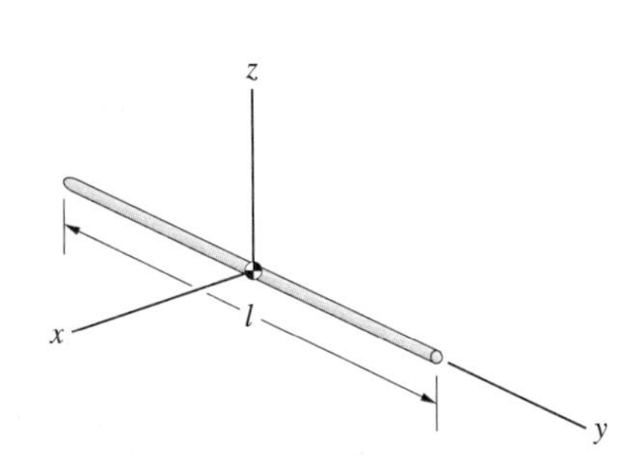

Figure 9.12
Aligning the y axis with the bar.

$$[I] = \begin{bmatrix} \frac{1}{12} ml^2 & 0 & 0 \\ 0 & 0 & 0 \\ 0 & 0 & \frac{1}{12} ml^2 \end{bmatrix}$$

Thin Plates Suppose that a homogeneous plate of uniform thickness T, area A, and unspecified shape lies in the x-y plane (Figure 9.13(a)). We can express its mass moments of inertia in terms of the moments of inertia of its cross-sectional area.

By projecting an element of area dA through the thickness T of the plate (Figure 9.13(b)), we obtain a differential element of mass $dm = \rho T\, dA$. We neglect the plate's thickness in calculating the moments of inertia, so the

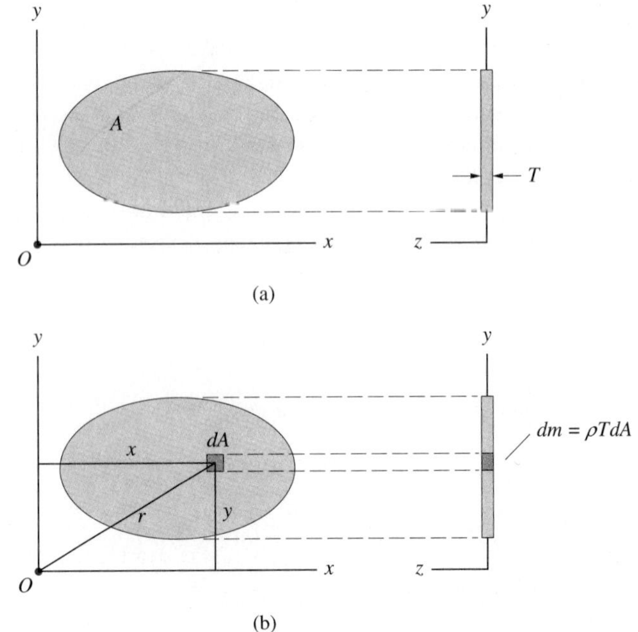

Figure 9.13
(a) A thin plate lying in the x-y plane.
(b) Obtaining a differential element of mass by projecting an element of area dA through the plate.

coordinates of the element dm are $(x, y, 0)$. The plate's moment of inertia about the x axis is

$$I_{xx} = \int_m (y^2 + z^2)\, dm = \rho T \int_A y^2\, dA = \rho T I_x$$

where I_x is the moment of inertia of the plate's cross-sectional area about the x axis. Since the mass of the plate is $m = \rho T A$, the product $\rho T = m/A$, and we obtain the moment of inertia in the form

$$I_{xx} = \frac{m}{A} I_x$$

The moment of inertia about the y axis is

$$I_{yy} = \int_m (x^2 + z^2)\, dm = \rho T \int_A x^2\, dA = \frac{m}{A} I_y$$

where I_y is the moment of inertia of the cross-sectional area about the y axis. The moment of inertia about the z axis is

$$I_{zz} = \int_m (x^2 + y^2) dm = \frac{m}{A} J_O$$

where $J_O = I_x + I_y$ is the polar moment of inertia of the cross-sectional area. The product of inertia I_{xy} is

$$I_{xy} = \int_m xy\, dm = \frac{m}{A} I_{xy}^A$$

where

$$I_{xy}^A = \int_A xy\, dA$$

is the product of inertia of the cross-sectional area. (We use a superscript A to distinguish the product of inertia of the plate's cross-sectional area from the product of inertia of its mass.) If the cross-sectional area A is symmetric about either the x or the y axis, $I_{xy}^A = 0$.

Because the z coordinate of dm is zero, the products of inertia I_{xz} and I_{yz} are zero. The inertia matrix for the thin plate is

$$[I] = \begin{bmatrix} \frac{m}{A} I_x^A & -\frac{m}{A} I_{xy}^A & 0 \\ -\frac{m}{A} I_{xy}^A & \frac{m}{A} I_y^A & 0 \\ 0 & 0 & \frac{m}{A} J_O \end{bmatrix} \tag{9.13}$$

If you know, or can determine, the moments of inertia and products of inertia of the plate's cross-sectional area, you can use these expressions to obtain the moments and products of inertia of its mass.

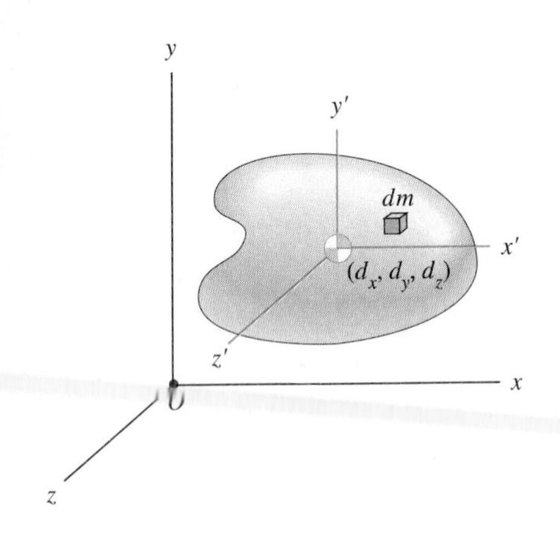

Figure 9.14
A coordinate system $x'y'z'$ with its origin at the centre of mass and a parallel coordinate system xyz.

Parallel-Axis Theorems

Suppose that we know an object's inertia matrix $[I']$ in terms of a coordinate system $x'y'z'$ with its origin at the centre of mass, and we want to determine the inertia matrix $[I]$ in terms of a parallel coordinate system xyz (Figure 9.14). Let (d_x, d_y, d_z) be the coordinates of the centre of mass in the xyz coordinate system. The coordinates of a differential element of mass dm in the xyz system are given in terms of its coordinates in the $x'y'z'$ system by

$$x = x' + d_x \qquad y = y' + d_y \qquad z - z' + d_z \tag{9.14}$$

By substituting these expressions into the definition of I_{xx}, we obtain

$$\begin{aligned} I_{xx} &= \int_m [(y')^2 + (z')^2]\, dm + 2d_y \int_m y'\, dm \\ &= 2d_z \int_m z'\, dm + (d_y^2 + d_z^2) \int_m dm \end{aligned} \tag{9.15}$$

The first integral on the right is the object's moment of inertia about the x' axis. We can show that the second and third integrals are zero by using the definitions of the object's centre of mass expressed in terms of the $x'y'z'$ coordinate system:

$$\bar{x}' = \frac{\int_m x'\, dm}{\int_m dm} \qquad \bar{y}' = \frac{\int_m y'\, dm}{\int_m dm} \qquad \bar{z}' = \frac{\int_m z'\, dm}{\int_m dm}$$

The object's centre of mass is at the origin of the $x'y'z'$ system, so $\bar{x}' = \bar{y}' = \bar{z}' = 0$. Therefore the second and third integrals on the right of Equation (9.15) are zero, and we obtain

$$I_{xx} = I_{x'x'} + (d_y^2 + d_z^2)m$$

where m is the mass of the object. Substituting Equation (9.14) into the definition of I_{xy}, we obtain

$$\begin{aligned} I_{xy} &= \int_m x'y'\, dm + d_x \int_m y'\, dm + d_y \int_m x'\, dm + d_x d_y \int_m dm \\ &= I_{x'y'} + d_x d_y m \end{aligned}$$

Proceeding in this way for each of the moments and products of inertia, we obtain the **parallel-axis theorems**:

$$\begin{aligned} I_{xx} &= I_{x'x'} + (d_y^2 + d_z^2)m \\ I_{yy} &= I_{y'y'} + (d_x^2 + d_z^2)m \\ I_{zz} &= I_{z'z'} + (d_x^2 + d_y^2)m \\ I_{xy} &= I_{x'y'} + d_x d_y m \\ I_{yz} &= I_{y'z'} + d_y d_x m \\ I_{zx} &= I_{z'x'} + d_z d_x m \end{aligned} \tag{9.16}$$

If you know an object's inertia matrix in terms of a particular coordinate system, you can use these theorems to determine its inertia matrix in terms of any parallel coordinate system. You can also use them to determine the inertia matrices of composite objects.

Moment of Inertia About an Arbitrary Axis

If we know a rigid body's inertia matrix in terms of a given coordinate system with origin O, we can determine its moment of inertia about an arbitrary axis through O. Suppose that the rigid body rotates with angular velocity $\boldsymbol{\omega}$ about an arbitrary fixed axis L_O through O, and let $\mathbf{e}$ be a unit vector with the same direction as $\boldsymbol{\omega}$ (Figure 9.15). In terms of the moment of inertia I_O about L_O, the rigid body's angular momentum about L_O is

$$H_O = I_O|\boldsymbol{\omega}|$$

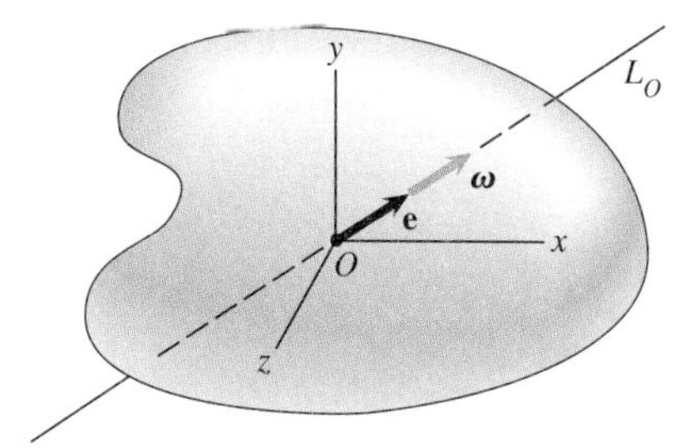

Figure 9.15
Rigid body rotating about L_O.

We can express the angular velocity vector as

$$\boldsymbol{\omega} = |\boldsymbol{\omega}|(e_x\,\mathbf{i} + e_y\,\mathbf{j} + e_z\,\mathbf{k})$$

so that $\omega_x = |\boldsymbol{\omega}|e_x$, $\omega_y = |\boldsymbol{\omega}|e_y$, and $\omega_z = |\boldsymbol{\omega}|e_z$. Using these expressions and Equations (9.5), the angular momentum about L_O is

$$\begin{aligned} H_O = \mathbf{H}_O \cdot \mathbf{e} &= (I_{xx}|\boldsymbol{\omega}|e_x - I_{xy}|\boldsymbol{\omega}|e_y - I_{xz}|\boldsymbol{\omega}|e_z)e_x \\ &\quad + (-I_{yx}|\boldsymbol{\omega}|e_x + I_{yy}|\boldsymbol{\omega}|e_y - I_{yz}|\boldsymbol{\omega}|e_z)e_y \\ &\quad + (-I_{zx}|\boldsymbol{\omega}|e_x - I_{zy}|\boldsymbol{\omega}|e_y + I_{zz}|\boldsymbol{\omega}|e_z)e_z \end{aligned}$$

Equating our two expressions for H_O, we obtain

$$I_O = I_{xx}e_x^2 + I_{yy}e_y^2 + I_{zz}e_z^2 - 2I_{xy}e_xe_y - 2I_{yz}e_ye_z - 2I_{zx}e_ze_x \quad (9.17)$$

Notice that the moment of inertia about an arbitrary axis depends on the products of inertia, in addition to the moments of inertia about the coordinate axes. If you know an object's inertia matrix, you can use Equation (9.17) to determine its moment of inertia about an axis through O whose direction is specified by the unit vector $\mathbf{e}$.

Principal Axes

For *any* object and origin O, at least one coordinate system exists for which the products of inertia are zero:

$$[I] = \begin{bmatrix} I_{xx} & 0 & 0 \\ 0 & I_{yy} & 0 \\ 0 & 0 & I_{zz} \end{bmatrix} \tag{9.18}$$

These coordinate axes are called **principal axes**, and the moments of inertia are called the **principal moments of inertia**.

If you know the inertia matrix of a rigid body in terms of a coordinate system $x'y'z'$ and the products of inertia are zero, $x'y'z'$ is a set of principal axes. Suppose that the products of inertia are not zero, and you want to find a set of principal axes xyz and the corresponding principal moments of inertia (Figure 9.16). It can be shown that the principal moments of inertia are roots of the cubic equation

$$\begin{aligned} &I^3 - (I_{x'x'} + I_{y'y'} + I_{z'z'})I^2 \\ &+ (I_{x'x'}I_{y'y'} + I_{y'y'}I_{z'z'} + I_{z'z'}I_{x'x'} - I_{x'y'}^2 - I_{y'z'}^2 - I_{z'x'}^2)I \\ &- (I_{x'x'}I_{y'y'}I_{z'z'} - I_{x'x'}I_{y'z'}^2 - I_{y'y'}I_{x'z'}^2 - I_{z'z'}I_{x'y'}^2 - 2I_{x'y'}I_{y'z'}I_{z'x'}) = 0 \end{aligned} \tag{9.19}$$

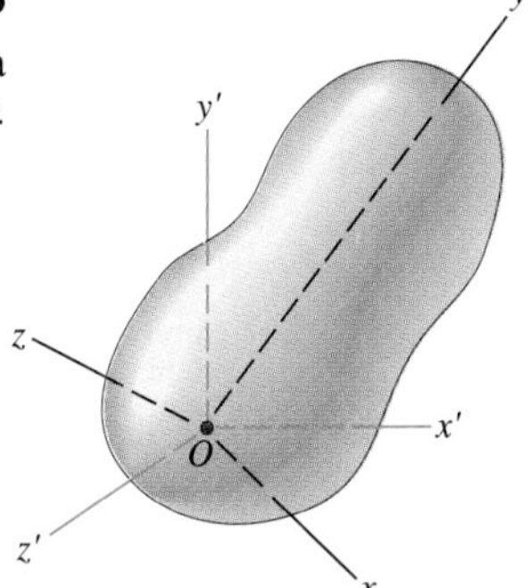

Figure 9.16
The $x'y'z'$ system with its origin at O and a set of principal axes xyz.

For each principal moment of inertia I, the vector $\mathbf{V}$ with components

$$\begin{aligned} V_{x'} &= (I_{y'y'} - I)(I_{z'z'} - I) - I_{y'z'}^2 \\ V_{y'} &= I_{x'y'}(I_{z'z'} - I) + I_{x'z'}I_{y'z'} \\ V_{z'} &= I_{x'z'}(I_{y'y'} - I) + I_{x'y'}I_{y'z'} \end{aligned} \tag{9.20}$$

is parallel to the corresponding principal axis.

To determine the principal moments of inertia, you must obtain the roots of Equation (9.19). Then substitute one of the principal moments of inertia into Equations (9.20) to obtain the components of a vector parallel to the corresponding principal axis. By repeating this step for each principal moment of inertia, you can determine the three principal axes. If you don't obtain a solution from Equations (9.20), try one of the other principal moments of inertia. You can choose the axes you identify as x, y and z arbitrarily, although you must make sure your coordinate system is right-handed. See Example 9.5.

Axes through O about which an object's moment of inertia is a minimum or maximum are principal axes. If the three principal moments of inertia are equal, any coordinate system with its origin at O is a set of principal axes, and the moment of inertia has the same value about any axis through O. This is the case, for example, if the object is a homogeneous sphere and the origin is at its centre (Figure 9.17(a)). If two of the principal moments of inertia are equal, you can determine a unique principal axis from the third one, and any axes perpendicular to the unique principal axis are principal axes. This is the case when an object has an axis of rotational symmetry and the origin is a point on the axis (Figure 9.17(b)). The axis of symmetry is the unique principal axis.

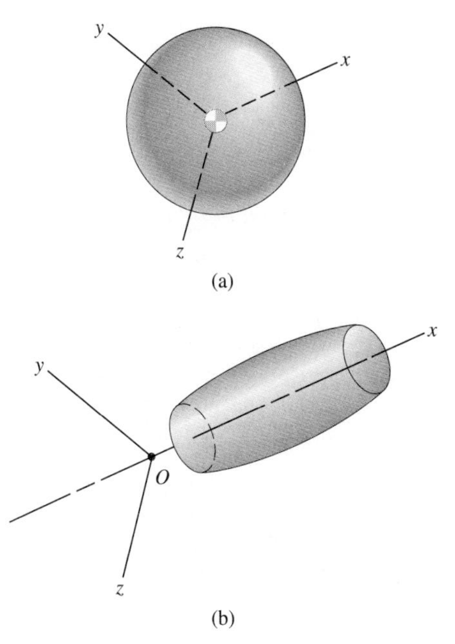

Figure 9.17

(a) A homogeneous sphere. Any coordinate system with its origin at the centre is a set of principal axes.
(b) A rotationally symmetric object. The axis of symmetry is a principal axix, and any perpendicular axes are principal axes.

In the following examples we determine moments and products of inertia of simple objects, apply the parallel-axis theorems, and evaluate the angular momenta of rigid bodies.

Example 9.3

The boom AB of the crane in Figure 9.18 has a mass of 4800 kg, and the boom BC has a mass of 1600 kg and is perpendicular to boom AB. Modelling each boom as a slender bar and treating them as a single object, determine the moments and products of inertia of the object in terms of the coordinate system shown.

Figure 9.18

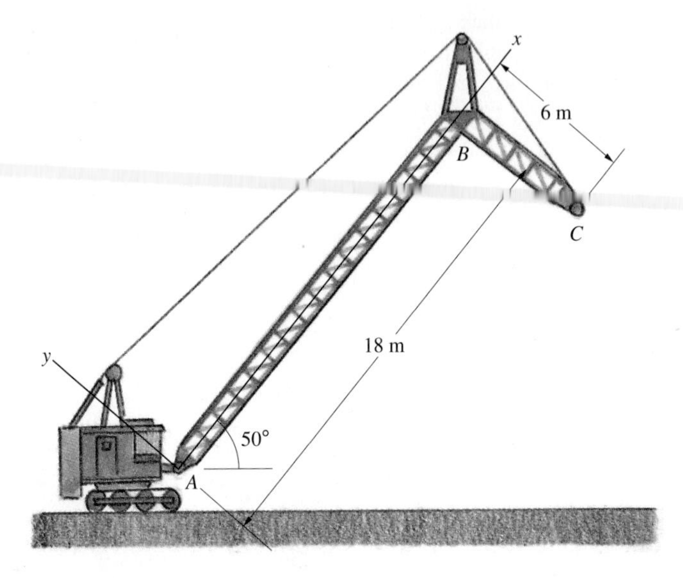

STRATEGY

We can apply the parallel-axis theorems to each boom to determine its moments and products of inertia in terms of the given coordinate system. The moments and products of inertia of the combined object are the sums of those for the two booms.

SOLUTION

Boom AB In Figure (a) we introduce a parallel coordinate system $x'y'z'$ with its origin at the centre of mass of boom AB. In terms of the $x'y'z'$ system, the inertia matrix of boom AB is

$$[I'] = \begin{bmatrix} 0 & 0 & 0 \\ 0 & \frac{1}{12}ml^2 & 0 \\ 0 & 0 & \frac{1}{12}ml^2 \end{bmatrix}$$

$$= \begin{bmatrix} 0 & 0 & 0 \\ 0 & \frac{1}{12}(4800)(18)^2 & 0 \\ 0 & 0 & \frac{1}{12}(4800)(18)^2 \end{bmatrix} \text{kg.m}^2$$

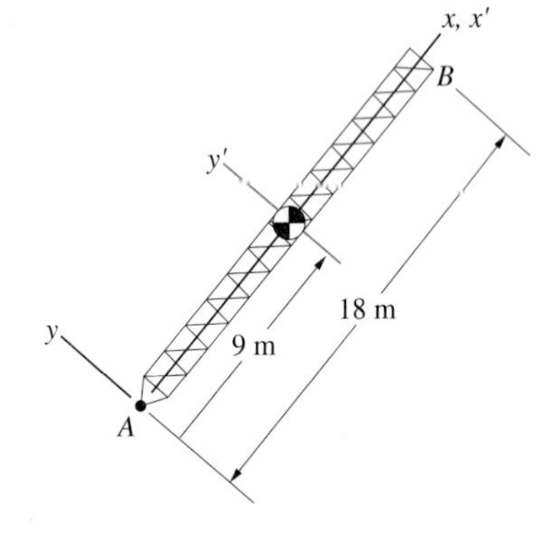

(a) Applying the parallel-axis theorems to boom AB.

The coordinates of the origin of the $x'y'z'$ system relative to the xyz system are $d_x = 9\,\text{m}$, $d_y = 0$, $d_z = 0$. Applying the parallel-axis theorems, we obtain

$$I_{xx} = I_{x'x'} + (d_y^2 + d_z^2)m = 0$$

$$I_{yy} = I_{y'y'} + (d_x^2 + d_z^2)m = \frac{1}{12}(4800)(18)^2 + (9)^2(4800)$$

$$= 518\,400\,\text{kg.m}^2$$

$$I_{zz} = I_{z'z'} + (d_x^2 + d_y^2)m = \frac{1}{12}(4800)(18)^2 + (9)^2(4800)$$

$$= 518\,400\ \text{kg.m}^2$$

$$I_{xy} = I_{x'y'} + d_x d_y m = 0$$

$$I_{yz} = I_{y'z'} + d_y d_z m = 0$$

$$I_{zx} = I_{z'x'} + d_z d_x m = 0$$

Boom BC In Figure (b) we introduce a parallel coordinate system $x'y'z'$ with its origin at the centre of mass of boom *BC*. In terms of the $x'y'z'$ system, the inertia matrix of boom *BC* is

$$[I'] = \begin{bmatrix} \frac{1}{12}ml^2 & 0 & 0 \\ 0 & 0 & 0 \\ 0 & 0 & \frac{1}{12}ml^2 \end{bmatrix}$$

$$= \begin{bmatrix} \frac{1}{12}(1600)(6)^2 & 0 & 0 \\ 0 & 0 & 0 \\ 0 & 0 & \frac{1}{12}(1600)(6)^2 \end{bmatrix} \text{kg.m}^2$$

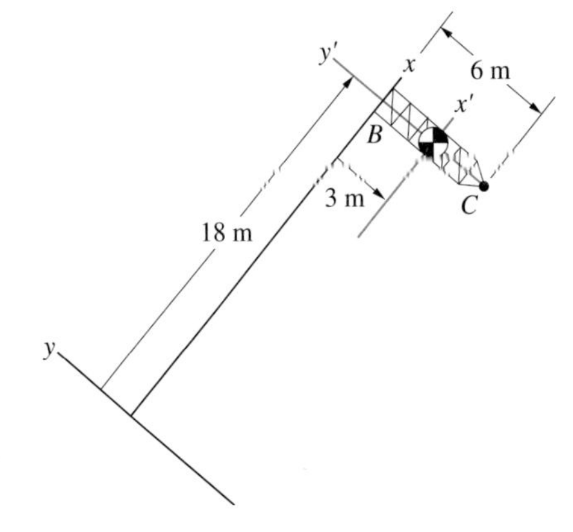

(b) Applying the parallel-axis theorems to boom *BC*.

The coordinates of the origin of the $x'y'z'$ system relative to the xyz system are $d_x = 18$ m, $d_y = -3$ m, $d_z = 0$. Applying the parallel-axis theorems, we obtain

$$I_{xx} = I_{x'x'} + (d_y^2 + d_z^2)m = \frac{1}{12}(1600)(6)^2 + (-3)^2(1600)$$

$$= 19\,200\ \text{kg.m}^2$$

$$I_{yy} = I_{y'y'} + (d_x^2 + d_z^2)m = 0 + (18)^2(1600) = 518\,400\ \text{kg.m}^2$$

$$I_{zz} = I_{z'z'} + (d_x^2 + d_y^2)m = \frac{1}{12}(1600)(6)^2 + [(18)^2 + (-3)^2](1600)$$

$$= 537\,600\ \text{kg.m}^2$$

$$I_{xy} = I_{x'y'} + d_x d_y m = 0 + (18)(-3)(1600) = -86\,400\ \text{kg.m}^2$$

$$I_{yz} = I_{y'z'} + d_y d_z m = 0$$

$$I_{zx} = I_{z'x'} + d_z d_x m = 0$$

Summing the results for the two booms, we obtain the inertia matrix for the single object:

$$[I] = \begin{bmatrix} 19\,200 & -(-86\,400) & 0 \\ -(-86\,400) & 518\,400 + 518\,400 & 0 \\ 0 & 0 & 518\,400 + 537\,600 \end{bmatrix}$$

$$= \begin{bmatrix} 19\,200 & 86\,400 & 0 \\ 86\,400 & 1\,036\,800 & 0 \\ 0 & 0 & 1\,056\,000 \end{bmatrix} \text{kg.m}^2$$

Example 9.4

The 4 kg rectangular plate in Figure 9.19 lies in the *x*-*y* plane of the body-fixed coordinate system.
(a) Determine the plate's moments and products of inertia.
(b) Determine the plate's moment of inertia about the diagonal axis L_O.
(c) If the plate is rotating about the fixed point O with angular velocity $\omega = (4\mathbf{i} - 2\mathbf{j})$ rad/s, what is the plate's momentum about O?

Figure 9.19

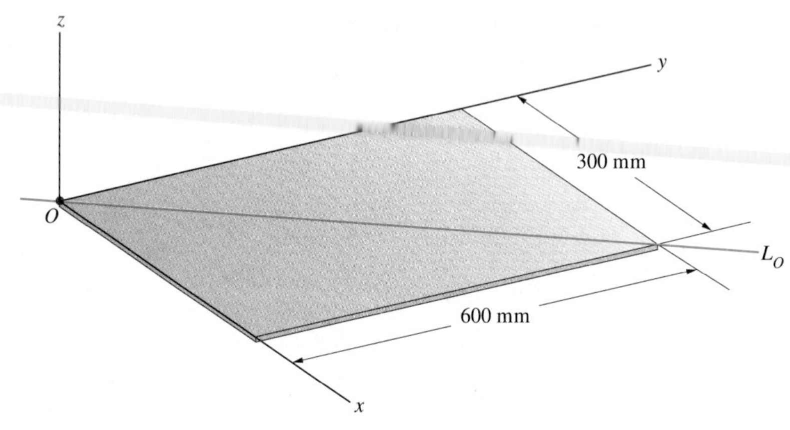

STRATEGY

(a) We can obtain the moments of inertia and products of inertia of the plate's rectangular area from Appendix B and use Equations (9.13) to obtain the moments and products of inertia.
(b) Once we know the moments and products of inertia, we can use Equation (9.17) to determine the moment of inertia about L_O.
(c) The angular momentum about O is given by Equation (9.8).

SOLUTION

(a) From Appendix B, the moments of inertia of the plate's cross-sectional area are (Figure (a)):

$$I_x = \frac{1}{3}bh^3 \qquad I_y = \frac{1}{3}hb^3$$

$$I_{xy}^A = \frac{1}{4}b^2h^2 \qquad J_O = \frac{1}{3}(bh^3 + hb^3)$$

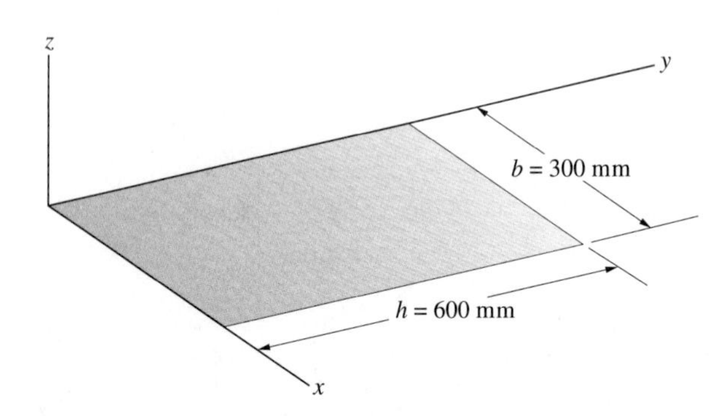

(a) Determining the moments of inertia of the plate's area.

Therefore the moments and products of inertia are

$$I_{xx} = \frac{m}{A} I_x = \frac{(4)}{(0.3)(0.6)} \left(\frac{1}{3}\right)(0.3)(0.6)^3 = 0.48\,\text{kg.m}^2$$

$$I_{yy} = \frac{m}{A} I_y = \frac{(4)}{(0.3)(0.6)} \left(\frac{1}{3}\right)(0.6)(0.3)^3 = 0.12\,\text{kg.m}^2$$

$$I_{xy} = \frac{m}{A} I_{xy}^A = \frac{(4)}{(0.3)(0.6)} \left(\frac{1}{4}\right)(0.3)^2(0.6)^2 = 0.18\,\text{kg.m}^2$$

$$I_{zz} = \frac{m}{A} J_O = \frac{(4)}{(0.3)(0.6)} \left(\frac{1}{3}\right)[(0.3)(0.6)^3 + (0.6)(0.3)^3] = 0.60\,\text{kg.m}^2$$

$$I_{xz} = I_{yz} = 0$$

(b) To apply Equation (9.17), we must determine the components of a unit vector parallel to L_O:

$$\mathbf{e} = \frac{300\,\mathbf{i} + 600\,\mathbf{j}}{|300\,\mathbf{i} + 600\,\mathbf{j}|} = 0.447\,\mathbf{i} + 0.894\,\mathbf{j}$$

The moment of inertia about L_O is

$$\begin{aligned} I_O &= I_{xx}e_x^2 + I_{yy}e_y^2 + I_{zz}e_z^2 - 2I_{xy}e_xe_y - 2I_{yz}e_ye_z - 2I_{zx}e_ze_x \\ &= (0.48)(0.447)^2 + (0.12)(0.894)^2 - 2(0.18)(0.447)(0.894) \\ &= 0.048\,\text{kg.m}^2 \end{aligned}$$

(c) The plate's angular moment about O is

$$\begin{aligned} \begin{bmatrix} H_{Ox} \\ H_{Oy} \\ H_{Oz} \end{bmatrix} &= \begin{bmatrix} I_{xx} & -I_{xy} & -I_{xz} \\ -I_{yx} & I_{yy} & -I_{yz} \\ -I_{zx} & -I_{zy} & I_{zz} \end{bmatrix} \begin{bmatrix} \omega_x \\ \omega_y \\ \omega_z \end{bmatrix} \\ &= \begin{bmatrix} 0.48 & -0.18 & 0 \\ -0.18 & 0.12 & 0 \\ 0 & 0 & 0.6 \end{bmatrix} \begin{bmatrix} 4 \\ -2 \\ 0 \end{bmatrix} \\ &= \begin{bmatrix} 2.28 \\ -0.96 \\ 0 \end{bmatrix} \text{kg.m}^2/\text{s} \end{aligned}$$

Example 9.5

In terms of a coordinate system $x'y'z'$ with its origin at the centre of mass, the inertia matrix of a rigid body is

$$[I'] = \begin{bmatrix} 4 & -2 & 1 \\ -2 & 2 & -1 \\ 1 & -1 & 3 \end{bmatrix} \text{kg.m}^2$$

Determine the principal moments of inertia and the directions of a set of principal axes relative to the $x'y'z'$ system.

SOLUTION

Substituting the moments and products of inertia into Equation (9.19), we obtain the equation

$$I^3 - 9I^2 + 20I - 10 = 0$$

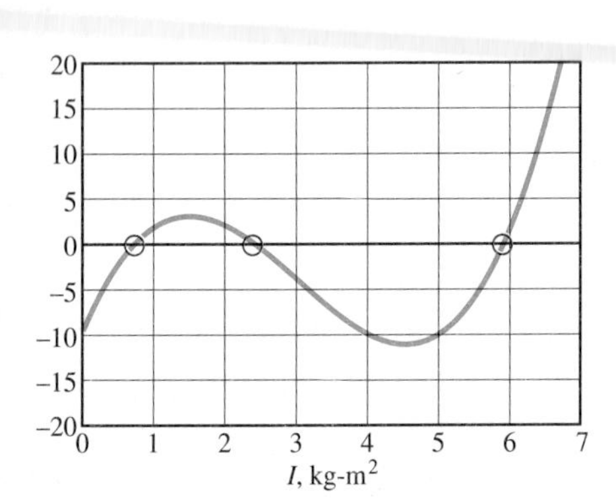

Figure 9.20
Graph of $I^3 - 9I^2 + 20I - 10$.

We show the value of the left side of this equation as a function of I in Figure 9.20. The three roots, which are the values of the principal moments of inertia in kg.m^2, are $I_1 = 0.708$, $I_2 = 2.397$ and $I_3 = 5.895$.

Substituting the principal moment of inertia $I_1 = 0.708\,\text{kg.m}^2$ into Equations (9.20) and dividing the resulting vector $\mathbf{V}$ by its magnitude, we obtain a unit vector parallel to the corresponding principal axis:

$$\mathbf{e}_1 = 0.473\,\mathbf{i} + 0.864\,\mathbf{j} + 0.171\,\mathbf{k}$$

Substituting $I_2 = 2.397\,\text{kg.m}^2$ into Equations (9.20), we obtain the unit vector

$$\mathbf{e}_2 = -0.458\,\mathbf{i} + 0.076\,\mathbf{j} + 0.886\,\mathbf{k}$$

and substituting $I_3 = 5.895\,\text{kg.m}^2$ into Equations (9.20), we obtain the unit vector

$$\mathbf{e}_3 = 0.753\,\mathbf{i} - 0.497\,\mathbf{j} + 0.432\,\mathbf{k}$$

We have determined the principal moments of inertia and the components of unit vectors parallel to the corresponding principal axes. In Figure 9.21 we show the principal axes, arbitrarily designating them so that $I_{xx} = 5.895\,\text{kg.m}^2$, $I_{yy} = 0.708\,\text{kg.m}^2$ and $I_{zz} = 2.397\,\text{kg.m}^2$.

Figure 9.21
The principal axes. Our choice of which ones to call x, y and z is arbitrary.

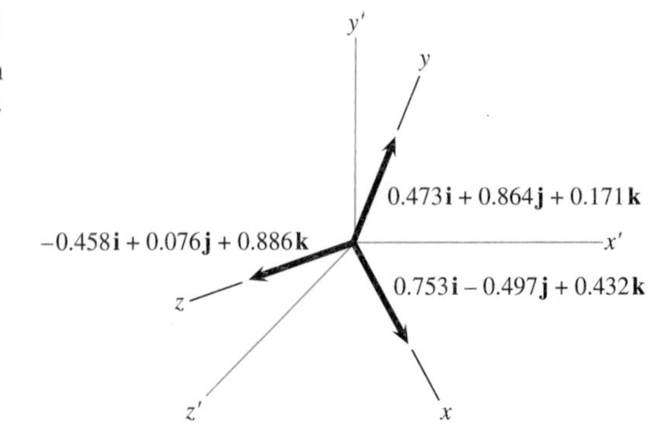

Problems

9.18 The inertia matrix of a rigid body in terms of a body-fixed coordinate system with its origin at the centre of mass is

$$[I] = \begin{bmatrix} 4 & 1 & -1 \\ 1 & 2 & 0 \\ -1 & 0 & 6 \end{bmatrix} \text{kg.m}^2$$

If the rigid body's angular velocity is $\boldsymbol{\omega} = (10\,\mathbf{i} - 5\,\mathbf{j} + 10\,\mathbf{k})$ rad/s, what is its angular momentum about its centre of mass?

9.19 What is the moment of inertia of the rigid body in Problem 9.18 about the axis that passes through the origin and the point $(4, -4, 7)$ m?

Strategy: Determine the components of a unit vector parallel to the axis and use Equation (9.17).

9.20 A rigid body rotates about a fixed point O. Its inertia matrix in terms of a body-fixed coordinate sytem with its origin at O is

$$[I] = \begin{bmatrix} 1 & -1 & 0 \\ -1 & 5 & 1 \\ 0 & 1 & 7 \end{bmatrix} \text{kg.m}^2$$

If the rigid body's angular velocity is $\boldsymbol{\omega} = (6\,\mathbf{i} + 6\,\mathbf{j} - 4\,\mathbf{k})$ rad/s, what is its angular momentum about O?

9.21 What is the moment of inertia of the rigid body in Problem 9.20 about the axis that passes through the origin and the point $(-1, 5, 2)$ m?

9.22 The mass of the homogeneous slender bar is 6 kg. Determine its moments and products of inertia in terms of the coordinate system shown.

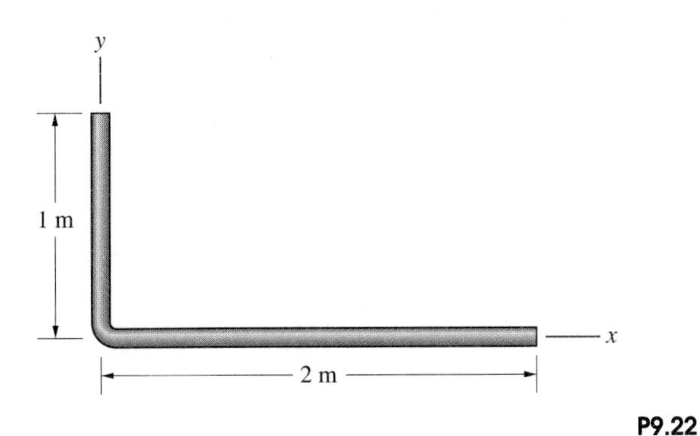

P9.22

9.23 Consider the slender bar in Problem 9.22.
(a) Determine its moments and products of inertia in terms of a parallel coordinate system $x'y'z'$ with its origin at the bar's centre of mass.
(b) If the bar is rotating with angular velocity $\boldsymbol{\omega} = 4\,\mathbf{i}$ rad/s, what is its angular momentum about its centre of mass?

9.24 The 4 kg thin rectangular plate lies in the x-y plane. Determine its moments and products of inertia in terms of the coordinate system shown.

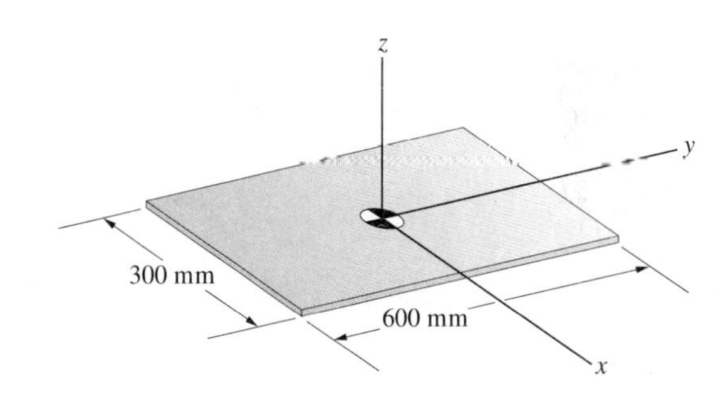

P9.24

9.25 If the plate in Problem 9.24 is rotating with angular velocity $\boldsymbol{\omega} = (6\,\mathbf{i} + 4\,\mathbf{j} - 2\,\mathbf{k})$ rad/s, what is its angular momentum about its centre of mass?

9.26 The 30 kg thin triangular plate lies in the x-y plane. Determine its moments and products of inertia in terms of the coordinate system shown.

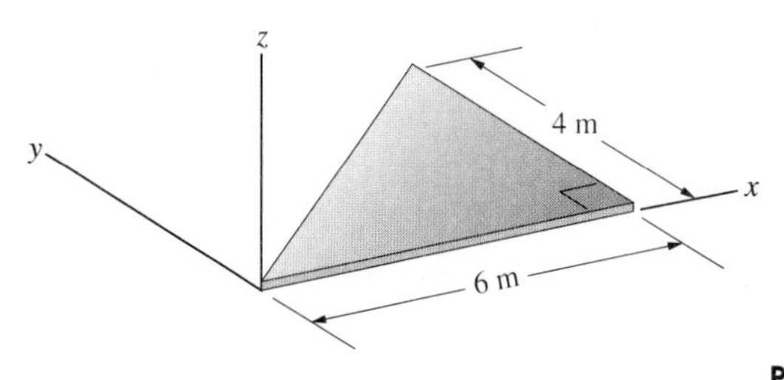

P9.26

9.27 Consider the triangular plate in Problem 9.26.
(a) Determine its moments and products of inertia in terms of a parallel coordinate system $x'y'z'$ with its origin at the plate's centre of mass.
(b) If the plate is rotating with angular velocity $\boldsymbol{\omega} = (20\,\mathbf{i} - 12\,\mathbf{j} + 16\,\mathbf{k})$ rad/s, what is its angular momentum about its centre of mass?

9.28 The slender bar of mass m rotates about the fixed point O with angular velocity $\boldsymbol{\omega} = \omega_y\,\mathbf{j} + \omega_z\,\mathbf{k}$. Determine its angular momentum (a) about its centre of mass; (b) about O.

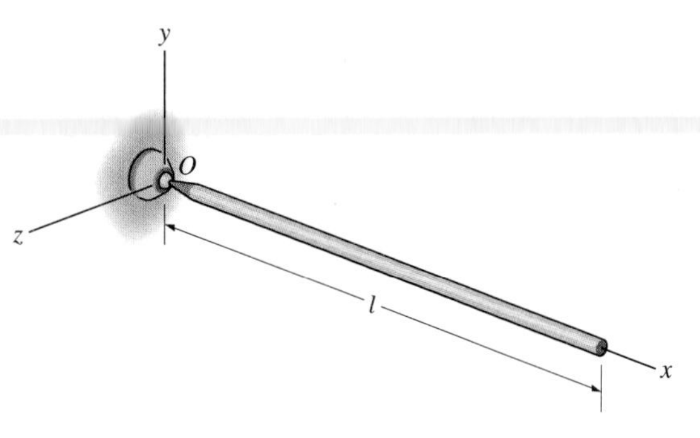

P9.28

9.29 The slender bar of mass m is parallel to the x axis. If the coordinate system is body-fixed and its angular velocity about the fixed point O is $\boldsymbol{\omega} = \omega_y\,\mathbf{j}$, what is the bar's angular momentum about O?

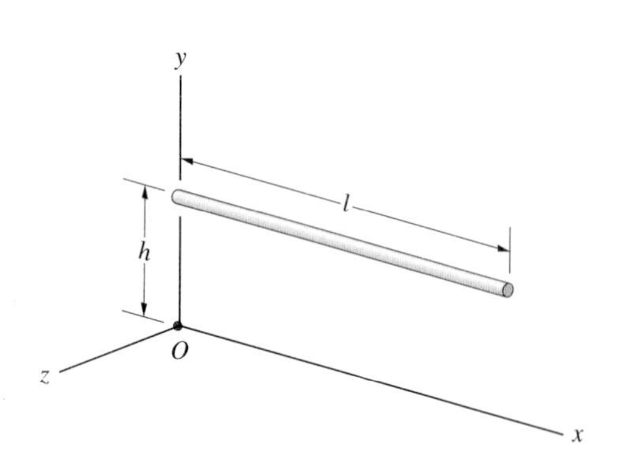

P9.29

9.30 Determine the inertia matrix of the 10 kg thin plate in terms of the coordinate system shown.

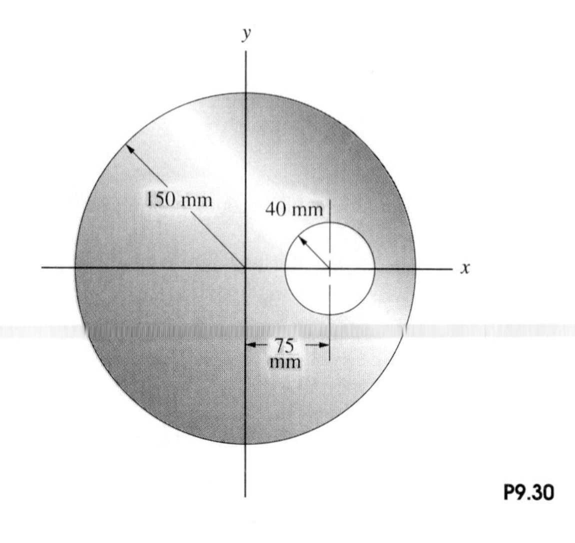

P9.30

9.31 The 10 kg thin plate in Problem 9.30 has angular velocity $\boldsymbol{\omega} = (10\,\mathbf{i} + 10\,\mathbf{j})$ rad/s. What is its angular momentum about its centre of mass?

9.32 In Example 9.3 the moments and products of inertia of the object consisting of the booms AB and BC were determined in terms of the coordinate system shown in Figure 9.18. Determine the moments and products of inertia of the object in terms of a parallel coordinate system $x'y'z'$ with its origin at the centre of mass of the object.

9.33 Suppose that the crane described in Example 9.3 undergoes a rigid body rotation about the vertical axis at 0.1 rad/s in the counterclockwise direction when viewed from above.
(a) What is its angular velocity vector $\boldsymbol{\omega}$?
(b) What is the angular momentum of the object consisting of the booms AB and BC *about its centre of mass*?

9.34 A 3 kg slender bar is rigidly attached to a 2 kg thin circular disc. In terms of the body-fixed coordinate system shown, the angular velocity of the composite object is $\boldsymbol{\omega} = (100\,\mathbf{i} - 4\,\mathbf{j} + 6\,\mathbf{k})$ rad/s. What is the object's angular momentum about its centre of mass?

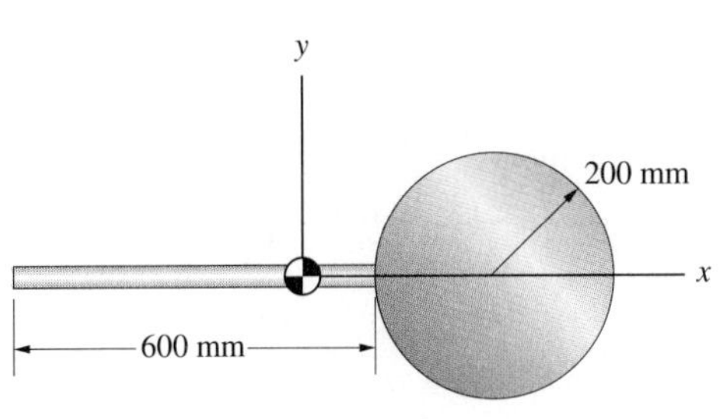

P9.34

9.35 The mass of the homogeneous slender bar is m. If the bar rotates with angular velocity $\omega = \omega_0(24\,\mathbf{i} + 12\,\mathbf{j} - 6\,\mathbf{k})$, what is its angular momentum about its centre of mass?

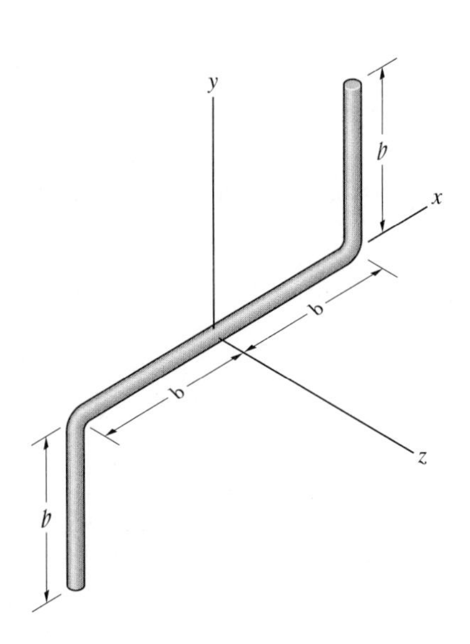

P9.35

9.36 The 8 kg homogeneous slender bar has ball and socket supports at A and B.
(a) What is the bar's moment of inertia about the axis AB?
(b) If the bar rotates about the axis AB at 4 rad/s, what is the magnitude of its angular momentum about its axis of rotation?

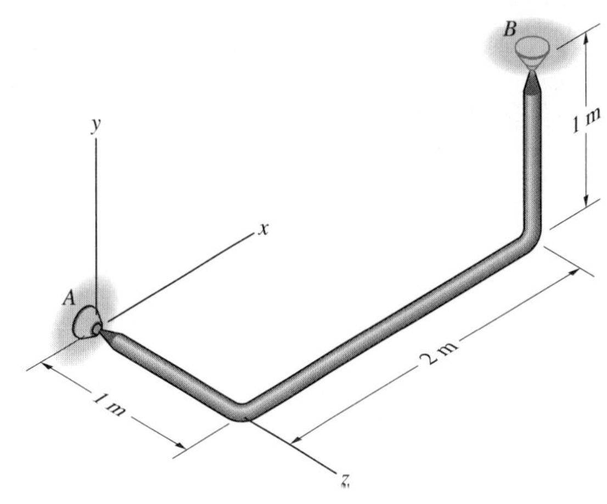

P9.36

9.37 The 8 kg homogeneous slender bar in Problem 9.36 is released from rest in the position shown. (The x-z plane is horizontal.) At that instant, what is the magnitude of the bar's angular acceleration about the axis AB?

9.38 In terms of a coordinate system $x'y'z'$ with its origin at the centre of mass, the inertia matrix of a rigid body is

$$[I'] = \begin{bmatrix} 20 & 10 & -10 \\ 10 & 60 & 0 \\ -10 & 0 & 80 \end{bmatrix} \text{kg.m}^2$$

Determine the principal moments of inertia and unit vectors parallel to the corresponding principal axes.

9.39 In terms of a coordinate system $x'y'z'$ with its origin at the centre of mass, the inertia matrix of a rigid body is

$$[I'] = \begin{bmatrix} 4 & -2 & 0 \\ -2 & 12 & 0 \\ 0 & 0 & 16 \end{bmatrix} \text{kg.m}^2$$

If the rigid body rotates, what is its angular momentum about its centre of mass?

9.40 The 1 kg, 1 m long slender bar lies in the *x*-*y* plane. Its moment of inertia matrix is

$$[I'] = \begin{bmatrix} \frac{1}{12}\sin^2\beta & -\frac{1}{12}\sin\beta\cos\beta & 0 \\ -\frac{1}{12}\sin\beta\cos\beta & \frac{1}{12}\cos^2\beta & 0 \\ 0 & 0 & \frac{1}{12} \end{bmatrix}$$

Use Equations (9.19) and (9.20) to determine the principal moments of inertia and unit vectors parallel to the corresponding principal axes.

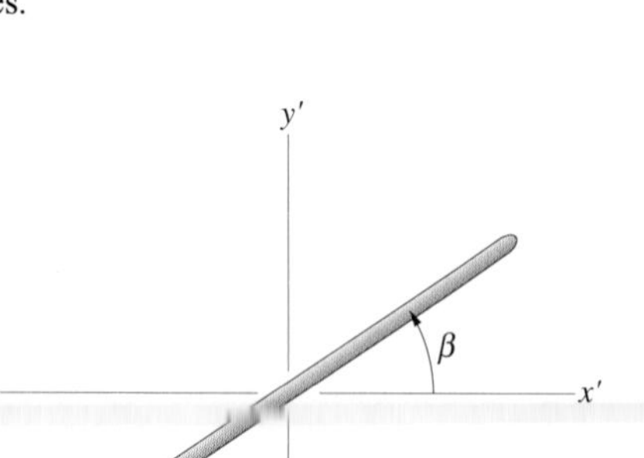

P9.40

9.41 The mass of the homogeneous thin plate is 140 kg. For a coordinate system with its origin at *O*, determine the principal moments of inertia and unit vectors parallel to the corresponding principal axes.

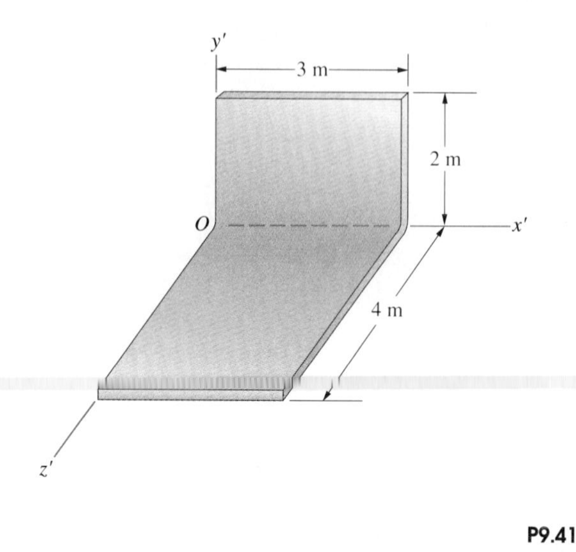

P9.41

9.4 Euler's Equations

The equations governing three-dimensional motion of a rigid body, which are known as Euler's equations, consist of Newton's second law

$$\Sigma \mathbf{F} = m\mathbf{a}$$

and equations of angular motion. In the following sections we derive the equations of angular motion, beginning with momentum principles for a system of particles developed in Chapter 7 and using our expressions for the angular momentum of a rigid body in three dimensions.

Rotation About a Fixed Point

If a rigid body rotates about a fixed point *O*, the sum of the moments about *O* due to external forces and couples equals the rate of change of the angular momentum about *O* (Equation 7.11):

$$\Sigma \mathbf{M}_O = \frac{d\mathbf{H}_O}{dt} \tag{9.21}$$

To obtain the equations of angular motion, we must substitute the components of the angular momentum given by Equations (9.5) into this equation. The coordinate system used to express these components is usually body-fixed and so rotates with the angular velocity $\boldsymbol{\omega}$ of the rigid body. In some situations, it is convenient to use a coordinate system that rotates but is not body-fixed. Let us denote the coordinate system's angular velocity vector by $\boldsymbol{\Omega}$, where $\boldsymbol{\Omega} = \boldsymbol{\omega}$

if the coordinate system is body-fixed. Expressing $\mathbf{H}_O$ in terms of its components,

$$\mathbf{H}_O = H_{Ox}\,\mathbf{i} + H_{Oy}\,\mathbf{j} + H_{Oz}\,\mathbf{k}$$

the rate of change of the angular momentum is

$$\frac{d\mathbf{H}_O}{dt} = \frac{dH_{Ox}}{dt}\,\mathbf{i} + H_{Ox}\frac{d\mathbf{i}}{dt} + \frac{dH_{Oy}}{dt}\,\mathbf{j} + H_{Oy}\frac{d\mathbf{j}}{dt} + \frac{dH_{Oz}}{dt}\,\mathbf{k} + H_{Oz}\frac{d\mathbf{k}}{dt}$$

By expressing the time derivatives of the unit vectors in terms of the coordinate system's angular velocity Ω,

$$\frac{d\mathbf{i}}{dt} = \Omega \times \mathbf{i} \qquad \frac{d\mathbf{j}}{dt} = \Omega \times \mathbf{j} \qquad \frac{d\mathbf{k}}{dt} = \Omega \times \mathbf{k}$$

we can write Equation (9.21) as

$$\Sigma\,\mathbf{M}_O = \frac{dH_{Ox}}{dt}\,\mathbf{i} + \frac{dH_{Oy}}{dt}\,\mathbf{j} + \frac{dH_{Oz}}{dt}\,\mathbf{k} + \Omega \times \mathbf{H}_O \tag{9.22}$$

Substituting the components of $\mathbf{H}_O$ from Equation (9.5) into this equation, we obtain the equations of angular motion:

$$\begin{aligned}
\Sigma M_{Ox} = {} & I_{xx}\frac{d\omega_x}{dt} - I_{xy}\frac{d\omega_y}{dt} - I_{xz}\frac{d\omega_z}{dt} \\
& - \Omega_z(-I_{yx}\omega_x + I_{yy}\omega_y - I_{yz}\omega_z) \\
& + \Omega_y(-1_{zx}\omega_x - I_{zy}\omega_y + I_{zz}\omega_z) \\[1ex]
\Sigma M_{Oy} = {} & -I_{yx}\frac{d\omega_x}{dt} + I_{yy}\frac{d\omega_y}{dt} - I_{yz}\frac{d\omega_z}{dt} \\
& + \Omega_z(I_{xx}\omega_x - I_{xy}\omega_y - I_{xz}\omega_z) \\
& - \Omega_x(-I_{zx}\omega_x - I_{zy}\omega_y + I_{zz}\omega_z) \\[1ex]
\Sigma M_{Oz} = {} & -I_{zx}\frac{d\omega_x}{dt} - I_{zy}\frac{d\omega_y}{dt} + I_{zz}\frac{d\omega_z}{dt} \\
& - \Omega_y(I_{xx}\omega_x - I_{xy}\omega_y - I_{xz}\omega_z) \\
& + \Omega_x(-I_{yx}\omega_x + I_{yy}\omega_y - I_{yz}\omega_z)
\end{aligned} \tag{9.23}$$

Since the components of the angular velocity, like the components of $\mathbf{H}_O$, are expressed in terms of a coordinate system rotating with angular velocity $\mathbf{\Omega}$, the rigid body's angular acceleration is

$$\alpha = \frac{d\boldsymbol{\omega}}{dt} = \frac{d\omega_x}{dt}\,\mathbf{i} + \frac{d\omega_y}{dt}\,\mathbf{j} + \frac{d\omega_z}{dt}\,\mathbf{k} + \Omega \times \boldsymbol{\omega} \tag{9.24}$$

If the coordinate system does not rotate or is body-fixed, the terms $d\omega_x/dt, d\omega_y/dt$ and $d\omega_z/dt$ are the components of the rigid body's angular acceleration. Otherwise, you must use Equation (9.24) to determine the angular acceleration.

We can write Equations (9.23) as the matrix equation

$$\begin{bmatrix} \Sigma M_{Ox} \\ \Sigma M_{Oy} \\ \Sigma M_{Oz} \end{bmatrix} = \begin{bmatrix} I_{xx} & -I_{xy} & -I_{xz} \\ -I_{yx} & I_{yy} & -I_{yz} \\ -I_{zx} & -I_{zy} & I_{zz} \end{bmatrix} \begin{bmatrix} d\omega_x/dt \\ d\omega_y/dt \\ d\omega_z/dt \end{bmatrix} + \begin{bmatrix} 0 & -\Omega_z & \Omega_y \\ \Omega_z & 0 & -\Omega_x \\ -\Omega_y & \Omega_x & 0 \end{bmatrix} \begin{bmatrix} I_{xx} & -I_{xy} & -I_{xz} \\ -I_{yx} & I_{yy} & -I_{yz} \\ -I_{zx} & -I_{zy} & I_{zz} \end{bmatrix} \begin{bmatrix} \omega_x \\ \omega_y \\ \omega_z \end{bmatrix} \tag{9.25}$$

General Motion

The sum of the moments about the centre of mass of a rigid body due to external forces and couples equals the rate of change of the angular momentum about the centre of mass (Equation 7.13):

$$\Sigma \mathbf{M} = \frac{d\mathbf{H}}{dt} \tag{9.26}$$

In terms of the components of the angular momentum vector, we can write this equation as

$$\Sigma \mathbf{M} = \frac{dH_x}{dt}\mathbf{i} + \frac{dH_y}{dt}\mathbf{j} + \frac{dH_z}{dt}\mathbf{k} + \mathbf{\Omega} \times \mathbf{H} \tag{9.27}$$

where $\mathbf{\Omega}$ is the angular velocity of the coordinate system. Substituting Equations (9.9) into this equation, we obtain the equations of angular motion

$$\begin{aligned} \Sigma M_x =& - I_{xx}\frac{d\omega_x}{dt} - I_{xy}\frac{d\omega_y}{dt} - I_{xz}\frac{d\omega_z}{dt} \\ & - \Omega_z(-I_{yx}\omega_x + I_{yy}\omega_y - I_{yz}\omega_z) \\ & + \Omega_y(-I_{zx}\omega_x - I_{zy}\omega_y + I_{zz}\omega_z) \\ \Sigma M_y =& - I_{yx}\frac{d\omega_x}{dt} + I_{yy}\frac{d\omega_y}{dt} - I_{yz}\frac{d\omega_z}{dt} \\ & + \Omega_z(I_{xx}\omega_x - I_{xy}\omega_y - I_{xz}\omega_z) \\ & - \Omega_x(-I_{zx}\omega_x - I_{zy}\omega_y + I_{zz}\omega_z) \\ \Sigma M_z =& - I_{zx}\frac{d\omega_x}{dt} - I_{zy}\frac{d\omega_y}{dt} + I_{zz}\frac{d\omega_z}{dt} \\ & - \Omega_y(I_{xx}\omega_x - I_{xy}\omega_y - I_{xz}\omega_z) \\ & + \Omega_x(-I_{yx}\omega_x + I_{yy}\omega_y - I_{yz}\omega_z) \end{aligned} \tag{9.28}$$

or

$$\begin{bmatrix} \Sigma M_x \\ \Sigma M_y \\ \Sigma M_z \end{bmatrix} = \begin{bmatrix} I_{xx} & -I_{xy} & -I_{xz} \\ -I_{yx} & I_{yy} & -I_{yz} \\ -I_{zx} & -I_{zy} & I_{zz} \end{bmatrix} \begin{bmatrix} d\omega_x/dt \\ d\omega_y/dt \\ d\omega_z/dt \end{bmatrix}$$
$$+ \begin{bmatrix} 0 & -\Omega_z & \Omega_y \\ \Omega_z & 0 & -\Omega_x \\ -\Omega_y & \Omega_z & 0 \end{bmatrix} \begin{bmatrix} I_{xx} & -I_{xy} & -I_{xz} \\ -I_{yx} & I_{yy} & -I_{yz} \\ -I_{zx} & -I_{zy} & I_{zz} \end{bmatrix} \begin{bmatrix} \omega_x \\ \omega_y \\ \omega_z \end{bmatrix} \tag{9.29}$$

The equations of angular motion for general motion, and the expressions for the moments and products of inertia, are identical in form to those we obtained for rotation about a fixed point. However, in the case of general motion the equations of angular motion are expressed in terms of the components of the moment about the centre of mass, and the moments and products of inertia are expressed in terms of a coordinate system with its origin at the centre of mass.

In the following examples we use the Euler equations to analyse three-dimensional motions of rigid bodies. This typically involves three steps:

(1) Choose a coordinate system – *If an object rotates about a fixed point O, it usually simplifies the equations of angular motion if you express them in terms of a coordinate system with its origin at O. Otherwise, you must use a coordinate system with its origin at the centre of mass. In either case, be sure to choose the coordinate system's orientation to simplify your determination of the moments and products of inertia.*

(2) Draw the free-body diagram – *Isolate the object and identify the external forces and couples acting on it.*

(3) Apply the equations of motion – *Use Newton's second law and the equations of angular motion to relate the forces and couples acting on the object to the acceleration of its centre of mass and its angular acceleration.*

Example 9.6

During an assembly process, the 4 kg rectangular plate in Figure 9.22 is held at O by a robotic manipulator. Point O is stationary. At the instant shown, the plate is horizontal, its angular velocity is $\boldsymbol{\omega} = (4\,\mathbf{i} - 2\,\mathbf{j})$ rad/s, and its angular acceleration is $\alpha = (-10\,\mathbf{i} + 6\,\mathbf{j})$ rad/s^2. Determine the couple exerted on the plate by the manipulator.

Figure 9.22

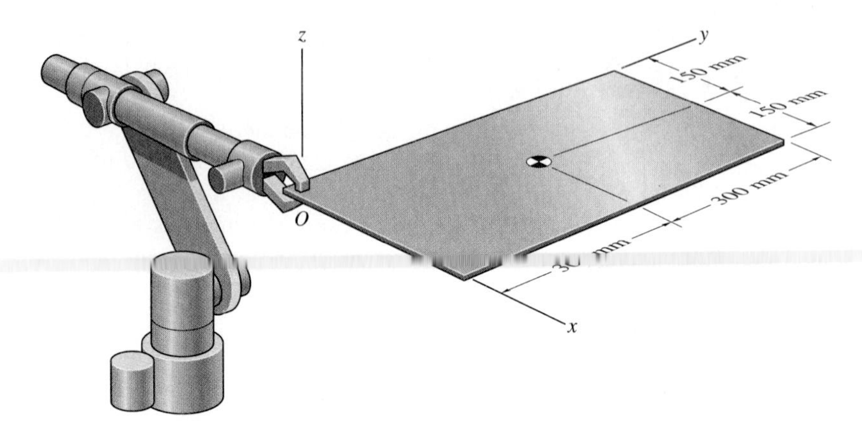

STRATEGY

The plate rotates about the fixed point O, so we can use Equation (9.25) to determine the total moment exerted on the plate about O.

SOLUTION

Draw the Free-Body Diagram We denote the force and couple exerted on the plate by the manipulator by **F** and **C** (Figure (a)).

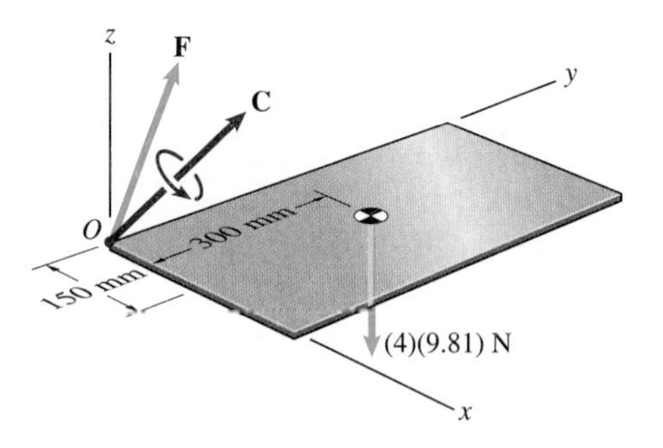

(a) Free-body diagram of the plate.

Apply the Equations of Motion The total moment about O is the sum of the couple exerted by the manipulator and the moment about O due to the plate's weight:

$$\begin{aligned}\Sigma \mathbf{M}_O &= \mathbf{C} + (0.15\,\mathbf{i} + 0.30\,\mathbf{j}) \times [-(4)(9.81)\,\mathbf{k}] \\ &= (\mathbf{C} - 11.77\,\mathbf{i} + 5.89\,\mathbf{j})\,\text{N-m}\end{aligned} \tag{9.30}$$

To obtain the unknown couple $\mathbf{C}$, we can determine the total moment about O from Equation (9.25).

We let the coordinate system be body-fixed, so its angular velocity $\boldsymbol{\omega}$ equals the plate's angular velocity $\boldsymbol{\Omega}$. We determine the plate's inertia matrix in Example 9.4, obtaining

$$[I] = \begin{bmatrix} 0.48 & -0.18 & 0 \\ -0.18 & 0.12 & 0 \\ 0 & 0 & 0.6 \end{bmatrix}$$

Therefore the total moment about O exerted on the plate is

$$\begin{aligned}\begin{bmatrix} \Sigma M_{Ox} \\ \Sigma M_{Oy} \\ \Sigma M_{Oz} \end{bmatrix} &= \begin{bmatrix} I_{xx} & -I_{xy} & -I_{xz} \\ -I_{yx} & I_{yy} & -I_{yz} \\ -I_{zx} & -I_{zy} & I_{zz} \end{bmatrix} \begin{bmatrix} d\omega_x/dt \\ d\omega_y/dt \\ d\omega_z/dt \end{bmatrix} \\ &\quad + \begin{bmatrix} 0 & -\omega_z & \omega_y \\ \omega_z & 0 & -\omega_x \\ -\omega_y & \omega_x & 0 \end{bmatrix} \begin{bmatrix} I_{xx} & -I_{xy} & -I_{xz} \\ I_{yx} & I_{yy} & -I_{yz} \\ I_{zx} & -I_{zy} & I_{zz} \end{bmatrix} \begin{bmatrix} \omega_x \\ \omega_y \\ \omega_z \end{bmatrix} \\ &= \begin{bmatrix} 0.48 & -0.18 & 0 \\ -0.18 & 0.12 & 0 \\ 0 & 0 & 0.6 \end{bmatrix} \begin{bmatrix} -10 \\ 6 \\ 0 \end{bmatrix} \\ &\quad + \begin{bmatrix} 0 & 0 & -2 \\ 0 & 0 & -4 \\ 2 & 4 & 0 \end{bmatrix} \begin{bmatrix} 0.48 & -0.18 & 0 \\ -0.18 & 0.12 & 0 \\ 0 & 0 & 0.6 \end{bmatrix} \begin{bmatrix} 4 \\ -2 \\ 0 \end{bmatrix} \\ &= \begin{bmatrix} -5.88 \\ 2.52 \\ 0.72 \end{bmatrix} \text{N.m}\end{aligned}$$

Substituting this result into Equation (9.30),

$$\Sigma \mathbf{M}_O = \mathbf{C} - 11.77\,\mathbf{i} + 5.89\,\mathbf{j} = -5.88\,\mathbf{i} + 2.52\,\mathbf{j} + 0.72\,\mathbf{k}$$

we determine the couple $\mathbf{C}$:

$$\mathbf{C} = (5.89\,\mathbf{i} - 3.37\,\mathbf{j} + 0.72\,\mathbf{k})\,\text{N.m}$$

Example 9.7

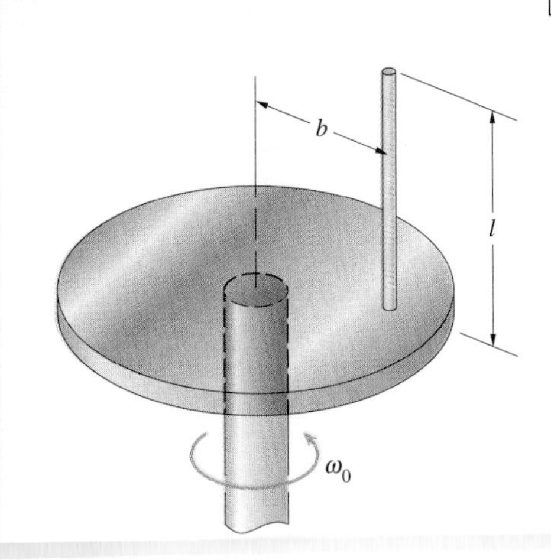

Figure 9.23

A slender vertical bar of mass m is rigidly attached to a horizontal disc rotating with constant angular velocity ω_0 (Figure 9.23). What force and couple are exerted on the bar by the disc?

STRATEGY

The external forces and couples on the bar are its weight and the force and couple exerted on it by the disc. The angular velocity and acceleration of the bar are given and we can determine the acceleration of its centre of mass, so we can use the Euler equations to determine the total force and couple.

SOLUTION

Choose a Coordinate System In Figure (a) we place the origin of a body-fixed coordinate system at the centre of mass with the y axis vertical and the x axis in the radial direction. With this orientation we will obtain simple expressions for the bar's angular velocity and the acceleration of its centre of mass.

Draw the Free-Body Diagram We draw the free-body diagram of the bar in Figure (a), showing the force $\mathbf{F}$ and couple $\mathbf{C}$ exerted by the disc.

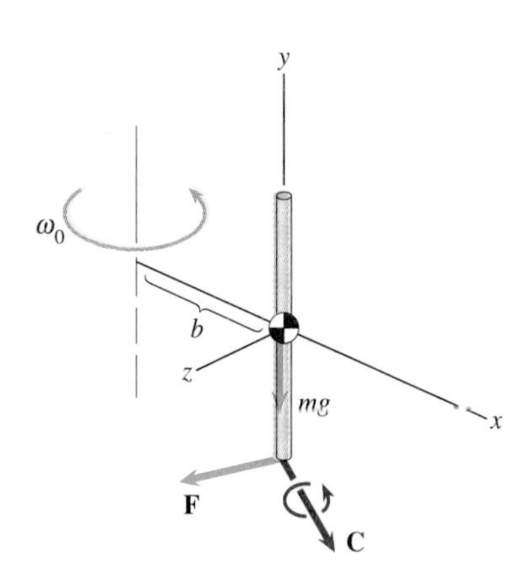

(a) Free-body diagram of the bar.

Apply the Equations of Motion The acceleration of the centre of mass of the bar due to its motion along its circular path is $\mathbf{a} = -\omega_0^2 b\,\mathbf{i}$. From Newton's second law,

$$\Sigma\,\mathbf{F} = \mathbf{F} - mg\,\mathbf{j} = m(-\omega_0^2 b\,\mathbf{i})$$

we obtain the force exerted on the bar by the disc:

$$\mathbf{F} = -m\omega_0^2 b\,\mathbf{i} + mg\,\mathbf{j}$$

The total moment about the centre of mass is the sum of the couple $\mathbf{C}$ and the moment due to $\mathbf{F}$:

$$\begin{aligned}\Sigma\,\mathbf{M} &= \mathbf{C} + \left(-\frac{1}{2}l\,\mathbf{j}\right) \times (-m\omega_0^2 b\,\mathbf{i} + mg\,\mathbf{j}) \\ &= C_x\,\mathbf{i} + C_y\,\mathbf{j} + \left(C_z - \frac{1}{2}mlb\omega_0^2\right)\mathbf{k}\end{aligned}$$

The bar's inertia matrix in terms of the coordinate system in Figure (a) is

$$[I] = \begin{bmatrix} \frac{1}{12}ml^2 & 0 & 0 \\ 0 & 0 & 0 \\ 0 & 0 & \frac{1}{12}ml^2 \end{bmatrix}$$

and its angular velocity, $\boldsymbol{\omega} = \omega_0\,\mathbf{j}$, is constant. The equation of angular motion, Equation (9.29), is

$$\begin{bmatrix} Cx \\ C_y \\ C_z - \frac{1}{2}mlb\omega_0^2 \end{bmatrix} = \begin{bmatrix} 0 & 0 & \omega_0 \\ 0 & 0 & 0 \\ -\omega_0 & 0 & 0 \end{bmatrix} \begin{bmatrix} \frac{1}{12}ml^2 & 0 & 0 \\ 0 & 0 & 0 \\ 0 & 0 & \frac{1}{12}ml^2 \end{bmatrix} \begin{bmatrix} 0 \\ \omega_0 \\ 0 \end{bmatrix}$$

The right side of this equation equals zero, so the components of the couple exerted on the bar by the disc are $C_x = 0$, $C_y = 0$, $C_z = \frac{1}{2} mlb\omega_0^2$.

Alternative Solution The bar rotates about a fixed axis, so we can also determine the couple **C** by using Equation (9.25). Let the fixed point O be the centre of the disc (Figure (b)), and let the body-fixed coordinate system be oriented with the x axis through the bottom of the bar. The total moment about O is

$$\Sigma \mathbf{M}_O = \mathbf{C} + (b\,\mathbf{i}) \times (-m\omega_0^2 b\,\mathbf{i} + mg\,\mathbf{j}) + \left(b\,\mathbf{i} + \frac{1}{2} l\,\mathbf{j}\right) \times (-mg\,\mathbf{j})$$

$$= \mathbf{C}$$

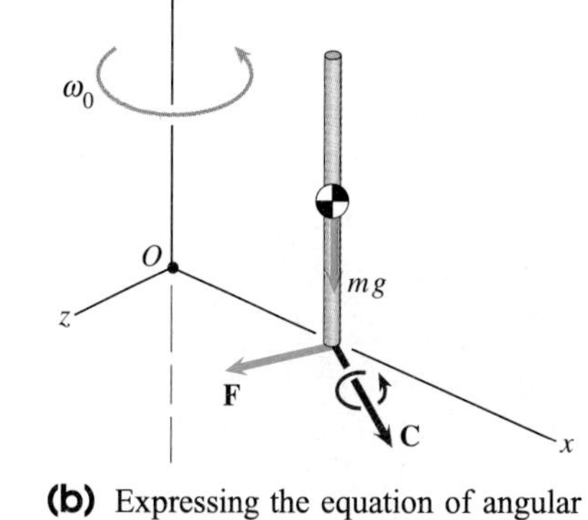

(b) Expressing the equation of angular motion in terms of the fixed point O.

Thus the only moment about O is the couple exerted by the disc. Applying the parallel-axis theorems, the bar's moments and products of inertia are (Figure (c))

$$I_{xx} = I_{x'x'} + (d_y^2 + d_z^2)m = \frac{1}{12} ml^2 + \left(\frac{1}{2} l\right)^2 m = \frac{1}{3} ml^2$$

$$I_{yy} = I_{y'y'} + (d_x^2 + d_z^2)m = mb^2$$

$$I_{zz} = I_{z'z'} + (d_x^2 + d_y^2)m = \frac{1}{12} ml^2 + \left[b^2 + \left(\frac{1}{2} l\right)^2\right] m = \frac{1}{3} ml^2 + mb^2$$

$$I_{xy} = I_{x'y'} + d_x d_y m = 0 + (b)\left(\frac{1}{2} l\right) m = \frac{1}{2} mbl$$

$$I_{yz} = I_{y'z'} + d_y d_z m = 0$$

$$I_{zx} = I_{z'x'} + d_z d_x m = 0$$

Substituting these results into Equation (9.25), we obtain

$$\begin{bmatrix} C_x \\ C_y \\ C_z \end{bmatrix} = \begin{bmatrix} 0 & 0 & \omega_0 \\ 0 & 0 & 0 \\ -\omega_0 & 0 & 0 \end{bmatrix} \begin{bmatrix} \frac{1}{3} ml^2 & -\frac{1}{2} mbl & 0 \\ -\frac{1}{2} mbl & mb^2 & 0 \\ 0 & 0 & \frac{1}{3} mb^2 + mb^2 \end{bmatrix} \begin{bmatrix} 0 \\ \omega_0 \\ 0 \end{bmatrix}$$

$$= \begin{bmatrix} 0 \\ 0 \\ \frac{1}{2} mlb\omega_0^2 \end{bmatrix}$$

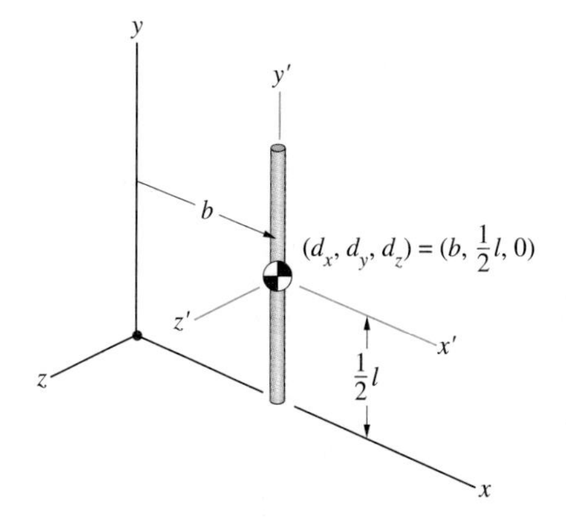

(c) Applying the parallel-axis theorem.

DISCUSSION

If the bar were attached to the disc by a ball and socket support instead of a built-in support, you can see that the bar would rotate outwards due to the disc's rotation. We have determined the couple that the built-in support exerts on the bar that prevents it from rotating outwards.

Example 9.8

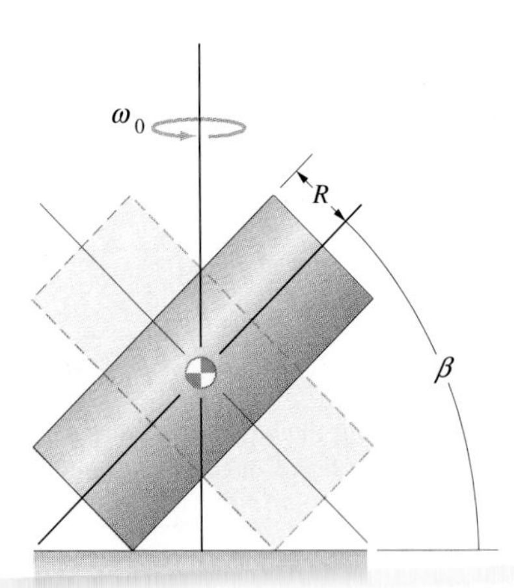

Figure 9.24

The tilted homogeneous cylinder in Figure 9.24 undergoes a steady motion in which one end rolls on the floor while its centre of mass remains stationary. The angle β between the cylinder axis and the horizontal remains a constant, and the cylinder axis rotates about the vertical axis with constant angular velocity ω_0. The cylinder has mass m, radius R, and length l. What is ω_0?

STRATEGY

By expressing the equations of angular motion in terms of ω_O, we can determine the value of ω_O necessary for the equations to be satisfied. Therefore our first task is to determine the cylinder's angular velocity $\boldsymbol{\omega}$ in terms of ω_0. We can simplify this task by using a coordinate system that is not body-fixed.

SOLUTION

Choose a Coordinate System We use a coordinate system in which the z axis remains aligned with the cylinder axis and the y axis remains horizontal (Figure (a)). The reason for this choice is that the angular velocity of the coordinate system is easy to describe – the coordinate system rotates about the vertical axis with the angular velocity ω_0 – and the rotation of the cylinder relative to the coordinate system is also easy to describe. The angular velocity vector of the coordinate system is

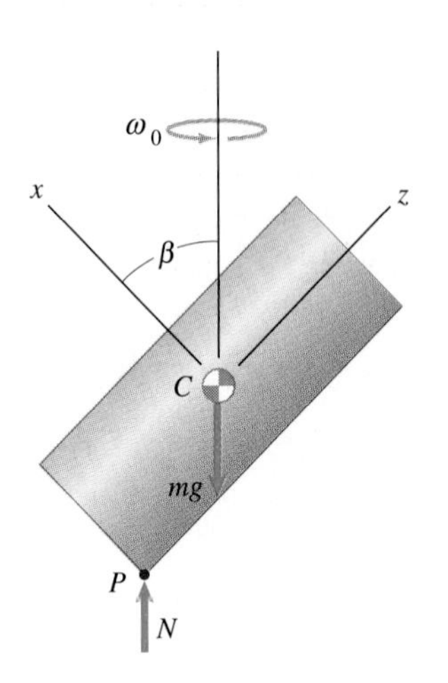

(a) Coordinate system with the z axis aligned with the cylinder axis and the y axis horizontal.

$$\Omega = \omega_0 \cos\beta\,\mathbf{i} + \omega_0 \sin\beta\,\mathbf{k}$$

Relative to the coordinate system, the cylinder rotates about the z axis. Writing its angular velocity relative to the coordinate system as $\omega_{\text{rel}}\,\mathbf{k}$, the angular velocity vector of the cylinder is

$$\boldsymbol{\omega} = \Omega + \omega_{\text{rel}}\,\mathbf{k} = \omega_0 \cos\beta\,\mathbf{i} + (\omega_0 \sin\beta + \omega_{\text{rel}})\,\mathbf{k}$$

We can determine ω_{rel} from the condition that the velocity of the point P in contact with the floor is zero. Expressing the velocity of P in terms of the velocity of the centre of mass C, we obtain

$$\mathbf{v}_P = \mathbf{v}_C + \boldsymbol{\omega} \times \mathbf{r}_{P/C}:$$

$$0 = 0 + [\omega_0 \cos\beta\,\mathbf{i} + (\omega_0 \sin\beta + \omega_{\text{rel}})\,\mathbf{k}] \times \left[-R\,\mathbf{i} - \frac{1}{2}l\,\mathbf{k}\right]$$

$$= \left[\frac{1}{2}l\omega_0 \cos\beta - R(\omega_0 \sin\beta + \omega_{\text{rel}})\right]\mathbf{j}$$

Solving for ω_{rel}, we obtain

$$\omega_{\text{rel}} = \left[\frac{1}{2}(l/R)\cos\beta - \sin\beta\right]\omega_0$$

Therefore the cylinder's angular velocity vector is

$$\boldsymbol{\omega} = \omega_0 \cos\beta\,\mathbf{i} + \frac{1}{2}(l/R)\omega_0 \cos\beta\,\mathbf{k}$$

Draw the Free-Body Diagram We draw the free-body diagram of the cylinder in Figure (a), showing its weight and the normal force exerted by the floor. Because the centre of mass is stationary, we know that the floor exerts no horizontal force on the cylinder and the normal force is $N = mg$.

Apply the Equations of Motion The moment about the centre of mass due to the normal force is

$$\Sigma \mathbf{M} = \left(mgR \sin\beta - \frac{1}{2} mgl \cos\beta \right) \mathbf{j}$$

From Appendix C, the inertia matrix is

$$\begin{bmatrix} \frac{1}{4}mR^2 + \frac{1}{12}ml^2 & 0 & 0 \\ 0 & \frac{1}{4}mR^2 + \frac{1}{12}ml^2 & 0 \\ 0 & 0 & \frac{1}{2}mR^2 \end{bmatrix}$$

Substituting our expressions for $\mathbf{\Omega}, \boldsymbol{\omega}, \Sigma \mathbf{M}$, and the moments and products of inertia into the equation of angular motion, Equation (9.29), and evaluating the matrix products, we obtain the equation

$$mg\left(R\sin\beta - \frac{1}{2} l \cos\beta \right) = \left(\frac{1}{4}mR^2 + \frac{1}{12}ml^2 \right) \omega_0^2 \sin\beta\cos\beta - \frac{1}{2}\left(\frac{1}{2}mR^2 \right) \omega_0^2 (l/R)\cos^2\beta$$

We can solve this equation for ω_0^2:

$$\omega_0^2 = \frac{g(R\sin\beta - \frac{1}{2}l\cos\beta)}{(\frac{1}{4}R^2 + \frac{1}{12}l^2)\sin\beta\cos\beta - \frac{1}{4}lR\cos^2\beta} \qquad (9.31)$$

DISCUSSION

If our solution yields a negative value for ω_0^2 for a given value of β, the assumed steady motion of the cylinder is not possible. For example, if the cylinder's diameter is equal to its length, $2R = l$, we can write Equation (9.31) as

$$\frac{R\omega_0^2}{g} = \frac{\sin\beta - \cos\beta}{\frac{7}{12}\sin\beta\cos\beta - \frac{1}{2}\cos^2\beta}$$

In Figure 9.25 we show the graph of this equation as a function of β. For values of β from approximately 40° to 45°, there is no real solution for ω_0. Notice that at $\beta = 45°$, $\omega_0 = 0$, which means that the cylinder is stationary and balanced with the centre of mass directly above point P.

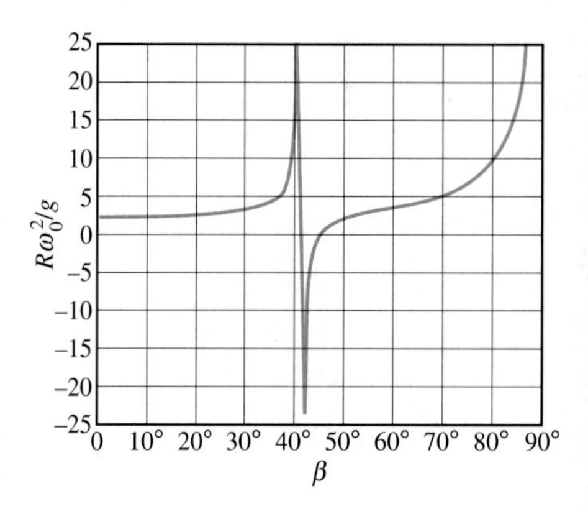

Figure 9.25
Graph of $R\omega_0^2/g$ as a function of β.

Problems

9.42 The inertia matrix of a rigid body in terms of a body-fixed coordinate system with its origin at the centre of mass is

$$[I] = \begin{bmatrix} 4 & 1 & -1 \\ 1 & 2 & 0 \\ -1 & 0 & 6 \end{bmatrix} \text{kg.m}^2$$

If the rigid body's angular velocity is $\boldsymbol{\omega} = (10\,\mathbf{i} - 5\,\mathbf{j} + 10\,\mathbf{k})$ rad/s and its angular acceleration is zero, what are the components of the total moment about its centre of mass?

9.43 If the total moment about the centre of mass of the rigid body in Problem 9.42 is zero, what are the components of its angular acceleration?

9.44 A rigid body rotates about a fixed point O. Its inertia matrix in terms of a body-fixed coordinate system with its origin at O is

$$[I] = \begin{bmatrix} 1 & -1 & 0 \\ -1 & 5 & 1 \\ 0 & 1 & 7 \end{bmatrix} \text{kg.m}^2$$

If the rigid body's angular velocity is $\boldsymbol{\omega} = (6\,\mathbf{i} + 6\,\mathbf{j} - 4\,\mathbf{k})$ rad/s and its angular acceleration is zero, what are the components of the total moment about O?

9.45 If the total moment about O due to the forces and couples acting on the rigid body in Problem 9.44 is zero, what are the components of its angular acceleration?

9.46 At $t = 0$, the stationary rectangular plate of mass m is subjected to the force F perpendicular to the plate. No other external forces or couples act on the plate. What is the magnitude of the acceleration of point A at $t = 0$?

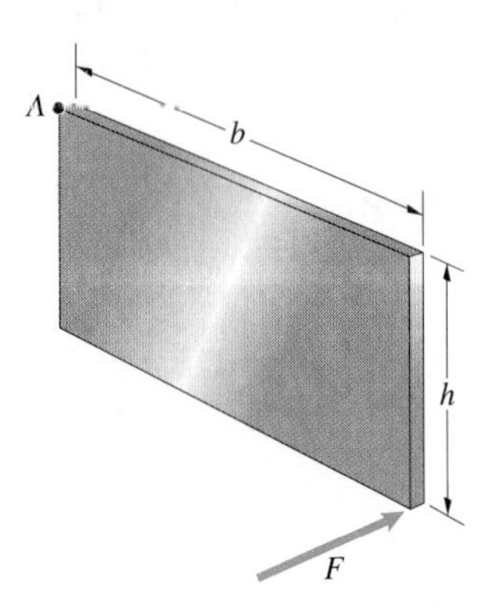

P9.46

9.47 The mass of the homogeneous slender bar is 6 kg. At $t = 0$, the stationary bar is subjected to the force $\mathbf{F} = 12\,\mathbf{k}$ N at the point $x = 2$ m, $y = 0$. No other external forces or couples act on the bar.
(a) What is the bar's angular acceleration at $t = 0$?
(b) What is the acceleration of the point $x = 2$ m, $y = 0$ at $t = 0$?

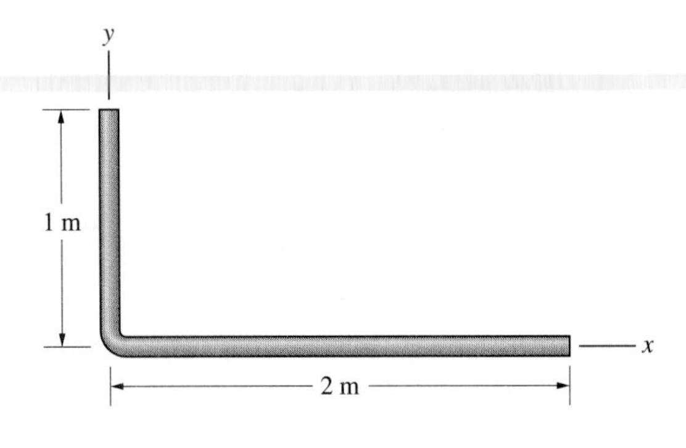

P9.47

9.48 The mass of the homogeneous slender bar is 1.2 kg. At $t = 0$, the stationary bar is subjected to the force $\mathbf{F} = (2\,\mathbf{i} + 4\,\mathbf{k})$ N at the point $x = 1$ m, $y = 1$ m. No other forces or couples act on the bar.
(a) What is the bar's angular acceleration at $t = 0$?
(b) What is the acceleration of the point $x = -1$ m, $y = -1$ m at $t = 0$?

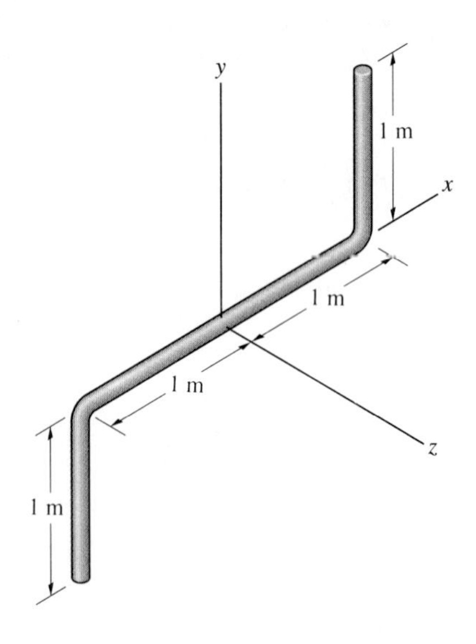

P9.48

9.49 The 10 kg thin plate has angular velocity $\omega = (10\,\mathbf{i} + 10\,\mathbf{j})$ rad/s, and its angular acceleration is zero. What are the components of the total moment exerted on the plate about its centre of mass?

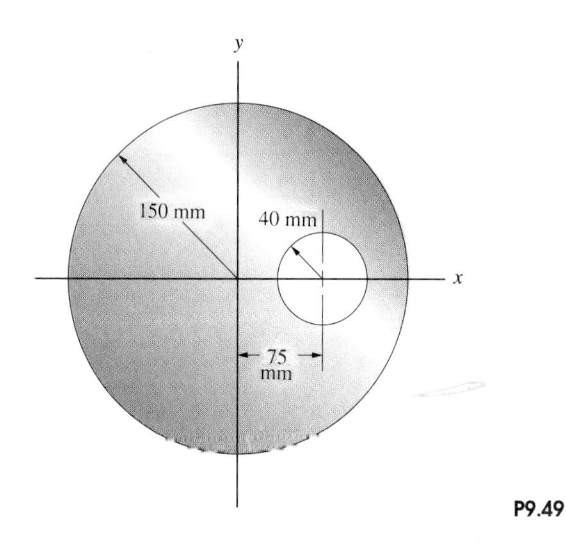

P9.49

9.50 At $y = 0$, the plate in Problem 9.49 has angular velocity $\omega = (10\,\mathbf{i} + 10\,\mathbf{j})$ rad/s and is subjected to the force $\mathbf{F} = -40\,\mathbf{k}$ N acting at the point (0, 150, 0) mm. No other forces or couples act on the plate. What are the components of its angular acceleration at that instant?

9.51 A 3 kg slender bar is rigidly attached to a 2 kg thin circular disc. In terms of the body-fixed coordinate system shown, the angular velocity of the composite object is $\omega = (100\,\mathbf{i} - 4\,\mathbf{j} + 6\,\mathbf{k})$ rad/s and its angular acceleration is zero. What are the components of the total moment exerted on the object about its centre of mass?

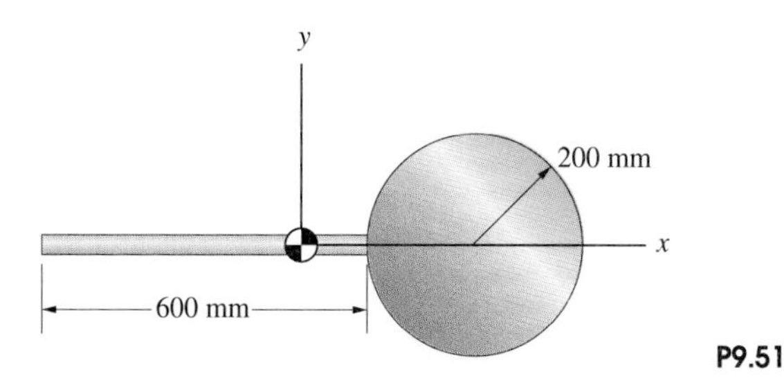

P9.51

9.52 At $t = 0$, the composite object in Problem 9.51 is stationary and is subjected to the moment $\Sigma\,\mathbf{M} = (-10\,\mathbf{i} + 10\,\mathbf{j})$ N.m about its centre of mass. No other forces or couples act on the object. What are the components of its angular acceleration at $t = 0$?

9.53 The base of the dish antenna is rotating with a constant angular velocity of 1 rad/s. The angle $\theta = 30°$, $d\theta/dt = 20°/\text{s}$, and $d^2\theta/dt^2 = -40°/\text{s}^2$. The mass of the antenna is 280 kg, and its moments and products of inertia in kg.m^2 are $I_{xx} = 140$, $I_{yy} = I_{zz} = 220$, $I_{xy} = I_{yz} = I_{zx} = 0$. Determine the couple exerted on the antenna by its support A at this instant.

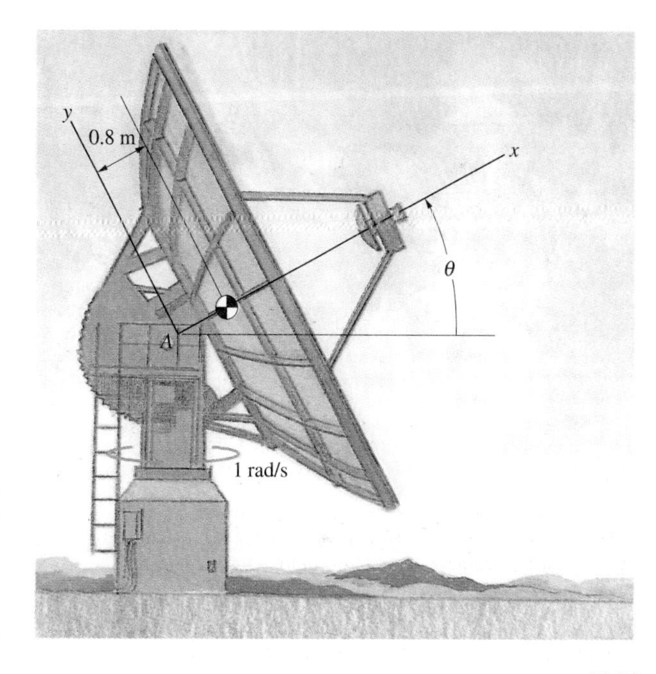

P9.53

9.54 A thin triangular plate of mass m is supported by a ball and socket at O. If it is held in the horizontal position and released from rest, what are the components of its angular acceleration at that instant?

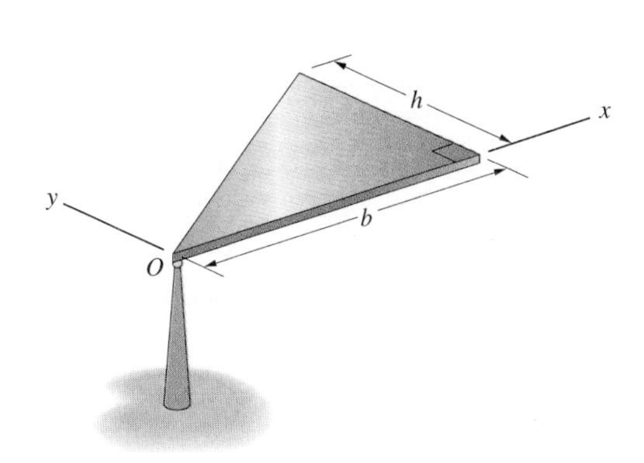

P9.54

9.55 Determine the force exerted on the triangular plate in Problem 9.54 by the ball and socket support at the instant of release.

9.56 In Problem 9.54, the mass of the plate is 5 kg, $b = 900$ mm, and $h = 600$ mm. If the plate is released in the horizontal position with angular velocity $\boldsymbol{\omega} = 4\mathbf{i}$ rad/s, what are the components of its angular acceleration at that instant?

9.57 A subassembly of a space station can be modelled as two rigidly connected slender bars, each of mass 5 Mg. The subassembly is not rotating at $t = 0$ when a reaction control motor exerts a force $\mathbf{F} = 400\,\mathbf{k}$ N at B. What is the acceleration of point A at that instant?

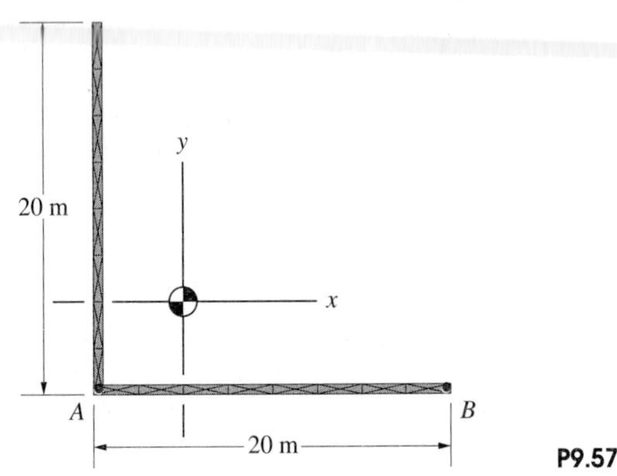

P9.57

9.58 If the subassembly described in Problem 9.57 rotates about the x axis at a constant rate of one revolution every 10 minutes, what is the magnitude of the couple that its reaction control system must exert on it?

9.59 The thin circular disc of radius R and mass m is attached rigidly to the vertical shaft. The disc is slanted at an angle β relative to the horizontal plane. The shaft rotates with constant angular velocity ω_0. What is the magnitude of the couple exerted on the disc by the shaft?

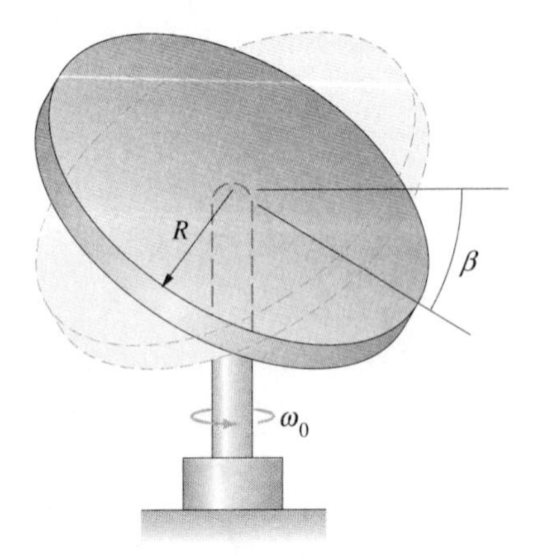

P9.59

9.60 A slender bar of mass m and length l is welded to a horizontal shaft that rotates with constant angular velocity ω_0. Determine the magnitudes of the force $\mathbf{F}$ and couple $\mathbf{C}$ exerted on the bar by the shaft. (Write the equations of angular motion in terms of the body-fixed coordinate system shown.)

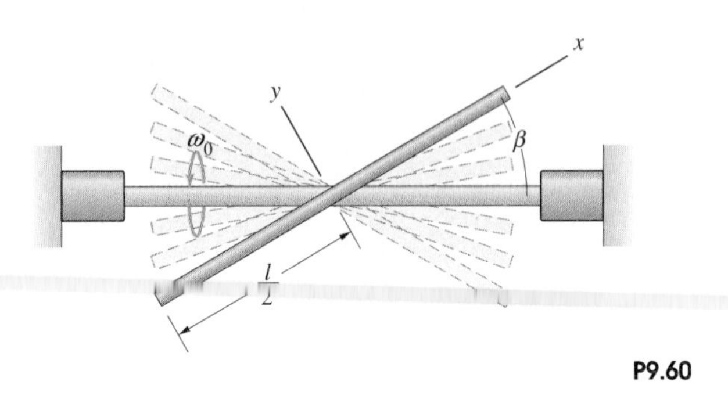

P9.60

9.61 A slender bar of mass m and length l is welded to a horizontal shaft that rotates with constant angular velocity ω_0. Determine the magnitude of the couple $\mathbf{C}$ exerted on the bar by the shaft. (Write the equation of angular motion in terms of the body-fixed coordinate system shown.)

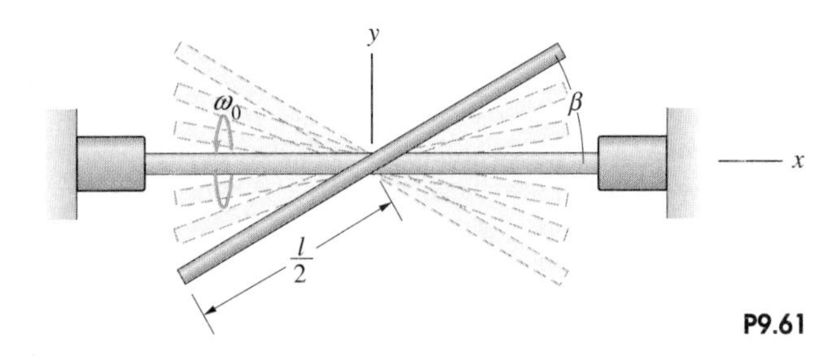

P9.61

9.62 The slender bar of length l and mass m is pinned to the vertical shaft at O. The vertical shaft rotates with a constant angular velocity ω_0. Show that the value of ω_0 necessary for the bar to remain at a constant angle β relative to the vertical is

$$\omega_0 = \sqrt{3g/(2l\cos\beta)}$$

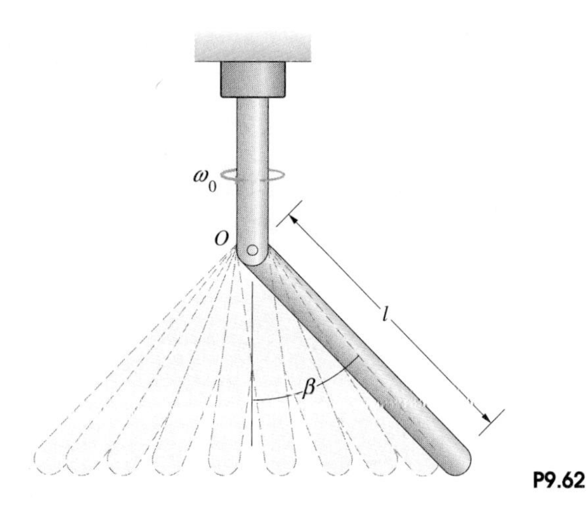

P9.62

9.63 The vertical shaft rotates with constant angular velocity ω_0. The 35° angle between the edge of the 10 kg thin rectangular plate pinned to the shaft and the shaft remains constant. Determine ω_0.

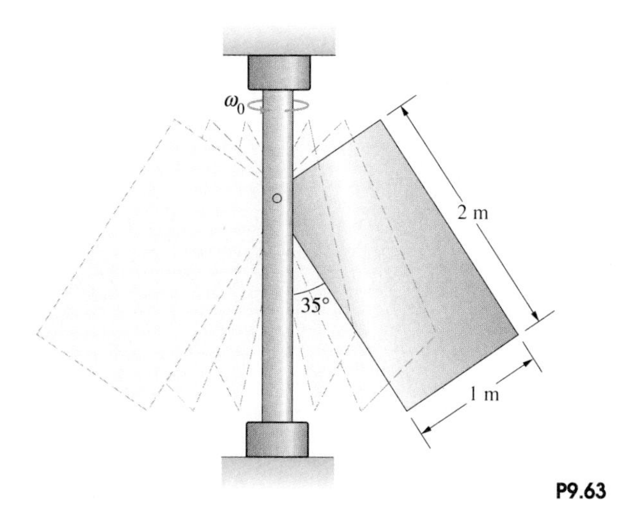

P9.63

9.64 A thin circular disc of mass m mounted on a horizontal shaft rotates relative to the shaft with constant angular velocity ω_d. The horizontal shaft is rigidly attached to a vertical shaft rotating with constant angular velocity ω_0. Determine the magnitude of the couple exerted on the disc by the horizontal shaft.

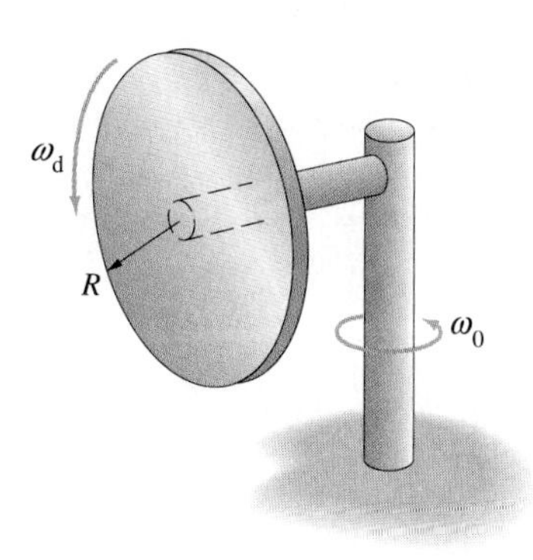

P9.64

9.65 The thin triangular plate has ball and socket supports at A and B. The y axis is vertical. If the plate rotates with constant angular velocity ω_0, what are the horizontal components of the reactions on the plate at A and B?

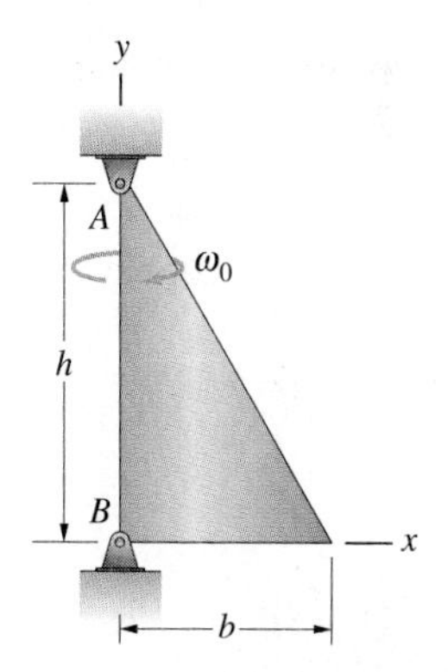

P9.65

9.66 The 5 kg thin circular disc is rigidly attached to the 6 kg slender horizontal shaft. The disc and horizontal shaft rotate about the axis of the shaft with constant angular velocity $\omega_d = 20\,\text{rad/s}$. The entire assembly rotates about the vertical axis with constant angular velocity $\omega_0 = 4\,\text{rad/s}$. Determine the components of the force and couple exerted on the horizontal shaft by the disc.

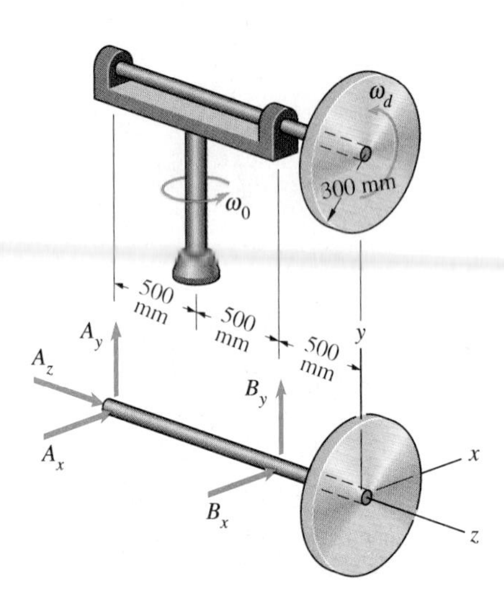

P9.66

9.67 In Problem 9.66, determine the reactions exerted on the horizontal shaft by the two bearings.

9.68 The thin rectangular plate is attached to the rectangular frame by pins. The frame rotates with constant angular velocity ω_0. Show that

$$\frac{d^2\beta}{dt^2} = -\omega_0^2 \sin\beta\cos\beta$$

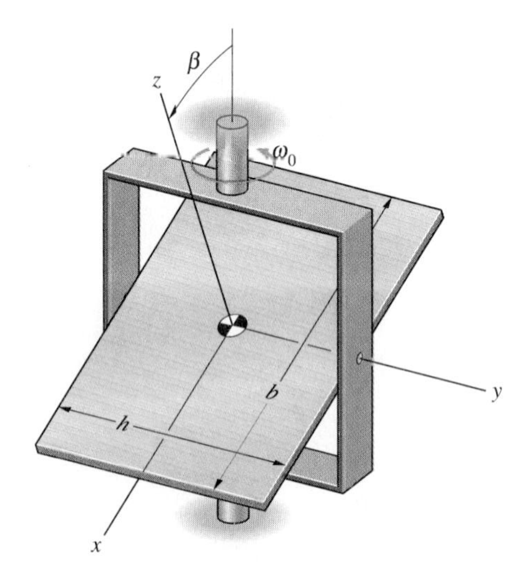

P9.68

9.69 The axis of the right circular cone of mass m, height h, and radius R spins about the vertical axis with constant angular velocity ω_0. Its centre of mass is stationary and its base rolls on the floor. Show that the angular velocity ω_0 necessary for this motion is $\omega_0 = \sqrt{10g/3R}$.

Strategy: Let the z axis remain aligned with the axis of the cone and the x axis remain vertical.

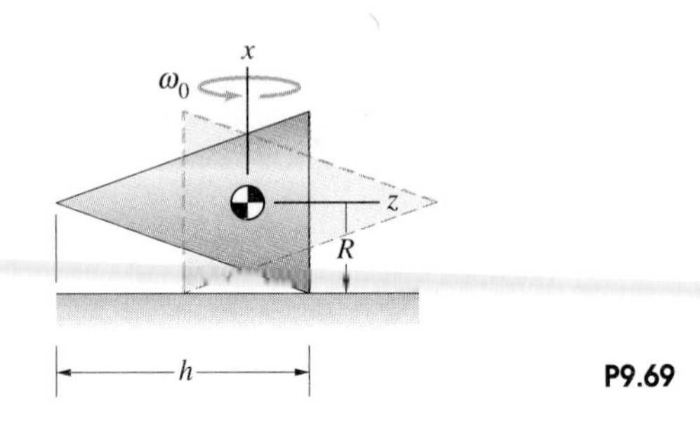

P9.69

9.70 A thin circular disc of radius R and mass m rolls along a circular path of radius r. The magnitude v of the velocity of the centre of the disc and the angle θ between the disc's axis and the vertical are constants. Show that v satisfies the equation

$$v^2 = \frac{\frac{2}{3}g\cot\theta(r - R\cos\theta)^2}{r - \frac{5}{6}R\cos\theta}$$

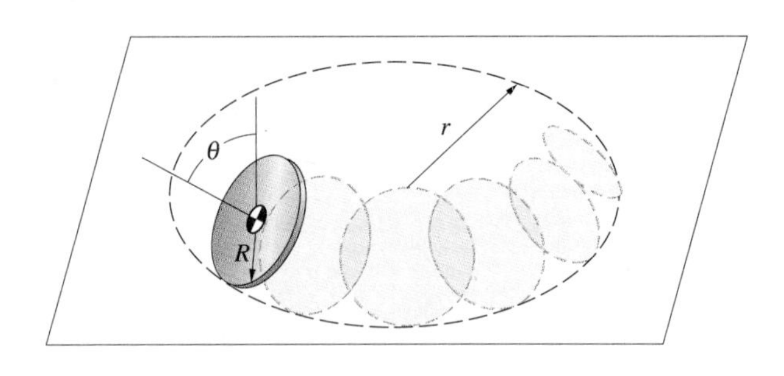

P9.70

9.71 The vertical shaft rotates with constant angular velocity ω_0, causing the grinding mill to roll on the horizontal surface. Assume that point P of the mill is stationary at the instant shown, and that the force N exerted on the mill by the surface is perpendicular to the surface and acts at P. The mass of the mill is m, and its moments and products of inertia in terms of the coordinate system shown are I_{xx}, $I_{yy} = I_{zz}$, and $I_{xy} = I_{yz} = I_{zx} = 0$. Determine N.

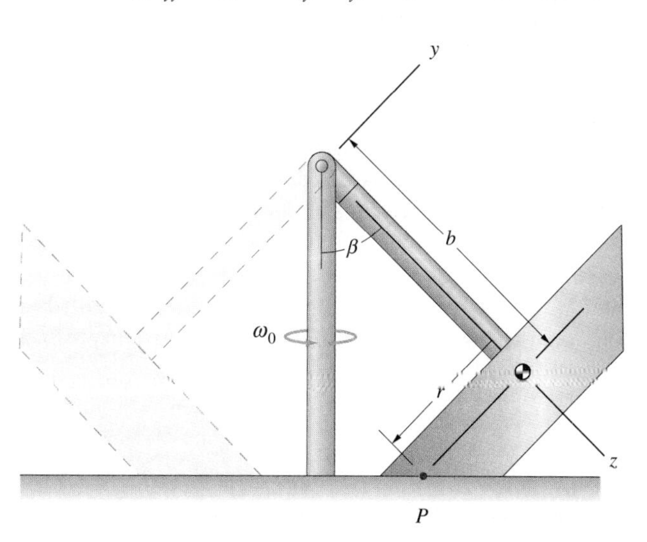

P9.71

9.72 The view of an aeroplane's landing gear looking from behind the aeroplane is shown in Figure (a). The radius of the wheel is 300 mm, and its moment of inertia is $2\,\text{kg.m}^2$. The aeroplane takes off at 30 m/s. After takeoff, the landing gear retracts by rotating towards the right side of the aeroplane as shown in Figure (b). Determine the magnitude of the couple exerted by the wheel on its support. (Neglect the aeroplane's angular motion.)

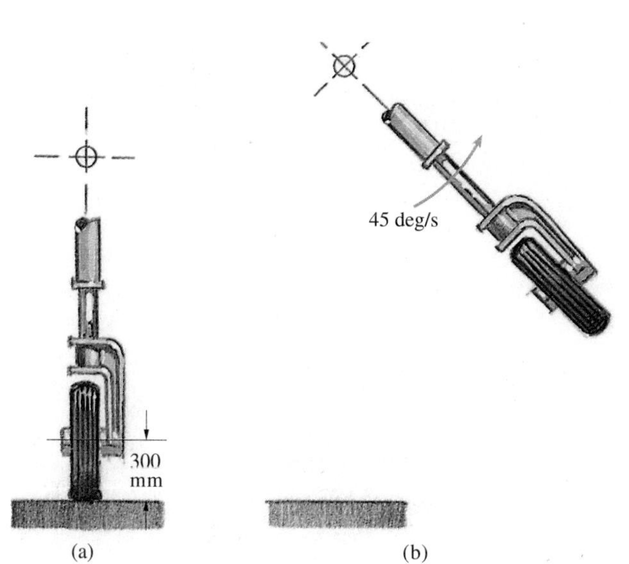

P9.72

9.73 If the rider turns to his left, will the couple exerted on the motorcycle by its wheels tend to cause the motorcycle to lean towards the rider's left side or his right side?

P9.73

9.74 By substituting the components of $\mathbf{H}_O$ from Equations (9.5) into Equation (9.22), derive Equations (9.23).

9.5 Eulerian Angles

The equations of angular motion relate the total moment acting on a rigid body to its angular velocity and acceleration. If we know the total moment and the angular velocity, we can determine the angular acceleration. But how can we use the angular acceleration to determine the rigid body's angular position, or orientation, as a function of time? To explain how this is done, we must first show how to specify the orientation of a rigid body in three dimensions.

You have seen that describing the orientation of a rigid body in planar motion requires only the angle θ that specifies the body's rotation relative to some reference orientation. In three-dimensional motion, three angles are required. To understand why, consider a particular axis that is fixed relative to the rigid body. Two angles are necessary to specify the direction of the axis, and a third angle is needed to specify the rigid body's rotation about the axis. Although several systems of angles for describing the orientation of a rigid body are commonly used, the best known system is the one called the **Eulerian** angles. In this section we define these angles and express the equations of angular motion in terms of them.

Objects with an Axis of Symmetry

We first explain how the Eulerian angles are used to describe the orientation of an object with an axis of rotational symmetry, because this case results in simpler equations of angular motion.

Definitions We assume that an object has an axis of rotational symmetry and introduce two coordinate systems: xyz, with its z axis coincident with the object's axis of symmetry, and XYZ, an inertial **reference coordinate system**. We begin with the object in a reference position in which xyz and XYZ are superimposed (Figure 9.26(a)).

Our first step is to rotate the object and the xyz system together through an angle ψ about the Z axis (Figure 9.26(b)). In this intermediate orientation, we denote the coordinate system by $x'y'z'$. Next, we rotate the object and the xyz system together through an angle θ about the x' axis (Figure 9.26(c)). Finally, we rotate the object relative to the xyz system through an angle ϕ about its axis of symmetry (Figure 9.26(d)). Notice that the x axis remains in the XY plane.

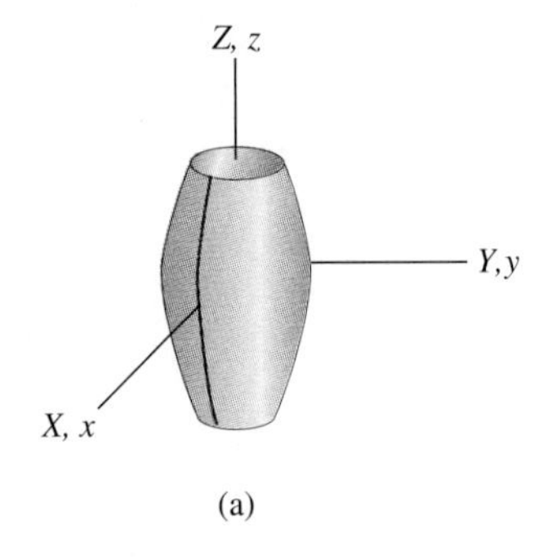

(a)

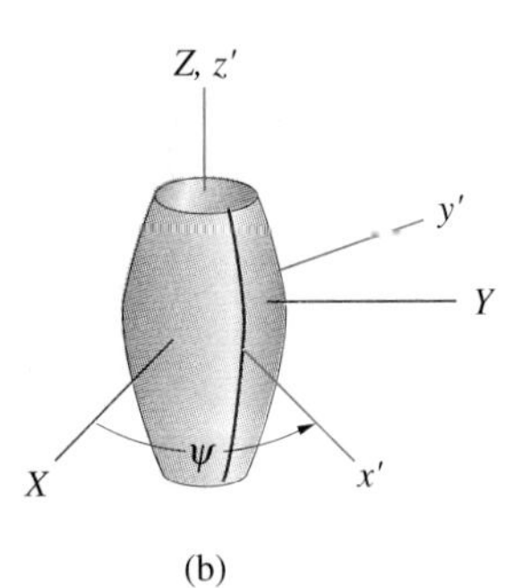

(b)

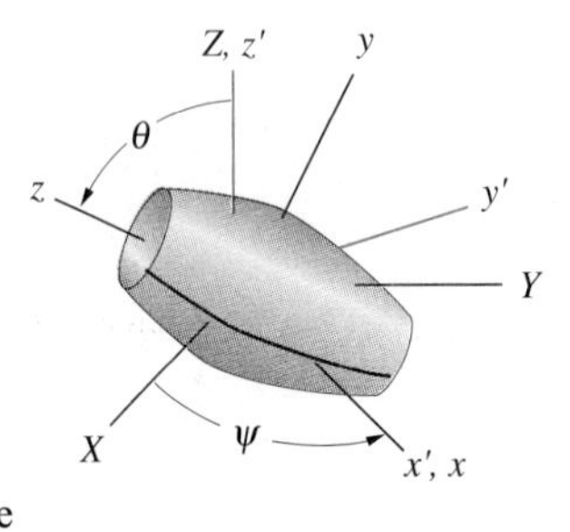

(c)

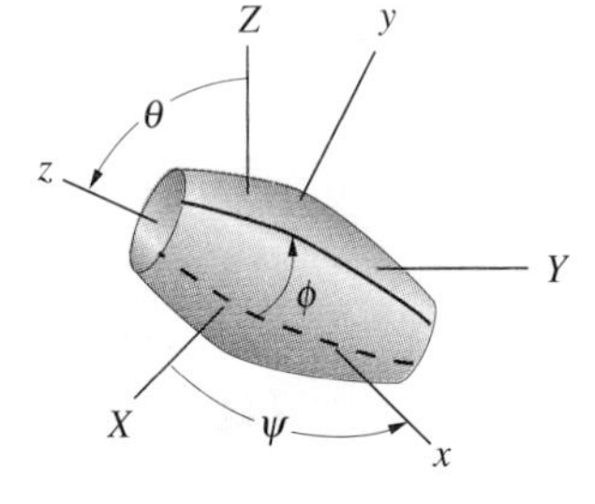

(d)

Figure 9.26
(a) The reference position.
(b) The rotation ψ about the Z axis.
(c) The rotation θ about the x' axis.
(d) The rotation ϕ of the object relative to the xyz system.

The angles ψ and θ specify the orientation of xyz system relative to the reference XYZ system. ψ is called the **precession angle**, and θ is called the **nutation angle**. The angle ϕ specifying the rotation of the rigid body relative to the xyz system is called the **spin angle**. These three angles specify the orientation of the rigid body relative to the reference coordinate system and are called the **Eulerian angles**. We can obtain any orientation of the object relative to the reference coordinate system by appropriate choices of these angles: we choose ψ and θ to obtain the desired direction of the axis of symmetry, then choose ϕ to obtain the desired rotational position of the object and its axis of symmetry.

Equations of Angular Motion To analyse an object's motion in terms of the Eulerian angles, we must express the equations of angular motion in terms of them. Figure 9.27(a) shows the rotation ψ from the reference orientation of the xyz system to its intermediate orientation $x'y'z'$. We represent the angular velocity of the coordinate system due to the rate of change of ψ by the angular velocity vector $\dot{\psi}$ pointing in the z' direction. (We use a dot to denote the derivative with respect to time.) Figure 9.27(b) shows the second rotation θ. We represent the angular velocity due to the rate of change of θ by the vector $\dot{\theta}$ pointing in the x direction. We also resolve the angular velocity vector $\dot{\psi}$ into components in the y and z directions. The components of the angular velocity of the xyz system relative to the reference coordinate system are

$$\Omega_x = \dot{\theta}$$
$$\Omega_y = \dot{\psi}\sin\theta \qquad (9.32)$$
$$\Omega_z = \dot{\psi}\cos\theta$$

In Figure 9.27(c), we represent the angular velocity of the rigid body relative to the xyz system by the vector $\dot{\phi}$. Adding this angular velocity to the angular velocity of the xyz system, we obtain the components of the angular velocity of the rigid body relative to the XYZ system:

$$\omega_x = \dot{\theta}$$
$$\omega_y = \dot{\psi}\sin\theta \qquad (9.33)$$
$$\omega_z = \dot{\phi} + \dot{\psi}\cos\theta$$

Taking the time derivatives of these equations, we obtain

$$\frac{d\omega_x}{dt} = \ddot{\theta}$$
$$\frac{d\omega_y}{dt} = \ddot{\psi}\sin\theta + \dot{\psi}\dot{\theta}\cos\theta \qquad (9.34)$$
$$\frac{d\omega_z}{dt} = \ddot{\phi} + \ddot{\psi}\cos\theta - \dot{\psi}\dot{\theta}\sin\theta$$

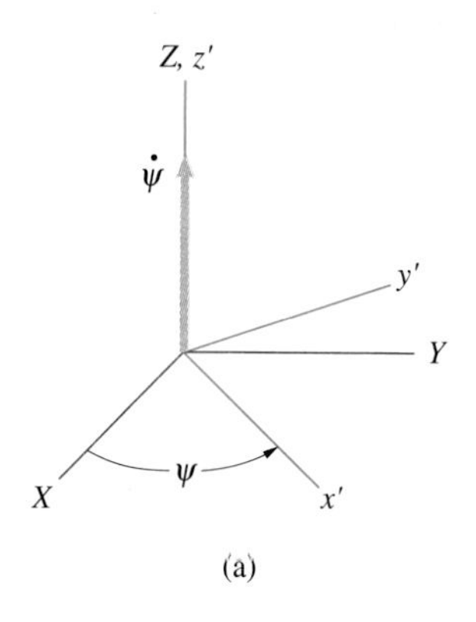

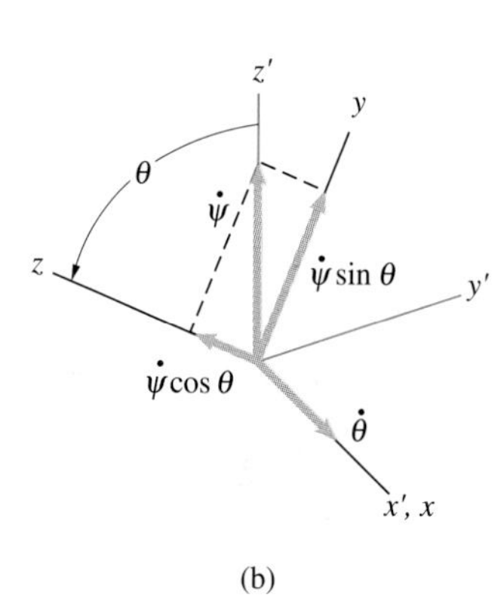

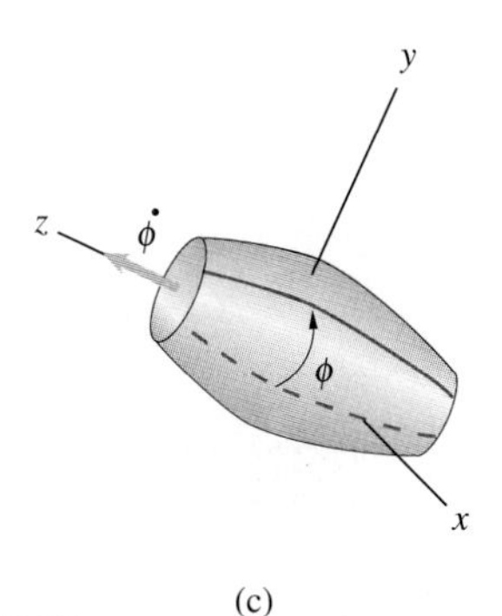

Figure 9.27
(a) The rotation ψ and the angular velocity $\dot{\psi}$.
(b) The rotation θ, the angular velocity $\dot{\theta}$, and the components of the angular velocity $\dot{\psi}$.
(c) The rotation ϕ and the angular velocity $\dot{\phi}$.

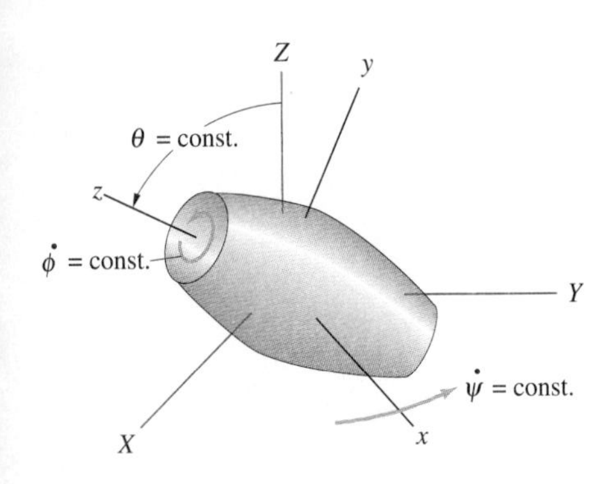

Figure 9.28
Steady precession.

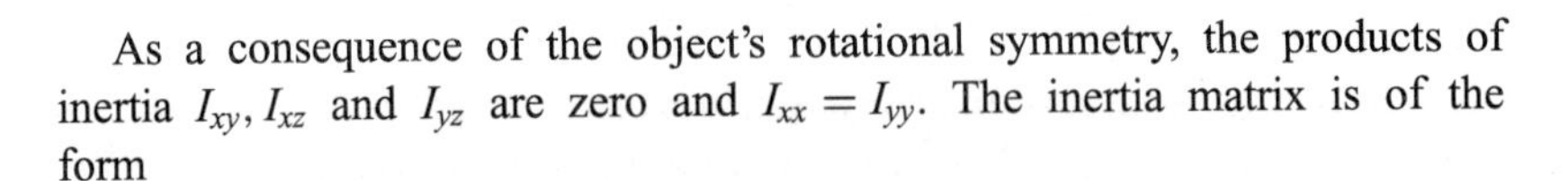
As a consequence of the object's rotational symmetry, the products of inertia I_{xy}, I_{xz} and I_{yz} are zero and $I_{xx} = I_{yy}$. The inertia matrix is of the form

$$[I] = \begin{bmatrix} I_{xx} & 0 & 0 \\ 0 & I_{xx} & 0 \\ 0 & 0 & I_{zz} \end{bmatrix} \tag{9.35}$$

Substituting Equations (9.32)–(9.35) into Equations (9.28), we obtain the equations of angular motion in terms of the Eulerian angles:

$$\Sigma M_x = I_{xx}\ddot{\theta} + (I_{zz} - I_{xx})\dot{\psi}^2 \sin\theta \cos\theta + I_{zz}\dot{\phi}\dot{\psi}\sin\theta \tag{9.36}$$

$$\Sigma M_y = I_{xx}(\ddot{\psi}\sin\theta + 2\dot{\psi}\dot{\theta}\cos\theta) - I_{zz}(\dot{\phi}\dot{\theta} + \dot{\phi}\dot{\theta}\cos\theta) \tag{9.37}$$

$$\Sigma M_z = I_{zz}(\ddot{\phi} + \ddot{\psi}\cos\theta - \dot{\psi}\dot{\theta}\sin\theta) \tag{9.38}$$

To determine the Eulerian angles as functions of time when the total moment is known, we must usually solve these equations by numerical integration. However, we can obtain an important class of closed-form solutions by assuming a specific type of motion.

Steady Precession The motion called **steady precession** is commonly observed in tops and gyroscopes. The object's rate of spin $\dot{\phi}$ relative to the xyz coordinate system is assumed to be constant (Figure 9.28). The nutation angle θ, the inclination of the **spin axis** z relative to the Z axis, is assumed to be constant, and the **precession rate** $\dot{\psi}$, the rate at which the xyz system rotates about the Z axis, is assumed to be constant. The last assumption explains the name given to this motion.

With these assumptions, Equations (9.36)–(9.38) reduce to

$$\Sigma M_x = (I_{zz} - I_{xx})\dot{\psi}^2 \sin\theta\cos\theta + I_{zz}\dot{\phi}\dot{\psi}\sin\theta \tag{9.39}$$

$$\Sigma M_y = 0 \tag{9.40}$$

$$\Sigma M_z = 0 \tag{9.41}$$

We discuss two examples: the steady precession of a spinning top and the steady precession of an axially symmetric object that is free of external moments.

(a)

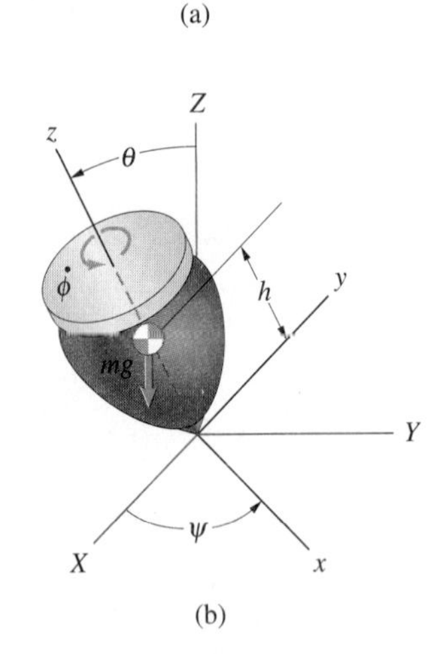

(b)

Figure 9.29
(a) A spinning top seems to defy gravity. (b) The precession angle ψ and nutation angle θ specify the orientation of the spin axis.

Precession of a Top. The peculiar behaviour of a top (Figure 9.29(a)), inspired some of the first analytical studies of three-dimensional motions of rigid bodies. When the top is set into motion, its spin axis may initially remain vertical, a motion called **sleeping**. As friction reduces the spin rate, the spin axis begins to lean over and rotate about the vertical axis. This phase of the top's motion approximates steady precession. (The top's spin rate continuously decreases due to friction, whereas in steady precession we assume the spin rate to be constant.)

To analyse the motion, we place the reference system XYZ with its origin at the point of the top and the Z axis upwards. Then we align the z axis of the xyz system with the spin axis (Figure 9.29(b)). We assume that the top's point rests in a small depression so that it remains at a fixed point on the floor. The precession angle ψ and nutation angle θ specify the orientation of the spin axis, and the spin rate of the top relative to the xyz system is $\dot{\phi}$.

The top's weight exerts a moment $M_x = mgh \sin\theta$ about the origin, and the moments $M_y = 0$ and $M_z = 0$. Substituting $M_x = mgh \sin\theta$ into Equation (9.39), we obtain

$$mgh = \dot{\psi}(I_{zz} - I_{xx})\dot{\psi} \cos\theta + I_{zz}\dot{\phi} \tag{9.43}$$

and Equations (9.40) and (9.41) are identically satisfied. Equation (9.42) relates the spin rate, nutation angle and rate of precession. For example, if we know the spin rate $\dot{\phi}$ and nutation angle θ, we can solve for the top's precession rate $\dot{\psi}$.

Moment-Free Steady Precession. A spinning axisymmetric object that is free of external moments, such as an axisymmetric satellite in orbit, can exhibit a motion similar to the steady precessional motion of a top. We observe this motion when an American football is thrown in a 'wobbly' spiral. To analyse it, we place the origin of the xyz system at the object's centre of mass (Figure 9.30(a)). Equation (9.39) becomes

$$(I_{zz} - I_{xx})^2_{\psi} \cos\theta + I_{zz}\dot{\phi} = 0 \tag{9.42}$$

and Equations (9.40) and (9.41) are identically satisfied. For a given value of the nutation angle, Equation (9.43) relates the object's rates or precession and spin.

We can interpret Equation (9.43) in a way that enables you to visualize the motion. Let's look for a point in the y-z plane (other than the centre of mass) at which the object's velocity relative to the XYZ system is zero at the current instant. We want to find a point with coordinates $(0, y, z)$ such that

$$\begin{aligned}\boldsymbol{\omega} \times (y\,\mathbf{j} + z\,\mathbf{k}) &= [(\dot{\psi} \sin\theta)\,\mathbf{j} + (\dot{\phi} + \dot{\psi}\cos\theta)\,\mathbf{k}] \times [y\,\mathbf{j} + z\,\mathbf{k}] \\ &= [z\dot{\psi}\sin\theta - y(\dot{\phi} + \dot{\psi}\cos\theta)]\,\mathbf{i} = \mathbf{0}\end{aligned}$$

This equation is satisfied at points in the y-z plane such that

$$\frac{y}{z} = \frac{\dot{\psi}\sin\theta}{\dot{\phi} + \dot{\psi}\cos\theta}$$

This relation is satisfied by points on the straight line at an angle β relative to the z axis in Figure 9.30(b), where

$$\tan\beta = \frac{y}{z} = \frac{\dot{\psi}\sin\theta}{\dot{\phi} + \dot{\psi}\cos\theta}$$

Solving Equation (9.43) for $\dot{\phi}$ and substituting the result into this equation, we obtain

$$\tan\beta = \left(\frac{I_{zz}}{I_{xx}}\right)\tan\theta$$

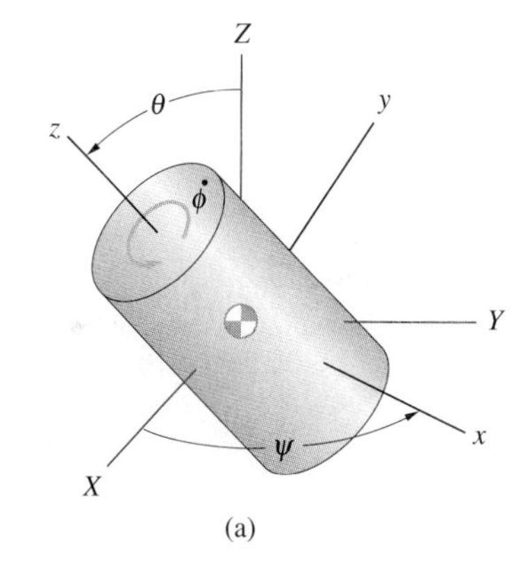

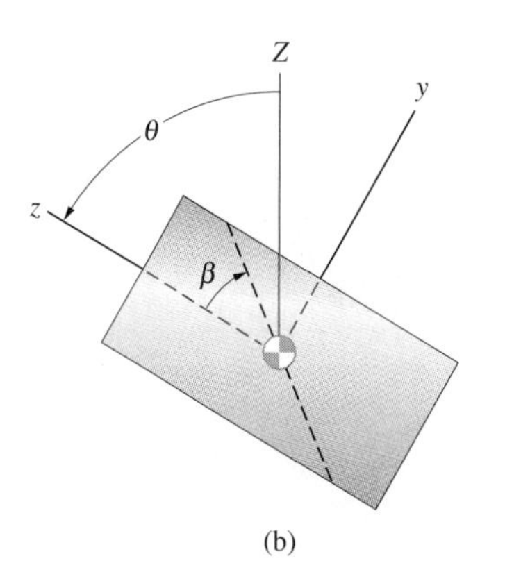

Figure 9.30
(a) An axisymmetric object.
(b) Points on the straight line at an angle from the z axis are stationary relative to the XYZ coordinate system.

If $I_{xx} > I_{zz}$, the angle $\beta < \theta$. In Figure 9.30(c), we show an imaginary cone of half-angle β, called the **body cone**, whose axis is coincident with the z axis. The body cone is in contact with a fixed cone, called the **space cone**, whose axis is coincident with the Z axis. If the body cone rolls on the curved surface of the space cone as the z axis precesses about the Z axis (Figure 9.30(d)), the points of the body cone lying on the straight line in Figure 9.30(b) have zero velocity relative to the XYZ system. That means that *the motion of the body cone is identical to the motion of the object.* You can visualize the object's motion by visualizing the motion of the body cone as it rolls around the outer surface of the space cone. This motion is called **direct precession**.

If $I_{xx} < I_{zz}$, the angle $\beta > \theta$. In this case you must visualize the *interior* surface of the body cone rolling on the fixed space cone (Figure 9.30(e)). This motion is called **retrograde precession**.

Arbitrary Objects

In this section we show how the equations of angular motion can be expressed in terms of the Eulerian angles for an arbitrary object. The initial steps are similar to our treatment of an axially symmetric object, but in this case we assume that the xyz coordinate system is body-fixed.

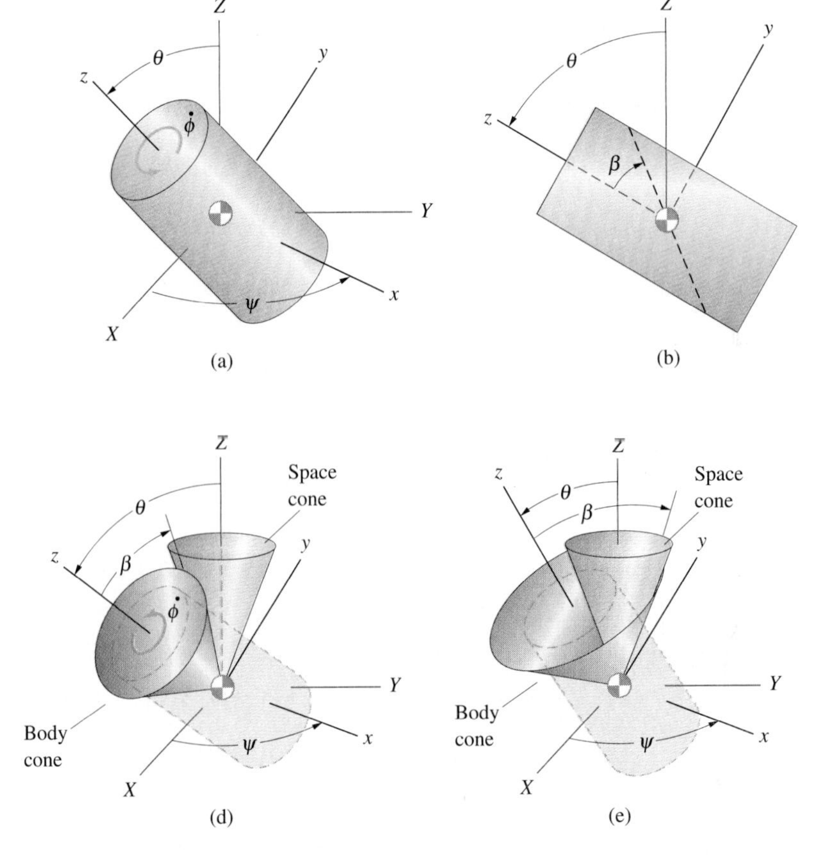

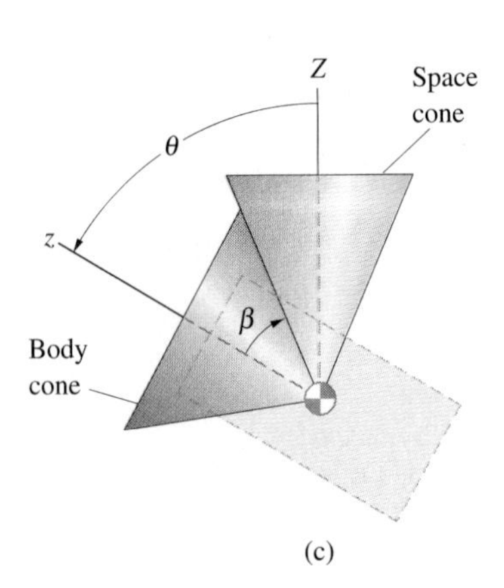

Figure 9.30

(c), (d) The body and space cones. The body cone rolls on the stationary space cone.
(e) When $\alpha > \theta$, the interior surface of the body cone rolls on the stationary space cone.

Definitions We begin with a reference position in the xyz and XYZ systems are superimposed (Figure 9.31(a)). First, we rotate the xyz system through the precession angle ψ about the Z axis (Figure 9.31(b)) and denote it by $x'y'z'$ in this intermediate orientation. Then we rotate the xyz system through the nutation angle θ about the x' axis (Figure 9.31(c)), denoting it by $x''y''z''$. We obtain the final orientation of the xyz system by rotating it through the angle ϕ about the z'' axis (Figure 9.31(d)). Notice that we have used one more rotation of the xyz system than in the case of an axially symmetric object.

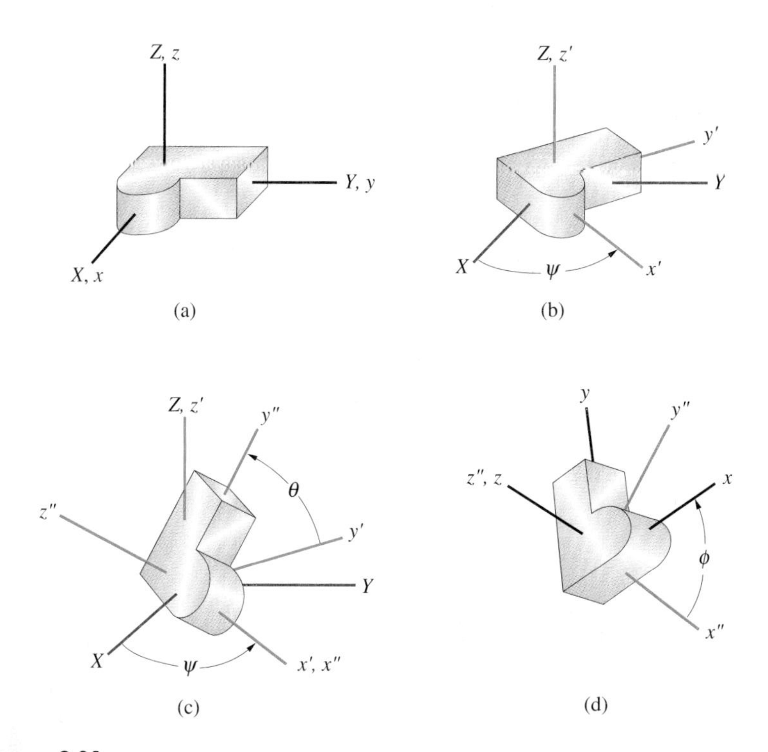

Figure 9.31
(a) The reference position.
(b) The rotation ψ about the Z axis.
(c) The rotation θ about the x' axis.
(d) The rotation ϕ about the z'' axis.

We can obtain any orientation of the body-fixed coordinate system relative to the reference coordinate system by these three rotations. We choose ψ and θ to obtain the desired direction of the z axis, then choose ϕ to obtain the desired orientation of the x and y axes.

Just as in the case of an object with rotational symmetry, we must express the components of the rigid body's angular velocity in terms of the Eulerian angles to obtain the equations of angular motion. Figure 9.32(a) shows the rotation ψ from the reference orientation of the xyz system to the intermediate orientation $x'y'z'$. We represent the angular velocity of the body-fixed coordinate system due to the rate of change of ψ by the vector $\dot{\psi}$ pointing in the z' direction. Figure 9.32(b) shows the next rotation θ that takes the body-fixed coordinate system to the intermediate orientation $x''y''z''$. We represent the angular velocity due to the rate of change of θ by the vector $\dot{\theta}$ pointing in the x'' direction. In this figure we also show the components of the angular velocity vector $\dot{\psi}$ in the y'' and z'' directions. Figure 9.32(c) shows the third rotation ϕ that takes the body-fixed coordinate system to its final orientation defined by the three Eulerian angles. We represent the angular velocity due to the rate of change of ϕ by the vector $\dot{\phi}$ pointing in the z direction.

To determine ω_x, ω_y and ω_z in terms of the Eulerian angles, we need to determine the components of the angular velocities shown in Figure 9.32(c) in the x, y and z directions. The vectors $\dot{\phi}$ and $\dot{\psi}\cos\theta$ point in the z axis direction. In Figures 9.32(d) and (e), which are drawn with the z axis pointing out of the page, we determine the components of the vectors $\dot{\psi}\sin\theta$ and $\dot{\theta}$ in the x and y axis directions.

By summing the components of the angular velocities in the three coordinate directions (Figure 9.32f), we obtain

$$\begin{aligned} \omega_x &= \dot{\psi}\sin\theta\sin\phi + \dot{\theta}\cos\phi \\ \omega_y &= \dot{\psi}\sin\theta\cos\phi - \dot{\theta}\sin\phi \\ \omega_z &= \dot{\psi}\cos\theta + \dot{\phi} \end{aligned} \tag{9.44}$$

Figure 9.32

(a) The rotation ψ and the angular velocity $\dot{\psi}$.
(b) The rotation θ, the angular velocity $\dot{\theta}$, and the components of $\dot{\psi}$ in the $x''y''z''$ system.
(c) The rotation ϕ and the angular velocity $\dot{\psi}$.
(d), (e) The components of the angular velocities $\dot{\psi}\sin\theta$ and $\dot{\theta}$ in the xyz system.
(f) The angular velocities ω_x, ω_y and ω_z.

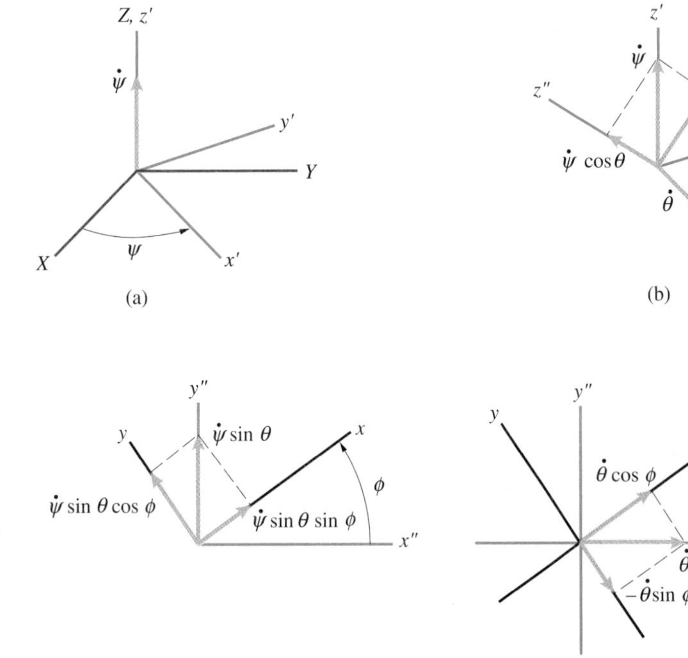

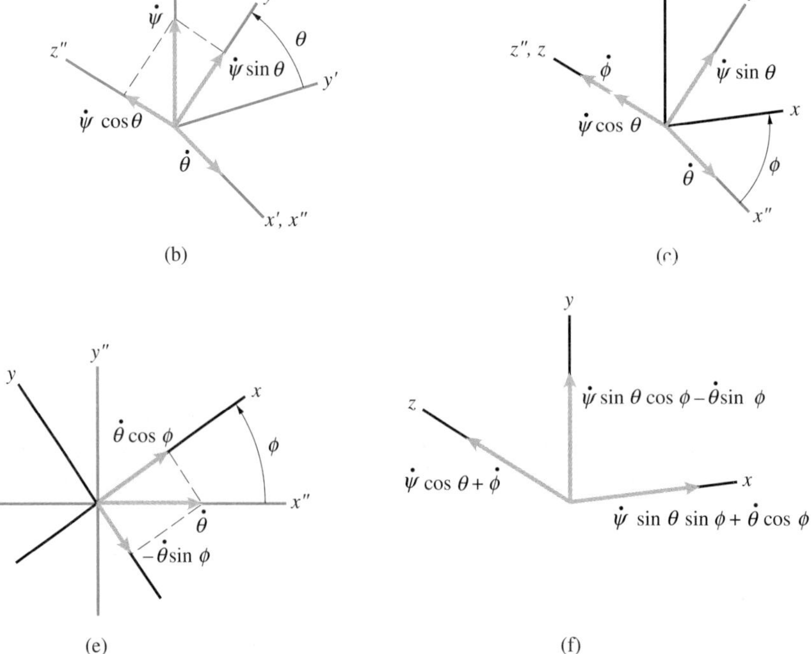

The time derivatives of these equations are

$$\begin{aligned}
\frac{d\omega_x}{dt} &= \ddot{\psi}\sin\theta\sin\phi + \dot{\psi}\dot{\theta}\cos\theta\sin\phi + \dot{\psi}\dot{\phi}\sin\theta\cos\phi \\
&\quad + \ddot{\theta}\cos\phi - \dot{\theta}\dot{\phi}\sin\phi \\
\frac{d\omega_y}{dt} &= \ddot{\psi}\sin\theta\cos\phi + \dot{\psi}\dot{\theta}\cos\theta\cos\phi - \dot{\psi}\dot{\phi}\sin\theta\sin\phi \\
&\quad - \ddot{\theta}\sin\phi - \dot{\theta}\dot{\phi}\cos\phi \\
\frac{d\omega_z}{dt} &= \ddot{\psi}\cos\theta - \dot{\psi}\dot{\theta}\sin\theta + \ddot{\phi}
\end{aligned} \tag{9.45}$$

Equations of Angular Motion With Equations (9.44) and (9.45), we can express the equations of angular motion in terms of the three Eulerian angles. To simplify the equations, we assume that the body-fixed coordinate system xyz is a set of principal axes. Then the equations of angular motion, Equations (9.28), are

$$\begin{aligned}
\Sigma M_x &= I_{xx}\frac{d\omega_x}{dt} - (I_{yy} - I_{zz})\omega_y\omega_z \\
\Sigma M_y &= I_{yy}\frac{d\omega_y}{dt} - (I_{zz} - I_{xx})\omega_z\omega_x \\
\Sigma M &= I_{zz}\frac{d\omega_z}{dt} - (I_{xx} - I_{yy})\omega_x\omega_y
\end{aligned}$$

Substituting Equations (9.44) and (9.45) into these equations, we obtain the equations of angular motion in terms of Eulerian angles:

$$\begin{aligned}
M_x &= I_{xx}\ddot{\psi}\sin\theta\sin\phi + I_{xx}\ddot{\theta}\cos\phi \\
&\quad + I_{xx}(\dot{\psi}\dot{\theta}\cos\theta\sin\phi + \dot{\psi}\dot{\phi}\sin\theta\cos\phi - \dot{\theta}\dot{\phi}\sin\phi) \\
&\quad - (I_{yy} - I_{zz})(\dot{\psi}\sin\theta\cos\phi - \dot{\theta}\sin\phi)(\dot{\psi}\cos\theta + \dot{\phi}) \\
M_y &= I_{yy}\ddot{\psi}\sin\theta\cos\phi - I_{yy}\ddot{\theta}\sin\phi \\
&\quad + I_{yy}(\dot{\psi}\dot{\theta}\cos\theta\cos\phi - \dot{\psi}\dot{\phi}\sin\theta\sin\phi - \dot{\theta}\dot{\phi}\cos\phi) \\
&\quad - (I_{zz} - I_{xx})(\dot{\psi}\cos\theta + \dot{\phi})(\dot{\psi}\sin\theta\sin\phi + \dot{\theta}\cos\phi) \\
M_z &= I_{zz}\ddot{\psi}\cos\theta + I_{zz}\ddot{\phi} - I_{zz}\dot{\psi}\dot{\theta}\sin\theta \\
&\quad - (I_{xx} - I_{yy})(\dot{\psi}\sin\theta\sin\phi + \dot{\theta}\cos\phi)(\dot{\psi}\sin\theta\cos\phi - \dot{\theta}\sin\phi)
\end{aligned} \tag{9.46}$$

If you know the Eulerian angles and their first and second time derivatives, you can solve these equations for the components of the total moment. Or, if you know the total moment, the Eulerian angles, and the first time derivatives of the Eulerian angles, you can determine the second time derivatives of the Eulerian angles. You can use these equations to determine the Eulerian angles as functions of time when you know the total moment, but numerical integration is usually necessary.

In the following example we analyse the motion of an object in steady precession. By aligning the coordinate system as shown in Figure 9.28, you can use Equation (9.39) to relate the total moment about the x axis to the nutation angle θ, precession rate $\dot{\psi}$, and spin rate $\dot{\phi}$.

Example 9.9

The thin circular disc of radius R and mass m in Figure 9.33 rolls along a horizontal circular path of radius r. The angle θ between the disc's axis and the vertical remains constant. Determine the magnitude v of the velocity of the centre of the disc as a function of the angle θ.

Figure 9.33

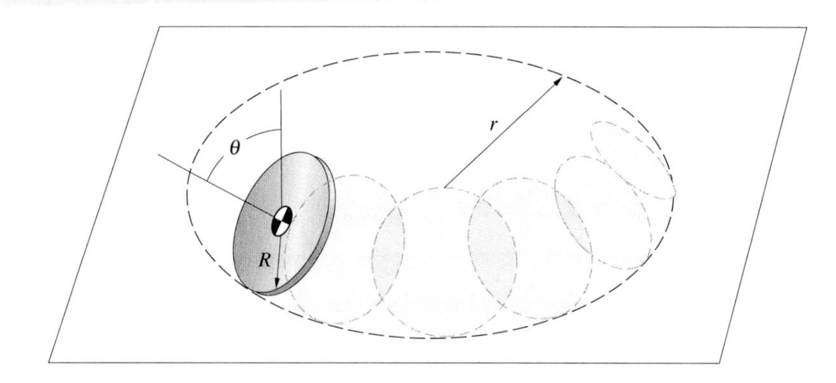

STRATEGY

We can obtain the velocity of the centre of the disc by assuming that the disc is in steady precession and determining the conditions necessary for the equations of motion to be satisfied.

SOLUTION

In Figure (a) we align the z axis with the disc's spin axis and assume that the x axis remains parallel to the surface on which the disc rolls. The angle θ is the nutation angle. The centre of mass moves in a circular path of radius $r_G = r - R\cos\theta$. Therefore the precession rate, the rate at which the x axis rotates in the horizontal plane, is

$$\dot{\psi} = \frac{v}{r_G}$$

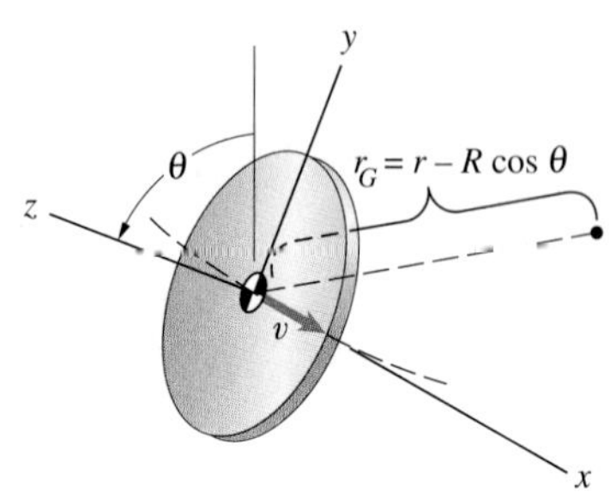

(a) Aligning the z axis with the spin axis. The x axis is horizontal.

From Equations (9.33), the components of the disc's angular velocity are

$$\omega_x = \dot{\theta} = 0$$

$$\omega_y = \dot{\psi}\sin\theta = \frac{v}{r_G}\sin\theta$$

$$\omega_z = \dot{\phi} + \dot{\psi}\cos\theta = \dot{\phi} + \frac{v}{r_G}\cos\theta$$

where $\dot{\phi}$ is the spin rate. To determine $\dot{\phi}$, we use the condition that the velocity of the point of the disc in contact with the surface is zero. In terms of the velocity of the centre the velocity of the point of contact is

$$\mathbf{0} = v\,\mathbf{i} + \boldsymbol{\omega} \times (-R\,\mathbf{j}) = v\,\mathbf{i} + \begin{vmatrix} \mathbf{i} & \mathbf{j} & \mathbf{k} \\ 0 & \dfrac{v}{r_G}\sin\theta & \dot{\phi} + \dfrac{v}{r_G}\cos\theta \\ 0 & -R & 0 \end{vmatrix}$$

Expanding the determinant and solving for $\dot{\phi}$, we obtain

$$\dot{\phi} = -\frac{v}{R} - \frac{v}{r_G}\cos\theta$$

We draw the free-body diagram of the disc in Figure (b). Because the centre of mass moves in the horizontal plane, its acceleration in the vertical direction is zero. Therefore the normal force $N = mg$. The acceleration of the centre of mass in the direction perpendicular to its circular path is $a_n = v^2/r_G$. Newton's second law in the direction perpendicular to the circular path is

$$T = m\frac{v^2}{r_G}$$

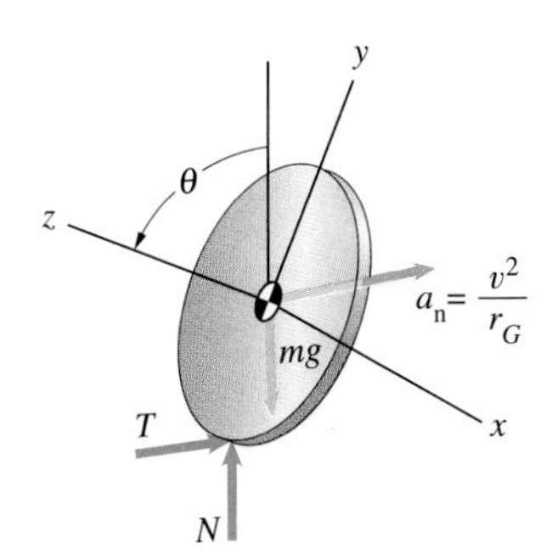

(b) Free-body diagram of the disc showing the normal acceleration of the centre of mass.

Therefore the components of the total moment about the centre of mass are

$$\Sigma M_x = TR\sin\theta - NR\cos\theta = m\frac{v^2}{r_G}R\sin\theta - mgR\cos\theta$$

$$\Sigma M_y = 0$$

$$\Sigma M_z = 0$$

Substituting our expressions for $\dot{\psi}$, $\dot{\phi}$ and ΣM_x into the equation of angular motion for steady precession, Equation (9.39), and solving for v, we obtain

$$v = \sqrt{\frac{\frac{2}{3}g\cot\theta(r - R\cos\theta)^2}{r - \frac{5}{6}R\cos\theta}}$$

Problems

9.75 A ship has a turbine engine. The spin axis of the axisymmetric turbine is horizontal and aligned with the ship's longitudinal axis. The turbine rotates at 10 000 rpm (revolutions per minute). Its moment of inertia about its spin axis is 1000 kg.m^2. If the ship turns at a constant rate of 20 degrees per minute, what is the magnitude of the moment exerted on the ship by the turbine?

Strategy: Treat the turbine's motion as steady precession with nutation angle $\theta = 90°$.

P9.75

9.76 The centre of the car's wheel A travels in a circular path about O at 24 km/hr. The wheel's radius is 0.3 m and its moment of inertia about its axis of rotation is 0.9 kg.m^2. What is the magnitude of the total external moment about the wheel's centre of mass?

Strategy: Treat the wheel's motion as steady precession with nutation angle $\theta = 90°$.

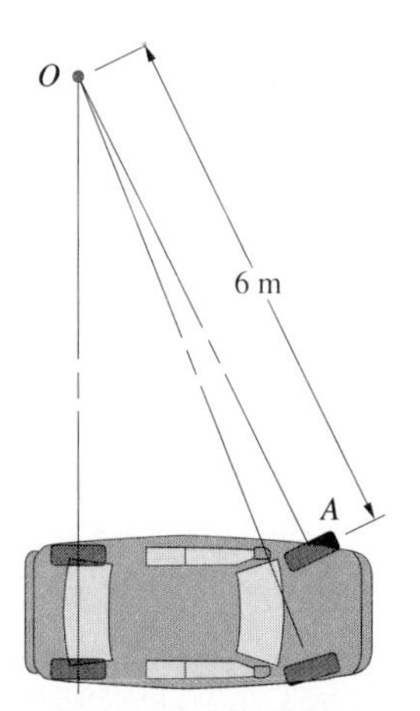

P9.76

9.77 Solve Problem 9.64 by treating the motion as steady precession.

9.78 Solve Problem 9.69 by treating the motion as steady precession.

9.79 A thin circular disc undergoes moment-free steady precession. The z axis is perpendicular to the disc. Show that the disc's precession rate is $\dot{\psi} = -2\dot{\phi}/\cos\theta$. (Notice that when the nutation angle is small, the precession rate is approximately two times the spin rate.)

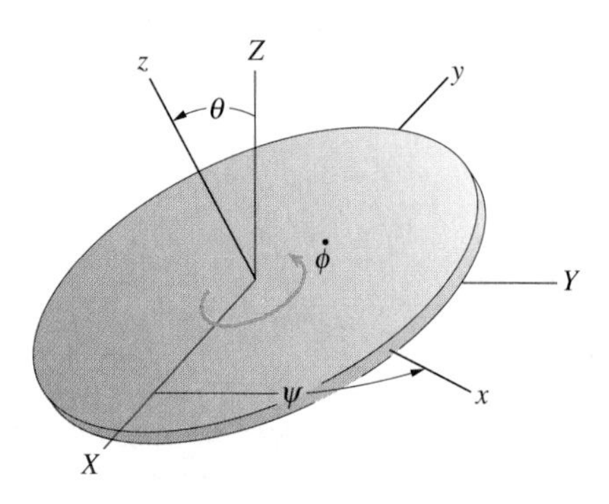

P9.79

9.80 The rocket is in moment-free steady precession with nutation angle $\theta = 40°$ and spin rate $\dot{\phi} = 4$ revolutions per second. Its moments of inertia are $I_{xx} = 10\,000$ kg.m^2 and $I_{zz} = 2000$ kg.m^2. What is the rocket's precession rate $\dot{\psi}$ in revolutions per second?

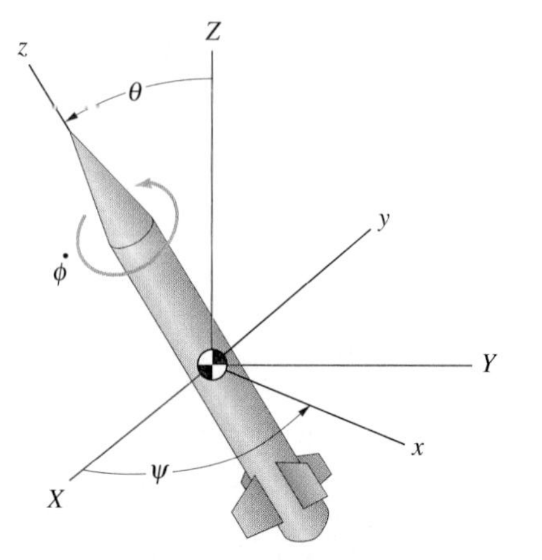

P9.80

9.81 Sketch the body and space cones for the motion of the rocket in Problem 9.80.

9.82 The top is in steady precession with nutation angle $\theta = 15°$ and precession rate $\dot{\psi} = 1$ revolution per second. The mass of the top is 12×10^{-3} kg, its centre of mass is 25 mm from the point, and its moments of inertia are $I_{xx} = 6 \times 10^{-6}$ kg.m^2 and $I_{zz} = 2 \times 10^{-6}$ kg.m^2. What is the spin rate $\dot{\phi}$ of the top in revolutions per second?

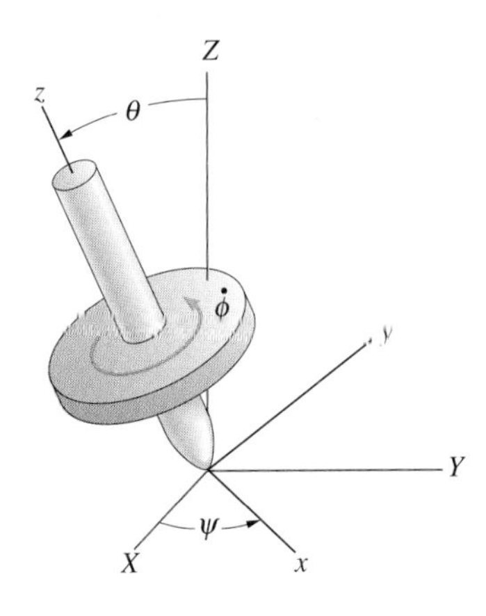

P9.82

9.83 The top described in Problem 9.82 has a spin rate $\dot{\phi} = 15$ revolutions per second. Draw a graph of the precession rate (in revolutions per second) as a function of the nutation angle θ for values of θ from zero to 45°.

9.84 The rotor of a tumbling gyroscope can be modelled as being in moment-free steady precession. Its moments of inertia are $I_{xx} = I_{yy} = 0.04$ kg.m^2, $I_{zz} = 0.18$ kg.m^2. Its spin rate is $\dot{\phi} = 1500$ rpm and its nutation angle is $\theta = 20°$.
(a) What is its precession rate in rpm?
(b) Sketch the body and space cones.

9.85 A satellite can be modelled as an 800 kg cylinder 4 m in length and 2 m in diameter. If the nutation angle is $\theta = 20°$ and the spin rate $\dot{\phi}$ is one revolution per second, what is the satellite's precession rate $\dot{\psi}$ in revolutions per second?

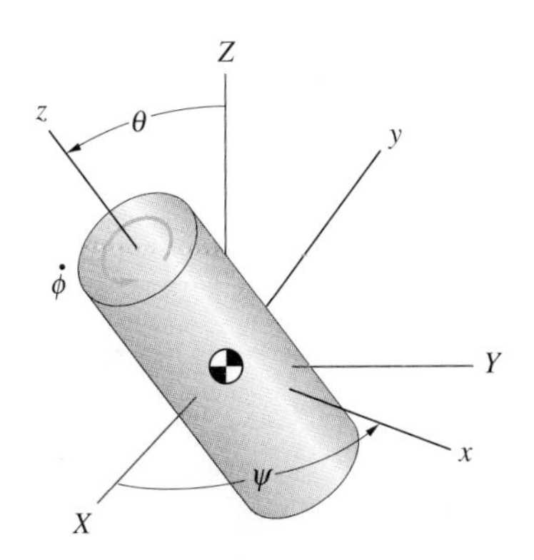

P9.85

9.86 Solve Problem 9.71 by treating the motion as steady precession.

9.87 Solve Problem 9.72 by treating the motion as steady precession.

9.88 Solve Problem 9.73 by treating the motion as steady precession.

9.89 Suppose that you are testing a car and use accelerometers and gyroscopes to measure its Eulerian angles and their derivatives relative to a reference coordinate system. At a particular instant, $\psi = 15°, \theta = 4°, \phi = 15°$, the rates of change of the Eulerian angles are zero, and their second time derivatives are $\ddot{\psi} = 0, \ddot{\theta} = 1$ rad/s^2 and $\ddot{\phi} = -0.5$ rad/s^2. The car's principal moments of inertia in kg.m^2 are $I_{xx} = 2200, I_{yy} = 480$ and $I_{zz} = 2600$. What are the components of the total moment about the car's centre of mass?

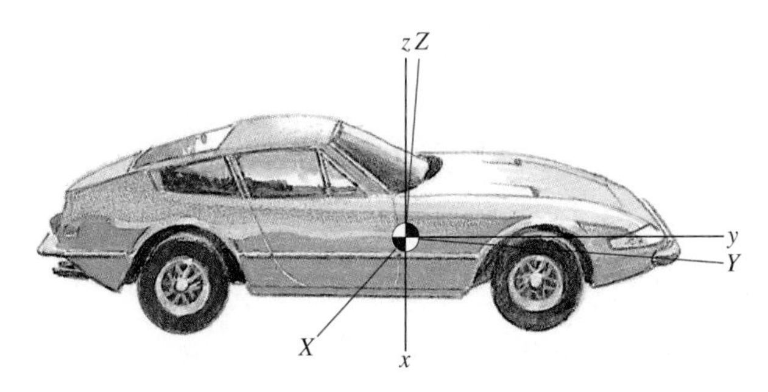

P9.89

9.90 If the Eulerian angles and their second derivatives of the car described in Problem 9.89 have the given values but their rates of change are $\dot{\psi} - 0.2$ rad/s, $\dot{\theta} = -2$ rad/s and $\ddot{\phi} = 0$, what are the components of the total moment about the car's centre of mass?

9.91 Suppose that the Eulerian angles of the car described in Problem 9.89 are $\psi = 40°, \theta = 20°$ and $\phi = 5°$, their rates of change are zero, and the components of the total moment about the car's centre of mass are $\Sigma M_x = -400$ N.m, $\Sigma M_y = 200$ N.m and $\Sigma M_z = 0$. What are the x, y and z components of the car's angular acceleration?

Chapter Summary

Kinematics

A rigid body undergoing any motion other than translation has an instantaneous axis of rotation. The direction of this axis at a particular instant and the rate at which the rigid body rotates about the axis are specified by its angular velocity vector $\boldsymbol{\omega}$.

The velocity of a point A of a rigid body is given in terms of the angular velocity vector and the velocity of a point B by

$$\mathbf{v}_A = \mathbf{v}_B + \boldsymbol{\omega} \times \mathbf{r}_{A/B} \qquad \textbf{Equation (9.1)}$$

The acceleration of a point A of a rigid body is given in terms of the angular velocity vector, the angular acceleration vector $\boldsymbol{\alpha} = d\boldsymbol{\omega}/dt$, and the acceleration of a point B by

$$\mathbf{a}_A = \mathbf{a}_B + \boldsymbol{\alpha} \times \mathbf{r}_{A/B} + \boldsymbol{\omega} \times (\boldsymbol{\omega} \times \mathbf{r}_{A/B}) \qquad \textbf{Equation (9.2)}$$

Let $\boldsymbol{\Omega}$ be the angular velocity of a rotating coordinate system xyz relative to a fixed reference frame, and let $\boldsymbol{\omega}_{\text{rel}}$ be the angular velocity of a rigid body *relative to* the xyz system. The rigid body's angular velocity and angular acceleration relative to the fixed reference frame are

$$\boldsymbol{\omega} = \boldsymbol{\Omega} + \boldsymbol{\omega}_{\text{rel}}$$

$$\boldsymbol{\alpha} = \frac{d\boldsymbol{\Omega}}{dt} + \frac{d\omega_{\text{rel}\,x}}{dt} + \frac{d\omega_{\text{rel}\,y}}{dt} + \frac{d\omega_{\text{rel}\,z}}{dt} + \boldsymbol{\Omega} \times \omega_{\text{rel}} \qquad \textbf{Equation (9.3)}$$

Angular Momentum

If a rigid body rotates about a fixed point O with angular velocity $\boldsymbol{\omega}$ (Figure (a)) the components of its angular momentum about O are given by

$$\begin{bmatrix} H_{Ox} \\ H_{Oy} \\ H_{Oz} \end{bmatrix} = \begin{bmatrix} I_{xx} & -I_{xy} & -I_{xz} \\ -I_{yx} & I_{yy} & -I_{yz} \\ -I_{zx} & -I_{zy} & I_{zz} \end{bmatrix} \begin{bmatrix} \omega_x \\ \omega_y \\ \omega_z \end{bmatrix} \qquad \textbf{Equation (9.8)}$$

This equation also gives the components of the rigid body's angular momentum about the centre of mass in general three-dimensional motion (Figure (b)). In that case the moments of inertia are evaluated in terms of a coordinate system with its origin at the centre of mass.

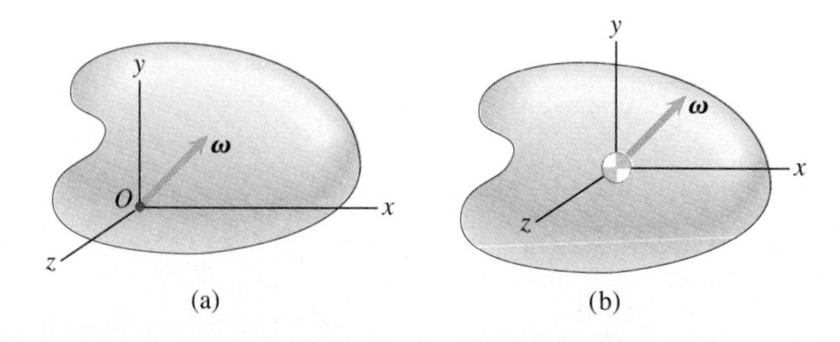

(a) (b)

Euler Equations

The equations governing three-dimensional motion of a rigid body include Newton's second law and equations of angular motion. For a rigid body rotating about a fixed point O (Figure (a)), the equations of angular motion are expressed in terms of the components of the total moment about O:

$$\begin{bmatrix} \Sigma M_{Ox} \\ \Sigma M_{Oy} \\ \Sigma M_{Oz} \end{bmatrix} = \begin{bmatrix} I_{xx} & -I_{xy} & -I_{xz} \\ -I_{yx} & I_{yy} & -I_{yz} \\ -I_{zx} & -I_{zy} & I_{zz} \end{bmatrix} \begin{bmatrix} d\omega_x/dt \\ d\omega_y/dt \\ d\omega_z/dt \end{bmatrix} + \begin{bmatrix} 0 & -\Omega_z & \Omega_y \\ \Omega_z & 0 & -\Omega_x \\ -\Omega_y & \Omega_x & 0 \end{bmatrix} \begin{bmatrix} I_{xx} & -I_{xy} & -I_{xz} \\ -I_{yx} & I_{yy} & -I_{yz} \\ -I_{zx} & -I_{zy} & I_{zz} \end{bmatrix} \begin{bmatrix} \omega_x \\ \omega_y \\ \omega_z \end{bmatrix} \qquad \textbf{Equation (9.25)}$$

where $\mathbf{\Omega}$ is the angular velocity of the coordinate system. If the coordinate system is body-fixed, $\mathbf{\Omega} = \boldsymbol{\omega}$. In the case of general three-dimensional motion (Figure (b)), the equations of angular motion are identical except that they are expressed in terms of the components of the total moment about the centre of mass.

The rigid body's angular acceleration is related to the derivatives of the components of $\boldsymbol{\omega}$ by

$$\alpha = \frac{d\boldsymbol{\omega}}{dt} = \frac{d\omega_x}{dt}\mathbf{i} + \frac{d\omega_y}{dt}\mathbf{j} + \frac{d\omega_z}{dt}\mathbf{k} + \mathbf{\Omega} \times \boldsymbol{\omega} \qquad \textbf{Equation (9.24)}$$

If the coordinate system does not rotate or is body-fixed, the terms $d\omega_x/dt$, $d\omega_y/dt$ and $d\omega_z/dt$ are the components of the angular acceleration.

Moments and Products of Inertia

In terms of a given coordinate system xyz, the **inertia matrix** of an object is defined by [Equation (9.11)]

$$[I] = \begin{bmatrix} I_{xx} & -I_{xy} & -I_{xz} \\ I_{yx} & I_{yy} & -I_{yz} \\ -I_{zx} & -I_{zy} & I_{zz} \end{bmatrix}$$

$$= \begin{bmatrix} \int_m (y^2 + z^2)\, dm & -\int_m xy\, dm & -\int_m xz\, dm \\ -\int_m yx\, dm & \int_m (x^2 + z^2)\, dm & -\int_m yz\, dm \\ -\int_m zx\, dm & -\int_m zy\, dm & \int_m (x^2 + y^2)\, dm \end{bmatrix}$$

where x, y and z are the coordinates of the differential element of mass dm. The terms I_{xx}, I_{yy} and I_{zz} are the moments of inertia about the x, y and z axes, and I_{xy}, I_{yz} and I_{zx} are the **products of inertia**.

If $x'y'z'$ is a coordinate system with its origin at the centre of mass of an object and xyz is a parallel system, the **parallel-axis theorems** state that

$$I_{xx} = I_{x'x'} + (d_y^2 + d_z^2)m$$

$$I_{yy} = I_{y'y'} + (d_x^2 + d_z^2)m$$

$$I_{zz} = I_{z'z'} + (d_x^2 + d_y^2)m$$

Equation (9.16)

$$I_{xy} = I_{x'y'} + d_x d_y m$$

$$I_{yz} = I_{y'z'} + d_y d_z m$$

$$I_{zx} = I_{z'x'} + d_z d_x m$$

where (d_x, d_y, d_z) are the coordinates of the centre of mass in the xyz coordinate system.

The moment of inertia about an axis through the origin parallel to a unit vector **e** is given by

$$I_O = I_{xx}e_x^2 + I_{yy}e_y^2 + I_{zz}e_z^2 - 2I_{xy}e_x e_y - 2I_{yz}e_y e_z - 2I_{zx}e_z e_x$$ **Equation (9.17)**

For any object and origin O, at least one coordinate system exists for which the products of inertia are zero. The coordinate axes are called **principal axes**, and the moments of inertia are called the **principal moments of inertia**. If the inertia matrix is known in terms of a coordinate system $x'y'z'$, the principal moments of inertia are roots of the cubic equation

$$I^3 - (I_{x'x'} + I_{y'y'} + I_{z'z'})I^2$$
$$+ (I_{x'x'}I_{y'y'} + I_{y'y'}I_{z'z'} + I_{z'z'}I_{x'x'} - I_{x'y'}^2 - I_{y'z'}^2 - I_{z'x'}^2)I$$ **Equation (9.19)**
$$- (I_{x'x'}I_{y'y'}I_{z'z'} - I_{x'x'}I_{y'z'}^2 - I_{y'y'}I_{x'z'}^2 - I_{z'z'}I_{x'y'}^2 - 2I_{x'y'}I_{y'z'}I_{z'x'}) = 0$$

For each principal moment of inertia I, the vector **V** with components

$$V_{x'} = (I_{y'y'} - I)(I_{z'z'} - I) - I_{y'z'}^2$$

$$V_{y'} = I_{x'y'}(I_{z'z'} - I) + I_{x'z'}I_{y'z'}$$ **Equation (9.20)**

$$V_{z'} = I_{x'z'}(I_{y'y'} - I) + I_{x'y'}I_{y'z'}$$

is parallel to the corresponding principal axis.

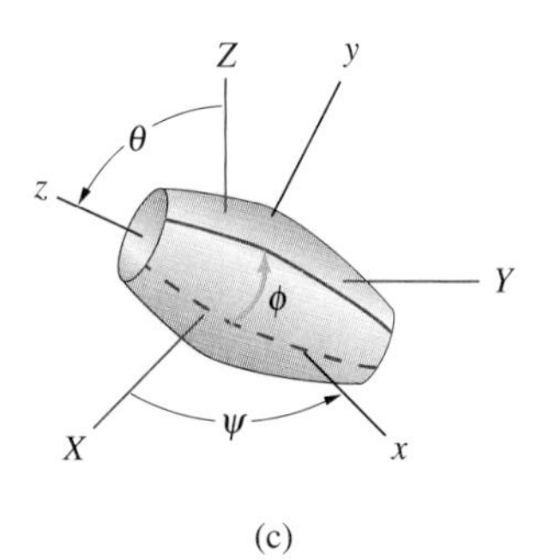

(c)

Eulerian Angles: Axisymmetric Objects

In the case of an object with an axis of rotational symmetry, the orientation of the xyz system relative to the reference XYZ system is specified by the **precession angle** ψ and the **nutation angle** θ (Figure (c)). The rotation of the object relative to the xyz system is specified by the **spin angle** ϕ.

The components of the rigid body's angular velocity relative to the XYZ system are given by

$$\omega_x = \dot{\theta}$$
$$\omega_y = \dot{\psi} \sin\theta \qquad \textbf{Equation (9.33)}$$
$$\omega_z = \dot{\phi} + \dot{\psi} \cos\theta$$

The equations of angular motion expressed in terms of these **Eulerian angles** are [Equations (9.36)–(9.38)]:

$$\Sigma M_x = I_{xx}\ddot{\theta} + (I_{zz} - I_{xx})\dot{\psi}^2 \sin\theta \cos\theta + I_{zz}\dot{\phi}\dot{\psi} \sin\theta$$
$$\Sigma M_y = I_{xx}(\ddot{\psi} \sin\theta + 2\dot{\psi}\dot{\theta} \cos\theta) - I_{zz}(\dot{\phi}\dot{\theta} + \dot{\psi}\dot{\theta} \cos\theta)$$
$$\Sigma M_z = I_{zz}(\ddot{\phi} + \ddot{\psi} \cos\theta - \dot{\psi}\dot{\theta} \sin\theta)$$

In **steady precession** of an axisymmetric spinning object, the **spin rate** $\dot{\phi}$, the **nutation angle** θ and the **precession rate** $\dot{\psi}$ are assumed to be constant. With these assumptions, the equations of angular motion reduce to [Equations (9.39)–(9.41)]

$$\Sigma M_x = (I_{zz} - I_{xx})\dot{\psi}^2 \sin\theta \cos\theta + I_{zz}\dot{\phi}\dot{\psi} \sin\theta$$
$$\Sigma M_y = 0$$
$$\Sigma M_z = 0$$

Eulerian Angles: Arbitrary Objects

In the case of an arbitrary object, the orientation of the body-fixed xyz system relative to the reference XYZ system is specified by the precession angle ψ, the nutation angle θ and the spin angle ϕ (Figure (d)).

The components of the rigid body's angular velocity relative to the XYZ system are given by

$$\omega_x = \dot{\psi} \sin\theta \sin\phi + \dot{\theta} \cos\phi$$
$$\omega_y = \dot{\psi} \sin\cos\phi - \dot{\theta} \sin\phi \qquad \textbf{Equation (9.44)}$$
$$\omega_z = \dot{\psi} \cos\theta + \dot{\phi}$$

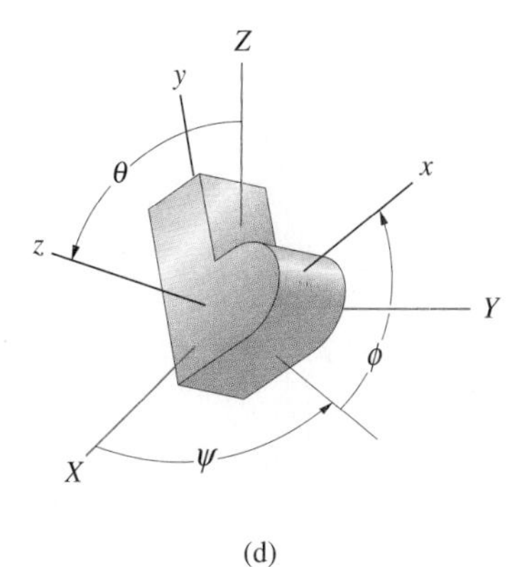

(d)

If xyz is a set of principal axes, the equations of angular motion in terms of Eulerian angles are given by Equations (9.46).

Review Problems

9.92 The slender bar of length l and mass m is pinned to the L-shaped bar at O. The L-shaped bar rotates about the vertical axis with a constant angular velocity ω_0. Determine the value of ω_0 necessary for the bar to remain at a constant angle β relative to the vertical.

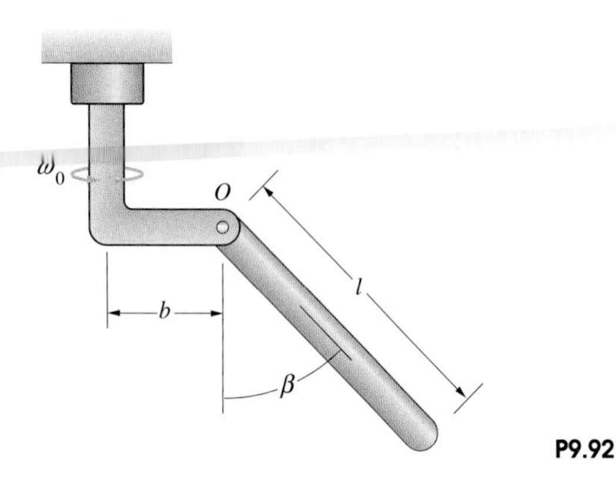

P9.92

9.93 A slender bar of length l and mass m rigidly attached to the centre of a thin circular disc of radius R and mass m. The composite object undergoes a motion in which the bar rotates in the horizontal plane with constant angular velocity ω_0 about the centre of mass of the composite object and the disc rolls on the floor. Show that $\omega_0 = 2\sqrt{g/R}$.

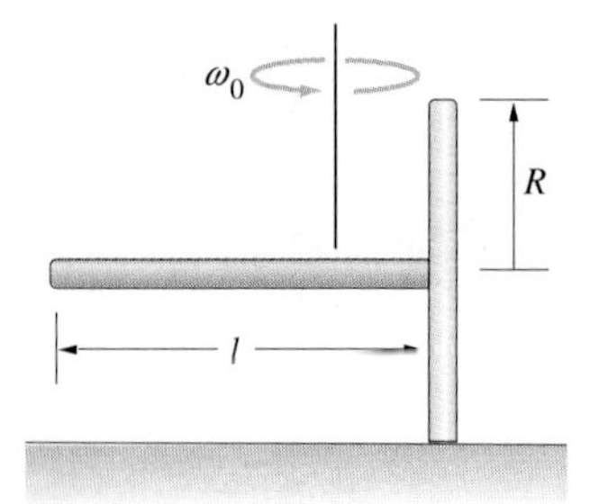

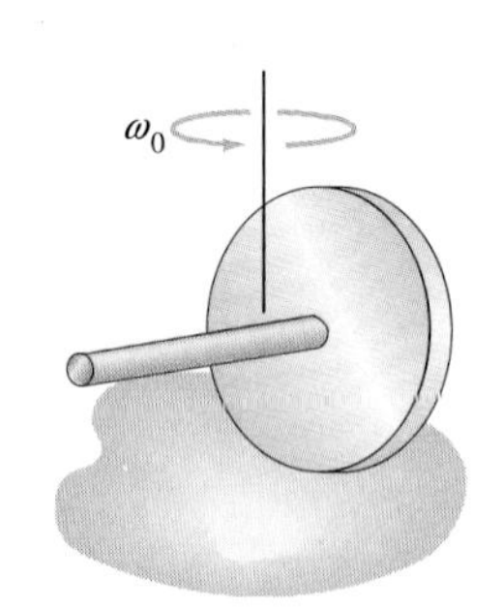

P9.93

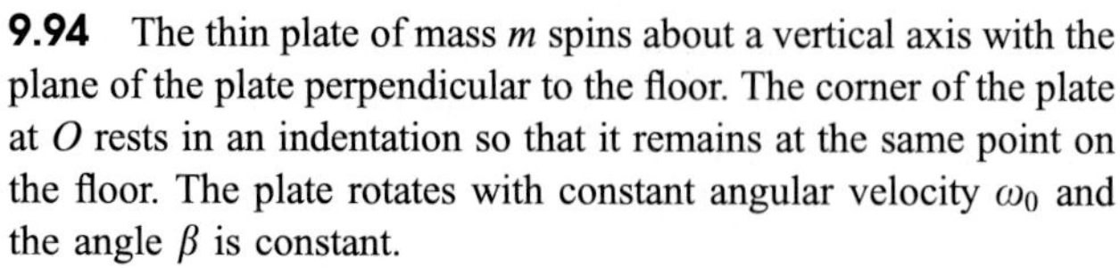

9.94 The thin plate of mass m spins about a vertical axis with the plane of the plate perpendicular to the floor. The corner of the plate at O rests in an indentation so that it remains at the same point on the floor. The plate rotates with constant angular velocity ω_0 and the angle β is constant.

(a) Show that the angular velocity ω_0 is related to the angle β by

$$\frac{h\omega_0^2}{g} = \frac{2\cos\beta - \sin\beta}{\sin^2\beta - 2\sin\beta\cos\beta - \cos^2\beta}$$

(b) The equation you obtained in part (a) indicates that $\omega_0 = 0$ when $2\cos\beta - \sin\beta = 0$. What is the interpretation of this result?

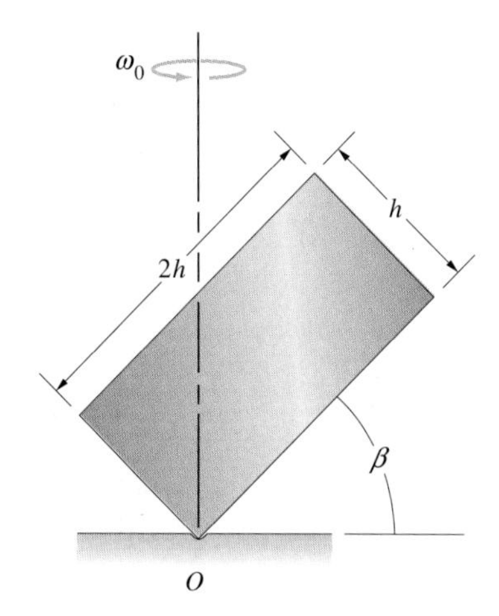

P9.94

9.95 In Problem 9.94, determine the range of values of the angle β for which the plate will remain in the steady motion described.

9.96 Arm BC has a mass of 12 kg, and its moments and products of inertia in terms of the coordinate system shown are $I_{xx} = 0.03\,\text{kg.m}^2$, $I_{yy} = I_{zz} = 4\,\text{kg.m}^2$, $I_{xy} = I_{yz} = I_{xz} = 0$. At the instant shown, arm AB is rotating in the horizontal plane with a constant angular velocity of 1 rad/s in the counterclockwise direction viewed from above. Relative to arm AB, arm BC is rotating about the z axis with a constant angular velocity of 2 rad/s. Determine the force and couple exerted on arm BC at B.

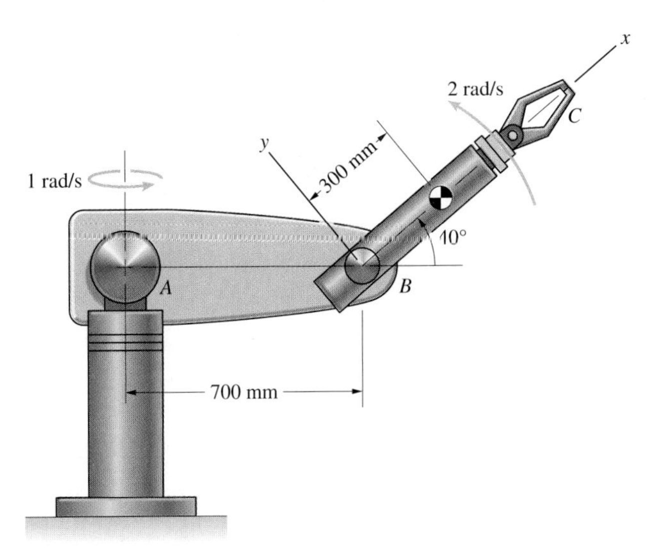

P9.96

9.97 Suppose that you throw a football in a wobbly spiral with a nutation angle of 25°. The football's moments of inertia are $I_{xx} = I_{yy} = 0.003\,\text{kg.m}^2$ and $I_{zz} = 0.001\,\text{kg.m}^2$. If the spin rate is $\dot{\phi} = 4$ revolutions per second, what is the magnitude of the precession rate (the rate at which it wobbles)?

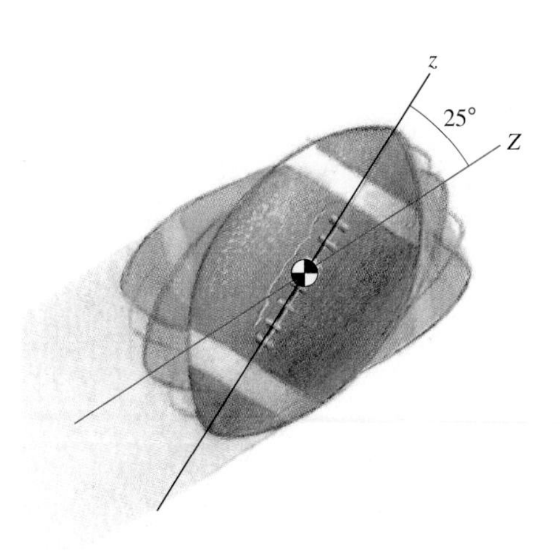

P9.97

9.98 Sketch the body and space cones for the motion of the football in Problem 9.97.

9.99 The mass of the homogeneous thin plate is 1 kg. For a coordinate system with its origin at O, determine the principal moments of inertia and the directions of unit vectors parallel to the corresponding principal axes.

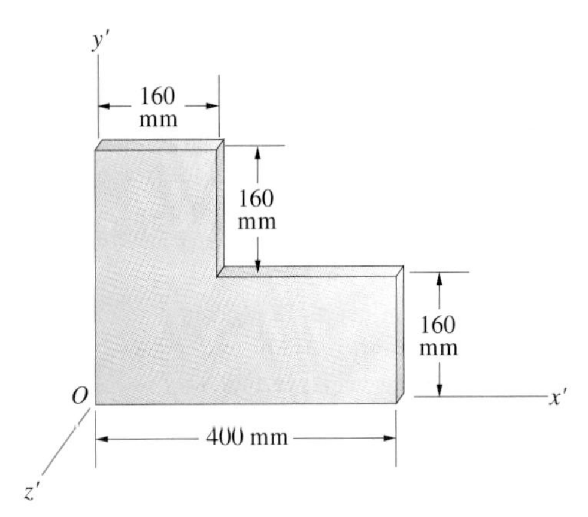

P9.99

9.100 The aeroplane's principal moments of inertia in kg.m^2 are $I_{xx} = 8000$, $I_{yy} = 48\,000$ and $I_{zz} = 50\,000$.
(a) The aeroplane begins in the reference position shown and manoeuvres into the orientation $\psi = \theta = \phi = 45°$. Draw a sketch showing its orientation relative to the XYZ system.
(b) If the aeroplane is in the orientation described in part (a), the rates of change of the Eulerian angles are $\dot{\psi} = 0$, $\dot{\theta} = 0.2$ rad/s and $\dot{\phi} = 0.2$ rad/s, and their second time derivatives are zero, what are the components of the total moment about the aeroplane's centre of mass?

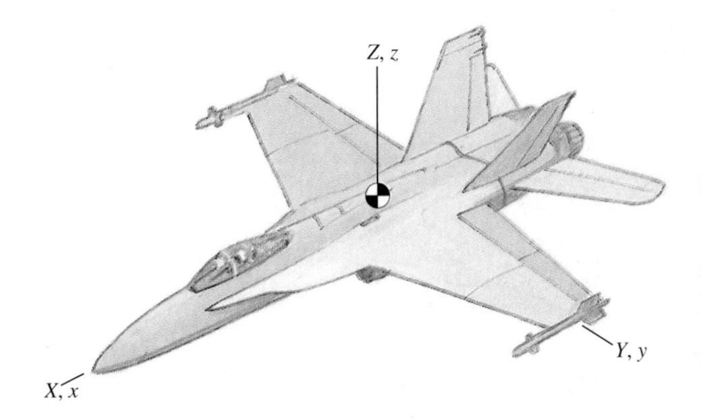

P9.100

9.101 What are the x, y and z components of the angular acceleration of the aeroplane described in Problem 9.100?

9.102 If the orientation of the aeroplane in Problem 9.100 is $\psi = 45°$, $\theta = 60°$, $\phi = 45°$, the rates of change of the Eulerian angles are $\dot{\psi} = 0$, $\dot{\theta} = 0.2$ rad/s and $\dot{\phi} = 0.1$ rad/s, and the components of the total moment about the centre of mass are $\Sigma M_x = 400$ N.m, $\Sigma M_y = 1200$ N.m and $\Sigma M_z = 0$, what are the x, y and z components of the aeroplane's angular acceleration?

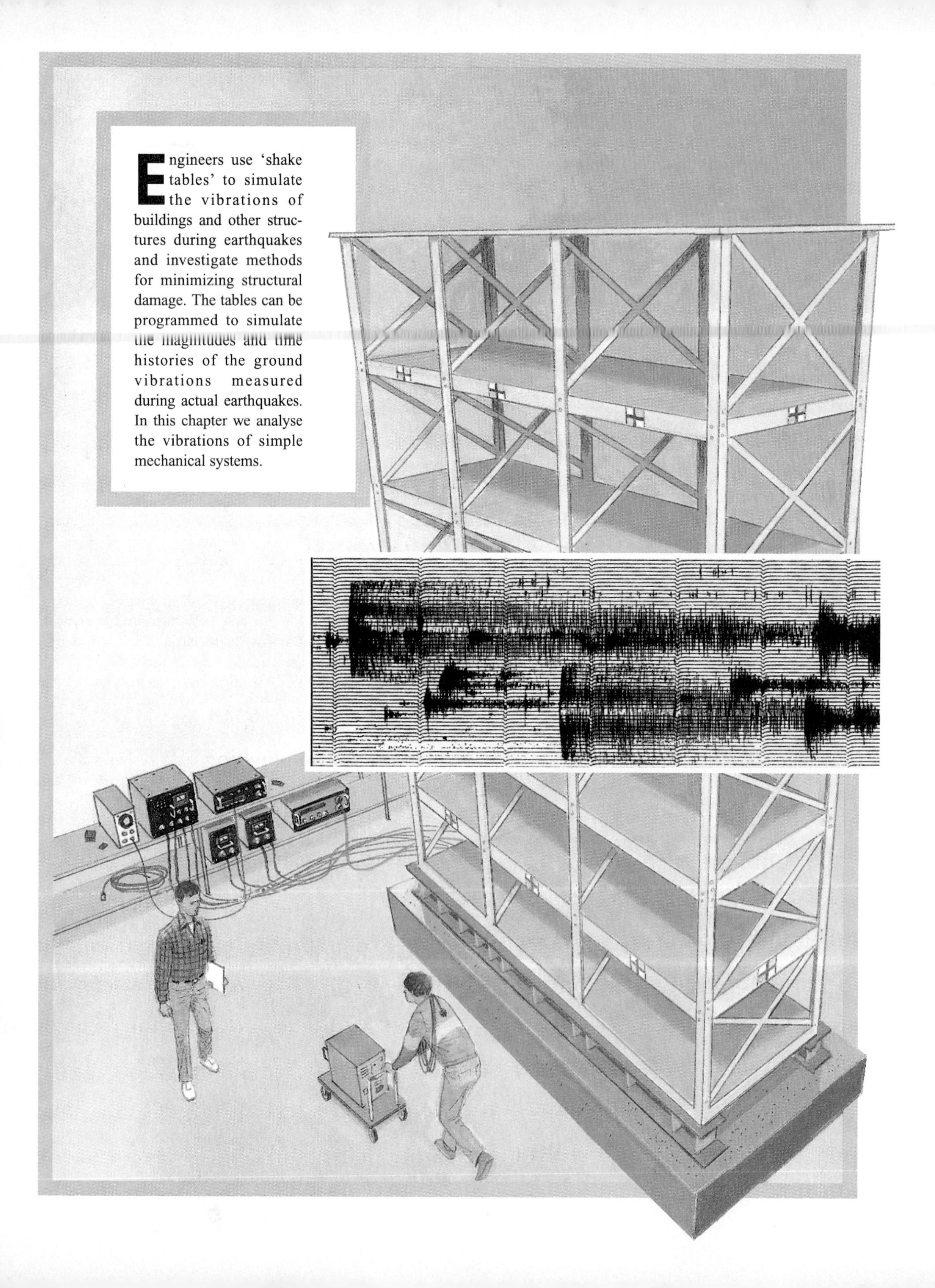

Engineers use 'shake tables' to simulate the vibrations of buildings and other structures during earthquakes and investigate methods for minimizing structural damage. The tables can be programmed to simulate the magnitudes and time histories of the ground vibrations measured during actual earthquakes. In this chapter we analyse the vibrations of simple mechanical systems.

Chapter 10

Vibrations

VIBRATIONS have been of concern in engineering at least since the beginning of the industrial revolution. The oscillatory motions of rotating and reciprocating engines subject their parts to large loads that must be considered in their design. Operators and passengers of vehicles powered by these engines must be isolated from their vibrations. Beginning with the development of electromechanical devices capable of creating and measuring mechanical vibrations, such as loudspeakers and microphones, engineering applications of vibrations have included the various areas of acoustics, from architectural acoustics to earthquake detection and analysis.

In this chapter we consider vibrating systems with one degree of freedom; that is, the position, or configuration, of each system is specified by a single variable. Many actual vibrating systems either have only one degree of freedom or their motions can be modelled by a one-degree-of-freedom system in particular circumstances. We discuss fundamental concepts, including amplitude, frequency, period, damping and resonance, that are also used in the analysis of systems with multiple degrees of freedom.

10.1 Conservative Systems

We begin by presenting different examples of one-degree-of-freedom systems subjected to conservative forces, demonstrating that their motions are described by the same differential equation. We then examine solutions of this equation and use them to describe the vibrations of one-degree-of-freedom conservative systems.

Examples

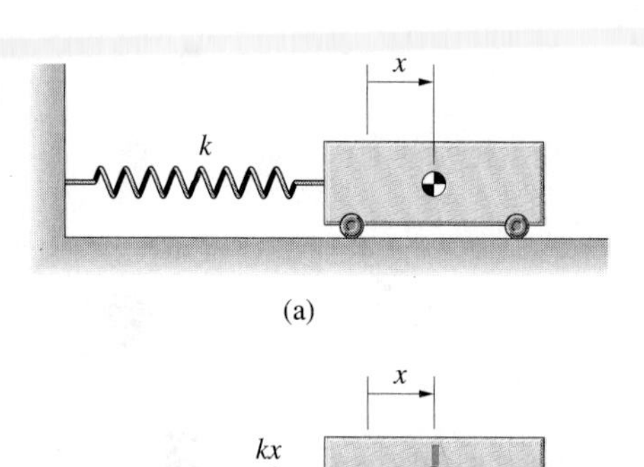

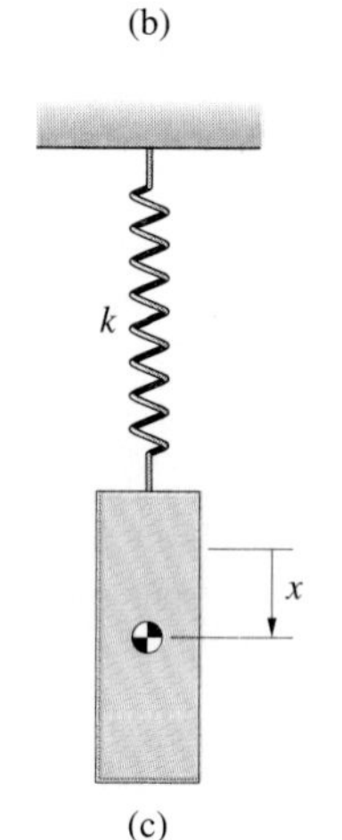

Figure 10.1
(a) The spring-mass oscillator has one degree of freedom.
(b) Free-body diagram of the mass.
(c) Suspending the mass.

The spring-mass oscillator (Figure 10.1(a)) is the simplest example of a one-degree-of-freedom vibrating system. A single coordinate x measuring the displacement of the mass relative to a reference point is sufficient to specify the position of the system. We draw the free-body diagram of the mass in Figure 10.1(b), neglecting friction and assuming that the spring is unstretched when $x=0$. Applying Newton's second law, we can write the equation describing horizontal motion of the mass as

$$\frac{d^2x}{dt^2}+\frac{k}{m}x=0 \tag{10.1}$$

We can also obtain this equation by using a different method that you will find very useful. The only force that does work on the mass, the force exerted by the spring, is conservative, which means that the sum of the kinetic and potential energies is constant:

$$\frac{1}{2}m\left(\frac{dx}{dt}\right)^2+\frac{1}{2}kx^2=\text{constant}$$

Taking the time derivative of this equation, we can write the resulting equation as

$$\left(\frac{dx}{dt}\right)\left(\frac{d^2x}{dt^2}+\frac{k}{m}x\right)=0$$

again obtaining Equation (10.1).

Suppose that the mass is suspended from the spring, as shown in Figure 10.1(c), and it undergoes vertical motion. If the spring is unstretched when $x=0$, you can easily confirm that the equation of motion is

$$\frac{d^2x}{dt^2}+\frac{k}{m}x=g$$

If the suspended mass is stationary, the magnitude of the force exerted by the spring must equal the weight, $kx=mg$, so the equilibrium position is $x=mg/k$. (Notice that we can also determine the equilibrium position by setting the acceleration equal to zero in the equation of motion.) Let us introduce a new variable $\tilde{x}$ that measures the position of the mass relative to its equilibrium position: $\tilde{x} = x - mg/k$. Writing the equation of motion in terms of this variable, we obtain

$$\frac{d^2\tilde{x}}{dt^2} + \frac{k}{m}\tilde{x} = 0$$

which is identical to Equation (10.1). The vertical motion of the mass in Figure 10.1(c) *relative to its equilibrium position* is described by the same equation that describes the horizontal motion of the mass in Figure 10.1(a) relative to its equilibrium position.

Let's consider a different one-degree-of-freedom system. If we rotate the slender bar in Figure 10.2(a) through some angle and release it, it will oscillate back and forth. (An object swinging from a fixed point is called a **pendulum**.) There is only one degree of freedom, since θ specifies the bar's position.

Drawing the free-body diagram of the bar (Figure 10.2(b)) and writing the equation of angular motion about A, we obtain

$$\frac{d^2\theta}{dt^2} + \frac{3g}{2l}\sin\theta = 0 \qquad (10.2)$$

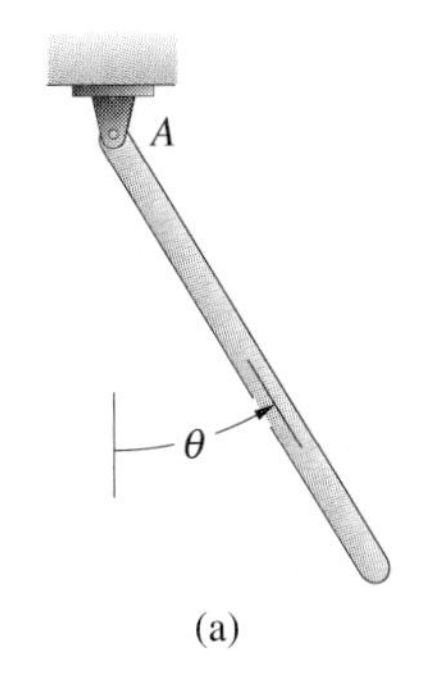

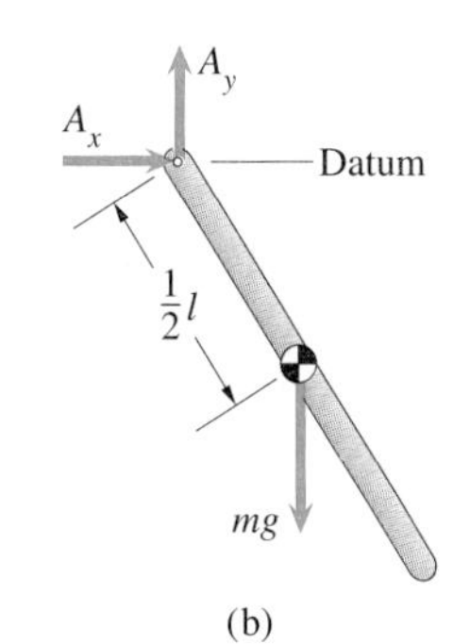

Figure 10.2
(a) A pendulum consisting of a slender bar.
(b) Free-body diagram of the bar.

We can also obtain this equation by using conservation of energy. The bar's kinetic energy is $T = \frac{1}{2}I_A(d\theta/dt)^2$. If we place the datum at the level of point A (Figure 10.2(b)), the potential energy associated with the bar's weight is $V = -mg(\frac{1}{2}l\cos\theta)$, so

$$T + V = \frac{1}{2}\left(\frac{1}{3}ml^2\right)\left(\frac{d\theta}{dt}\right)^2 - \frac{1}{2}mgl\cos\theta = \text{constant}$$

Taking the time derivative of this equation and writing the result in the form

$$\left(\frac{d\theta}{dt}\right)\left(\frac{d^2\theta}{dt^2} + \frac{3g}{2l}\sin\theta\right) = 0$$

we obtain Equation (10.2).

Equation (10.2) does not have the same form as Equation (10.1). However, if we express $\sin\theta$ in terms of its Taylor series,

$$\sin\theta = \theta - \frac{1}{6}\theta^3 + \frac{1}{120}\theta^5 + \cdots$$

and assume that θ remains small enough to approximate $\sin\theta$ by θ, then Equation (10.2) becomes identical in form to Equation (10.1):

$$\frac{d^2\theta}{dt^2} + \frac{3g}{2l}\theta = 0 \tag{10.3}$$

Our analyses of the spring-mass oscillator and pendulum resulted in equations of motion that are identical in form. To accomplish this in the case of the suspended spring-mass oscillator, we had to express the equation of motion in terms of displacement relative to the equilibrium position. In the case of the pendulum, we needed to assume that the motions are small. But within those restrictions, you will see that the equation we obtained describes the motions of many one-degree-of-freedom conservative systems.

Solutions

Let us consider the differential equation

$$\boxed{\frac{d^2x}{dt^2} + \omega^2 x = 0} \tag{10.4}$$

where ω is a constant. You have seen that with $\omega^2 = k/m$, this equation describes the motion of a spring-mass oscillator, and with $\omega^2 = 3g/2l$, it describes small motions of a suspended slender bar. Equation (10.4) is an **ordinary differential equation**, because it is expressed in terms of ordinary (not partial) derivatives of the dependent variable x with respect to the independent variable t. It is **linear**, meaning there are no non-linear terms in x or its derivatives, and it is **homogeneous**, meaning that each term contains x or one of its derivatives. The general solution of this differential equation is

$$\boxed{x = A \sin \omega t + B \cos \omega t} \tag{10.5}$$

where A and B are arbitrary constants.

Although in practical problems you will usually find Equation (10.5) to be the most convenient form of the solution of Equation (10.4), we can describe the properties of the solution more easily by expressing it in the alternative form

$$x = E \sin(\omega t - \phi) \tag{10.6}$$

where E and ϕ are constants. To show that these two solutions are equivalent, we can use the identity

$$\begin{aligned} E \sin(\omega t - \phi) &= E(\sin \omega t \cos \phi - \cos \omega t \sin \phi) \\ &= (E \cos \phi) \sin \omega t + (-E \sin \phi) \cos \omega t \end{aligned}$$

This expression is identical to Equation (10.5) if the constants A and B are related to E and ϕ by

$$A = E \cos \phi \qquad B = -E \sin \phi \tag{10.7}$$

Equation (10.6) clearly demonstrates the oscillatory nature of the solution of Equation (10.4). Called **simple harmonic motion**, it describes a sinusoidal function of ωt. The constant ϕ determines the horizontal placement of the sinusoidal function relative to the origin $\omega t = 0$, which is called the **phase**. We define ϕ to be the distance to the right of $\omega t = 0$ at which the solution first crosses the horizontal axis with positive slope (Figure 10.3). The positive constant E is called the **amplitude** of the vibration. By squaring Equations (10.7) and adding them, we obtain a relation between the amplitude and the constants A and B:

$$E = \sqrt{A^2 + B^2} \tag{10.8}$$

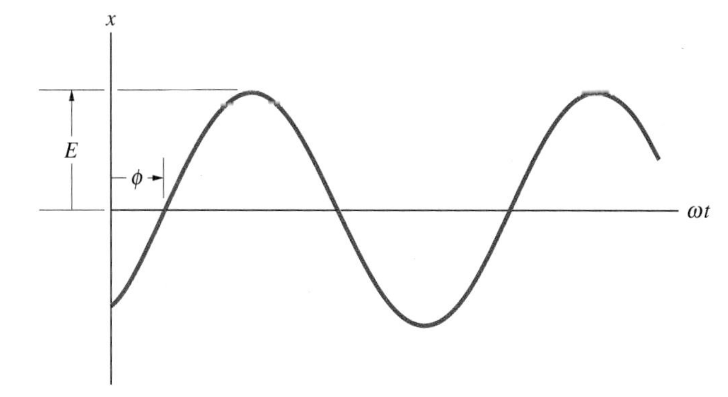

Figure 10.3
Graph of x as a function of ωt.

We can interpret Equation (10.6) in terms of the uniform motion of a point along a circular path. We draw a circle whose radius equals the amplitude (Figure 10.4) and assume that the line from O to P rotates in the counter-clockwise direction with constant angular velocity ω. If we choose the position of P at $t = 0$ as shown, the projection of the line OP onto the vertical axis is $E \sin(\omega t - \phi)$.

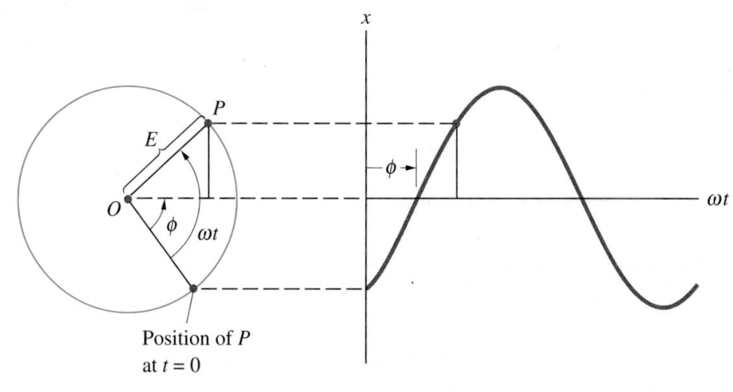

Figure 10.4
Correspondence of simple harmonic motion with circular motion of a point.

Thus there is a one-to-one correspondence between the circular motion of P and Equation (10.6). Point P makes one complete revolution, or **cycle**, during the time required for the angle ωt to increase by 2π radians. The time $\tau = 2\pi/\omega$ required for one cycle is called the **period** of the vibration. Since τ is the time required for one cycle, its inverse $f = 1/\tau$ is the number of cycles per unit time, or **natural frequency** of the vibration. The frequency is usually expressed in cycles per second, or hertz (Hz). We illustrate the effect of changing the period and frequency in Figure 10.5.

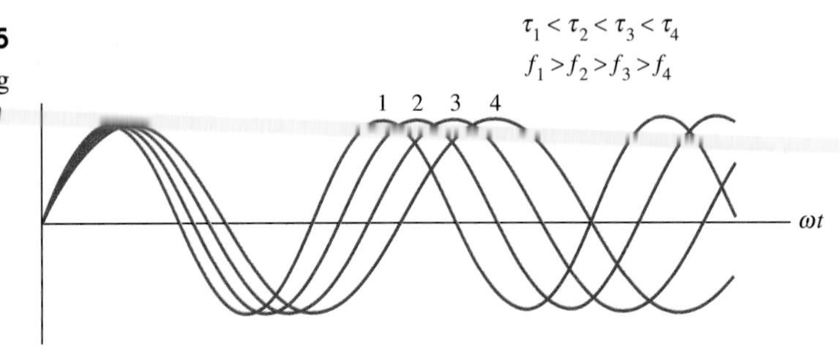

Figure 10.5
Effect of increasing the period (decreasing the frequency) of simple harmonic motion

In summary, the period and natural frequency are

$$\tau = \frac{2\pi}{\omega} \tag{10.9}$$

$$f = \frac{\omega}{2\pi} \tag{10.10}$$

A system's period and natural frequency are determined by its physical properties and do not depend on the functional form in which its motion is expressed.

The natural frequency f is the number of revolutions the point P moves around the circular path in Figure 10.4 per unit time, so $\omega = 2\pi f$ is the number of radians per unit time. Therefore ω is also a measure of the frequency and is expressed in rad/s. To distinguish it from f, ω is called the **circular natural frequency**. The term ωt, and the variable ϕ that specifies the phase, can be specified in either radians or degrees.

Suppose that Equation (10.6) describes the displacement of the spring-mass oscillator in Figure 10.1(a), so that $\omega^2 = k/m$. The kinetic energy of the mass is

$$T = \frac{1}{2}m\left(\frac{dx}{dt}\right)^2 = \frac{1}{2}mE^2\omega^2 \cos^2(\omega t - \phi)$$

and the potential energy of the spring is

$$V = \frac{1}{2}kx^2 = \frac{1}{2}mE^2\omega^2 \sin^2(\omega t - \phi)$$

The sum of the kinetic and potential energies, $T + V = \frac{1}{2}mE^2\omega^2$, is constant (Figure 10.6). As the system vibrates, its total energy oscillates between kinetic and potential energy. Notice that the total energy is proportional to the square of the amplitude and the square of the natural frequency.

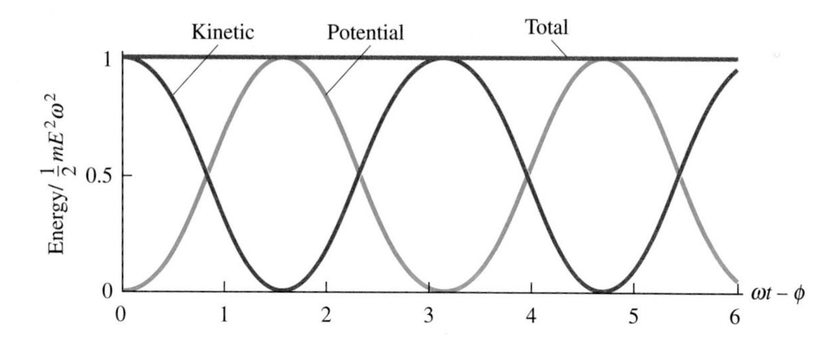

Figure 10.6

Kinetic, potential and total energies of a spring-mass oscillator.

In the following examples we analyse one-degree-of-freedom vibrating systems. Your first objective will be to determine the equation of motion of the system and express it in the form of Equation (10.4):

$$\frac{d^2x}{dt^2} + \omega^2 x = 0$$

To do so, you must write the equation of motion in terms of the displacement of the system relative to its equilibrium position. You may also need to linearize the equation by assuming that the displacement is small, as we did in obtaining Equation (10.3) from Equation (10.2). Once you have the equation of motion in this form, you know the value of ω for the system and can use it to obtain the period and natural frequency from Equations (10.9) and (10.10). You can also determine the motion of the system from Equation (10.5) or Equation (10.6) if you are given sufficient information to determine the arbitrary constants.

Example 10.1

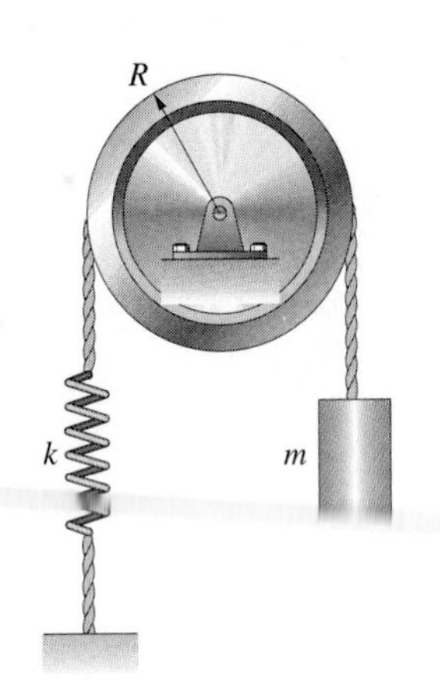

Figure 10.7

The pulley in Figure 10.7 has radius R and moment of inertia I, and the cable does not slip relative to the pulley. The mass m is displaced downwards a distance h from its equilibrium position and released from rest at $t = 0$.
(a) What is the natural frequency of the resulting vibrations?
(b) Determine the position of the mass relative to its equilibrium position as a function of time.

STRATEGY

A single coordinate specifying the vertical displacement of the mass specifies the positions of both the mass and pulley, so there is one degree of freedom. We obtain the equation of motion of the system both by writing the individual equations of motion of the mass and pulley and by using conservation of energy.

SOLUTION

Let x be the downward displacement of the mass relative to its position when the spring is unstretched. We draw the free-body diagrams of the pulley and mass in Figure (a), where T_C is the tension in the cable and α is the angular acceleration of the pulley. Applying Newton's second law to the mass, we obtain

$$mg - T_C = m\frac{d^2x}{dt^2} \qquad (10.11)$$

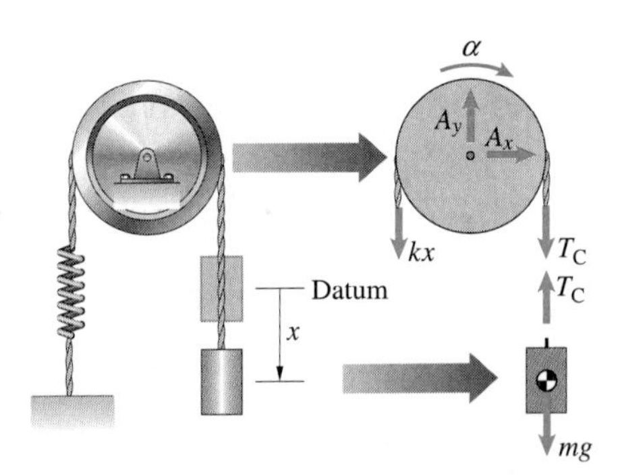

(a) Free-body diagrams of the pulley and mass.

The equation of angular motion for the pulley is

$$T_C R - (kx)R = I\alpha \qquad (10.12)$$

The angular acceleration of the pulley and the acceleration of the mass are related by $\alpha = (d^2x/dt^2)/R$, so we can write Equation (10.12) as

$$T_C - kx = (I/R^2)\frac{d^2x}{dt^2}$$

Summing this equation and Equation (10.11), we obtain the equation of motion

$$(m + I/R^2)\frac{d^2x}{dt^2} + kx = mg \qquad (10.13)$$

Alternative Solution In terms of the velocity of the mass, the angular velocity of the pulley is $(dx/dt)/R$. Therefore we can write the total kinetic energy of the mass and pulley as

$$T = \frac{1}{2}m\left(\frac{dx}{dt}\right)^2 + \frac{1}{2}I\left[\frac{1}{R}\left(\frac{dx}{dt}\right)\right]^2$$

Placing the datum for the potential energy associated with the weight of the mass at $x = 0$, the total potential energy is

$$V = -mgx + \frac{1}{2}kx^2$$

The sum of the kinetic and potential energies is constant:

$$T + V = \frac{1}{2}(m + I/R^2)\left(\frac{dx}{dt}\right)^2 - mgx + \frac{1}{2}kx^2 = \text{constant}$$

Taking the time derivative of this equation, we again obtain Equation (10.13).

By setting $d^2x/dt^2 = 0$ in Equation (10.13), we see that the equilibrium position is $x = mg/k$. By expressing Equation (10.13) in terms of a new variable $\tilde{x} = x - mg/k$ that measures the position of the mass relative to its equilibrium position, we obtain the equation

$$\frac{d^2\tilde{x}}{dt^2} + \omega^2\tilde{x} = 0$$

where

$$\omega^2 = \frac{k}{m + I/R^2}$$

(a) The natural frequency of vibration of the system is

$$f = \frac{\omega}{2\pi} = \frac{1}{2\pi}\sqrt{\frac{k}{m + I/R^2}}$$

(b) From Equation (10.5), we can write the general solution for $\tilde{x}$ in the form

$$\tilde{x} = A \sin \omega t + B \cos \omega t$$

When $t = 0$, $\tilde{x} = h$ and $d\tilde{x}/dt = 0$. The derivative of the general solution is

$$\frac{d\tilde{x}}{dt} = A\omega \cos \omega t - B\omega \sin \omega t$$

The initial conditions yield the equations

$$h = B \qquad 0 = A\omega$$

so the position of the mass relative to its equilibrium position is

$$\tilde{x} = h \cos \omega t$$

Example 10.2

The spring attached to the slender bar of mass m in Figure 10.8 is unstretched when $\theta = 0$. Neglecting friction, determine the natural frequency of small vibrations of the bar relative to its equilibrium position.

Figure 10.8

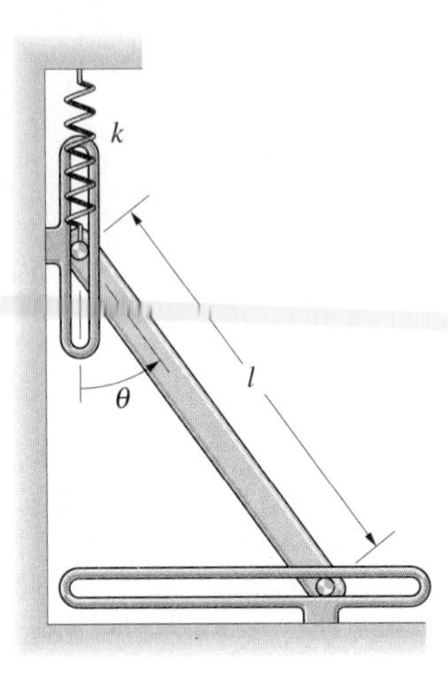

STRATEGY

The angle θ specifies the bar's position, so there is one degree of freedom. We can express the kinetic and potential energies in terms of θ and its time derivative and take the time derivative of the total energy to obtain the equation of motion.

SOLUTION

The kinetic energy of the bar is

$$T = \frac{1}{2}mv^2 + \frac{1}{2}I\left(\frac{d\theta}{dt}\right)^2$$

where v is the velocity of the centre of mass and $I = \frac{1}{12}ml^2$. The distance from the bar's instantaneous centre to its centre of mass is $\frac{1}{2}l$ (Figure (a)), so $v = (\frac{1}{2}l)(d\theta/dt)$, and the kinetic energy is

$$T = \frac{1}{2}m\left[\frac{1}{2}l\left(\frac{d\theta}{dt}\right)\right]^2 + \frac{1}{2}\left(\frac{1}{12}ml^2\right)\left(\frac{d\theta}{dt}\right)^2 = \frac{1}{6}ml^2\left(\frac{d\theta}{dt}\right)^2$$

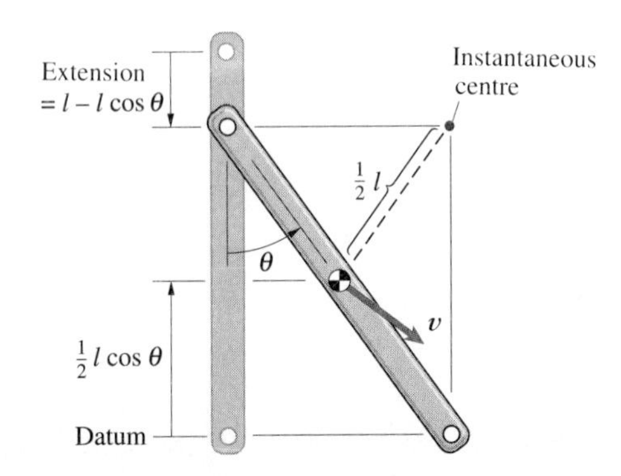

(a) Determining the velocity of the centre of mass, the extension of the spring, and the height of the centre of mass above the datum.

In terms of θ, the extension of the spring is $l - l\cos\theta$. We place the datum for the potential energy associated with the weight at the bottom of the bar (Figure (a)), so the total potential energy is

$$V = mg\left(\frac{1}{2}l\cos\theta\right) + \frac{1}{2}k(l - l\cos\theta)^2$$

The sum of the kinetic and potential energies is constant:

$$T + V = \frac{1}{6}ml^2\left(\frac{d\theta}{dt}\right)^2 + \frac{1}{2}mgl\cos\theta + \frac{1}{2}kl^2(1-\cos\theta)^2 = \text{constant}$$

Taking the time derivative of this equation, we obtain the equation of motion:

$$\frac{1}{3}ml^2\frac{d^2\theta}{dt^2} - \frac{1}{2}mgl\sin\theta + kl^2(1-\cos\theta)\sin\theta = 0 \qquad (10.14)$$

To express this equation in the form of Equation (10.4), we need to write it in terms of small vibrations relative to the equilibrium position. Let θ_e be the value of θ when the bar is in equilibrium. By setting $d^2\theta/dt^2 = 0$ in Equation (10.14), we find that θ_e must satisfy the relation

$$\cos\theta_e = 1 - \frac{mg}{2kl} \qquad (10.15)$$

We define $\tilde{\theta} = \theta - \theta_e$, and expand $\sin\theta$ and $\cos\theta$ in Taylor series in terms of $\tilde{\theta}$:

$$\sin\theta = \sin(\theta_e + \tilde{\theta}) = \sin\theta_e + \cos\theta_e\tilde{\theta} + \cdots$$

$$\cos\theta = \cos(\theta_e + \tilde{\theta}) = \cos\theta_e - \sin\theta_e\tilde{\theta} + \cdots$$

Substituting these expressions into Equation (10.14), neglecting terms in $\tilde{\theta}$ of second and higher orders, and using Equation (10.15), we obtain

$$\frac{d^2\tilde{\theta}}{dt^2} + \omega^2\tilde{\theta} = 0$$

where

$$\omega^2 = \frac{3g}{l}\left(1 - \frac{mg}{4kl}\right)$$

From Equation (10.10), the natural frequency of small vibrations of the bar is

$$f = \frac{\omega}{2\pi} = \frac{1}{2\pi}\sqrt{\frac{3g}{l}\left(1 - \frac{mg}{4kl}\right)}$$

Problems

10.1 Confirm that $x = A \sin \omega t + B \cos \omega t$, where A and B are arbitrary constants, satisfies Equation (10.4).

10.2 Confirm that $x = E \sin(\omega t - \phi)$, where E and ϕ are arbitrary constants, satisfies Equation (10.4).

10.3 (a) Show that $x = G \cos(\omega t - \psi)$, where G and ψ are arbitrary constants, satisfies Equation (10.4). (b) Determine the constants A and B in the form of the solution given by Equation (10.5) in terms of the constants G and ψ.

10.4 The position of a vibrating system is

$$x = [(1/\sqrt{2}) \sin \omega t - (1/\sqrt{2}) \cos \omega t] \text{ m}$$

(a) Determine the amplitude E of the vibration and the angle ϕ in degrees.
(b) Draw a sketch of x for values of ωt from zero to 4π radians, showing the angle ϕ.

10.5 The position of a vibrating system is

$$x = [-\sqrt{2} \sin \omega t + \sqrt{2} \cos \omega t] \text{ m}$$

(a) Determine the amplitude E of the vibration and the angle ϕ in degrees.
(b) Draw a sketch of x for values of ωt from zero to 4π radians, showing the angle ϕ.

10.6 The mass $m = 10$ kg and $k = 90$ N/m. The coordinate x measures the displacement of the mass relative to its equilibrium position. At $t = 0$, the mass is released from rest in the position $x = 0.1$ m.
(a) Determine the period and natural frequency of the resulting vibrations.
(b) Determine x as a function of time.

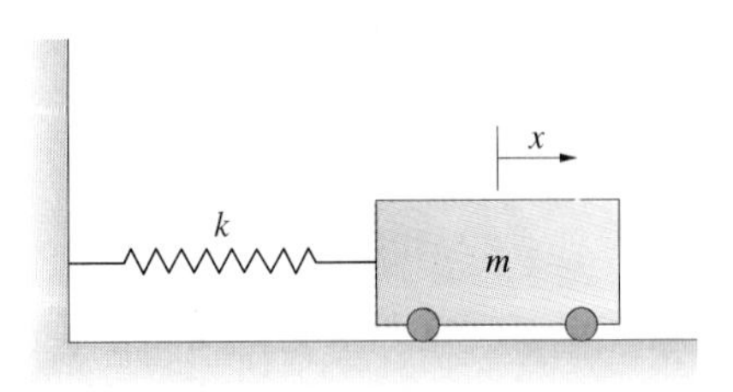

P10.6

10.7 The suspended object weighs 130 N and $k = 300$ N/m. At $t = 0$, the displacement of the object *relative to its equilibrium position* is $\tilde{x} = 75$ mm and it is moving downwards at 450 mm/s.
(a) Determine the period and natural frequency of the resulting vibrations.
(b) Determine $\tilde{x}$ as a function of time.
(c) Draw a graph of $\tilde{x}$ as a function of time from $t = 0$ to $t = 3$ s.

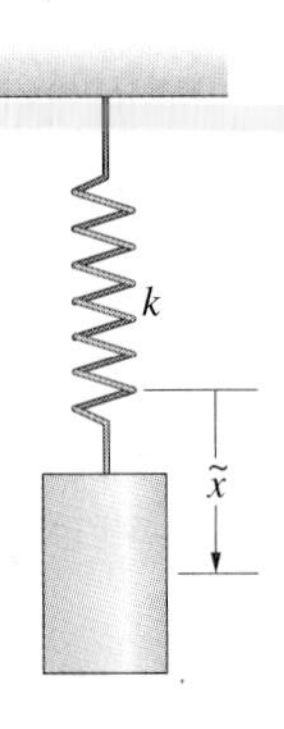

P10.7

10.8 The coordinate x measures the displacement of the mass relative to the position in which the spring is unstretched. The mass is given the initial conditions

$$t = 0 \begin{cases} x = 0.1 \text{ m} \\ \dfrac{dx}{dt} = 0 \end{cases}$$

(a) Determine the position of the mass as a function of time.
(b) Draw graphs of the position and velocity of the mass as functions of time for the first 5 s of motion.

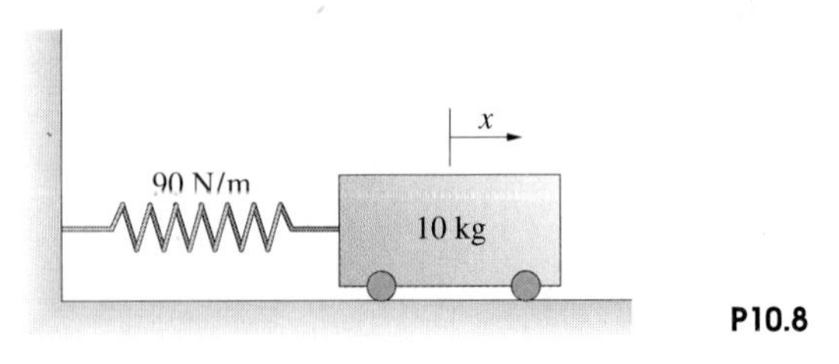

P10.8

10.9 When $t = 0$, the mass in Problem 10.8 is in the position in which the spring is unstretched and has a velocity of 0.3 m/s to the right. Determine the position of the mass as a function of time and the amplitude of the vibration: (a) by expressing the solution in the form given by (10.5); (b) by expressing the solution in the form given by (10.6).

10.10 Determine the natural frequency of vibration of the mass relative to its equilibrium position.

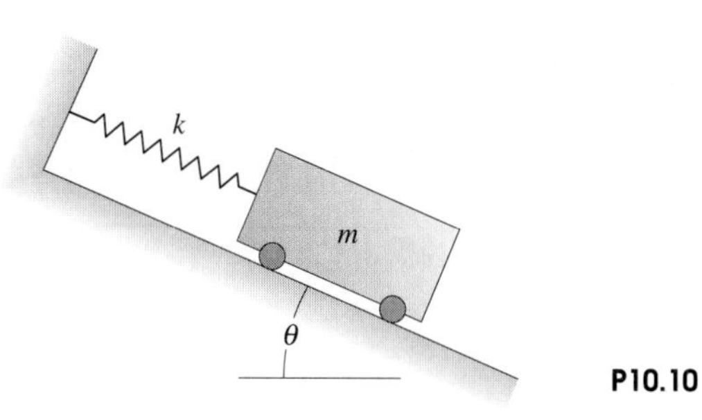

P10.10

10.11 A 90 kg 'bungee jumper' jumps from a bridge above a river. The bungee cord has an unstretched length of 18 m and it stretches an additional 12 m before he rebounds. If you model the cord as a linear spring, what is the period of his vertical oscillations?

P10.11

10.12 A homogeneous disc of mass m and radius R rotates about a fixed shaft and is attached to a torsional spring with constant k. (The torsional spring exerts a restoring moment of magnitude $k\theta$, where θ is the angle of rotation of the disc relative to its position in which the spring is unstretched.) Show that the period of rotational vibrations of the disc is

$$\tau = \pi R\sqrt{2m/k}$$

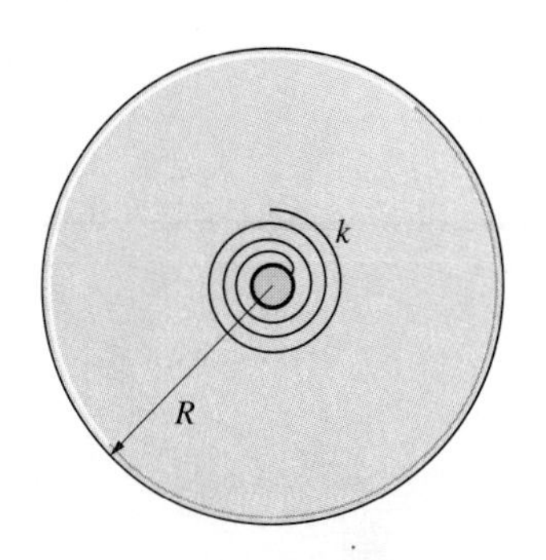

P10.12

10.13 Assigned to determine the moments of inertia of astronaut candidates, an engineer attaches a horizontal platform to a vertical steel bar. The moment of inertia of the platform about L is $7.5\ \text{kg.m}^2$, and the natural frequency of torsional oscillations of the unloaded platform is 1 Hz. With an astronaut candidate in the position shown, the natural frequency of torsional oscillations is 0.520 Hz. What is the candidate's moment of inertia about L?

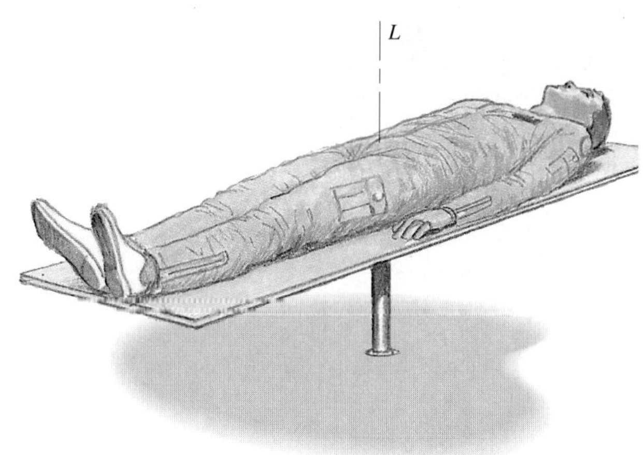

P10.13

10.14 The pendulum consists of a homogeneous 1 kg disc attached to a 0.2 kg slender bar. What is the natural frequency of small vibrations of the pendulum?

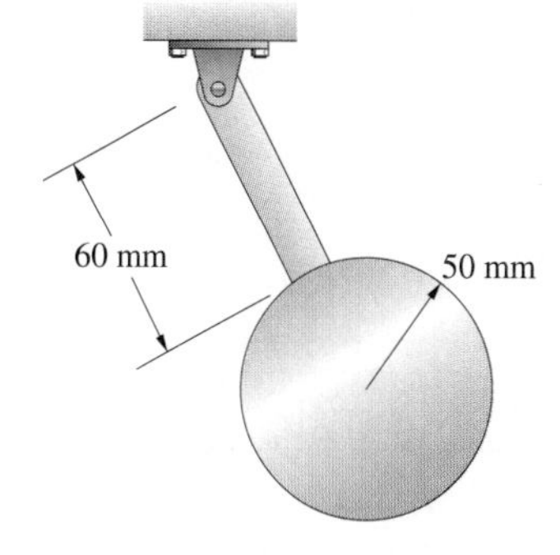

P10.14

10.15 The homogeneous disc weighs 445 N and its radius is $R = 0.3$ m. It rolls on the plane surface. The spring constant is $k = 1460$ N/m.
(a) Determine the natural frequency of vibrations of the disc relative to its equilibrium position.
(b) At $t = 0$, the spring is unstretched and the disc has a clockwise angular velocity of 2 rad/s. What is the amplitude of the resulting vibrations of the disc and what is the angular velocity of the disc when $t = 3$ s?

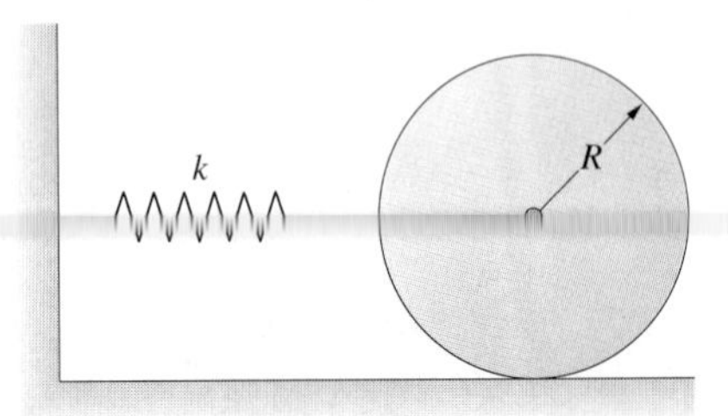

P10.15

10.16 The radius of the disc is $R = 100$ mm, and its moment of inertia is $I = 0.1$ kg.m^2. The mass $m = 5$ kg, and the spring constant is $k = 135$ N/m. The cable does not slip relative to the disc. The coordinate x measures the displacement of the mass relative to the position in which the spring is unstretched.
(a) What are the period and natural frequency of vertical vibrations of the mass relative to its equilibrium position?
(b) Determine x as a function of time if the system is released from rest with $x = 0$.

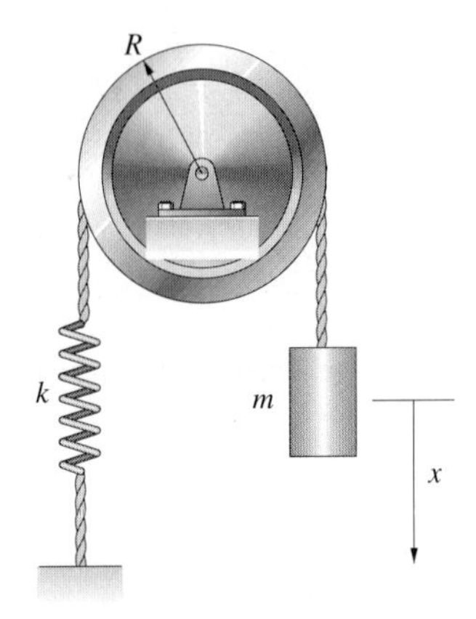

P10.16

10.17 The 22 kg platen P rests on four roller bearings. The roller bearings can be modelled as 1 kg homogeneous cylinders with 30 mm radii, and the spring constant is $k = 900$ N/m. What is the natural frequency of horizontal vibrations of the platen relative to its equilibrium position?

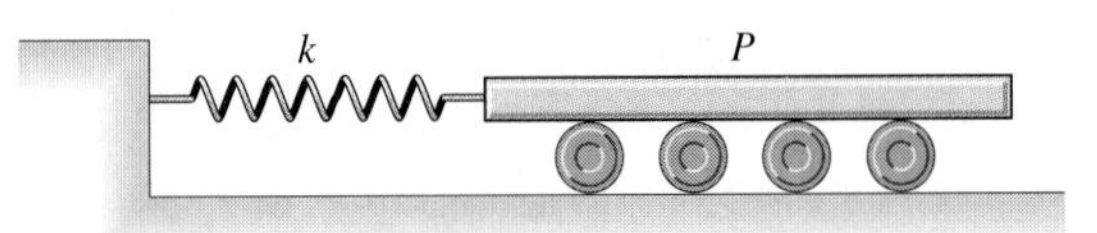

P10.17

10.18 At $t = 0$, the platen described in Problem 10.17 is 0.1 m to the left of its equilibrium position and is moving to the right at 2 m/s. What are the platen's position and velocity at $t = 4$ s?

10.19 A homogeneous disc of mass m and radius r rolls on a curved surface of radius R. Show that the natural frequency of small vibrations of the disc relative to its equilibrium position is

$$f = \frac{1}{\pi}\sqrt{\frac{g}{6(R - r)}}$$

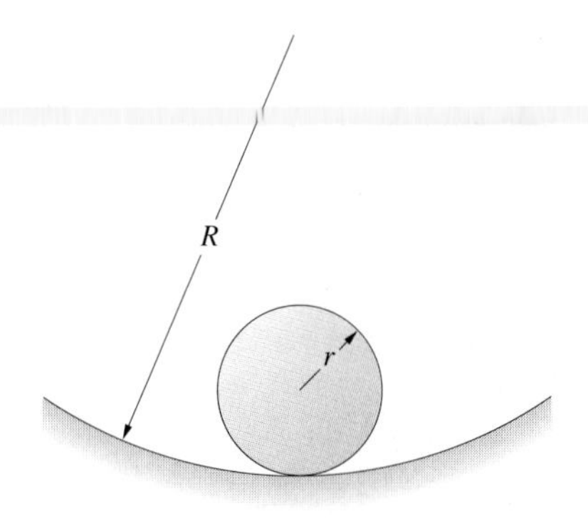

P10.19

10.20 The slender bar has roller supports at its ends and is at rest in a circular depression with an 8 m radius. What is the frequency of small vibrations of the bar relative to its equilibrium position?

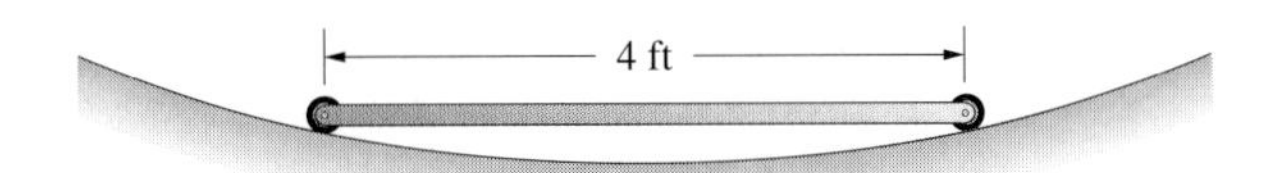

P10.20

10.21 A slender bar of mass m and length l is pinned to a fixed support as shown. A torsional spring of constant k attached to the bar at the support is unstretched when the bar is vertical. Show that the equation governing small vibrations of the bar from its vertical equilibrium position is

$$\frac{d^2\theta}{dt^2} + \omega^2\theta = 0 \quad \text{where} \quad \omega^2 = \frac{(k - \frac{1}{2}mgl)}{\frac{1}{3}ml^2}$$

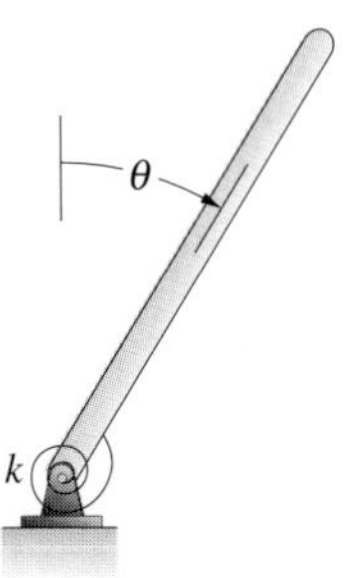

P10.21

10.22 The initial conditions of the slender bar in Problem 10.21 are

$$t=0\begin{cases}\theta=0\\ \dfrac{d\theta}{dt}=\dot{\theta}_0\end{cases}$$

(a) If $k>\frac{1}{2}mgL$, show that θ is given as a function of time by

$$\theta=\frac{\dot{\theta}_0}{\omega}\sin\omega t \quad \text{where} \quad \omega^2=\frac{(k-\frac{1}{2}mgl)}{\frac{1}{3}ml^2}$$

(b) If $k<\frac{1}{2}mgL$, show that θ is given as a function of time by

$$\theta=\frac{\dot{\theta}_0}{2h}(\mathrm{e}^{ht}-\mathrm{e}^{-ht}) \quad \text{where } h^2=\frac{(\frac{1}{2}mgl-k)}{\frac{1}{3}ml^2}$$

Strategy: To do part (b), seek a solution of the equation of motion of the form $x=C\mathrm{e}^{\lambda t}$, where C and λ are constants.

10.23 A floating sonobuoy (sound measuring device) is in equilibrium in the vertical position shown. (Its centre of mass is low enough that it is stable in this position.) It is a 10 kg cylinder 1 m in length and 125 mm in diameter. The water density is 1025 kg/m^3, and the buoyancy force supporting the buoy equals the weight of the water that would occupy the volume of the part of the cylinder below the surface. If you push the sonobuoy slightly deeper and release it, what is the natural frequency of the resulting vertical vibrations?

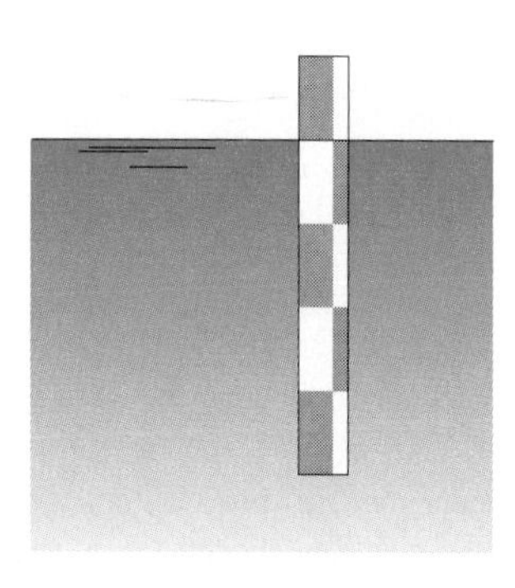

P10.23

10.24 A disc rotates about a fixed *vertical* axis with constant angular velocity Ω. (The plane of the disc is horizontal.) A mass m slides in a smooth slot in the disc and is attached to a spring with constant k. The distance from the centre of the disc to the mass when the spring is unstretched is r_0. Show that if $k/m>\Omega^2$, the natural frequency of vibration of the mass is $f=(1/2\pi)\sqrt{k/m-\Omega^2}$.

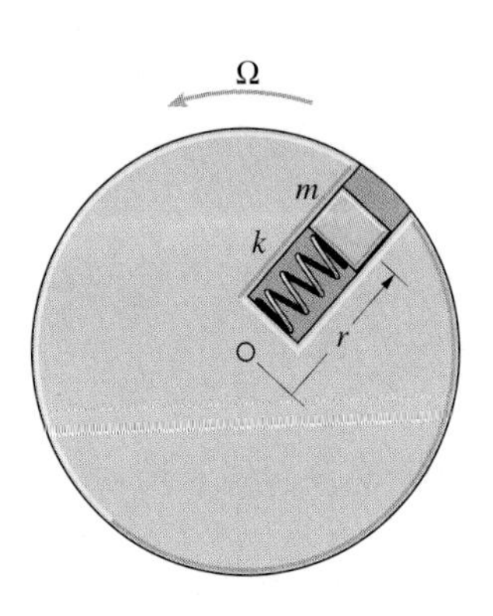

P10.24

10.25 Suppose that at $t=0$, the mass described in Problem 10.24 is located at $r=r_0$ and its radial velocity is $dr/dt=0$. Determine the position r of the mass as a function of time.

10.26 A homogeneous 100 kg disc with radius $R=1$ m is attached to two identical cylindrical steel bars of length $L=1$ m. The relation between the moment M exerted on the disc by one of the bars and the angle of rotation θ of the disc is

$$M=\frac{GJ}{L}\theta$$

where J is the polar moment of inertia of the cross-section of the bar and $G=80$ GN/m^2 is the shear modulus of the steel. Determine the required radius of the bars if the natural frequency of rotational vibrations of the disc is to be 10 Hz.

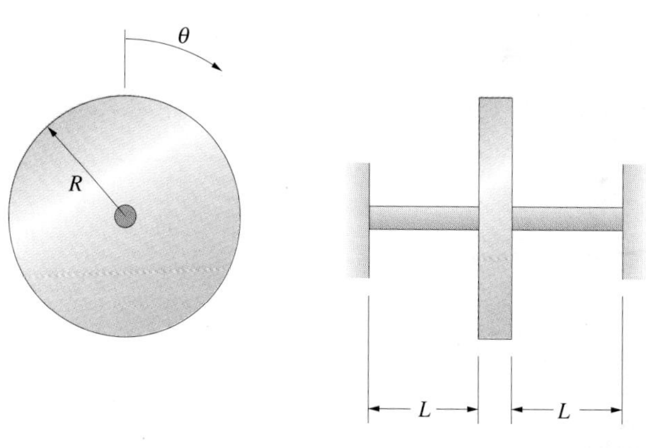

P10.26

10.27 The moments of inertia of gears A and B are $I_A = 0.025\,\text{kg.m}^2$ and $I_B = 0.100\,\text{kg.m}^2$. Gear A is connected to a torsional spring with constant $k = 10\,\text{N.m/rad}$. What is the natural frequency of small angular vibrations of the gears?

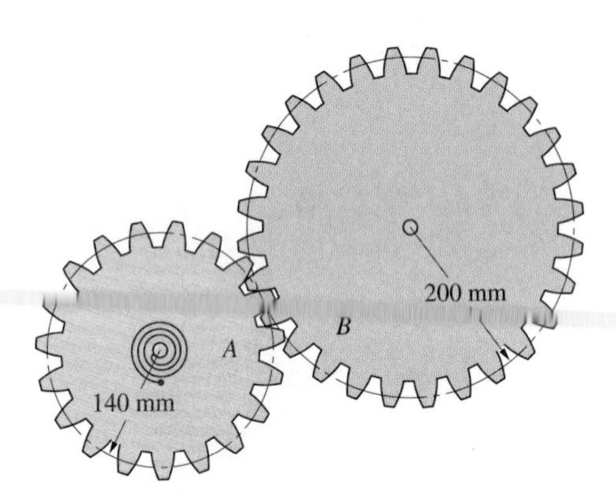

P10.27

10.28 At $t = 0$, the torsional spring in Problem 10.27 is unstretched and gear B has a counterclockwise angular velocity of 2 rad/s. Determine the counterclockwise angular position of gear B relative to its equilibrium position as a function of time.

10.29 The moments of inertia of gears A and B are $I_A = 0.014\,\text{kg.m}^2$ and $I_B = 0.100\,\text{kg.m}^2$. Gear A is connected to a torsional spring with constant $k = 2\,\text{N.m/rad}$. What is the natural frequency of small angular vibrations of the gears relative to their equilibrium position?

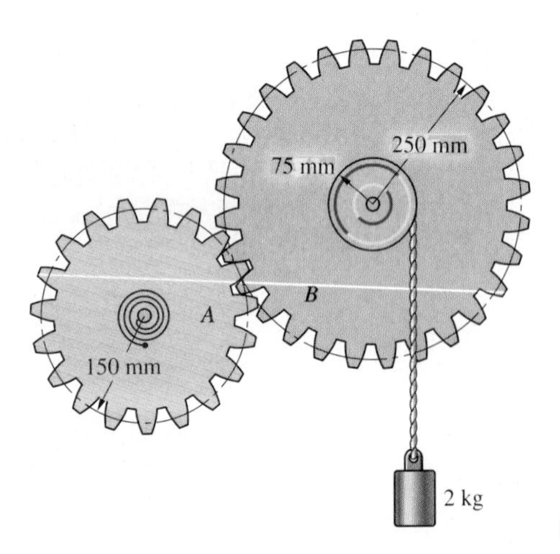

P10.29

10.30 The 2 kg weight in Problem 10.29 is raised 10 mm from its equilibrium position and released from rest at $t = 0$. Determine the counterclockwise angular position of gear B relative to its equilibrium position as a function of time.

10.31 Each slender bar is of mass m and length l. Determine the natural frequency of small vibrations of the system.

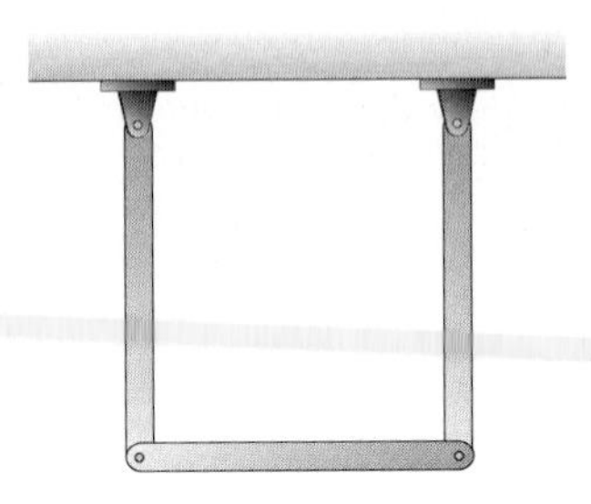

P10.31

10.32 The mass of each slender bar is 1 kg. If the natural frequency of small vibrations of the system is 0.935 Hz, what is the mass of the object A?

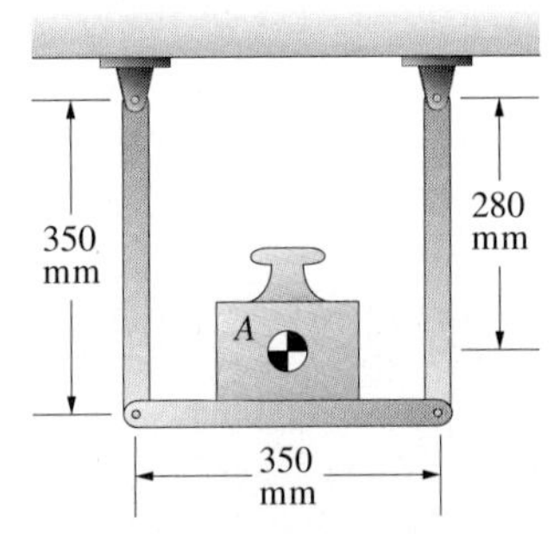

P10.32

10.33 The slender bar of mass m and length l is held in equilibrium in the position shown by a torsional spring with constant k. The spring is unstretched when the bar is vertical. Determine the natural frequency of small vibrations relative to the equilibrium position shown.

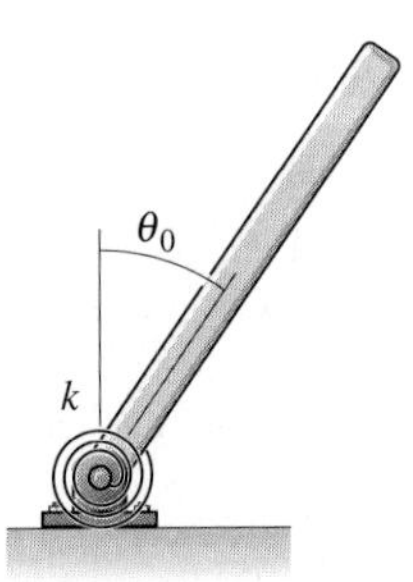

P10.33

10.34 The masses of the slender bar and the homogeneous disc are m and m_d, respectively. The spring is unstretched when $\theta = 0$. Assume that the disc rolls on the horizontal surface.
(a) show that the motion is governed by the equation

$$\left(\frac{1}{3}+\frac{3m_d}{2m}\cos^2\theta\right)\frac{d^2\theta}{dt^2}-\frac{3m_d}{2m}\sin\theta\cos\theta\left(\frac{d\theta}{dt}\right)^2 -\frac{g}{2l}\sin\theta+\frac{k}{m}(1-\cos\theta)\sin\theta=0$$

(b) If the system is in equilibrium at the angle $\theta = \theta_e$ and $\tilde{\theta} = \theta - \theta_e$, show that the equation governing small vibrations relative to the equilibrium position is

$$\left(\frac{1}{3}+\frac{3m_d}{2m}\cos^2\theta_e\right)\frac{d^2\tilde{\theta}}{dt^2} +\left[\frac{k}{m}(\cos\theta_e-\cos^2\theta_e+\sin^2\theta_e)-\frac{g}{2l}\cos\theta_e\right]\tilde{\theta}=0$$

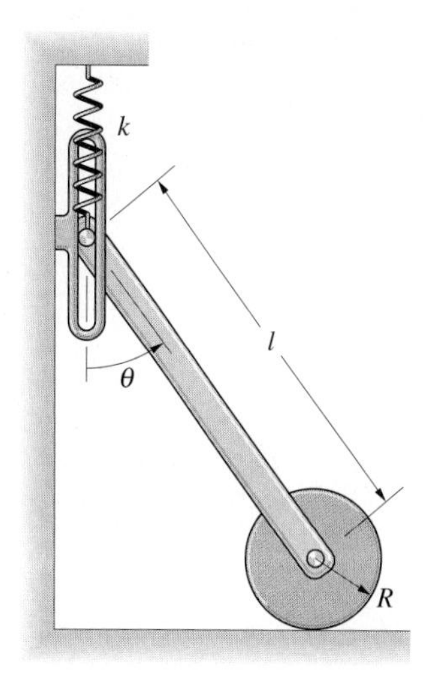

P10.34

10.35 The masses of the bar and disc in Problem 10.34 are $m = 2$ kg and $m_d = 4$ kg, respectively. The dimensions $l = 1$ m and $R = 0.28$ m, and the spring constant is $k = 70$ N/m.
(a) Determine the angle θ_e at which the system is in equilibrium.
(b) The system is at rest in the equilibrium position, and the disc is given a clockwise angular velocity of 0.1 rad/s. Determine θ as a function of time.

10.2 Damped Vibrations

If you displace the mass of a spring-mass oscillator and release it, you know that it won't continue to vibrate indefinitely. It will slow down and eventually stop as a result of frictional forces, or **damping mechanisms**, acting on the system. Damping mechanisms damp out, or **attenuate**, the vibration. In some cases, engineers intentionally include damping mechanisms in vibrating systems. For example, the shock absorbers in a car are designed to damp out vibrations of the suspension relative to the frame. In the previous section we neglected damping, so the solutions we obtained describe motions of systems only over periods of time brief enough that the effects of damping can be neglected. We now discuss a simple method for modelling damping in vibrating systems.

The spring-mass oscillator in Figure 10.9(a) has a **damping element**. The schematic diagram for the damping element represents a piston moving in a cylinder of viscous fluid, which is called a **dashpot**. The force required to lengthen or shorten a damping element is defined to be the product of a constant c, the **damping constant**, and the rate of change of its length (Figure 10.9(b)). Therefore the equation of motion of the mass is

$$-c\frac{dx}{dt}-kx=m\frac{d^2x}{dt^2}$$

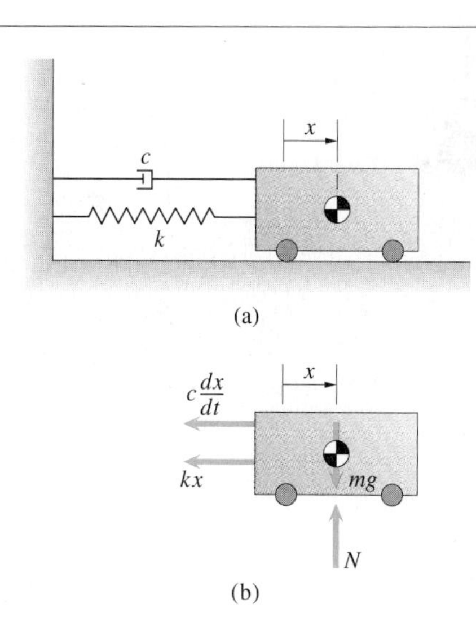

Figure 10.9
(a) Damped spring-mass oscillator.
(b) Free-body diagram of the mass.

By defining $\omega = \sqrt{k/m}$ and $d = c/(2m)$, we can write this equation in the form

$$\frac{d^2x}{dt^2} + 2d\frac{dx}{dt} + \omega^2 x = 0 \tag{10.16}$$

This equation describes the vibrations of many damped, one-degree-of-freedom systems. The form of its solution, and consequently the character of the predicted behaviour of the system, depends on whether d is less than, equal to or greater than ω. We discuss these cases in the following sections.

Subcritical Damping

If $d < \omega$, a system is said to be **subcritically damped**. By assuming a solution of the form

$$x = Ce^{\lambda t} \tag{10.17}$$

where C and λ are constants, and substituting it into Equation (10.16), we obtain

$$\lambda^2 + 2d\lambda + \omega^2 = 0$$

This quadratic equation yields two roots for the constant λ that we can write as

$$\lambda = -d \pm i\omega_d$$

where $i = \sqrt{-1}$ and

$$\omega_d = \sqrt{\omega^2 - d^2} \tag{10.18}$$

Because we are assuming that $d < \omega$, the constant ω_d is a real number. The two roots for λ give us two solutions of the form of Equation (10.17). The resulting general solution of Equation (10.16) is

$$x = e^{-dt}(Ce^{i\omega_d t} + De^{-i\omega_d t})$$

where C and D are constants. By using the identity $e^{i\theta} = \cos\theta + i\sin\theta$, we can express the general solution in the form

$$x = e^{-dt}(A \sin \omega_d t + B \cos \omega_d t) \tag{10.19}$$

where A and B are constants. Equation (10.19) is the product of an exponentially decaying function of time and an expression identical in form to the solution we obtained for an undamped system. The exponential function causes the expected effect of damping: the amplitude of the vibration attenuates with time. The coefficient d determines the rate at which the amplitude decreases.

Damping has an important effect in addition to causing attenuation. Because the oscillatory part of the solution is identical in form to Equation

(10.5) except that the circular natural frequency ω is replaced by ω_d, it follows from Equations (10.9) and (10.10) that the period and natural frequency of the damped system are

$$\tau = \frac{2\pi}{\omega_d} \qquad f = \frac{1}{\tau} = \frac{\omega_d}{2\pi} \tag{10.20}$$

From Equation (10.18) we see that $\omega_d < \omega$, so *the period of the vibration is increased and its natural frequency is decreased as a result of subcritical damping.*

The rate of damping is often expressed in terms of the **logarithmic decrement** δ, which is the natural logarithm of the ratio of the amplitude at a time t to the amplitude at time $t+\tau$. Since the amplitude is proportional to e^{-dt}, we can obtain a simple relation between the logarithmic decrement, the coefficient d, and the period:

$$\delta = \ln\left[\frac{e^{-dt}}{e^{-d(t+\tau)}}\right] = d\tau$$

Critical and Supercritical Damping

When $d \geqslant \omega$, the character of the solution of Equation (10.16) is very different from the case of subcritical damping. Suppose that $d > \omega$. When this is the case, the system is said to be **supercritically damped**. We again substitute a solution of the form

$$x = Ce^{\lambda t} \tag{10.21}$$

into Equation (10.16), obtaining

$$\lambda^2 + 2d\lambda + \omega^2 = 0 \tag{10.22}$$

We can write the roots of this equation as

$$\lambda = -d \pm h$$

where

$$h = \sqrt{d^2 - \omega^2} \tag{10.23}$$

The resulting general solution of Equation (10.16) is

$$x = Ce^{-(d-h)t} + De^{-(d+h)t} \tag{10.24}$$

where C and D are constants.

When $d = \omega$, the system is said to be **critically damped**. The constant $h = 0$, so Equation (10.22) has a repeated root, $\lambda = -d$, and we obtain only one solution of the form (10.21). In this case it can be shown that the general solution of Equation (10.16) is

$$x = Ce^{-dt} + Dte^{-dt} \tag{10.25}$$

where C and D are constants.

Equations (10.24) and (10.25) indicate that the motion of the system is not oscillatory when $d \geqslant \omega$. They are expressed in terms of exponential functions and do not contain sines and cosines. The condition $d = \omega$ defines the minimum amount of damping necessary to avoid oscillatory behaviour, which is why it is referred to as the critically damped case. Figure 10.10 shows the effect of increasing amounts of damping on the behaviour of a vibrating system.

The concept of critical damping has important implications in the design of many systems. For example, it is desirable to introduce enough damping into a car's suspension so that its motion is not oscillatory, but too much damping would cause the suspension to be too 'stiff'.

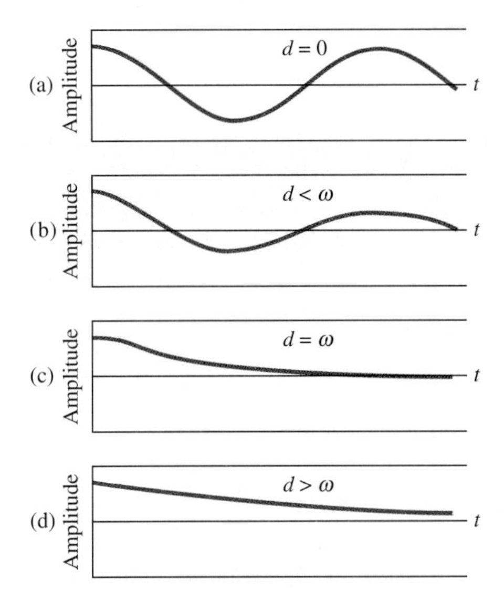

Figure 10.10
Amplitude history of a vibrating system that is (a) undamped; (b) subcritically damped; (c) critically damped; (d) supercritically damped.

In the following examples we analyse damped one-degree-of-freedom systems. By expressing the equation of motion of the system in the form of Equation (10.16), you can determine d ***and*** ω***. Their values tell you whether the damping is subcritical, critical or supercritical, which indicates the form of solution you should use:***

	Type of Damping	Solution
$d < \omega$:	Subcritical	Equation (10.19)
$d = \omega$:	Critical	Equation (10.25)
$d > \omega$:	Supercritical	Equation (10.24)

Example 10.3

The damped spring-mass oscillator in Figure 10.9(a) has mass $m = 2$ kg, spring constant $k = 8$ N/m and damping constant $c = 1$ N.s/m. At $t = 0$, the mass is released from rest in the position $x = 0.1$ m. Determine its position as a function of time.

SOLUTION

The constants $\omega = \sqrt{k/m} = 2$ rad/s and $d = c/2m = 0.25$ rad/s, so the damping is subcritical and the motion is described by Equation (10.19). From Equation (10.18),

$$\omega_d = \sqrt{\omega^2 - d^2} = 1.98 \text{ rad/s}$$

From Equation (10.19),

$$x = e^{-0.25t}(A \sin 1.98t + B \cos 1.98t)$$

and the velocity of the mass is

$$\frac{dx}{dt} = -0.25e^{-0.25t}(A \sin 1.98t + B \cos 1.98t)$$
$$+ e^{-0.25t}(1.98A \cos 1.98t - 1.98B \sin 1.98t)$$

From the conditions $x = 0.1$ m and $dx/dt = 0$ at $t = 0$, we obtain $A = 0.0126$ and $B = 0.1$ m, so the position of the mass is

$$x = e^{-0.25t}(0.0126 \sin 1.98t + 0.1 \cos 1.98t) \text{ m}$$

The graph of x for the first 10 s of motion in Figure 10.11 clearly exhibits the attenuation of the amplitude.

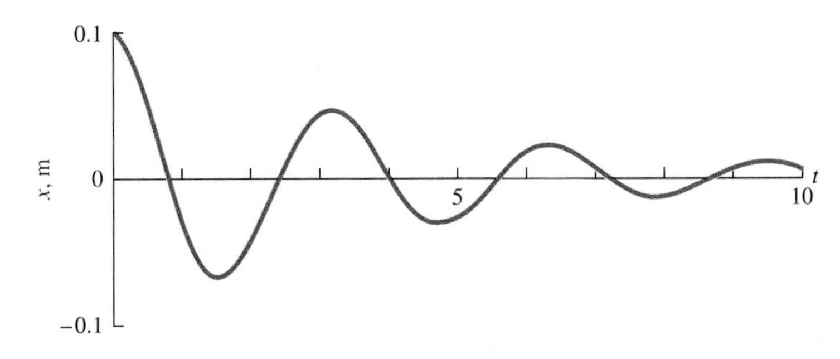

Figure 10.11

Position of the mass as a function of time.

Example 10.4

The 20 kg stepped disc in Figure 10.12 is released from rest with the spring unstretched. Determine the position of the centre of the disc as a function of time if $R = 0.3$ m, $k = 161$ N/m, $c = 64.4$ N.s/m and the moment of inertia expressed in terms of the mass m of the disc is $I = 3mR^2$.

Figure 10.12

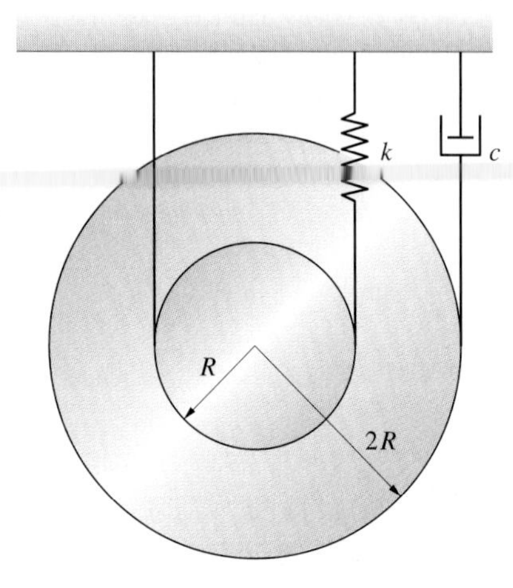

SOLUTION

Let x be the downward displacement of the centre of the disc relative to its position when the spring is unstretched. From the position of the disc's instantaneous centre (Figure (a)), we can see that the rate at which the spring is stretched is $2(dx/dt)$ and the rate at which the damping element is lengthened is $3(dx/dt)$. When the centre of the disc is displaced a distance x, the extension of the spring is $2x$.

We draw the free-body diagram of the disc in Figure (b), showing the forces exerted by the spring, the damping element and the tension in the cable. Newton's second law is

$$mg - T - 2kx - 3c\frac{dx}{dt} = m\frac{d^2x}{dt^2}$$

and the equation of angular motion is

$$RT - R(2kx) - 2R\left(3c\frac{dx}{dt}\right) = (3mR^2)\alpha$$

The angular acceleration is related to the acceleration of the centre of the disc by $\alpha = (d^2x/dt^2)/R$. Adding the equation of angular motion to Newton's second law, we obtain the equation of motion

$$4m\frac{d^2x}{dt^2} + 9c\frac{dx}{dt} + 4kx = mg$$

By setting d^2x/dt^2 and dx/dt equal to zero in this equation, we determine that the equilibrium position of the disc is $x = mg/4k$. Rewriting the equation of motion in terms of the variable $\tilde{x} = x - mg/4k$, we obtain

$$\frac{d^2\tilde{x}}{dt^2} + \left(\frac{9c}{4m}\right)\frac{d\tilde{x}}{dt} + \frac{k}{m}\tilde{x} = 0$$

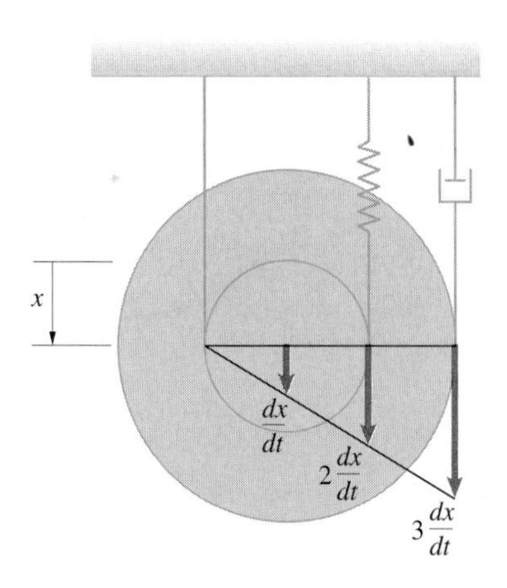

(a) Using the instantaneous centre to determine the relationships between the velocities.

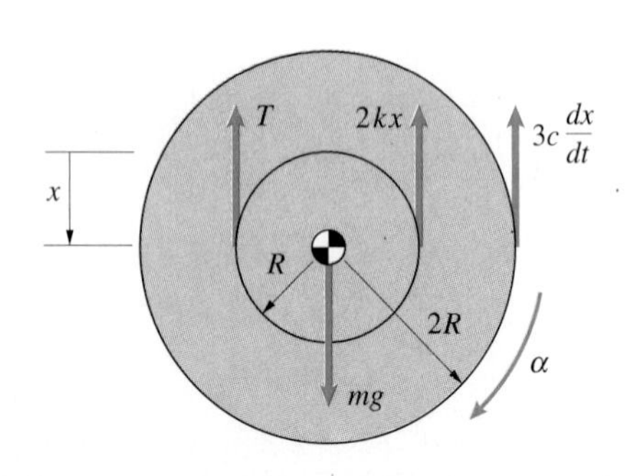

(b) Free-body diagram of the disk.

This equation is identical in form to Equation (10.16), where the constants d and ω are

$$d = \frac{9c}{8m} = \frac{(9)(64.4)}{(8)(20)} = 3.62 \text{ rad/s}$$

$$\omega = \sqrt{\frac{k}{m}} = \sqrt{\frac{161}{20}} = 2.84 \text{ rad/s}$$

The damping is supercritical, so the motion is described by Equation (10.24) with $h = \sqrt{d^2 - \omega^2} = 2.25$ rad/s:

$$\tilde{x} = Ce^{-(d-h)t} + De^{-(d+h)t} = Ce^{-1.37t} + De^{-5.87t}$$

The velocity is

$$\frac{d\tilde{x}}{dt} = -1.37Ce^{-1.37t} - 5.87De^{-5.87t}$$

At $t = 0$, $\tilde{x} = -mg/4k = -0.3$ m and $d\tilde{x}/dt = 0$. From these conditions, we obtain $C = -0.384$ m and $D = 0.084$ m, so the position of the centre of the disc relative to its equilibrium position is

$$\tilde{x} = -0.384e^{-1.37t} + 0.084e^{-5.87t} \text{ m}$$

We show the graph of the position for the first four seconds of motion in Figure 10.13.

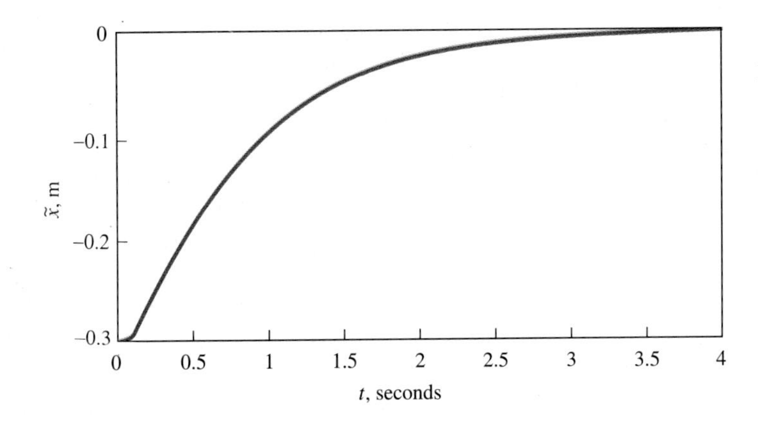

Figure 10.13

Position of the centre of the disc as a function of time.

Problems

10.36 (a) What are the natural frequency and period of the spring-mass oscillator described in Example 10.3? (b) What are the natural frequency and period if the damping element is removed?

10.37 (a) What value of c is necessary for the stepped disc in Example 10.4 to be critically damped? (b) If c equals the value you determined in part (a) and the disc is released from rest with the spring unstretched, determine the position of the centre of the disc relative to its equilibrium position as a function of time.

10.38 The damping constant of the damped spring-mass oscillator is $c = 20$ N.s/m. What are the period and natural frequency of the system? Compare them with the period and natural frequency when the system is undamped.

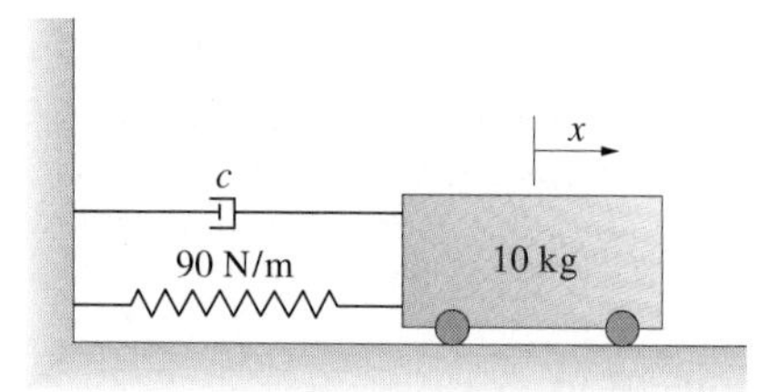

P10.38

10.39 At $t = 0$, the position of the mass in Problem 10.38 relative to its equilibrium position is $x = 0$ and its velocity is 1 m/s to the right. Determine x as a function of time.

10.40 In Problem 10.38, what value of the damping constant c will cause the amplitude of vibration of the system to decrease to one-half of its initial value in 10 s?

10.41 At $t = 0$, the position of the mass in Problem 10.38 is $x = 0$ and it has a velocity of 1 m/s to the right. Determine x as a function of time if c has twice the value necessary for the system to be critically damped.

10.42 The homogeneous slender bar is 1 m long and weighs 40 N. Aerodynamic drag and friction at the support exert a resisting moment on the bar of magnitude $0.5(d\theta/dt)$ N.m, where $d\theta/dt$ is the angular velocity of the bar in rad/s.
(a) What are the period and natural frequency of small vibrations of the bar?
(b) How long does it take for the amplitude of vibration to decrease to one-half of its initial value?

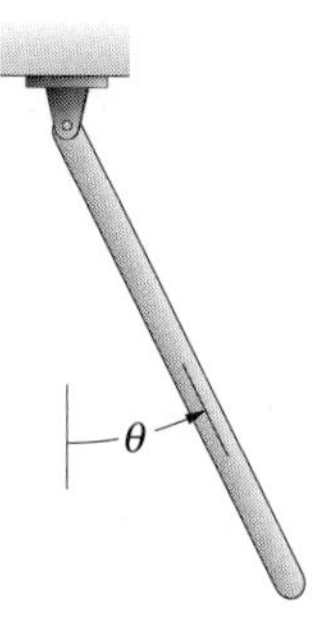

P10.42

10.43 If the bar in Problem 10.42 is displaced a small angle θ_0 and released from rest at $t = 0$, what is θ as a function of time?

10.44 The radius of the pulley is $R = 100$ mm and its moment of inertia is $I = 0.1$ kg.m^2. The mass $m = 5$ kg, and the spring constant is $k = 135$ N/m. The cable does not slip relative to the pulley. The coordinate x measures the displacement of the mass relative to the position in which the spring is unstretched. Determine x as a function of time if $c = 60$ N.s/m and the system is released from rest with $x = 0$.

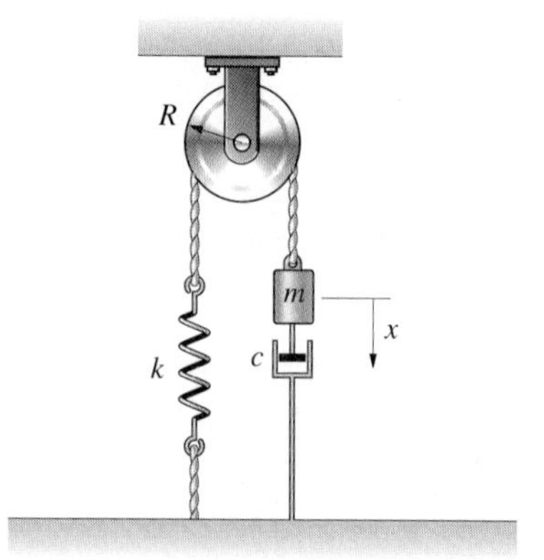

P10.44

10.45 For the system described in Problem 10.44, determine x as a function of time if $c = 120$ N.s/m and the system is released from rest with $x = 0$.

10.46 For the system described in Problem 10.44, choose the value of c so that the system is critically damped and determine x as a function of time if the system is released from rest with $x = 0$.

10.47 The homogeneous disc weighs 450 N and its radius is $R = 0.3$ m. It rolls on the plane surface. The spring constant is $k = 1500$ N/m and the damping constant is $c = 45$ N.s/m. Determine the natural frequency of small vibrations of the disc relative to its equilibrium position.

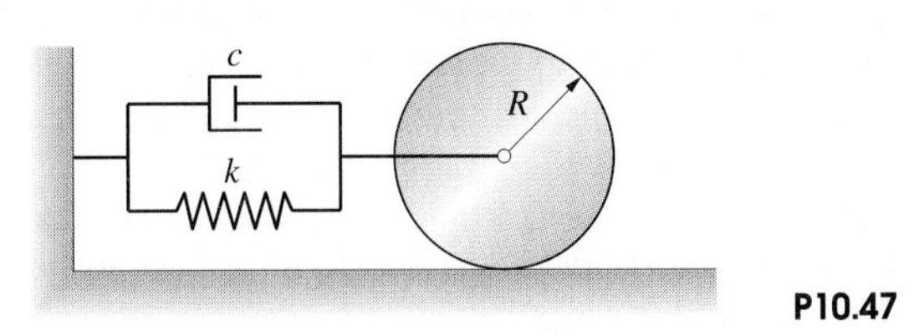

P10.47

10.48 In Problem 10.47, the spring is unstretched at $t = 0$ and the disc has a clockwise angular velocity of 2 rad/s. What is the angular velocity of the disc when $t = 3$ s?

10.49 The moment of inertia of the stepped disc is I. Let θ be the angular displacement of the disc relative to its position when the spring is unstretched. Show that the equation governing θ is identical in form to Equation (10.16), where

$$d = \frac{R^2 c}{2I} \quad \text{and} \quad \omega^2 = \frac{4R^2 k}{I}$$

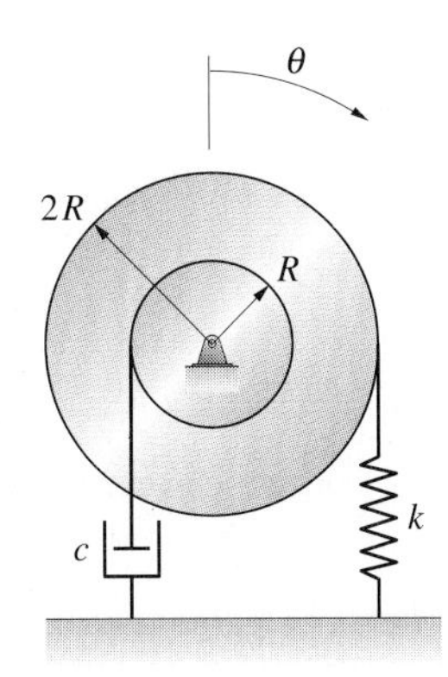

P10.49

10.50 In Problem 10.49, the radius $R = 250$ mm, $k = 150$ N/m and the moment of inertia of the disc is $I = 2$ kg.m^2.
(a) What value of c will cause the system to be critically damped?
(b) At $t = 0$, the spring is unstretched and the clockwise angular velocity of the disc is 10 rad/s. Determine θ as a function of time if the system is critically damped.
(c) Using the result of part (b), determine the maximum resulting angular displacement of the disc and the time at which it occurs.

10.51 The moments of inertia of gears A and B are $I_A = 0.025$ kg.m^2 and $I_B = 0.100$ kg.m^2. Gear A is connected to a torsional spring with constant $k = 10$ N.m/rad. The bearing supporting gear B incorporates a damping element that exerts a resisting moment on gear B of magnitude $2(d\theta_B/dt)$ N.m, where $d\theta_B/dt$ is the angular velocity of gear B in rad/s. What is the frequency of small angular vibrations of the gears?

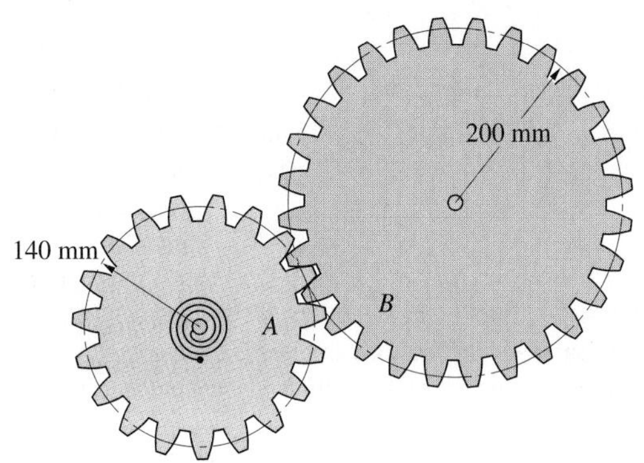

P10.51

10.52 At $t = 0$, the torsional spring in Problem 10.51 is unstretched and gear B has a counterclockwise angular velocity of 2 rad/s. Determine the counterclockwise angular position of gear B relative to its equilibrium position as a function of time.

10.53 The moments of inertia of gears A and B are $I_A = 0.014$ kg.m^2 and $I_B = 0.100$ kg.m^2. Gear A is connected to a torsional spring with constant $k = 2$ N.m/rad. The bearing supporting gear B incorporates a damping element that exerts a resisting moment on gear B of magnitude $1.5(d\theta_B/dt)$ N.m, where $d\theta_B/dt$ is the angular velocity of gear B in rad/s. What is the frequency of small angular vibrations of the gears?

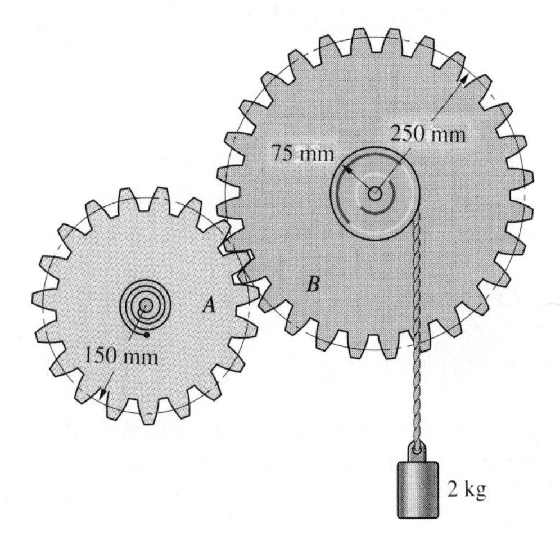

P10.53

10.54 The 2 kg mass in Problem 10.53 is raised 10 mm from its equilibrium position and released from rest at $t = 0$. Determine the counterclockwise angular position of gear B relative to its equilibrium position as a function of time.

10.55 For the case of critically damped motion, confirm that the expression

$$x = Ce^{-dt} + Dte^{-dt}$$

is a solution of Equation (10.16).

10.3 Forced Vibrations

The term **forced vibrations** means that external forces affect the vibrations of a system. Until now, we have discussed **free vibrations** of systems, vibrations unaffected by external forces. For example, during an earthquake, a building undergoes forced vibrations induced by oscillatory forces exerted on its foundations. After the earthquake subsides, the building vibrates freely until its motion damps out.

The damped spring-mass oscillator in Figure 10.14(a) is subjected to a horizontal time-dependent force $F(t)$. From the free-body diagram of the mass (Figure 10.14(b)), its equation of motion is

$$F(t) - kx - c\frac{dx}{dt} = m\frac{d^2x}{dt^2}$$

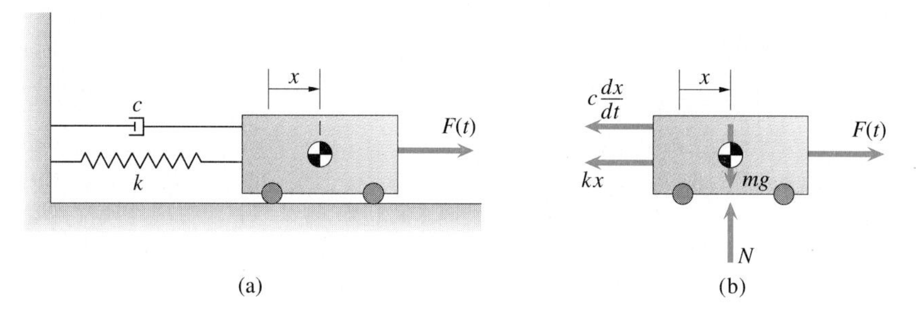

Figure 10.14
(a) A damped spring-mass oscillator subjected to a time-dependent force.
(b) Free-body diagram of the mass.

Defining $d = c/2m$, $\omega^2 = k/m$ and $a(t) = F(t)/m$, we can write this equation in the form

$$\boxed{\frac{d^2x}{dt^2} + 2d\frac{dx}{dt} + \omega^2 x = a(t)} \qquad (10.26)$$

We call $a(t)$ the **forcing function**. Equation (10.26) describes the forced vibrations of many damped, one-degree-of-freedom systems. It is non-homogeneous, because the forcing function does not contain x or one of its derivatives. Its general solution consists of two parts, the homogeneous and particular solutions:

$$x = x_\mathrm{h} + x_\mathrm{p}$$

The **homogeneous solution** x_h is the general solution of Equation (10.26) with the right side set equal to zero. Therefore the homogeneous solution is the general solution for free vibrations, which we described in Section 10.2. The **particular solution** x_p is a solution that satisfies Equation (10.26). In the following sections we discuss the particular solutions for two types of forcing functions that occur frequently in applications.

Oscillatory Forcing Function

Unbalanced wheels and shafts exert forces that oscillate at their frequency of rotation. When your car's wheels are out of balance, they exert oscillatory forces that cause vibrations you can feel. Engineers design electromechanical devices that transform oscillating currents into oscillating forces for use in testing vibrating systems. But the principal reason we are interested in this type of forcing function is that nearly any forcing function can be represented as a sum of oscillatory forcing functions with several different frequencies or with a continuous spectrum of frequencies.

By studying the motion of a vibrating system subjected to an oscillatory forcing function, we can determine its response as a function of the frequency of the force. Suppose that the forcing function is an oscillatory function of the form

$$u(t) = a_0 \sin \omega_0 t + b_0 \cos \omega_0 t \tag{10.27}$$

where a_0, b_0 and the circular frequency of the forcing function ω_0 are given constants. We can obtain the particular solution to Equation (10.26) by seeking a solution of the form

$$x_p = A_p \sin \omega_0 t + B_p \cos \omega_0 t \tag{10.28}$$

where A_p and B_p are constants we must determine. Substituting this expression and Equation (10.27) into Equation (10.26), we can write the resulting equation as

$$(-\omega_0^2 A_p - 2d\omega_0 B_p + \omega^2 A_p - a_0) \sin \omega_0 t + (-\omega_0^2 B_p + 2d\omega_0 A_p + \omega^2 B_p - b_0) \cos \omega_0 t = 0$$

Equating the coefficients of $\sin \omega_0 t$ and $\cos \omega_0 t$ to zero and solving for A_p and B_p, we obtain

$$A_p = \frac{(\omega^2 - \omega_0^2)a_0 + 2d\omega_0 b_0}{(\omega^2 - \omega_0^2)^2 + 4d^2\omega_0^2} \qquad B_p = \frac{-2d\omega_0 a_0 + (\omega^2 - \omega_0^2)b_0}{(\omega^2 - \omega_0^2)^2 + 4d^2\omega_0^2} \tag{10.29}$$

Substituting these results into Equation (10.28), the particular solution is

$$x_p = \left[\frac{(\omega^2 - \omega_0^2)a_0 + 2d\omega_0 b_0}{(\omega^2 - \omega_0^2)^2 + 4d^2\omega_0^2}\right] \sin \omega_0 t + \left[\frac{-2d\omega_0 a_0 + (\omega^2 - \omega_0^2)b_0}{(\omega^2 - \omega_0^2)^2 + 4d^2\omega_0^2}\right] \cos \omega_0 t \tag{10.30}$$

The amplitude of the particular solution is

$$E_p = \sqrt{A_p^2 + B_p^2} = \frac{\sqrt{a_0^2 + b_0^2}}{\sqrt{(\omega^2 - \omega_0^2)^2 + 4d^2\omega_0^2}} \tag{10.31}$$

We showed in Section 10.2 that the solution of the equation describing free vibration of a damped system attenuates with time. For this reason, the particular solution for the motion of a damped vibrating system subjected to an oscillatory external force is also called the **steady-state solution**. The motion approaches the steady-state solution with increasing time. (See Example 10.5.)

We illustrate the effects of damping and the frequency of the forcing function on the amplitude of the particular solution in Figure 10.15. We plot the non-dimensional expression $\omega^2 E_p/\sqrt{a_0^2 + b_0^2}$ as a function of ω_0/ω for several values of the parameter d/ω. When there is no damping ($d = 0$), the amplitude of the particular solution approaches infinity as the circular frequency ω_0 of the forcing function approaches the circular natural frequency ω. When the damping is small, the amplitude of the particular solution approaches a finite maximum value at a value of ω_0 that is smaller than ω. The frequency at which the amplitude of the particular solution is a maximum is called the **resonant frequency**.

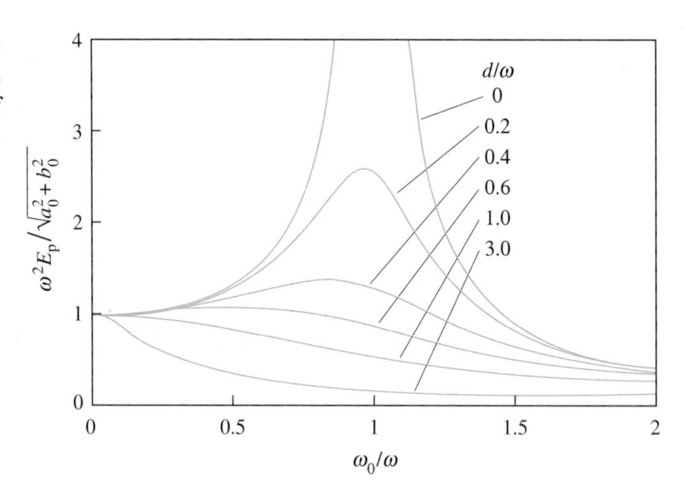

Figure 10.15
Amplitude of the particular (steady-state) solution as a function of the frequency of the forcing function.

The phenomenon of resonance is a familiar one in our everyday experience. For example, when a wheel of your car is out of balance, you notice the resulting vibrations when the car is moving at a certain speed. At that speed, the wheel rotates at the resonant frequency of your car's suspension. Resonance is of practical importance in many applications, because relatively small oscillatory forces can result in large vibration amplitudes that may cause damage or interfere with the functioning of a system. The classic example is that of soldiers marching across a bridge. If their steps in unison coincide with one of the bridge's resonant frequencies, they may damage it even though the bridge can safely support their weight.

Polynomial Forcing Function

Suppose that the forcing function $a(t)$ in Equation (10.26) is a polynomial function of time:

$$a(t) = a_0 + a_1 t + a_2 t^2 + \cdots + a_N t^N$$

where $a_1, a_2, \ldots, a_N$ are given constants. This forcing function is important in applications because you can approximate many smooth functions by polynomials over a given interval of time. In this case, we can obtain the particular solution to Equation (10.26) by seeking a solution of the same form:

$$x_p = A_0 + A_1 t + A_2 t^2 + \cdots + A_N t^N \tag{10.32}$$

where A_0, A_1, $A_2, \ldots,$ A_N are constants we must determine.

For example, if $a(t) = a_0 + a_1 t$, Equation (10.26) is

$$\frac{d^2x}{dt^2} + 2d\frac{dx}{dt} + \omega^2 x = a_0 + a_1 t \tag{10.33}$$

and we seek a particular solution of the form $x_p = A_0 + A_1 t$. Substituting this solution into Equation (10.33), we can write the resulting equation as

$$(2dA_1 + \omega^2 A_0 - a_0) + (\omega^2 A_1 - a_1)t = 0$$

This equation can be satisfied over an interval of time only if

$$2dA_1 + \omega^2 A_0 - a_0 = 0$$

and

$$\omega^2 A_1 - a_1 = 0$$

Solving these two equations for A_0 and A_1, we obtain the particular solution:

$$x_p = (a_0 - 2da_1/\omega^2 + a_1 t)/\omega^2$$

You should confirm that this is a solution by substituting it into Equation (10.33).

In the following examples we analyse forced vibrations of one-degree-of-freedom systems. After expressing the equation of motion of the system in the form

$$\frac{d^2x}{dt^2} + 2d\frac{dx}{dt} + \omega^2 x = a(t)$$

you must usually determine the homogeneous and particular solutions. The forms of the homogeneous solution are given in Section 10.2.

Example 10.5

An engineer designing a vibration isolation system for an instrument console models the console and isolation system by the damped spring-mass oscillator in Figure 10.14(a) with mass $m = 2$ kg, spring constant $k = 8$ N/m and damping constant $c = 1$ N.s/m. To determine the system's response to external vibration, she assumes that the mass is initially stationary with the spring unstretched, and at $t = 0$ a force

$$F(t) = 20 \sin 4t \text{ N}$$

is applied to the mass.
(a) What is the amplitude of the particular (steady-state) solution?
(b) What is the position of the mass as a function of time?

STRATEGY

The forcing function is $a(t) = F(t)/m = 10 \sin 4t$ m/s^2, which is an oscillatory function of the form of Equation (10.27) with $a_0 = 10$ m/s^2, $b_0 = 0$ and $\omega_0 = 4$ rad/s. The amplitude of the particular solution is given by Equation (10.31), and the particular solution is given by Equation (10.30). We must also determine whether the damping is subcritical, critical or supercritical and choose the appropriate form of the homogeneous solution.

SOLUTION

(a) The circular natural frequency of the undamped system is $\omega = \sqrt{k/m} = 2$ rad/s and the constant $d = c/(2m) = 0.25$ rad/s. Therefore the amplitude of the particular solution is

$$E_p = \frac{a_0}{\sqrt{(\omega^2 - \omega_0^2)^2 + 4d^2\omega_0^2}} = \frac{10}{\sqrt{[(2)^2 - (4)^2]^2 + 4(0.25)^2(4)^2}} = 0.822 \text{ m}$$

(b) Since $d < \omega$, the system is subcritically damped and the homogeneous solution is given by Equation (10.19). The circular frequency of the damped system is $\omega_d = \sqrt{\omega^2 - d^2} = 1.98$ rad/s, so the homogeneous solution is

$$x_h = e^{-0.25t}(A \sin 1.98t + B \cos 1.98t)$$

From Equation (10.30), the particular solution is

$$x_p = -0.811 \sin 4t - 0.135 \cos 4t$$

and the complete solution is

$$\begin{aligned} x &= x_h + x_p \\ &= e^{-0.25t}(A \sin 1.98t + B \cos 1.98t) - 0.811 \sin 4t - 0.135 \cos 4t \end{aligned}$$

At $t = 0$, $x = 0$ and $dx/dt = 0$. Using these conditions to determine the constants A and B, we obtain $A = 1.651$ m and $B = 0.135$ m. The position of the mass as a function of time is

$$\begin{aligned} x = {} & [e^{-0.25t}(1.651 \sin 1.98t + 0.135 \cos 1.98t) \\ & - 0.811 \sin 4t - 0.135 \cos 4t] \text{ m} \end{aligned}$$

Figure 10.16 shows the homogeneous, particular and complete solutions for the first 25 seconds of motion. The complete solution has an initial 'transient' phase due to the homogeneous part of the solution. As the homogeneous solution attenuates, the complete solution approaches the particular, or steady-state, solution.

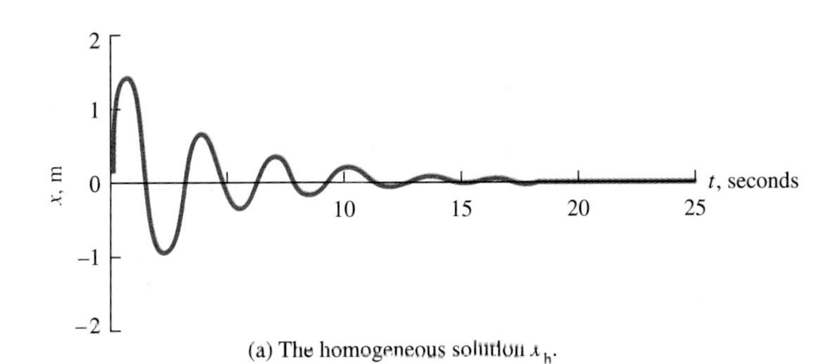

(a) The homogeneous solution x_h.

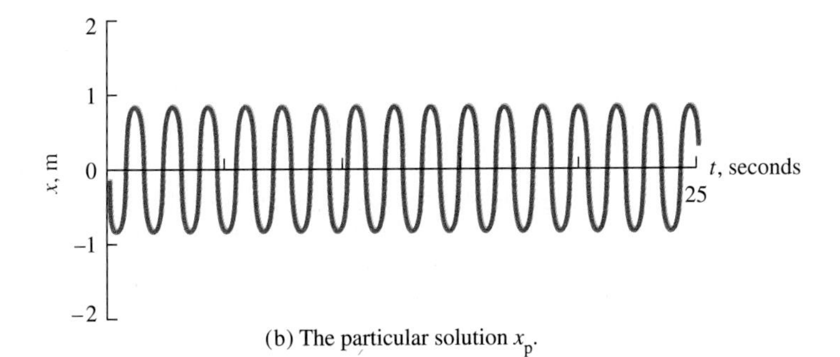

(b) The particular solution x_p.

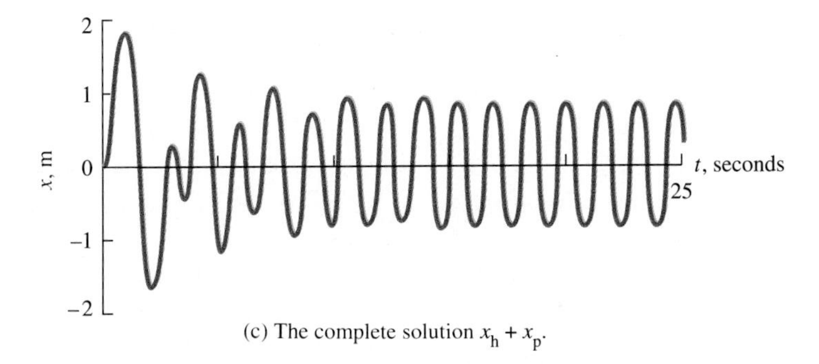

(c) The complete solution $x_h + x_p$.

Figure 10.16

The homogeneous, particular and complete solutions.

Example 10.6

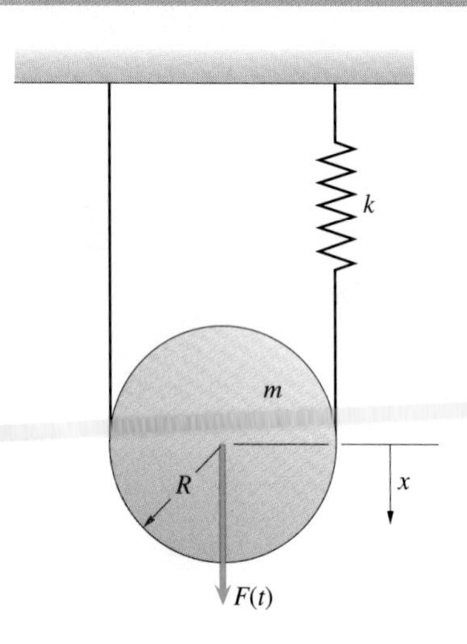

Figure 10.17

The homogeneous disc in Figure 10.17 has radius $R = 2$ m and mass $m = 4$ kg. The spring constant is $k = 30$ N/m. The disc is initially stationary in its equilibrium position, and at $t = 0$ a downward force $F(t) = (12 + 12t - 0.6t^2)$ N is applied to the centre of the disc. Determine the position of the centre of the disc as a function of time.

STRATEGY

The force $F(t)$ is a polynomial, so we can seek a particular solution of the form of Equation (10.32).

SOLUTION

Let x be the displacement of the centre of the disc relative to its position when the spring is unstretched. We draw the free-body diagram of the disc in Figure (a), where T is the tension in the cable on the left side of the disc. From Newton's second law,

$$F(t) + mg - 2kx - T = m\frac{d^2x}{dt^2} \tag{10.34}$$

The angular acceleration of the disc in the clockwise direction is related to the acceleration of the centre of the disc by $\alpha = (d^2x/dt^2)/R$. Using this expression, we can write the equation of angular motion of the disc as

$$\Sigma M = I\alpha:$$

$$TR - 2kxR = \left(\frac{1}{2}mR^2\right)\left(\frac{1}{R}\frac{d^2x}{dt^2}\right)$$

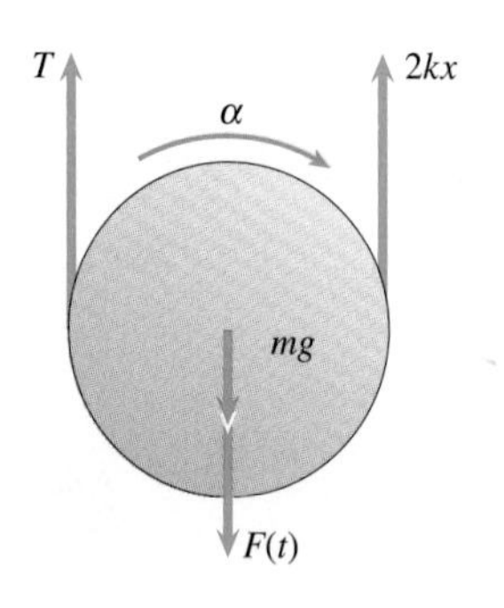

(a) Free-body diagram of the disk.

Solving this equation for T and substituting the result into Equation (10.34), we obtain the equation of motion

$$\frac{3}{2}m\frac{d^2x}{dt^2} + 4kx = F(t) + mg \tag{10.35}$$

Setting $d^2x/dt^2 = 0$ and $F(t) = 0$ in this equation, we find that the equilibrium position of the disc is $x = mg/4k$. In terms of the position of the centre of the disc relative to its equilibrium position, $\tilde{x} = x - mg/4k$, the equation of motion is

$$\frac{d^2\tilde{x}}{dt^2} + \frac{8k}{3m}\tilde{x} = \frac{2F(t)}{3m}$$

This equation is identical in form to Equation (10.26). Substituting the values of k and m and the polynomial function $F(t)$, we obtain

$$\frac{d^2\tilde{x}}{dt^2} + 20\tilde{x} = 2 + 2t - 0.1t^2 \tag{10.36}$$

Comparing this equation with Equation (10.26), we see that $d = 0$ (there is no damping) and $\omega^2 = 20$ (rad/s)^2. From Equation (10.19), the homogeneous solution is

$$\tilde{x}_{\rm h} = A \sin 4.472t + B \cos 4.472t$$

To obtain the particular solution, we seek a solution in the form of a polynomial of the same order as $F(t)$:

$$\tilde{x}_p = A_0 + A_1 t + A_2 t^2$$

where A_0, A_1 and A_2 are constants we must determine. We substitute this expression into Equation (10.36) and collect terms of equal powers in t:

$$(2A_2 + 20A_0 - 2) + (20A_1 - 2)t + (20A_2 + 0.1)t^2 = 0$$

This equation is satisfied if the coefficients multiplying each power of t equal zero:

$$2A_2 + 20A_0 = 2$$

$$20A_1 = 2$$

$$20A_2 = -0.1$$

Solving these three equations for A_0, A_1 and A_2, we obtain the particular solution:

$$\tilde{x}_p = 0.101 + 0.100t - 0.005t^2$$

The complete solution is

$$\begin{aligned}\tilde{x} &= \tilde{x}_h + \tilde{x}_p \\ &= A \sin 4.472t + B \cos 4.472t + 0.101 + 0.100t - 0.005t^2\end{aligned}$$

At $t=0$, $\tilde{x}=0$ and $d\tilde{x}/dt = 0$. Using these conditions to determine A and B, we obtain the position of the centre of the disc as a function of time:

$$\tilde{x} = -0.022 \sin 4.47t - 0.100 \cos 4.47t + 0.101 + 0.100t - 0.005t^2$$

The position is shown for the first 25 s of motion in Figure 10.18. You can see the undamped oscillatory homogeneous solution superimposed on the slowly varying particular solution.

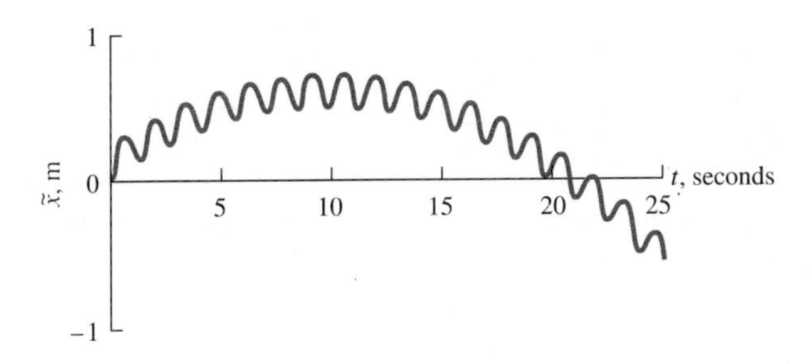

Figure 10.18

Position of the centre of the disc as a function of time.

Example 10.7

Application to Engineering

Displacement Transducers

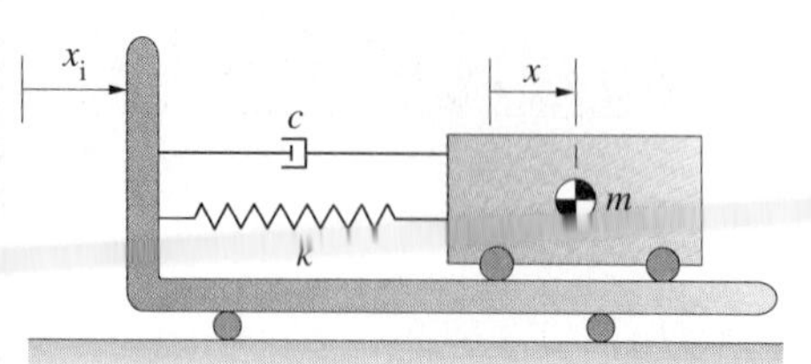

Figure 10.19

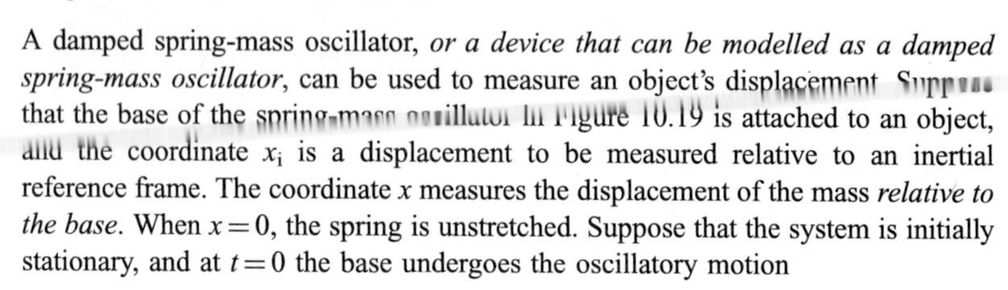

A damped spring-mass oscillator, *or a device that can be modelled as a damped spring-mass oscillator*, can be used to measure an object's displacement. Suppose that the base of the spring-mass oscillator in Figure 10.19 is attached to an object, and the coordinate x_i is a displacement to be measured relative to an inertial reference frame. The coordinate x measures the displacement of the mass *relative to the base*. When $x=0$, the spring is unstretched. Suppose that the system is initially stationary, and at $t=0$ the base undergoes the oscillatory motion

$$x_i = a_i \sin \omega_i t + b_i \cos \omega_i t \tag{10.37}$$

If $m=2$ kg, $k=8$ N/m, $c=4$ N.s/m, $a_i=0.1$ m, $b_i=0.1$ m and $\omega_i=10$ rad/s, what is the resulting steady-state amplitude of the displacement of the mass relative to the base?

SOLUTION

The acceleration of the mass relative to the base is d^2x/dt^2, so its acceleration relative to the inertial reference frame is $(d^2x/dt^2)+(d^2x_i/dt^2)$. Newton's second law for the mass is

$$-c\frac{dx}{dt} - kx = m\left(\frac{d^2x}{dt^2} + \frac{d^2x_i}{dt^2}\right)$$

We can write this equation as

$$\frac{d^2x}{dt^2} + 2d\frac{dx}{dt} + \omega^2 x = a(t)$$

where $d=c/2m=1$ rad/s, $\omega = \sqrt{k/m} = 2$ rad/s and the function $a(t)$ is

$$a(t) = -\frac{d^2x_i}{dt^2} = a_i\omega_i^2 \sin \omega_i t + b_i\omega_i^2 \cos \omega_i t \tag{10.38}$$

Thus we obtain an equation of motion identical in form to that for a spring-mass oscillator subjected to an oscillatory force. Comparing Equation (10.38) to Equation (10.27), we can obtain the amplitude of the particular (steady-state) solution from Equation (10.31) by setting $a_0 = a_i\omega_i^2$, $b_0 = b_i\omega_i^2$ and $\omega_0 = \omega_i$:

$$E_p = \frac{\omega_i^2\sqrt{a_i^2 + b_i^2}}{\sqrt{(\omega^2 - \omega_i^2)^2 + 4d^2\omega_i^2}}$$

Therefore the steady-state amplitude of the displacement of the mass relative to its base is

$$E_p = \frac{(10)^2\sqrt{(0.1)^2 + (0.1)^2}}{\sqrt{[(2)^2 - (10)^2]^2 + 4(1)^2(10)^2}} = 0.144 \text{ m} \tag{10.39}$$

DESIGN ISSUES

A microphone transforms sound waves into a varying voltage that can be recorded or transformed back into sound waves by a loudspeaker. A device that transforms a mechanical input into an electromagnetic output, or an electromagnetic input into a mechanical output, is called a **transducer**. Transducers can be used to measure displacements, velocities and accelerations by transforming them into measurable voltages or currents.

In the case of the spring-mass oscillator in Figure 10.19, the coordinate x_i is the displacement to be measured, the input. A transducer can be used to measure the displacement x of the mass relative to the base, the output. If the relationship between the input and output is known, the displacement x_i can be determined. Some seismographs (Figure 10.20) measure motions of the earth in this way.

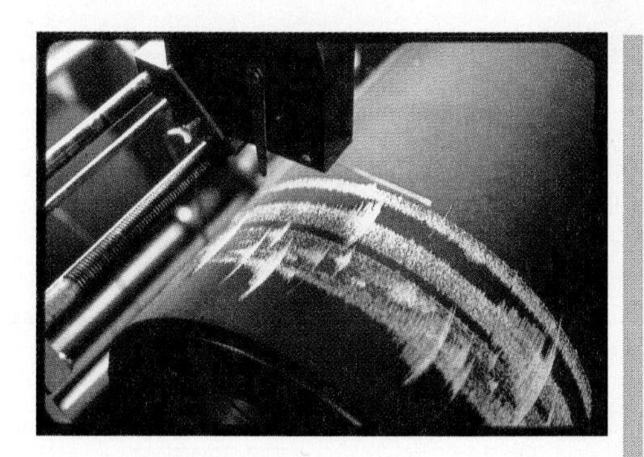

Figure 10.20

A seismograph that measures the local displacement of the earth.

If the input is an oscillatory displacement given by Equation (10.37), the amplitude of the output is given by Equation (10.39). We can write the latter equation as

$$\frac{E_p}{E_i} = \frac{(\omega_i/\omega)^2}{\sqrt{[1-(\omega_i/\omega)^2]^2 + 4(d/\omega)^2(\omega_i/\omega)^2}}$$

where $E_i = \sqrt{a_i^2 + b_i^2}$ is the amplitude of the input. In Figure 10.21 we show the ratio E_p/E_i as a function of the ratio of the input frequency to the natural frequency of the undamped system, ω_i/ω, for several values of d/ω. If the parameters of the spring-mass oscillator are known, you can use a graph of this type to determine the amplitude of the input by measuring the amplitude of the output.

In practice, the input displacement does not usually have a single frequency, but consists of a combination of different frequencies or even a continuous *spectrum* of frequencies. For example, the displacements resulting from earthquakes have a spectrum of frequencies. In that case, it is desirable for the ratio of the output amplitude to the input amplitude to be approximately equal to 1 over the range of the input frequencies. The response of the instrument is said to be 'flat'. In Figure 10.21 you can see that the response is approximately flat for frequencies ω_i greater than about 2ω if the damping of the system is chosen so that d/ω is in the range 0.6-0.7. Also, notice that making the natural frequency ω small increases the range of input frequencies over which the response of the instrument is flat. For that reason, seismographs are often designed with large masses and relatively weak springs.

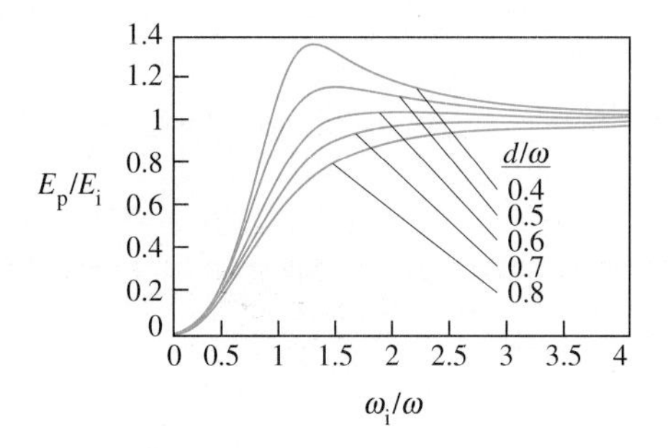

Figure 10.21

Ratio of the output amplitude to the input amplitude.

Problems

10.56 The mass $m=2$ kg and $k=200$ N/m. Let x be the position of the mass relative to its position when the spring is unstretched. The force $F(t)=36 \sin 8t$ N.
(a) Determine the particular solution.
(b) At $t=0$, $x=1$ m and the velocity of the mass is zero. Determine x as a function of time.

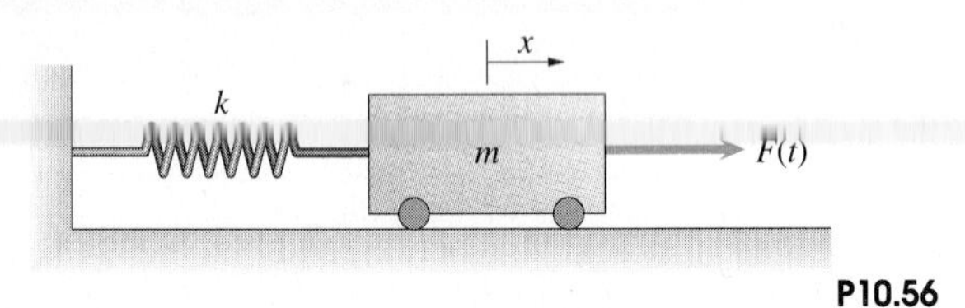

P10.56

10.57 The damped spring-mass oscillator is initially stationary with the spring unstretched. At $t=0$, a constant 1.2 N force is applied to the mass.
(a) What is the steady-state (particular) solution?
(b) Determine the position of the mass as a function of time.

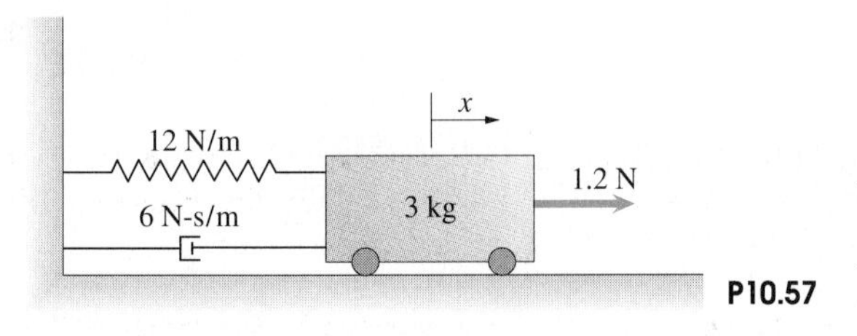

P10.57

10.58 A disc with moment of inertia I rotates about a fixed shaft and is attached to a torsional spring with constant k. The angle θ measures the angular position of the disc relative to its position when the spring is unstretched. The disc is initially stationary with the spring unstretched. At $t=0$, a time-dependent moment $M(t) = M_0(1 - e^{-t})$ is applied to the disc, where M_0 is a constant. Show that the angular position of the disc as a function of time is:

$$\theta = \frac{M_0}{I}\left[-\frac{1}{\omega(1+\omega^2)}\sin \omega t - \frac{1}{\omega^2(1+\omega^2)}\cos \omega t + \frac{1}{\omega^2} - \frac{1}{(1+\omega^2)}e^{-t}\right]$$

Strategy: To determine the particular solution, seek a solution of the form

$$\theta_p = A_p + B_p e^{-t}$$

where A_p and B_p are constants that you must determine.

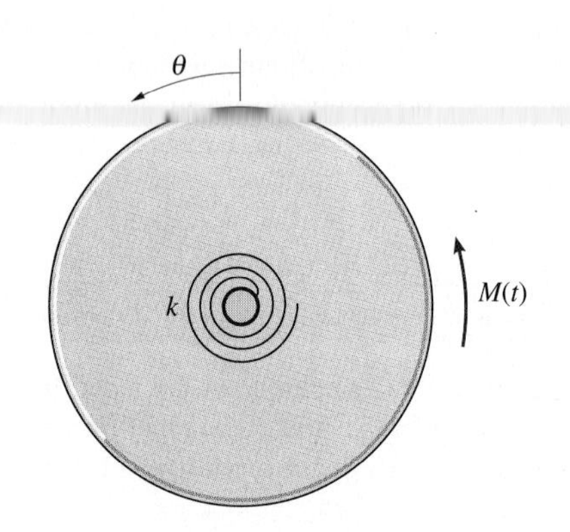

P10.58

10.59 The stepped disc weighs 90 N and its moment of inertia is $I=0.8$ kg.m^2. It rolls on the horizontal surface. The disc is initially stationary with the spring unstretched, and at $t=0$ a constant force $F=45$ N is applied as shown. Determine the position of the centre of the disc as a function of time.

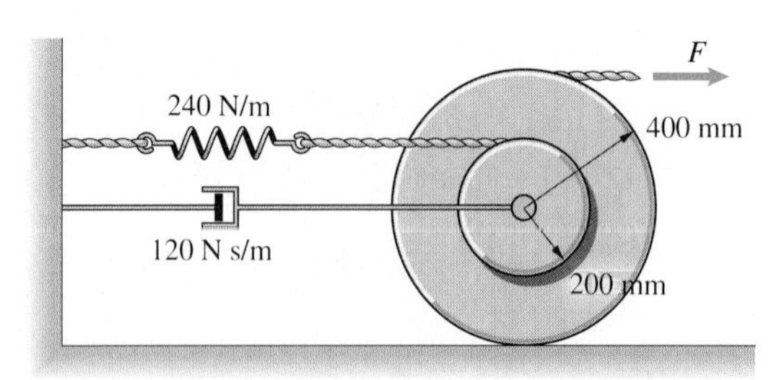

P10.59

10.60 An electric motor is bolted to a metal table. When the motor is on, it causes the tabletop to vibrate horizontally. Assume that the legs of the table behave like linear springs and neglect damping. The total weight of the motor and the tabletop is 650 N. When the motor is not turned on, the frequency of horizontal vibration of the tabletop and motor is 5 Hz. When the motor is running at 600 rpm (revolutions per minute), the amplitude of the horizontal vibration is 0.25 mm. What is the magnitude of the oscillatory force exerted on the table by the motor at this speed?

P10.60

10.61 The moments of inertia of gears A and B are $I_A = 0.014\,\text{kg.m}^2$ and $I_B = 0.100\,\text{kg.m}^2$. Gear A is connected to a torsional spring with constant $k = 2\,\text{N.m/rad}$. The system is in equilibrium at $t = 0$ when it is subjected to an oscillatory force $F(t) = 20 \sin 3t\,\text{N}$. What is the downward displacement of the 2 kg mass as a function of time?

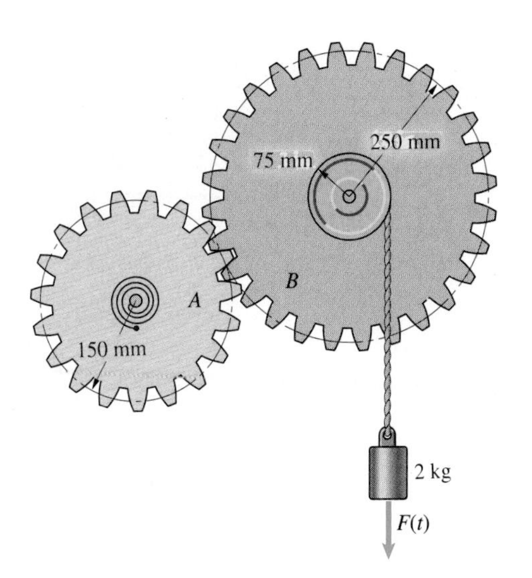

P10.61

10.62 A 1.5 kg cylinder is mounted on a 'sting' in a wind tunnel with the cylinder axis transverse to the flow direction. When there is no flow, a 10 N vertical force applied to the cylinder causes it to deflect 0.15 mm. When air flows in the wind tunnel, vortices subject the cylinder to alternating lateral forces. The velocity of the air is 5 m/s, the distance between vortices is 80 mm, and the magnitude of the lateral forces is 1 N. If you model the lateral forces by the oscillatory function $F(t) = (1.0) \sin \omega_0 t\,\text{N}$, what is the amplitude of the steady-state lateral motion of the sphere?

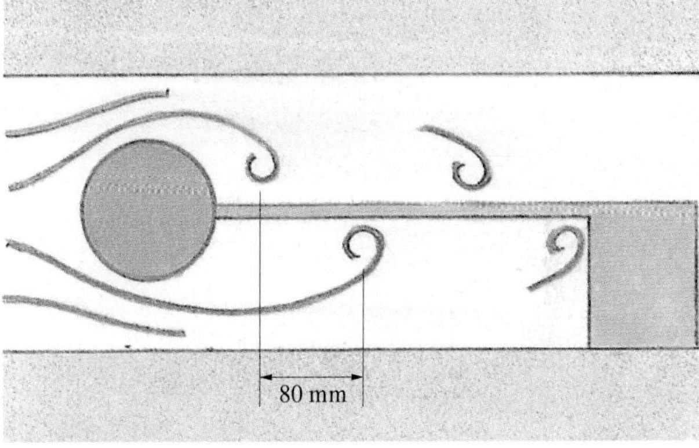

P10.62

10.63 Show that the amplitude of the particular solution given by Equation (10.31) is a maximum when the frequency of the oscillatory forcing function is $\omega_0 = \sqrt{\omega^2 - 2d^2}$.

Problems 10.64–10.67 are related to Example 10.7.

10.64 The mass in Figure 10.19 is 25 kg. The spring constant is $k = 3\,\text{kN/m}$ and $c = 150\,\text{N.s/m}$. If the base is subjected to an oscillatory displacement x_i of amplitude 250 mm and circular frequency $\omega_i = 15\,\text{rad/s}$, what is the resulting steady-state amplitude of the displacement of the mass relative to the base?

10.65 The mass in Figure 10.19 is 100 kg. The spring constant is $k = 4\,\text{N/m}$, and $c = 24\,\text{N.s/m}$. The base is subjected to an oscillatory displacement of circular frequency $\omega_i = 0.2\,\text{rad/s}$. The steady-state amplitude of the displacement of the mass relative to the base is measured and determined to be 200 mm. What is the amplitude of the displacement of the base?

10.66 A team of engineering students builds the simple seismograph shown. The coordinate x_i measures the local horizontal ground motion. The coordinate x measures the position of the mass relative to the frame of the seismograph. The spring is unstretched when $x = 0$. The mass $m = 1$ kg, $k = 10$ N/m and $c = 2$ N.s/m. Suppose that the seismograph is initially stationary and at $t = 0$ it is subjected to an oscillatory ground motion $x_i = 10 \sin 2t$ mm. What is the amplitude of the steady-state response of the mass?

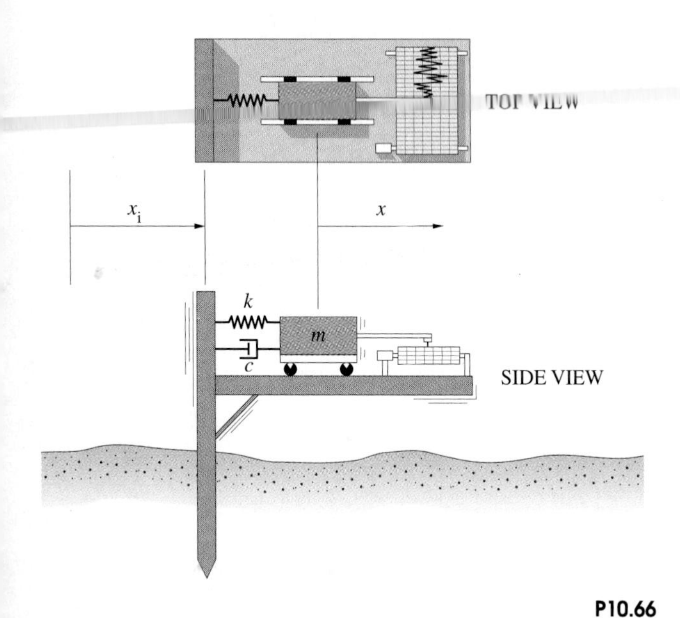

P10.66

10.67 In Problem 10.66, determine the position x of the mass relative to the base as a function of time.

10.68 A sonobuoy (sound-measuring device) floats in a standing wave tank. It is a cylinder of mass m and cross-sectional area A. The water density is ρ, and the buoyancy force supporting the buoy equals the weight of the water that would occupy the volume of the part of the cylinder below the surface. When the water in the tank is stationary, the buoy is in equilibrium in the vertical position shown in Figure (a). Waves are then generated in the tank, causing the depth of the water at the sonobuoy's position *relative to its original depth* to be $d = d_0 \sin \omega_0 t$. Let y be the sonobuoy's vertical position relative to its original position. Show that the sonobuoy's vertical position is governed by the equation

$$\frac{d^2y}{dt^2} + \left(\frac{A\rho g}{m}\right)y = \left(\frac{A\rho g}{m}\right)d_0 \sin \omega_0 t$$

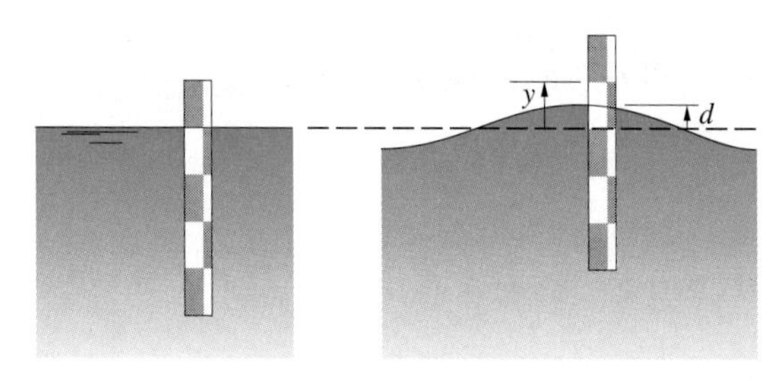

P10.68

10.69 Suppose that the mass of the sonobuoy in Problem 10.68 is $m = 10$ kg, its diameter is 125 mm, and the water density is $\rho = 1025$ kg/m^3. If the water depth is $d = 0.1 \sin 2t$ m, what is the magnitude of the steady-state vertical vibrations of the sonobuoy?

Computational Mechanics

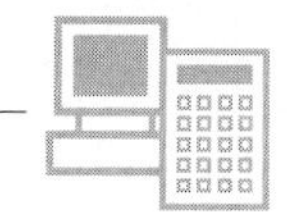

The material in this section is designed for the use of a programmable calculator or computer.

In Section 10.1 we derived the equation of motion for a pendulum consisting of a slender bar (Figure 10.22), obtaining

$$\frac{d^2\theta}{dt^2} + \frac{3g}{2l} \sin\theta = 0 \tag{10.40}$$

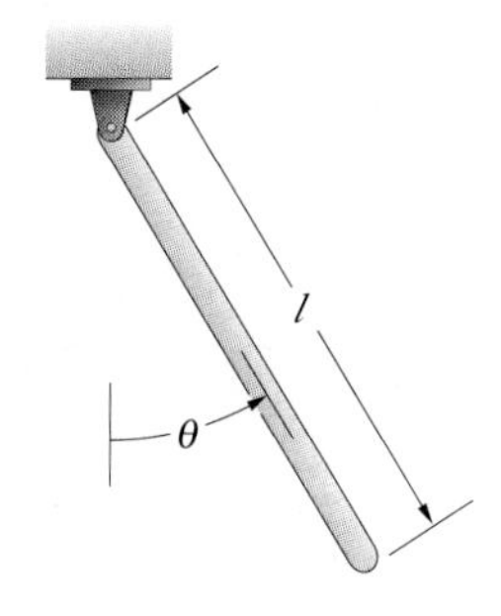

Figure 10.22
A pendulum in the form of a slender bar.

By assuming that the vibrations were sufficiently small to make the approximation $\sin\theta = \theta$, we obtained the equation for a spring-mass oscillator. But suppose that the vibrations are not small, and we cannot linearize the equation of motion. How can we determine the pendulum's motion?

When systems undergo large-amplitude vibrations, the differential equations describing their motions are usually non-linear. Although an analytical solution of Equation (10.40) exists, in most instances you cannot obtain closed-form solutions to such equations. Even when an analytical solution is possible for the unforced system, it may be subjected to external forces that are too complicated to permit a closed-form solution.

We can obtain approximate solutions to such problems by numerical integration. Let's consider an equation of motion that is sufficiently general to include most one-degree-of-freedom vibrating systems. Suppose that the acceleration is a function of the time, position and velocity:

$$\frac{d^2x}{dt^2} = f\left(t, x, \frac{dx}{dt}\right)$$

By defining $v = dx/dt$, we can express the equations governing the system as two first-order differential equations:

$$\frac{dx}{dt} = v$$
$$\frac{dv}{dt} = f(t, x, v) \tag{10.41}$$

We can approximate the values of x and v as functions of time by using Euler's method. Assuming that at a certain time t_0 we know the values $x(t_0)$ and $v(t_0)$, we determine their values at $t_0 + \Delta t$ by approximating the derivatives in Equation (10.41) by forward differences:

$$x(t_0 + \Delta t) = x(t_0) + v(t_0)\Delta t$$

$$v(t_0 + \Delta t) = v(t_0) + f(t_0, x(t_0), v(t_0))\Delta t$$

We then repeat the procedure, using $x(t_0 + \Delta t)$ and $v(t_0 + \Delta t)$ as initial conditions to determine the values of x and v at $t_0 + 2\Delta t$, and so on. In this way, we can determine the position and its rate of change as functions of time for a one-degree-of-freedom system.

Example 10.8

The pendulum in Figure 10.22 is 1 mm long and is released from rest with an initial displacement θ_0. Determine θ as a function of time from $t = 0$ to $t = 4$ s for the cases $\theta_0 = 2°$, $45°$ and $90°$.

SOLUTION

We can express the pendulum's equation of motion, Equation (10.40), as two first-order equations of the same forms as Equation (10.41):

$$\frac{d\theta}{dt} = \omega$$

$$\frac{d\omega}{dt} = -\frac{3g}{2l} \sin\theta$$

Let us consider the case $\theta_0 = 45°$, and let $\Delta t = 0.01$ s. At the initial time $t_0 = 0$, $\theta(t_0) = 45° = 0.7854$ rad and $\omega(t_0) = 0$. At time $t_0 + \Delta t = 0.01$ s, the angle is

$$\theta(t_0 + \Delta t) = \theta(t_0) + \omega(t_0)\Delta t :$$

$$\theta(0.01) = \theta(0) + \omega(0)\Delta t$$

$$= 0.7854 + (0)(0.01) = 0.7854 \text{ rad}$$

The angular velocity is

$$\omega(t_0 + \Delta t) = \omega(t_0) + \left[-\frac{3g}{2l} \sin(\theta(t_0))\right]\Delta t :$$

$$\omega(0.1) = 0 - \frac{(3)(9.81)}{(2)(1)} \sin(0.7854)(0.01) = -0.1041 \text{ rad/s}$$

Using these values as the initial conditions for the next time step, the angle and angular velocity at $t = 0.02$ s are

$$\theta(0.02) = 0.7854 + (-0.1041)(0.01) = 0.7844 \text{ rad}$$

$$\omega(0.02) = -0.1041 - \frac{(3)(9.81)}{(2)(1)} \sin(0.7854)(0.01) = -0.2081 \text{ rad/s}$$

Continuing in this way, we obtain the following values for the first five time steps:

Time, s	θ, rad	ω, rad/s
0.00	0.7854	0.0000
0.01	0.7854	−0.1041
0.02	0.7844	−0.2081
0.03	0.7823	−0.3120
0.04	0.7792	−0.4158
0.05	0.7750	−0.5192

In Figure 10.23, we compare the numerical solutions for the angle (normalized by its initial value $\theta_0 = 45°$) obtained using $\Delta t = 0.01$ s, 0.001 s and 0.0005 s. Trials with smaller time intervals indicate that $\Delta t = 0.0005$ s closely approximates the exact solution over this interval of time. Using $\Delta t = 0.0005$ s, we obtain the solutions for $\theta_0 = 2°$, 45° and 90° shown in Figure 10.24. The normalized results for the different initial amplitudes are qualitatively similar, but the period of the vibrations increases with increasing amplitude – the bar takes longer to make larger swings.

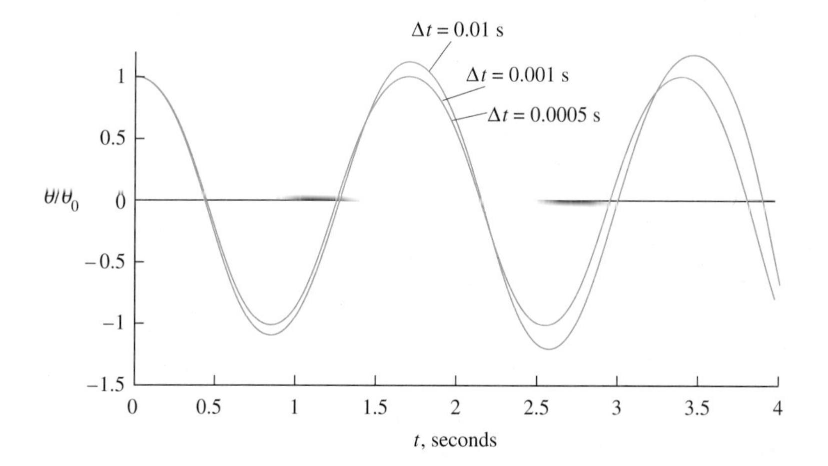

Figure 10.23

Numerical solutions for $\theta_0 = 45°$ using different values of Δt.

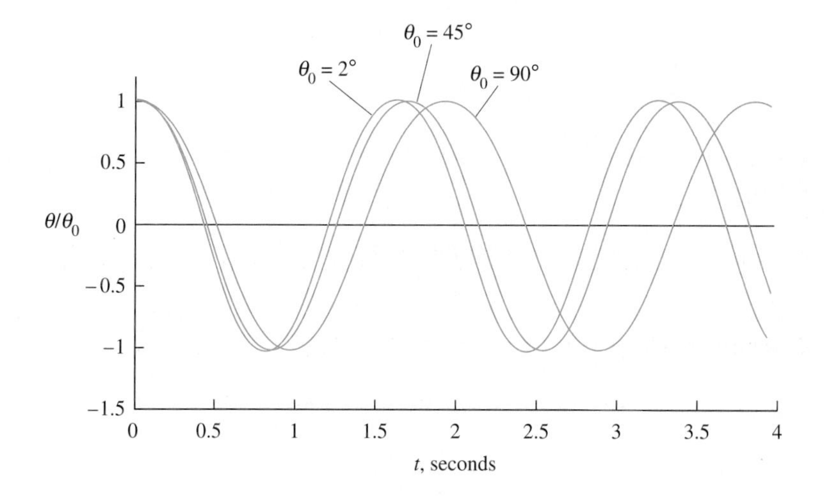

Figure 10.24

Position of the bar as a function of time for different initial displacements.

Problems

10.70 The pendulum described in Example 10.8 is released from rest with $\theta_0 = 2°$. Using $\Delta t = 0.01$ s, calculate the values of the angle and angular velocity for the first five time steps.

10.71 The pendulum described in Example 10.8 is released from rest with $\theta_0 = 2°$. By linearizing the equation of motion, obtain the closed-form solution for the angle as a function of time. Draw a graph comparing your solution for the angle in degrees from $t = 0$ to $t = 4$ s with the numerical solution obtained using $\Delta t = 0.01$ s.

10.72 In his initial design of a wall to protect workers in the event of explosion of a chemical reactor, an engineer assumes that the wall behaves like a spring-mass oscillator with a mass of 4380 kg and a spring constant of 146 kN/m. He models the blast force exerted on the wall by the function $F = 292te^{-4t}$ kN. Using $\Delta t = 0.001$ s, determine the displacement and velocity of the wall for the first five time steps.

P10.72

10.73 Using $\Delta t = 0.001$ s, obtain graphs of the displacement and velocity of the wall described in Problem 10.72 from $t = 0$ to $t = 2$ s.

Chapter Summary

Conservative Systems

Small vibrations of many one-degree-of-freedom conservative systems relative to an equilibrium position are governed by the equation

$$\frac{d^2x}{dt^2} + \omega^2 x = 0 \qquad \textbf{Equation (10.4)}$$

where ω is a constant determined by the properties of the system. Its general solution is

$$x = A \sin \omega t + B \cos \omega t \qquad \textbf{Equation (10.5)}$$

where A and B are constants. Its general solution can also be expressed in the form

$$x = E \sin (\omega t - \phi) \qquad \textbf{Equation (10.6)}$$

where the constants E and ϕ are related to A and B by

$$A = E \cos \phi \qquad B = -E \sin \phi \qquad \textbf{Equation (10.7)}$$

The **amplitude** of the vibration is

$$E = \sqrt{A^2 + B^2} \qquad \textbf{Equation (10.8)}$$

The **period** τ of the vibration is the time required for one complete oscillation, or **cycle**. The **natural frequency** f is the number of cycles per unit time. The period and natural frequency are related to ω by

$$\tau = \frac{2\pi}{\omega} \qquad \textbf{Equation (10.9)}$$

$$f = \frac{\omega}{2\pi} \qquad \textbf{Equation (10.10)}$$

The term $\omega = 2\pi f$ is called the **circular natural frequency**.

Damped Vibrations

Small vibrations of many damped one-degree-of-freedom systems relative to an equilibrium position are governed by the equation

$$\frac{d^2x}{dt^2} + 2d\frac{dx}{dt} + \omega^2 x = 0 \qquad \textbf{Equation (10.16)}$$

Subcritical Damping If $d < \omega$, the system is said to be subcritically damped. In this case, the general solution of Equation (10.16) is

$$x = \mathrm{e}^{-dt}(A \sin \omega_d t + B \cos \omega_d t) \qquad \textbf{Equation (10.19)}$$

where A and B are constants and ω_d is defined by

$$\omega_d = \sqrt{\omega^2 - d^2} \qquad \textbf{Equation (10.18)}$$

The period and frequency of the damped vibrations are

$$\tau = \frac{2\pi}{\omega_d} \qquad f = \frac{\omega_d}{2\pi} \qquad \textbf{Equation (10.20)}$$

Critical and Supercritical Damping If $d > \omega$, the system is said to be supercritically damped. The general solution is

$$x = C\mathrm{e}^{-(d-h)t} + D\mathrm{e}^{-(d+h)t} \qquad \textbf{Equation (10.24)}$$

where C and D are constants and h is defined by

$$h = \sqrt{d^2 - \omega^2} \qquad \textbf{Equation (10.23)}$$

If $d = \omega$, the system is said to be critically damped. The general solution is

$$x = C\mathrm{e}^{-dt} + Dt\mathrm{e}^{-dt} \qquad \textbf{Equation (10.25)}$$

where C and D are constants.

Forced Vibrations

The forced vibrations of many damped, one-degree-of-freedom systems are governed by the equation

$$\frac{d^2x}{dt^2} + 2d\frac{dx}{dt} + \omega^2 x = a(t) \qquad \textbf{Equation (10.26)}$$

where $a(t)$ is the **forcing function**. The general solution of Equation (10.26) consists of the homogeneous and particular solutions:

$$x = x_h + x_p$$

The **homogeneous solution** x_h is the general solution of Equation (10.26) with the right side set equal to zero, and the **particular solution** x_p is a solution that satisfies Equation (10.26).

Oscillatory Forcing Function If $a(t)$ is an oscillatory function of the form

$$a(t) = a_0 \sin \omega_0 t + b_0 \cos \omega_0 t$$

where a_0, b_0 and ω_0 are constants, the particular solution is (Equation 10.30)

$$x_p = \left[\frac{(\omega^2 - \omega_0^2)a_0 + 2d\omega_0 b_0}{(\omega^2 - \omega_0^2)^2 + 4d^2\omega_0^2}\right] \sin \omega_0 t + \left[\frac{-2d\omega_0 a_0 + (\omega^2 - \omega_0^2)b_0}{(\omega^2 - \omega_0^2)^2 + 4d^2\omega_0^2}\right] \cos \omega_0 t$$

and its amplitude is (Equation 10.31)

$$E_p = \sqrt{A_p^2 + B_p^2} = \frac{\sqrt{a_0^2 + b_0^2}}{\sqrt{(\omega^2 - \omega_0^2)^2 + 4d^2\omega_0^2}}$$

The particular solution for the motion of a damped vibrating system subjected to an oscillatory external force is also called the **steady-state solution**. The motion approaches the steady-state solution with increasing time.

Polynomial Forcing Function If $a(t)$ is a polynomial of the form

$$a(t) = a_0 + a_1 t + a_2 t^2 + \cdots + a_N t^N$$

where $a_1, a_2, \ldots, a_N$ are constants, the particular solution can be obtained by seeking a solution of the same form:

$$x_p = A_0 + A_1 t + A_2 t^2 + \cdots + A_N t^N \qquad \textbf{Equation (10.32)}$$

where $A_1, A_2, \ldots, A_N$ are constants that must be determined.

Review Problems

10.74 The mass of the slender bar is m. The spring is unstretched when the bar is vertical. The light collar C slides on the smooth vertical bar so that the spring remains horizontal. Determine the natural frequency of small vibrations of the bar.

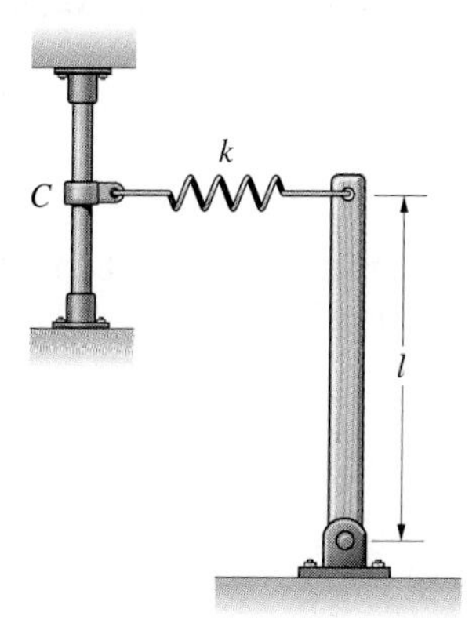

P10.74

10.75 A homogeneous hemisphere of radius R and mass m rests on a level surface. If you rotate the hemisphere slightly from its equilibrium position and release it, what is the natural frequency of its vibrations?

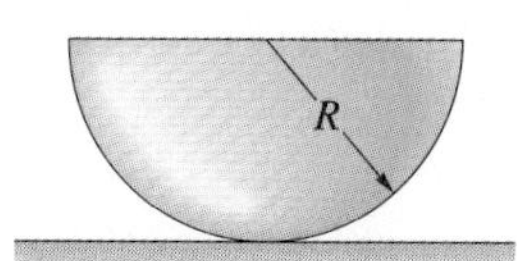

P10.75

10.76 The frequency of the spring-mass oscillator is measured and determined to be 4.00 Hz. The spring-mass oscillator is then placed in a barrel of oil, and its frequency is determined to be 3.80 Hz. What is the logarithmic decrement of vibrations of the mass when the oscillator is immersed in oil?

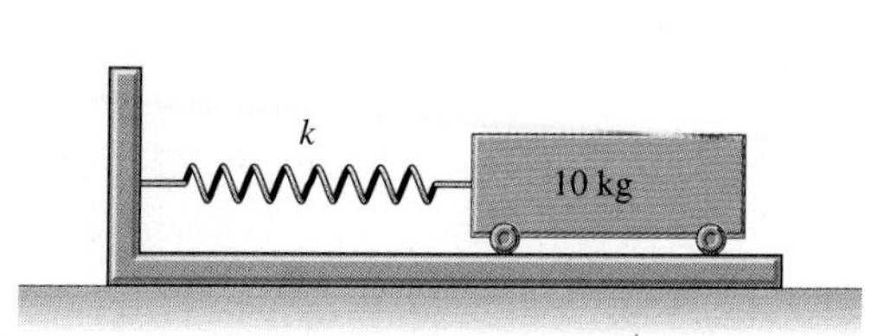

P10.76

10.77 Consider the oscillator immersed in oil described in Problem 10.76. If the mass is displaced 0.1 m to the right of its equilibrium position and released from rest, what is its position relative to the equilibrium position as a function of time?

10.78 The stepped disc weighs 90 N and its moment of inertia is $I = 0.8\ \text{kg.m}^2$. It rolls on the horizontal surface. If $c = 120$ N.s/m, what is the frequency of small vibrations of the disc?

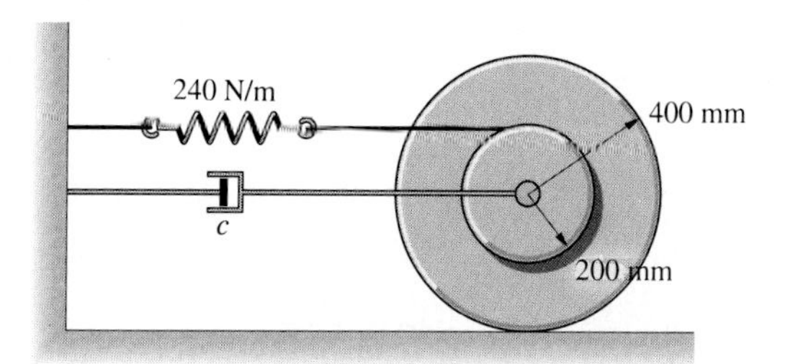

P10.78

10.79 The stepped disc described in Problem 10.78 is initially in equilibrium, and at $t = 0$ it is given a clockwise angular velocity of 1 rad/s. Determine the position of the centre of the disc relative to its equilibrium position as a function of time.

10.80 The stepped disc described in Problem 10.78 is initially in equilibrium, and at $t = 0$ it is given a clockwise angular velocity of 1 rad/s. Determine the position of the centre of the disc relative to its equilibrium position as a function of time if $c = 240$ N.s/m.

10.81 The 22 kg platen P rests on four roller bearings. The roller bearings can be modelled as 1 kg homogeneous cylinders with 30 mm radii, and the spring constant is $k = 900$ N/m. The platen is subjected to a force $F(t) = 100 \sin 3t$ N. What is the magnitude of the platen's steady-state horizontal vibration?

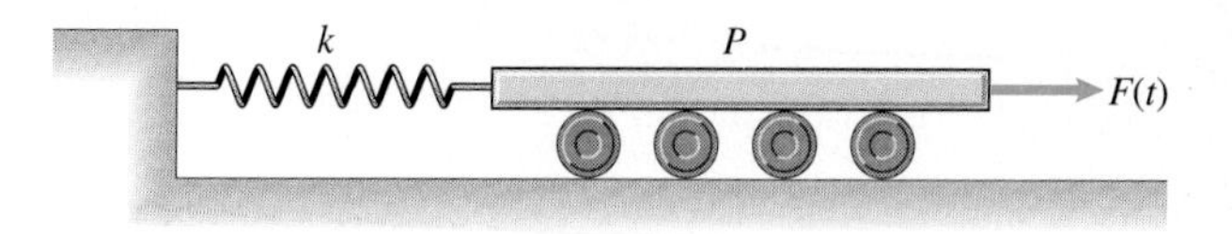

P10.81

10.82 At $t=0$, the platen described in Problem 10.81 is 0.1 m to the right of its equilibrium position and is moving to the right at 2 m/s. Determine the platen's position relative to its equilibrium position as a function of time.

10.83 The base and mass m are initially stationary. The base is then subjected to a vertical displacement $h \sin \omega_i t$ relative to its original position. What is the magnitude of the resulting steady-state vibration of the mass m *relative to the base*?

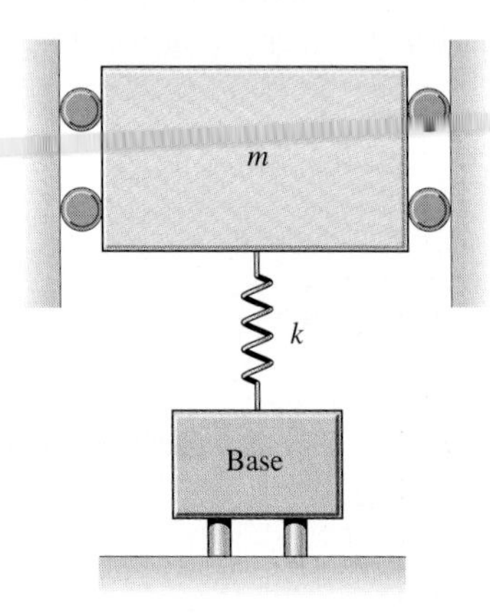

P10.83

10.84 The mass of the trailer, not including its wheels and axle, is m, and the spring constant of its suspension is k. To analyse the suspension's behaviour, an engineer assumes that the height of the road surface relative to its mean height is $h \sin(2\pi x/\lambda)$. Assume that the trailer's wheels remain on the road and its horizontal component of velocity is v. Neglect the damping due to the suspension's shock absorbers.

(a) Determine the magnitude of the trailer's vertical steady-state vibration *relative to the road surface*.

(b) At what velocity v does resonance occur?

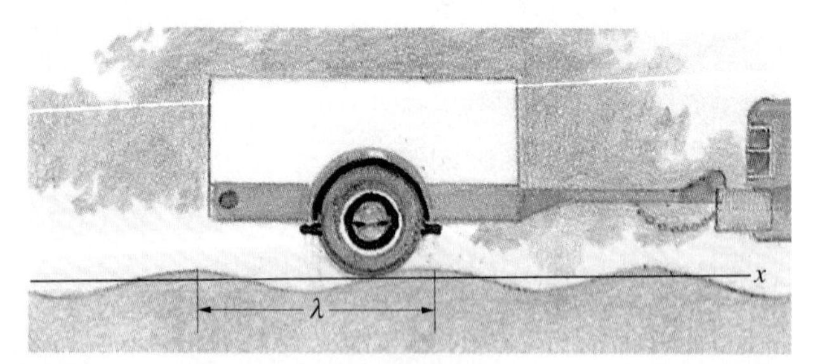

P10.84

10.85 The trailer in Problem 10.84, not including its wheels and axle, weighs 4450 N. The spring constant of its suspension is $k=35$ kN/m, and the damping coefficient due to its shock absorbers is $c=2920$ N.s/m. The road surface parameters are $h=50$ mm and $\lambda=2.44$ m. The trailer's horizontal velocity is $v=1.25$ m/s. Determine the magnitude of the trailer's vertical steady-state vibration relative to the road surface: (a) neglecting the damping due to the shock absorbers; (b) not neglecting the damping.

3.74 $\mu_s = 0.406$. The mass slips towards O.

3.76 $(-51\,\mathbf{e}_r - 11\,\mathbf{e}_\theta)$ N.

3.78 $k = 2m\omega_0^2$.

3.82

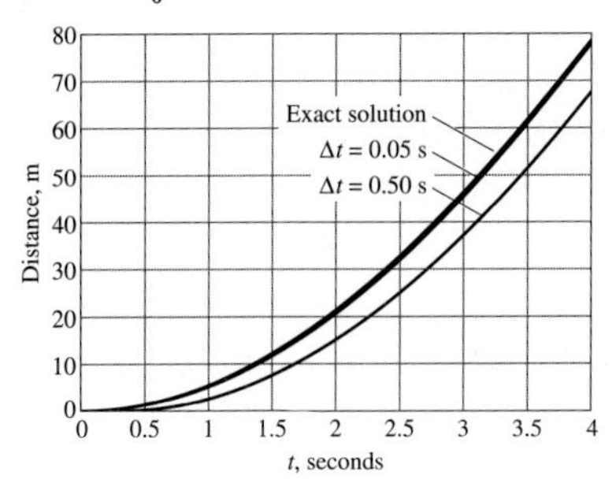

3.84

Time, s	Position, m	Velocity, m/s
0.00	1.0000	0.0000
0.01	1.0000	−0.0100
0.02	0.9999	−0.0200
0.03	0.9997	−0.3000
0.04	0.9994	−0.0400
0.05	0.9990	−0.0500

3.86 $x = 7.55\,m$, $v_x = 8.1$ m/s.

3.88

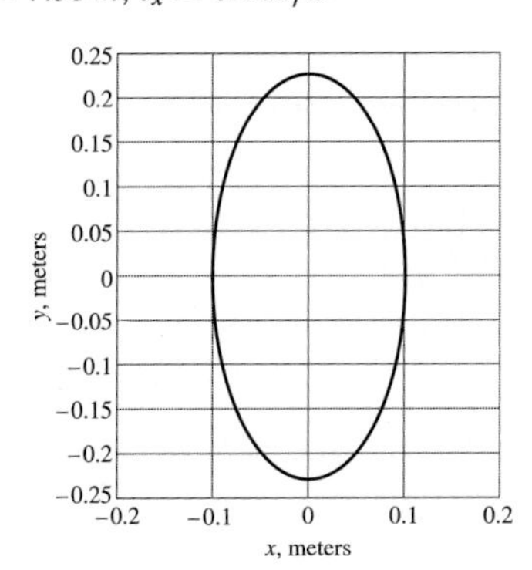

3.90 $C = 0.0217$.

3.92 (a) $F_1 = 63.1$ kN, $F_2 = 126.1$ kN, $F_3 = 189.2$ kN.
(b) $F_1 = 75.1$ kN, $F_2 = 150.1$ kN, $F_3 = 225.2$ kN.

3.94 4364 m.

3.96 47.1 N.

3.98 $a_A = 1.226\,\text{m/s}^2$, $T = 262.5$ N.

3.100 8.96 m.

3.102 $\tan\alpha = v^2/(\rho g)$.

3.104 $5.15\,\text{m/s}^2$.

3.106 9.30 N.

3.108 $\Sigma\mathbf{F} = (-10.7\,\mathbf{e}_r + 2.55\,\mathbf{e}_\theta)$ N.

3.110 $1.62\,\text{m/s}^2$.

Chapter 4

4.2 (a) 1.728×10^6 tonnes. (b) 275×10^6 kg.

4.4 (a) 8.889×10^6 N.m. (b) 22.2 kN.

4.6 2963 kW.

4.8 129 kN.

4.10 (a) 14.4 N.m. (b) 6.15 m/s.

4.12 $U = \frac{1}{2}m[(v_0 - cs_f/m)^2 - v_0^2]$.

4.14 $v = 21.8$ m/s.

4.16 66.7 N.

4.18 10 kN/m.

4.20 1.542 m/s.

4.22 $k = 4288$ N/m.

4.24 (a) 1.55 m. (b) 1.74 m/s.

4.26 (a), (b) Work = 588.6 N.m, $v = 6.26$ m/s.

4.28 $v = 1.72$ m/s.

4.30 $v = 1.14$ m/s.

4.32 (a) 932.9 N. (b) Work = 89.6 N.m, $v = 1.06$ m/s.

4.34 21.0 m/s.

4.36 16 m/s or 57.6 km/hr

4.38 $T = 3mg\sin\alpha$.

4.40 2.33 kN.

4.42 (a), (b), (c) 63.4 m/s.

4.44 $\alpha = \arccos(2/3)$.

4.46 0.327 m.

4.48 0.267 m.

4.50 $v_2 = 7.03$ m/s.

4.52 5.77 m/s.

4.54 4.90 m/s.

4.56 36.85 m/s.

4.58 $v_A = 5.87$ m/s.

4.60 1442 m/s.

4.62 2.25×10^{10} N.m.

4.64 $P = -180.6$ kW

4.66 $P_{av} = 10.58$ kW.

4.68 245 kW.

4.70 (a) 7.72 m. (b), (c) 6.48 m.

4.72 60°.

4.74 (a) 686.7 N. (b) 882.9 N.

4.76 $W[1 + \sqrt{1 + 4C/W}]$.

4.78 $V = \frac{1}{2}kS^2 + \frac{1}{4}qS^4$.

4.80 $v = 2.30$ m/s.

4.82 $v = 3.764$ m/s.

4.84 (a) $V = p_0 s_0^\gamma A s^{1-\gamma}/(\gamma - 1)$.
(b) $v = \sqrt{2p_0 s_0^\gamma A(s_0^{1-\gamma} - s^{1-\gamma})/[m(\gamma - 1)]}$.

4.86 $v = 11.0$ km/s.

4.88 $v_C = 2887$ m/s.

4.90 (a) $\mathbf{F} = (4x\,\mathbf{i} - \mathbf{j})$ N. (b) The work is -1 N.m for any path from position 1 to position 2.

4.92 $\mathbf{F} = -[k(r - r_0) + q(r - r_0)^3]\,\mathbf{e}_r$.

4.94 (a) $\mathbf{F} = (\sin\theta - 2r\cos^2\theta)\mathbf{e}_r + (\cos\theta + 2r\sin\theta\cos\theta)\mathbf{e}_\theta$.

(b) The work is 2 N.m for any path from 1 to 2.

4.98 (a) and (c).

4.100 552.5 mm.

4.102 58.7 mm, 1277.6 N.m/s (watts).

4.104 2.15 m.

4.106 $v = 6.32$ m/s.

4.108 $P = -11.84$ MW.

4.110 (a), (b) 61.9 m.

4.112 He should choose strategy (b). Impact velocity is 11.8 m/s, work is −251.3 kN.m. In strategy (a), impact velocity is 13.9 m/s, work is −119.2 kN.m.

4.114 $v = 2.08$ m/s.

4.116 (a) 2.08×10^8 N.m. (b) $v = 100(1 - e^{-(F_0/100\,m)t})$ km/hr.

4.118 $k = 2408$ N/m.

4.120 (a) $k = 809.0$ N/m. (b) $v_2 = 6.29$ m/s.

4.122 $h = 0.179$ m.

4.124 1.56 m/s.

4.126 (a) $\theta = 26.3°$. (b) 704 N. (c) 946.8 N.

4.128 $v_1 = 2.65$ m/s.

4.130 1.02 m.

4.132 9.45 m/s.

4.134 (a) $v = 0$. (b) $v = \sqrt{gR_E/2} = 5590$ m/s or 20 123 km/hr.

4.136 (a) 11.34 MW (megawatts). (b) 9.45 MW.

Chapter 5

5.2 (a) 120 kN.s. (b) 20 kN.

5.4 (a) 4.4 kN.s. (b) 32 m/s or 115 km/hr.

5.6 (a) $(180\mathbf{i} - 120\mathbf{j} + 40\mathbf{k})$ N.s. (b) $(10\mathbf{i} - 6\mathbf{j} + 4\mathbf{k})$ m/s.

5.8 (a) 375 kN.s. (b) 8 s.

5.10 10.2 m/s.

5.12 (a) 221.7 N.s. (b) 1.85 m/s.

5.14 0.192 s.

5.16 3.27 m/s.

5.18 $(-7.07\mathbf{i} + 7.26\mathbf{j})$ m/s.

5.20 (a) $-490\mathbf{j}$ N.s. (b) $\mathbf{v} = (6\mathbf{i} - 9.23\mathbf{j})$ m/s.

5.22 1450 N.

5.24 (a) 8467 N. (b) 6.67 m/s^2.

5.26 333 N (approximately 600 times the watch's weight).

5.28 Horizontal force is 0.234 N, vertical force is 0.364 N.

5.30 $(-0.35\mathbf{i} + 122.96\mathbf{j})$ N.

5.32 (a) 0.000 57 m/s. (b) 0.003 43 m.

5.34 (a), (b) 1.63 m/s.

5.36 Velocity is 6.96 km/s, $\beta = 0.174°$.

5.38 $v = \sqrt{2\mu_k g D}(1 + m_B/m)$.

5.40 683 m/s.

5.42 (a) $30\mathbf{j}$ mm/s. (b) $(6\mathbf{i} + 6\mathbf{j})$ m.

5.44 $e = 0.2$, $v'_A = 0.4 v_A$.

5.46 0.75 m/s to the right.

5.48 A: 1.05 m/s to the left; B: 2.55 m/s to the right.

5.50 A: 33.3 m/s^2; B: 20 m/s^2.

5.52 $v_A = 7.70$ m/s.

5.54 $e = 0.77$.

5.56 Helmet: 1.09 m/s to the right; head: 1.07 m/s to the left.

5.58 $e = 2\sqrt{(1 - \cos\beta)/(1 - \cos\theta)} - 1$.

5.60 $\mathbf{v}'_A = (\mathbf{i} + \mathbf{j})$ m/s, $\mathbf{v}'_B = (-\mathbf{i} + \mathbf{j})$ m/s.

5.64 $e = 0.304$.

5.66 2.30 m.

5.68 $v_S = 35.3$ m/s.

5.70 0.65 m/s.

5.72 $\mathbf{v} = (120\mathbf{i} + 40\mathbf{j} - 30\mathbf{k})$ m/s or $\mathbf{v} = (-120\mathbf{i} - 40\mathbf{j} + 30\mathbf{k})$ m/s.

5.74 (a) $-1440\mathbf{k}$ kg.m^2/s. (b) 1.2 m/s.

5.76 $\mathbf{v} = -v_0\,\mathbf{e}_r + \dfrac{[(1/2)r_0 t^2 - (1/3)v_0 t^3]C}{(r_0 - v_0 t)m}\mathbf{e}_\theta$.

5.78 2.31 m.

5.80 0.111 revolutions per minute.

5.82 $|v_r| = 3378$ m/s, $|v_\theta| = 2828$ m/s.

5.86 919.3 N.m.

5.88 750 N.

5.90 (a) $(351\mathbf{i} - 849\mathbf{j})$ N. (b) $(1200\mathbf{i} - 1200\mathbf{j})$ N. (c) $2400\mathbf{i}$ N.

5.92 $0.753\mathbf{k}$ N.m.

5.94 $(51.9\mathbf{i} - 282.2\mathbf{j})$ N.

5.96 1.435 km/s.

5.98 (a) 1.870 km/s. (b) 1.946 km/s. (c) 1.797 km/s.

5.100 (a) $F = (45s + 0.85)$ N. (b) 33.4 J.

5.102 $v = v_0/[1 + (\rho_L/2m)s]$.

5.104 6815 N.

5.106 18.4 kN.

5.108 (a) 5. (b) 193 kN. (c) 3.4 m/s^2.

5.110 876.9 kN.

5.112 Mass = 2.0×10^{13} kg. Radius = 880.1 m.

5.114 (a) 10.85 m/s. (b) 21.7 kJ.

5.116 $0.775 \le e \le 0.828$.

5.118 (a) 3.45 m/s. (b) 18.7 kN.

5.120 (a) 8.23 kN. (b) 16.5 kN.

5.122 $v_A\sqrt{mk/2}$.

5.124 1.36 m.

5.126 $x = -0.32$ m, $y = 0.24$ m.

5.128 3.90 kN (including the weight of the drum).

5.130 (a) 9.15 m/s. (b) 14.11 m/s.

Chapter 6

6.2 (a) 20 rad/s^2 clockwise. (b) 1.59 revolutions.

6.4 8 rad/s.

6.6 $\mathbf{v}_A = (-0.2\,\mathbf{i} - 0.2\,\mathbf{j})$ m/s, $\mathbf{a}_A = (2\,\mathbf{i} - 2\,\mathbf{j})$ m/s^2.

6.8 (a) $\mathbf{v}_B = 2.4\,\mathbf{j}$ m/s. (b) $\mathbf{v}_B = (-0.34\,\mathbf{i} - 2.38\,\mathbf{j})$ m/s.

6.10 $|\mathbf{v}_B| = 4.5$ m/s, $|\mathbf{a}_B| = 90$ m/s^2.

6.12 $\boldsymbol{\omega} = 30\,\mathbf{i}$ rad/s.

6.14 $\boldsymbol{\omega}_{AB} = 10\,\mathbf{k}$ rad/s, $\boldsymbol{\omega}_{BC} = -10\,\mathbf{k}$ rad/s, $\boldsymbol{\omega}_{CD} = 10\,\mathbf{k}$ rad/s.

6.16 (a) $\boldsymbol{\omega} = \omega\,\mathbf{k}$. (b) $\mathbf{v}_{A/B} = r_{A/B}\omega\,\mathbf{j}$.

6.18 (a) $\boldsymbol{\omega} = 5\,\mathbf{k}$ rad/s. (b) $\mathbf{v}_A = (-5\,\mathbf{i} + 8.66\,\mathbf{j})$ m/s.

6.20 (a) 92.6 rad/s clockwise. (b) The top; 200 km/hr, or 55.6 m/s.

6.22 Angular velocity = 17.95 rad/s, or 171.4 rpm clockwise; velocity = 3.14 m/s.

6.24 (a) $\boldsymbol{\omega} = 2.13\,\mathbf{k}$ rad/s, (b) $\mathbf{v}_A = -0.73\,\mathbf{j}$ m/s.

6.26 8 rad/s clockwise, $\mathbf{v}_D = (4\,\mathbf{i} - 0.8\,\mathbf{j})$ m/s.

6.28 $\omega_{CD} = 10$ rad/s counterclockwise.

6.30 $\omega_{AB} = 2$ rad/s clockwise, $\omega_{BC} = 3.22$ rad/s, counterclockwise

6.32 $\mathbf{v}_E = -12.3\,\mathbf{j}$ m/s.

6.34 $|\omega_{AB}| = 4.91$ rad/s, $|\omega_{DE}| = 6.77$ rad/s.

6.36 $\omega_{AB} = 2.31$ rad/s clockwise, $v_B = 3.15$ m/s to the left.

6.38 $\mathbf{v}_C = 0.628$ m/s.

6.40 $\mathbf{v}_A = (1.2\,\mathbf{i} + 1.2\,\mathbf{j})$ m/s.

6.42 $\omega_{BC} = 2.61$ rad/s, $\mathbf{v}_C = -9.10\,\mathbf{i}$ m/s.

6.44 0.95 m/s.

6.46 $\mathbf{v}_C = (0.272\,\mathbf{i} + 0.188\,\mathbf{j})$ m/s.

6.48 $v_W = 0.2$ m/s, 4 rad/s counterclockwise.

6.50 Angular velocity = 52.06 rad/s, or 497.1 rpm, $|\mathbf{v}_A| = 5.21$ m/s.

6.52 $x_C = 3$m, $y_C = 0$, $v_B = 10$ m/s.

6.54 $x = 0.105$ m, $y = -0.45$ m.

6.56 $\mathbf{v}_G = (1.00\,\mathbf{i} - 0.36\,\mathbf{j})$ m/s.

6.58 (a) (1.04, 1.2) m. (b) $v_A = 0.6\,\mathbf{i}$ m/s.

6.60 0.164 m/s.

6.62 $\mathbf{v}_E = -12.25\,\mathbf{j}$ m/s.

6.64 $\omega_{BC} = 5.33$ rad/s counterclockwise, $\omega_{CD} = 4.57$ rad/s clockwise.

6.68 (a) $\mathbf{a}_A = (-50\,\mathbf{e}_r + 60\,\mathbf{e}_\theta)$ m/s^2. (b), (c) $\mathbf{a}_A = (-50\,\mathbf{i} + 60\,\mathbf{j})$ m/s^2.

6.70 (a) $\mathbf{a}_B = -800\,\mathbf{i}$ m/s^2. (b) $\mathbf{a}_A = -400\,\mathbf{i}$ m/s^2.

6.72 $\omega_{AC} = 0$, $\alpha_{AC} = 1.13$ rad/s^2 clockwise.

6.74 $\mathbf{a}_A = -0.125\,\mathbf{j}$ m/s^2, $\mathbf{a}_B = 0.075\,\mathbf{j}$ m/s^2.

6.76 Acceleration = 3.35 m/s^2.

6.78 Velocity is 1.4 m/s to the right; acceleration is 22.7 m/s^2 to the left.

6.80 $\omega_{BD} = 2.67$ rad/s clockwise, $\alpha_{BD} = 6.22$ rad/s^2 counterclockwise.

6.82 $\mathbf{v}_C = 1.02\,\mathbf{i}$ m/s, $\mathbf{a}_C = 0.175$ m/s^2.

6.84 $\mathbf{a}_C = 5.15\,\mathbf{i}$ m/s^2.

6.86 $\omega_{AB} = -0.879$ rad/s, $\alpha_{AB} = -1.06$ rad/s^2 $\omega_{BC} = -1.15$ rad/s, $\alpha_{BC} = -2.14$ rad/s^2.

6.88 $\mathbf{a}_E = -12.25\,\mathbf{j}$ m/s^2.

6.90 $\mathbf{a}_C = (-7.78\,\mathbf{i} - 33.54\,\mathbf{j})$ m/s^2.

6.92 $\mathbf{a}_D = (-0.135\,\mathbf{i} - 0.144\,\mathbf{j})$ m/s^2.

6.94 $\omega_{AB} = 3.55$ rad/s clockwise, $\alpha_{AB} = 12.05$ rad/s^2 clockwise, $\omega_{BC} = 2.36$ rad/s counterclockwise, $\alpha_{BC} = 16.53$ rad/s^2 counterclockwise.

6.96 $\alpha_{\text{planet}} = 41.43$ rad/s^2 clockwise, $\alpha_{\text{sun}} = 82.86$ rad/s^2 counterclockwise.

6.98 $\alpha_{CD} = 26.45$ rad/s^2 clockwise, $\alpha_{DE} = 31.14$ rad/s^2 counterclockwise.

6.100 $\mathbf{a}_A = (-200\,\mathbf{i} + 80\,\mathbf{j})$ m/s^2.

6.102 $\mathbf{a}_C = (-8.80\,\mathbf{i} + 5.60\,\mathbf{j})$ m/s^2.

6.104 $\omega_{AC} = 3$ rad/s counterclockwise, $\alpha_{AC} = 6$ rad/s^2 clockwise.

6.106 0.549 m/s to the left.

6.108 $\omega_{BC} = 1.16$ rad/s clockwise, 0.385 m/s towards C.

6.110 $\omega_{BC} = 6.17°$ per second counterclockwise; rate of extension is 0.109 m/s.

6.112 $\omega_{AC} = 8.66$ rad/s counterclockwise, and bar AC slides through the sleeve at 5 m/s towards A.

6.114 $\omega_{AC} = 0.293$ rad/s clockwise, $\mathbf{v}_C = (-0.221\,\mathbf{i} - 0.411\,\mathbf{j})$ m/s.

6.116 It is extending at 0.324 m/s.

6.118 (a) $\omega_{AB} = 2$ rad/s clockwise. (b) $v_{B\text{rel}} = 2$ m/s towards C.

6.120 $\omega_{AB} = 3.88$ rad/s counterclockwise.

6.122 $\omega_{\text{plate}} = 2$ rad/s counterclockwise, and the velocity at which the pin slides relative to the slot is 0.2 m/s downwards.

6.126 (a) $\mathbf{v}_A = R\omega\,\mathbf{j}$, $\mathbf{a}_A = -R\omega^2\,\mathbf{i}$. (b) $\mathbf{v}_A = 2R\omega\,\mathbf{j}$, $\mathbf{a}_A = -2R\omega^2\mathbf{I}$.

6.128 $\mathbf{v}_A = (-3.5\,\mathbf{i} + 0.5\,\mathbf{j} + 4.0\,\mathbf{k})$ m/s, $\mathbf{a}_A = (-10\,\mathbf{i} - 6.5\,\mathbf{j} - 19.25\,\mathbf{k})$ m/s^2.

6.130 $-2\,\mathbf{i}$ m/s.

6.132 $\mathbf{a}_A$ rel $= (-4.2\,\mathbf{i} - 0.6\,\mathbf{j})$ m/s^2.

6.134 (a) $\mathbf{v}_{A\,\text{rel}} = v\,\mathbf{j}$, $\mathbf{a}_{A\,\text{rel}} = -(v^2/R_{\text{E}})\,\mathbf{i}$. (b) $\mathbf{v}_A = v\,\mathbf{j} - \omega_{\text{E}}R_{\text{E}}\cos L\,\mathbf{k}$, $\mathbf{a}_A = (-v^2/R_{\text{E}} - \omega_{\text{E}}^2R_{\text{E}}\cos^2 L)\,\mathbf{i} + \omega_{\text{E}}^2R_{\text{E}}\sin L\cos L\,\mathbf{j} + 2\omega_{\text{E}}v\sin L\,\mathbf{k}$.

6.136 (a) $-9.81\,\mathbf{k}$ m/s^2. (b) $(3.29\,\mathbf{i} - 9.81\,\mathbf{k})$ m/s^2.

6.138 (a) $(0.1\,\mathbf{i} + 0.1\,\mathbf{j})$ m/s^2. (b) $(0.125\,\mathbf{i} + 0.085\,\mathbf{j} + 0.106\,\mathbf{k})$ m/s^2.

6.142 $\mathbf{v}_C = 9.50\,\mathbf{j}$ m/s.

6.144 $\alpha_{AB} = 13.60 \times 10^3$ rad/s^2 clockwise, $\alpha_{BC} = 8.64 \times 10^3$ rad/s counterclockwise.

6.146 $\mathbf{a}_D = -87.2$ m/s^2.

6.148 $\mathbf{a}_G = (-0.20\,\mathbf{i} - 0.66\,\mathbf{j})$ m/s^2.

6.150 $\mathbf{v}_C = (-1.48\,\mathbf{i} + 0.79\,\mathbf{j})$ m/s.

6.152 $\mathbf{a}_C = (-2.99\,\mathbf{i} - 1.40\,\mathbf{j})\,\text{m/s}^2$.

6.154 $\omega_{BD} = 0.733$ rad/s counterclockwise.

6.156 $\omega_{AB} = 0.261$ rad/s counterclockwise, $\omega_{BC} = 2.80$ rad/s counterclockwise.

6.158 $\omega_{BC} = 1.22$ rad/s clockwise, 17.96 m/s from B towards C.

6.160 Velocity = 2.07 m/s upwards; acceleration = 50.6 m/s^2 upwards.

6.162 $T_{BC} = 0$.

6.164 (a) $\mathbf{v}_{A\,\text{rel}} = 5\,\mathbf{i}$ m/s, $\mathbf{a}_{A\,\text{rel}} = 0$.
(b) $\mathbf{v}_{A/B} = (5\,\mathbf{i} + 2\,\mathbf{j})$ m/s, $\mathbf{a}_{A/B} = (-4\,\mathbf{i} + 20\,\mathbf{j})\,\text{m/s}^2$.

Chapter 7

7.2 $F = (b/2c)mg$.

7.4 Time = 0.980 s, distance = 2.94 m.

7.6 $F = W(b - ca/g)/h$.

7.8 $T_A = 3681.1$ N, $T_B = 4137.1$ N.

7.10 6.78 m/s^2, $N_A = 1356.4$ N, $N_B = 213.2$ N.

7.12 $\omega = 7.5$ rad/s.

7.14 49.6 N.m.

7.16 (a) $\alpha = 27$ rad/s^2 clockwise, $T = 90$ N.
(b) $\alpha = 19.1$ rad/s^2 clockwise, $T = 63.7$ N.

7.18 Time = 1.15 s.

7.20 (a), (b) $\alpha = 10.4$ rad/s^2 counterclockwise, $A_x = 0$, $A_y = 20.2$ N.

7.22 $x = l/\sqrt{12}$, $\alpha_{\max} = \sqrt{3}g/l$.

7.24 $A_x = -10.28$ kN, $A_y = -7.04$ kN.

7.26 $M_B = 27.1$ N.m counterclockwise, $B_x = -11.0$ N, $B_y = 108.5$ N.

7.28 $t_{\text{ring}}/t_{\text{disc}} = \sqrt{4/3}$.

7.30 $v_{\max} = \frac{2}{7}R\omega_0$, time $= \frac{2}{7}R\omega_0/(\mu_k g)$.

7.32 (a) It doesn't slip, $\alpha = 22.2$ rad/s^2 clockwise. (b) It does slip, $\alpha = 53.6$ rad/s^2 clockwise.

7.34 (a) $\alpha = 1.84$ rad/s^2 counterclockwise. (b) 112.3 N.

7.36 9.54 rad/s^2 clockwise.

7.38 61.3 rad/s^2 clockwise.

7.40 0.170 m/s^2 to the right.

7.42 $\mu_s = 0.0339$.

7.44 $M = 405.8$ N.m.

7.46 $B_x = -296$ N, $B_y = -506$ N, $C_x = 56.3$ N, $C_y = 556$ N.

7.48 $\alpha_{BC} = 9.7$ rad/s^2 counterclockwise, normal force = 10.8 N.

7.50 (a) 9.34 m/s^2. (b) 515.5 N.m.

7.54 $\alpha_P = 0.174$ rad/s^2 clockwise, force = 45.4 N.

7.56 $|P| = 44.1$ kN.

7.58

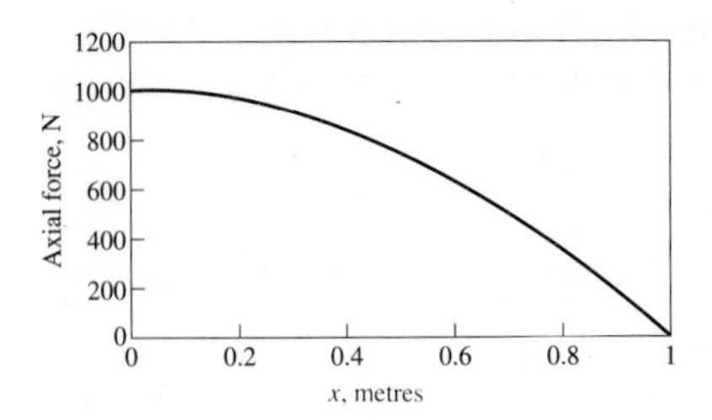

7.60

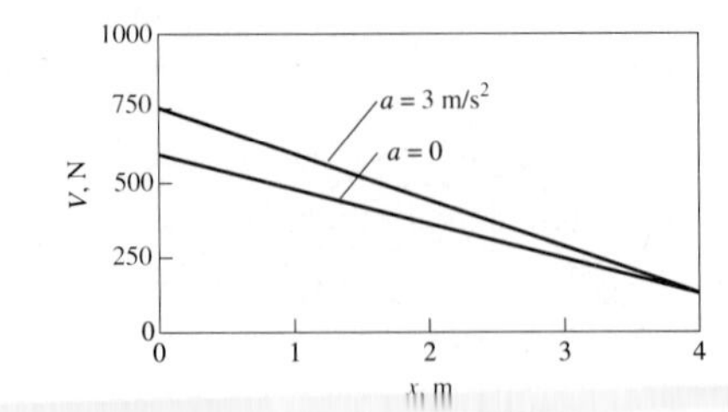

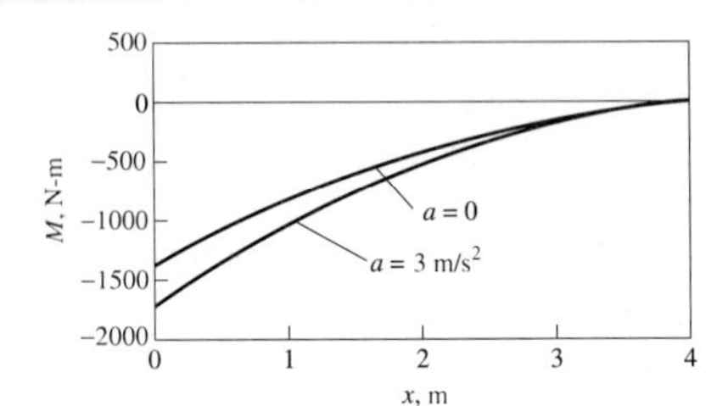

7.62

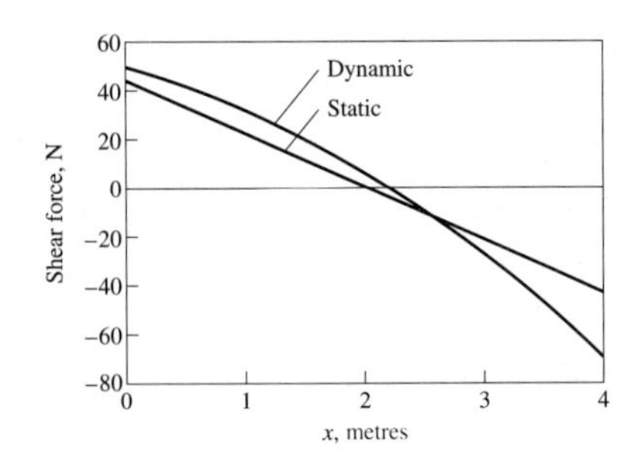

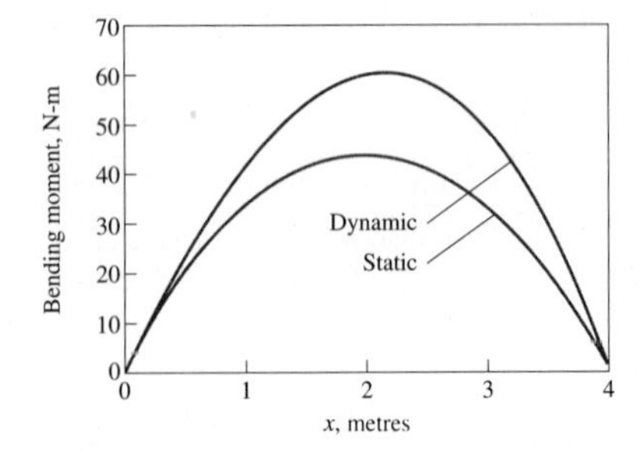

7.64

Time, s	θ, rad	ω, rad/s	Closed-form ω, rad/s
0.0	0.000	0.000	0.000
0.2	0.000	0.250	0.250
0.4	0.050	0.499	0.498
0.6	0.150	0.747	0.744
0.8	0.299	0.991	0.987
1.0	0.498	1.232	1.225

7.64

Time, s	θ, rad	ω, rad/s
0.0	0.000	0.000
0.1	0.000	1.472
0.2	0.147	2.943
0.3	0.441	4.399
0.4	0.881	5.729
0.5	1.454	6.665

7.68

7.70 $I_0 = \frac{1}{3}ml^2$.

7.72 $I = \frac{1}{12}ml^2 \sin^2\theta$.

7.74 $I_{(x\,\text{axis})} = \frac{1}{12}mh^2$, $I_{(y\,\text{axis})} = \frac{1}{12}mb^2$, $I_{(z\,\text{axis})} = \frac{1}{12}m(b^2 + h^2)$.

7.76 $I_{(y\,\text{axis})} = 32.6\,\text{kg.m}^2$.

7.78 $20.8\,\text{kg.m}^2$.

7.80 $I_0 = \frac{17}{12}ml^2$.

7.82 $I_{(z\,\text{axis})} = 47.0\,\text{kg.m}^2$.

7.84 $4119.3\,\text{kg.m}^2$.

7.86 $I_{(z\,\text{axis})} = 9.00\,\text{kg.m}^2$.

7.88 $I_{(y\,\text{axis})} = 0.130\,\text{kg.m}^2$.

7.90 $I_0 = 0.0188\,\text{kg.m}^2$.

7.92 $I_{(x\,\text{axis})} = I_{(y\,\text{axis})} = m(\frac{3}{20}R^2 + \frac{3}{5}h^2)$.

7.94 $I_{(x\,\text{axis})} = I_{(y\,\text{axis})} = m(\frac{1}{20}\omega^2 + \frac{3}{5}h^2)$.

7.96 $I = 0.5815\,\text{kg.m}^2$.

7.98 $I_0 = 0.003\,67\,\text{kg.m}^2$.

7.100 $I_{(z\,\text{axis})} = 0.902\,\text{kg.m}^2$

7.102 (a) 12.27 s. (b) 0.647 kN.

7.104 (a) $20\,\text{m/s}^2$. (b) $c \le 49.1$ mm.

7.106 $I = 2.05\,\text{kg.m}^2$.

7.108 40.2 kN.

7.110 $\alpha = -0.420\,\text{rad/s}^2$, $F_x = 336.3$ N, $F_y = 1709.7$ N.

7.112 $\alpha = \dfrac{[3(1-\mu^2)\sin\theta - 6\mu\cos\theta]g}{(2-\mu^2)l}$ counterclockwise.

7.114 $B_x = -1959$ N, $B_y - 1238$ N, $C_x = 2081$ N, $C_y = -922$ N.

7.116 $\alpha_{OA} = 0.425\,\text{rad/s}^2$ counterclockwise, $\alpha_{AB} = 1.586\,\text{rad/s}^2$ clockwise.

7.118 $\alpha_{HP} = 5.37\,\text{rad/s}^2$ clockwise.

7.120 $208.2\,\text{m/s}^2$ to the left.

Chapter 8

8.2 504.3 rad/s (4816 rpm).

8.4 10.05 revolutions.

8.6 0.103 revolution per second.

8.8 (a) 21.7 rad/s. (b) 17.3 rad/s.

8.10 (a), (b) 2.63 rad/s.

8.12 (a) $\omega = \sqrt{2gx/(\frac{1}{12}l^2 + x^2)}$. (b) $x = l/\sqrt{12}$.

8.14 139.0 rad/s counterclockwise.

8.16 (a) 0.397 m. (b) 0.382 m.

8.18 $U = 5634$ N.m.

8.20 $\omega = \sqrt{\dfrac{(mg\sin\beta + M/R)2b}{I + R^2m}}$.

8.22 16.7 rad/s clockwise.

8.24 $v = 0.413$ m/s.

8.28 $P_{max} = 810$ W, $P_{av} = 405$ W.

8.30 $|\omega| = 4.32$ rad/s.

8.32 0.518 m/s.

8.34 3.69 rad/s clockwise.

8.36 2.48 m/s.

8.38 4.52 rad/s counterclockwise.

8.40 2.80 rad/s counterclockwise.

8.42 $\omega_{AB} = 4.67$ rad/s clockwise, $\omega_{BC} = 5.74$ rad/s counterclockwise.

8.44 3 s.

8.46 (a) 126.4 rad/s. (b) 200 rad/s.

8.48 $\omega = 685.6$ rad/s.

8.50 $h = 1.33$ m, $\omega = 3.75$ rad/s.

8.52 $\omega_2 = 4.55$ revolutions per second.

8.54 $\omega' = 0.721$ rad/s.

8.56 14.5°.

8.58 4.72 rad/s counterclockwise.

8.60 0.006 02 rad/s counterclockwise.

8.62 1.389 rad/s counterclockwise.

8.64 3.75 rad/s counterclockwise.

8.66 (a) 50 N.m before, 31.3 N.m after. (b) 50 N.m before, 50 N.m after.

8.70 Energy lost is $\frac{1}{6}mgl$.

8.72 1.8 m/s.

8.74 Velocity= $(1.72\,\mathbf{i} - 0.63\,\mathbf{j})$ m/s, angular velocity = 3.13 rad/s counterclockwise.

8.76 $\omega = \dfrac{6v_0}{7l}$ counterclockwise.

8.78 $\omega = 0.0997$ rad/s = 5.71 deg/s counterclockwise.

8.80 $\omega'_A = 0.380$ rad/s clockwise, $\omega'_B = 0.122$ rad/s clockwise, $\mathbf{v}'_A = 0.174\,\mathbf{i}$ m/s, $\mathbf{v}'_B = 0.567\,\mathbf{i}$ m/s.

8.82 0.003 36 rad/s clockwise.

8.84 $v = 2.05$ m/s.

8.86 $v = \sqrt{Fb/[\frac{1}{2}m_c + 2(m + I/R^2)]}$.

8.88 $\omega_s = 12.33\,\text{rad/s}$.

8.90 $\omega_{AB} = 11.6\,\text{rad/s}$ clockwise, $v_C = 2.7\,\text{m/s}$.

8.92 11.07 rad/s.

8.94 1.77 rad/s counterclockwise.

8.96 $N = (373/283)mg$.

8.98 $\omega' = (\frac{1}{3} + \frac{2}{3}\cos\beta)\omega$.

8.100 $b = 2\,\text{m}$.

8.102 10.5 rad/s counterclockwise.

8.104 $0.0822\,\mathbf{k}\,\text{rad/s}$.

8.106 4.45 rad/s counterclockwise.

8.108 (a) 3.90 rad/s. (b) 0.162 m/s.

Chapter 9

9.2 $\mathbf{a}_A = (358.8\,\mathbf{i} + 24.0\,\mathbf{j} + 149.6\,\mathbf{k})\,\text{km/s}^2$.

9.4 $\mathbf{a}_A = (-255.2\,\mathbf{i} + 199.3\,\mathbf{j} - 209\,\mathbf{k})\,\text{m/s}^2$.

9.6 $\mathbf{a}_C = (25\,\mathbf{i} - 100\,\mathbf{j} - 25\,\mathbf{k})\,\text{m/s}^2$.

9.8 (a) $\alpha = (77.8\,\mathbf{i} + 44.4\,\mathbf{j} + 44.4\,\mathbf{k})\,\text{rad/s}^2$.
(b) $(160.0\,\mathbf{i} - 140.0\,\mathbf{j} - 140.0\,\mathbf{k})\,\text{km/s}^2$.

9.10 (a) $\omega = \omega_d\,\mathbf{i} + \omega_0\,\mathbf{j}$.
(b) $\mathbf{v}_A = -R\omega_0\cos\theta\,\mathbf{i} + R\omega_d\cos\theta\,\mathbf{j}$
$+(R\omega_d\sin\theta - b\omega_0)\,\mathbf{k}$.

9.12 $\mathbf{v}_A = 80\,\mathbf{i}\,\text{mm/s}$, $\mathbf{v}_B = (-103\,\mathbf{i} + 282\,\mathbf{j} - 56\,\mathbf{k})\,\text{mm/s}$.

9.14 (a) $\omega_{BC} = (0.1\,\mathbf{j} + 0.4\,\mathbf{k})\,\text{rad/s}$.
(b) $\mathbf{v}_C = (-0.315\,\mathbf{i} + 0.085\,\mathbf{j} - 0.131\,\mathbf{k})\,\text{m/s}$.

9.16 (a) $\omega = (20\,\mathbf{i} - 5\,\mathbf{j})\,\text{rad/s}$.
(b) $\mathbf{v}_A = (0.25\,\mathbf{i} + 1.00\,\mathbf{j} + 3.73\,\mathbf{k})\,\text{m/s}$.

9.18 $\mathbf{H} = (25\,\mathbf{i} + 50\,\mathbf{k})\,\text{kg.m}^2/\text{s}$.

9.20 $\mathbf{H}_O = (20\,\mathbf{j} - 22\,\mathbf{k})\,\text{kg.m}^2$.

9.22 $I_{xx} = 0.67\,\text{kg.m}^2$, $I_{yy} = 5.33\,\text{kg.m}^2$, $I_{zz} = 6\,\text{kg.m}^2$,
$I_{xy} = I_{yz} = I_{zx} = 0$.

9.24 $I_{xx} = 0.12\,\text{kg.m}^2$, $I_{yy} = 0.03\,\text{kg.m}^2$,
$I_{zz} = 0.15\,\text{kg.m}^2$, $I_{xy} = I_{yz} = I_{zx} = 0$.

9.26 $I_{xx} = 80\,\text{kg.m}^2$, $I_{yy} = 540\,\text{kg.m}^2$, $I_{zz} = 620\,\text{kg.m}^2$,
$I_{xy} = 180\,\text{kg.m}^2$, $I_{yz} = I_{zx} = 0$.

9.28 (a) $\mathbf{H} = \frac{1}{12}ml^2(\omega_z\,\mathbf{k})$. (b) $\mathbf{H}_O = \frac{1}{3}ml^2(\omega_y\,\mathbf{j} + \omega_z,\mathbf{k})$.

9.30 $I_{xx} = 0.0603\,\text{kg.m}^2$, $I_{yy} = 0.0560\,\text{kg.m}^2$,
$I_{zz} = 0.1162\,\text{kg.m}^2$, $I_{xy} = I_{yz} = I_{zx} = 0$.

9.32 $I_{x'x'} = 15\,600\,\text{kg.m}^2$, $I_{y'y'} = 226\,800\,\text{kg.m}^2$, $I_{z'z'} =$
$242\,400\,\text{kg.m}^2$, $I_{x'y'} = -32\,400\,\text{kg.m}^2$, $I_{yz} = I_{zx} = 0$.

9.34 $\mathbf{H} = (2.00\,\mathbf{i} - 1.64\,\mathbf{j} + 2.58\,\mathbf{k})\,\text{kg.m}^2/\text{s}$.

9.36 (a) $I = 3.56\,\text{kg.m}^2$. (b) $14.22\,\text{kg.m}^2/\text{s}$.

9.38 $I_1 = 16.15$, $I_2 = 62.10$, $I_3 = 81.75\,\text{kg.m}^2$, $\mathbf{e}_1 = 0.964\,\mathbf{i} -$
$0.220\,\mathbf{j} + 0.151\,\mathbf{k}$, $\mathbf{e}_2 = -0.204\,\mathbf{i} - 0.972\,\mathbf{j} - 0.114\,\mathbf{k}$,
$\mathbf{e}_3 = 0.172\,\mathbf{i} + 0.079\,\mathbf{j} - 0.982\,\mathbf{k}$.

9.40 $I_1 = 0$, $I_2 = 1/12$, $I_3 = 1/12\,\text{kg.m}^2$, $\mathbf{e}_1 = \cos\beta\,\mathbf{i} +$
$\sin\beta\,\mathbf{j}$, $\mathbf{e}_2 = -\sin\beta\,\mathbf{i} + \cos\beta\,\mathbf{j}$, $\mathbf{e}_3 = \mathbf{k}$.

9.42 $\Sigma\mathbf{M} = (-250\,\mathbf{i} - 250\,\mathbf{j} + 125\,\mathbf{k})\,\text{N.m}$.

9.44 $\Sigma\mathbf{M}_O = (-52\,\mathbf{i} + 132\,\mathbf{j} + 120\,\mathbf{k})\,\text{N.m}$.

9.46 $|\mathbf{a}_A| = 5F/m$.

9.48 (a) $(\alpha = (28.57\,\mathbf{i} + 5.71\,\mathbf{j} - 2.00\,\mathbf{k})\,\text{rad/s}^2$.
(b) $(-0.33\,\mathbf{i} + 2.00\,\mathbf{j} - 19.54\,\mathbf{k})\,\text{m/s}^2$.

9.50 $\alpha = (-99.5\,\mathbf{i} + 3.9\,\mathbf{j} + 3.7\,\mathbf{k})\,\text{rad/s}^2$.

9.52 $\alpha = (-500.0\,\mathbf{i} + 24.4\,\mathbf{j})\,\text{rad/s}^2$.

9.54 $\alpha = \frac{4}{3}(g/b)\,\mathbf{j}$.

9.56 $\alpha = (14.53\,\mathbf{j} + 4.65\,\mathbf{k})\,\text{rad/s}^2$.

9.58 27.4 N.m.

9.60 $|\mathbf{F}| = mg$, $|\mathbf{C}| = \frac{1}{12}ml^2\omega_0^2|\sin\beta\cos\beta|$.

9.64 $\frac{1}{2}mR^2\omega_0\omega_d$.

9.66 $(-49\,\mathbf{j} + 80\,\mathbf{k})\,\text{N}$, $-18\,\mathbf{i}\,\text{N.m}$.

9.72 157 N.m.

9.76 22.2 N.m.

9.80 $\dot{\psi} = 1.305\,\text{rev/s}$.

9.82 39.2 rpm.

9.84 (a) $\dot{\psi} = -2052\,\text{rpm}$.
(b)

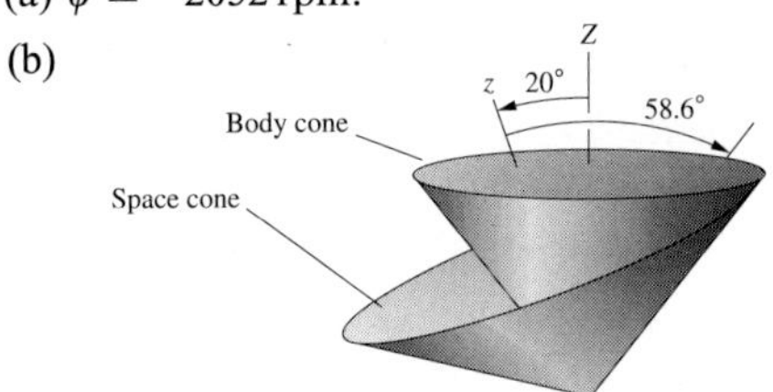

9.86 $N = \omega_0^2[(b/r)I_{zz}\sin^2\beta - I_{xx}\sin\beta\cos\beta]/\,(b\sin\beta -$
$r\cos\beta)$.

9.88 His right side.

9.90 $\Sigma\mathbf{M} = (2122.5\,\mathbf{i} - 155.4\,\mathbf{j} - 534.0\,\mathbf{k})\,\text{N.m}$.

9.92 $\omega_0 = \sqrt{g\sin\beta/(\frac{2}{3}l\sin\beta\cos\beta + b\cos\beta)}$.

9.94 (b) If $\omega_0 = 0$, the plate is stationary. The solution of the equation $2\cos\beta - \sin\beta = 0$ is the value of β for which the centre of mass of the plate is directly above point O; the plate is balanced on one corner.

9.96 $\mathbf{B} = (52.72\,\mathbf{i} + 97.35\,\mathbf{j} + 9.26\,\mathbf{k})\,\text{N}$,
$\mathbf{M}_B = (0.05\,\mathbf{i} - 10.25\,\mathbf{j} + 30.63\,\mathbf{k})\,\text{N.m}$.

9.98

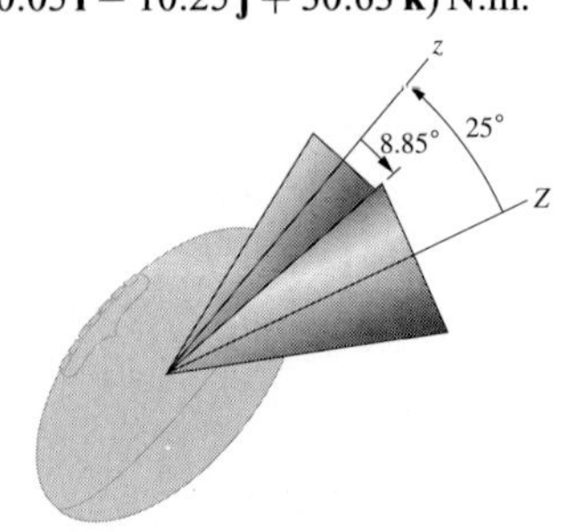

9.100 $\Sigma\mathbf{M} = (-282.8\,\mathbf{i} - 2545.6\,\mathbf{j} - 800\,\mathbf{k})\,\text{N.m}$.

9.102 $\alpha = (0.0535\,\mathbf{i} + 0.0374\,\mathbf{j} + 0.0160)\,\text{rad/s}^2$.

Chapter 10

10.4 (a) $E = 1, \phi = 45°$.

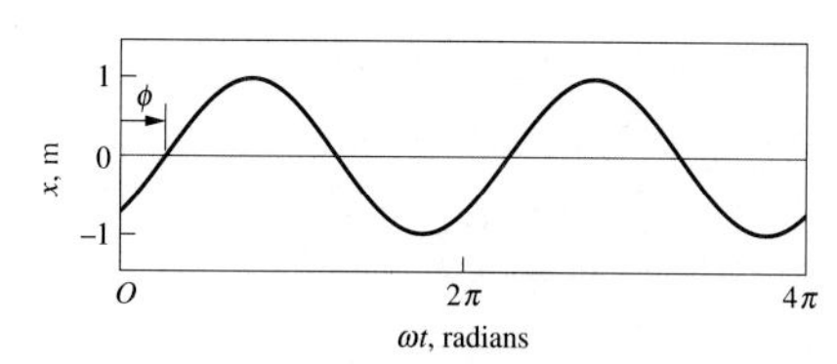

10.6 (a) $\tau = 2.09$ s, $f = 0.477$ Hz. (b) $x = (0.1)\cos 3t\, m$.

10.8 (a) $x = (0.1)\cos 3t$ m.

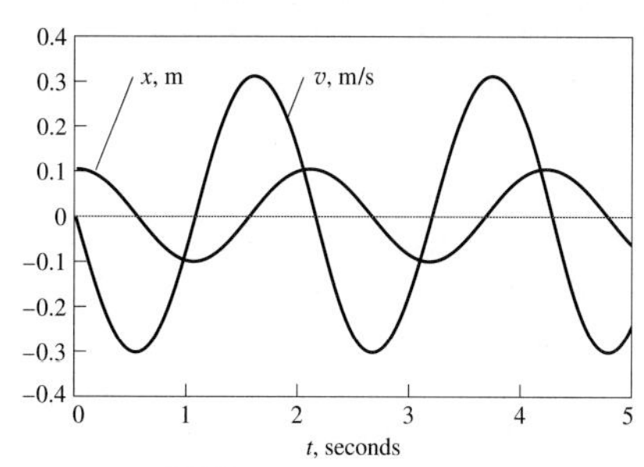

10.10 $f = (1/2\pi)\sqrt{k/m}$.

10.14 1.46 Hz.

10.16 (a) $\tau = 2.09$ s, $f = 0.48$ Hz. (b) $x = 0.36(1 - \cos 3t)$ m.

10.18 0.212 m to the left, 1.630 m/s towards the right.

10.20 $f = 0.177$ Hz.

10.26 30 mm.

10.28 $\theta_B = 0.172 \sin 11.62t$ rad.

10.30 $\theta_B = 0.133 \cos 6.08t$ rad.

10.32 $m_A = 4.38$ kg.

10.36 (a) $\tau = 3.17$ s, $f = 0.316$ Hz. (b) $\tau = 3.14$ s, $f = 0.318$ Hz.

10.38 $\tau = 2.22$ s, $f = 0.450$ Hz. (The undamped values are $\tau = 2.09$ s, $f = 0.477$ Hz.)

10.40 $c = 1.39$ N.s/m.

10.42 (a) $\tau = 1.64$ s, $f = 0.61$ Hz. (b) 3.768 s.

10.44 $x = [e^{-2t}(-0.325 \sin 2.24t - 0.363 \cos 2.24t) + 0.363]$ m.

10.46 $x = [-(0.363 + 1.090t)e^{-3t} + 0.363]$ m.

10.48 0.0708 rad/s clockwise.

10.50 (a) 277.1 N.s/m. (b) $\theta = 10te^{-4.33t}$ rad. (c) $\theta_{max} = 0.850$ rad at $t = 0.231$ s.

10.52 $\theta_B = 0.209e^{-6.62t} \sin 9.55t$ rad.

10.54 $\theta_B = e^{-5.05t}(0.244 \sin 3.45t + 0.167 \cos 3.45t)$ rad.

10.56 (a) $x_p = 0.5 \sin 8t$ m. (b) $x = (-0.4 \sin 10t + \cos 10t + 0.5 \sin 8t)$ m.

10.60 49 N.

10.62 0.113 mm.

10.64 0.407 m.

10.66 5.5 mm.

10.70

Time, s	θ, rad	ω, rad/s
0.00	0.0349	0.0000
0.01	0.0349	−0.0051
0.02	0.0349	−0.0103
0.03	0.0348	−0.0154
0.04	0.0346	−0.0205
0.05	0.0344	−0.0256

10.72

Time, s	Displacement, m $\times 10^6$	Velocity, m/s $\times 10^3$
0.000	0.0000	0.0000
0.001	0.0000	0.0000
0.002	0.0000	0.0664
0.003	0.0664	0.1987
0.004	0.2651	0.3963
0.005	0.6614	0.6587

10.74 $f = (1/2\pi)\sqrt{3[(k/m) - (g/2l)]}$.

10.76 $\delta = 2.07$.

10.78 $f = 0.714$ Hz.

10.80 $x = 0.0345(e^{-2.672t} - e^{-14.26t})$ m.

10.82 $x = (0.253 \sin 6.19t + 0.100 \cos 6.19t + 0.145 \sin 3.00t)$ m.

10.84 (a) $E_p = (2\pi v/\lambda)^2 h/[k/m) - (2\pi v/\lambda)^2]$. (b) $v = \lambda\sqrt{k/m}/2\pi$.

Index

A

B

C

D

E

F

G

H

O

P

R

Unit Conversion Factors

TIME

$1\ \text{min} = 60\ \text{s}$
$1\ \text{hr} = 60\ \text{min} = 3600\ \text{s}$
$1\ \text{day} = 24\ \text{hr} = 86\,400\ \text{s}$

LENGTH

$1\ \text{m} = 3.281\ \text{ft} = 39.37\ \text{in}$
$1\ \text{km} = 0.6214\ \text{mi}$
$1\ \text{in} = 0.083\,33\ \text{ft} = 0.025\,40\ \text{m}$
$1\ \text{ft} = 12\ \text{in.} = 0.3048\ \text{m}$
$1\ \text{mi} = 5280\ \text{ft} = 1.609\ \text{km}$
$1\ \text{nautical mile} = 1852\ \text{m} = 6000\ \text{ft}$

ANGLE

$1\ \text{rad} = 180/\pi\ \text{deg} = 57.30\ \text{deg}$
$1\ \text{deg} = \pi/180\ \text{rad} = 0.017\,45\ \text{rad}$
$1\ \text{revolution} = 2\pi\ \text{rad} = 360\ \text{deg}$
$1\ \text{rev/min (rpm)} = 0.1047\ \text{rad/s}$

AREA

$1\ \text{mm}^2 = 1.550 \times 10^{-3}\ \text{in}^2 = 1.076 \times 10^{-5}\ \text{ft}^2$
$1\ \text{m}^2 = 10.76\ \text{ft}^2$
$1\ \text{in}^2 = 645.2\ \text{mm}^2$
$1\ \text{ft}^2 = 144\ \text{in}^2 = 0.0920\ \text{m}^2$

VOLUME

$1\ \text{mm}^3 = 6.102 \times 10^{-5}\ \text{in}^3 = 3.531 \times 10^{-8}\ \text{ft}^3$
$1\ \text{m}^3 = 6.102 \times 10^4\ \text{in}^3 = 35.31\ \text{ft}^3$
$1\ \text{in}^3 = 1.639 \times 10^4\ \text{mm}^3 = 1.639 \times 10^{-5}\ \text{m}^3$
$1\ \text{ft}^3 = 0.028\,32\ \text{m}^3$

VELOCITY

$1\ \text{m/s} = 3.281\ \text{ft/s}$
$1\ \text{km/hr} = 0.2778\ \text{m/s} = 0.6214\ \text{mi/hr} = 0.9113\ \text{ft/s}$
$1\ \text{mi/hr} = (88/60)\ \text{ft/s} = 1.609\ \text{km/hr} = 0.4470\ \text{m/s}$
$1\ \text{knot} = 1\ \text{nautical mile/hr} = 0.5144\ \text{m/s} = 1.689\ \text{ft/s}$

ACCELERATION

$1\ \text{m/s}^2 = 3.281\ \text{ft/s}^2 = 39.37\ \text{in/s}^2$
$1\ \text{in/s}^2 = 0.083\,33\ \text{ft/s}^2 = 0.025\,40\ \text{m/s}^2$
$1\ \text{ft/s}^2 = 0.3048\ \text{m/s}^2$
$1\ g = 9.81\ \text{m/s}^2 = 32.2\ \text{ft/s}^2$

MASS

$1\ \text{kg} = 0.0685\ \text{slug} = 2.205\ \text{lbm}$
$1\ \text{slug} = 14.59\ \text{kg} = 32.2\ \text{lbm}$
$1\ \text{t (metric tonne)} = 10^3\ \text{kg} = 68.5\ \text{slug}$
$1\ \text{ton} = 2240\ \text{lbm} = 1016\ \text{kg}$

FORCE

$1\ \text{N} = 0.2248\ \text{lb}$
$1\ \text{lb} = 4.448\ \text{N}$
$1\ \text{kip} = 1000\ \text{lb} = 4448\ \text{N}$
$1\ \text{ton} = 2240\ \text{lb} = 9967\ \text{N}$
$1\ \text{ton (US)} = 2000\ \text{lb} = 8896\ \text{N}$

WORK AND ENERGY

$1\ \text{J} = 1\ \text{N.m} = 0.7376\ \text{ft.lb}$
$1\ \text{ft.lb} = 1.356\ \text{J}$

POWER

$1\ \text{W} = 1\ \text{N.m/s} = 0.7376\ \text{ft.lb/s} = 1.340 \times 10^{-3}\ \text{hp}$
$1\ \text{ft.lb/s} = 1.356\ \text{W}$
$1\ \text{hp} = 550\ \text{ft.lb/s} = 746\ \text{W}$

PRESSURE

$1\ \text{Pa} = 1\ \text{N/m}^2 = 0.0209\ \text{lb/ft}^2 = 1.451 \times 10^{-4}\ \text{lb/in}^2$
$1\ \text{bar} = 10^5\ \text{Pa}$
$1\ \text{lb/in}^2\ \text{(psi)} = 144\ \text{lb/ft}^2 = 6891\ \text{Pa}$
$1\ \text{lb/ft}^2 = 6.944 \times 10^{-3}\ \text{lb/in}^2 = 47.85\ \text{Pa}$